PO Box   ,
05
Limpopo Province
orders@SirNewtonsfraud
The Absolute Relevancy of Singularity in terms of Cosmology.
Addressed to the Dean of Physics Professor……………………………………………

I0405084

Dear Professor……………………….,
I am sending my work (again) in the hope of finding an informed audience that is in an intellectual position to appreciate the information I divulge in these books. In that line of thought I hope that…. The work I send you I (might) have sent to your academic University address in the past or may not have sent or that you might not have been filling the office at the time and therefore might not have been the recipient. My approach might have been somewhat different at the time I sent it in the past or my line of introduction might have been different in view of how I now approach the introduction. The work now might be more defined when compared to as it was in the past. Whichever way, the work I send to you now is the same as it was previously should I have sent it to you.

In this work I have abandoned previous politeness, as it then served me not and by trying to accommodate what I do believe is only unproven rhetoric every one loses through diplomacy and correct protocol. I no longer apologise for being correct. My work was definitely not as straight to the point and forthcoming in my disagreeing with the conduct in teaching science as I now aim to achieve. Previously I was afraid to step on toes whereas now I try to break knuckles because my niceties got me little attention and even less notoriety in the past. I now call a spade a spade without beating about the bush and those that don't like it I dare them to prove me wrong. By teaching non-existing, unproven and unsupported ideas just because the ideas support a culture of teaching principles coming from the past is brainwashing and misconduct. All evidence shows mass is not present in allocating planet positioning and yet because Newton envisaged it, this upholds the past concept culture that is still taught as if confirmed. How can academics presently teaching physics still support this method of promoting disinformation?

This work **The Absolute Relevancy of Singularity in terms of Cosmology** contains a new theory holding an entirely new approach to the way gravity forms the cosmos and the construction of the Universe entirely. I take a very new approach to cosmic physics and show that cosmic physics diverts completely from normal physics. Whereas normal physics has a factor such as mass which is a substitute (I found no viable reason why) for weight and it therefore plays some significance, in cosmology mass is completely absent. I found cosmology places singularity in the absolute centre stage of cosmic physics and I show that I found in cosmology it is the presence of singularity that provides time in space. Studying the tables Kepler left we find clear evidence of space moving. The evidence from studying Kepler supports the idea that material only moves by the space in which it is that moves. There is no force or pulling of anything closer and because of that reason all cosmic objects circle specific centres.

Space a^3 with or without material moves T^2k. This is the formula Kepler brought into the minds of man and is without doubt correct. Singularity charges gravity by forming movement in space of space. Therefore I removed mass as a major factor and replace mass with singularity. This concept is new. This entire idea must come across to Newtonian minded physicist as blasphemous scandal. I expect total rejection but then I challenge any person to prove where the cosmos applies mass other than in the imagination of physicists. I totally reject mass as forming any influencing factor in the cosmos and moreover I question its existence in any form used in the cosmos by nature applying physics. In short: There is no mass anywhere. I base my work entirely on the findings of Kepler. I reject Newton's deforming of Kepler's work. I use and explore Kepler's $a^3 = T^2k$ and reject $a^3 = T^2$ that obviously has no validation since Kepler gave values to k^{-1} by $T^2 / a^3 = k^{-1}$ and that shows space retracts towards the sun. The sun contracts space holding matter or not holding matter it makes no difference. That is where the gravity shown as the Titius Bode law finds significance in forming Π and all the consequences arriving from validating Π to form orbiting circles without planets moving towards the sun. By establishing a different position to Π every time the position the planet holds is thereby confirmed. That explains the Titius Bode law and planets do not follow mass as the rule in allocating individual positions. Therefore I discard mass because nowhere could I detect a factor such as mass in cosmic physics. In upholding that view I challenge any person to prove one place where any person could find mass in any form playing any part in cosmology except in the imagination of Newton and his followers, the Newtonian thinkers. Mass has a role in forming weight in normal physics but that can't transfer to the cosmos because no where is there any evidence proving mass in the cosmos but that misinterpretation is an accepted fact. The speed by which planets rotate around the sun is about equal which removes mass as a factor.

I prove that gravity comes as a cosmic factor by forming Π in relation to singularity. In forming Π the movement that forms space goes through four principles on which the entirety of the Universe rests. These principles are the **Titius Bode law**; the **Roche limit**; the **Lagrangian points** all combining as the **Coanda effect**. You might be under the impression that these principles are well explained but Newtonian science has not began to understand the principles that are in place, why the principles are in place and how massive a role these principles form in what forms the Universe. Everything in the Universe moves in relation to everything else also moving and this forms the Universe. Everything that moves, how small and insignificant or how big and overwhelming, does so conducting the process that goes through upholding Pythagoras in line with the four principles combing to produce a circle that moves in a straight line, and that forms gravity. Only through applying the law of Pythagoras does Π form gravity and that forms the Universe. Saying this I know it would impress you little but that is because you have never been introduced to the overall picture. Every one of the mentioned four principles plays a part in forming Π by measure of Pythagoras. Gravity started the Universe and by the same token it still starts movement as gravity starts movement by introducing the law of Pythagoras. In that manner Π forms and that is how gravity forms. Gravity forms for the same reason as why the straight line, half circle and triangle all are equal in measure but not in form. It is all part of the relevancy that singularity applies to cosmic form by exercising control on movement and it is the movement that forms gravity by repositioning space, not the mass that pulls to form gravity.

All that science at present has to use as a basis is mass pulling mass and there is no evidence of that happening. If it did the Universe must contract and proving that would be the Moon coming closer to the Earth. That is not happening because the Moon is drifting away from the Earth by plus minus four centimetres per year. Whichever way one applies Newton's gravitational formula indicating mass reduces the radius it is a failure. Everything connected to material including all material expands at a predetermined value and ratio and in that this removes mass pulling mass and black matter hiding in obscurity for no obvious reason. The circumference of the earth is increasing and that can only be because the earth is expanding in size. The Universe including the earth and including every atom on earth is expanding, not contracting. If you remove mass from Newtonian science there is nothing left to support any idea Newton envisaged but that is only as far as cosmology is concerned. I should know because I remove mass and placed in its stead singularity and I discovered an entire new cosmos made up of logic and a correctness science has never experienced before, but on the condition of removing of all fabricated forces. Should any one feel shocked by my declaring the forces fabricated then prove there are forces? Space moves in relation to material but to call such movement a force is incorrect since I prove it is the space that moves.

I go against the institution of physics' believes and in that is my downfall in the past but also that I cannot avoid to have. All I ask is prove me wrong, that is all. My work might have been too laborious in the past to validate your attention, as it does require a lot of concentration due to a lot of new information never before discovered, because if anything, it is comprehensive and it demands the full concentration of an intellectual reader. However, think of the work you receive as but a condensed summery of the overall totality of what my work contains. It is merely a sample-summery of what there is, compressed for your benefit and informing you by almost providing highlights. In the work I send now I try to keep the complexity to a minimum as much as my efforts can manage.

The entirety of the work has no connection to any work or part of any other person's work. Without dramatising or over playing any point I try to make I must stress that this work does require intellectual attention in order to bring comprehensive understanding about the concept. This I say because in the past my work was given to pre-graduates and postgraduate teenagers that did not even have the ability to read the language of the work much less than understand any concept explained in the work and in that the effort was lost completely. In view of my past experience with Universities and academics in physics the past twelve years while trying to find a channel or a medium whereby I can introduce my work with some degree of success forthcoming from any positive reaction I am not holding my breath. However, for my personal interest I am this time keeping record of documentation sent to established institutions of learning, because I just wish to establish where does my effort always go wrong?

The written book sent to you contains
Book 7 The Absolute Relevancy of Singularity in terms of The Thesis (reduced)
Book 1 The Absolute Relevancy of Singularity in terms of Applying Cosmic Physics,
Book 2 The Absolute Relevancy of Singularity in terms of the Sound Barrier
Book 3 The Absolute Relevancy of Singularity in terms of the Four Cosmic Phenomena and
Book 4 The Absolute Relevancy of Singularity in terms of The Cosmic Code

On CD and in PDF I sent **The Absolute Relevancy of Singularity The Dissertation**
Book 0 The Absolute Relevancy of Singularity in terms of Newton
Book 1 The Absolute Relevancy of Singularity in terms of Cosmic Physics
Book 2 The Absolute Relevancy of Singularity in terms of The Sound Barrier
Book 3 The Absolute Relevancy of Singularity in terms of The Four Cosmic Phenomena
Book 4 The Absolute Relevancy of Singularity in terms of The Cosmic Code
Book 5 The Absolute Relevancy of Singularity in terms of Life
Book 6 The Absolute Relevancy of Singularity in terms of Investigating Kepler
Book 7 The Absolute Relevancy of Singularity in terms of The Thesis
Read my work and go on a journey of discovery you never thought was possible.
I am sending a copy to each of the following institutions: The University of Pretoria, The University of South Africa, The University of the Witwatersrand, The University of Cape Town.
Best wishes

P.S.J. Schutte

A More Complex Commercial Science Book

ISBN-13: 978-1539133803

ISBN-10: 153913380X

Written by P.S.J. (Peet) Schutte

© KOSMOLOGIESE EN ASTRONOMIESE TEGNIKA

All rights are reserved.
No part, parts or the entirety of this book may be reproduced by publishing, electronically copied, duplicated by whatever means that form reproduction or duplication, without the prior written consent of the copy rite owner.

This is the book showing everyone that there is

A Conspiracy in Science in Progress

ISBN 978-1-920430-05-4 Written by P.S.J. (Peet) Schutte

© KOSMOLOGIESE EN ASTRONOMIESE TEGNIKA

mailto: orders@singularityrelevancy.com

All rights are reserved. Publishing, or alternatively reproducing electronically or using any other means of duplicating without the consent of the copyright owner is strictly prohibited. No parts of this book may be reproduced in any way that is possible.

If you aren't into science this might just be the reason why you chose not to study science. It is The Ultimate Conspiracy Theory.

I do find much pride in my status as being Afrikaner and would like to have my names used by pronouncing it in the manner Afrikaans dictates…therefore I would sincerely appreciate the courtesy when readers will take note that my name and last name are pronounced in Afrikaans, which is originally from Dutch and must be pronounced that way. Peet one would pronounce "here" which is the closest English to the pronouncing of the "ee". The "Sch" in Schutte is pronounced exactly as school is where both actually are pronounced Skutte or "skool". By pronouncing my name in Afrikaans you do me the utmost courtesy any one can.

Being an Afrikaner is what I am most proud of. I submit article to well known physics magazines but my articles are rejected on the most unappeasable grounds and for the most outrageously ridiculous reasons the Newtonians can think of. I explain how gravity forms but I am rejected because they are of the opinion that my work does not meet. One such an article I may use because I said I was going to use the material as an open letter I gladly show.

This book was done with a $25 ⁰⁰ scanner and a $35 ⁰⁰ printer and the reason I explain inside. For the same reason this book was not edited or linguistically checked. I could not because that does not work because I am in the writing business and not the spelling business and while I check spelling the writing gets more and so does the spelling and grammar errors. I had a choice; doing the books with no funding or not doing it at all because while I rubbish Newtonian science and show it is the fake it is, they will never publish my work because I trash Newton. Not having funds and trying to fight science for the truth with the truth was a fight that physically broke my health and still I am not published except in this manner. I apologise for the spelling and language but in poverty that was the best I could do under the prevailing circumstances in which I find myself…. This book is a first in every sense… it unites science and religion because science and religion was separated by human stupidity

Please take note that I sell information and not words or books and therefore the information takes priority and not the spelling or words used to inform the readers. This represents the work of God and not the word of God and so there is no interpretations applying and versus you can learn and sound intellectual but only cold facts you will have to understand.

It is the Conspiracy Concerning Physics

THE WELL INTENDED WARNING

I have to congratulate you; you are one of the very first persons ever to find out how the cosmos layout works using the true way. It is not the hoax you all were told at school about mass forming a structure that holds the solar system in place and the idea that mass is the way planets line up. No, you are about to learn the true way, the way in which the Titius Bode law forms and this no Physics Academic knows. Now you will know what they never knew before and you know what they should've known even before you know but don't because before they can know they first have to admit I am correct about mass not forming the layout of the solar system. Admitting that I am correct makes every book or theses they've ever written on cosmology a fake and science fantasy and therefore they refuse to read what I reveal. I reveal how the Universe is truly built by proving how the laws on cosmic nature in reality apply. It does not stroke with what the Newtonian concepts are in which Mainstream Science believes. No Academic would trash his or her work in favour of admitting my work is correct and so my work is ignored by Mainstream physics as if it carries the plague. The Universe and the solar system apply the Titius Bode law to form space and not one Academic in physics ever attempted to admit this reality on cosmology. Now you are about to find out what the reality is about how the solar system forms and therefore I have to congratulate you.

Reading this book will **intellectually** be **very challenging** to any person since what I say was never yet published. I disagree in principle with science's accepted principles on even the very basic issues and that fact is undeniably true. I also propose new principles and what you read is very new to everyone alike. Yet, the Super-Educated-Masters have preset conditions they prescribe to information and they can't break their mould. As my views are new the Super-Educated-Masters only use information stored by culture and if they can't bring the information to mind by recognising they fail to understand new science concepts. So although you will see that I am absolutely correct yet they fail to understand my work. You are going to read evidence of this in a letter that was sent to me proving just that. You will learn that I call physicists that studied for so many years just to have learned nothing in the end Super-Educated-Masters because they think of their positions as Super-Educated-Masters but don't realise the most fundamental foundation of physics is completely wrong and yet they build fantasy castles on presumptions that does not exist. Where the Super-Educated-Masters fail to see my arguments and the obvious correctness thereof they also fail to see that I found that the ordinary persons with a scholastic physics background cope with the difficult explaining much better than does Super-Educated-Masters. The ordinary person do not have a culture to defend and a work ethic that might be compromised whereas those in Academic office have a lot of Academic pride and years of study material they will lose. Therefore, the purpose with which I wrote this book is to get around the network by which Super-Educated-Masters strangle any form of science that does not fit their views or match their liking. If what anyone says does not stroke with what the Physicist says control physics and agree with "Mainstream Science" or echo their thinking, they just smother all intellectual publication on the grounds that it is not fitting their profile on science. I disagree most strongly but I do also supply proof thereof. Still Mainstream Science blocks the publishing of my views on science that does not compliment their views. If you believe science is more accurate than God don't read further and live out your fantasy. If you want to know the truth about **how students** and the public **are brainwashed** by **mind control** in **science** this will wake you from your slumber. Read this and wake from the culture you believe in; that which science has lulled you into and made you accept science as the absolute undeniable rock fast truth. They instated this culture concept of science only working with the truth as a religiosity. Should you believe that then stop reading or get your tranquillising anti depressants next to you with a large bowl of water. "Mainstream Science" hides behind maths. You will find some mathematical equations, if you are not familiar with it ignore it because it only shows the silliness of "Mainstream Science". If you don't read the mathematical equations because you find it tiresome you will still understand the linguistic explaining when reading the language whereby I explain it. I need help to fight their fraud. I need you to help me fight them.

For more about the conspiracy also visit

www.questionablescience.net or
http://www.singularityrelavancy.com/
and www.sirnewtonsfraud.com

In this book I prove that what you think of as Science, is only ideology based on rumoured disinformation made up by those making up facts as they hide what they don't know, hide what they don't want to know and conceal what they don't want others to know. By proclaiming they know everything while pretending to know all there is to know they preach what they falsify to sustain their religiosity in the form of unproven dogma. Science tells what they wish and reserve what they wish not to tell and hide what they don't wish others to know that they don't know. My biggest undertaking is to prove that what science presents as credible facts is unfounded and incorrect. Science is believable because culture sustains Science. My problem us busting a culture people trust as Science and be a whistleblower about accepted religiosity people unconditionally believe in as being trustworthy. The public never once thought to question what they were taught as science. This forms a problem since what is presented as science is falsified information and that which we need instead is a new approach that is truthful about science. Therefore no one realises what big a problem there is in the level of correctness there is in the approach to reality science now presents. Science never shows the support or proof in what they present as ultimately credible facts. Bluntly put; Newton was never proven to be correct but only accepted. I firstly need to promote the need we have for the truth to replace what we have as science before I can promote what the truth is that we need to form science. Science hides their flaws under a profile that they present as the uncompromising truth while in fact what they present is lies covered by a veneer of truthful facts and this then neglects the urgent need for the truth by denying there is a problem. If you don't believe this statement I make on the first page that you read then I challenge you to read on and challenge me on any fact I present about the untruthfulness you think of as science and if you disagree with me send me a e-mail, but if you believe me then help me to challenge the establishment about their scandalous disinformation that they hide to commit the conspiracy they call science. Everyone is unbelievably naïve while Everyone is getting hoaxed and your children are brainwashed to unconditionally believe in science that is totally without credit. Read what it is that science hides to put in place this notion that they know everything and what they know are without blemish.

Many years ago, while I was teaching, I couldn't live with my conscience and get paid to betray my scholars by lying to then about what accuracy the science I was teaching them had. I had to quite teaching. It was better for my sole to live on the brink of starvation as I have been doing and experience complete poverty than to lie to the children that look at me with those trusting eyes. To them I am completely trustworthy, even more that they believe in the knowledge their parents represents because I am the teacher and the teacher they believe. …And then I look into their eyes and tell the about science I know is totally fictitious. How can I take money and in the end meet my Father in Heaven and tell my

Father in Heaven I sold not only my sole, but also the soles of children I had to teach for money. I had to live an entire lifetime with lies and betrayal not to one or two persons but to scholars year after year.

I had to teach science and I did not believe a word about what I was teaching the scholars. Then I had to force my innocent students to believe what I don't believe How morally uplifting is that to sit with questions that need answers while giving information that you see is completely false. Ask any mathematics teacher to explain the Newton formula on gravity $F = G \dfrac{M_1 M_2}{r^2}$ and the mathematics teacher will tell you there is no chance in hell that this formula can apply as Newton said it does. According to this formula there are trillion times more gravity between my feet that there are between the earth and the sun because the smaller the radius gets, the more it increases the influence the mass has.

With all those science professors all having many doctoral degrees in science and mathematics across the world on every continent there is they never saw the error of this formula let alone all the other bullshit they were propagating...and still they believe there is no Science Conspiracy Network in place to fool the public about the correctness and the honesty in science? You wish to tell me not one could question Newton on the merits of this formula...and then they all act as if there is no conspiracy going on. Well I got to a point that I could not be part of such a conspiracy any longer and I chose rather to live in poverty than to drape my sole with hogwash. Never is there one teacher that makes an effort to explain the formula. I will never accept in all this time that not one sat back and saw this was a load of rubbish.

It is not what science declares that is important but it is always what scientists don't declare that holds prominence and more so the reason why science keeps a silence about the information they do not disclose. It is never about what they say but it is why they don't say other things they keep quiet about. You will read how they never disclose the entire truth because science is about promoting one-sided and selectively opinionated information forming fraud no less. I have been per suiting a new cosmic theory that I partly present in a six part theses, of which the investigating research began in 1977. In 1999 I compiled my theory and searched for a publisher.

First I located what was wrong in physics then formed a correct approach. I compiled my presentation of it in a theses that I call The Absolute Relevancy of Singularity and then six separate thesis parts forming the theses published through LULU.com which I saw as the only manner whereby I could generate funding by which I would be able to have the twenty seven books I already wrote linguistically edited and then to have the books published on a Print-On-Demand basis.

I compiled a new cosmic theory by which I eliminated all the incorrectness that Newton has burdened science with but with this being my opinion I did not find a garage full of academics supporters waiting to applaud me and to uphold my views on the matter. Yet still I was not going to be ambushed by their relentless stonewalling my efforts and blocking my efforts in introducing both the incorrectness and the new cosmic theorem I concluded. Their mannerism in blocking and frustrating my opinion when showing the mistakes in science convinced me about a Conspiracy in Science in Progress and this spurred me on to tell the entire world about their brainwashing students minds. By the manner they selectively withhold information when teaching science, amounts to deliberate brainwashing of students in physics by "normal" education practises.

Trying to convey my message kept me busy for the past going on to twelve years on full time basis whereby I was trying to introduce my findings to many academics without finding much joy from my efforts. This past eleven years plus saw me go without any income as I tried to get my theorem recognised as well as get my warning noted.

Going without a steady income left me almost destitute and in order to find a manner to get my theory across to the attention of influential readers, I decided to publish a theses of six books electronically as to try and get around the stranglehold of Newtonian bias controlling science at present worldwide. I decided to publish electronically which those in power do not control. However to get people to believe me is to change science that everyone believes as culture.

With my first language not English and the books not linguistically checked by an expert there are bound to be language errors that readers will notice. In the past I tried to check my work myself but after checking say one hundred and fifty pages for language corrections, then after days of toiling instead of having corrected work I ended having four hundred pages of newly written information which is still not linguistically corrected but holds a lot more information. The language and spelling errors compiled instead of reduced. This is because my priorities lie elsewhere. I aim to spend money on correcting the work as far as language goes, as I receive money in the selling of my theses and in the hope that I will receive money. I will have all my work including the one you are reading edited professionally and corrected as I find money to do so...But first I have to get the public aware of the problem to get the academics to appreciate the problem. In everyone's mind science is more perfect than religion is.

In the event of any readers who may have questions concerning more facts as it is presented in this book, please feel free to contact me, PEET SCHUTTE. All information divulged came about through independent self-study during the past thirty-two years or so. I have to warn the readers that the topics are showing a very new approach with no quick answers abstaining from proof or holding just a few lines and the information is new in nature but not hard to grasp.

Should anyone desire to contact me about raising an opinion, or sharing your opinion about my views on science then feel free to do so by using mailto:info@questionablescience.net or the e-mail address mailto:info@singularityrelavancy.com but please do so when coming to the end of this book at least and then see if you understand the entire concept that I introduced. In most cases all your questions are answered further down the line and I am not fond of repeating what is already said in the book. The book forms a line in explaining and the concepts are best understood if the reader follows the designated line. However do share your views with me or with a Physics professor, I'll be glad to hear from you and he/she will be less pleased to read my work.

How the Solar System Forms: An Academic Presentation by Peet (P.S.J.) Schutte
ISBN-13: 978-1523217021 (CreateSpace-Assigned)
ISBN-10: 1523217022

A Cosmic Birth as an Academic Presentation Book 1 by Peet (P.S.J.) Schutte
ISBN-13: 978-1517066970 (CreateSpace-Assigned)
ISBN-10: 1517066972

A Cosmic Birth...as a Special Presentation Book 2 by Peet (P.S.J.) Schutte
ISBN-13: 978-1517525460 (CreateSpace-Assigned)
ISBN-10: 1517525462

An Academic Introducing to The Titius Bode Law Book 1 by (P.S.J.) Peet Schutte
ISBN-13: 978-1507845851 (CreateSpace-Assigned)
ISBN-10: 1507845855

An Academic Introducing to The Titius Bode Law Book 2 by Peet (P.S.J.) Schutte
ISBN-13: 978-1507853788 (CreateSpace-Assigned)
ISBN-10: 1507853785

An Academic Introducing to The Titius Bode Law Book 3 by Peet (P.S.J.) Schutte
ISBN-13: 978-1505874884 (CreateSpace-Assigned)
ISBN-10: 1505874882

How the Solar System Forms: a Pre- Script by Peet (P.S.J.) Schutte
ISBN-13: 978-1503023895 (CreateSpace-Assigned)
ISBN-10: 1503023893

Relevant applying literature Go to Google Amazon.com: Peet Schutte: Books
http://www.amazon.com/s?ie=UTF8&page=1&rh=n%3A283155%2Cp_27%3APeet%20Schutte.
Oxford dictionary of Astronomy web site naturescosmicconcept

The Following books are all available from CreateSpace web site.
The Absolute Relevance of Singularity The Journal
The Absolute Relevance of Singularity The Unpublished Article
The Absolute Relevance of Singularity The Dissertation
The Absolute Relevance of Singularity in terms of Newton Book 0
The Absolute Relevance of Singularity in terms of Cosmic Physics Book 1
The Absolute Relevance of Singularity in terms of The Sound Barrier Book 2
The Absolute Relevance of Singularity in terms of The Four Cosmic Phenomena Book 3
The Absolute Relevance of Singularity in terms of The Cosmic Code Book 4
The More Colute Relevancal of Singularity in terms of Life Book 5
The Absolute Relevance of Singularity in terms of Investigating Kepler Book 6
The Absolute Relevance of Singularity in terms of The Thesis Book 7
The Absolute Relevance of Singularity in terms of The Cosmic Creation Book 8

peet@naturescosmicconcept.co.za   mail.naturescosmicconcept.co.za

What you read I prove even in this book and I dare any one to prove otherwise or reprimand me. However when you do reprimand me don't come up with silly nonessential remarks such as Larz Law commented in example (commenting on another book I published that has much less information available)<u>: The only comment I received was from a person going by the name of Larz Law if that is his name. Larz Law DISLIKES my book and he says: This book is in concise and displays no knowledge about what is being argued. The Books first page is grossly wrong and trying to say gravity is wrong is the worst way to start a book like this. I am as of yet undecided as to my view of the world and what forces are at play (divine or other wise).</u> This comment only says that Larz Law of no specific e-mail address that Larz Law (if that is in fact his name) has been brainwashed beyond recognising any consent that would not stroke with his what his teaching about physics would allow him to believe or to recognise. If that is your mentality stop reading because this book is far too complicated for your mental development. This says nothing but that his mind had been altered and his mind is controlled beyond having a free will in thought or creating a valid opinion about what he reads. I disprove most of the views science has on cosmology, which is the science on stars and galactica, and how the Universe formed. I start by introducing a new view on how gravity applies and from there dismiss most of popular claims on science. Any person that challenges me on the legitimacy of content given within this book, please do so any time. However, do so on the grounds of all the information provided in this book. What I say is don't be opinionated and then reject on what is culture but inform yourself with what I provide and then gauge what I say in relation to the half truths mainstream physics provide.

The content of this book **Science: A Conspiracy To Misrepresent The Truth** AIMS TO SHOW ORDINARY GOD FEARING MEN AND WOMEN, THE TRUTH ABOUT SCIENCE BUT IT IS NOT A RELIGIOUS BOOK AT ALL. Moreover it shows how students are mentally abused and brainwashed to believe physics. The book is not linguistically edited and there might be a lot of linguistic errors within the pages. Please look at the meaning of the content and not the accuracy of language used because English is my second language with Afrikaans being my first. The content however is more accurate than any printing you can purchase that is underwritten by mainstream science and that I guarantee. You are going to read some **mathematics in equations** and expressions in **mathematical formulas** placed to defend my position but if you **don't like it** then just **skip** the mathematics because the content and grounds the mathematics proof or disprove is not important and it is there to disprove the Brainy Bunch. It is there not to scare readers away but to silence the Brainy Bunch critics by showing them the foolishness of their arguments. By using mathematics the Brainy Bunch have been cheating the public and have been brainwashing students for centuries. That cheating is how they do it and I have to show and uncover the dishonesty in mathematics.

Take these questions very seriously because I do not ask this lightly! Are you ready for the truth? Can you stomach the trut after studied for years to find out your studies are a farce? Can you face the truth without firstly knowing what you will meet? Before you say yes that quickly ask yourself if you have the perception or the insight or moreover the understanding to relate to what the truth might hold. It is quick to agree to the request but I have yet to find one professor in physics that will respond to the truth! Should you face up to the truth you are stronger than any one of almost two thousand physicists in academic posts I have contacted over twelve years. Moreover what will you do when you know the truth or are you satisfied to be told what to believe is the truth without ever investigating by personal standards as to finding the truth about science. Then go on reading and confront the truth, as you never had ever before.

<u>I am Petrus Stephanus Jacobus Schutte going by the nickname of Peet and I am a married male, with a sane mind and I hold very sober habits being a lifelong teetotaller, therefore my mind is clear. When you read what you are about to read it is not the hallucinations of a drunkards mind.</u>

I see myself as very responsible and am not well liked because of my straightforward personality where I say what needs to be said to maintain honesty above friendship and what I say is always not infusive or congenial to make friends. I wrap no feelings in cotton wool for the sake of peace. I mention these facts to establish beforehand that I am not a danger to society. Although I am the only one in the entire world that disagrees with modern mainstream science I am not criminally insane. I would not attack your dog before your dog will bark at me just because I hate your dog. I would rather attack you and not your dog after he barks at me just because I don't like you but I do like dogs. I say this to convince you that I have never been jailed in all my life on grounds that I attempted to mislead or fool or tried to mislead anyone, as you can see. I try hard to be honest

The proof of my honesty is that I am one of the poorest people in society. I write books that don't sell because I try to convince people about a mistake no one on earth is aware of and therefore no body bloody cares except me. I am the only one that can see a mistake and seeing it in the full consequence thereof by what it holds to the entire human race it is frustrating me senseless. Take it from me that not selling books is not a very profitable enterprise and is frustrating at any level.

If I was less honest and went about bullshitting everyone about the accuracy that Newtonian science portrays in the modern era it would at least sell some books because some dinosaur Newtonian physicist would be pleased about it but then I was a cheat although I would have been richer but being as poor as I am I in favour of being as honest in what I am, I can't be a cheat.

I studied physics and in my first year of studies I came upon a mistake concerning physics. My roommate told me it was Newton that said what I disputed and I remember telling him I don't care who this Newton fellow is but he is wrong and ever since then I stuck to what I said. That discovery made me disagree with the establishment forming principles in physics. Later in time I have detected much more than a mistake. I have discovered a Pandora's box of mind manipulation and brainwashing going on to force students to accept science notwithstanding.

If you feel I am exaggerating or that I am a mentally impaired asylum escapee that found a writing pad, ink and paper and that I now start to scrabble senseless suggestions to while away my social frustrations then I challenge you to prove that students are **NOT** being **brainwashed** to **believe** what they are taught in **science**. I started off being likable and being nice but the deeper I delved into the conspiracy the less I could care about who thought what about what I said because I am going for the truth as hard as I can. If you are a student you better read on…and if you are a teacher I advise you to read and get wise in realising what you do. I am going to give facts that will support me in the claim that science is all about falsifying facts to prove what is crooked in order to make those that should know science look good while in fact those in physics and teaching physics know less about physics than does a cat know about opera.

This mistake is about the cosmic phenomena called gravity. Detecting the mistake is simple because it is uncomplicated to understand. Academics in Science say that a feather will fall with the same speed as what a large hammer would fall. It not so much what they say but it is about what they hide about physics. If you are a student in physics then this information you are about to read is most important. In the classes you attend in physics has any one confirmed a location where one might find the centre of the Universe? Have you been told how mass causes gravity to pull by force? Have you as a student in terms of the fact that you are being a student been informed how mass confirms gravity? What evokes the force that establishes the pulling that confirms the mass that produces the gravity? If no one went to the trouble to tell you why, then isn't it about time that someone exerts himself to do the honours? On the other hand have you been asking what evokes the force that establishes the pulling that confirms the mass that produces the gravity? Why have you not gone to the trouble and just ask this simple question, its science.

It would be most interesting to hear the answers those lecturers will come up with since these questions I now put to you and as basic as these questions are in science, these questions have not found answers, up to now that is. I wrote a six part theses in which I found a means to define gravity. I did accomplish this for the first time ever since the time Newton introduced gravity. This is more than what Newton achieved and it is more than what the whole lot of Newtonians achieved in three hundred and fifty years. I could do that by accomplishing one thing all others thought not to be possible! Before I achieved finding what gravity is, I first had to find the centre of the Universe because it is there that anyone can locate gravity. I can now show how gravity forms because I detected the centre of the Universe to where gravity moves. Indecently it is also the very first spot where the Universe stated at the very first moment when it started.

Is there any Newtonian applauding my effort and congratulating me on my achievement? If there is one such a Newtonian that Newtonian still awaits birth. I couldn't find one Newtonian even being prepared to read what I have to say about what they have nothing to say about. I could therefore not locate one publisher that was prepared to publish my work because before publishing they first have to read my work and no one was prepared to even glance at my work let alone read it intensively with publishing in mind. But I need to get the information out to everyone to get anyone to read my work. In achieving that I had to resort to private publishing because from the nature of my work I take Mainstream science head on and am confrontational on most aspects of astronomy including astrophysics and the founding principle guarding the authenticity of physics. If I say physics is a hoax made up of corrupted facts you then will stop reading and therefore would you blame any publisher that is not willing to come near my work?

To have a publisher backing me in order to publish my book the publisher had to find an academic prepared to back up my statements that Newton is a criminal that committed extensive scientific and mathematical fraud! In that sense there does not seem to be any publisher that wants to go head bashing with the Physics Custodian establishment of science on official science principles, which I have to do to convey my message in no uncertain language.

I argue that if it is the correct practise to have gravity to move to a centre then it should be such a centre one would have to locate. If gravity is what moves space towards the centre of the earth as the law of gravitational principle demands gravity must flow to the centre of the Universe. In that sense I followed where gravity leads and I located singularity, the point where gravity pulls the Universe "flat". Einstein was the one that said or proved that gravity pulls the Universe flat. If you don't believe me I located the place then explain what he says in this statement.

By merely putting gravity in the Universe by telling everyone that gravity is acting as a mysterious FORCE that is pulling towards a common point in an allocated general centre is rather avoiding the question with simplicity because the question about how and why remains unanswered. Not knowing the answer to where the centre of the Universe is, will leave any Academic physicist empty and not knowing is the same as suicide on a mental level. That is why you are primarily a student. We all are always students. Being a student is being in search of information and knowing you might never achieve the prime information in physics must be devastating to eager minds. Ask yourself the following: If gravity pulls towards a centre and gravity holds the Universe attached the question arising from that simplistic answer is then ... where is the centre of the Universe? Newton never found the answer. Newtonians took all of three hundred and fifty years not to find the answer. Do you wish to spend a lifetime searching and never find the answer? Then become the next generation of Newtonian Masters. However if you discard the falsifying of facts that I charge Newton with, physics will present you with an answer as we follow Kepler's lead. Should you decide to download and read this book, it will bring along a new perception about Kepler. Science sees to it that Kepler stays the least appreciated Cosmologist where as in truth Kepler proved gravity, proved singularity, proved space-time, proved the Big Bang, proved every dynamic most of the wise persons afterwards thought about. Yet no one gave Kepler any recognition up to now because science denies Kepler his limelight. All they can see is the way Newton raped Kepler by falsifying everything Kepler introduced. All I had to do was correct Newton's changes to Kepler to find the answers.

Many years ago, while I was teaching, I couldn't live with my conscience and get paid to betray my scholars by lying to them about what accuracy the science I was teaching them had. I had to quit teaching. It was better for my soul to live on the brink of starvation as I have been doing and experience complete poverty than to lie to the children that look at me with those trusting eyes. To them I am completely trustworthy, even more that they believe in the knowledge their parents represents because I am the teacher and the teacher they believe. ...And then I look into their eyes and tell them about science I know is totally fictitious. How can I take money and in the end meet my Father in Heaven and tell my Father in Heaven I sold not only my sole, but also the souls of children I had to teach for money. I had to live an entire lifetime with lies and betrayal not to one or two persons but to scholars year after year.

I had to teach science and I did not believe a word about what I was teaching the scholars. Then I had to force my innocent students to believe what I don't believe. How morally uplifting is that to sit with questions that need answers while giving information that you see is completely false. Ask any mathematics teacher to explain the Newton formula on gravity $F = G \frac{M_1 M_2}{r^2}$ and the mathematics teacher will tell you there is no chance in hell that this formula can apply as Newton said it does. According to this formula there are trillion times more gravity between my feet and the earth that there are between the earth and the sun because the smaller the radius gets, the more it increases the influence the mass has.

With all those science professors all having many doctoral degrees in science and mathematics across the world on every continent there is they never saw the error of this formula let alone all the other bullshit they were propagating...and still they believe there is no Science Conspiracy Network in place to fool the public about the correctness and the honesty in science? You wish to tell me not one could question Newton on the merits of this formula...and then they all act as if there is no conspiracy going on. Well I got to a point that I could not be part of such a conspiracy any longer and I chose rather to live in poverty than to drape my soul with hogwash. Never is there one teacher that makes an effort to explain the formula. I will never accept in all this time that not one sat back and saw this was a load of rubbish.

It is not what science declares that is important but it is always what scientists don't declare that holds prominence and more so the reason why science keeps a silence about the information they do not disclose. It is never about what they say but it is why they don't say other things they keep quiet about. You will read how they never disclose the entire truth because science is about promoting one-sided and selectively opinionated information forming fraud no less. I have been pursuing a new cosmic theory that I partly present in a six part theses, of which the investigating research began in 1977. In 1999 I compiled my theory and searched for a publisher.

First I located what was wrong in physics then formed a correct approach. I compiled my presentation of it in a thesis that I call The Absolute Relevancy of Singularity and then six separate theses parts forming the thesis published through LULU.com which I saw as the only manner whereby I could generate funding by which I would be able to have the twenty seven books I already wrote linguistically edited and then to have the books published on a Print-On-Demand basis.

I compiled a new cosmic theory by which I eliminated all the incorrectness that Newton has burdened science with but with this being my opinion I did not find a garage full of academics supporters waiting to applaud me and to uphold my views on the matter. Yet still I was not going to be ambushed by their relentless stonewalling my efforts and blocking my efforts in introducing both the incorrectness and the new cosmic theorem I concluded. Their mannerism in blocking and frustrating my opinion when showing the mistakes in science convinced me about a Conspiracy in Science in Progress and this spurred me on to tell the entire world about their brainwashing students' minds. By the manner they selectively withhold information when teaching science, amounts to deliberate brainwashing of students in physics by "normal" education practises.

Trying to convey my message kept me busy for the past going on to twelve years on full time basis whereby I was trying to introduce my findings to many academics without finding much joy from my efforts. This past eleven years plus saw me go without any income as I tried to get my theorem recognised as well as get my warning noted.

Going without a steady income left me almost destitute and in order to find a manner to get my theory across to the attention of influential readers, I decided to publish a theses of six books electronically as to try and get around the stranglehold of Newtonian bias controlling science at present worldwide. I decided to publish electronically which those in power do not control. However to get people to believe me is to change science that everyone believes as culture.

In the event of any readers who may have questions concerning more facts as it is presented in this book, please feel free to contact me, PEET SCHUTTE. All information divulged came about through independent self-study during the past thirty-two years or so. I have to warn the readers that the topics are showing a very new approach with no quick answers abstaining from proof or holding just a few lines and the information is new in nature but not hard to grasp.

Should anyone desire to contact me about raising an opinion, or sharing your opinion about my views on science then feel free to do so by using mailto: info@questionablescience.net or the e-mail address mailto: info@singularityrelavancy.com but please do so when coming to the end of this book at least and then see if you understand the entire concept that I introduced. In most cases all your questions are answered further down the line and I am not fond of repeating what is already said in the book. The book forms a line in explaining and the concepts are best understood if the reader follows the designated line. However do share your views with me or with a Physics professor, I'll be glad to hear from you and he/she will be less pleased to read my work.

Part 1
THE START TO

Science is A Conspiracy 2 Misrepresent The Truth

This is the book showing everyone that

Science A Conspiracy 2 Misrepresent The Truth

ISBN 978-1-920430-56-6

Written by P.S.J. (Peet) Schutte

© KOSMOLOGIESE EN ASTRONOMIESE TEGNIKA

mailto: info@singularityrelevancy.com or   mailto: info@questionablescience.net.

All rights are reserved. Publishing, or alternatively reproducing electronically or using any other means of duplicating without the consent of the copyright owner is strictly prohibited. No parts of this book may be reproduced in any way that is possible.

I WISH TO DEFINE THE CATOGORISING I USE AS PART OF THE BOOK.
I have the utmost admiration for Scientists and I shall never dream of placing me in the same category as academics mainly because of their intellect and achievements. They pushed their corrupt conspiracy of a hoax they present as science and which they further by brutal brainwashing through over 300 years of never getting detected and that in itself is an achievement unheard of in human history. I will not be able to bullshit anybody for 300 seconds because I am not built to be dishonest and Newtonians went about for far longer than 300 years fooling even the brightest among us on earth. That achievement is most brilliant and no religion of magical mysteries in the past could ever match the Newtonians. Every time I go against Mainstream Science which is another name for upholding Newtonian blindness I am told I do not seem to have the intellect or mental capacity to *"understand Newton's classical mechanics"* and then because of my limited vision on physics I should know my place and retire to a dark corner where I would then silently and quietly vanish from earthly records. They forever tell me there are two positions on earth: those with the mental capacity **to understand Newton** and then there are those in my sector **that is mindless to the point of not understanding Newton.** In that sense there are two classes, the clever ones that **understand Newton** and then me, the mindless that just cant **understand Newton.**

To substantiate this segregation I use some referring to place distinction between the highly schooled super trained academics that spent most if not all of their lives in preparing to further their minds, filling it with the same void they fill the Universe and calling it "nothing". When I asked where is more nothing: Between Pluto and the sun because Pluto is the furthers from the sun holding the most "nothing" between it and the sun or in the centre of the sun because there is nothing standing between the sun's centre and the centre of the sun, I was discredited as incoherent and irrational. I tip the opposite of the scale as I spent little time repeating the brainwashing they subdue every student with to believe in the norms taught as the official policy in learning and education I have to be on the "other end". I don't believe their crap and tell it as I see it and therefore I am dumber than a pig, or that is their opinion. From where I stand and admire those in science, I can only see intellect as they fooled every person on earth for centuries non-stop: and moreover that achievement is presented as the academic's common denominator. If that is the common denominator used on the one side, fooling everyone by using unsubstantiated rumours and gossip and putting that as the joining factor, then on the other side, which has to be *"my side"* must then be the class of stupidity. To those forming the brilliance in science and their class such a remark would be an insult but to me (and therefore my class) it rings truth and that makes it not an insult but a norm we should except and learn to live with. I would rather be stupid and not **understand Newton** than be **Brainy** and believe I **understand Newton...** how stupid must I be before I would be able to **understand Newton.**

It is rather a pity that while the SUPER CLASS will never say it to our faces; the SUPER CLASS is strongly of such opinion that we on the other side of the Universe have no minds to think in any way, and it is therefore our duty as much as it is our absolute privilege to except what the SUPER–EDUCATED, the ones occupying the informed side of the Universe inform us to what we should accept and the SUPER–EDUCATED live by that idea. As I said I have to live with it too and if I am the ill literate, then the SUPER CLASS must be the SUPER–EDUCATED; where I am the class amounting to stupidity the SUPER CLASS must be the Brainy Bunch. It all comes from the fact that there is such a huge differentiation between us. Those that **understand Newton** is therefore Superior and I, that don't **understand Newton** are of the lesser blessed. To distinctly point to grouping or class or whatever the readers wish to consider the division there are between the SUPER–EDUCATED and me I refer to the SUPER–EDUCATED side of the Universe by the names I use above. Further more when I refer to mistakes that I do prove to be mistakes in the book as we go along I refer to it as Xepted mistakes to clear another distinction of necessity. In short I don't **understand Newton** and therefore I am stupid and they **understand Newton** and therefore they are brilliant and what I present must hold the categories in such class divisions.

If you read on you are going to Read About The Biggest Conspiracy ever enlisted, it's more than just the next conspiracy because it is A Science Conspiracy Network. I prove everything I say about the conspiracy I announce…read and be shocked.

This Network which I denounce Engulfs the Entire Civil Human Race in Every Aspect and willing or not, you reading this are participating without knowing about what you support with every part of the mental intellect you have. It involves the most trusted members in the part of our civil society and we are deceived by the ones EVERYBODY trust the most, the teachers and Professors teaching at academic institutions everywhere. For over a decade I have been knocking on doors with the information I present in this book and lots more evidence, just to be ignored and to be turned away. This behaviour of physicists ignoring me or even attacking me when I point out a very legitimate case of mistakes in science made me realise there is a conspiracy going on. Science has been getting away with this conspiracy since the Dark Ages. What you are about to read is not for the simple minded because science requires much intelligence and in that physicists get away with turning Creation into a joke. By turning the facts that form the conspiracy into what you think of and you accept as natural and as culture, man has been cheated and mislead for three hundred years because those teaching science consider everyone as representing stupidity. To them everyone else are simple-minded peasants who are not worth having independent minds who could be able to think on their very exclusive superior level that they are able to think. That is how they get away with the conspiracy. They pretend to be intellectually superior and to understand what seems senseless to the rest of us and then look down their noses on the rest of the world. I will show you what it is they can't see and then miss. They underestimate the public as much as the public are lazy too think and the misleading carry on. Those in science think their mathematical abilities make them Gods–on–earth while we others are mindless human fodder and for that reason they brainwash our children at school by applying mind control as their mathematical understanding makes them gods. What they say all must believe because if they speak using mathematics then God spoke.

THIS IS THE WEBSITE THAT WHISTLE BLOWS ON A SCIENCE CONSPIRACY NETWORK. I AM GOING TO SHOW YOU THEIR UNBELIEVABLY POOR MATHEMATICAL SKILLS!

This conspiracy is so widely active that every person on earth teaching physics participates, most probably without knowing, some reluctant and others well knowingly to set out to brainwash by thought control the minds of innocent students forcing them to believe in dark aged ritual beliefs in the ritual practicing of forces and allow the conspiracy to be enlisted with all the vigour they could muster. People would argue religious concepts but everybody accepts science on face value. The guilty party that I refer to betray us in every way possible by teaching us lies that never could be true and all the while we believe them with all our hearts and entrust them with our entire future… They take money from parents with the sole aim to brainwash the students who are entrusted to their care and demolish their intellectual understanding …and before you think my claims are exaggerated I say again I prove every word I say. I show what is true and what is wrong in science and as a thank you for all the concern and devoting care I show, the establishment of science frustrates me, ignores me or runs me down, while I challenge everyone and any one to disprove me. Decide after reading this book weather what you read is not a conspiracy…however, familiarise yourself first with what I present and then decide.

Those I refer to are allowed the powers to take as much tax dollars as they wish and endeavour on all sorts of projects that they could fantasize about. Their budgets are limitless and they never face questions because they are the intellectuals that we dare not question. With the same powers we give them they brainwashed your child making science a religiosity. The methods they use on your child are to make your child believe in science and that is just methodical corruption of their thinking. Physics teachers take money from parents with the sole aim to brainwash the students that are entrusted to their care so that students will believe in science unequivocally and regardless. When teachers teach falsified information then they are worse than the mafia, that operates openly by crooked intention and in that the Mafia maintains honesty, which the physics teachers don't have. If you are a parent then mind what you do and beware and read on! If you think I am going one step too far in accusing the academics teaching physics and I slander their names, I challenge you to ask this question again after you have read this entire book and thought about everything and considered every aspect.

Read on and you are about to read About The Biggest Conspiracy ever enlisted, it's more than just the next conspiracy because it is A Science Conspiracy Network. I prove everything I say about the conspiracy I announce…read and be shocked.

I call science a scam and if many readers if not most will completely disagree with me about science living as a scam then the next medical scam I wish to bring to every one's attention will support me. This scam is called humanity as it involves the medical science more than physics. It covers not what you are told or what you are taught but what science remains silent about. It is what is never said that is of serious importance. It is never what the research reveals but what science hides from getting revealed. It lurks in the profits of the pharmaceutical companies or the doctors or the hospitals the industry never whish to reveal. Euthanasia is not about the absolute sanctity of life but it is keeping the patient alive when the patient is going to need the most expensive drugs and the most expensive intensive care. I saw a patient with a knife in his head wheeled out of hospital to find some other place to die because that patient did not have medical care to fund his desperate situation. Because there was no money for treatment there was no treatment and this goes on in every private hospital in every city around the world. If that person had a good medical care and the medical insurer was willing to pay then a team of doctors would fight day and night to keep the person alive and no cost will be spared. It is during the last few months / weeks/ days / hours that most care and medicine is needed. During that time when the end of the life approaches and the fatal disease is going to take its final toll that the pharmaceutical company, the hospital and its staff, and every person sanctioned with keeping this individual alive will fill the bill as fast and as hard as possible to get the last money from the dying sucker. You have a lawyer walk in there holding your last testament that states you are no longer presuming responsibilities for the costs and you have the lawyer tell them if they don't guarantee success with the treatment payment will stop on that minute and the costs for keeping you alive will be on the hospital and its staff that treat you and then you see how quickly they stop machines keeping you breathing. If you start legalising euthanasia you kill the part that serves the highest profits and they would rather see you suffer than they would agree that you might die cheaply. If you are not prepared to pay for their effort of holding your life so ever dearly they will lose all interest and let you die as quickly as the machines are switched off. Your life and all life have value when there is someone prepared to pay the bill and those doctors are the worst criminals there mathematically is. But yet again behind this medical industry there is insurers and behind the insurers there are bankers. Fighting euthanasia is a conspiracy because euthanasia will kill the huge profits the medical industry makes. If we are all so against euthanasia lets take the profits of these pharmaceutical companies, the hospitals and the doctors and spend their money to pay for every one dying that can't afford treatment. Now it is a case that they would rather see people suffer and die in agony when such persons can't pay for the treatment because after all being part of science makes them equal to God and therefore others must suffer so that they can reap the award of the money they invest in promoting and furthering science. THIS IS THE WEBSITE THAT WHISTLE BLOWS ON A SCIENCE CONSPIRACY NETWORK.

I AM GOING TO SHOW YOU THEIR MATHEMATICAL SKILLS! This conspiracy is so widely active that every person on earth participates, most probably without knowing, some reluctant and others well knowingly pursuing the goals of the conspiracy with all the vigour they could muster. People would argue religious concepts but everybody accepts science on face value. The guilty party that I refer to betray us in every way possible by teaching us lies that never could be true and all the while we believe them with all our hearts and entrust them with our entire future… They take money from parents with the sole aim to brainwash the students that are entrusted to their care and demolish their intellectual understanding …and before you think my claims are exaggerated I say again I prove every word I say. I show how the establishment of science frustrate me, ignore me or run me down and I challenge everyone and any one to disprove me. Those I refer too are allowed the powers to take as much tax dollars as they wish and endeavour on all sorts of projects that they could fantasize about. Their budgets are limitless and they never face questions because they are the intellectuals that we dare not question. With the same powers we give them they brainwashed your child making science a religiosity. The methods they use on your child are to make your child believe in science and that is just methodical corruption of their thinking.

They take money from parents with the sole aim to brainwash the students that are entrusted to their care so that students will believe in science unequivocally and regardless. If you are a parent then mind what you do and beware and read on! If you think I am going one step too far in accusing the academics teaching physics and I slander their names, I challenge you to ask this question again after you have read this entire book and thought about everything and considered every aspect.

In this book I prove that what you think of as Science, is only ideology based on rumoured disinformation made up by those making up facts as they hide what they don't know, hide what they don't want to know and hide what they don't want others to know by preaching what they falsify to sustain unproven dogma. Science tells what they wish to reserve what they wish to tell and hide what they don't wish to know.

My biggest undertaking is to prove that what science presents as credible facts is unfounded and incorrect. In other words my problem Is breaking the culture people have used for three centuries to unconditionally believe in science.

The public never once though to question what they were taught to be science.
This forms a problem since what they present as science is falsified information and that which we need instead is a new approach that is truthful about science.

Therefore no one realises what big a problem there is in the level of correctness there is in the approach to reality science now presents. Science never shows the support in proof about what they present as completely credible facts. I firstly need to promote the need we have for the truth to replace what we have as science before I can promote what the truth we need forming science.

Science hides their flaws under a profile that they present as the uncompromising truth while promoting in fact what they present is lies covered by a veneer of truthful facts and neglects the urgent need for the truth by denying there is a problem.

If you don't believe this statement I challenge you to read on and challenge me on any fact I present about the untruthfulness you think of as science and if you disagree with me send me an e-mail, but if you believe me then help me to challenge the establishment about their scandalous disinformation that they hide to commit the conspiracy they call science.

You are unbelievably naïve while you are getting hoaxed and your children are brainwashed to believe in science that is totally without credit.

Read what it is that science hides to put in place this notion that they know everything and what they know are without blemish.

What you read I prove even in this book and I dare any one to prove otherwise or reprimand me.

I disprove most of the views science has on cosmology, which is the science on stars and galactica, and how the Universe formed. I start by introducing a new view on how gravity applies and from there dismiss most of popular claims on science. Any person that challenges me on the legitimacy of content given within this book, please do so any time. However, do so on the grounds of all the information provided in this book. What I say is don't be opinionated and then reject on what is culture but inform yourself with what I provide and then gauge what I say in relation to the half truths mainstream physics provide.

THE CONTENT OF THIS BOOK **A Conspiracy in Science in Progress** AIM TO SHOW ORDINARY GOD FEARING MEN AND WOMEN, THE TRUTH ABOUT SCIENCE BUT IT IS NOT A RELIGIOUS BOOK AT ALL. Moreover it shows how students are mentally abused and brainwashed to believe physics. The book is not linguistically edited and there might be a lot of linguistic errors within the pages. Please look at the meaning of the content and not the accuracy of language used because English is my second language with Afrikaans being my first. The content however is more accurate than any printing you can purchase that is underwritten by mainstream science and that I guarantee.

You will find I never compromise truth for friendship and in that light I say what they needed to hear and not what is wanted to please who ever should feel pleased. Meet the Newtonian physicist. In this book I try to introduce the reader to the brilliant Newtonian conspirator that has been dragging all of intelligent man by the nose for three centuries on a string.
The more the conspirator pretends to be an intellectual physicist the better a fool those conspirators become.
He looks sheepish because he acts sheepish because as he follows he never questions what he believes and brainwash students to do the same. Read how clever the physicists are in hiding their stupidity from

students and the public alike. By enlisting thought control those teaching physics force students to believe in science and to believe science

This book started off as a website to inform about a science conspiracy but grew into a book that serves much more information than what I first intended to supply. It grew into a comprehensive study on cosmology. At times you may observe while reading this book that it seems as if my frustration will ring through like the chiming of the Big Ben Bell. For that there is a reason. At times my frustration and anger will boil over drowning my politeness and that is true, which I admit. For twelve years I have had the answer that would correct the philosophy that has a stranglehold on cosmological science. I discovered the building blocks of nature where my discovery puts all other cosmic aspects of science into science fiction.

Those who force-feed non-existing dogma do so to brainwash students to hide the incompetence of "modern science" so they can rule supreme while ignoring the truth that they deliberately hide by concocting a conspiracy. To keep everyone unguarded they practise a conspiracy by which they perform an accepted practise of thought control on students to further the false dogma presently in place. I try to blow the whistle on such a practise but accepting my resolution makes every thesis any person that is part of "Mainstream Science" ever wrote becomes science fiction.

Therefore no one in science dare to read my work leave alone appreciates the revolutionary nature thereof. Any association with my work must condemn the own work of such an associating person. Whatever is deemed now to be accepted science would by accepting my work then become what is the past tense and belonging to the past in science. My work unmasks the flaws that those in power of concocted science principles kept in place and in thought for centuries on end as forming the untouchable truth. It will then be rejected to show the holes formed as accepted science! They try to silence me but surely somehow somewhere I have to break through with my massage! I bring you a true form of science as never seen before in all of history. I do that by disposing of the conspiracy that hides the incorrectness and the failures that haunts science today. They use mathematical interpretations that take the explaining away from logical facts.

In the event of any readers who may have questions concerning more facts as it is presented in this book, please feel free to contact me, PEET SCHUTTE. All information divulged came about through independent self-study during the past thirty-two years or so. I have to warn the readers that the topics are showing a very new approach with no quick answers abstaining from proof or holding just a few lines and the information is new in nature but not hard to grasp. Should anyone wish to confront me or wish to contact me then do so by E-MAIL AT: e-mail mailto:info@questionablescience.net

There is a conspiracy in science and about this fact there can be no doubt. Once you finished this book and you still disagree about the conspiracy going on it then will be because you are a professional physicist who are paid a huge income to spread the falsified science and contribute to the well being of the conspiracy. However, you will be convinced either way. At the end of this book you will believe in the incorrectness in mainstream science and be convinced that the incorrectness runs as deep as science goes and involves all person that studied science, worked with science, teaches science and thought about science. In short it touches every person that lives. Everyone works with fake science and conspires to promote science fiction passing it off as Newton's science.

I wish to tell the truth about science, but what is the truth about science?
Are you ready for the truth?   Better still…
Those making a living from science, can you stomach the truth?
Can you face the truth without firstly knowing what you will be presented with?
The truth is going to have what you believe Is the truth come tumbling down on you.
So, before you say yes that quickly, ask yourself if you have the perception or the insight or moreover the understanding to relate to what the truth might hold.
It is easy to agree to the request but I have yet to find one professor in physics that will respond to the truth! Should you face up to the truth you are stronger than any one of almost two thousand physicists in academic posts I have contacted over twelve years.

Moreover what will you do when you know the truth or are you satisfied to be told what to believe is the truth without ever investigating by personal standards as to finding the truth about science.
I managed my discoveries because I DISCOVERED A MISTAKE IN SCIENCE.
Then as a result of finding the mistake I DISCOVERED THE TRUTH ABOUT SCIENCE

What you are about to read is the mother of all the conspiracies in science. This that you will read is the root of the conspiracy. You are going to read about how science applies a system of mind control with thought processing. Every conspiracy that Science ever thought up such as the Critical Density Conspiracy or the Dark Energy Conspiracy, or any conspiracy connected to science is in place to protect this theoretical conspiracy. By this conspiracy those teaching science hides the truth from becoming known to anyone outside physics. The conspiracies in place hides the Mother Conspiracy so effective that in three centuries no one outside science got a sniff about what the Mother Conspiracy entails.

The Mother Conspiracy is in place so that students in physics are **brainwashed** by instigating the deliberate sanctioning of **mind control on students** through their practising of **enforcing thought control** that they unleash on students. I prove that the Mother Conspiracy is in place. I reveal the Mother Conspiracy and I show why it is in place. I show how every student including you reading this has been brainwashed to believe science is unquestionable believable. Read this book to get the nastiest surprise of your life. See how much you are brainwashed. Then also read how the cosmos truly applies science and read explanations of what was never explained before and how the explained factors interlink.

In the event of any readers who may have questions concerning more facts as it is presented in this book, please feel free to contact me, PEET SCHUTTE. All information divulged came about through independent self-study during the past thirty-two years or so. I have to warn the readers that the topics are showing a very new approach with no quick answers abstaining from proof or holding just a few lines and the information is new in nature but not hard to grasp. Should anyone wish to confront me or wish to contact me to ask questions about the New Cosmic Principles I put forward then do so by **E-MAIL AT:** e-mail mailto:info@questionablescience.net.

Read how clever the physicists are in hiding their stupidity from students and the public alike. By enlisting thought control those teaching physics force students to believe in science and to believe science as being the only correct mindset there are. This idea of science being the ultimate in correctness is totally ridiculous.

There are those brainwashed to the point they believe physicist are above reprimand and should be treated as if they are God by never criticizing their actions. I wish to provide one such example and you can determine what halfwits and brainwashed non-thinkers fills the faculty of physics. To them everything is correct as long as it is part of physics because although they cannot think, they "can understand physics". That they have no idea about what they understand is apparent to everybody with a brain but still they wish to show the world they "understand physics" just because they believe they are so superior that "they understand Newton"

One such an example I can show is a person going by some name such as Symeof.

Symeof of no known e-mail address dislikes this book. According to him this is why ...He says

Pathetic material. The author has no legitimacy in disputing physics: he is merely an amateur trying to disprove what the greatest minds of our times have taken centuries to prove (by the way, you need to have studied a subject before you can criticize it, which the author didn't). In my opinion, the author is so narrow-minded that when he came to scientists to talk about his material, the scientists must have told him that it was nonsensical. Then, of course, he couldn't imagine he might have been wrong, so he rationalized his failure to be accepted by creating a conspiracy theory: this is probably the reason why this e-book exists. This is no scientific material.

Please study Symeof's criticism. If you are not prepared for the shock awaiting you about the conspiracy in science, you should not continue. It will be as big a shock to you as it apparently is to Symeof. You will as he did realise that so many years of study has gone wasted because what you learned was one big hoax, covered by the mother conspiracy in science and as a result of this awakening you will feel to lash out but don't try to kill the messenger. There is the mother conspiracy, which I expose. If you are unprepared for the shock then reading this book will be more harmful than never knowing the truth. To find all your professional knowledge came to nothing must be unbearable and insufferable. The mother conspiracy is in place but you decide if you wish to know about it or not, the choice are yours to make.

What you are going to read is the mother of all the conspiracies in science, which is about how science applies mind control by processing thought control. Every conspiracy ever linked to science is in place to protect that conspiracy from becoming known. I prove that there is a mother conspiracy in place. The mother conspiracy is in place whereby students are brainwashed through the instigation of mind control through enforcing the acceptance of dogma on students. I also introduce a new cosmic vision with the entirety called the Universe, which is formed by singularly taking on every shape, and space that we know. Singularity is the point where the Universe first started according to Einstein.

Those making such remarks as Symeof does must remember that persons on a higher level of education can immediately gauge the level of your education development when it is so inferior. This person might seem highly educated in his own eyes but it is clear he is on a very low level of understanding physics. This person clearly never heard of the cosmic laws named **1) the Lagrangian system 2) the Roche limit 3) the Titius Bode law 4) the Coanda affect,** which I explain by delivering mathematical proof as to how they fit into the overall picture of gravity and which I mention and explain in much detail.

Reading this remark it is evident that Symeof never came as far as the explaining of the four cosmic laws or such explaining as I give went past this reader without him noticing the explaining as it was too far advanced and much above his level of understanding. That indicates that the person never understood the explaining of the laws and therefore has a very small insight and a low level of understanding physics. It would be much wiser to shut up and get wise than to advertise your uneducated stupidity to the world as you did. When you say "*The author has no legitimacy in disputing physics: he is merely an amateur trying to disprove what the greatest minds of our times have taken centuries to prove*" with that remark you are trying to dispute what is apparently very much above your limited understanding of physics and that which I prove you did not even come to read. It is a pity but then again we can't be all intellectual.

When I refer to brainwashing students into believing this is the exact example I refer to. I could not have asked for a better example even if I ordered it myself. This is the typical learn by repeat and never question those teaching you because questions will uncover the mother conspiracy and that the teachers avoid. The mother conspiracy is in place to teach about what is not present in the cosmos like telling that things such as mass is positioning planets while never mention what forms the true basis of cosmology which is **1) the Lagrangian system 2) the Roche limit 3) the Titius Bode law 4) the Coanda affect** Those who are on a low education level such as Symeof so clearly is, would never have heard of these laws. I explain for the first time in the history of man how these laws apply and how to read gravity from their applications. However he never came to know such information is in place and thus informing him and those such as he is almost impossible. When explaining these laws I have to discredit what Newtonian science uses because what they use has no validation or credit, as I will show very soon.

This is how the interpretation of physics would apply if Newton was correct and mass did position planets. I am, going to show everyone with even a childlike understanding of reality that there is not even a remote chance that the positioning of the planets go in accordance with mass or $4\pi^2 a^3 = P^2 G (M + m)$. This is the formula Newton introduced and it is the formula that all those studying physics still are taught to accept as the valid formula. Do you realise there is much more "gravity produced by mass" in the space your feet has contact with the earth than there could ever be between Jupiter and the sun when using this formula $F = G \frac{M_1 M_2}{r^2}$? But this is part of what Symeof never can understand. It shows that those who study physics never tested what they were told apply in cosmic physics. They never put the correct mass as applying numbers into $F = G \frac{M_1 M_2}{r^2}$ and put in place the radius between the sun and whatever planet and come to a mathematical conclusion. Any person trying to put mathematical values in this equation that is supposed to be in place to show how solar bodies are positioned will realise it has no value.

Symeof, those that you admire and follow without question is just like you a lot of sheep following their leaders' lead without even thinking… and then you criticize me for thinking! I mentioned the four Pillars namely **1) the Lagrangian system 2) the Roche limit 3) the Titius Bode law 4) the Coanda affect** **Have you ever heard of these laws… no you did not because I worked out how they generate gravity while you an d you flock of sheep known as physicist run on e behind the other that is in front while chasing mindless un proven hogwash science dish up as uncompromising facts and truth.**

Have you ever questioned the sheep you follow to explain how mass pulls mass to form gravity. No but in your ignorance you think it wise to criticize me. Read the following page and get wise!

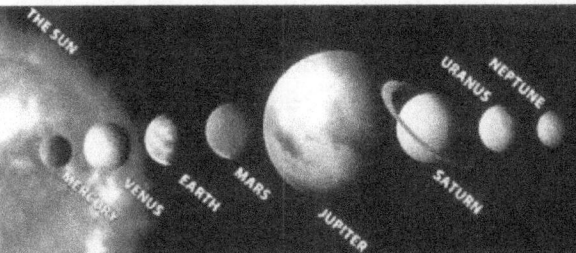

The Titius Bode law is how planets distribute places in the solar system. There is a precise sequence and order in which planets hold places and this annihilates the idea that mass plays any part in this sequence. In contrast the Newtonian system that science promotes currently says that gravity "pulls" according to mass. This is fictional. The mass of the planets is totally random and Newton holds no theoretical basis that nature supports. Look at the arrangement and you can visually see using your eyes that the distribution by mass is a hoax and is an invention Newton concocted and science validates for the past 300 years. Explaining the Titius Bode law for the first time ever breaks the 300-year-old myth and brings truth. Now we know how nature forms the smallest particles to build the cosmos.

The Roche limit is the law that says stars do not collide ever in spite of Newton's ridiculous idea that stars can or do collide. If two stars are in each other's atmosphere the law reads that when the minor of the two stars is closer than 2.4674 of the diameter of the major star, then the major star will liquefy the minor star into a gas plume that it then treats as more atmosphere. This is most significant in the cosmic principles I now put forward. There is not one instance or any evidence that shows where two stars do collide. When two or more stars are evenly matched by gravity the two stars do not collide either but become binary stars that spin around each other.

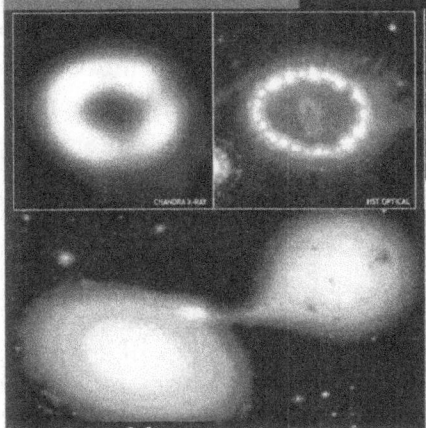

Stars or planets never collide and when a meteor enters the atmosphere like at Tunguska in Russia in 1908, the earth vaporises and liquefies the meteor and the meteor becomes fragments as well as more dust clouds in the atmosphere. This totally annihilates Newton's idea that a radius between structures diminishes by the gravitational attraction of mass. This book shows what happens in truth in nature.

The Coanda effect connects with the previous law. Gravity is a ratio that forms between what is solid space and what is liquid space. This applies when the movement of the solid (star or planet or atom) turns and by turning 7° it reduces the value of the circle of the space surrounding the solid from $\Pi = 21.991/7$ to the compressed value $\Pi = 3.142/1$ which connects to singularity. There is no pulling of material but only reducing of space when gravity contracts gas into liquid such as what the atmosphere is.

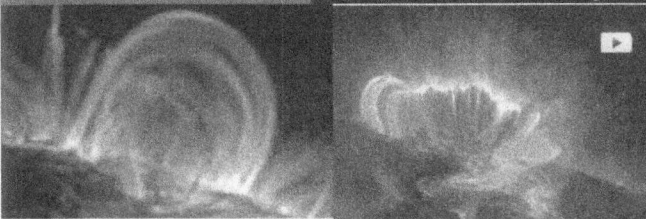

The Lagrangian Points This law proves that science is committed to deception for centuries. It is known that sattelites turn around planets in the same fashion as do planets orbit the sun. Saturn does not "pull" the satelites into its atmosphere or even "pull" the satellites closer and this knowleidge never drove science to question the validity of Newton ot his concept about "mass" pulling "mass". Even the rings form as debris spins in a circle that always remains constant and science should have rejected Newton's ideas on the grounds that nature never not once supports newton and Newtonian science in any way thinkable.

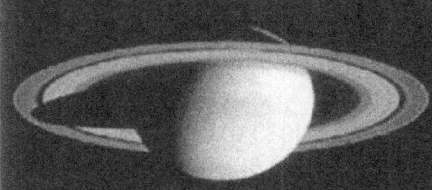

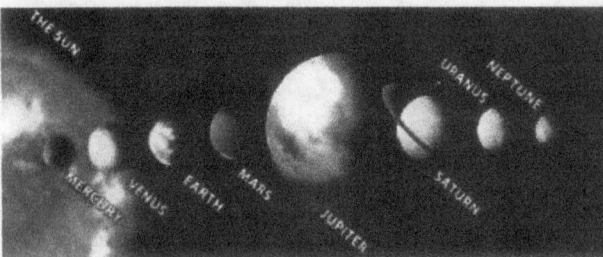

This is the Titius Bode law
The Titius Bode law proves that mass has no place in science. See in the picture how random mass is and with such randomness how can mass place planets in the positions they hold. By my effort to solve the mystery of the Titius Bode law I prove that gravity forms not by mass but gravity forms by Π forming in movement Π^2. Solving the Titius Bode law and proving from that how gravity works opens up a new view on the cosmos.

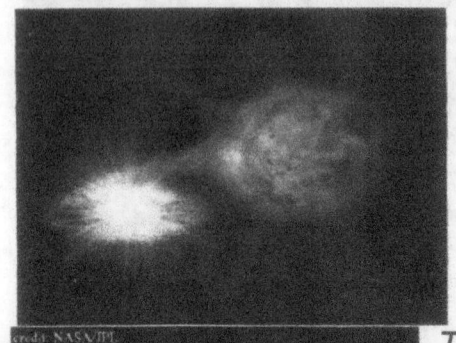

This is the Roche limit.
The Roche limit proves amongst others how the sound barrier applies and works. It also proves that cosmic structures with an atmosphere can never collide because the Roche limit that produces the atmosphere prevent foreign object from moving faster that $\Pi^2 / 4$ within the boundary limitation of that atmosphere. The Roche limit brings further proof that using the truth about gravity in physics the answer is simple; it is that gravity is Π.

This is the Lagrangian points
The Lagrangian points have been known to science for centuries and with all the mathematical splendour available not one calculation could ever explain why this event is taking place. The satellites form precise locations positioned around the major planet and never comes closer while remaining the their positions.

This is the Coanda effect

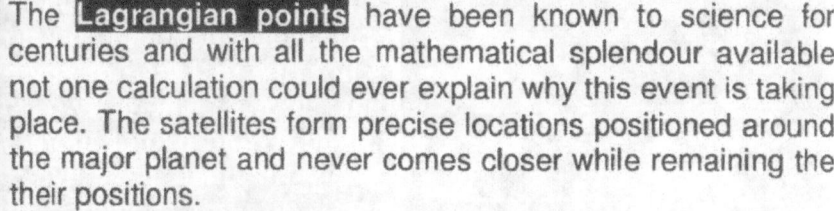

The Coanda effect has powered turbine engines and aeroplanes in flight for almost a century and with all the mathematical splendour available to design the most terrific aircraft, not one engineer could mathematically compute one fact to show understanding why this takes place.
How sad it is that those claiming of much superior intellect in physics remain just no more than having computing power. The understanding is not complex.

I have to warn the readers that the topics are showing a very new approach with no quick answers. Understanding is in the proof and that does not come by reading just a few lines and then forming conclusions. The information is new but not hard to grasp. I did not put these phenomena in place and these phenomena nullifies Newton's correctness and that proof I bring goes beyond any doubt. I prove the Titius Bode law. Go to the Internet and see how science doubt the Titius Bode law and the correctness thereof while to solve the problem you add 3 plus 4 to get 7 that is if you want to find a solution. I have published the Titius Bode law in four already published books but in this one I go deeper than the four already published. In each of the books I present I disclose how the Titius Bode law forms gravity

Everyone is so taken by the accuracy of Einstein's formula that $E=mC^2$ but this is exactly Kepler's formula where $E^3=mC^2$ is taken from Kepler's formula when accurately used as **$a^3=T^2k$.** There is no $a^3 = P^2$.

This is so typical Newtonian in every sense there is in science. The Newtonians gave the Titius Bode law a formula and that explains the lot. To they're under achieving standards that is very satisfactory. Now it is written in mathematics then what more do we need to know? The fact that the distance that Mercury has from the sun is doubled by that which Venus has from the sun is completely ignored. In cosmic reality mass plays no part. Then again the distance that Venus has from the sun is doubled by that which the earth has. This clearly has nothing to do with the size or mass of the planets. Explaining that part is completely ignored. Then again the distance that the earth has from the sun is doubled by that which Venus has and inexplicably this forms the layout of all planets in the solar system. Where do Newton and his idea of mass fit into what truly applies in outer space. Moreover, why does science never mention this? This is my formulated explanation about how the Titius Bode law forms.

If you think my accusations are baseless or the ravings of a madman then go on and download what you have opened and read for yourself. What you download is free and I do not benefit financially from this explanation I present to you. There is no such a thing as mass anywhere in the cosmos. If there were a factor such as mass every planet would orbit distinctly positioned according to mass, but they don't. Should you think of the size of a body containing more or containing less material and put that in terms of mass that forms gravity then the orbital layout of the Universe or solar system would very distinctly NOT be the way it is. The cosmos shows no mass as a factor and we can either regard the cosmos as correct or Newtonian science as correct as Newtonian science diverts totally from the physics that the cosmos displays. The choice to make is do we believe science or do we believe the cosmos you choose?

Show that the square of the mass in relation to the gravitational common factor puts planets in their allocated positions! Put the orbit of Jupiter in relation to the mass of Jupiter and in relation to the position Jupiter holds. Forget getting swept away by the fancy Mathematics; just get to the task of putting the mass of any of the planets in relation or ratio of that particular planet has and then in connection with the position that any of the planets hold. Don't come up with the argument that science works and therefore Newton works. Please then show as he put it: _The author has no legitimacy in disputing physics: he is merely an amateur trying to disprove what the greatest minds of our times have taken centuries to prove (by the way, you need to have studied a subject before you can criticize it, which the author didn't). In my opinion, the author is so narrow-minded that when he came to scientists to talk about his material, the scientists must have told him that it was nonsensical._ That is a lame excuse to hide incompetence. If you are unable to do it your physics is a giant fraud, which is based on a century old lie and a hoax. It then shows your small-mindedness because you have never put physics as it is taught to the test. This is taken from the idea that Newton had when Newton changed Kepler's formula from $a^3 = (T^2k)$ to $a^3 = T^2$ because without any legal mathematical backing Newton said the third dimension is equal to anything holding a second dimension or a cube a^3 is equal to a **square T^2**.

Mathematical reality is that any person in the third dimension a^3 having three sides can climb into a mirror T^2 being absolutely flat and then climb back to the third dimension because $a^3 = T^2$. This is the garbage those slow witted person's such as Symeof failed to see when he was taught that $a^3 = T^2$ because Newton said so.

Reading this mathematically encrypted coded formula of the cosmos given to Kepler and keeping it removed from Newton it reads as being that the space a^3 is equal to = the motion T^2 of the space a^3 in ratio k to a centre k^0, which is relevant to the positioning of k. If we bring in the full equation it will be $k^0 = a^3 \div (T^2k)$ which means half of space is solid $k = a^3 \div T^2$ and half of space is liquid $k^{-1} = T^2 \div a^3$ where liquid is moving. However, it is also true that everything through movement defines a value in relation to one point holding singularity k^0 and that is what the formula $k^0 = a^3 \div (T^2k)$ underwrites. What this proves is that gravity is the motion of space provided by time being the liquid. Please allow me to explain. In the formula $a^3 = T^2 k$ the space forms as the space is in motion. Newton suggested that $\frac{dJ}{dt} = 0$ where he stopped time to have the motion of the circle demolish the work that the circle does. It is because the chambers of physics are filled with brainwashed victims such as Symeof that physics is the joke it is.

There is a conspiracy of conducting fraud by claiming non- existing forces but such claims are utterly fraudulent. I have been trying for twelve years to introduce the true forming of gravity but all Physicists I have encountered prevented me of doing so. They stop me because my work makes Newtonian science look like the farce it is and when removing the notion that a pulling force of gravity works by the value of mass, most of their work becomes science fiction that falls apart in substance. Read this and see how **students in physics** are methodically **brainwashed** to get the students to believe in the absolute accuracy of science. Professors and teachers participate knowingly or unknowingly in this thought manipulation process by means of conducting **mind control.** By applying this **mind-altering process** those teaching physics ensure they subdue students into becoming mind-altered zombies.

It is a process going on for centuries and which without science would have no foot to stand on in the modern environment. By presenting incorrect, falsified or unproven facts and other untruths as proven truth they exert **thought control** and thereby change the student's ability to appreciate what is correct and believable logic and then force students to discard such judgement ability in favour of accepting the institutionalised untested norms and values of science in order to unequivocally believe in science. The accepted teaching methods force students to comply by compromising their better judgment and then systematically to capitulate under teacher pressure by making their own what science prescribes what should be believed. I prove this and you get this for free so what have you got to lose…**but you can get wise to what forms a better understanding about science**! By using the building blocks that forms the Universe I take you back to the instant the Universe started and I show you how the Universe fits like a jigsaw puzzle.

This astronaut has "mass" and the earth also in the picture has "mass" and the pulling should be going on since there is no restraining that would prevent the man from falling to the earth so why is Newton's "mass" not applying? This astronaut is circling the earth at a certain speed notwithstanding "whatever mass" science contributes to the pilot. If "mass" was pulling then why is the man not falling. The man will fall depending on the speed by which the man rotates the earth. His spacecraft (not in the picture) has about a hundred times more mass that should entice about a hundred times more pulling but both float above the earth at a specific pre-calculated speed. It is the speed that determines the distance of circling and not "mass".

$$F = G \frac{M_1 M_2}{r^2}$$

If the astronaut were on earth his mass would be the same as his weight. Therefore on earth there is no distinction between the mass factor on earth and the weight factor on earth. It is clear that things change when this astronaut walked on the moon. If the astronaut were walking on the moon his weight would not be equal to his mass.

The astronaut would weigh less than the mass he has on earth. It is said that the gravity changes the numbers. The moon has less mass and that gives the man less weight while the mass remains the same. Say the man weighed 90 kg on earth he would have mass equal to 90kg.

When the same man walks on the moon he would weigh 30 kg while having a mass of 90 kg being on the moon. On the moon the machine may have the same weight as the man on earth but the mass of the machine stays what it was on earth and so does the mass of the man. That becomes science in fictional perspective. If the man could walk on the sun his weight would become a thousand times more but the mass of the man remains 90 kg. If I believe that bit then somewhere I am fooled just because I am that stupid and I deserve to be fooled that easy. Then when the astronaut is in space he still has a mass of 90 kg but when walking in air the man has no weight with a mass of 90 kg. The indication is that while the astronaut still has mass and so the earth does he circles around the earth by gravity. Gravity is circle and not a pulling by force. Gravity comes about in terms of density applying through motion exerted on any object. It is movement that determines the location of an object and the movement forms gravity. How stupid must any person be not to see and therefore realise this as a fact.

If you as reader feel I have no right to dispute physics then it is not I that dispute physics but it is physics that is in dispute with reality. Those "professionals Physicist" who are so superior educated as Symeof is, please use the formula in the formula $4\pi^2 a^3 = P^2 G(M + m)$ on which all Newtonian physics rests and prove that all planets use "mass" to position their allocations. Then try to use $4\pi^2 a^3 = P^2 G(M + m)$ to explain the Titius Bode law because the Titius Bode law is in place while "mass" positioning the planets are a hoax. What this formula says is that the mass of the sun and the specific mass of the planet would form the allocated position in which that specific planet is.

I explain again: that the mass of the sun and the specific mass of the planet $G(M + m)$ would form the allocated position P^2 in which that specific planet is. Those so professionals explain why Jupiter is where it is by putting in the mass of Jupiter and the mass of the sun and then prove that Mercury or Venus or the earth or whatever planet is in its specific location by putting in the distance $4\pi^2$ that is according to the mass a^3 because Newtonians say that $4\pi^2 a^3$ is equal to $P^2 G(M + m)$

This idea of mass being responsible for planets positions only science shares while it contradicts nature completely and yet we all have to embrace this concept notwithstanding that nature embraces the Titius Bode law. This now is the explanation of how the Universe forms and this is for the first time in human history that any correct explanation is presented of the formation of the solar system.

Bode's Law:

Planet	Mercury	Venus	Earth	Mars	Ceres	Jupiter	Saturn	Uranus	Neptune	Pluto
Bode's Law distance	4	7	10	16	28	52	100	196	-	388
Actual distance	3.9	7.2	10	15.2	28	52	95.4	191.8	300.7	394.6

The Titius Bode's law is a numerical sequence announced by J.E. Bode in 1772, which matches the distances from the Sun of the six planets then known. It is also known as the Titus-Bode law, as it was first pointed out by the German mathematician Johann Daniel Titius (1729-96) in 1766. It is formed from the sequence 0,3,6,12,24,48,96, and 192 by adding 4 to each number. The planets were seen to fit this sequence quite well – as did Uranus, discovered in 1781. However, Neptune and Pluto do not conform to the 'law'. Bode's Law stimulated the search for a planet orbiting between Mars and Jupiter that led to the discovery of the first asteroids. It is often said that the law has no theoretical basis, but it does show how orbital resonance can lead to commensurability.

The importance that becomes known is the sequence the Titius – Bode law saw in the number arrangement of 3; 6; 12; 24; 48; 96 etc. The incorrect application of the Titus Bode law lies in subtracting the figure of 3 from 10 leaving 7. The other way of reasoning is to add four each time to the firs value of three starting with 3 and so on. The true significance of the Titus-Bode law is that it points directly to a circular growth of 7 stages.

The 7 relating to 10 is a precise derogative of the Roche limit or the Roche limit is a precise derogative of the Titius Bode principle because he two systems interlink. This is how I mange to explain the Titius Bode law that is in the solar system by the ratio applying that really form the solar system in the way nature shows space growing by time. What you see on the next page was never been shown but on the other hand Physicist say this mathematics are too simple to apply as physics!

To find the mean distances of the planets, beginning with the following simple sequence of numbers:
0 3 6 12 24 48 96 192 384

With the exception of the first two, the others are simple twice the value of the preceding number.

Add 4 to each number:
4 7 10 16 28 52 100 196 388

Then divide by 10:
0.4 0.7 1.0 1.6 2.8 5.2 10.0 19.6 38.8

The resulting sequence is very close to the distribution of mean distances of the planets from the Sun:
Body   Actual distance (A.U.)   Bode's Law <A.U.)< td>

Mercury	0.39	0.4
Venus	0.72	0.7
Earth	1.00	1.0
Mars	1.52	1.6
		2.8
Jupiter	5.20	5.2
Saturn	9.54	10.0
Uranus	19.19	19.6

The Titius Bode law proves that in the Universe laws apply that positions objects in terms of other rules that mass. That means the Newtonians hides their lack of understanding behind mass that they invent. The Newtonians gave the Titius Bode law a formula and that explains the lot. To they're under achieving standards that is very satisfactory. Now it is written in mathematics then what more do we need to know. The fact that the distance that Mercury has from the sun is doubled by that which Venus has from the sun is completely ignored. In cosmic reality mass plays no part. Then again the distance that Venus has from the sun is doubled by that which the earth has. This clearly has nothing to do with the size or mass of the planets. Explaining that part is completely ignored. Then again the distance that the earth has from the sun is doubled by that which Venus has and inexplicably this forms the layout of all planets in the solar system. Where does Newton's idea of mass fit into what truly applies in outer space. Moreover, why does science never mention this? This is my formulated explanation about how the Titius Bode law forms. The numbers we need to find the key to the mystery of the Titius Bode law is 3, 4, 7, and 10.

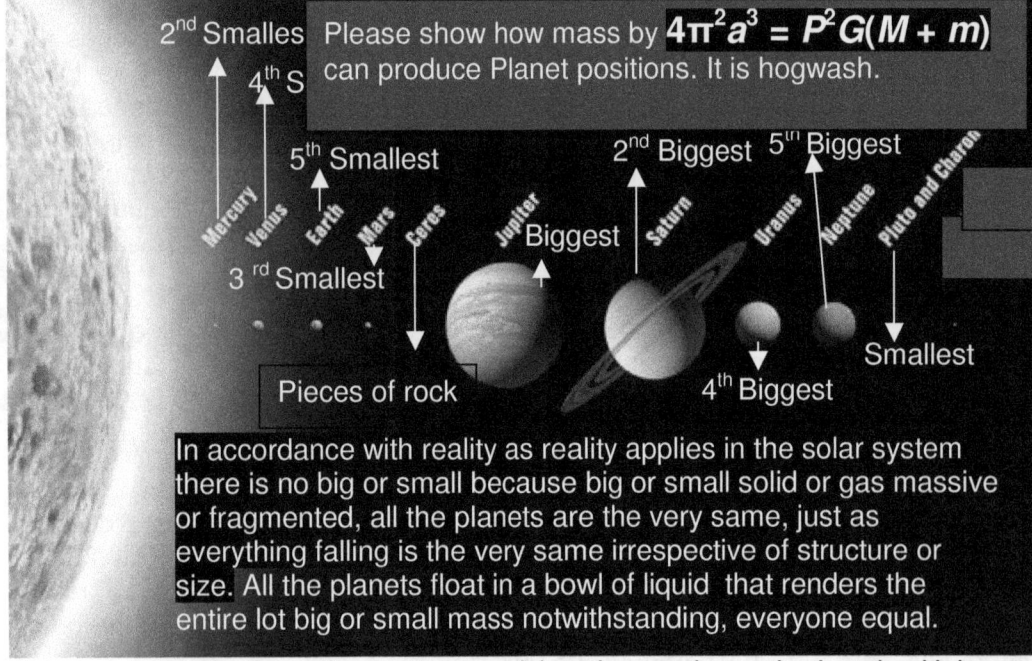

By depicting the solar system in such a presentation with the distances so inaccurately presented as Newtonians normally do such as the picture next this form of presenting the layout without providing correct spacing purposely corrupts the entire structure formation by which the solar system develops.

This is what is in the solar system applying the serving ratio that the Universe uses. It is not the fake Newtonian $4\pi^2 a^3 = P^2 G (M + m)$ that has no basis except for Newtonians brainwashing students into believing that which otherwise can't be accommodated by the Universe. These are ratio values that are there…used by the Universe as actual factors forming space. Use this picture below to show me where the planets are positioned according to mass or where the orbit going around the sun goes according to mass. The entire Newtonian idea of mass creating gravity by pulling is the complete misrepresentation of the truth. I show what principles are in place and do give the reason why. It is easy to talk about "mass" and never get "mass" part of reality when hiding the truth.

It then purposely hides the essence that forms the solar system. In the Universe all things are equal in size because Neptune spins around the sun equal to mercury's time and Jupiter floats around the sun equal to mars or Neptune. Notwithstanding what "size" or "mass" they grant the planet to have the rotation happens equal and without mass bringing any favouring in positioning or in speed of movement. So where the hell is mass a factor in gravity forming? Ask a physicist to explain the fact that if mass pulls mass by gravity pulling, then why do the planets in the solar system not position their allocated places in relation to the sun by applying mass as the nominating and defining factor?

If Symeof and all those others who stand in line to correct me are so certain of his and their position why don't he correct me by showing how mass does put the planets in the order that they are. This is so typical of the brainwashed zombie who learned science where instead I studied science and to him and those there is a world of difference between the two concepts. Those brainwashed into learning science never tested on concept that which they were told to learn. The following picture on the next page shows the applying size and size ratio and the allocated places they have. Are those defending physics really that blind that they are unable to see what even small children see or are they so dumbfounded that 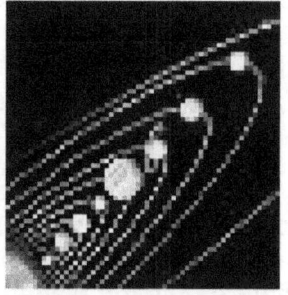 they will rather attack me to defend a senseless conception that can't be defended by logic or intellect. To those physicist brainwashed into stupidity this is not very serious. They truly can't see what the fuss is about as you can see with the reaction of the ever-so-wise Symeof and his gang of wise that so fiercely protect the honour and legitimacy of physics as a whole that is in a hole of deception. To them the fact that there is no evidence that mass is responsible for planet positioning is not a big issue although that forms the fundamental basis of their cosmological concept they put forward as the truth and the only truth. It is said that science only works with truth so how much truth is there in this concept?

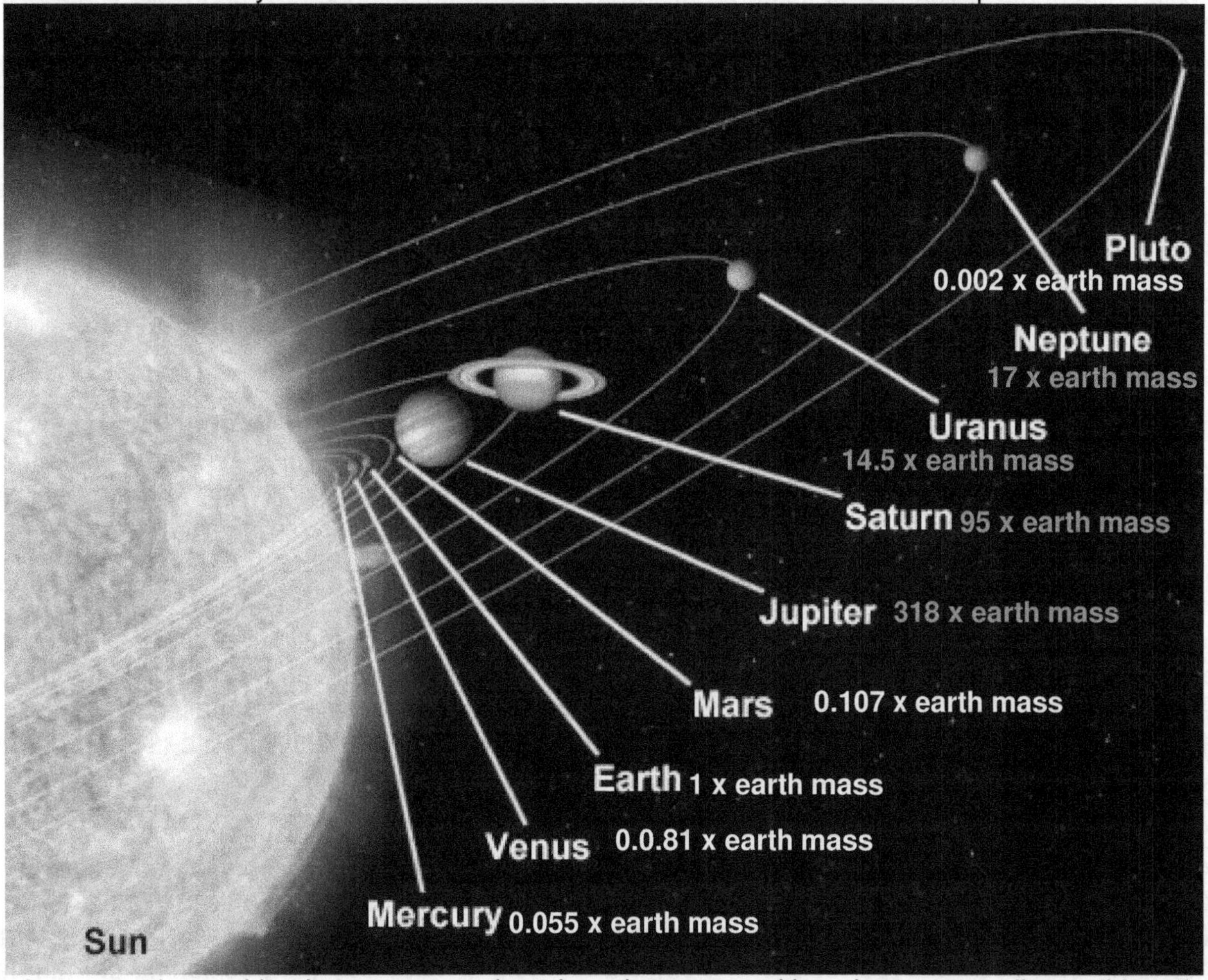

Pluto 0.002 x earth mass
Neptune 17 x earth mass
Uranus 14.5 x earth mass
Saturn 95 x earth mass
Jupiter 318 x earth mass
Mars 0.107 x earth mass
Earth 1 x earth mass
Venus 0.0.81 x earth mass
Mercury 0.055 x earth mass
Sun

This shows a ratio used by the cosmos and not by science to position planets according to this ratio that works independent of mass or size. That is why Mercury as the smallest planet are inside next to the sun and Jupiter being the largest is in the centre. If mass did apply then Jupiter should be inside and Mercury at the very end. Where Symeof of no known e-mail address holds the opinion that I as the author has no right to dispute physics I wish to ask him to explain why mass then holds planets in place as the Newton formula says and as used by the honourable in science since they are the ones who declare mass applies as $F = G \dfrac{M_1 M_2}{r^2}$. Symeof have you or any forming your league ever seen any explaining about the **Lagrangian system** 2) the **Roche limit** 3) the **Titius Bode law** 4) the **Coanda affect** or know that these laws exist? You say I prove nothing and yet you understand nothing about what I prove because you have such a little understanding. Do you realise the fools you appear in the eyes of the more intelligent amongst the many in society? In your case I can't figure out which is more, your stupidity, your arrogance or you pathetic self worth through which you believe in your absolute intelligence. To Symeof and all the others such as he is my advise to you lot is to go and ask your money back from the institution that you thought educated you as you should clearly see you have not been educated but you have been conned and tricked. Moreover, read on and see the deception you have undergone by those you trust in such high regard.

Why do I say there is a conspiracy? This is why I say there is a conspiracy in science. Science says the planets are in location because of the mass they have and this opinion stems from Newton. However look at the picture I provide on the next page and see how the mass applies. Then tell me you still think that nature puts the planets according to mass with the larger planets in the centre and the smallest on either end of the solar system. Is there any body that still say I am not aloud to criticize science with this blatant misrepresentation of the truth? Are there still those that think like Symeof and say I have no right to criticize the most honourable persons in science for misrepresenting the truth? This is but one of so many cases I bring to your mind, and yet the most clever say *"Then, of course, he couldn't imagine he might have been wrong, so he rationalized his failure to be accepted by creating a conspiracy theory: this is probably the reason why this e-book exists. This is no scientific material..* Tell me which part of this have I got wrong and about which part of this am I misrepresenting what science say incorrectly? Is the wrong part that I thought about what I was taught and the rest of you lot did not?

The cosmos however, does not apply mass in any form and the cosmos has four other principles in place. What Newton show that should place planets according to mass is not used by nature. It is the most incorrect idea ever put forward as the truth. I show what nature uses namely the Titius Bode law, The Roche limit, The Lagrangian Points and the Coanda effect and how this forms gravity as well as place the positions of the planets in accordance with singularity. Because I trash Newton's rubbish that does not fit and that can't apply no publisher of science books or science magazines will publish my work I show what goes on in nature while Newton's contribution of mass applying is total rubbish. Because I call it rubbish and I rubbish Newton I am ignored. How can any person believe mass forms gravity that puts planets in position in accordance with mass when viewing planets, in the Universe?

This picture shows the hoax the Newtonian conspiracy pampers to keep the rest of Newtonian physics believable. They never mention the Titius Bode law and try to explain the Titius Bode law while it is the Titius Bode law that is really in place in the solar system. I*f you wish to learn the truth then think again.* With all this evidence in place for centuries and no evidence of mass pulling in the Universe science still upholds $F = G \dfrac{M_1 M_2}{r^2}$ because in the known field of physics they have nothing else to show for centuries of labour. They have to be content with Newton because they only have Newton and while Newton never made sense it was all there was to cling to... Newtonians uphold their version of the law of physics they advocate as the truth applying without showing mercy to anybody that dares to challenge Newton. The very first things the Newtonians use to beat us into submission are to blast us with incomprehensible mathematical formulas.

If only the equations they say applies were truthful…Incomprehensible they are not truthful but still those Newtonian philosophies are used to scare anyone with the idea that the mathematical equations will get everyone hiding as it has done for hundreds of years. They bewilder students and the public alike with equations that put the fear of God into you; it used simply to make you feel inferior so that they can feel superior and frown down on your inferiority from a dizzy height. They say the positioning of planets are according to this formula [illegible formula].
This is pathetic, not my book you are reading. I challenge any person on earth to put in the mass of Jupiter and then from that show how it is arriving at the position any of the planets hold or how by mass any planet's allocation or position is derived. Or do it with any other planet. This is one big hoax that is a small part of the conspiracy science tries to hide. To make sense we have to look for Π in this because Π forms the value of gravity because of the circle nature that we find in cosmic space. We have 3 and 4 adding to be 7. Then we have 10 forming the other factor number. If we wish to stop pretending to make science a hoax we must come to some realistic conclusion that will prove what is working in the cosmos.

The cosmos works on a distinct ratio put in place by a distinct pattern. This is the ratio. It is a ratio that puts every planet in a precise location that finds a value in sequence in accordance with the turning movement of the sun. Mass never comes into the picture. Mass plays no part. Science never announces this fact with trumpets blowing because then science has to admit they know less about physics than their dog knows about British or American politics.

To find the mean distances of the planets, beginning with the following simple sequence of numbers:
0 3 6 12 24 48 96 192 384

With the exception of the first two, the others are simple twice the value of the preceding number.

Add 4 to each number:
4 7 10 16 28 52 100 196 388
Then divide by 10:
0.4 0.7 1.0 1.6 2.8 5.2 10.0 19.6 38.8
The resulting sequence is very close to the distribution of mean distances of the planets from the Sun:

Body	Actual distance (A.U.)	Bode's Law <A.U.)< td>
Mercury	0.39	0.4
Venus	0.72	0.7
Earth	1.00	1.0
Mars	1.52	1.6
Asteroids		2.8
Jupiter	5.20	5.2
Saturn	9.54	10.0
Uranus	19.19	19.6

First to find the ratio one puts the earth distance to the sun in as a cog in ratio to 1. This eliminates the human factor in presenting the positions and puts cosmic ratios in practise. The distance from the sun to the earth is 1 and that is the dependence of the ratio of the solar system applying.

In that ratio we have Mercury at 0.39 Astronomical Units from the sun and according to the Titius Bode law it should be 0.4. 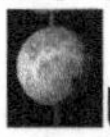 Using the solar ratio of placing the earth to the sun at 1 we have Venus at 0.72 Astronomical Units from the sun whereas according to the Titius Bode law it should be at 0.7. 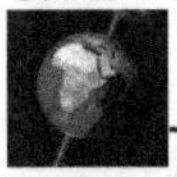 Then we have the earth at 1 Astronomical Units from the sun according to the ratio that the solar system applies.  Mars is at 1.52 Astronomical Units from the sun whereas its allocated position should be at 1.6 when the ratio is applied. At 2.8 Astronomical Units from the sun we find a huge lot of planetary debris that also holds an allocated ratio position although the lot is fragmented into small pieces. Yet, all the pieces orbit at 2.8 in ratio to the sun / solar system. 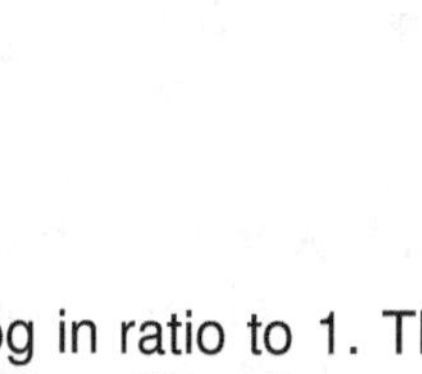 Jupiter the "giant" stands at 5.2 Astronomical Units from the sun in accordance with the ratio the solar system provides.  Saturn is at a cosmic ratio of 9.54 Astronomical Units from the sun and the mathematical ratio is 10. 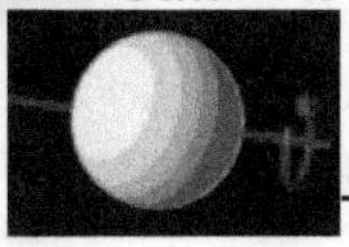 The next planet is at 19.19 and mathematically according to the solar ratio its allocated position should be 19 .6. This goes on and by using this ratio planets were discovered in the past. Yet science has no explanation as to why this is true and I am the very first person ever and alive that explain why this is the case.

This is a fixed motion driving the movement and forming time by placing the planets in relation to certain places the planets will maintain according to movement of time as it forms space by the movement in time. Using evidence that the solar system provides mass has no place but we are forced not to use the evidence provided by nature because nature clashes with Newton. Subsequently we must toss out what is there and what nature supplies us with as evidence and instead we have to apply Newton's vision of mass because God can be wrong by placing a ratio in Nature but when nature goes against Newton then God and Nature is wrong. This is the official position of science and you are going to read about this attitude many times more as this book develops. We will rather blot out what is in Nature than question the correctness of Newton and the way science view physics. If we do question Newton then we will learn that the "oh so clever" were a bunch of arrogant fools that did not know anything about what nature in reality offers. Then in this ratio another factor is present:

When reading the distance that mercury is from the sun this distance doubles from mercury to Venus and the distance from Venus to the sun doubles from Venus to the earth. Then this distance from the earth doubles to Mars as it doubles again from the sun to Jupiter. This goes on up to where the solar system ends. This shows no indication of mass forming any part in the planetary display. Science's claims since the time of Newton that by the Kepler laws which is what Newton fabricated everything depends on mass and the cosmos shows that mass has nothing to do with the planet positioning. It is folly that according to Kepler the planets hold individual positions according to mass. Science never put this formation in place when explaining the forming of cosmic science because then science has to admit they have no clue how the cosmos works and for that admission they are not ready because everybody must think that they know more than God Almighty about how science works. They can't come across as the ignorant fools they obviously are!

The cosmos shows no tendency to use the Kepler laws or any fabrication that science may propose that those laws claim. In nature the formula or $F = G \frac{M_1 M_2}{r^2}$ is completely absent. There is the Titius Bode law that shows a positioning according to a ratio and the distance doubles every time a new position arrives in relation to the inside planet of that planet. Without any proof of this formula having any legitimacy whatsoever science continues to use this formula when explaining gravity and the way the Universe applies.

Notwithstanding all these facts available to science it still is propagated that mass forms the allocated positions that the planets hold according to the sun in . This conspiracy is in place to hide the fact that science has no idea what puts the planets in position and rather than admitting they have no idea they keep teaching unsuspected students a lot of unsupported hogwash just to cover the fact that science is clueless about the most basic knowledge about gravity.

I am the first person ever to explain WHY the Titius Bode is in space functioning as it is. But since those such as Symeof and the "Small-brain-brigade" such as he can't fathom any explanation going this far they are of the opinion that: "*The author has no legitimacy in disputing physics: he is merely an amateur trying to disprove what the greatest minds of our times have taken centuries to prove (by the way, you need to have studied a subject before you can criticize it, which the author didn't).*

In my opinion, the author is so narrow-minded that when he came to scientists to talk about his material, the scientists must have told him that it was nonsensical." In their smallness they have not the ability to understand that planets are not allocated according to size or mass and this they do not understand notwithstanding all the planets' sizes or individual mass values proving distinctly that Newtonian science is senselessly wrong. Now I will explain it again in detail for those "professionals" such as Symeof and his band of "Super Educated Professionals" who "understand science" because they are so informed about physics and the application thereof and is well informed about physics. There is a ratio in place keeping time as if it runs on cogs. It is not connected to mass in any way. Science never come out and says this because they hide this. If they didn't hide this they then must admit they have been wrong all along about mass positioning planets.

Symeof who are so typical of the "brainwashed class of stupidity" and the "Small-brain-brigade" has the opinion that: *Then, of course, he couldn't imagine he might have been wrong, so he rationalized his failure to be accepted by creating a conspiracy theory: this is probably the reason why this e-book exists. This is no scientific material.*

If I show that in mathematics according to science the planets' positions are in accordance with Symbolically being: $P^2 \propto a^3$ and therefore $a^3 = P^2$ in position of P and therefore $a^3 / (P^2 P)$. This is taken from the idea that "Kepler said", which is totally fabricated that $a^3 = T^2$ where Kepler in fact said $a^3 = T^2 k$ and this is a big difference because $a^3 = T^2 k$ is the same as $E = mC^2$. *Then, of course, he couldn't imagine he might have been wrong, so he rationalized his failure by creating a baseless and unfounded theory: this is probably the reason why this e-book exists. This is no scientific material.*

PLANET	PERIOD (Years) (T)	MOVEMENT (T^2)	DISTANCE	SPACE (a^3)	RATIO k
Mercury	0.241	0.058	0.39	0.059	0.983
Venus	0.615	0.378	0.728	0.381	0.992
Earth	1.000	1.000	1.000	1.000	1.000
Mars	1.881	3.54	1.524	3.54	1.000
Jupiter	11.86	140.66	5.20	140.6	1.000
Saturn	29.46	867.9	9.54	868.25	0.999
Uranus	84.008	7069	19.19	7067	1.000
Neptune	164.8	27159	30.07	27189	0.999
Pluto	248.4	61703	39.46	61443	1.004

In this equation **k** has a defined value where **k** holds a value. The Newton ratio of $a^3 = T^2$ put forward can't have any substance. Kepler said $a^3 = T^2 k$ and that nullifies Newton's presumed ratio of $a^3 = T^2$

KEPLER'S LAW OF PERIODS FOR THE SOLAR SYSTEM

PLANET	SEMIMAJOR AXIS $a \, (10^{10} m)$	PERIOD T (y)	T^2 / a^3 $(10^{-34} \, y^2 / m^3)$
Mercury	5.79	0.241	k^{-1} = 2.99
Venus	10.8	0.615	k^{-1} = 3.00
Earth	15.0	1.00	k^{-1} = 2.96
Mars	22.8	1.88	k^{-1} = 2.98
Jupiter	77.8	11.9	k^{-1} = 3.01
Saturn	143	29.5	k^{-1} = 2.98
Uranus	287	84.0	k^{-1} = 2.98
Neptune	450	165	k^{-1} = 2.99
Pluto	590	248	k^{-1} = 2.99

With a ratio that Kepler calculated using the formula $k^{-1} = T^2/a^3$ the negative value of **k** proves that space moves in recline towards the sun. The sun contracts space from the outposts of the sun's influence field.

These are Kepler's actual numbers before Newton got hold of those numbers and altered it into something completely fraudulent. Science at present accepts that $a^3 = T^2$. Then Kepler had a specific column for k where if you multiply **T^2 with k** it then is equal to a^3. Kepler said if you divide T^2 with a^3 it **has a value of say** 2.96 as it is in the case of the earth. Mathematics teaches us that if you have numbers telling you that $a^3 = T^2 k$ then those numbers will give you a mathematical result of $T^2 \div a^3 = k^{-1}$. **What should then dawn on me that I could be wrong as he suggests in his remark** *of course, he couldn't imagine he might have been wrong, so he rationalized his failure to be accepted by creating a conspiracy theory: this is probably the reason why this e-book exists. This is no scientific material*? ÷

You brainwashed stupid zombie who tries to teach others what your simple mind do not even understand, where am I wrong? Do go and study what science says because in this part a few pages back I have given you the claims why science makes that **$T^2 \div a^3 = 0$**. Now with your little underdeveloped intellect, please show me that mathematically $T^2 \div a^3 = 0$. Where I contacted more that 2000 Academics worldwide in physics showing them just this anomaly and that their practising of mathematics are as incorrect as this, where no one answers me but either belittle me in precisely the same manner as you do without ever telling me where am I wrong but just telling me that I am wrong just as you do and NEVER substantiate my incorrectness with facts of where I go wrong, then yes, you are part of the conspiracy and I thank you for supplying me with such a splendid unmistakably example of how brainwashed students in physics become. I hold your example in such high esteem I even open this book with your writing.

So why do I think of this as a conspiracy. I have uncovered the formula by which the Titius Bode law works. The Titius Bode law is in place. It is not mass positioning planets but it is the Titius Bode law that holds the key and I have solved the riddle that unlocks the explanation about the Titius Bode law. Any person doubting my accusation is very much welcome to show the world exactly how mass works when all the planets are randomly sized. The Titius Bode law says one should apply the ratio according to the astronomical unit, which has the distance of the earth and the sun at a position of 1.

Then in ratio Mercury is .39 where it should be four. Then add four and divide by 10. That is the ratio but why that is the ratio has never been understood because it has never been explained or had any attempt by a living person to explain…that is up to now where I explain it in detail. That the ratio also has the distance double between planets next to each other I explain. This too has never been understood because it has never been explained or had any attempt by a living person to explain…and again that is up to now where I explain it in detail. This forms one of the four foundations on which the entire order of physics rests.

The others are the **Roche limit**, which I explain in detail, the **Lagrangian numbers**, which again I explain in much detail and the **Coanda effect** that forms gravity and that too I explain in excruciating detail. Then you get these arrogant half-wits such as Symeof and all the other "highly regarded Physicist of the world" who are unable to explain gravity but likes to belittle me because the evidence I show belittles them to the point where they are not even having a microscopic brain to work with. I show they cover fraud and then they turn around and declare I don't know physics because I show what they think they know is total shit.

Do you reading this or any person on earth wish to tell me that in science no one in almost four hundred years saw that the planets do not show arranging in accordance to mass and therefore ALL COSMIC OBJECTS show total indifference to mass positioning or allocating positions? With this evidence do you believe there is no conspiracy to remain quit and keep facts under cover?

All this detail is so small part of the entire web forming the conspiracy that I could use it as an opening to show how my critics work. Go back and read the entire criticism of Symeof and see what he says I do not understand or why he says I never studied physics when I know more about physics that what his little mind can hold. Go and see where he indicates one single piece of evidence that I do not understand. It is always rumours and thought thrown intro the air just as his idea of facts in science are. This is the rubbish I have to deal with. I have spent forty years on research and I can now explain what no one ever could explain and that is how gravity works.

Never as even once would science say they have no idea how the solar system holds positions in the cosmos. When admitting that they will have to forfeit their veneer of godliness about how enlightened they are on matters about how the cosmos works. I have made a study of almost thirty-five years to uncover how gravity works and I have concluded how the process of physics fit into the cosmos. I have showed what is wrong with physics and how to correct it. I am so grateful for suckers such as Symeof and his clan who so feverishly defend the science they believe in and thereby provide me with the opportunity to use this as a speaking example as to show how the brainwashing can apply. Symeof was brainwashed to believe that the formula he never could test did apply and he would die in the conviction that the formula he defends is correct and this formula carries more truth than does the Holy Bible. So all those who wants to crucify me when I prove what you lot studied was a waste of time, first apply this formula with the adding of the distances as well as the mass and prove to yourself why the planets hold their allocated positions according to mass. If you can't then A conspiracy is in place to prevent the truth from leaking. For thirteen years everyone in science prevents me from explaining the four comic principles that are there where they placed mass and they prevent me because then their image goes down the toilet because then they must admit of being wrong and they know nothing about what really applies in science.

By the way, those who learned physics and feel so proud about it, one first have to study physics to realize where there are incorrect adaptations in the reality of the concept when forming judgement on the entire matter as a whole. This is a drop in the ocean and those who studied physics if you are not prepared to live through the shock of awakening stop at this point. As you can well see with the example I provide, if you are so mind conditioned that you are unable to see the anomalies I provide, then stop reading because your mental faculties are at risk and you may find much mental strain when reading this and thereby coming to terms with what you never studied but thought you did. Prevent yourself the agony awaiting you.

WHOM IT MAY CONCERN,

I have already introduced my credentials but I do so again: I am P.S.J Schutte, nicknamed Peet. Being a white South African the language of my mother tongue is Afrikaans and my second language is English. I present a new cosmic theory that I partly present in a six part theses, of which the investigating research began in 1977. First I located what was wrong in physics. I compiled my presentation of The theses called The Absolute Relevancy of Singularity and then six separate thesis parts forming the theses published through LULU.com which I saw as the only manner whereby I could generate funding by which I would be able to have the twenty seven books I already wrote linguistically edited and then to have the books published on a Print-On-Demand basis. I compiled a new cosmic theory by which I eliminated all the incorrectness that Newton has burdened science with but with this being my opinion I did not find a garage full of academics supporters waiting to applaud me and to uphold my views on the matter. Yet still I was not going to be ambushed by their relentless stonewalling my efforts and blocking my efforts in introducing both the incorrectness and the new cosmic theorem I concluded. Their blocking convinced me about a Conspiracy in Science in Progress and this spurred me on to tell the entire world about their brainwashing of the minds of students. This kept me busy for the past going on to twelve years on full time basis whereby I was trying to introduce my findings to many academics without finding much joy from my efforts.

This past eleven years plus saw me go without any income as I tried to get my theorem recognised as well as get my warning noted. Going without a steady income left me almost destitute and in order to find a manner to get my theory across to the attention of influential readers, I decided to publish a thesis of six books electronically as to try and get around the stranglehold of Newtonian bias controlling science at present worldwide. I decided to publish electronically which those in power do not control.

With my first language not English and the books not linguistically checked by an expert there are bound to be language errors that readers will notice. In the past I tried to check my work myself but after checking say one hundred and fifty pages for language corrections, then after days of toiling instead of having corrected work I ended having four hundred pages of newly written information which is still not linguistically corrected but holds a lot more information. The language and spelling errors compiled instead of reduced. This is because my priorities lie elsewhere. I aim to spend money on correcting the work as far as language goes, as I receive money in the selling of my theses and in the hope that I will receive money. I will have all my work including the one you are reading edited professionally and corrected as I find money to do so...But first I have to get the public aware of the problem to get the academics to appreciate the problem. The public are thinking that only science is the truth and it is proven above all and is the only truth. Science is completely fabricated at the base where all science begins.

This is very important to me and in that light I repeat it again. I do not wish to have my names pronounced in English because it will remove my Afrikaans identity and I would rather die than to be thought of as being English. Therefore I repeat again please note that I do find much pride in my status as being Afrikaner and would like to have my names used by pronouncing it in the manner Afrikaans dictates...therefore I would sincerely appreciate the courtesy when readers will take note that my name and last name are pronounced in Afrikaans, which is originally from Dutch and must be pronounced that way. Peet one would pronounce "here" which is the closest English to the pronouncing of the "ee". The "Sch" in Schutte is pronounced exactly as school is where both actually are pronounced Skutte or "skool". By pronouncing my name in Afrikaans you do me the utmost courtesy any one can. Being an Afrikaner is what I am most proud of. If you have been a scholar at school you would know how to pronounce Schutte.

In order to explain why I am so vocal in criticizing the teaching profession of the day is because they brought on me so much poverty and hardship and all the time I was portrayed as the cunning knifing hooligan out to mesmerise the innocent by blackening the characters of the highly appraises physicists. I chose poverty and therefore had to quite teaching because I could not betray children in my class any longer. I decided on poverty for I thought it would be better for my soul that I live on the brink of starvation as I have been doing and experience complete poverty than to cheat the children who trust me and look at me with those trusting eyes. To them as a teacher I am completely trustworthy, even more than they believe in the knowledge their parents represents because I am the teacher and the teacher they believe. …And then I look into their eyes and tell them to believe in what I teach while I know the science I know is totally fictitious. How can I take money and in the end meet my Father in Heaven and tell my Father in Heaven I sold not only my sole, but also their soles for money. I had to live an entire lifetime with lies and betrayal not to one or two persons but to scholars year after year while I know every word I say on science is corrupt. I have to teach what I know is unproven dogma and I am forced to teach if I wish to earn a living. I brought much hardship on my children by forsaking Newtonian standpoints and searched for most of my life to find the truth and then those living the lie would thrust my finding aside because they would cling on to the lie that feed them, keep them famous and by which they can pretend to be wise. How can any person with a conscience, with moral standards and with dignity spread untruths and still feel unremorseful about the dishonesty they teach. If I call them scoundrels it is because I believe it.

In my effort to bring change I wrote many articles and books, amongst which I wrote an article, which I call the Absolute Relevancy of Singularity in which I go much further than Einstein did on the relevancy of singularity in the general application of all types of physics. In doing that I had to reject Newton. I prove the entirety called the Universe is made up of singularity and contains only singularity in many forms thereof. I prove singularity is not a general phenomenon but is the concept forming the absolute basis of physics and this foundation is transferred to the beginning of mathematics. The Universe is singularity coming in many forms and this is simple to prove; it is the other concepts flowing from this that complicates things. Modern Science can't explain any of the modern findings such as the Big Bang because of a lack of proficiency. After finding the building blocks of the Universe I now can show where is the centre of the Universe. Have you thought where is the centre of the Universe? I can answer that…and why is the Universe still grows since the Big Bang…why did the Universe start so small…why did the Universe fit into a neutron at one time…how did everything expand from fitting into a neutron…why does space grow from small to large…where is it going while it is growing…why was the Universe any specific size...what was everything before the Big bang…if you read this you will know all of this!

My discovery is the fact that singularity presents the total and complete control of everything there is in all forms of everything there ever could form in the Universe and discovering this fact led me to prove that gravity is the application of Π and from extending Π further into a six dimensional sphere that is spinning in a six-sided cube, is what makes the Universe form three dimensions using time thereby the cosmos enlists the multi dimensional sphere we live in. However, at the stage where the Universe holds gravity as Π, which is at the point before it forms multi-dimensional space, the Universe still is flat although the Universe then is already accepting form by going round in shape-forming when extending from forming singularity as a value when retaining form in the principle of a sphere. The foundation of physics is gravity forming Π by allowing Π forming. This explanation comes across, as simple but believe me no academic this far was able to understand to entire context of such an explanation.

My findings are not congenial with Newtonian views and therefore it means that no one in Mainstream science could dare to look at my work. If they do read my work and confess to the truth they then have to confess that Newtonian science is a corrupt conspiracy that is hiding issues most important and portraying a fantasy as the truth. Then I decided to put the cat amongst the pigeons and see what flew out of the bush into the sunlight. I wrote seven articles after the first article that was (according to my page set-up) in total 15 pages that I sent to Annalen Der Physics. This first article and the seven afterwards I sent to the addresses of Dr. Ulrich Eckern as well as Dr. Fredrich-Wilhelm Hehl and at the bottom of this page I give their responses. But as to getting a reaction about the other seven I am waiting for many, many months but as to date still never received any correspondence as to confirm the acceptance / rejecting of the seven articles explaining the four cosmic principles. It is true that I could never give sufficient proof substantiating my work, but as I put it to Dr Ulrich Eckern, one cannot prove a complete new trend in thought about science going using completely new principles about gravity when only devoting 15 pages to the concept. I showed that singularity is time forming space and where to to locate singularity. I always submit articles to well known physics magazines but my articles are rejected on the most unappeasable grounds and for the most outrageously ridiculous reasons the Newtonians can think of. I explain how gravity forms but I am rejected because they are of the opinion that my work does not meet an acceptable level of standard since I am at odds with the way science in the present think about gravity.

I say and I prove there is no such a thing as "mass" with the ability to "pull" anything. I do not say a person does not have weight but I say no person has "mass" that brings about a force that pulls anything closer by using gravity. I explained in the article the Coanda effect and the way the Coanda effect applies according to gravity. I also showed that this explanation forms the fundamental and basis of physics and how it forms the foundation of physics. It is the heartbeat of what no one knows to explain in physics.

In this book I prove that what you think of as Science, is only ideology based on rumoured disinformation made up by those making up facts, as they hide what they don't know, hide what they don't want to know and conceal what they don't want others to know. By proclaiming they know everything while pretending to know all there is to know, they preach what they falsify to sustain their religiosity in the form of unproven dogma. Science tells what they wish and reserve what they wish not to tell and hide what they don't wish others to know that they don't know. My biggest undertaking is to prove that what science presents as credible facts is unfounded and incorrect. Science is believable because culture sustains Science. My problem is busting a culture people trust as Science and be a whistleblower about accepted religiosity people unconditionally believe in as being trustworthy. The public never once thought to question what they were taught as science. This forms a problem since what is presented as science is falsified information and that which we need instead is a new approach that is truthful about science. Therefore no one realises what big a problem there is in the level of correctness there is in the approach to reality science now presents. Science never shows the support or proof in what they present as ultimately credible facts. Bluntly put; Newton was never proven to be correct but only accepted. I firstly need to promote the need we have for the truth to replace what we have as science before I can promote what the truth is that we need to form science. Science hides their flaws under a profile that they present as the uncompromising truth while in fact what they present are lies covered by a veneer of truthful facts and this then neglects the urgent need for the truth by denying there is a problem. If you don't believe this statement I make on on the fact of mass forming gravity then I challenge you to read on and challenge me on any fact I present about the untruthfulness you think of as science and if you disagree with me send me a e-mail, but if you believe me then help me to challenge the establishment about their scandalous disinformation that they hide to commit the conspiracy they call science. Everyone that has contact with physics is unbelievably naïve while everyone is getting hoaxed and your children are being brainwashed to unconditionally believe in science, which is totally without credit. Everyone accepts without questioning the science because they believe in the persons teaching them and not necessarily in the credit of the information they are taught. Read what it is that science hides to put in place this notion that they know everything and what they know are without blemish.

There is an obvious conspiracy in science that everyone misses and about this fact there is no doubt about it. It runs as deep as it can go and involves every person that studied science worked with science teaches science and thought about science. In short it touches very person that lives notwithstanding culture, ethnicity, race or religion. Everyone on earth believes the science taught at schools and everyone is working with fake science and conspires to promote science fiction passing it off as Newton's science.

Moreover what will you do when you know the truth or are you satisfied to be told what to believe is the truth without ever investigating by personal standards as to finding the truth about science. Then go on reading and confront the truth, as you never had ever before. The truth about a conspiracy is that everyone involved with the conspiracy will fight tooth and nail to stop the conspiracy to get leaked and leave those behind the conspiracy and all those feeding from the corruption of the conspiracy exposed.

Those that have the power to maintain the conspiracy would keep the waters as still as possible as to draw no attention to such conspiracy.

A true conspiracy has to be as quiet and as unseen as it could be.

A true conspiracy must involve everyone without anyone detecting even a hint of what the conspirers hide.

There are many conspiracies going on such as the banks involvement with crime and the bankers profiting from gamble rackets and drug selling.

The same goes for the Insurance business profiting from lenient sentencing of courts holding very merciful judges in office so that the crime cases and the burglaries, car theft, hi-jacking and all other forms of crime will shoot through the roof every year and grow by thousand percentage points from which insurance will sell ever-more cover-policies.

The more crime is about and committed daily, the more people need insurance covering and thereby the more money insurers bank giving bankers profit to spend on the stock exchange by controlling the economy.

Bankers buy democracies with money they give politicians to write laws that protect the rich against the poor. In time I might write about this but now I cover another conspiracy, a bigger case file about a conspiracy everyone on earth participates in...it is about...

Should anyone desire to contact me with an opinion about my views on science then please do by using mailto:info@questionablescience.net or mailto:info@singularityrelavancy.com but please do so when coming to the end of this book at least and then see if you understand the entire concept that I introduced. In most cases all your questions are answered further down the line and I am not fond of repeating what is already said in the book. The book forms a line in explaining and the concepts are best understood if the reader follows the designated line.

This book started off as a website to inform about a science conspiracy but although reduced still it grew into a book that serves much more information than what I first intended to supply. You will see many new aspects about gravity please make sure you understand what you read. It grew into a comprehensive study on cosmology. At times you may observe while reading this book that it seems as if my frustration will ring through like the chiming of the Big Ben Bell.

For that there is a reason.

At times my frustration and anger will boil over drowning my politeness and that is true, which I admit. For twelve years I have had the answer that would correct the philosophy that has a stranglehold on cosmological science.

I discovered the building blocks of nature where my discovery puts all other cosmic aspects of science into science fiction.

Those who force-feed non-existing dogma do so to brainwash students to hide the incompetence of "modern science" so they can rule supreme while ignoring the truth that they deliberately hide by concocting a conspiracy.

To keep everyone unguarded they practise a conspiracy by which they perform an accepted practise of thought control on students to further the false dogma presently in place.
I try to blow the whistle on such a practise but accepting my resolution makes every thesis ever written science fiction.

Therefore no one in science dare to read my work leave alone appreciate the revolutionary nature thereof. Whatever now is deemed to be accepted science would then become what is the past tense in science because the flaws that those in power of science principles kept coated for centuries on end as untouchable truth will then be rust that breaks the surface to show the holes!

They try to silence me but surely somehow somewhere I have to break through with my message! I bring you a true form of science as never seen before in all of history and I do that when I dispose of the conspiracy that hides all the incorrectness and the failures that haunts science today. Science is accepted as the most righteous information available to man and that is a scam.

go to mailto:info@questionablescience.net.

You are going to read about a conspiracy but people think of a conspiracy in many terms. Let us define not by definition but by interpretation to what a conspiracy constitutes. What do you think is a conspiracy?

All the conspiracies you know about is known about because someone somewhere makes money by allowing the revealing of the conspiracy. Silencing the conspiracy does not make money but informing a suspicious public loosens the flow of money. If it were a true conspiracy no one would know about the conspiracy because the powerful would make money from not revealing the conspiracy. The revealing of the facts about any conspiracy would be stopped before it leaked because it would kill the flow of money.

A conspiracy is thought to be a gossip story that makes money and by not revealing it or revealing it goes in line with making money or not making money. You can download this book free of charge because I don't make money by revealing the conspiracy. I truly want to find an audience to divulge the truth. I want to make money but it is by showing how I can correct the flaws in science, not by hiding it in a conspiracy. People put a conspiracy in the same realms as a gossip story, an old wives tail, which is going about but does not intend to harm and mostly serves as amusement to many. Hearing about a conspiracy tests your intellectual comprehension. It is some quiz that you match your truth against the truth that the conspiracy reveals. It is a funny, but it is not funny until you catch the funny part hiding behind the conspiracy and only when you measure the catch behind the conspiracy are you treated to be amused.

If the conspiracy does not touch the person directly then no harm is felt and no harm is intended. Every one holds this view that a conspiracy is on a slightly higher level than gossip. It is a gossip story about someone living in the neighbouring village known only to some people next door but has no direct linking to me or has no threat to the safety of others directly associated to me. Everyone treats a conspiracy as if it is something amusing that holds no threat at all. It is something that goes around as a joke of sorts.

We all live in a bubble we call civilisation and we all run after a dream called peace and we all preach a fantasy we named democracy and every one excluding no one lives a fantasy we love to believe but never can. Those that say they lead us are programming us in an assortment of ways to have us believe that we are happy content with our fate and we are lucky to be as prosperous as we are and this they do by a process of programming us mentally. They call it advertising. They call it politically socialising. They even call it following teaching procedure. The teachers tell us what we must believe and then force us to write examinations about what we must believe and according to how we believe what they say we have to believe in, they set our future.

A hundred or two years ago it was believed that one of the unhealthiest practises any one can indulge in was to have a bath. In the reign of Henry 14th of France it was believed to bathe was to progress the aging process. I presume that they saw our skin in our hands wrinkle from being in the water for too long and when the skin wrinkles it is a clear sign of quick aging. With the wrinkling process so evident the thing to stay away was to have contact with water. It ages you enormously. Should you wish to die young and miserably old at the same time then you have to needlessly bathe regularly. It was believed at the time that it was bathing that made you old before your time or so it was believed throughout the Western Civilised world. One should much rather drown yourself in perfume if you were rich and smell like a cheap modern whore would from a mile off. The status symbol at the time was to smell like a perfume factory that has gone bust and you could get perfume for free. If you smelled like that you had the same aroma as the King did and his entire entourage of the royal upper class. In that period drowning your body in perfume was the fashion of the moment and the statement that you had money. It was the same as in modern times carry a hundred credit cards in many wallets for all to see how credit worthy you really are.

It is playing mind games with the public and makes those with simple minds feel important if they were fashionable and trendy. Then someone found a way of advertising and what was advertised first was soap. If you look at films and photos about hundred and fifty to hundred years ago was with soap advertisements that was filling the background in these true to life photos and films. The soap barons got in on the act first and through scrupulous mind control started to program the mind and the mindset of the everyday persons.

Even today and more so today every advertisement are programming you to believe that the subject advertised will be most unique and if you do not have it your soul would be much poorer.

Let's take soap advertisements.

One hundred years ago soap started with a program to get people to use soap more often. Look at the very old pictures of the turn of the century going to the 20th century and you will find a soap advertisement looming in the back ground when a film was shot showing a scene of everyday life. At the time of the turn of the 20th century the people were bathing once a week. One day in the week was set aside for having a thorough bath, shave and hair wash. The cleansing process took all night and all night was from sunset to about seven or going very late then about half-past eight. The gentlemen shaved with a cut-thought razor and the woman tool about an hour or more to dry their hair. This was a lengthy process and was done with most care in mind.

There was no advertisement prompting the public to buy a horse because that was all you could buy in order to travel without using trains as transport. You either had a horse and you rode on it or you had no horse and walked. So a horse you needed and therefore advertising to use a horse was senseless. But soap was something else. They had to get people to use more soap and they had to open a need for people to need more soap. Therefore if you were modern you didn't need a horse because you needed a horse in any case so no one is going to pay you to believe you needed a horse. But they had to tell you that you needed soap so that you would realise a need to purchase soap you never needed before.

How many houses had indoor plumbing during the American civil war in the 1860's or the Anglo Boere War at the turn of the 20th century? And were those persons a dirty barbaric bunch of lunatics only craving to plunder and destroy everything they came across…well frankly yes, those on the British side were because under the orders of a mad murder called Lord Roberts they killed thirty thousand unarmed woman, old-folk and children by starving them and by lack of nourishment they killed in the Boer-concentration camps the thirty thousand mentioned from a population of a hundred thousand all included while only forty thousand Boer fighters died all told on the battle field. …But all the murdering and the desecration on the side of the British, well that was not due to their habits of having a bath once a month or so. That was because excluding Portugal, the British warred with every other nation on earth and that too had nothing to do with the bathing habits of the day. As the soldiers came home they were not frowned upon as dirty disease ridden dirty pigs but notwithstanding going for months without bathing they were accepted as those representing the normal population. But then came movies and advertising soap in papers and advertising soap on boards many times the size of reality and by making money advertising changed the thinking of man. To buy soap was the in-thing and that liberated the modern age from the depressing past that the older generation was burdened with. Wherever your eyes would take you, there was some commercial scripture telling you that cleanliness was next to Godliness and therefore having a bath was presumably equal to with having a conversation God Almighty. Doing this mental rebalancing was done systematically to get the public trendy at the time advertising started getting the world moving the way the Mammonites planned and being clean began being very fashionable indeed. You had to wash more often to be civilised and cultured. The educated was clean and the want-to-bees were a dirty lot. If you had a tip pens worth of brains and a thought of decency you will use more soap more often.

If you walked past someone and you did not smell like a bar of soap and did not leave an aroma like a lump of soap behind you were a dirty gutter boy who was uncivilised, crude and uncultured. So you had to bathe as often as you could to smell like a walking brick of soap. You had to bath very regularly otherwise you were a peasant without reading skills or any cultural morality. All and everyone forming all looked down their noses at your indecency because not bathing often made you immoral and disgusting. If you had a bath you were well educated, very intellectual and very modern and the sun shone from your arse. The downside of this endeavour of bathing was that the world was not yet accommodating this process of utter cleanliness because back then in the living quarters of the time then the process of having a bath was one of labour. The water in which you bathed was boiled on a Dover stove in a massive pot. The boiling was done after the nighttime meal was cooked and the bathing started after everyone had eaten. The water was carried to a galvanised bath outside the house because no room was set aside for bathing. I know this personally because I grew up on a Transvaal farm where electricity was a commodity that belonged to town folk.

To get the water boiled on the pot took about half an hour and this water boiling only happened after dinner because before dinner food was cooked and a meal was prepared. This was a waiting process you had to go through every night. In times before the war, (there was always a time before the war because there was always a time during a war because there was a country such as Britain always going to war with everyone only excluding the Portuguese and the someone at present is Iraq because they have oil and having something the Brits and Yanks can steal, Iraq are declared the Axis of Evil that has to be destroyed and therefore they are experiencing a War with the Anglo American Allies at present) the tediousness of the process of bathing forced everybody to pick a day during every week that was the day that that particular family member had a turn to bath. Because of necessity it was impossible for every one to bath every day so it was primarily a full nights bother to get your hair washed and your body sanitised and cleansed. Except this did not apply to me during my boyhood days where I became part of this bath as a dayly process while enduring circumstances befitting a bath a week conditions. I had to boil water not on a Dover stove but on a stove called the Ellis de Lux but while the stove was special for cooking food it was as tedious as a Dover was when it came to boiling water. It had a boiler at the back but that was for hot water in the kitchen and therefore I was not spared this pot with the boiling water every night of the week. Then when this water boiled in the large pot I had to carry this large pot filled with boiling water. This meant I had to move a few gallons of boiling water to a bath that was either in some room (if it was winter) or outside in a covering shelter that guaranteed our privacy. For a little boy less than ten carrying this pot with a few gallons of hot water was laborious to say the least and spilling only a drop was a painful affair. Even a drop of boiling water dripping on bear feet had you jump but not for joy. However, notwithstanding and nevertheless you had to take your bath before you took to bed.

That was a rule set in stone. It was like reading your Bible by candle light after having you bath then praying and then going to sleep. That was also set in stone. Then again I was part of society where the town folk had running water and two taps built into the wall that was serving a large bath. This was in the late fifty and early sixties. Those in town had a bath plug you could pull out and the water would run into a drain. I was not that lucky. This system they had in town was not the system I was used to on the farm so I had to drain the water with a bucket and that was bucket by bucket. I was part of the bath-in-the-house era while I was going about in the previous era where bathing was done once a week. Before the war everybody lived in a house that had bathing commodities outside. I grew up in a dual culture and had experience on both sides of the war. While my friend also had a bath I had to bath outside every night in a lamp while carrying my water

When we were living on the farm back when we were seen as the unsophisticated because we had to shit in a long-drop situated far outside the house and we ate inside the house in the dining room. Then when we moved to town where we became very sophisticated so therefore we had to shit inside in a lavatories while we ate outside calling the process Bar-B – Queue or as we in South Africa call it having a "Braai" coming from the Afrikaans word "braaivleis" that translates directly as "roasting meat". This remark is quite beside the point but duly valid. In my days of growing up if you are old fashioned you shit outside and eat inside and if you are modern you shit inside and eat outside.

Going back to soap; after the war there was money to build new accommodation and building new accommodation brought about that there was money to be made by those already floating in money. Electricity and water was brought to every house by cable and pipe. Water was heated by electricity. Every house no longer had a need for a bathroom because there were three in every house. This got the soap advertisement selling and the soap advertising got even bigger after Television invaded all houses and destroyed civilised privacy. Then a new range of story telling by camera brought about the name for it as it was called "TV-soapies". The Movie-soapies were dim and mindless enough to make every housewife cry because their life brought no opportunities or reasons to cry and so crying became entertainment while they were told to use soap to bath so that their morals could also be washed away.

Nevertheless it was promoting soap with making a large profit in the selling thereof that got the human race to reject the natural smell of humans in favour of smelling like soap. It was rejecting what God gave to man as a sent in favour of smelling like perfumed cleaning material. When a man smells like sweat and manly flavoured work resulting after smell such as he would smell when finishing twelve hours of toil that no woman on earth lesbian or otherwise could do, he then is thought to be a scruffy beast to be avoided. This ridiculous but quite acceptable scenario is now in such a common practise it is done in all aspects of life. If you smoke you will die. If you eat fatty red meat you will die. If you eat oily chips you will die. If you run or if you don't run you will die. If you eat liver you will die. If you drink wine your baby will be a misfit. If you eat pork meat you will land in hell. If you eat turkey instead of fish you will die.

Everyone wishes to be God and force you not to eat so that you can live forever. Any old arsehole that has written a paper on whatever wishes to put the fear of God into you by telling you that if you don't do exactly as he scares then you will die. If you do not fear his finding you are going to die.

I have news for you…you are going to die whatever you do. The only thing that is as sure as tomorrow arriving by sunlight is that in the event of you that went through a process we call birth then you are going to die. If your mother gave birth to you then you are going to die and whatever you do might slow the process down but if you do that you will die of old age fading away and that is more horrible than any other death I can think of. She could not even go to relieve herself for years and in the end she looked at a sealing for days at a time being to weak to sit up straight. I saw my mother disappearing into a vial of demeaning loss of self respect and I saw my father getting a heart attack at the age of fifty one dying in seconds flat and if I had a choice in the matter, I will choose to go like my father did at the age of fifty one. I can't die at that age any longer because I am way past that age but still my father had a good life and spared his health very little as he enjoyed everything he was told by his medical doctor to avoid. Then how correct are all these arseholes trying to chase Godly admiration and eternal fame compared to the likes of me telling you that you are going to die whatever you do and I do not try to claim fame with this announcement. Every researcher portrays what he found as if it is going to put the earth in another orbit.

If you wish to be clean you wash three times a day. If you wish to be healthy you eat crackers with water and hope you will not last long because it is horrible to live healthy and enjoy life simultaneously. On all levels everyone is doomed to be a robot by those thinking they are the wise and they can help you not to die while the only thing you are eventually and finally going to do is die. It is not what they tell you that holds significance but it is what they prevent you form finding out that becomes critical to the way you live. Now hear this and come to terms with what you don't know and find out why you don't know the following.

Reading this book will **intellectually** be **very challenging** to any person since what I say was never yet published. That I in principle disagree with science's accepted principles on very basic issues is a fact that is undeniably true. What you read about the principles I propose is new to everyone alike. However, I found that ordinary persons with a scholastic physics background cope with the difficult explaining much better than does Super-Educated-Masters. The Super-Educated-Masters have information stored by culture and if they can't bring the information to mind by recognition they fail to understand new science concepts. You are going to read this in a letter that was sent to me. The Super-Educated-Masters have preset conditions they prescribe to information and they can't break their mould. The purpose with which I wrote this book is to get around the network of Super-Educated-Masters who strangle any form of science that does not fit their views or match their liking. If what anyone says does not stroke with what the Brainy Bunch says who controls physics and agree with "Mainstream Science" or echo their thinking, they just smother all intellectual publication on the grounds that it is not fitting their profile on science. I disagree most strongly but I do also supply proof thereof where Mainstream Science blocks the publishing of my views on science that does not compliment their views. If you believe science is more accurate than God, then live your fantasy out and don't read further. If you want to know the truth about how students and the public are brainwashed by mind control in science this will wake you from your slumber. Read this and wake from the culture you believe in; that which science has lulled you into and made you accept science as the absolute undeniable rock fast truth by instating it as a religiosity then stop reading or get your tranquillising anti depressants next to you with a large bowl of water. You will find some mathematical equations, if you are not familiar with it ignore it because it shows the silliness of "Mainstream Science". If you don't read it you will still understand the explaining by reading the language where I explain it. "Mainstream Science" hides behind maths. I need help to fight their fraud and I need you to help me fight them. What you read I prove even in this book and I dare any one to prove otherwise or reprimand me.

I disprove most of the views science has on cosmology, which is the science on stars and galactica, and how the Universe formed. I start by introducing a new view on how gravity applies and from there dismiss most popular claims on science. Any person who challenges me on the legitimacy of content given within this book, please do so any time. However, do so on the grounds of all the information provided in this book. What I say is don't be opinionated and then reject on what is culture but inform yourself with what I provide and then gauge what I say in relation to the half truths mainstream physics provide.

THE CONTENT OF THIS BOOK A Conspiracy in Science in Progress AIM TO SHOW ORDINARY GOD FEARING MEN AND WOMEN, THE TRUTH ABOUT SCIENCE BUT IT IS NOT A RELIGIOUS BOOK AT ALL. Moreover it shows how students are mentally abused and brainwashed to believe physics. The book opens facts that have been covered up by everyone teaching science for the past three hundred years or more. Please look at the meaning of the content and the accuracy of what this brings to mind and see what they hide to dupe all of us. It is where the entire defrauding of everyone starts. They are pushing the button of the Pope when he dismissed the work of Galileo Galilee but they are still busy with the same technique only this time it is to save their hides and not that of the Roman Catholic Church. The content however is more accurate than any printing you can purchase that is underwritten by mainstream science and that I guarantee.

You are going to read some **mathematics in equations** and expressions in **mathematical formulas** placed **to defend my position** but if you <u>don't like it</u> then just <u>skip</u> the **mathematics** because the content and grounds the mathematics prove or disprove is not important and it is there to disprove the Brainy Bunch. It is there not to scare readers away but to silence the Brainy Bunch critics by showing them the foolishness of their arguments. By using mathematics those professing intellectual splendour have been cheating the public and have been brainwashing students for centuries. They use mathematics to mislead every person that is part of the public. Should you not believe me take what you read and challenge any physicist you come across and let him or her give you a satisfactory argument. That cheating is how they do it and I have to show and uncover the dishonesty in mathematics.

I am going to show that there is no such a thing in the Universe as mass. No that would be incorrect. I am going to prove that mass is an invention Newton created to fool every person since Newton. It includes all those that pretended to be intelligent. It is a fairy storey coming true. You've heard about the King and the magic invisible clothes, apparently magical because only the wise and the clever in society could see it. Everyone was fascinated by the beauty of the clothes The clothes was only visible to the very wise and because no one wanted to be seen as being a fool in the community, and least of all the King wanted not to be seen as a fool that couldn't see his own clothes, the lot in the Kingdom saw the fantastic magic clothes. As the King paraded naked through the streets a young girl shouted: "The king is naked!" Then everyone realised the King was naked and even the King realised he was the biggest fool in his entire Kingdom. How does this have any connection to a book that is devoted to science in any way you might ask. Isn't this a famous story about a bunch of fools pretending, well it is not a story in science but it applies to everyone in science. Everyone since Newton that knew physics was able to see mass because no one wanted to be the mindless fool that couldn't see mass. So much does this story apply to physics that we can only change the name of the King to Newton and everything else in the story is in place. Newton created mass by magic and everybody since then could see mass. This was the case for three hundred years where everyone in science saw mass as clear as daylight until I came along in 1977 an I was unable to see mass as a factor when seeing all the evidence that disproves mass as a cosmic factor. I was the simpleminded student that couldn't see mass and I still can't see mass. As you read this you feel agitated by my remarks, don't you? At a later chapter I give you the entire story behind this letter that I received from a Journal in response to an article in which I showed how gravity comes together by forming the Coanda effect. The Coanda effect is when liquid of any form and solid of any form interact.

The following is the response I received when I submitter a paper on how gravity forms by singularity forming.

Dear Dr. Schutte,
You submitted an article of 15 pages to the Annalen. The content of this paper doesn't constitute a theory in physics. With a lot of words and some simple algebraic relations, there is no way to "explain" the world of physics. You seem to be out of touch with modern developments. This is also shown by the fact that you don't quote any relevant literature.

I am sorry to say, but the Annalen is not able to publish your work. I am sorry for having no better news for your.
Best regards,
Friedrich Hehl

Co-Editor Annalen der Physik (Berlin)
--
Friedrich W. Hehl, Inst. Theor. Physics
* University of Cologne, 50923 Koeln _____/\/_____ Germany
fon +49-221-470-4200 or -4306, fax -5159
hehl@thp.uni-koeln.de, http://www.thp.uni-koeln.de/gravitation
* Univ. of Missouri, Dept. Phys. & Astr., Columbia, MO, USA

Dear Prof Friedrich W. Hehl, I have received your e-mail reply and I wish to respond on your letter. The article of 15 pages to the Annalen had in mind to introduce a very wide-ranging concept contained in many books. I wish to promote books in which I introduce a much larger and much more detailed cosmic picture. It is six books that actually form seven volumes of one theme supporting **The New Cosmic Theory**. I wish to unveil a totally new approach to the thinking in cosmology. The concept is proposed in the article I sent to you which is "revealing" **The New Cosmic Theory**

In the article as much as the theme I wish to go where no one ever attempted to go before. I introduce the Universe of singularity, a state in which the Universe still is because it is a state from which the Universe grows. It is where material in a dimensional dynamic does not apply because it is where Einstein said, "the Universe goes "flat"". I show you how and where the Universe goes "flat" I will guide you to the point where I go…so that you may see where my books and the article lead you. It is in the domain of singularity

When you read work about the Big Bang you have to go right down the development (in reverse order) to the point where the Theory of the Big Bang points at a spot named singularity. It shows the very start from where all material developed. At that point one will find The Absolute Relevancy of Singularity and there has never been any attempt by any person ever to venture beyond the dimensional birth of the cosmos, which is called the Big Bang by going into the era where singularity prevailed. I take you there in my books as well as the unpublished article.

However, going there requires a very high degree of concentration and calls for understanding that a very little number of persons are capable to show. I try to show how the Universe goes "flat" as Einstein said the Universe goes "flat". Even by completing this unimpressive letter you will also know how the Universe goes "flat". Even where you failed to read the article I sent you, then by just reading this letter you will be able to find where singularity takes the Universe "flat". But it requires a mental capacity to understand because where I venture no one ever in the history of mankind reached into before.

I do not speculate but even in the unpublished article I show with pictures and sketches as well as "some simple algebraic relations" where to go to where the Universe starts, but you failed to read that because you are opinionated as to what conditions should the Universe have before the Universe will allow any one into physics.

That is a pity.

One should learn from the cosmos and not tell the cosmos what it must be to qualify as the cosmos. Then in the article I show you by almost taking your finger to the spot, the very point where the Universe ends and that too I qualify. You might dispute my arguments and show me about what you disagree, but it shows very little understanding of reason on your part about qualities man should have before understanding the Universe.

I go into a Universe that was in place before light was in place in the Universe and only darkness prevailed because light calls for space and in that era of singularity space was not even a thought yet. I show why the Universe goes "flat" and in a "flat" Universe only the value of 1 holds value since singularity is 1. If you can understand 1 or $5^0 \times 7^0 \times 3^0 = 1^1$ you have all the mathematical skills required to understand the applying concepts. To reach a value of 1 does not require big mathematical equations but to reach singularity requires 1.

The collection I named The Absolute Relevancy of Singularity: The Theses and the collection as such forms a small introduction to the thirty-two or so books I wrote on various matters concerning physics with gravity in mind, but **The Theses** as such in the entirety of the four books does not officially even start to introduce the spectrum of every aspect of my work. I have been in contact with numerous Academics and about one in one hundred reply.

When the one in a hundred reply, the academic always uses a most aggressive tone which I came to accept as what I receive from academics, and because of that I was most delighted to find some kind remarks from you as a practicing academic, and might I add, the first such kind remark in ten years of my trying to contact any person in physics that would take note of what I have to say about a new line of thought, because the few others that replied were extremely aggressive about me confronting Newton. I only began to submit books to publishers after twenty-seven years of studying Newton and the role Newton plays in cosmology and thereafter which was ten years ago I began promoting these ideas. The New Cosmic Theory is a process wherein I try to introduce a study that is ongoing for about thirty-seven years, give or take a few and I did not jump into the frying pan having my first thought about the matter published as an article when I sent the article to the address of Annalen der physics.

This is the ridiculous concepts that Professor Doctor Friedrich W. Hehl, Inst. Theor. Physics supports. Because I don't conduct tainted mathematics and highly suspicious Newtonian concepts covered by completely ridiculous malfunctioning concepts, I am despised as the ridiculous small-minded novice who is openly very feeble in thought. I do not support complete rubbish as this formula indicates the Newtonian vision of physics represents, but for my failing to go along with total trash my work gets rejected time and again by every science publisher I approach.

I was told, "With a lot of words and some simple algebraic relations, there is no way to "explain" the world of physics". Now I am going to go as public as possible and to use "words" to explain physics and to show how incredibly untrustworthy the world of Newtonian physics is. All you who think Newtonian physics is in reliability equal to God professing through the mouth of Newton don't be cowards by not reading my work as you always do but confront the conspiracy all of you are part of.

What you should answer is "how did this formula $F = \dfrac{r^2}{M_1 M_2}$ become equal in every respect to this formula $F = G \dfrac{M_1 M_2}{r^2}$ and never change any aspect of its characteristics or identity. You go and do this swapping of relevancies in your mathematic class and then see what your mathematics teacher thinks of your modus operandi. This is what physics believe is true and because I question this I am wrong.

Let us test Newton's attraction theory in practice.

Let us now proceed and test the basis on which all physics are founded

Let us then test the truthfulness and believability of the foundation on which the entirety of Newton's physics principles rests. I show you the absolute accuracy of $F = G \dfrac{M_1 M_2}{r^2}$

According to the Big Bang principle everything is growing and moving "further apart" and therefore there is no "mass" pulling anything closer according to "body mass" because nothing has "body mass". But my views are rejected because my views clash with the wisdom of **the Masters forming the principles who direct the thinking of science in science according to science**.

If you want to read how corrupt the thinking is in science then read on but do so with the sole purpose to prove me wrong. That is one thing not one of the **Masters forming the principles who direct the thinking of science in science according to science** carrying all the wisdom they can manage could ever do! Modern Science still can't explain the following questions because of a lack of proficiency.

Have you ever thought about…where is the centre of the Universe?

That I can answer…and I also can answer…why is the Universe still growing since the Big Bang…why did the Universe start so very small…why did the Universe fit into a neutron at one time…how did everything expand from fitting into a neutron…why does space grow from small to large…where is it going while it is growing…why was the Universe any specific size…what was everything before the Big bang…if you read this you will know all of this! No one in science dares to read my work leave alone appreciates the revolutionary nature thereof. Any association with my work must condemn the own work of such an associating person. Whatever is deemed now to be accepted science would by accepting my work then become what is the past tense and belonging to the past in science. My work unmasks the flaws that those in power of concocted science principles kept in place and in thought for centuries on end as forming the untouchable truth. It will then be rejected to show the holes formed as accepted science! They try to silence me but surely somehow somewhere I have to break through with my message! I bring you a true form of science as never seen before in all of history. I do that by disposing of the conspiracy that hides the incorrectness and the failures that haunts science today. They use mathematical interpretations that take the explaining away from logical facts.

Knowing my credentials better would lead to understanding my motivation behind my effort in trying to bring about total change to cosmic physics and science. You will find I never compromise truth for friendship and in that light I say what those in science needs to hear and not what they want to hear which then in that will please who ever should feel pleased.

Meet the Newtonian physicist. In this book I try to introduce the reader to the brilliant Newtonian conspirator that has been dragging all of intelligent man by the nose for three centuries on a string.

The more the conspirator pretends to be an intellectual physicist the better a fool those conspirators become. This is the picture that paints the modern Physicist in full colour.

Meet the the modern Newtonian physicist who hides the truth about everything concerning the science the Newtonian do not know and do not wish to know and do not want anybody else to know. Isn't he cute with the woolly fluff and all? Don't you think the brilliant Newtonian Physicist looks rather sheepish? For him earning that look there is a reason! Look at him for he looks sheepish because he acts sheepish because as he follows he never questions in what he believes and then brainwash students to do the same. He follows Newton like a mindless sheep follows the rest of the flock of sheep. Read how clever the physicists are hiding their stupidity from students and the public alike by pretending to be clever. By enlisting thought control those teaching physics force students to believe in science and to believe science. Those who force-feed non-existing dogma do so to brainwash students to hide the incompetence of "modern science" so they can rule supreme while ignoring the truth that they deliberately hide by concocting a conspiracy. To keep everyone unguarded they practise a conspiracy by which they perform an accepted practise of thought control on students to further the false dogma presently in place. I try to blow the whistle on such a practise but if anyone accepts my resolution that admission makes every thesis any person that is part of "Mainstream Science" ever wrote as his or her thesis on physics then becomes **science fiction.**

In this book I prove that what you think of as Science, is only ideology based on rumoured disinformation made up by those making up facts, as they hide what they don't know, hide what they don't want to know and conceal what they don't want others to know. By proclaiming they know everything while pretending to know all there is to know, they preach what they falsify to sustain their religiosity in the form of unproven dogma. Science tells what they wish and reserve what they wish not to tell and hide what they don't wish others to know that they don't know. My biggest undertaking is to prove that what science presents as credible facts is unfounded and incorrect. Science is believable because culture sustains Science. My problem us busting a culture people trust as Science and be a whistleblower about accepted religiosity people unconditionally believe in as being trustworthy. The public never once thought to question what they were taught as science. This forms a problem since what is presented as science is falsified information and that which we need instead is a new approach that is truthful about science. Therefore no one realises what big a problem there is in the level of correctness there is in the approach to reality science now presents. Science never shows the support or proof in what they present as ultimately credible facts. Bluntly put; Newton was never proven to be correct but only accepted. I firstly need to promote the need we have for the truth to replace what we have as science before I can promote what the truth is that we need to form science. Science hides their flaws under a profile that they present as the uncompromising truth while in fact what they present is lies covered by a veneer of truthful facts and this then neglects the urgent need for the truth by denying there is a problem. If you don't believe this statement I make on the first page that you read then I challenge you to read on and challenge me on any fact I present about the untruthfulness you think of as science and if you disagree with me send me a e-mail, but if you believe me then help me to challenge the establishment about their scandalous disinformation that they hide to commit the conspiracy they call science. Everyone that has contact with physics is unbelievably naïve while everyone is getting hoaxed and your children are being brainwashed to unconditionally believe in science which is totally without credit. Everyone accepts without questioning the science because they believe in the persons teaching them and not necessarily in the credit of the information they are taught. Read what it is that science hides to put in place this notion that they know everything and what they know are without blemish.

There is an obvious conspiracy in science that everyone misses and about this fact there is no doubt. It runs as deep as it can go and involves every person who studied science worked with science teaches science and thought about science. In short it touches every person who lives notwithstanding culture, ethnicity, race or religion. Everyone on earth believes the science taught at schools and everyone is working with fake science and conspires to promote science fiction passing it off as Newton's science.

Moreover what will you do when you know the truth or are you satisfied to be told what to believe is the truth without ever investigating by personal standards as to finding the truth about science. Then go on reading and confront the truth, as you never had ever before.

The truth about a conspiracy is that everyone involved with the conspiracy will fight tooth and nail to stop the conspiracy to get leaked and leave those behind the conspiracy and all those feeding from the corruption of the conspiracy exposed.

Those who have the power to maintain the conspiracy would keep the waters as still as possible as to draw no attention to such conspiracy.

A true conspiracy has to be as quiet and as unseen as it could be.

A true conspiracy must involve everyone without anyone detecting even a hint of what the conspirers hide.

There are many conspiracies going on such as the banks involvement with crime and the bankers profiting from gamble rackets and drug selling.

The same goes for the Insurance business profiting from lenient sentencing of courts holding very merciful judges in office so that the crime cases and the burglaries, car theft, hi-jacking and all other forms of crime will shoot through the roof every year and grow by thousand percentage points from which insurance will sell ever-more cover-policies.

The more crime is about and committed daily, the more people need insurance covering and thereby the more money insurers bank giving bankers profit to spend on the stock exchange by controlling the economy.

Bankers buy democracies with money they give politicians to write laws that protect the rich against the poor. In time I might write about this but now I cover another conspiracy, a bigger case file about a conspiracy everyone on earth participates in…it is about…

Should anyone desire to contact me with an opinion about my views on science then please do by using mailto:info@questionablescience.net or mailto:info@singularityrelavancy.com but please do so when coming to the end of this book at least and then see if you understand the entire concept that I introduced. In most cases all your questions are answered further down the line and I am not fond of repeating what is already said in the book. The book forms a line in explaining and the concepts are best understood if the reader follows the designated line.

This book started off as a website to inform about a science conspiracy but although reduced still it grew into a book that serves much more information than what I first intended to supply. You will see many new aspects about gravity please make sure you understand what you read. It grew into a comprehensive study on cosmology. At times you may observe while reading this book that it seems as if my frustration will ring through like the chiming of the Big Ben Bell.

For that there is a reason.

At times my frustration and anger will boil over drowning my politeness and that is true, which I admit. For twelve years I have had the answer that would correct the philosophy that has a stranglehold on cosmological science.

I discovered the building blocks of nature where my discovery puts all other cosmic aspects of science into science fiction.

Those who force-feed non-existing dogma do so to brainwash students to hide the incompetence of "modern science" so they can rule supreme while ignoring the truth that they deliberately hide by concocting a conspiracy.

To keep everyone unguarded they practise a conspiracy by which they perform an accepted practise of thought control on students to further the false dogma presently in place.

I try to blow the whistle on such a practise but accepting my resolution makes every thesis ever written science fiction.

Therefore no one in science dare to read my work leave alone appreciate the revolutionary nature thereof. Whatever now is deemed to be accepted science would then become what is the past tense in science because the flaws that those in power of science principles kept coated for centuries on end as untouchable truth will then be rust that breaks the surface to show the holes!

They try to silence me but surely somehow somewhere I have to break through with my message! I bring you a true form of science as never seen before in all of history and I do that when I dispose of the conspiracy that hides all the incorrectness and the failures that haunts science today. Science is accepted as the most righteous information available to man and that is a scam.

Read The Absolute Relevancy of Singularity The Theses that is written as the first introduction to introduce singularity forming gravity in the new The Absolute Relevancy of Singularity The Website offering a theorem that for the first time explains gravity in a founded manner.

The Absolute Relevancy of Singularity The (proposed) Article Free of Charge from Lulu
http://www.lulu.com/content/e-book/the-absolute-relevancy-of-singularity-the-unpublished-article/7747133]

The Absolute Relevancy of Singularity The Dissertation
http://www.lulu.com/content/e-book/the-absolute-relevancy-of-singularity-the-dissertation/5994478]

The Absolute Relevancy of Singularity in terms of Newton
http://www.lulu.com/content/e-book/book-0-the-absolute-relevancy-of-singularity-in-terms-of-newton/7190018]

The Absolute Relevancy of Singularity in terms of Cosmic Physics
http://www.lulu.com/content/e-book/book-1-the-absolute-relevancy-of-singularity-in-terms-of-cosmic-physics/6624181]

The Absolute Relevancy of Singularity in terms of The Four Cosmic Phenomena
http://www.lulu.com/content/e-book/book-2-the-absolute-relevancy-of-singularity-in-terms-of-the-four-cosmic-pillars/7181003]

The Absolute Relevancy of Singularity in terms of The Sound Barrier
http://www.lulu.com/content/e-book/book-3-the-absolute-relevancy-of-singularity-in-terms-of-the-sound-barrier/6621856]

The Absolute Relevancy of Singularity in terms of The Cosmic Code
http://www.lulu.com/content/e-book/book-4-the-absolute-relevancy-of-singularity-in-terms-of-the-cosmic-code/6625975]

The Absolute Relevancy of Singularity in terms of Life
http://www.lulu.com/content/e-book/book-5-the-absolute-relevancy-of-singularity-in-terms-of-life/6626316]

The Absolute Relevancy of Singularity in terms of Investigating Kepler
http://www.lulu.com/content/e-book/book-6-the-absolute-relevancy-of-singularity-investigating-kepler/7179110]

The Absolute Relevancy of Singularity in terms of The Theses
Project URL: http://www.lulu.com/content/e-book/book-7-the-absolute-relevancy-of-singularity-the-thesis/7587475]

The Absolute Relevancy of Singularity in terms of A Cosmic Creation
[Project URL: http://www.lulu.com/content/e-book/book-8-the-absolute-relevancy-of-singularity-in-terms-of-a-cosmic-creation/7846786]

mailto:info@singularityrelavancy.com or
mailto:info@questionablescience.net

This is the book showing everyone that

Science A Conspiracy 2 Misrepresent The Truth

ISBN 978-1-920430-54-2

Written by P.S.J. (Peet) Schutte. ©KOSMOLOGIESE EN ASTRONOMIESE TEGNIKA

All rights are reserved. Publishing, or alternatively reproducing electronically or using any other means of duplicating without the consent of the copyright owner is strictly prohibited. No parts of this book may be reproduced in any way that is possible.

For more about the conspiracy also visit
www.questionablescience.net or http://www.singularityrelavancy.com/
and www.sirnewtonsfraud.com

This is a simplified or reduced version of

A Conspiracy in Science in Progress

ISBN 978-1-920430-05-4

http://www.lulu.com/content/e-book/a-conspiracy-in-science-in-progress/8352010]

I wish to again reaffirm my commitment and motivation behind my writing the book that you are reading. It is, believe it or not, but it is about modern science never fully revealing the truth and as part of physics hide the truth by doing what those who don't understand physics but still teach physics do at schools and Universities.

mailto:info@questionablescience.net

Please do take note of the following warning:
The content of this book might seem to be intellectually challenging and reading it might require much concentration at some points because the ideas I put forward will be very new to any person reading it and will sometimes seem to call on intense detailed analysis of some of the new information I present in this book. One example for instance is reading how I prove the Universe started off at the very beginning before the beginning those in science refer to as the Big Bang.

I am able to introduce this breakthrough that I claim I achieved because I found the building blocks used in the building of the Universe and by applying the four cosmic principles I am able to go as far back as before the Big Bang event. The Big Bang only happened recently when atoms formed the known Universe. In this book I do not touch on this aspect but I go to the period predating the Big Bang era as I inform the reader how the start took place when the Universe first came about. I found the working principles of the cosmic principles called the Roche limit, the Lagrangian points, the Titius Bode law and the Coanda effect that are in place and forms that which forms the construction layout of what builds the entirety of the Universe. The horror in this is that notwithstanding the importance these principles hold almost no one knows about them.

I ensure you that what you read even in this book was never written before and absorbing new information is difficult. You are confronted with facts you never heard of before. So therefore notwithstanding the level of the person's academic qualifications or how well developed background the person has in the field of science, this is not your average garden-variety storybook. However, in the end after you have come to realise the conspiracy in science, you will be a lot wiser than you were before. The information might ask for a lot of your attention span at times, but then again challenging the mind to think is what makes life that more interesting!

I found a mistake and then realised I had to search for a solution to the mistake. First I had to located what was wrong in physics and from there formed a correct approach. I compiled my presentation of it in a theses that I call The Absolute Relevancy of Singularity and then six separate thesis parts forming the theses published through LULU.com which I saw as the only manner whereby I could generate funding by which I would be able to have the twenty seven books I already wrote linguistically edited and then to have the books published on a Print-On-Demand basis. I compiled a new cosmic theory by which I eliminated all the incorrectness that Newton has burdened science with but with this being my opinion I did not find a garage full of academics supporters waiting to applaud me and to uphold my views on the matter. Yet still I was not going to be ambushed by their relentless stonewalling my efforts and blocking my efforts in introducing both the incorrectness and the new cosmic theorem I concluded. Their mannerism in blocking and frustrating my opinion when showing the mistakes in science convinced me about a Conspiracy in Science in Progress and this spurred me on to tell the entire world about their brainwashing students minds. By the manner they selectively withhold information when teaching science, amounts to deliberate brainwashing of students in physics by "normal" education practises.

Trying to convey my message kept me busy for the past going on to twelve years on full time basis whereby I was trying to introduce my findings to many academics without finding much joy from my efforts. This past eleven years plus saw me go without any income as I tried to get my theorem recognised as well as get my warning noted. Going without a steady income left me almost destitute and in order to find a manner to get my theory across to the attention of influential readers, I decided to publish a theses of six books electronically as to try and get around the stranglehold of Newtonian bias controlling science at present worldwide. I decided to publish electronically which those in power do not control. However to get people to believe me is to change science that everyone believes as culture.

With my first language not English and the books not linguistically checked by an expert there are bound to be language errors that readers will notice. In the past I tried to check my work myself but after checking say one hundred and fifty pages for language corrections, then after days of toiling instead of having corrected work I ended having four hundred pages of newly written information which is still not linguistically corrected but holds a lot more information. The language and spelling errors compiled instead of reduced. This is because my priorities lie elsewhere. I aim to spend money on correcting the work as far as language goes, as I receive money in the selling of my theses and in the hope that I will receive money. I will have all my work including the one you are reading edited professionally and corrected as I find money to do so...But first I have to get the public aware of the problem to get the academics to appreciate the problem. In everyone's mind science is more perfect than religion is.

In the event of any readers who may have questions concerning more facts as it is presented in this book, please feel free to contact me, PEET SCHUTTE. All information divulged came about through independent self-study during the past thirty-two years or so. I have to warn the readers that the topics are showing a very new approach with no quick answers abstaining from proof or holding just a few lines and the information is new in nature but not hard to grasp. Should anyone desire to contact me with an opinion about my views on science then please do by using the worldwide web addresses mailto:info@questionablescience.net or mailto:info@singularityrelavancy.com but please do so when coming to the end of this book at least and then see if you understand the entire concept that I introduced. In most cases all your questions are answered further down the line and I am not fond of repeating what is already said in the book. The book forms a line in explaining and the concepts are best understood if the reader follows the designated line. This book started off as a website to inform about a science conspiracy but although reduced still it grew into a book that serves much more information than what I first intended to supply. You will see many new aspects about gravity please make sure you understand what you read. It grew into a comprehensive study on cosmology. At times you may observe while reading this book that it seems as if my frustration will ring through like the chiming of the Big Ben Bell. For that there is a reason. At times my frustration and anger will boil over drowning my politeness and that is true, which I admit. For twelve years I have had the answer that would correct the philosophy that has a stranglehold on cosmological science.

I discovered the building blocks of nature where my discovery puts all other cosmic aspects of science into science fiction. Those who force-feed non-existing dogma do so to brainwash students to hide the incompetence of "modern science" so they can rule supreme while ignoring the truth that they deliberately hide by concocting a conspiracy. To keep everyone unguarded they practise a conspiracy by which they perform an accepted practise of thought control on students to further the false dogma presently in place. I try to blow the whistle on such a practise but accepting my resolution makes every thesis ever written science fiction. Therefore no one in science dare to read my work leave alone appreciate the revolutionary nature thereof. Whatever now is deemed to be accepted science would then become what is the past tense in science because the flaws that those in power of science principles kept coated for centuries on end as untouchable truth will then be rust that breaks the surface to show the holes! They try to silence me but surely somehow somewhere I have to break through with my massage! I bring you a true form of science as never seen before in all of history and I do that when I dispose of the conspiracy that hides all the incorrectness and the failures that haunts science today. Science is accepted as the most righteous information available to man and that is a scam.

There is a conspiracy in science and about this fact there can be no doubt. Once you finished this book and you still disagree about the conspiracy going on it then will be because you are a professional physicist that are paid a huge income to spread the falsified science and contribute to the well being of the conspiracy. However, you will be convinced either way. At the end of this book you will believe convinced that it runs as deep as science goes and involves all person that studied science, worked with science, teaches science and thought about science. In short it touches every person that lives. Everyone works with fake science and conspires to promote science fiction passing it off as Newton's science.

I am Petrus Stephanus Jacobus Schutte going by the nickname of Peet and I am a married male, with a sane mind and I hold very sober habits being a lifelong teetotaller, therefore my mind is clear. When you read what you are about to read it is not the hallucinations of a drunkards mind.

I see myself as very responsible and am not well liked because of my straightforward personality where I say what needs to be said to maintain honesty above friendship and what I say is always not infusive or congenial to make friends. I wrap no feelings in cotton wool for the sake of peace. I mention these facts to establish beforehand that I am not a danger to society. Although I am the only one in the entire world that disagrees with modern mainstream science I am not criminally insane. I would not attack your dog before your dog will bark at me just because I hate your dog. I would rather attack you and not your dog after he barks at me just because I don't like you but I do like dogs. I say this to convince you that I have never been jailed in all my life on grounds that I attempted to mislead or fool or tried to mislead anyone, as you can see. I try hard to be honest

The proof of my honesty is that I am one of the poorest people in society. I write books that don't sell because I try to convince people about a mistake no one on earth is aware of and therefore no body bloody cares except me. I am the only one that can see a mistake and seeing it in the full consequence thereof by what it holds to the entire human race it is frustrating me senseless. Take it from me that not selling books is not a very profitable enterprise and is frustrating at any level.

If I was less honest and went about bullshitting everyone about the accuracy that Newtonian science portrays in the modern era it would at least sell some books because some dinosaur Newtonian physicist would be pleased about it but then I was a cheat although I would have been richer but being as poor as I am I in favour of being as honest in what I am, I can't be a cheat.

I studied physics and in my first year of studies I came upon a mistake concerning physics. My roommate told me it was Newton that said what I disputed and I remember telling him I don't care who this Newton fellow is but he is wrong and ever since then I stuck to what I said. That discovery made me disagree with the establishment forming principles in physics. Later in time I have detected much more than a mistake. I have discovered a Pandora's box of mind manipulation and brainwashing going on to force students to accept science notwithstanding.

If you feel I am exaggerating or that I am a mentally impaired asylum escapee that found a writing pad, ink and paper and that I now start to scrabble senseless suggestions to while away my social frustrations then I challenge you to prove that students are NOT being brainwashed to believe what they are taught in science. I started off being likable and being nice but the deeper I delved into the conspiracy the less I could care about who thought what about what I said because I am going for the truth as hard as I can. If you are a student you better read on…and if you are a teacher I advise you to read and get wise in realising what you do. I am going to give facts that will support me in the claim that science is all about falsifying facts to prove what is crooked in order to make those that should know science look good while in fact those in physics and teaching physics know less about physics than does a cat know about opera.

There are so many conspiracy theories going around, floating like a bad smelling odour in the room, but most of those have no more factual substance or evidential backing by truthful facts than a James Bond novel has. I also have a science conspiracy theory but mine has all the proof and all the substance of authenticity that any argument can ever hold. It involves **physics** as **physics are taught** at institutions worldwide. If you now don't believe me, read on and learn the truth yourself, it is free of charge.

Mine is the most unbelievable one that has most truth. This book is not about any conspiracy ever mentioned before but takes into consideration what worries people.

When you see what science achieves and you hear what I accuse science of, then it seems as if I am a nutcase who escaped from the loony house and I must be the one having very dangerous tendencies to harm the innocent.

I should be locked up and taken from society.

However, read my story and you'll see that I don't harm the innocent. It is those we all trust without reservations that brainwash to forge deceit through corruption and malice.

The blame is with those admirable, yet astute members in our society, the educated leaders in physics that we trust without even thinking about their trustworthiness or question their sincerity regarding their blameless honesty.

Do you believe in the absolute unquestionable correctness of science and that everything that forms science is truthful being far beyond suspicion?

Do you believe that those furthering the science of physics works with no less than absolute proven facts and the facts are beyond any doubt at all.

If you do then I have very bad news for you, **news that will change the grain of what you believe is truthful.** I am about to shock you into reality!

This is going to reveal the biggest conspiracy ever concocted by the human mind, ever since the time a human had a mind to use.

It is A Conspiracy about Fundamental Science. It started when Newton gave us the idea that gravity is the result of mass and in gravity mass is the biggest contributing factor.

I say that mass as a factor in physics and not the value of weight, but mass with the pulling power only pulls stupidity over the eyes of incompetent idiots raving in their personal sublimation. Let see what I say. Would those in science cheat you and withhold the truth from you?

www.singularityrelavancy.com that is telling everyone about
An Ultimate Science Conspiracy, the biggest conspiracy ever concocted.

NATURE put TIME into place TO PREVENT Everything FROM HAPPENING all at once.

This proverb sounds silly and yet, silly it is not.

While this proverb is the truth, science say they don't know what time is…and this is ignorance they hold while living with time for millennia all the while Newtonians say they uphold Galileo and everything Galileo implemented. Gravity drives the Universe as the earth spins around.

In other words what science can't see is that there is a time –connected plan in the form of gravity driving the Universe along. Time is in the form of gravity and the cosmos is driven by time. Time and gravity are the same, which is precisely what science is unable to see.
If gravity drives the Universe, the Universe is powered by time.

When Galileo's pendulum can measure the flow of gravity and record time to a precise science while swinging in space, then the flow of gravity in space has to be measure of the pendulum swinging in air while reading time. The pendulum measures time when swinging through the air that surrounds us and what surrounds us is there because of gravity.

This obvious conclusion about gravity forming time, which is what the pendulum reads, any child can draw but as simple as it is, this went past Newtonian understanding of science for centuries.

They never could realise the flow of gravity and time is one and the very same since it then conflicts with Newton's ideas. Still the pendulum proves that reading time and measuring the flow of gravity is the same. Galileo used gravity to measure time accurately, yet for as long as the pendulum recorded time, this fact eluded science in a way that science doesn't even know about it. It is what science doesn't know that becomes important to man. If gravity measures time by applying a swinging pendulum this becomes the biggest fact in physics and it still went past science because their cleverness brought on their rigged stupidity.

This religious belief in the infallibility of their Godly insight becomes their stupidity. Their ignorance is what science hides. It is that what they hide which is what you don't see. Science hides what science doesn't know which they cover under the larger pretence of their cleverness.

Also science hides from everyone's view what science doesn't know so that students will never realise Newtonians are covering their stupidity and ignorance behind a curtain of arrogance. This conspiracy to not reveal their stupidity then becomes the conspiracy they nurture for centuries. Science conspires to hide by covering their stupidity from open view. The main thing is that science is clueless about time and time is the driving cosmic plan so science hasn't a plan to see the truth.

I have been in fruitless conversation mainly going one way about what I see as their stupidity in defending flawed science. I told them this but with having no response. Still notwithstanding their arrogance about ignoring my showing right from wrong, still those physicists filling high academic office are so infatuated by their superiority and their personal righteousness they can only see the malice that the Pope showed towards Galileo when he differed from the general science views and yet they do the very same today. They are overwhelmed by their correctness.

you will see how those hypocrites condemning the pope having the holier than though attitude because they think they are cleverer than all other intellectually lesser sub-humans. They accuse the Pope of constraining science progress while back then but they torment me in much crueller ways than the pope did to Galileo. At least the Pope allowed Galileo to print a book while they did everything to stifle my efforts when I just tried to be heard by others about my views!

Whenever anyone show those in science that their religiosity called Newtonian science is wrong, they act in exactly the same way as the Catholic Church did to Galileo five hundred years ago, in no less a manner while they uphold in public that they maintain their stand on condemning the Church for not excepting Galileo's principles. The crude thing is that I prove all those filling academic office in science have also still not accepted Galileo to the letter and I challenge anyone to prove me wrong. That is one of the parts of the big conspiracy I see and call it the Mother Conspiracy!

By the swinging pendulum in space while connecting it to the earth Galileo proved gravity is time and that principle goes unnoticed as it diverts from Newton's idea that "mass" producers a magical "force" of a "pulling nature" he called "gravity". This view they can never explain while this is what they will always defend as the truth.

Those oh, so clever Dark Aged wizards pronouncing the upholding of free speech and science liberty for all of mankind while hiding behind their superior positions will befall the same memory as what befell the Pope in the days of Galileo. They will be remembered for being the last of those that were stupid enough to believe that an inexplicable magical "force" of "gravity" "pulled" the Universe by "mass" while E.P. Hubble showed the Universe is not contracting but otherwise expanding.

Their sublimation covering their stupidity causing their ignorance will outlast them for all time to come. As the Pope five hundred years ago is remembered for his ignorant stupidity, so those currently in office would also befall the same fait just because they were as stubbornly arrogant and ignorant as the Pope was back then when ignoring mistakes.

They conspire to conceal the dishonesty that they **don't wish to reveal.**

Part 2

THIS IS THE BOOK THAT

WHISTLE BLOWS
ON A SCIENCE CONSPIRACY NETWORK

You think their abilities are mathematically superhuman: Then you hold on to that thought because I AM GOING TO SHOW YOU THE EXTENT OFTHEIR MATHEMATICAL SKILLS! **This conspiracy is so widely active that every person on earth participates, most probably without knowing, some reluctant and others well knowingly pursuing the goals of the conspiracy with all the vigour they can muster.**

I am going to ask you this again later on but be honest not to me but to your own mind as you answer yes to the following questions. See how gullible you are in terms of your surroundings.

I can tell from personal experience that you purchased this book because you are a person who likes to be on top of events as they happen and you insist on being well informed. You buy the paper every day to read about world affairs going on around you. You take everything as it happens while it happens being as a way to be as awake as any civil person could dream to be. Being this well informed and on top of present matters: Ask yourself if you are truthful about your answering the following questions: Will your government lie to you to deliberately deceive you and misinform you by virtue of a media set - up…you know for sure they not only would but they do and they will stipulate how honest and trustworthy the media is because they control what the media says when the believable media misinform you as best they can. The media will report everything not to inform you but to misinform as the government tells them how to do that. The media is no longer the extension of the people but it is used as a political extension.

You believe the papers and the Television news would tell you events as it unfolds and they would report what happens as honest to God as if they are in prayer about what they inform you. You think that you are learning the truth from the news media as if you are busy with Gospel. If that is your view I do not wish to wake you because where ignorance is blessed it is a sin to be wise. I do not wish to shock you because as far as I am concerned you already died and crossed over to the other side while your body slumbers on our side waiting for the date of completing the dying ceremony. You are not sleeping no you are already dead. The news media reports what the Government tells them to report and no more and no less. The Government tells the media to say what, when and where and if any reporter crosses this line that would be the end of such a reporter. You can ask Peter Arnett of CNN what happens to such a reporter who steps out of line and reports the truth on matters of war and not what the Government tells reporters they should show. The media will spoon-feed you and you will digest what they whish you should know and you will be in an information cocoon as if you are a small baby because they wish to keep you thinking like a small baby. It is the Mammonites and the money barons who own the banks and because the banks are holding the money-tycoon's investment the banks dutifully tell the politicians what the money moguls' want and the Mammonites who owns the banks tell the banks that own the media what fits where and by owning the media they control what the public should know about events and thereby they are controlling the politician in that they enslave the public. Best part of it all is that it worthless for members of the public to become dissatisfied because they control even the dissatisfaction you may display. All with selected uprisings and rampaging is staged and controlled by them since when the public feel like revolting that action they may display the Mammonites take charge of the revolt. By launching a protest this action make the public feel they are in control of the negative ness and by staging and paying for such an uprising they stage a protest that aims to release the tension. To take control of the negative revolt they get thugs to kick around garbage cans and burn vehicles and staging this on the news on the television. The rioting and burning of vehicles and buildings are done in front of television cameras to gain the most effect they wish for. With all the burning and throwing of missiles taken into consideration think of this question; whenever did one such uprising lead to any government turning around on a made decision? The government enforces the will of the bankers which is the will of the Mammonites, those who own the money of the world, onto the public and the public has to take what the Money barons dish out whether they can stomach it or not because they will do what the elected official force them to do and that is what the Mammonites tell the politicians to do. Those who control the money are the same ones who regulate the negative protests. All the great protests and uprising are paid by the Mammonites because then while they push ahead their will, they control the bad feeling in the public's domain and allow the steam to be released with little derailing to their plans. It is great to control a bunch of fools who believe in democracy because this is democracy at work in all its splendour.

In line with this then ask yourself if you are truthful about the following questions:
Will your government lie to you to deliberately deceive you…for sure they not only would lie to you but in comparison to the Mafia we find the Mafia much more trustworthy than your Government. See this is why they are so paranoid about Wikileaks because suddenly for the first time they don't control the media. They wish you to believe they are about protecting you and in that they are less truthful than what the Mafia is. If the mafia tells you they are going to slit your throat it is going to happen and you can count on it but the Government will never tell you they are going to slit your throat because they only tell the CIA to do it and then your throat is slit. The government hold tabs on every person on that is not because they feel responsible for your well-being, no they will kill you and make you disappear if you step out of line or if they see you are becoming some threat to them. The government is partly formed by administrators and partly by politicians. Will the politician you so dutifully re-elect bring you harm? Yes because they made you a brainwashed zombie who believes in the power of democracy. Will the politician you so dutifully re-elect ever lie to you? By God I have never heard one single one tell the truth in all my life. To them the truth is rhetoric that will get them re-elected and democracy is about telling you what you wish to hear but never what the Mammonites tell them they must do. If you believe your Politician will only speak the truth about everything he governs…you better not read further because this earth is so corrupt that it does will destroy innocence such as you have. But to all others who got wise from getting bitter, you know they are the biggest bunch of thugs going around and yet you vote for them because they deceive you again and again. You are about to find out about THE ORIGIN OF ALL SCIECE CONSPIRACIES. The conspiracy <u>That Proves Science is An Art to Further Opinionated Incorrectness</u> by Withholding, Misleading or misinforming one and all…In the event of you answering yes to all the questions I put to you, then by god get back into your carrycot and wait for your Mommy to bring your bottle. Your naiveties must render you no more than six months of being on this earth.

In this book that is in principle also a Theses I show you how science suckers-punches everybody including you and science is a stitched up corrupted dogma that is in place because science as it currently is (whether you believe what I say or not) is running on so many phoney principles of which those in charge of science holding the highest office in science benefit richly from and therefore discard the utter flaws on which science in the present way is founded. Their job is to ignore the flaws there are. By discarding my work they secure their work. If they agree about the flaws, their papers on science and their degrees in science becomes something for the paper wastebasket.

People would argue religious concepts but everybody accepts science on face value. The guilty party that I refer to betray us in every way possible by teaching us lies that never could be true and all the while we believe them with all our hearts and entrust them with our entire future… They take money from parents with the sole aim to brainwash the students that are entrusted to their care and demolish their intellectual understanding …and before you think my claims are exaggerated I say again I prove every word I say. I show how the establishment of science frustrate me, ignore me or run me down and I challenge everyone and any one to disprove me.

Those I refer to are allowed the powers to take as much tax dollars as they wish and endeavour on all sorts of projects that they could fantasize about. Their budgets are limitless and they never face questions because they are the intellectuals whom we dare not question. With the same powers we give them they brainwashed your child making science a religiosity. The methods they use on your child are to make your child believe in science and that is just methodical corruption of their thinking.

They take money from parents with the sole aim to brainwash the students who are entrusted to their care so that students will believe in science unequivocally and regardless. If you are a parent then mind what you do and beware and read on! If you think I am going one step too far in accusing the academics teaching physics and I slander their names, I challenge you to ask this question again after you have read this entire book and thought about everything and considered every aspect.

For more about the conspiracy also visit **http://www.questionablescience.net/** **http://www.singularityrelavancy.com/** and **http://www.sirnewtonsfruad.com/**
If you read on you are going to Read About The Biggest Conspiracy ever enlisted, it's more than just the next conspiracy because it is A Science Conspiracy Network. I prove everything I say about the conspiracy I announce…read and be shocked.

This Network which I denounce Engulfs the Entire Civil Human Race in Every Aspect and willing or not, you reading this are participating without knowing about what you support with every part of the mental intellect you have.

It involves the most trusted members in the part of our civil society and we are deceived by the ones EVERYBODY trust the most, the teachers and Professors teaching at academic institutions everywhere. For over a decade I have been knocking on doors with the information I present in this book and lots more evidence, just to be ignored and to be turned away.

This behaviour of physicists ignoring me or even attacking me when I point out a very legitimate case of mistakes in science made me realise there is a conspiracy going on. Science has been getting away with this conspiracy since the Dark Ages. What you are about to read is not for the simple minded because science requires much intelligence and in that physicists get away with turning Creation into a joke.

By turning the facts that form the conspiracy into what you think of and you accept as natural and as culture, man has been cheated and mislead for three hundred years because those teaching science consider everyone as representing stupidity. To them everyone else are simple-minded peasants who are not worth having independent minds who could be able to think on their very exclusive superior level that they are able to think. That is how they get away with the conspiracy.

They pretend to be intellectually superior and to understand what seems senseless to the rest of us and then look down their noses on the rest of the world. I will show you what it is they can't see and then miss.

They underestimate the public as much as that the public are too lazy to think and the misleading carry on. Those in science think their mathematical abilities make them Gods–on–earth while we others are mindless human fodder and for that reason they brainwash our children at school by applying mind control as their mathematical understanding makes them gods. What they say all must believe because if they speak using mathematics then it is as if God spoke.

THIS IS THE BOOK THAT **WHISTLE BLOWS** ON A SCIENCE CONSPIRACY NETWORK.
You might think physicists are mathematical Masters knowing every aspect of whatever part of mathematics one may stumble upon but I AM GOING TO SHOW YOU THEIR UNBELIEVABLY POOR MATHEMATICAL SKILLS!

This conspiracy is so widely active that every person on earth teaching physics participates, most probably without knowing, some reluctant and others well knowingly to setting out to brainwash by thought control the minds of innocent students forcing them to believe in dark aged ritual beliefs in the ritual practicing of forces and allow the conspiracy to be enlisted with all the vigour they could muster.

People would argue religious concepts but everybody accepts science on face value. The guilty party that I refer to betray us in every way possible by teaching us lies that never could be true and all the while we believe them with all our hearts and entrust them with our entire future… They take money from parents with the sole aim to brainwash the students who are entrusted to their care and demolish their intellectual understanding …and before you think my claims are exaggerated I say again I prove every word I say. I show what is true and what is wrong in science and as a thank you for all the concern and devoting care I show, the establishment of science frustrates me, ignores me or runs me down, while I challenge everyone and any one to disprove me. Decide after reading this book whether what you read is not a conspiracy…however, familiarise yourself first with what I present and then decide.

Those I refer to are allowed the powers to take as much tax dollars as they wish and endeavour on all sorts of projects that they could fantasize about. Their budgets are limitless and they never face questions because they are the intellectuals who we dare not question. With the same powers we give them they brainwashed your child making science a religiosity. The methods they use on your child are to make your child believe in science and that is just methodical corruption of their thinking.

Physics teachers take money from parents with the sole aim to **brainwash** the students who are entrusted to their care so that students will believe in science unequivocally and regardless. When teachers teach falsified information then they are worse than the mafia, that operates openly by crooked intention and in that the Mafia maintains honesty, which the physics teachers don't have. If you are a parent then mind what you do and beware and read on! If you think I am going one step too far in accusing the academics teaching physics and I slander their names, I challenge you to ask this question again after you have read this entire book and thought about everything and considered every aspect.

Read About The Biggest Conspiracy ever enlisted, it's more than just the next conspiracy because it is A Science Conspiracy Network. I prove everything I say about the conspiracy I announce…read and be shocked. I call science a scam and if many readers if not most will completely disagree with me about science living as a scam then the next medical scam I wish to bring to every one's attention will support me. This scam is called humanity as it involves the medical science more than physics. It covers not what you are told or what you are taught but what science remains silent about. It is what is never said that is of serious importance. It is never what the research reveals but what science hides from getting revealed.

It lurks in the profits of the pharmaceutical companies or the doctors or the hospitals the industry never whishes to reveal. Euthanasia is not about the absolute sanctity of life but it is keeping the patient alive when the patient is going to need the most expensive drugs and the most expensive intensive care. I saw a patient with a knife in his head wheeled out of hospital to find some other place to die because that patient did not have medical care to fund his desperate situation. Because there was no money for treatment there was no treatment and this goes on in every private hospital in every city around the world. If that person had a good medical care and the medical insurer was willing to pay, then a team of doctors would fight day and night to keep the person alive and no cost will be spared. It is during the last few months / weeks/ days / hours that most care and medicine is needed.

During that time when the end of life approaches and the fatal disease is going to take its final toll that the pharmaceutical company, the hospital and its staff, and every person sanctioned with keeping this individual alive will fill the bill as fast and as hard as possible to get the last money from the dying sucker. You have a lawyer walk in there holding your last testament that states you are no longer presuming responsibilities for the costs and you have the lawyer tell them if they don't guarantee success with the treatment payment will stop on that minute and the costs for keeping you alive will be on the hospital and its staff that treat you and then you see how quickly they stop machines keeping you breathing. If you start legalising euthanasia you kill the part that serves the highest profits and they would rather see you suffer than they would agree that you might die cheaply.

If you are not prepared to pay for their effort of holding your life so ever dearly they will lose all interest and let you die as quickly as the machines are switched off. Your life and all life have value when there is someone prepared to pay the bill and those doctors are the worst criminals there mathematically could be. But yet again behind this medical industry there are insurers and behind the insurers there are bankers.

Fighting euthanasia is a conspiracy because euthanasia will kill the huge profits the medical industry makes. If we are all so against euthanasia lets take the profits of these pharmaceutical companies, the hospitals and the doctors and spend their money to pay for every one dying that can't afford treatment. Now it is a case that they would rather see people suffer and die in agony when such persons can't pay for the treatment because after all being part of science makes them equal to God and therefore others must suffer so they can reap the reward of the money they invest in promoting and furthering science.

Money for space travel goes straight to NASA while money for poor black children that goes to Africa is first skimmed and siphoned and tapped to fund the wealthy who feed the banks from which the politician gets funded under handed. Then how long will it take to get the politician to get the money going to hungry black children in Africa that then is siphoned to feed the rich and the bankers and the politicians and while everyone gets a share worth fighting for the hungry children do get the left - over that Europe and America was going to dump into the sea in any way.

In the sixties the Russians were so far ahead with the "space race" that Kennedy the then President had to say something to hide his embarrassment with the lack of credibility and progress that the Americans had in outer space. He said the Americans would be on the moon at the turn of the decade and the speech would have been lost if it was not for a fatal intervention. Everybody that was somebody had a finger in the pie with the death of Kennedy the then president and instead of making a martyr that was killed and with a million questions to have answered they made Kennedy a hero with a dream. The Americans who was in power either had something to do with the assassination or knew somebody that had something to do with the assassination. In order to divert the attention away from questions needing serious answers they upheld Kennedy's dream of landing on the moon at the turn of the decade. So whether the Russians were in on it or not, the Russians participated in a race to the moon except the Russians never had any intention of this race. The Russians saw money would be much better spent on establishing a laboratory in outer space and they proceeded with that gaol. But the Americans had the Russians in this race that the Russians never took part in. So America was going to the moon at all cost to hide the accountability of those that had ordered or was in place to order the death of Kennedy.

The Kennedy's were arrogant brats that Joe raised to believe they were above end beyond whoever else was sharing the earth with the Kennedy's. The Dad Joe did make his money on the collapse of Wall Street in 1929 during the time when everyone was losing he was gaining the millions that others lost. He was so big on insider trading Joe Kennedy could buy the stock exchange after the collapse just as he did and got rich from that. This I am telling you to show there were no scruples in the house that raised the Kennedy's as the cream of America only fit enough to fill the high profile in History as they did. The two Kennedy's were making enemies a lot faster than they were making friends and they almost dumped the world into a nuclear war because JF was so high on drugs he did not know what he was doing. Both of them were on a witch-hunt and persecuted every person who had fame or influence as part of the Surge "to drive out communists". This venture had them way down on the popularity list and they both had their personal assignation coming because they were acting tough. Why am I telling you this? I tell you this because I don't want to make a hero out of a drug enslaved, womanising sex crazed, spoilt brat that thought he could put millions maybe billions of lives at risk in a nuclear war just because he wanted to show Castro he was tough. He attacked Castro and made an arse of America's "war machinery" and so Castro ran for help. In that he made himself the most unpopular person in the world and had his people rid themselves of him. So with this I wish to put the first moon race in perspective. It was employed to draw attention away from who did it and who was involved because I think the list was never-ending to begin with. I mention this to show that every aspect of life holds some form of a conspiracy and wherever you may tread, you are going to step on many conspiracies forming part of one conspiracy.

But then the first moon race was in place as a front forming a conspiracy to cover very important, very rich and very influential persons from becoming more known and more famous than they were before Kennedy's death. So let us all not get too ambiguous about the event and put it into the correct perspective because this must play a part in the second planned moon race. Then the moon race was in place from the American point of view and guess what; America won the race mainly because the Russians never got in on the act. I am not criticizing the event because for once this did further Human achievement and played out for the better thereof. I hail what Nasa did on behalf of America.

After hitting the moon with a big ho - ha and all the excitement going with it in the seventies the excitement went down to becoming cold. Kennedy was dead and Kennedy was made a notional hero and everybody hated Lee Harvey Oswald for being killed before he could be stoned for something he was convicted of before he was even charged in a court of justice. Now came the seventies and the fire went cold. Then came the eighties and traces of the fire disappeared. Then the nineties brought questions that no one seemingly was able to answer. Were "we" on the moon…this eventually is all part of the conspiracy. Why don't NASA defend their stance and prove that "we" were on the moon! The newsreels showed how the astronauts danced the moon dance and flew like moon men. They show the moon buggy twisting rough the moon rocks. The fact that everyone misses is that the dust comes down much faster when the moon buggy or the astronauts kick up dust. For something that never were lifted from the moon surface the dust came down as fast as it would on earth with earth gravity. This big news that then became news that became ordinary news and ended up as boring stuff with no news. The seventies went past and the moon had no interest to John and Jane public in America.

The simmer of Global warming loomed in every news item. If a dog went for a shit it was because of Global warming and if a dog did not go for a shit it had something to do with Global warming. This was an attempt too scare the daylights out of John and Jane public and to get every person on a roll to discuss Global warming. If there was a haunting ghost you then called it Global warming because the only threatening thing you could endure was Global warming. Global warming is supposed to happen when the earth goes warmer. It happens when the surface temperature of the earth rises by some significant number of degrees. Go back some 31 thousand years and you will find the earth was in the midst of an ice age. The ice was so bad it covered England and the entire Europe. The only place habitable was the desert regions because where there now is dry desert there then was water that was not snowed over. If it was not covered with snow it was environmentally liveable. The ice age is a repeating process that comes and goes every thirty one thousand years when the polarity of the earth switches. It holds a link with gravity and gravity is the result of pi forming.

What you see on this map as brown now is desert but back then was an area with water and life during the ice age. What you see as green was the same as what the snow covered Antarctic regions now is. The green parts had ice and ice had fields that were lifeless. Because Newtonians think with their arses they put the centre of the Universe in the centre of their arse. Then from there they observe the Universe through the one hole they use as a tool with which to think.

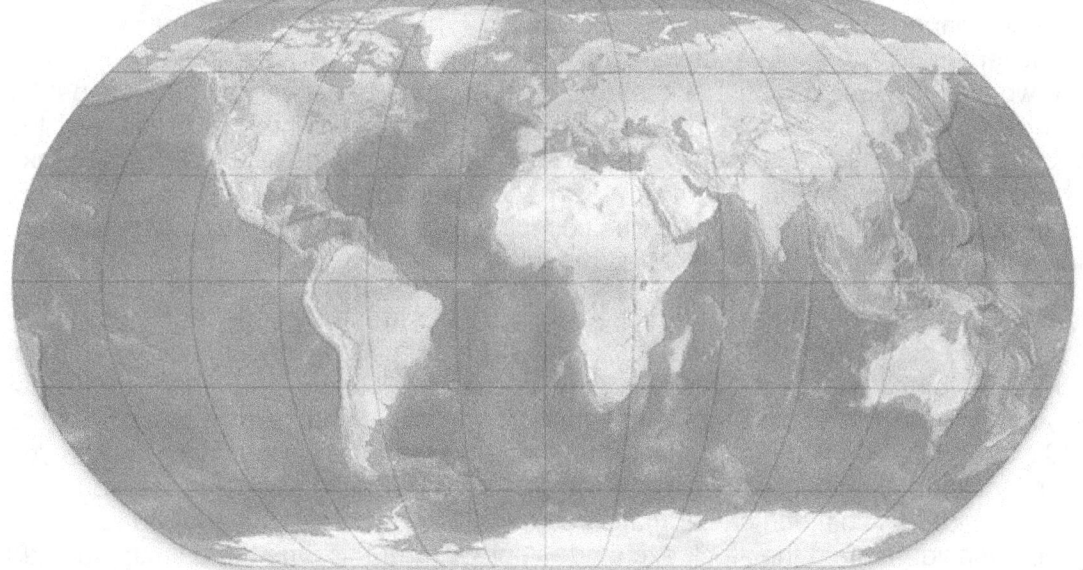

Global warming is not an earth thing but it is a territorial thing. In Antarctica and Arnica at the tip of the top of the earth the ice age never dissipated. In Britain the ice age went away about five or six thousand years ago and in Italy it went away I guess about ten thousand years ago. During the ice age the Sahara desert brimmed with life and all evidence is there for all to see. There had to be big grassy patches because every known animal was found roaming in the desert. We have human rock paintings all over the desert showing these animals grazing lively. That is where civilisation began, in the centre of those deserts. This happened because Europe and Asia were covered in ice that was miles thick. As the earth grows hotter the ice age reclines further towards the poles. This will end as soon as the polarity switches and the ice age will begin again. Therefore I call the one we are now approaching the liquid age and I call the "ice age" the solid age because in one part fresh water is mostly liquid and in the other age fresh water is frozen solid in ice. But telling me something is going to end the earth whereas the evidence the earth provides shows that evidently this eventuality happens every thirty one years but I now am forced to believe it is something I now have to shout about in fear is one of the conspiracies they force upon mankind. The earth has been through droughts we can't even imagine and it has been flooded in ways they can't even think of and yet I have to be scared that I am ruining the earth with their petrol I have to burn to survive the world they created. All of a sudden I am part of the end of the earth that is upon us and I have to pay them or I shall die the most awful death they can imagine. If that is not conspiring to defraud the earth then what is it? I tell you it is normal science behaviour because they have been defrauding the earth for centuries in the past. Science is nothing but fraud conspiring to fool everyone.

I mention this to put a warning to the conspiracy of scaring everyone into the state of Global madness, as everyone is scared of everything going hot. Go and look at rocks formed in the desert and see how water made the rock surfaces smooth. That means there was water once that flowed as a liquid and that went on for many generations, maybe not all the time but it was definitely tidal and concurrent. The evidence is there but conveniently missed by science. Ever wondered why they choose not to see it? They are said to be worried about some supposed carbon footprint. The Mammonites pump oil to the surface at the tiniest cost because they get the product for free. It is there and because they have the oil in the ground below then they own the oil. They force me to live in an oil dependant world where I have to commute or starve. I have to pay to travel to earn a living while they own the factories that build the cars that run on oil by-products that they pump from the earth for free. Now they want me to pay for some f%@#$en mad crazy scheme to pump the gas back into the ground where they removed the oil from. In order to do what? It is so that they get paid by John and Joe public to cool the earth. And they keep the human monkeys they force feed garbage to pay for the expenses. We have to buy the cars and buy the fuel and pay them to remove the gas from their moneymaking madness.

They print the money and tell me what it is worth. They only pay for paper and ink and a printing press and then they have some printed-paper that they determine the worth thereof by forcing the public to work for this much money, labouring hours to get this printed paper they call money in return. The have what I produced with sweat. That which has true worth belongs to them but is a product of my labour and in return I get some worthless paper I have to give back to them in return to make a living and feed my family. Then they have the shops that we have to buy food from and they decide how much paper I am going to get and how much paper I have to give to be able to live. They decide the worth of worthless paper and force me to accept that a piece of paper with some number printed on it has the buying power of that amount and me working my arse off must accept that for that piece of paper they print almost for free I have to labour for so many hours as they decide it will be. Now they will take even more of the worthless paper they force us to earn to pay for a scheme we will be unable to monitor. We have to take their word on what it will cost to pump unseen gas into a hole in the ground we don't even know exists but only have to believe their say-so. But through all the callus conspiring we again must trust those bunch of criminals as we are forced to work as the slaves of the Mammonites.

If the stock exchange crashes the money the stock represented can't just simply disappear. If it was worth money the money must still be somewhere because no banker will loan money without finding sufficient collateral. If you paid $10 000 for ten shares and the next day the shares fell to a $ 1 000 then you did not lose $9 000 because you have ten shares and that is as much as you had. Whether you ten shares are worth $ 10 000 or $ 100 or $ 1000 000 you have ten shares and only when selling the shares would you have gained money or lost money. But if you thought the ten shares was going to repay you your investment of $ 10 000 it would then so much easier make you $ 1000 and you can get a far better return on your investment. Every ten years or so the stock exchange implodes but this time the banks imploded and that loss was (so they say) much bigger than the stock exchange losses. The banks made losses by weak investments but never do we hear what the bad investments were. The all the banks in all the countries made these losses and now all the taxpayers have to fit the bill of the losses. In every country it is the taxpayer that pays for money the government bailed to the banks. Where is the money going?

Every country has to sacrifice health care and government benefits for whatever causes. The students get less money and has to pay back more. The schools get less money and the staff is reduced so where does all this saving of money go? The Hospitals must do more with much less and the hospital staff working in the hospitals is reduced in numbers working while the pay the staff earns are also reduced. Where is al these billions and trillions of savings going? The money was there before to pay for the lot and suddenly out of the blue the money disappeared. Whereto did the money go because there was money before the savings were announced and that money that was there must still be there but that money is now channelled to other places? Who is getting the savings that before went to welfare and the payment of staff that since became redundant? A quarter on average is no longer earning while the three quarters remaining is doing a quarter more work for a quarter less pay. Where are the beneficiaries of these savings? The losses of bad loans and the losses of carbon cleaning and the losses of stock exchange rampaging and the savings of redundant welfare did not vaporise the money so the money in terms of being must still be but where is it? I shall get back to this so just keep this conspiracy of making money disappear in the back of your mind. Those in power are investing in things we have no part of in the planning thereof. We only pay for the investment with our effort and sweat but later on we will get to that. Let's start a song…where has all the money gone so liberally taken from us by force? Money can't just vanish no matter how fraudulent the idea of money is. Money can be stolen, money can be hidden but money can't just be and not be. If they take 25 % of the money of the economy for so called and debt this money has to be paid to someone and it must serve a purpose, but what would that purpose be?

The oil they pump they get for free. There is no cost hikes involved that will pump up the price. They have to get it from the bottom where the oil is located to the top where they remove it and that is not so costly. As a farmer I pumped a lot of water from a borehole and my pumping was much less than that but the price of pumping would have been steady if not for the electricity supply company fiddling the prices that much. To hike the price of oil is illegal because it is creating monetary currency. It is the same as printing money. There is a commodity that you have a selling price of say $10.00. Then you jack the price to $20.00 thereby creating $10.00 that never existed. You have printed $10.00 that was not there before the jacking of the oil price. The persons or company that did this falsified as many $10.00's as the barrels there are by which they sell oil.

Where does the extra money go? The oil is the same oil but they charge more for the oil. It is not the oil that became more expensive but it is the paper they use that became less valued. The worthless printed rubbish we work for just became less valued and thereby more worthless than it was before because now I have to work more to do less with this worthless ink printed paper than what I could do with the worthless rubbish beforehand. If I make cars in a factory they don't pay me with a part of the car as Bugatti did in the thirties when he on occasion ran out of currency to pay his workers. They print as much money as they wish and they give me the pulp but never the product that I could go and sell. The product they keep and the pulp is mine. Then the government raises taxes so the money I get is even more worthless. I have even less purchasing currency that I had before. But this becomes profit to someone somewhere because my labour in hours remained the same and the produce I generate remains in quantity the same and the house I then have to pay more for because they raised the interest rates remained the same and in the end I am far worse off but the money they steal from me in this manner has to go to someone's pocket.

If the money is in vaults they must realise as I do that the paper is worthless and to have more of the senseless commodity stuck in larger vaults are doing them as little good as paying more by using more money to receive less goods. If it was that simple they could print money by cleaning the trees of Brazil and it will do them no good. If the money do have a value and they increase the value then what do they gain from receiving more or paying less for my labour and forcing me to pay more for receiving less goods. What do they do with the profits they make? Where are the extra profits going? Why am I being ripped off? Now put this into the context of their ambitions suddenly raising by their dreams of going to the moon and then onto mars. They are suddenly making my slave labour much more where I receive much less and if I am not satisfied I am going to be one of those who are walking around and looking for a job. So now I have to be great fill for being allowed to have a job because I could be one of those who does not have a job. When they start a new factory they tell us so many new jobs are created but in the background there are vultures who make ten times more from my labour than they pay me for my labour. If that hyena pays me $5.00 for my hour of work that vulture will get $50.00 for the things I deliver to the company during the elapse for that hour. I have to be thankful I can be a slave and make that Mammonite richer just so that the pig will allow my family and me to live. Where do all the profits go and why the urge to go to mars? Why would the urge coincide with the scheming of the profits of every country that serves the cause to enrich the Mammonites even further? We will come back to this when we delved into some other part of the conspiracy. But first let's get back to the moon where we were or where we were not...

We accept what science accepts and we reject whatever science rejects. We put our faith in the hands of science as we humans put our faith in the hands of the clergy five hundred years ago. On what grounds do we rest our infallible belief in science, where we never question the legality? The conspiracy in science is summarised by the question that a philosopher once asked when he asked the most decisive question that he ever could ask: "In order for us to know anything, we first have to question everything we know." By not asking questions science conspired to conceal. The conspiracy I detected in science about science is that science never had questioned everything known to science. I found a conspiracy in science that is in place keeping it alive by not to question everything in science that science knows and hiding the knowledge no one ever questions; that is the conspiracy I wish to inform every person about. The conspiracy is to never question and by never questioning, science hides that what science does not wish to have others to know about. The "not questioning" became the obsession that became a conspiracy amongst all person working with and in science. By not questioning that which science knows they hide that which they should question. By the hiding of the questioning the conspiracy comes alive and not questioning maintains the conspiracy in science about science!

Apollo 11 moon landing

Are you convinced about the outcome in the search to uncover the possible conspiracy about NASA really being on the moon or faking all the evidence they show?

Chapter 2 MOON CRASHING.

Do you think NASA was on the moon in the sixties and seventies? Do you think they ever landed and if they did then how many times did they land?

Have you got this eerie feeling that somehow there is somewhere something about science that comes across as not that believable and you in your mind have your personal limbering suspicion, concerns and doubt but you cant see it?

Do you think there is a conspiracy in the unknown regions of science hiding in the dark avoiding the obvious detection of everyone but no one can point a finger as to where the itch is coming from? Is it not how we all feel about science!

Politicians mislead the public, bankers fake figures and mislead the public, and warring Generals fake facts to mislead the always-trusting Public.

Now it is the turn where we find science that seems to go wanting for the truth!

The question on everyone's lips is "were we there or were we not there?" Suddenly everyone questions the moon landing and everything that is connected to the moon landing in the sixties and the seventies. Why would everybody have an urge to distrust the loyalties of science?

"We" were never there to begin with. NASA was there and to demean the efforts of NASA by questioning the integrity of what those intellectual giants achieved is despicable to say the least.

"We" were not there because "we" have not the guts to land on the moon. It is always a small number of giants, a handful of wise men who carry a nation or in this case that carries a human population on earth. However, to question or not to question is the conspiracy that science hides.

Was NASA on the moon…well yes, I think for sure they were on the moon? Did NASA at the time and during the event make the movies that they then showed as if broadcasting live about the event and as it took place on the moon that very minute? Don't be simple minded to think or expect that such film making on the moon was possible because the television technology allowing that was not in place yet and what the public saw most probably was a pre-recorded program.

I truly can't see how the computer facilities at the time could handle such a broadcast transmission about what took place on the moon at that moment in time. There was no transmission possible to broadcast from the moon back then, live in the instant frame-by-frame photography. There was no facilities in those days to broadcast live events from the moon to the public, but that is what the public wanted and if the public wants to be fooled then the public get what they expect to please them.

Think of what NASA achieved and see how small the related accusations are when compared. Moreover we know the American public is the most demanding people there is and if they wanted to see the events unfold while having a live broadcast that was what they would get. Think of the cabbage doll outcry that happened some Christmas some time later but during the same decade and how the American public got nasty over a silly doll. Can you then blame NASA for going undercover about some of the truth?

Everyone tries to prostitute everything for money without showing scruples about it. No on shows a sign of having a conscience because there is an all-conquering lust for money and this has no limits or boundaries. Everyone out there tries to make money by selling improvable suggestions or toxic rumours filled with ridiculous defamation and defamation it is because they can't prove what they declare is true.

I suspect the public thought by paying billions of tax dollars for the entertainment that the least NASA could do was to oblige by delivering the expectation that was the buzz at the time. Does the live broadcasting or not deter the importance from what the event is in history… sure as the earth is round the event remains as important and as real as if the broadcast was coming from some live sport event. But they used sliding rulers to calculate on board the spacecraft. I used sliding rulers to calculate at the time because hand held take-away computers were science fiction and with no computers to broadcast the entire idea was fiction. No one had the facility to put a computer in place that was capable of relating the broadcast from the moon back to earth. The technology was not in place but everyone wanted it to be in place. I don't say that it happened in this manner but from my perspective it could have lead to the NASA reaction. But this doubt that is evoked comes as a result of other doubt that is lingering in suspicion. The smallest question of doubt has another origin coming from another source in science that I found but was apparently never yet detected. The doubt that these questions raise could be the part of the conspiracy that science tries to hide. By looking at these issues, everyone is looking in the wrong direction and looking away from where the conspiracy is and then to what is not important. Of all the enterprises in the last century, the achievements of NASA rates bigger than any of the other industrial-related achievement, coming only second to what we gained from using electricity as man's main energy.

The achievements of science must leave all of us with awe. Their efforts are frightening as much as it is mind-boggling and goes beyond what anyone could ever have dreamed of in a fairy tale say two hundred years prior. If those generations that came before us could come alive today and could see this that man now achieve they would think of man as God. If those generations that preceded us could see what is taking place today, they would run from us and hide in fear of our achievements and our abilities in living

as we do today! Look at the picture and see the engineering effort portrayed in this image of tranquillity. Not long ago in human history people would have fallen on their knees and would start to worship what they saw if they saw this and whoever witnessed this would father a brand new religion just there and then on the spot. Still unmistakably their effort remains incredible and the picture reminds us of Godly inspired wonders. If ever anything man made proved intellect way above the ordinary and into the realms of superhuman thoughts, then this is proof of such achievement. Yet it took so many man-hours of superhuman labour, planning and concentration to bring everything in this picture together. With the entire marvel this picture holds and all the greatness it portrays of man's ability to achieve, out of view it hides a dark side of the brilliance of man and more particular of science. The conspiracy comprises science as much as it involves the entire faculty of science. It is about what science conceals hiding underneath everything science reveals.

With all the amazing achievements accounted for and when recognising all that science changed our way of living on the earth and what was achieved by scientists developing this super mentality and in that also giving science all the admiration dually admitted, notwithstanding I am about to dump on you the biggest conspiracy that has ever been presented and that was ever undertaken by any group of persons in the entire human race. Think of anything you might think is big or outlandish by nature and that dwarfs in comparison to what I am about to reveal. It is so large that there is nothing in the past history of man with which one could compare it to. It involves every aspect of the life of every human being and this shadow in our midst covers the darkest secret that was ever hidden from intellectual human view. It is perpetrated by those we absolutely unconditionally trust in all aspects, It touched on every individual walking the surface of the earth and that excludes no person of any status albeit it an infant or someone in old age.

Do you fully believe in science as if everything about science is proven fact and is truthful, never questioning or rethinking one question in having the minutes doubt just because this story has been repeated for centuries? Are you so confident in science up to the point you will put your life on the line to prove the accuracy of trusting the facts we have in physics? Are you one hundred percent sure about the honesty of science and are you sure about the trust we put in the honesty with which we regard science.

We know that scientists claim to work with facts and truth and the commitment to further the truth science presents as facts. This trust we have in the truthful accuracy of science goes beyond any and all suspicion of any kind. We all know that physics can never lie and physics represent the truth as no other form of knowledge could ever have. The presence of God or the absence of God becomes doubtful and is scientifically debatable but the accuracy of physics can never be doubted in having any suspicion. When you then worry about NASA being on the moon or not, I have to warn you to forget about NASA going to the moon and to start thinking about the moon coming to us when we are taking Newton's physics principles under review. What we see in this picture is that heat can release objects from the gravity that confines everything to the earth. Yet this says nothing to the physicist about the nature of physics and that surprises me as much as it disappoints me. No one ever gave a thought about the question that if a hot air balloon can lift "mass" up into the air and break the shackles of gravity, why is that feature in physics possible when it is "mass" that forms the confining gravity.

With raw heat streaming from engines as we see in the picture it is flames that convert fuel into fire to form the power that is so great it releases so many tonnage of "mass" into outer space. We can lift so much "mass" into orbit and by doing that then overcome the burden of gravity where that gravity burden is the result of mass attracting all other mass to produce a pulling force.

Yet, by releasing heat from engines, this power of gravity can be overcome and send many hundreds of tons of mass into outer space. That means heat can overcome gravity and gravity is a force. But heat will only react to heat and a force will be neutral to heat expanding. However, in a hot air balloon it is heat that also produces lift and we know that heat produces expanding space and therefore gravity shows more inclination to heat increasing and reducing than chasing after Dark Age "forces" fighting contraction. Holding this evidence in the front of my mind I have formed a new concept concerning physics and in particular gravity in which I attach other rules to the forming of gravity. By trying to introduce these concepts aligning heat and cold to gravity I ran into the biggest stonewall formed by resistance. Then with many decades of research and studying facts science avoids I came to form a conclusion I am about to share with you. I am about to shock your socks from your feet with the accusation I am about to make!

If you are looking at such concepts as the moon landing or no moon landing issues you are looking at a conspiracy in science from a totally wrong area. I am going to show you where the conspiracy in science hides. There are far greater issues that science avoids than the fact of landing on the moon or not. The question should be why is nobody informing the human race about the biggest disaster that man could ever imagine. Why is there a deliberate silence about a pending but certain collision that is coming and that would end everything that man made including the life form "man"?

It is, believe it or not, but it is about modern Physics as they teach physics at schools and Universities by those who don't understand physics. If you aren't into science this might just be why you chose not to study science. You most probably were smart enough not to understand Newton. What kept you out of science is The Ultimate Conspiracy Theory; it is the Conspiracy Concerning Physics.

If there were no reason why NASA would tolerate this nonsense of not being to the moon they would have silenced this Mickey Mouse as science has been silencing me for many years.

They use this rubbish as a lightning conductor. They draw attention away by having much debate over no existing enterprising arguments. But from what are they drawing attention away and why are they drawing attention away? Why would NASA not just come out and settle the argument in the direction they want. Why keep the tongs clicking? Could it be that they are not deflating the situation because they wish to draw all the attention away from a reality that forms a conspiracy and anything will do to deflect the attention from where it would bring harm?

Do you think a conspiracy has to be as dramatic as Dan Brown's conspiracy he called *The Da Vinci Code* and *The Lost Symbol,* no it is not because what Dan Brown wrote about is entirely fiction and therefore it **must be** dramatized to **gain maximum publicity** in order to sell. It is not intended to be silenced because by siliencing the conspiracy then the conspiracy would remain a conspiracy and the conspiracy would not unveil the billions of dollars it must make. The entire idea is to blurr out as loud as much gossip minded hogwash to sell as much as it can. Conspiracies are kept silent and Dan Brown's is anything but quiet and that makes it no conspiracy.

The best seller of "factual mystery" is one big hoax and every person even my wife is included (except me and maybe a minute number of others I don't even know) bought it and fell for this devious prank, gun, barrel and flintlock. They ate it like it was a green banana and got choked on the pornography of lies and clever deceit. They wanted it to be true and if you ask me why, I would have to admit it is one of the rarities I can't solve but I think it has to do with their utter F&c@en stupidity and half a brain not working in most people that read it (including my wife and my daughter in-law that enjoyed the book) and was fascinated by the garbage. The problem I see is that we live in a society that brainwash the paupers such as I who forms a part of the rest of us money less scum of the earth, or that is how the rich and powerful feel about us drifting in the hogwash they create. In the past many decades ago before television and advertising put us into stupidity, the suppressing by the upper class was bad but now it has become shear undiluted slavery. On television they target persons with a mighty effort to keep people thoughtless in order to make the advertising slogans work. Make those looking at the advertising think less and they get impressed easier and quicker. Everyone knows there is a problem that everyone endures but in the comatose state they keep the public captured no one can see clearly because of the comatose state in which the general public are.

Everyone became a robotic sleepwalker fed with wild sporting events making billions on the go and movie fantasies less realistic than anyone's life is paid entertainment while doping the wits. This is the way society stitches people together. From the churches to politicians to solicitors to doctors to bankers to the army/navy/air force including the police and teachers all vow to keep whoever they can under a control as to keep everyone in slavery. Since we all know there is a problem but we all can't see the problem since we have become most of the problem, everyone grasps for a solution to bring an end to the mental zombie-state in which we all are. Now we are all grasping for air and start to look for conspiracies. Most of those making money from detecting conspiracies play along in keeping us intoxicated with brainwashing.

Even when Dan Brown said in court his books were entirely just a fantasy, a flimsy hoax put together by compiling no evidence, he was ignored. That evidence he gave in court about the absence in the truthful research he supposedly brought to light statement was lost. The entire story is founded on the Leonardo Da Vinci's painting portraying something that is totally non-biblical and false. The way they portray this hoax is as if Leonardo Da Vinci was present at the last supper and he had a stand-in portraying of the actual event. It's as if Da Vinci was eye witnessing while documenting the event as one being present.

On this interpretation the entire idea is vested. And all the mindless masses indulge in the carnage of mindlessness while feasting on bullshit and getting stuffed by mental stupidity that feeds on the replacing of their miserable existence as to give them a view into the higher working class. They pretend Da Vinci was present in the room and saw the diners sitting around the table. That is as far from the truth as that idea that the Pope can forgive sins but this last comment again is the way religion of all kind put the fear of God into you by presenting those clergy as having a better connection to God and can converse with God on a far elevated level than we, the low class "Grass Roots" commoners can. …And this made Dan Brown a multi millionaire …by selling the truth short and being a conspirer in the effort to make money.

The media batters the public to become mentally instable in an all out war to make money and to create slaves. **Or do you think a true conspiracy theory should be as devilish looking as the Illuminate. Are you convinced one must see the evil of the devil being portrayed in every symbol that represents the illuminate? No, those in power would silence all. It's just what happened in the Kennedy conspiracy.** Those with the power hide the blame so effectively by law no less that the guessing will remain in place during the next century. People who have the power to control our coming and going can't just be nameless. They must be named and what better name as naming them the enlighten

one. They are the ones who can see in the dark and they are the ones who can see through your bedroom window and they can visit your wife when you sleep and they can farther your child when you are not looking because they have the power of God on their side by practising dark Magic. We know we are cheated hands down by government but it can't be because of our own impotent stupidity. We know we are cheated politically but they must have an evil Satan Devil on their side to play with us so ruthlessly.

Everyone wants change. George Bush being the President in office promises change during an election campaign while he serves the rich. He sits in office knowing the public is unsatisfied with the way things are but instead of changing while being in office he promises to bring change. No one wants what is coming their way and still they want to be part of the process that rule them, so they opt to go democratic. We think their power they derive by which they control all aspects of our lives can't be because we are brainwashed into thinking democracy gives us power. No, if that is the case then we must take the blame for being that stupid and we are clever since everyone in power tells us we can fight for what we want because we are committed to democracy. We think that it is by democracy that we govern the country by putting government in power and we tell the politicians how to govern because we can vote, and we are the ones living in the 21 first century thinking we came to be the most intellectuals of all times? In Britain there are 44 million voters and of the 44 million each one can vote once. I have to stand in line to make a cross next to some political party and that turns me into Superman while in fact that makes me stupider than Ape-man during the time before he could think. If you can jump off a cliff and I tell you every once in 44 million someone survives the fall to tell his or her tale, will you jump? Still you do think being one of 44 million, that honour gives you the power to rule the country by democracy? Can we still find I people that stupid, yes there are and the western civilisation is filled to the brim with such mindless masses, begging for the privilege to be able to vote. No wonder those in power can keep so many on as slaves and never pay a dime to have slaves to fill the sole purpose of making those in power also rich at the same time.

If I cast my vote I can change the world. What a bunch of dumb bastards and mindless idiots would fall for that. To think if I was a Britain I have one chance in 44 million and still they want me to be an idiot to think that the one vote I have counts! No, I can't be that stupid in believing I am plaid like a drunken chicken! No there has to be an evil-eye bunch of Devil worshipers who call on the power of Satan by formulating dark magic rituals and then they get the power to influence my life so that I become my worst nightmare. I become so stupid that I believe it is my right as it is my duty to send my children to be killed in a war they wanted and planned and that they put in place so that they can profit from and to protect their billions.

Hey liquid brain, you are not one in a million making you special, you are one of 44 million making you anonymous. No matter whatever the party is that you vote for, the Bankers got to the politicians first. To thank Tony Blair for giving the oil fields of Iraq to the British oil barons, they awarded Tony Blair with $30 million this far just to go around and make speeches. He gave the bankers the oil and while he is alive he is forever a rich individual. It paid to allow the Liberal Democrats to be against the war to present a democratic resistance and pay them handsomely for being negative about the war, while the Liberals and the Conservatives believed whole heartedly the photos of the weapons of mass destruction was real and that piped out to be demolished war debris laying in the desert. So both the main parties voted for the war that would enrich the bankers and industrialists much more then they are already rich and the politicians on all fronts were paid generously to believe pipes in old mines could be atomic missile launchers. All they had to do was to brainwash the public about the urgency of going to war to save the country!

No matter for what party you vote, you will keep the money-mighty Mammonites in power and in turn they will keep the Politicians in party as long as the politicians serve their cause well. The public will still have the privilege to send their children to death to become national war heroes and die for national pride while when they get back, there is not even money to supply the war cripples with artificial limbs or to give the crippled a living pension. The penniless have the fortune to send their offspring to die and to murder so that the rich and powerful can become richer. When will Tony Blair's boy go to Iraq or Afghanistan to serve his country? Are Gordon Brown's two boys going to enlist to fight for their "Queen and country" No it is the " grass roots" scum who only has a purpose of keeping the rich safe that would have the honour to be killed for Queen and Country or to kill as many Muslims in Iraq as possible for "Queen and country".

Turning our heads to America things are much worse because in America I think you as an American will compete with something like 160 million to one to get any change in the two houses Governing or to get the President to go look for a new job. Think of what the chance is to fight a battle against things you don't care for and playing a 160 million to one role in the outcome. You lot wgho believe in democracy are simpleminded. With one vote you think you can change the government. The government uses your simple-mindedness to get their illegal governing legalised.

The money power even control the anti-whatever protesting with fabricating a protest and getting the mood that was in favour swinging against the riot. This is how they do it. We as civilised persons get tired of petrol prises risings through the roof. We want to get our message across. Then over the news we learn that there are going to be protest meetings in several cities. That is what we need. We need the government to see we are deeply unsatisfied with the way things are going under their government. That night we see TV cameras bringing us the riots in long sections in the news. Wee see a bunch of hooligans burn cars and beat the daylights out of the Police. We think of the Police being those who protect us from such hooligans. We as civil people don't want to see the police getting kicked and beaten and we reject that behaviour completely. We see the rioters burning cars and demolishing shops and going on like madmen and we can't associate with that type of behaviour. We have to distance ourselves from such madness and we turn our backs on the protest. The protest was about how we felt but those with money hired a bunch of criminals to hijack the protest and after seeing what the protesters were we don't want even to think we were acknowledging anything they stood for. Immediately we go against the protesters but going against the protesters we then agree with the raising of the fuel price because that was what the protest was all about. By distancing ourselves from the protest we come on the side of the moderates and the moderates accept a new fuel price like all citizens do who believe in "democracy". So for a few pound and pennies the Mammonites bout their right to raise the price of petrol and by me being civilised they remove my protest in hiring criminals to make the protest unacceptable and hijacking my feelings in the process. My rejection of such behaviour as those they hired is much higher than my rejection of the new fuel prices and they buy my silence and my rejection about what I believe in by hiring thugs to throw police around and to attack descent people's cars and burn it. Since I can not associate with such behaviour I distance myself from what I first believe in and I become part of the silent majority that holds my values higher than my frustrations and my feelings of pure contempt. That was the reason why the Mammonites had Hoover in charge of the FBI for so many years. With a smile on his face the biggest crook in America (that is according to my values) was the head of the biggest law enforcement office in the land. If he could kill of every person he found undesirable, just because he found them undesirable killing Kennedy would bring him the ultimate challenge and would leave him the most powerful man on earth, the man that could kill the most powerful man on Earth.

We know Hoover was involved by his involvement of the FBI in the cover up because Hoover had to hide his gambling debt and his homosexuality. It was the Mafia that involved Hoover and the FBI. If you wanted someone executed from behind, Hoover was your man. If you wanted anyone executed you could entrust Hoover with the job unless your name was Fidel Castro and you lived in Cuba working as the Cuban president. Other than that Hoover and his FBI squad of men had their jobs cut out as legal assassins. The truth never mattered to J Edgar Hoover and this we saw in the way he shot and killed Ma Baker and her boys and so many other brutal criminals that the Media paraded as already convicted and then made them enemy number one. You ask Ma Barker- head of the Barker-Karpis Gang that supposedly committed a spree of robberies, kidnappings, and other crimes between 1931 and 1935, but in contrast to the popular image of her as the gang's leader and its criminal mastermind has been found to be fictitious. Fictitious or not they were slain by the FBI without ever been convicted in a court and America stood for it. Another one was George Nelson, was a bank robber and murderer in the 1930s. Gillis was known as Baby Face Nelson, a nickname given to him due to his youthful appearance and small stature. Usually referred to by criminal associates as "Jimmy", Nelson partnered with John Dillinger, helping him escape from prison in the famed "wooden pistol" escape, and was later labelled along with the remaining gang members as public enemy number one. These were but a few that was killed by Hoover without facing a trial but was shot because of the say so crimes Hoover charged them with without ever going to court. Being public enemy number one made you convicted without trial and already dead. The Kansas City Massacre, or Union Station Massacre, occurred on June 17, 1933. Frank "Jelly" Nash, a convicted mail train robber and sometime member of the Barker-Karpis Gang, was being returned to Leavenworth, from which he had escaped three years earlier, when would-be underworld rescuers attacked the two carloads of lawmen guarding him with machine guns. A federal agent, Raymond J. Caffrey, two Kansas City police detectives, Frank Hermanson and William J. Grooms, and Orrin H. (Ott) Reed, Chief of Police of McAlester, Oklahoma, were all killed in the attack, as was Nash himself. It was later proven that the "massacre was the doing of the totally incompetent FBI field agents that couldn't handle their firearms and they killed their own people. This evidence was swept under the carpet by Hoover that played the public like a fiddle when he convicted the criminals without having a judge present and then had them executed on site. He was so good with this with all the practice he had through the years he was quite capable of getting the Kennedy assassination successfully completed, that would or should I say also be with much help from other partners.

Hoover's involvement in the mafia was so blatant it shook the foundations of the Kennedy boys. Hoover denied there was organised crime because organised crime or better known as the mafia had him wrapped up in horse racing gambling debts and therefore he had to keep the FBI off of the heels of the Mafia. It is known that Hoover was a notorious gambler and had a craving for horse racing. He must have owed the mafia lots and lots of money because the mafia was in control of horse racing gambling and he had a free hand when it came to punting on horses. Hoover couldn't go for a piss without the permission of the mafia because of his love for horses and betting on them as well as his soft spot he had for men. He had a very open homosexual relationship with a number of men and this is information that the mafia used to their advantage. I wouldn't be surprised if the two Kennedy boys also started threatening Hoover with this information to get him out of the FBI office.

Hoover saw his position as the man that brought civilisation to America by creating the G-men, those who rid America of crime albeit by blatant distortions of the truth and using courts and media spectacle to kill those he blamed for crimes he wanted solved. He J Edgar Hoover, saw J Edgar Hoover as equipped to judge and qualified to convict any person out there which just by being in his office J Edgar Hoover then could decide there in his office that the person was guilty of everything J Edgar Hoover charged that person with and had the execution executed from his office without having a nights sleep loss. If ever there was a maniac overflowing with self-righteous narcissism it was J Edgar Hoover. If J Edgar Hoover put a man on trial he did it by allowing the media to sell papers and then get the public in general to become the jury and convict enemy number one that month by giving them selected well rehearsed propaganda and let the public decide the person's guilt long before any court got to the person.

I would have loved to read their minds when J. Edgar Hoover, JF Kennedy and Robert Kennedy shared space in one room. I believe with their ego's filling the room furniture would have flown out the window due to lack of space. I am sure Hoover knew how Kennedy got to office because Hoover had a file an everyone in America that had any importance and the Kennedy family surely knew who they were dealing with when they took on the FBI boss. **The death of President John F. Kennedy is a true conspiracy and that we can see from** the fact that there are no results coming from any inquest launched at even a Presidential level.

Then when the Kennedy's came to office it was by the father Joe Kennedy rigging an election outcome in Chicago and when he did this he called in the help of the mafia allowing the Mafia to swing the votes in his son's favour. But as any business deal a favour required another favour and Kennedy had a promise to keep. The mafia did this because Kennedy had to bring the lucrative Cuban gambling casinos back to the mob and the mob did their part. It was choosing between the mafia and the Russians and the mafia expected Kennedy to deliver on his promise because the mafia tends to be bad losers. Kennedy had to get Cuba back to mob rule and the gambling houses back on behalf of the mob and when JFK turned on his election promise to the mob they showed that not even a President is above their retribution. The presidency was their part of the deal. Cuba was part of the Kennedy's end of the arrangement. But Kennedy thought that being the President made him beyond any approach and I guess this would have been true if he and his brother did not make such an effort to step on every important and influential toe in America. In the process they united parties that would never have befriended one another but with a common enemy the worst opponents for the time that the enemy unites them in a common cause.

Suddenly his boss was a thirty something rich kid that was on a crusade to fight crime and that placed Hoover in a tight spot. Hoover was the one that for decades decided what was the crime of the day and who was to be fought and now this rich kid that never fought one crime case in court was going to turn the apple cart upside down. Hoover himself had not the power to do anything to Bobby because Bobby had a Big Brother in office going by the name of JFK so Bobby could tell Hoover what to do and where to go because JFK was the president and this did not sit very well with Hoover. Hoover was all so happy to oblige by involving the FBI because removing Kennedy also removed the little pest called Bobby and he could do well without the Kennedy brothers becoming problem number one in the life of J Edgar Hoover. No one realised this but J Edgar Hoover alone was a criminal institution deciding who and what is the criminal institution of the month to fight.

Kennedy had one promise to keep and that was to deliver Cuba back to the mafia and that Kennedy failed to do. Kennedy thought the Russians could not touch the President of America but he lost sight of the influence of the Texas oil Barons, the mafia and the FBI and many Senators that did not sit well with Kennedy trying to get people out of Vietnam or then better said, some President that was not getting America involved in Vietnam fast enough. Think of the money that was made by the rich and the powerful that was selling and supplying war material. It was almost twenty years after the Second World War and the ammunition stockpiles as well as the feeding supply was grinding to a halt. The supply was going anywhere and those making money saw drought coming if America did not enter the East in a war effort. Korea was almost a disaster with the North fighting back and with America almost getting their arse kicked by the Chinese. That was one trap the money machine did not want to enter but relieving France from the Vietnamese burden was one way to sell arms to America. Vietnam was such a pathetic little country and to bomb them into submission or just to pacify them with half-ton bombs was a sure way to go to keep the war machine supplied with arms and to feed the warmongering arms industry.

All this had a part in the tragic demise of the President but not one of those alone had the influence to get the job done smartly. It was down to the Dallas police, the FBI and the mafia and they all had a lot invested in whatever Kennedy was threatening. Then Kennedy made a compromising deal with Russia that said America was in Future going to leave Castro and his Cuban Island alone. This affair did not sit well with the Mafia because the Mafia had Hoover deciding there was no organised crime such as the Mafia and when the Mafia thought they bought a president by swinging votes they got a witch hunter going after their security. These two rich kids became a menace and everyone in crime had something to lose with the Kennedy's in office and nothing to lose with the Kennedy's out of office.

We know the Mafia was involved because they silenced Lee Harvey Oswald the innocent man that was openly framed because it was the mob that got Kennedy elected by faking the election results against Nixon. The mob by the hand of Jack Ruby silenced the investigation afterwards as they had other interested parties also in their pocket and those in high office helped afterwards getting JFK out of the way. Jack was a dying man and I guess owed the mafia quite a bit.

All the facts point to the death of President John F. Kennedy as a true conspiracy because as I said and the conspiracy is what we could see from the fact that the investigation was a shamble from the start. The Dallas police had a very good idea who to look for before the shots were fired but the location that they went to look for the lame duck did not deliver the groomed suspect because the fool was watching a movie and that he was not supposed to have done. If anybody was groomed to take the fall for anything years before the event then this was Lee Harvey Oswald. He was sent to Russia to get a Russian wife and to seem as a person that switched sides openly. He was sent to Cuba to seem as a Castro collaborator. He was ordered to get (no less) a mail ordered rifle with which he was to have assassinated the president of the United States. This man must have been a complete idiot to line up all the evidence against him so conspicuously or he was a sitting duck in place for the conspirers to poach whenever they felt like it. If this international tale in the complot was so evident, we then also can see the hand of the CIA touching the overall planning in this plot. They are the ones able to send agents to Russia and then to Cuba and get them to seem like communists as Lee Harvey Oswald did. One question no one ever asked is how did Oswald know about the route the President was supposed to have followed the day of the parade. How did Oswald know that he was in the correct building where he could keep a loaded gun and wait for the president to parade past this very building, in an open limo and that is if he did not have intense help from very informed persons close to the presidents entourage. You reading this, would you know when to bring a gun to work so that you will be in a position from where you could kill the President as he passed your office window? ...And Oswald knew when and where the President was going to pass during the Dallas parade so that he could order the correct rifle through the post no less and keep it loaded in the event that the President was coming past the building in which he worked so that he could kill the President of America. If you believe such bullshit, then my friend I believe that you are gullible enough to believe in democracy. Oswald knew he could kill the president from his office window as the president passed in an open top vehicle at that very instant that the president did. Then he went and killed a cop just to draw more attention to him and after that he went to watch a movie and make it easy for the cops to get him. Good God I pray for you lot because this method of thinking is so lame only Americans could believe that! ...And the Americans need advertisements to know how to cross the road.

We know Hoover was involved by his involving of the FBI in the cover up because Hoover had to hide his gambling debt and his homosexuality. It was the Mafia that involved Hoover and the FBI. Hoover denied there was organised crime because organised crime or better known as the mafia had him wrapped up in horse racing gambling debt and therefore he had to keep the FBI off of the heels of the Mafia. In spite of the mafia being able to remove any persons not withstanding their social positions. You take the case of Bugsy Siegel who was one of the prominent mafia bosses and has to take the credit for Las Vegas now being what Las Vegas became. Being the head of the FBI he would know what happened to even mafia bosses when they cost the firm money and he also knew not to investigate the matter since it was a family affair. But owing the mafia money for gambling debt Hoover also knew that made him part of the mafia family where the only family love was that if you owed the family you better pay up, not necessarily with money, but you did what the family needed you to do. Hoover was a notorious horse gambler and did this "sport" openly and frequently. Everyone knew he had as much of an eye for choosing a winner as Jimmy Hoffa had for choosing secure sites to meet the mafia. Hoover lost money as if he knew he would never repay in bank transferable currency because he did not earn that much even being the FBI boss. Hoover was in a commanding position and had debt to pay not only with his engagement with the mafia but there was a score to settle with the Kennedy boys that stepped on his toes and frightened his mafia friends. This plot had good riddance written all over it from every way you looked as far as Hoover went.

We know the Mafia was involved because they silenced Lee Harvey Oswald and Lee Harvey Oswald was the innocent man who was openly framed because it was the mob that got Kennedy elected by faking the election results against Nixon. The condition of being elected was that Kennedy had to get Cuba back to mob rule and the gambling houses back on behalf of the mob and when JFK turned on his election promise to the mob they showed that not even a President is above their retribution. There were many not as big celebrities as the Kennedy president but celebrities non the less that had money arrangements and went bad on their word and got hacked for not fulfilling a promise made to the mafia. Sonny Liston the boxer and Mario Lanza the singer went to their graves because they owed the mob gambling money and couldn't pay up. Sonny Liston was a paid fighter and the mafia was writing his cheque on the condition that Sonny Liston had to fight and people had to place bets and the mafia had to make money from the bleeding nose of Sonny Liston. When the punch-drunk boxer thought he was above and beyond the reach of the mafia, the mafia had their approach in place and that was the end of Sonny Liston. Hey, this is so common knowledge that even I know what comes from not meeting your obligations to the mob and I have ever even seen one mobster in my life. Mario Lanza owed money and had to sing in Las Vegas to repay. He could have had much more than what he owed if he only sang but he had stage fright and the mafia showed him his stage fright was nothing in comparison with how much he should have had to be scared of the mafia's retaliation techniques. If they could get to Jimmy Hoffa and Sonny Listen and Mario Lanza the Hoover knew they could get to J Edgar Hoover without even breaking a sweat. I think even Hoover met his end via the mafia because his death had all the trademarks and the lack of investigating the strange scenario has one big ring to it, the ring of the mafia buying with purchasing power that money alone as a currency to bribe can't buy.

Hoover knew the rules by which the mob enforces bad debt. So he allowed the mob by the hand of Jack Ruby to silence the investigation afterwards as they had other interested parties also in their pocket and those in high office helped afterwards getting JFK out of the way. Did the police look for powder residue on the face, hands and arms of Lee Harvey Oswald? Did the police determine how many shots were fired from the rifle of Lee Harvey Oswald that they found in the Texas School Book Depository store and how many cartridges were retrieved? Where are the cartridges now and if not what happened with the lost cartridges.

Those are the question to address, but no investigation ever asked the obvious questions. We know that someone ordered the conspiracy from very high up being in cahoots with the mob and the FBI because the investigation was screwed even before the assassination took place. The conspiracy does not end its involvement there, because Congress and the Senate also had to be involved. With Kennedy so quickly reaching for the Nuclear button, a lot of Bankers got stiff with anxiety because a few Nuclear bombs might not reach the Kennedy family in their bunkers, but it will write off billions of dollars of property all belonging to a lot of banks, either by mortgage or by ownership.

Then there were other contributing factors that removed so many influential persons from the adoring friends – list Kennedy had. With Kennedy so quickly reaching for the Nuclear button, a lot of Bankers got stiff with anxiety because a few Nuclear bombs might not reach the Kennedy family in their bunkers, but it will write off billions of dollars of property all belonging to a lot of banks, either by mortgage or by ownership. Going into a nuclear bomb contest is not what the bankers in America had in mind when finding a resolution to the Cuba problem. It is now bragged that America had more nuclear missiles than Russia had at the time but that does sooth the threat to the Bankers who had the buildings in the major cities. Say only one missile landed in New York and destroyed only one high rising building. It is not the building that goes bust but it is the rest of New York that goes contaminated that forms the problem. No one wishes to live in Chernobyl, not because of the building not standing any longer but it is because of the nuclear contaminated atmosphere that does not come across that inviting any more. Now think of the entire New York or Los Angeles or any major city being in the same state of contamination as Chernobyl and see how those bankers will be able to fill the buildings with tenants. One entire city going spooky is not what any banker requires for a future vision and the best insurance policy to prevent this strategy is to go anti-Kennedy in a big way. They all knew Kennedy was going to be re-elected because if a Chicago Democrat liberal as can be can get a cheer in Dallas in the Lone Star State as Kennedy received that tour in Dallas then Kennedy was going to be re – elected the very next election. That confidence of re-election was securing Kennedy in his office as far as democracy goes but that too got Kennedy riding the barrel of a cannon through a street surrounded by people on either side.

With a trigger-happy cowboy set and secured in the American Presidential office someone that is very influential somewhere might just lose billions and in that it is cheaper to get a few million flowing in the direction of politicians that will help the conspirers get Kennedy from the office to a coffin literally with a bang…of a gun. That many in the Senate and High Court Judges had a common interest in the demise of the Kennedy boys was apparent, not in Kennedy dying but the other Kennedy dying a few years on. This also is so apparent in the manner that the Warren inquest was so self-servingly righteous and maliciously opinionated which all shows an open cover-up and while every one on the Warren commission was greatly rewarded during the next decade, each in their own time one cannot draw another conclusion but that the outcome was decided beforehand. The commission rushed to a conclusion and the lot had very little interest in what other factors could bring about. They had one guilty man that was dead and one mafia man that turned hero just because he silenced the guilty man before the guilty man could defend his position with a not guilty plea.

This case too shows the true science of a conspiracy and even the brother of JFK got done in when he got to close to solving the mystery. But investigate as much as you like, you will keep on guessing for the rest of your life! Are you under the impression that a conspiracy theory cover-up must be as old, as wide spread and as powerfully involved as the Free Masons, which represents those that have the power to alter the destiny of the human race? If that were true then they would have the influence we find in the Kennedy case to keep their secrets covered under a veil of total silence and never to reveal any information to anyone! When there is a real conspiracy, even those forming the establishment that reveals the conspiracy guards the secret, and then it is well guarded.

So then you want a conspiracy theory to be as blatant and as staged, in your face and unreliable as the UFO landings where the sites are very vaguely concealed and as revealing as it can be with the most pathetic idiots becoming star witnesses. The media then blasts their idiotic untested testimonies over the air around the world as Gospel. It will be better to believe in ghosts. I don't ever say there are ghosts that

can haunt us but I can't use physics to prove there are not ghosts while I can prove with physics there just can't be aliens. Where would aliens come from and where do they return? A visit from a "nearby" galactica will take them tens of thousands of years to reach us and for what purpose, to have sex with the biggest idiots they find? You never think why those Aliens would choose to meet the most simple-minded idiots with a single figure IQ after travelling for many a century non-stop to reach us? Are you impressed with aliens coming to have intercourse with the most greasy-looking wench on earth that by her own admission in accordance with her intellect and her appearance could not find one male in three billion to inseminate her with wasted sperm, and so the alien got lucky? There is no one on earth in flesh wanting her for sex because no one is that desperate so she invents someone craving her from outer space.

She is so desperate for sex that she invents some one even more desperate than she is. Do you fall for her stories that her ten mindless children has a father many galactica away and did not even leave her alimony to bring up his very earthly looking brats. If you have that mentality, then go for every conspiracy theory you can read because facts of life you will find too challenging to understand and comprehend fully. If you think conspiracies must be as mind baffling and logic eluding as crop circles that are not in the fields when the sun sets and are there when the sun rises, then you are a sucker living for bullshit and fantasy, a monkey to be made of and you have the mentality a child has and you then are living in a fantasy-world only you can believe. Then this world is a place far to harsh for a person as gentle minded as you are. It would seriously be in your best interest to commit suicide very gently and even more quietly for this earth is far too complicated for your likes. You would be better off not being with us in this harsh unforgiving environment and we would be better off being teary by missing you than having you around.

Newtonian science can't disprove such mad ideas because firstly of incompetence as a result of flawed principles applying and secondly Newtonian science would rather deflect any scrutiny away from science and in the direction of something as mad as Newtonian science because then Newtonian science would not get scrutinised. They would rather keep quite and stand defenceless about mad claims that persons from outer space came visiting than to attack such madness head on. While people are preoccupied with statements made by loony cucumber heads trying to get one second of fame albeit only getting the attention of those even lamer brained than the attention seeker is, this prevents anyone starting to investigate science in the manner in which I have. To travel to the "nearest" galactica is 26 light years away. If any person travelled at $\Pi \times 10^3$ km / sec it will take thousands and thousands of years to get there or from there get here if it could but it can't be done and that part Newtonian science doesn't even realise. Much of the so-called preparing for space travel is another conspiracy hiding illicit international banned research

For us to break free from the gravity of the sun we have to travel faster than the gravity is within the very centre of the sun where the sun annuls all captured space-time. The sun is the centre and with gravity reducing space it is moving space towards the centre of the sun in the spot holding singularity, Therefore the movement such an object must apply must be faster and stronger than the movement within the very centre of the sun. If you keep on reading that is what I will show you; I will show you what gravity is. Within the very centre of the sun a fledgling Black Hole is forming gravity and in the end eventually the Sun will grow into a Black Hole because this centre Black Hole is growing progressively. If the something that tries to escape from the sun's gravity and then travels to another star, that something must travel faster than the attraction is within the centre of the sun or else it will become an orbiting missile at the distance of the debris probably at Oord's cloud or further but it will never be able to escape. That reason is the why comets circle around the sun and never crushes into the sun or escape and leave their orbit around the sun. Once captured by and being within the gravity of the sun there is no escaping possible and that is the biggest fact of gravity. That means that object trying to escape must be able to free itself as if it is escaping from the centre of the sun which means it must be stronger than the gravity is in the very centre of the sun. Only a person that knows nothing about physics and has an IQ of about 75 or less will claim that there can be space-travel outside the solar system and the Newtonians are unable to say why such travelling is not possible. Then people more stupid and with less brains will believe it. Travelling at that point within the sun is travelling at the speed of light and no persons or object can travel at the speed of light just because travelling exceeds the peed of sound and Newtonian science has no way to tell the difference through physics.

Human travelling would not exceed beyond Mars and go even beyond since gravity would not commit life to function at that point. Life developed on earth in the gravity of earth and on the long haul can't function fluently outside earth. Secondly it would take too long and we cannot travel faster than $\Pi \times 10^3$ km / sec after which electronics will start to malfunction. This has to do with the earth gravity and our natural material that formed on earth within the earth gravity. However, science likes to tell how we can travel at the speed of light because that way they bullshit everyone who has fantasies and not reality about physics. Newtonian physics are so poorly informed they do believe that it is possible to travel at the speed of light. In this way they prevent people from getting fed-up with them throwing money into a pit of mockery which is what their science is. So to create a fantasy world where science is on top of all aspects and controls Creation they don't distance their views from madness but also don't complement the outrageousness in the press. This plus the deceit they use to swindle and brainwash all forms part of the most elaborate conspiracy ever devised by any form of human deception ever thought out by man in any way possible. For three hundred years they were teaching about mass and never, not once proved mass.

The truth about a conspiracy is that everyone involved with the conspiracy will fight tooth and nail to stop the conspiracy to get leaked and leave those behind the conspiracy and all those feeding from the corruption of the conspiracy exposed. Those who have the power to maintain the conspiracy would keep the waters as still as possible as to draw no attention to such conspiracy. A true conspiracy has to be as quiet and as unseen and going as unnoticed as it could be. A true conspiracy must involve everyone without anyone detecting even a hint of what the conspirers hide. There are many conspiracies going on such as the banks involvement with crime and the bankers profiting from gamble rackets and drug selling. The same goes for the insurance business profiting from lenient sentencing of courts holding very merciful judges in office so that the insurance will sell cover-policies. Our social system in the Western World is a conspiracy. Democracy is held up as a saviour but politicians rule only to favour the rich.

The conspiracy of Government is to Govern the population with fear and the fear must be directed at a general enemy that can be pin pointed but never can be caught. Create a terrorist conspiracy and you have a conspiracy that can be directed at the population you wish to control and subdue. Many more people are killed by "crime" than by terrorism and yet there are billions more spent on fighting terrorism than eradicating murder and crime. The crime helps to subdue the people by fear and pressing the terrorism button does precisely the same so why not put fear about by being killed through terrorism in action in order to do undercover maintenance of fighting the poor to keep the rich safe and secure when it works? If a terrorist is caught they don't have to divulge the information of the danger of the conspirator because of national security. In most cases it is one man with a feeble mind having malicious intentions, that's all. The Government can control and spy on the population with CCTV cameras and secret police under the cover of fighting terrorists and the population will be so grateful losing their liberty and privacy because by doing that they don't help to fight crime which is the main enemy but a fictitious creation of government.

Burglars, robbers, car thieves, drug smugglers all are the main income of Bankers and Insurance firms. If there is one chance in two million that your house would be burgled or your car stolen would you take out a very expensive insurance, never in your life. You would take the chance to go without expensive insurance and take the risk yourself. Therefore if you have a chance of one in one hundred that you will become victimised in that your house will be burgled or your car will be stolen, then you better take the insurance notwithstanding the premium. In England the Bankers already got to the politicians to draft a law that makes it illegal not to have insurance. By enforcing insurance payment through law, now the Bankers and their money-Brothers can rob you blind legally and take you to court if you would not oblige.

They set the terms and the rates of how you will be robbed. The Insurers would never have the insurance premium holders pay the equal amount of what the amount is of cars being stolen. They would set the premiums at least at three times the number of crimes committed to cover their arses and cut a tidy profit on the run. Therefore we can assume that for every one car that is stolen the premium holder would pay three times what the crime rate asked because to the insurer it has to be worthwhile. That makes the insurance firms the biggest robbers, car thieves and crime hustlers in the world because they rob three times more from the public and take three times the cars or furniture or whatever was stolen than that which the criminals did. They steal more from you than thieves do and they have to if they want to make a profit. Then they use crime to keep us defenceless and vulnerable and easy prey, just as the villains prefer it. The more defenceless we are the better victims we could be and good victims are unarmed persons.

They use anticrime propaganda to get us to become complacent, as the criminals prefer us. It is so easy as taking candy from a baby… Unarm the law-abiding civilian to protect the criminal. Use antiviolence propaganda to promote violence flourishing, and achieving it is so simple they let the public do it. They allow the mindlessness of the public to get the better of public safety and then get the do-gooders to finish the task. When a madman goes on a killing spree and starts killing everyone and everybody the first thing that is mentioned is removing firearms. It was not the firearm that killed the say 30 something people or whatever, it is the madman that got hold of the gun. They blame everything on guns. A hundred years ago guns were much more common and accessible and there were no mass murdering of innocent people where a madman goes running down the street killing as far as he goes. Or some boys played a video game once too many times then go to school shooting from class to class. That is not a gun issue but it is a behavioural issue. It is society becoming crazy and show there are those on the fringes that goes that one step further. Guns left alone are harmless and guns used for hunting or sport in safety is harmless. But getting the young public hero-worshipping the blood drunken maniacs on screen helps.

The children sit and view Rambo type movies showing how Sylvester Stallone or van Dam or Vin Diesel runs through town killing and mutilating with all forms of explosives while being the hero. There were no such mass killing a hundred years back while there were firearms a plenty especially after the two wars. There were no such mass murdering or thoughts entertaining such killing sprees but there are now and now there are violence on the TV and in movies and displayed where you can participate in such brutal killings on a TV in the form of a video game. However they only need one professor to claim these games has nothing to do with the murders and then it is Gospel. It is pure science or in other words a lot of bullshit to say these violent pictures and this brutal bloodthirsty entertainment does no harm in the minds of the masses. They pay some University and one professor in that University to profess how unscientifically it is to blame these games or movies and science again conspires to pollute the minds of the public. What is wrong with this? The fact is that Science does this all the time with everything so what is so horrible in doing it once more, after all the entire lot is a bunch of cheats in any way. Ban Rambo who is a hero and a mass killer. Ban these movies but that won't happen because the banks bought the Politicians and the moviemakers and video game owners tell the banks what to tell the politicians. Stop this by banning Rambo who is a mass killing hero but there is too much money made from this form of entertainment so the Politicians are told to lay the blame elsewhere. Everyone with a lick of sense knows that the violence in Video games must corrupt some of the minds that play these bloodthirsty games.

Then after all it is so easy to blame guns and the democratic principle takes its course where the majority can immediately see the sense thereof. If there are no guns then the man can shoot no one. It is so simple as the simple minds believing the simple mess they create in the first place. In China there are no guns and therefore Chinese madmen do it with butcher knives. Are they going to ban all knives after they banned all the firearms because banning the firearms doesn't solve the problem? They have to remove violence from society in making violence and killing an everyday sport that entertains millions. That they won't do because then they will loose the billions the entertainment industry makes them each year!

All they ask for is that crime flourish so that no one could dare to be without insurance and remain alive. Then to get everyone scared and nippy about crime, they protect the child molesters and the rapists and arsonists just to keep John and Jane Dow and the rest of the public so scared of crime they haven't got time to become frustrated with the total rip off the public suffers under the "protection" of the Police and the Politicians. That is why the "law makers" make crime ever so humane and get parents to become the villains when they try to discipline their children. It is about making the law-abiding citizen feel guilty and vulnerable at the same time while allowing the criminal to victimise the innocent so that the public can be controlled. You see how fast they put a father in jail when he goes to castrate and remove the genitals of the rapist who raped his daughter. Having the rapist out in public will put the parents on high alert rendering to a condition where they are thankful just to have their children alive and moderately safe. When it does happen to the persons next door we pray on our knees thanking that it was not our turn.

The rapist gets a slap on the wrist with early release and the parent get tossed into the rough side of the jail for decades because no one in charge wants crime to get under control. We have gone as far as feeling grateful that a rapist did not kill his victim or that a burglar only took "earthly possessions" and that the victim wasn't physically harmed. We don't mind being robbed as long as we're not clubbed to death. That is how far these scoundrels wheeling money got into our safety. Those scoundrels are the ones governing you by employing "democracy" and bringing comfort to your door. Your credit cards, home loans, hire purchase, and all the credit available is paid by the blood of victims of the criminal enterprise the Bankers fight to uphold and even pay to maintain. To Bankers crime pays through insurance coverage.

The latest trend is the kidnapping of persons and taking ships for ransom. This became just one more way the bankers found a means to "tax" the public. If pirates in Somalia take an oil tanker, everybody sits back with a smile because it is not their problem since they are not involved. Wake up and smell the shit on you doorstep. We, the public, pay for the goods the cargo carriers carried. By paying the ransom to the pirates the price of the products went up. Not only that but now insurance are paid on all other products in the event of pirates grabbing ships. Insurers will now put a levy on every container carried on the high seas just in case of piracy. They up the price to ad profits on the go whether the cargo is carried through safe waters or not, it is money to be made by insurers and the Bankers get the benefit of the crime.

Everything on earth just went up with a few cents but at the end of the year it comes to billions of currency that was captured by the Bankers going into the pockets of the wealthy and out of the pockets of the poor. Crime pays so there has to be a way to boost the levels of crime so that the insurance firms and Bankers can benefit from the source of wealth. The trend of kidnapping will only rise and in the next decade or two everyone on earth will have a stipulation added that there is a certain premium put on the head of every policyholder. In the event of any person being kidnapped, the insurers of the policyholder would be willing to pay that sum of money that your policy covers so that you then are released. If you have not got such a clause in your contract, the kidnappers just get rid of your body and get the next victim to pay up. If you have not got insurance covering such an event, you will disappear without a trace, but if you do have such coverage, then your insurers would be kind enough to pay for your release. This is how the game is played. The Insurers would always see to it that they get many times more from crime than what the criminals could ever get. They can't insure one on one but will go ten to one in their favour. Ask yourself who steals the most from you? Illegal immigrants are another source of currency going into the pockets of the insurers and the Bankers. When an illegal immigrant lands in a country, that person has to eat and live and crime does it. That person didn't cross the shores to live on minimum wages and be the poorest part of society in that country. Yea sure, maybe a few percent are honest workers striving for better but the vast majority wanted more than what was on offer where they came from. These are perfect candidates because they are untraceable. The police have no names and no identity and the criminals are shadows coming out at night to harvest what is not theirs and then to live lavishly while the true criminals are the Insurance firms and Bankers that regard this affair as only being part of business.

Tony Blair as Prime Minster of Britain acknowledged he had no idea how many illegal immigrants there were in Britain and no Brit ever asked why he had no idea. He and his party were paid not to have any idea. Why is the Ministry of Home affairs the one department that is "not fit for purpose" to use the phrase they used. It is so that no one can trace anybody illegal even if they tried to do it. Criminals are renowned to walk out of jails with early release even when they were illegal immigrants. Now why would you think that would happen? The Bankers pay the Government to have the Home Office in such a state that no one knows arse from head and the bigger mess in this department is, the more illegal immigrants there are and the more the crime rises. Crime pays but only for the Bankers, the stockbrokers and the filthy rich that can buy their own security army. Therefore the rich benefit by deducting cost as tax deduction so that the poor must pay more tax. The poor remains part of the Rich's food chain notwithstanding democracy.

The illegal immigrants bring another sort of business with them. They bring girls they use as slaves for prostitution in the developed countries where there are money to be made. The girls get kidnapped and then their services get sold off. The price asked for one hour is not high but since the girls don't ask for much as they are drug slaves every man can afford to pay for the services and having many girls in a working position, the money eventually escalates into billions. The money is far too much to be kept outside bank vaults and so where will this money go…it goes to banks and the Bankers are the recipients of the cash in the end! Big drug busts don't stop the drug trade. It does not even influence the process by causing massive price hikes and that should be an indication of how much money goes to Banks.

Don't think of rescuing the girls in distress. You try and touch those girls to save them and see how you vanish from the scene either by getting jailed for interfering with police investigation or by the criminals who will make you disappear without a trace. You touch the girls and you touch the bankers who befriended the Politicians who are the bosses of the police and in that order you will annoy the command structure because the girls prostituted as slaves guarantee Bankers and other criminals much wealth. Those with money or with power are criminals or are criminal minded because that is the only way you can have money, steal from others to fill you pocket or use them as slaves to work for you. Don't believe this crap about only honest money going to Banks because with them in charge of the Political Government of all countries they will get the law to create loopholes through which they will work "legally" and with them having your elected government working just to please them they have nothing to worry about what is legal and correct and what is not because everything is just business.

The drug trade flourishes just because of one principle, the money has to end up in the coffers of Bankers. If the drug trade is worth a billion dollars a day, where does the money go? No one can walk around with that much money and therefore it has to go to bankers for safekeeping. I have not made a study on the subject since it is not my main interest to reveal crime statistics, but any one would see that printed money escalated by tens if not by hundred fold the past forty years. Why would they print that much more money? It is because that much more money is in circulation and the distribution thereof finally land in the banks. This gives banks the ultimate power they could ask for. Crime brings in cash and cash is their business, so crime becomes the business of banks. In order to further their campaign on behalf of criminals they get to politicians. Politicians not in favour of using "their election help "and aid by "election contributions and donations" don't get elected. The honest ones don't even get press publication. In this syndicate don't discard the TV stations and the printed press. Since the Bankers would own much of the press as shares legally held, the press is in place to further the cause of those in politics that would further the cause of Bankers and Bankers are in business to make money and crime pays good money. This is an animal that feeds off itself by consuming the consumer and enrich the rich.

There are many more "lobbyists" running down the corridors of the American House of Congress and the House of the Senate than there are politician official assistants in America. Why would that be? That is a smarter and more acceptable name for bribing the politicians to gang up against the electrets. It is a fancy way of bribing anyone and everyone so that the "democratic" process could be hijacked and rigged.

If Gordon Brown and the rest of the European leadership so full heartedly believed in democracy, why don't they put the death penalty to the system of democracy? Go on and give the public the choice, let the people decide... If they say they rule by democracy, then let the people decide on the main issues directly, call a referendum! Why don't they put it to the vote and disallow the one-sided advertisement funding of vote swinging so that the people can make a well-decided popular choice on the matter. They lie and cheat and bullshit before an election just to come afterwards and push the rule of the rich down the throat of the poor. The Police say they protect. If they say it let them prove it and then let them protect and if they don't solve crimes then put the Politicians in jail with the rest of the Bankers and the criminal mob. But it is a case of the quicker robbers and car thieves and rapist get out of jail, the quicker they get back to crime and the more crime is committed, the higher the premium get for insurers. Therefore the more the money is and the higher the profits are for insurance firms, the more bankers and industrialists gain. Also the higher the crime rate gets the more the people have to see that they are insured to cover their potential losses in crime. Therefore it pays to buy politicians to create laws that go soft on criminals.

Defend yourself in your own home against a burglar and you are going to jail! If you as a law-abiding citizen have a gun it is a crime, but is it because you make it difficult for criminals? They can't keep guns away from criminals so they take my gun away! You go and beat the holy shit out of a person that robbed you then you land in jail. If the robber gets away with the crime no single tear is shed but if you beat him blind you are the villain. This is to protect the Insurance firms that bank at the Bankers that invest on the Stock exchange that get the Rich richer so that the Poor must get poorer. The more crime is about and committed daily, the more people need insurance cover and thereby the more money insurers bank giving bankers profit to spend on the stock exchange by controlling the economy and the politics.

If the Government puts Police in charge of crime and disallows me as a citizen to take revenge on the criminal by going out and look for the robber myself, or get the rapist to swing on a lamp post because I have his finger prints in my house where he raped my wife, or cut the genitals out of the rapist who molested my four-year old daughter, then they must take the blame for unsolved crime and the high crime rate. Then the Politicians must see to it that when the rapist rapes again after being released from jail everyone involved with his release from the lenient judge down to my local politician that is responsible for crappy laws, must sit in jail with the rapist because they all participated in the crime being repeated again. Because of their weakness crime is committed, so the politician must guarantee my property or give myself the option to retrieve my belongings and then hang the criminal from the nearest lamppost.

It is time government takes responsibility for laws failing and because they favour the Bankers and Insurers it is my security that is compromised by them applying lenient laws that is about protecting criminals. If the government says their Police will protect my belongings and secure me then they take the responsibility. If I am being burgled, then the government must replace my lost property because it was they and their police and their laws that did not protect me and then if the crime is too high, the cabinet must resign and give other people more fit to do the job the chance to govern. Now at this moment we know that Politicians are the most crooked over paid under worked robbers of public funding there is on earth. Do like the Chinese do with crooked officials, those they catch with their hands in their cookie jar, take a crooked politician or official into the centre of town and put a bullet into his brain. Why would other politicians be scared of this practise when they are as innocent as newborn babies? That will get rid of the crooked politicians and place the genuine honest persons in public office. Crime pays for the elite who profit from the rate of crime becoming sky-high. The money is used to change laws in favour of crime so that crime can flourish and Bankers can profit. Bankers buy politicians and democracy through money donations. With the money they give politicians, laws are written that protect the rich against the poor. This is the way to run a conspiracy orderly. No one can trace it and no one suspects it because we all benefit. With this method the Bankers gain by inflating everyone's livelihood and eroding personal income and safety as to gain from that. In time I might write about this but now I cover another story, a much bigger story about a conspiracy everyone on earth has been brainwashed to hold a part of...it is about...

There are so many conspiracy theories going around, floating like a bad smelling odour in the room, but most of those have no more factual substance or evidential backing by truthful facts than a James Bond novel has. I also have a science conspiracy theory but mine has all the proof and all the substance of authenticity that any argument can ever hold. It involves **physics** as **physics are taught** at institutions worldwide. If you now don't believe me, read on and learn the truth yourself, it is free of charge.

My conspiracy I wish to share is the most unbelievable one that has most truth. This book is not about any conspiracy ever mentioned before but takes into consideration what worries people.

When you see what science achieves and you hear what I accuse science of, then it seems as if I am a nutcase who escaped from the loony house and I must be the one having very dangerous tendencies to harm the innocent.

I should be locked up and taken from society.

However, read my story and you'll see that I don't harm the innocent. It is those we all trust without reservations that brainwash to forge deceit through corruption and malice.

The blame is with those admirable, yet astute members in our society, the educated leaders in physics that we trust without even thinking about their trustworthiness or question their sincerity regarding their blameless honesty.

Do you believe in the absolute unquestionable correctness of science and that everything that forms science is truthful being far beyond suspicion?

Do you believe that those furthering the science of physics works with no less than absolute proven facts and the facts are beyond any doubt at all.

If you do then I have very bad news for you, **news that will change the grain of what you believe is truthful.** I am about to shock you into reality!

This is going to reveal the biggest conspiracy ever concocted by the human mind, ever since the time a human had a mind to use.

It is A Conspiracy about Fundamental Science. It started when Newton gave us the idea that gravity is the result of mass and in gravity mass is the biggest contributing factor.

I say that mass as a factor in physics and not the value of weight, but mass with the pulling power only pulls stupidity over the eyes of incompetent idiots raving in their personal sublimation. Let us now proceed and see what it is that I say is the conspiracy. Would those in science cheat you and withhold the truth from you?

Go to the website **www.singularityrelavancy.com** that is telling everyone about an Ultimate Science Conspiracy, the biggest conspiracy ever concocted.

NATURE put TIME into place TO PREVENT Everything FROM HAPPENING all at once.

This proverb sounds silly and yet, silly it is not.

While this proverb is the truth, science say they don't know what time is…and this is ignorance they hold while living with time for millennia all the while Newtonians say they uphold Galileo and everything Galileo implemented. Gravity drives the Universe as the earth spins around.

In other words what science can't see is that there is a time –connected plan in the form of gravity driving the Universe along. Time is in the form of gravity and the cosmos is driven by time. Time and gravity are the same, which is precisely what science is unable to see.

If gravity drives the Universe, the Universe is powered by time.

When Galileo's pendulum can measure the flow of gravity and record time to a precise science while swinging in space, then the flow of gravity in space has to be the measure of the pendulum swinging in air while reading time. The pendulum measures time when swinging through the air that surrounds us and what is it that surrounds us is there because of gravity.

This obvious conclusion about gravity forming time, which is what the pendulum reads, any child can draw but as simple as it is, this went past the Newtonian's understanding of science for centuries.

They never could realise the flow of gravity and time is one and the very same since it then conflicts with Newton's ideas. Still the pendulum proves that reading time and measuring the flow of gravity is the same. Galileo used gravity to measure time accurately, yet for as long as the pendulum recorded time, this fact eluded science in a way that science doesn't even know about it. It is what science doesn't know that becomes important to man. If gravity measures time by applying a swinging pendulum this becomes the biggest fact in physics and it still went past science because their cleverness brought on their rigged stupidity.

This religious belief in the infallibility of their Godly insight becomes their stupidity. Their ignorance is what science hides. It is that what they hide which is what you don't see. Science hides what science doesn't know which they cover under the larger pretence of their cleverness.

Also science hides from everyone's view what science doesn't know so that students will never realise Newtonians are covering their stupidity and ignorance behind a curtain of arrogance. This conspiracy to not reveal their stupidity then becomes the conspiracy they nurture for centuries. Science conspires to hide by covering their stupidity from open view. The main thing is that science is clueless about time and time is the driving cosmic plan so science hasn't a plan to see the truth.

I have been in fruitless conversation mainly going one way about what I see as their stupidity in defending flawed science. I told them this but with having no response. Still notwithstanding their arrogance about ignoring my showing right from wrong, still those physicists filling high academic offices are so infatuated by their superiority and their personal righteousness they can only see the malice that the Pope showed towards Galileo when he differed from the general science views and yet they do the very same today.

They are overwhelmed by their correctness.

You will see how those hypocrites condemning the pope having the holier than thou attitude because they think they are cleverer than all other intellectually lesser sub-humans. They accuse the Pope of constraining science progress while back then but they torment me in much crueller ways than the pope did to Galileo. At least the Pope allowed Galileo to print a book while they did everything to stifle my efforts when I just tried to be heard by others about my views!

Whenever anyone shows those in science that their religiosity called Newtonian science is wrong, they act in exactly the same way as the Catholic Church did to Galileo five hundred years ago, in no less a manner while they uphold in public that they maintain their stand on condemning the Church for not accepting Galileo's principles. The crude thing is that I prove all those filling academic office in science have also still not accepted Galileo to the letter and I challenge anyone to prove me wrong. That is one of the parts of the big conspiracy I see and call it the Mother Conspiracy!

By the swinging pendulum in space while connecting it to the earth Galileo proved gravity is time and that principle goes unnoticed as it diverts from Newton's idea that "mass" produces a magical "force" of a "pulling nature" he called "gravity". This view they can never explain while this is what they will always defend as the truth.

Those oh, so clever Dark Aged wizards pronouncing the upholding of free speech and science liberty for all of mankind while hiding behind their superior positions will befall the same memory as what befell the Pope in the days of Galileo. They will be remembered for being the last of those who were stupid enough to believe that an inexplicable magical "force" of "gravity" "pulled" the Universe by "mass" while E.P. Hubble showed the Universe is not contracting but otherwise expanding.

Their sublimation covering their stupidity causing their ignorance will outlast them for all time to come. As the Pope five hundred years ago is remembered for his ignorant stupidity, so those currently in office would also befall the same fate just because they were as stubbornly arrogant and ignorant as the Pope was back then when ignoring mistakes.

They conspire to conceal the dishonesty that they don't wish to reveal.

Remember we spoke about the money that governments have to pay back to repay everyone's bad debt all across the world? Where does all the bad debt money go that we suddenly are burdened?

If you have all the money you ever can have then all that money can only be worth in relation to what will you do with that much money. If you have all the gold money can buy what would the worth of the gold be. If you have all the oil there is what would be the worth of money that you trade for oil? Money has a worth in terms of what one could do with it. Sure having more must bring future security but after that and you still have more, what is the worth of having so much more?

But if there is a secret no one knows about and you are one of the few who can generate sufficient money to ensure your survival while billions upon billions of other will perish and will surely not survive the fact then that you can generate money at will to purchase your survival gives the money extra ordinary power as a currency. If you can buy a seat on a spacecraft that will take you to Mars from where you will sit and see how the moon destroys the earth and all others on the earth then generating more cash than any person ever knew existed will be worth the effort. If you know you buy a future for your loved ones while six and a half billion others that are not sufficiently provided for will die in misery, then you can find a worth for that much money that will pay for your place on Mars.

I guess all those who have a nuclear bunker will have a Martian address to go to and the rest of us that don't even know where the bunkers are will see the moon come by and give us tidal waves kilometres high.

Let us test Newton's attraction theory in practice. What everyone duly knows is that the mass of the earth pulls the mass of the moon closer and the fools who can't think that far is designated to perish.

Academics in physics insist that gravity is founded on the Newtonian gravitational principle of mass $F = G \frac{M_1 M_2}{r^2}$. This says the moon is pulling the earth at the same time that the earth is pulling the moon and keeping all this in mind, **then when is the inevitable collision between the earth and the moon coming?** When will the collision that will end all come? What do those in science that knows science say about the day the moon will hit the earth and destroy all forms of life in the Universe? **We have to know when is it the final doomsday when all life form will end.**

There are those in society that must know since all knows some information and some information is "classified" meaning the privilege knows and the rest are misinformed. Why is it always only the wealthy, the influential and the powerful that is informed on very important matters concerning the well being of all of us and not only those with money. With $F = G \frac{M_1 M_2}{r^2}$ being in place the earth and the moon is pulling on each other for longer than life is around. This is science at its core. This is fundamental physics. But all we get is conflicting messages coming from all over. It is as if someone is hiding something from everybody and this does not sit very favourable in my mind. Why give some information and then fail to give other formation. The formula $F = G \frac{M_1 M_2}{r^2}$ says the moon has mass pulling on the earth that also has mass and this mass pulling on each other is destroying the radius of the distance there is between the moon and the earth. The mass of both is pulling by force that forms gravity that pulls each other closer until the moon and the earth collides with a thump. This collision has the ferocity as was never seen before in the history of the solar system! But when is it coming with such force as nobody can anticipate? Humans had the formula for centuries and did nothing about figuring out when the moon and the earth will smash into each other...and having this formula that can produce the information not one person thought to find out when this event will happen? Science claims that the Newton formula of $F = G \frac{M_1 M_2}{r^2}$ forms the basis of all physics. That means gravity is there to reduce the value of the radius separating planets and indeed all cosmic objects so that all cosmic objects will finally one by one collide with one another until only one lump of structural graveyard holds all the star material in one place. This is called the Big Crunch theory that opposes the Big Bang theory. In this case the moon having mass is pulling the earth having mass and the distance between the two is reduced by the mass of the two as the pulling power or force reduces the radius between the earth and the moon that is parting the moon and the earth. Now this is an issue of diverting the attention in a way that brings much suspicion.

The suggestion in the image portrays a scenario where the radius of both celestial objects was reduced to nothing. The moon and the earth already collided. Mass forms a force that pulls both the earth as well as the moon by forming gravity. If the mass of the earth was next to the mass of the moon why did it move apart?

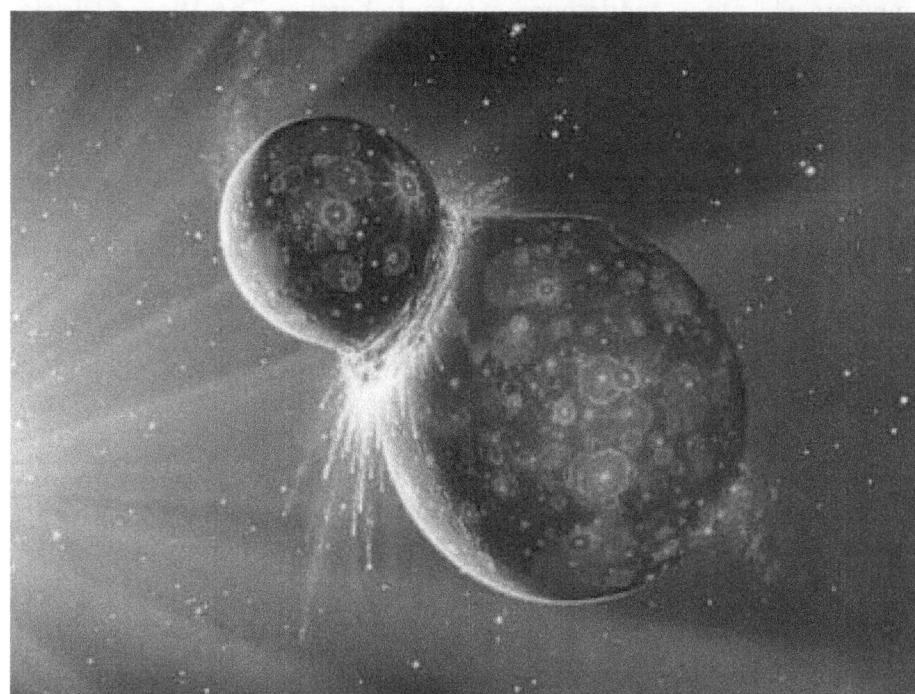

The big goal is to unite by having mass pull mass towards, not push mass away from mass. When this collision happened the deed was done, the puling was completed, the mass united with the mass and the lot waited for judgement day to come. Now someone is telling me that the moon was close and it took off and floated away. This does not meet with Newtonian principle requirements. This is a set-up.

The question is why would they come up with some fancy scheme that clashes with Newtonian wisdom head on? Why would they create some idea that totally contradicts the pulling of Newton's principles of forces pulling by the measure of mass? Why tell something that opposes all other things that confirms science's legitimacy. I can see how this story would divert attention away from what is truly happening. If Newton is correct and that is what everyone in physics believe unconditionally, then mass pulls pass by the force this pulling creates which is called gravity. With that being correct how can any one presume that the moon was part of the earth and then parted from the earth? This forms part of what I call www.questionablescience.net and the e-mail address is mailto:info@questionablescience.net.

Newtonian physics teach us the moon is coming while the theorists tell us the moon is leaving and what does that teach us; that there is total confusion in physics that should be the most sane science in the world. The confusion is deliberate and that made me very suspicious. Why come up with this deliberate confusing story and why does science not repudiate those spreading this contradicting message. If they say the moon is going while Newton says the moon is coming but they never mention the fact that the moon is coming it will leave everyone thinking that the moon is going. Why would they not wish to say the moon is coming to hit the earth instead of contradicting science by never saying anything about the obvious fact that the moon is coming to collide with the eart? Only those with power and those with money can buy this deceit. Only the politicians can persuade the scientists to obey what the bankers wish to allow the paupers to hear so we are kept in the dark and they feed us bullshit. As they make up science they make up science fiction and we have the task to see what is science and what forms science fiction and moreover separate the two as we part what is truth from the fiction.

This book is written (and so too is many of my other less complex books) to show how the science magazine Annalen Der Physics and all other intellectuals approach new science information that in the basis contradicts the science used in the modern era and from this I aim to promote my theses compiled by six parts thesis to introduce a new way of thinking in terms of science. Those masters of science practising science are up in arms about how the Roman Catholic Church treated Galileo Galilee while their conduct that clashes with their views receive no less condemnation from their ranks. As you will later read the condemnation by which I am treated for not agreeing with modern science **and proving them wrong no less** you will see those in Physics known as mainstream Physics are no better than Pope Paul III. They act as if they are holier than others while their condemnation is more harshly. The Theses I wrote

are in place because science as it currently is (whether you believe what I say or not) is running on so many phoney principles of which those in charge of science holding the highest office in science benefit richly from and therefore discard the utter flaws on which science in the present way is founded. Their job is to ignore the flaws there are. By discarding my work they secure their work. If they agree about the flaws that I show is in place then their Academic papers that they wrote on science and their research from which their degrees followed in science instantly becomes something for the paper wastebasket.

I wish to take your mind into the world of science and moreover physics. The physics I show is the basis of physics taught to children at the lowest level of science at schools worldwide. It should be simple enough for anybody to follow because it deals with the everyday interpretation of science as we think of science in terms of science. Every person dropped a glass during their lifetime and felt sorry and incompetent for being so clumsy. If you are able to drop a glass and break it on the floor then you are able to understand the following explanation about science. If you know how to fall you will be able to understand why you fall when you fall...or are you able to know why you fall when you fall...

To my thinking man became man when man saw shiny objects in the night sky and formed religious ideas about forces greater than that which man controls. Man saw the two large objects in the sky and found the two had most influence and then man. Then civilization came along as amongst many other ideas from which man grew man gave these two that apparently had prominence being the sun and the moon the most direct influence on human life and then with that then also the most importance. I say this because if anything makes man distinct from animal it is the fact that man see things inside the influence or outside the influence sphere of man's ability to control or to be controlled. As humans of all cultures we have from forever been fascinated by the noon and always tried to connect some purpose the moon would have on the destiny of life. From the beginning of whatever we connected to the idea of civilization, somewhere in the middle there was the moon playing a part that had a role to play in our destiny. If there were a bunch of gods in any collection that any culture or ethnic tribe formed you can bet your shirt on the fact that the moon had a name and the moon was an important part of the ruling class in the various gods' power constructions. There were lightening and thunder and the sun and the moon and some collection of stars where each had a group for every month. Nevertheless the moon had a prominence as part in religion and was no less influential or prominent than any of the others.

People attributed so many factors playing a part in their lives to the influence that the moon brought on them and many still do. Women that got pregnant became so with the influence coming from the moon. Women that did not get pregnant were also because of favours not coming forthwith from the moon. Gender was a result of the moon being either badly influenced or favouring some of the community or not granting favours. The moon was some influence to be respected and could form a menacing influence.

The moon played a most critical part of farming from the time when farming first began. In centuries gone by before the inventing of electricity and fossil fuel, ploughing was primarily done under the rise of the moon and therefore by the blessing of the moon. This idea is still very much alive in modern times. I don't think any person would have grounds to think of me, as backwards and as a farmer I know how intense we farmers gauge the moon to see rain coming or having extreme heat and cold spells, all conducive to the moon influencing changes in nature. I don't believe or deny this but many very intelligent farmers believe that if you plant crops such as corn or wheat or tomatoes that grow on top of the surface you plant when the moon is rising or better said getting fuller. When you plant crops that grow below the surface such as potatoes or sweet potatoes then you plant that when the moon is fading or reducing in volume. Do I know this works for sure, hell no I don't and yes I do because we all are planting in this way from the time many thousand years ago when farming became a science project and since that is how you plant therefore it is most scientific to believe in this method of planting. Since we all plant this way there is no evidence that this is not the way to go about farming and therefore with all the evidence pointing towards this method of planting it must work. Why take the risk of not planting in this manner if it is so well proven and tested. If you think I am silly and superstitious then wait till I get around in explaining Newton's gravity by mass forming forces all over the show, because that is superstition in the making.

Today in the present as I sit and write this information I know for sure that in about 90% of the rain that falls in this forbidding place of dissolution, the godforsaken drought stricken desert that God forgot to complete during the time of creation of the earth and then left alone to become the area which is the desert I chose to live in where the rain will fall only during a period from three days before full moon to three days after full moon in the summer months. It rains from three days before the event of full moon to three days after the event of full moon. It is a rule and if you think this has nothing to do with the moon then you come and farm here. In less than a year you will believe in this as much as we all believe in this rule. It might shower a few drops or be a cloud burst that month but it rains in that period of the month.

Then that confirms Newton because if the moon is full there is lots of moon pulling water into the air and releasing the water to fall as rain. Australia is flooding like never before. China is flooding like never before and South America is having entire towns covered by collapsing mountains as the result of drenching landslides. We see snow fall as we saw never before and this happens way before Christmas day even. You try to sell the idea of global warming to a town that had I meter of snow come down in one night one month before any snow was suppose to fall and I see you go home hungry because you will not do much selling.

We see how the gravity pulls the water up and how gravity pushes the water down in high and low tides when the moon's gravity comes into affect. If this is what you believe you are more Newtonian in science than most others are. The moon pulls the water up by gravity of the moon. The gravity of the earth releases the water to fall on the earth. The water that fall we think of and we call rain. One thing we never ever get is snow. Gravity is not strong enough to pull water to form snow.

There is always much talk of how the Universe will develop from how the Universe developed. But never do we find talking about our close companion. Some see the moon as the closest planet and others see the moon as an extension of the earth. There is much speculation how the moon got where it is but never is a word said about where and when will it hit the earth. If there is such a force as gravity then the cosmic question that needs an answer is … If this is true that the earth-mass pulls moon-mass as a pulling force, then when will the earth and the moon collide? This destruction of the earth and the moon must happen with the earth and moon being so close…

Is there anyone that ever read anywhere about when the moon will hit the earth? Why does no one ever refer to this doomsday that has to be part of our future!

Go on and hunt down as much research as you can but you will never see any researcher published anything about the day that the moon is going to hit the earth.

The moon has mass that pulls the earth and the earth has mass that pulls the moon, therefore this lot pulling each other has to come closer leaving a bang as the solar system never saw before…and with Newton's gravitational laws prevailing it has to come…but when is it coming and why does no one ever talk about this. Why is there absolute silence about this event that must take place? Have you reading this ever given this notion a thought and if not why did you not give it a thought before?

How far did the moon come closer this past say million years? We heard about animals that disappeared million of years ago and we see impact craters investigated of events gone to the past but no where is there any research offered on the moon coming towards the earth and the catastrophic end that will bring devastation to all forms of life. It will be the end, not to certain species or to some rainforests but to all forms of what can be called life! And yet as critical as this is, no one ever spared a thought about researching this doomsday looming in the future. This catastrophic end is coming as sure as Newton's mass is pulling to form gravity! There is not enough imagination in any one person's mind to replicate the disaster pending that will come when the earth and the moon collide. The impact will end any form of life there might be.

Ever since science broke "free" from the Dark Ages science insists on forces existing that control nature. They would not put these forces in the same category as witchcraft but a force to all purposes remains a force, a dark hidden unexplainable movement contracting and changing without finding any explanation as to what it is. The forces are hidden and the origin goes beyond what man can see or explain and yet the forces control our everyday as much as witchcraft did before we abolished witchcraft and made witchcraft a legal enterprise, Now we sit with four forces that no one can explain and the witchcraft behind it is illegal to think of as witchcraft. The force of mass pulls the force of mass and the result is a collision between cosmic bodies. This happened as often as science ran out of explanations about events that occurred in the past such as when the dinosaurs were killed off or when other tremendous unexplained occurrences happened. Yet every doomsday science connect with a force but the force does not connect with witchcraft because the force is a force and that is science.

$$F = G \; \frac{\text{Mass of the earth} \times \text{Mass of the moon}}{\text{radius between the two destroyed by the square thereof}}$$

Think how big the moon is and then think how big the earth is and then think how close the earth is from the moon … think of the force this must unleash and how this force has to increase as the moon is coming closer, speeding up the pace by increasing the speed of the moon closing the distance between the two solar bodies and where does this leave you? You being sandwiched in between the two!
Think how the sea will have high and low tides of thousands of meters pulling the waves to destroy the land long before the moon ever rushed into the earth with a bang. By the time the Moon appears this large in the sky, would there be human life remaining to see the event occurring or would all human life be seized by then? Think of Global Warming and this will bring us Global Warming as nothing did before.

Then again might this "global warming" not be an eye blind to cover up the real issue that is at hand, the moon coming towards the earth. To stop a panic attack that will destroy all forms of property and currency those in power then create the "global warming issue" to make us think that we are safe and that we are in charge of what we clearly never could be in charge of, and that is gravity destroying the moon and the earth? Why has no person ever thought even to mention this subject, ever only once?
Those in charge of putting the fear of God into us by arousing mass hysteria about Global Warming and carbon polluting or destruction looming such as the international press never even mention the possibility of such a collision! Why is that the case!

There are probes in place to detect renegade comets and asteroids that might inflict massive damage when it hits the earth. Science does show a tendency to locate this event beforehand because this brings a certain degree of worry to those charged with finding something to worry about. Billions in whatever currency is spent to detect some asteroid or comet that is heading our way with the sole purpose to destroy the earth, but when this happens the event is by chance. These events are not predictable and are avoidable and yet they are studied in anguish. There is a much more realistic desecration heading our way with the moon destroying the earth. This event has a better chance and more certainty, with Newton's principles applying the certainty factor and yet this possibility is never even mentioned by those charged with detecting concerns! Why would those doing research never mention the time that the earth and the moon are bound to collide? If it is billions of years away, then knowing that it is still in the very far future will soothe us and bring comfort. However, if it is a possibility that could happen and we could do nothing about it but commit suicide to escape this fate, then hiding this information from the public seems the way to go for those in power. The photos depict the Tunguska event in 1908 in Siberia when an Asteroid exploded above the earth. This was devastating.

See the trees that were giants seem as if they were matches strewn. The blast depicted by this event is comparable to a flash bulb lightning a room to take a photograph. Any person who thinks Tunguska was bad news to the earth should rethink his or her degree of worrying. Tunguska and its aftermath is a Christmas cracker side show compared to what is coming when the moon is coming to our part of the earth. Try the event when the moon comes hitting the earth and see what there truly is that could bring discomfort at bedtime when no one on earth would feel the need to sleep because the ultimate doomsday is with us and upon us.

The Tunguska event would in comparison seem to be a minor local inconvenience compared to when the moon collides with the earth and both objects becoming dust particles. In this light would it not be wiser for the rich and those in political power just to keep quiet about such looming catastrophic scenario and in keeping silence not arouse a hysteric human race going mad in fear and anxiety? Think how uncontrollably mad everyone would get knowing death is upon us? So hush the peoples concerns…

In late June of 1908, a fireball exploded above the remote Russian forests of Tunguska, Siberia, flattening more than 800 square miles of trees. At 7:17 AM on the morning of June 30, 1908, a mysterious explosion occurred in the skies over Siberia. Realistic pictures of the event are unavailable. I believe that we now know enough about large impacts to "decode" the subjective descriptions of the witnesses and create realistic views of this historic asteroid impact as seen from different distances.'

Previous speculation had ranged from comets to meteors. Noctilucent clouds are brilliant, night-visible clouds made of ice particles and only form at very high altitudes and in extremely cold temperatures. These clouds appeared a day after the Tunguska explosion and also appear following a shuttle mission.

The researchers contend that the massive amount of water vapour spewed into the atmosphere by the 1908 comet's icy nucleus was caught up in swirling eddies with tremendous energy by a process called two-dimensional turbulence, which explains why the noctilucent clouds formed a day later many thousands of miles away. Noctilucent clouds are the Earth's highest clouds, forming naturally in the mesosphere at about 55 miles over the Polar Regions during the summer months when the mesosphere is around minus 180 degrees Fahrenheit (minus 117 degrees Celsius).

The space shuttle exhaust plume, the researchers say, resembled the comet's action. A single space shuttle flight injects 300 metric tons of water vapour into the Earth's thermosphere, and the water particles have been found to travel to the Arctic and Antarctic regions, where they form the clouds after settling into the mesosphere. Following the Tunguska Event, the night skies shone brightly for several days across Europe, particularly Great Britain — more than 3,000 miles away. In both cases, water vapour was injected into the atmosphere. The scientists have attempted to answer how this water vapour travelled so far without scattering and diffusing, as conventional physics would predict. This "new" physics, the

researchers contend, is tied up in counter-rotating eddies with extreme energy. "Our observations show that current understanding of the mesosphere-lower thermosphere region is quite poor,"
Witnesses in the town of Kirensk and nearby towns at the same distance recollected the fireball flashing across the sky in the following terms: "A ball of fire appeared in the sky... "A flying star with a fiery tail; its tail disappeared into the air." After this object passed across the sky, it approached the horizon where it was consistently described from this distance of 400 km, as appearing like a "pillar of fire," then replaced by "a cloud of smoke rising from the ground," or "a cloud of ash...on the horizon," or "a huge cloud of black smoke. "It was called diffuse bright ball two or three times larger than the sun but not as bright; the trail was a "fiery-white band." Inconsistent colours were mentioned: white, red, flame-like, bluish-white.

Some minutes after the explosion, distant observers reported a column of smoke on the horizon. One observer said, "Where the body disappeared behind the horizon, a pillar of dark smoke rose up." I have wondered whether the dark colour could result from the smoke of the explosion containing black, sooty carbonaceous particles, in the same way that the explosion clouds on Jupiter from the impact of Comet Shoemaker-Levy 9 were very dark. Because the meteorite did not strike the ground or make a crater, early researchers thought the object might be a weak, icy fragment of a comet, which vaporized explosively in the air, and left no residue on the ground. However, modern planetary scientists have much better tools for understanding meteorite explosion in the atmosphere. Some of them dropped brick-sized fragments on the ground, but others, such as the one that hit Siberia, may produce primarily a fireball and cloud of fine dust and tiny fragments. In 1993 researchers studied the Siberian explosion and concluded it was of this type -- a stone meteorite that exploded in the atmosphere. This conclusion was supported when Russian researchers found tiny stony particles embedded in the trees at the collision site, matching the composition of common stone meteorites. The original asteroid fragment may have been roughly 50-60 meters (50-60 yards) in diameter.

Many steroidal fragments circle the Sun; the Siberian object was merely the largest to hit the Earth in the last century or so. Had it hit a populated area, devastation would have been enormous. If there are many asteroid fragments, why don't we see more hits? An Air Force satellite in the 1990s detected a smaller explosion over the Pacific. In 1972, a 1000-ton object skimmed tangentially through Earth's atmosphere over the Grand Tetons in Wyoming, and then skipped back out into space, like a stone skipping off water. Even larger objects have hit Earth, but they are more rare. Brick-sized interplanetary stones fall from the sky in various locations every year. Interplanetary space contains many small bodies of different sizes. Large enough bodies leave sizable craters on planets or satellites. If we continue to study asteroids and build more telescopes for detecting and tracking them, we will have better information about the frequency of such asteroid impact-explosions, and more chance to have warning about impending impacts. At 500 km (300 mi), observers reported "deafening bangs" and a fiery cloud on the horizon. At distances around 60 km, people were thrown to the ground or even knocked unconscious; windows were broken and crockery knocked off shelves.

In late June of 1908, a fireball exploded above the remote Russian forests of Tunguska, Siberia, flattening more than 800 square miles of trees. At 7:17 AM on the morning of June 30, 1908, a mysterious explosion occurred in the skies over Siberia. Realistic pictures of the event are unavailable. I believe that we now know enough about large impacts to "decode" the subjective descriptions of the witnesses and create realistic views of this historic asteroid impact as seen from different distances. Previous speculation had ranged from comets to meteors. Noctilucent clouds are brilliant, night-visible clouds made of ice particles and only form at very high altitudes and in extremely cold temperatures. These clouds appeared a day after the Tunguska explosion and also appear following a shuttle mission. The researchers contend that the massive amount of water vapour spewed into the atmosphere by the 1908 comet's icy nucleus was caught up in swirling eddies with tremendous energy by a process called two-dimensional turbulence, which explains why the noctilucent clouds formed a day later many thousands of miles away. Noctilucent clouds are the Earth's highest clouds, forming naturally in the mesosphere at about 55 miles over the Polar Regions during the summer months when the mesosphere is around minus 180 degrees Fahrenheit (minus 117 degrees Celsius).

The space shuttle exhaust plume, the researchers say, resembled the comet's action. A single space shuttle flight injects 300 metric tons of water vapour into the Earth's thermosphere, and the water particles have been found to travel to the Arctic and Antarctic regions, where they form the clouds after settling into the mesosphere. Following the Tunguska Event, the night skies shone brightly for several days across Europe, particularly Great Britain — more than 3,000 miles away. In both cases, water vapour was injected into the atmosphere. The scientists have attempted to answer how this water vapour travelled so far without scattering and diffusing, as conventional physics would predict. This "new" physics, the researchers contend, is tied up in counter-rotating eddies with extreme energy. "Our observations show that current understanding of the mesosphere-lower thermosphere region is quite poor,"

Witnesses in the town of Kirensk and nearby towns at the same distance recollected the fireball flashing across the sky in the following terms: "A ball of fire appeared in the sky... "A flying star with a fiery tail; its tail disappeared into the air." After this object passed across the sky, it approached the horizon where it was consistently described from this distance of 400 km, as appearing like a "pillar of fire," then replaced by "a cloud of smoke rising from the ground," or "a cloud of ash...on the horizon," or "a huge cloud of black smoke. "It was called diffuse bright ball two or three times larger than the sun but not as bright; the trail was a "fiery-white band." Inconsistent colours were mentioned: white, red, flame-like, bluish-white.

Some minutes after the explosion, distant observers reported a column of smoke on the horizon. One observer said, "Where the body disappeared behind the horizon, a pillar of dark smoke rose up." I have wondered whether the dark colour could result from the smoke of the explosion containing black, sooty carbonaceous particles, in the same way that the explosion clouds on Jupiter from the impact of Comet Shoemaker-Levy 9 were very dark. Because the meteorite did not strike the ground or make a crater, early researchers thought the object might be a weak, icy fragment of a comet, which vaporized explosively in the air, and left no residue on the ground. However, modern planetary scientists have much better tools for understanding meteorite explosion in the atmosphere. Some of them drop brick-sized fragments on the ground, but others, such as the one that hit Siberia, may produce primarily a fireball and cloud of fine dust and tiny fragments. In 1993 researchers studied the Siberian explosion and concluded it was of this type -- a stone meteorite that exploded in the atmosphere. This conclusion was supported when Russian researchers found tiny stony particles embedded in the trees at the collision site, matching the composition of common stone meteorites. The original asteroid fragment may have been roughly 50-60 meters (50-60 yards) in diameter.

Many steroidal fragments circle the Sun; the Siberian object was merely the largest to hit the Earth in the last century or so. Had it hit a populated area, devastation would have been enormous. If there are many asteroid fragments, why don't we see more hits? An Air Force satellite in the 1990s detected a smaller explosion over the Pacific. In 1972, a 1000-ton object skimmed tangentially through Earth's atmosphere over the Grand Tetons in Wyoming, and then skipped back out into space, like a stone skipping off water. Even larger objects have hit Earth, but they are more rare. Brick-sized interplanetary stones fall from the sky in various locations every year. Interplanetary space contains many small bodies of different sizes. Large enough bodies leave sizable craters on planets or satellites. If we continue to study asteroids and build more telescopes for detecting and tracking them, we will have better information about the frequency of such asteroid impact-explosions, and more chance to have warning about impending impacts. At 500 km (300 mi), observers reported "deafening bangs" and a fiery cloud on the horizon. At distances around 60 km, people were thrown to the ground or even knocked unconscious; windows were broken and crockery knocked off shelves.

With the formula as Newton introduced it being $F = G \dfrac{M_1 M_2}{r^2}$ any one with a little insight into reality could see that as the bottom part decreases (become smaller) this will affect the top part to become larger. I will show three values and you will see what I mean. $\dfrac{100}{100} = 1$ and $\dfrac{100}{1} = 100$ and then getting the bottom really small $\dfrac{100}{.001} = 100000$. The bottom part's decrease increases the speed of the top part or the gravitational movement.

By reducing the distance the force of gravity will grow exponentially faster!
I show you this to indicate that the closer the moon gets to the earth the faster it will come to the earth. The last minute the moon will travel about say one hundred million kilometres because the gravity pull will increase by a billion trillion times per second. During the last day's of the earth's existence the moon is coming closer to the earth by radius reduction to the square to the square per second and the reducing will be as quick as the last million years we now have had. The rushing moon will increase in speed by the square of the disappearing radius and we sit between the two as a human sandwich.

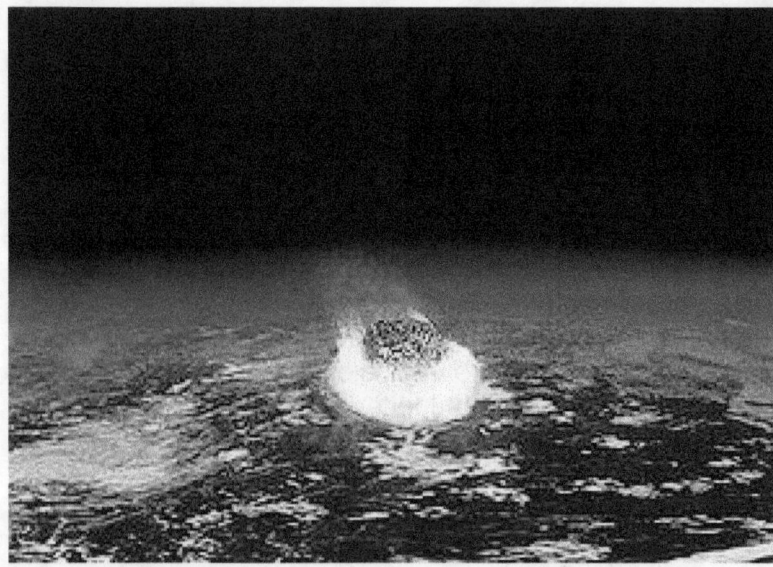

This is the tempo whereby the distance of the radius between the earth and the moon will shrink. Then these numbers still have to be divided by the squared of the reducing radius making the result incomprehensibly more devastating.

The smaller the radius becomes the faster the radius would become smaller and the more the tempo would be of the moon closing down on the earth. The one-year everything might seem perfectly under control and the next year everything will seem to go mad in haste.

The moon will seem to be on a runaway collision coarse going out of bounds the closer it gets.

The devastation it brings will increase exponentially and becoming so big no one would be able to realise the changes as the doomsday is increasing! No one would be able to keep track of the fatalities occurring. Everybody that died during the first and the Second World War plus the epidemic influenza that came after the First World War would be deaths happening in one day and during every day. And no one in science even thinks about this scenario while Newton's gravitational laws are in place? Can you think why not?

  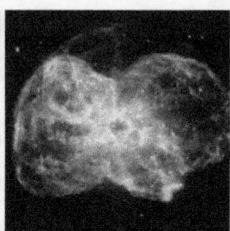 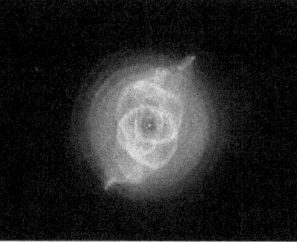

Those who should know better says when the sun runs out of its current fuel, the hydrogen starts to cool down, as it will expand into a red giant and could expand out to where Jupiter is circling or so they say it is going to be. These are the estimates I have seen quoted before and are not mine to quote.

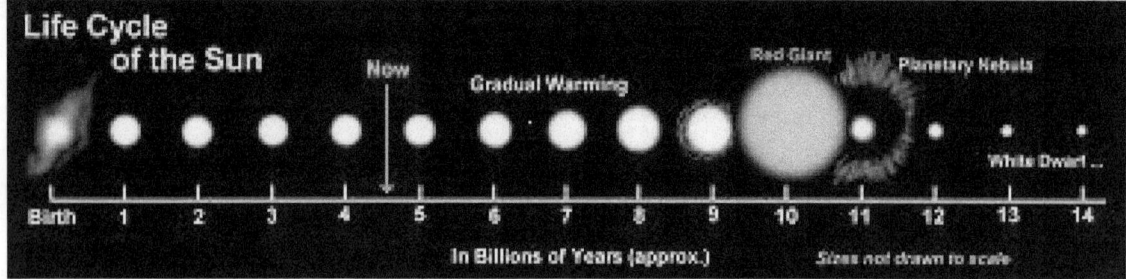

The end of the sun is billions of years away. It is a lot of hogwash but it shows someone somewhere got interested and started some form of thinking about how the sun and the solar system would end. Yet no one in science ever thought it worthwhile to conduct a study into when the earth and the moon will go into destruction.

Even if this is going to happen in a billion or five billion or a trillion years from now, still there are persons that find the end of the sun important enough to be concerned about. It shows that it is not because it is far into the future that there is no apparent interest into the collision between the moon and the earth. The moon would have hit the earth much sooner than the sun would explode, so why is there no investigative research done on the earth and moon contracting by the mass that Newton foresaw?

Chapter 2 MOON CRASHING.

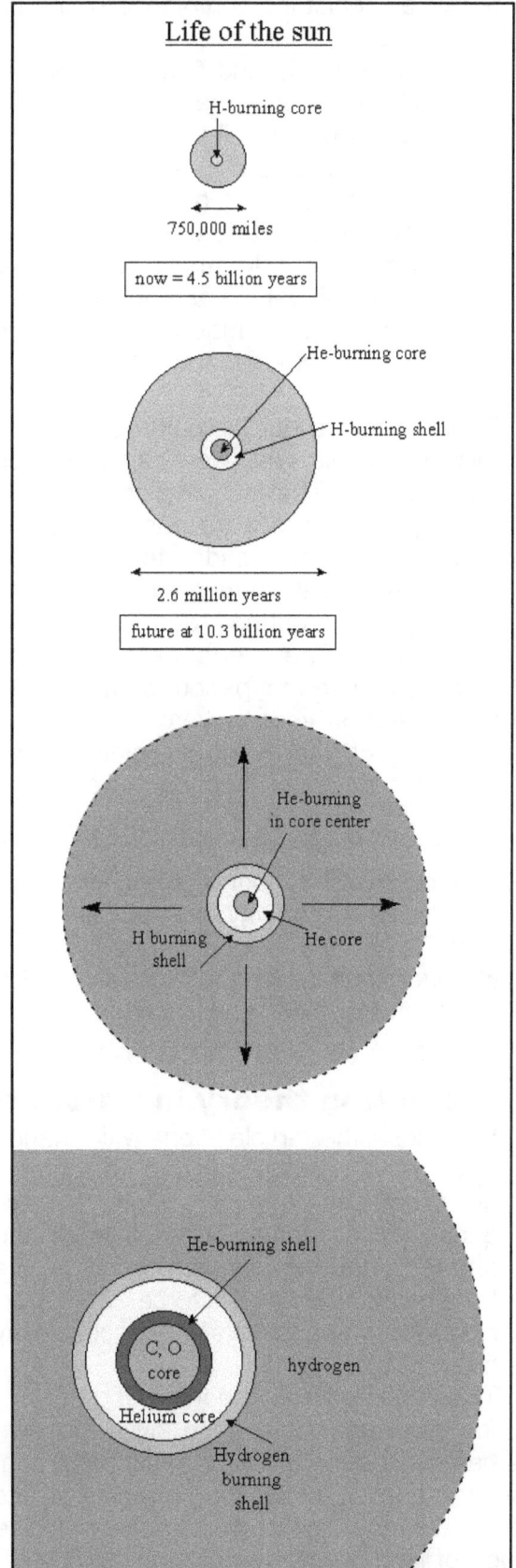

I don't for a minute say I agree with anything said about the sun, but at least it shows some investigation was done not looking at the correctness thereof, there was a thought spared about how the sun will end.

It is said using Newtonian vision about what the Universe holds install that the sun is going to blow up and burst and come to an end. Don't repeat me on this hogwash because all I wish to prove that in terms of even the sun coming to a closure of a life cycle there was research done in this line of thought. It is said that the sun will scourge the solar system out of existence but never does any one mention that the moon and the earth is having a get-together long before that formed on the grounds of Newton's vision on gravity coming from mass that pulls mass. If science can look this far ahead and find a doomsday waiting on us billions upon billions of years from now, why is everybody avoiding the apparent doomsday that is much closer to realising the ending of all of the earth and its moon including all (and the only) life that was ever present in the Universe.

With everyone in science saluting Newton's gravitational contracting there was an extended effort by Albert Einstein to find the critical density of the Universe. That is the backbreaking effort that science took with painstaking accuracy to find the level of interest in determining the end of the entire Universe. The critical density idea did not pan out and that left science high and dry for answers about science. They did not stop there, no Sir, they paddles on in darkness to find the answer. They left no stone untouched to come up with a conclusion…and yet not one person in all that time started to think that the contraction will be much better monitored by researching the moon reducing of the radius it has between it and the earth. Would it not have been much easier to study when the moon will splash into the earth and from there work out when Newton's attraction will have the worst collision we can think of happening in our backyard?

Then science goes further and tries to detect the untraceable that is invisible. They try to find dark matter in a shady Universe. They spend billions on dark material or dark energy research but not a single dime goes the way of finding out when the moon and the earth will have a gravitational self-destruction event. Do you not find that very odd…or is it just I wondering about what?

Building caves in mountains would not help because the massive waves will get us there in the final event. It is just a thought but why is there such a surge to investigate possible occupation of Mars? There is no spot on earth that will remain safe so where will those that hold the money and the power go when they need a safe haven. What would be the most obvious reason be to keep order and keep the money flowing and for what purpose would they wish to keep the money flowing? Think about it this way, if the rich and the powerful could take all of the tax money of all the lesser human beings and start to build on Mars under the disguise of research, then when the doomsday arrives those being the fortunate, the powerful, the influential and the law makers will have somewhere to go to escape, leaving the less worthy to take the moon head on! There was an urge to get to the moon and when they got there then afterwards the interest in the moon fell to an all time low. Then those in investigating investigated all the planets without showing the least interest in what the moon might hold. Then suddenly from the blue comes some hysteric research about Mars, just after someone discovered possible prehistoric semblance of life on Mars.

if we could trust Newton's physics of mass that pulls mass.

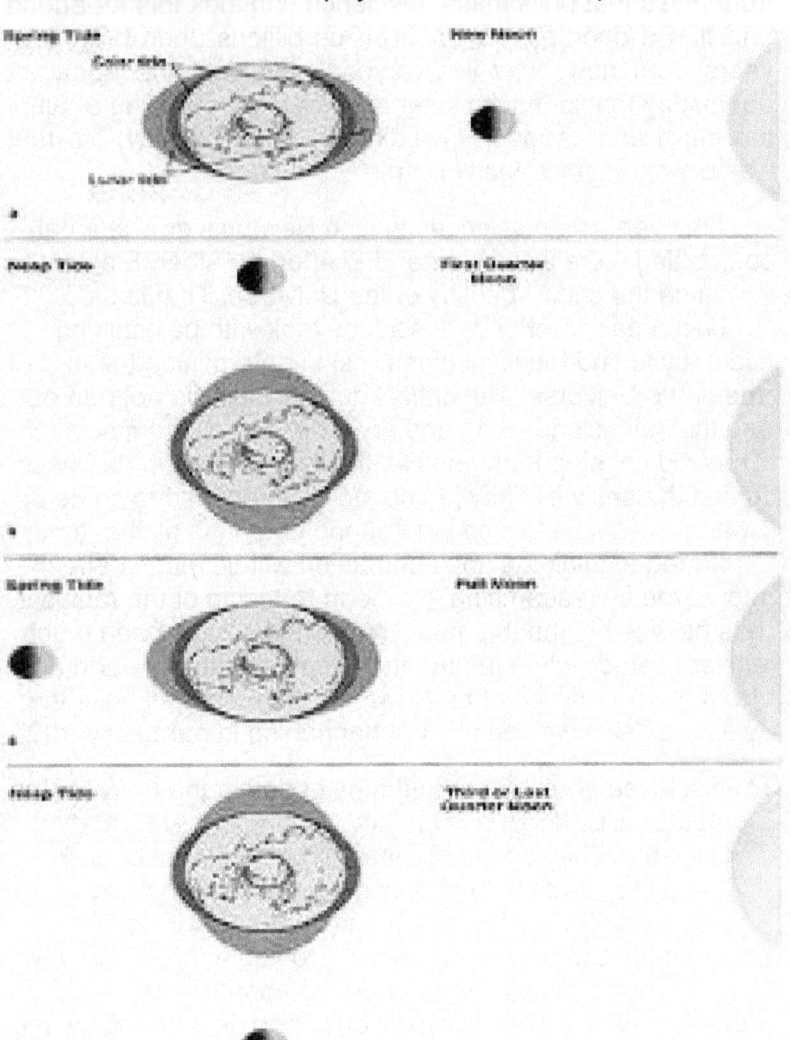

If they study and find this event is still far off into the future, then do the study and tell us. If they do the study and see it is going to happen in three or four generations from now, then include us into the plans of becoming Martians or say who will be left behind and on what merit is the choice made, but for God sake, do studies and tell us, that is

Those who are quick would observe the mistake in this very Newtonian science idea I made. The idea that it rains only six days a year is a true as anything in reason could be. The idea that the moon has more gravity during full moon and less gravity during dark moon is fabrications. The moon has the same mass all over the period of any given year going on year by year. So the idea of more of the being in place moon finding more mass forming more gravity that pulls more water is lame and yet it has a very Newtonian ring to the concept. Newton said the moon's gravity must pull and the earth's gravity must pull and then with tides the moon's gravity pulls stronger than the earth's gravity, which results in high tides and low tides.

Let us test Newton's attraction theory in practice

I suppose this article alone will change little in the onslaught of the conspiracy, but someone has to begin to ask questions at some point. I have the answers but the conspiracy stops me from being heard as the force of the most powerful Academic's silences my message.

This is quoted from the Internet

You know how the earth pulls the apple down? It pulls everything on or near its surface toward its own centre. And everything near enough to be pulled pulls back as hard as it can. The earth pulls the moon, and the moon pulls the earth. Although it is much smaller than the earth it is just the right size, for its distance away, to keep from falling into or away from the earth. We cannot see its pulling power on the solid parts of the earth. But the ocean is made of water. A slope of land, a brisk wind, many things set water in motion. It feels the pulling power of the moon. Whenever the moon rises over the ocean, it pulls the water that is just under it. So, a great wave, or tide, travels under the moon across the wide sea. When the shore is reached this wave rises higher against the rocks, or spreads over level sand beaches. When the moon sets, the wave goes back to the old level.

Right, this explanation is very Newtonian and very incoherent.
Newtonian science says, "Mass pulls to form gravity". Newtonian science says "gravity pulls" to form high and low tides.

Can anybody see the Newtonian double fork in this conflict of factor implications? If this is true that mass form gravity by creating pulling, then that means the moon earth mass relevancy must change twice in twenty-four hours. Either the moon must become more massive or less massive twice a day because the tide rises and the tide lowers in one circle rotation of the earth. It has nothing to do with the mass of the moon because if this "gravitational pull" forms gravity by mass, then either the earth or the moon has to increase the mass ratio to "draw" the water "up" cyclic or to "push" the water "down" in an earth rotation cycle.

Tidal waves again as everything else in nature proves Newton wrong and out of touch with cosmic reality. I formulate and I prove that gravity has nothing to do with mass and everything to do with the rotation movement of the earth and through the rational spin of the earth mass is created. The moon proves my theory as much as all of the solar system proves my thinking of how gravity forms as much as all the mentioned factors disprove Newton.

Yet Newton is accepted as if proven without question. Should you wish to find out how gravity does work then go to Lulu.com There is one question though? With the moon always at an even size and most often at an even distance, how does the moon pull the earth by 12-hour cycles?

The idea of proof comes automatically to the door of Newton although Newtonians will deny this fact as if they deny the honour of their Master Newton and that is what they have to do. Newton as a culture became a religiosity in which every human being believes unconditionally.

Is there anybody who would try to convince me that mass plays a part in the tidal wave of the moon? It is gravity pulling the tidal but that gravity is directly linked to the spin of the earth and has no connection to mass, although mass is a direct result of the spin of the earth. Is there anybody who would convenience me that Newton's idea of "mass forming gravity" whereby the moon pulls the sea level into high tide because if he or she might attempt then let's start with the mass of either the moon or the earth becoming bigger or smaller in every circle that the earth rotates.

Is there anybody gullible enough to believe that physicists in three hundred years never saw this tendency contradicting Newton's "mass that forms gravity" claim? There is no one that thought this should deserve a better look because this does not come across as very sane or believably accurate. This is the conspiracy. The conspiracy is to hide Newton's outlived ideas from all those standing outside physics because remove Newton's ideas from physics and you are left with nothing!

I prove with formulation and evidence of what applies in the cosmos that gravity is the result of the earth spinning. To do that I prove that the Titius Bode law, The Roche limit, the Lagrangian points and the Coanda effect combines to form gravity. These are the most important principles in physics and are in place used in the formation of the Universe but because they contradict Newton and make a fool of Newton they are not very pertinent. Because the layout and the reason why it is in place is not understood they make those Brainy-Bunch in science look incompetent, as incompetent as Newton's ideas are, and therefore it is very discretely hushed up. I challenge anyone to show where the cosmos uses mass but I can show where the cosmos uses the four principles I mention and that I prove, forms gravity.

However, there is nobody that is interested in what I have to say because there is too much money to be lost! If my work is accepted, it makes Newton part of science fiction and think of what that would do to the Powerful-In-Charge-Of Money-Matters? Every book about Newton then becomes fiction. Since the money in science depends on every person on earth believing in Newton just the sheer influence of money gone lost prevents the most powerful even to consider my views as long as I denounce Newton. But that I have to do because Newton is as far from the truth as a Christmas tree is from the Christian religion.

I prove that as the earth turns, by turning it compresses the space around the earth. The space we call atmosphere is thrust onto the earth and if anything circles the earth such an object then can leave the surface of the earth. By turning, the space around the earth compresses the space we call the atmosphere and by thrusting the atmosphere onto the earth the space holding a solid will then press onto the earth and thereby form mass. That's how simple it is.

Science says mass pulls mass. Then why is this astronaut floating in space? He still has mass and the earth still has mass and the mass must be pulling to form gravity. The man is floating on the same principle that the moon is floating. If the man is not going to fall, the moon and earth will never collide and that makes Newton's mass pulling mass a contradiction, which in turn makes science a joke. If Newton's "mass is pulling mass" is correct, then the moon and the earth must collide. So which is it, because it can't be both? The man has mass just like the moon has mass and the earth still has the mass it always had and the law of attraction between the human floating above the earth must be the same as the moon floating above the earth. If the man could float then this law of attraction is rather becoming suspicious. If that is the way you think then having those suspicions are very correct. Science confirms that there is always a force of attraction between all bodies. There must therefore be a force of attraction between the astronaut and the earth in, so why isn't the astronaut falling straight down in like with Newton's apple? The mass has gone nowhere but science cheats the answer by giving the man "micro gravity". This way they cheat by avoiding the truth without pulling a face.

If mass formed gravity, then Kepler's first law $r = \dfrac{p}{1 + \varepsilon \cos\theta}$ that Newton devised would not be possible because this shows deviation of the orbit while we all know that the mass remains totally constant and always the same. The why would the orbit circle in a loop coming "closer" at one point and going "wider" at the opposite side. Does anyone out there still wish to tell me there is no conspiracy about Newton and the truth going on in science?

If mass was responsible for forming gravity and it was the mass of both the sun and the planet, then a precise circle would form allowing a continuous radius that could never deviate. A child can see this but our Newtonian maters of conspiracy never once noticed this or gave any of their precious time to rethink this Newtonian anomaly…or did the lot just conspire for three centuries to look the other way because if they have not got Newton, then they have got nothing…and they forever tell they only deal in proven facts?

If Newton's presumed $F = G \dfrac{M_1 M_2}{r^2}$ was true then "Kepler's" second law could not apply because the Kepler's true formula, which is $a^3 = T^2 k$ would bring about again a perfect round circle because again mass being constant will sustain as perfect circle around the sun. Again no one in science in three hundred years saw this and did not consider it to be rather odd. Then every one outside physics thought those inside physics were so clever that those inside physics saw what those outside physics could not see and left it to be.

That gave those inside physics the chance to swindle with physics and was never challenged by those living outside physics for the fear of feeling inadequate and stupid.

All the while it is those inside physics that are inadequate and stupid because they never could understand that Newton can never apply because the cosmos disproves Newton at every chance there is.

Where Newton shows that mass is supposedly the factor that should place planets according to size in their respective positions, which is not true, I prove that mass is not used by nature, not anywhere and I am the one they frown upon because I don't accept Newton's and therefore I don't compromise by embracing the Newtonian conspiracy. I show how ridiculous it is to support Newton because Newton has no foundation or support from the cosmos, yet I am ridiculed as the one who is incoherent. I show what nature uses namely the **Titius Bode law**, **The Roche limit**, **The Lagrangian Points** and **the Coanda effect** and how this forms gravity as well as place the positions of the planets in accordance with singularity, not mass. Because I trash Newton's rubbish that does not fit and in that can't apply, no publisher of science books or science magazines will publish my work! I show what goes on in nature while Newton's contribution of mass applying is total rubbish. Because I call it rubbish and I rubbish Newton's ideas, I am ignored.

Let us now proceed and test the basis on which all physics are founded

Let us then test the truthfulness and believability of the foundation on which the entirety of Newton's physics principles rests. I show you the absolute accuracy of $F = G \dfrac{M_1 M_2}{r^2}$. This is Newton's formula on which the entire philosophy of gravity rests and it supposedly proves that mass forms gravity by force.. No one in science is capable of proving this formula worthy of being part of science and with this forming the basis of science it must be clear to what extent the brainwashing and the mind control and the thought manipulation is going on to force students to accept corrupted science that can never make sense.

Newtonian physics teaches that the mass of the earth and the mass of the moon is constantly contracting (reducing), which produces a force called gravity that reduces the distance between the two solar objects. The cosmic question that calls for an answer is … with this being true that the mass pulls mass, then when will the earth and the moon collide? This destruction of the earth and the moon must happen first since the earth and moon are the closest…

Academics in physics insist that gravity is founded on the Newtonian gravitational principle of $F = G \frac{M_1 M_2}{r^2}$, then when is the inevitable collision coming.

Now, everyone has to braise himself or herself for the inevitable impact that has to come, but when. When can we trust the conspirers and physics academics to tell us when this inevitable doomsday destruction is on the earth's doorstep?

Newton said this must come because of mass pulling mass to clash with each other and therefore so when is the day going to be when we will not be on earth any longer. We should know! For those not familiar with Newton's gravitational principles I will give you an update of science development that took place the past three hundred years, in the event you missed the latest findings.

Have you heard about the apple that fell from the tree that had a bigger influence on mankind than any religious intervention before the event or after the event? This incident with the apple changed the outlook of every person on earth notwithstanding religion, culture, ethnicity, or race. This apple that fell from a tree changed the lives of every individual on earth that remotely had anything to do with science.

Let's be serious. They say (they being those in science) the moon was a part of the earth and by some mystical collision the two came apart in some fashion no one can entirely understand. This does not make sense and is even more senseless where it does not stroke with science principles.

This is how you con people by diverting attention away from the place you question. In trickery and playing games of deception one uses a method of diverting the attention towards another place to commit a crime. One would bump hard into someone while picking the pocket while the person feels the thump of a trip when colliding with the victim. This is precisely what I see is happening.

This principle declares that a force of attraction by the value of mass of both solar objects is pulling by reducing the radius, therefore bringing the moon and the earth closer. This must lead to an inevitable collision that must end both the earth as well as the moon.

Let the Newtonians calculate when this event is due and inform the human race when our final day of doom will arrive by using Newton's law of attraction or $F = G \frac{M_1 M_2}{r^2}$. The day if (not when but if Newton is correct) this happens, Life in all forms will end and therefore knowing this event is crucially important for all of us having life on earth! We have to know! The future of life in the Universe depends on this knowledge. If we know then we have to find the ability to transplant life to Mars and beyond just to save life, as we know life is! Knowing when this will happen will give us humans a fighting chance to have a future somewhere safe!

Those Ever-So-Wisely-Educated cosmic Super-Brains always calculate the power that drives a Super Nova. Let them use those brilliant minds then to calculate the precise date when the earth and moon will destruct all life on Earth! Those Brainy- Mathematical-Masters always custom-design in detail space – whirls they invent with applied cosmic imagination. Let them bring such terrific astonishing human abilities closer to home. The Mathematical-Geniuses who can calculate the inside of a Black Hole should bring their splendour to a much better use in terms of where it concerns human future.

Instead of painting a Universe fit for Alice in wonder-world, and sprinkling it with the best mathematical formula that they think must put them on par with God, rather apply the same formula and show when the solar system, as big as it is, will collapse into the Sun. Ask them to not search for imaginary undetectable, unexplainable dark matter they can't even point out, but to find when the distance we have between the planets and the sun will dissolve and when the pull of the planets will have all the planets go crashing into the sun.

Those Ever-So-Wisely-Educated cosmic Super-Brains best art form is to deceive everyone by conspiring to hide the truth about Newtonian science. Those Mathematical-Masters will never even hint Newton's incorrectness of thought because it will reveal their fraud.

It is showing that those you think you can trust are those very persons that you dare not trust!

This website allows you to see what the degree of blatant corruption is presented as physics and what Academics in physics hide.

The law of gravity says that the mass of one object pulls by force on the mass of another object where the second object then pulls back also by force.

Newtonian physics says that the mass of the earth and the mass of the moon is constantly contracting (reducing) the distance they are apart, by forming a force called gravity, or so science believes. If there is such a force as gravity then the cosmic question that needs an answer is ... if this is true that the earth-mass pulls moon-mass as a pulling force, then when will the earth and the moon collide? This destruction of the earth and the moon must happen with the earth and moon being so close…

With the information at hand for centuries, then why not use it? Why don't they tell us when?
Is there anyone that ever read anywhere about when the moon will hit the earth? Why does no one ever refer to this doomsday that has to be part of our future! Go on and hunt down as much research as you can but you will never see any researcher published anything about the day that the moon is going to hit the earth. The moon has mass that pulls the earth and the earth has mass that pulls the moon, therefore this lot pulling each other has to come closer leaving a bang as the solar system never saw before…and with Newton's gravitational laws prevailing it has to come…but when is it coming and why does no one ever talk about this. These objects with the most mass are the closest and must form the biggest pulling power from our earthly perspective.

Why is there absolute silence about this event that must take place? Have you reading this ever given this notion a thought and why did you not give it a thought before? How far did the moon come closer this past say million years? We heard about animals that disappeared million of years ago and we see impact craters investigated of events gone to the past but no where is there any research offered on the moon coming towards the earth and the catastrophic end that will bring devastation to all forms of life. It will be the end, not to certain species or to some rainforests but to all forms of what can be called life! And yet as critical as this is, no one ever spared a thought about researching this doomsday looming in the future. This catastrophic end is coming as sure as Newton's mass is pulling to form gravity!

There is not enough imagination in any one person's mind to replicate the disaster pending that will come when the earth and the moon collide. The impact will end any form of life there might be.

Have you heard about the event about the falling apple that had such a concrete influence on mankind it changed the world more profoundly than Jesus Christ Did?

This book and what forms the content that I present has the dynamics to change science forever and that is not cheap exaggeration or promotional talk. You are about to find out how influential apples are in directing man's eternal destiny. According to the Bible it was a fruit that a man going by the name of Adam ate in the Garden of Eden and it is generally thought of as an apple that changed man's destiny. The fact that it was an apple is just surmising and is not a proven statement and that much I admit but such as it may be, correct or incorrect, the idea that it was an apple stuck. The correctness about the fruit of the Garden of Eden being an apple or not being an apple is disputed and is also never disputed but since it was never proven to be an apple or not to be an apple therefore it also was never disproved just as much as it was never proved. In the general accepted views of the broader view of person's that the concept of the apple was the forbidden fruit is widely accepted, although it is not widely proven and although widely accepted it is accepted without carrying strong doubt or religious rejection and therefore we are not going to dwell on the apple being the fruit which the Bible refers to. The point I wish to bring across is that even without proof that the apple was the fruit being illicit in the Garden of Eden is accepted almost without a second thought. That is the culture man has. Man accepts without question the small detail of what is accepted. In this where we go back to the beginning of Man it is accepted in that case that it took an apple falling to change the destiny of Man and the thought pattern of humanity, and everything about the idea that it was an apple that was the illicit fruit proves to be a lie!

Now you may think that I am referring to Adam and Eve and the probability that they ate from an illicit apple without permission. No this apple had far more reaching significance than Adam's apple had because although Adam and his Apple was important it had only had significance as far as man is concerned in context of sins man had and as far as some religions believe. The apple I am talking about changed the outlook man has on the entire cosmos. This apple brought the cosmos into a new light and everything in the cosmos became clearer all thanks to an apple that fell at the right place at the right time. Every time anything falls to the earth this should be your reminder that your final destiny with the moon is

eminent. As objects drop or fall, then by the same margin and token is the moon coming closer but quicker because the moon has a lot more mass favouring the speed by which it must then fall. Newton saw the apple fall and realised the moon was falling to the earth as the earth is falling to the sun. Newton was the one that concluded his principles of gravity by seeing an apple falling and then he extended this rather ordinary occurrence to become a Universally spectacle compelling as a cosmic connected event.

Newton as a student saw an apple fall while he was on his back as he was loitering under the English sun. As he saw an apple fall from a branch and he made this the most fundamental event in science as far as human's recollecting goes. Seeing the apple fall Newton concluded that the apple has mass and the earth has mass. He saw the apple fall down from the tree to the earth and recognised this as a spectacular cosmic spectacle with all the astronomical implications preceding the event. As great a mind as Newton had, he did not see an apple fall but he saw the way the moon is descending towards the earth. He realised there was more than one apple falling to the earth; no there was a force that was between the moon and the earth. The apple fell by the force of mass and whenever whatever falls, the falling is going by mass.

There are those who are atheists, which means they don't believe in this Garden of Eden apple incident and what was connected to such an event but they also attach their unproven and highly doubtful religion thought of as physics to another event also connecting an apple where in their religion the fact of the apple carries much less dispute than what was derived from that apple's falling. The apple in the Garden of Eden I refer to as the first apple incident and the Newton apple I therefore call the second apple controversy. Those who dispute the first apple incident normally are the same people who support the second apple incident and there is one person as far as I know that disputes the second apple incident. It is not the fact that an apple fell that I dispute but it is the outcome of the broader religion that formed due to this overall accepted incident that I dispute.

Chapter 2 MOON CRASHING.

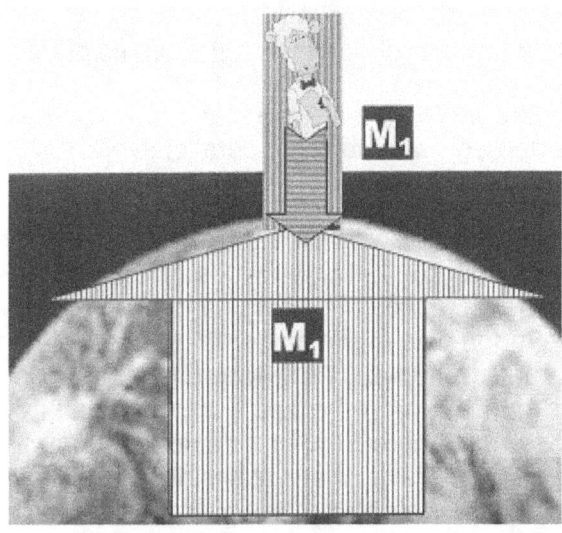

On the day in question a young student was idly wasting time in recreation while enjoying the shade of an apple tree close to a University in Britain and I believe moreover the University was in England. While loitering in the shade of this apple tree with no better things to do than feeling bored he witnessed an apple of general description break free from a branch on which this apple was hanging and fell to the ground. What grew from this event threw man in a change of destiny from which man never since could break free. Newton saw this apple coming to the Earth and apparently this filled him with inspiration that changed the world to larger degree than did all the teachings of Jesus Christ. I am not being sacrilegious but I am merely stating the honest truth. While everyone is disputing the religion of another person or the convictions of others there is only one person who looks on this second apple incident with a high degree of scepticism and that is me. There are on any specific second on Earth more persons preaching the physics of Newton that spawned from this vision Newton received than there are persons believing concepts about the teaching of the Gospel of Jesus Christ. The connection I see is that Jesus Christ is accepted as the new Adam where the old Adam failed his test by the incident involving the apple. Again I know the case of it being an apple or not being an apple is highly disputed but I shall return to this part since in this lies a large degree of the malice in physics.

After seeing an apple fall from an apple tree Newton was inspired. Newton was overcome by the magnificence of a Universe that opened the door to all physics at that moment. Newton saw an apple fall. What a divine revelation this turned out to be. This revelation equalled any revelation of Biblical proportions and then even topped any of those revelations coming from the Bible. This genius of a man hiding in the body of a young student in England going by the name of Isaac Newton was apparently sitting near or under an apple tree when he observed the now so famous apple falling from the tree. Seeing the apple land the student then made a brief calculation in relation to what he observed as a young man who saw an apple fall from a tree. He observed as a student (giving him a very special and rare quality to be able to be a student and to be able to observe with both abilities connected to the same individual) that from the observation he made the presumption that the apple fell by the weight that which the apple had when the apple was on the ground. Young Isaac thought the weight that which the apple had was responsible for having the apple falling. This sparked a cosmic connection never anticipated before. Isaac Newton was so inspired by an apple falling from an apple tree that from this a science was born.

Newton saw an apple hanging suspended from a tree. There was a distance between the apple and the earth and this distance he named the radius. Then he gave the earth mass, as he gave the apple mass. This he then subsequently formulated and from this formula everything connected to physics was born.

In his formula he envisaged $F = \dfrac{r^2}{M_1 M_2}$ he carried one part of the square $\dfrac{r^2}{}$ of the travelling distance he named as the radius $\dfrac{r}{}$, which he devoted to the earth $\dfrac{r}{M_1}$ leaving the other part of this radius $\dfrac{r}{}$, which is the falling distance $\dfrac{r^2}{}$, to what the apple $\dfrac{r}{M_2}$ when the apple had to travel to fall to the earth. He saw the mass of the apple was pulling the mass of the Earth along the radius as much as the mass of the Earth was pulling the mass of the apple along the same radius, and instantaneously the formula

$F = \dfrac{r^2}{M_1 M_2}$ became instant religiosity. This brainwave was the wave that shook the earth from its cradle of slumber in mindlessness. Isaac Newton was a Superstar before there was a superstar! He found the force containing the Universe into a unit. It was an all-time breakthrough with no other parallel ever.

However one can never keep a genius satisfied with one conclusion at a time. Because this was Isaac Newton and Isaac Newton was very clever it went further, much further when Isaac Newton saw the moon as the apple and the earth as the earth. He no longer saw an apple but saw cosmic structures forming their final destiny that drew to the last Universal conclusion. He saw the apple as the moon and he saw the earth, well I suppose as the earth and although the moon was much bigger than the apple it was also much further from the earth than the distance of the falling apple. …But according to this he then put his gravitational principle in place where he declared that between the apple and the earth a force called gravity is in place. He gave the apple a value in mass and he gave the earth a value in mass. Then he gave the mass the responsibility to create gravity and he gave gravity the responsibility to pull by the force the mass initiates. …And there are some educated person's who holds the opinion that I don't "understand Newton" or his physics principles…

The mass of the apple was pulling the earth but the mass the earth has made it impractical to expect that the apple would be able to pull the mass of the earth towards the apple.

The earth also pulled on the apple by the force the mass of the earth would enlist. The earth is pulling the apple with all of the mass the earth could apply while the apple was pulling the earth with all the mass the apple had about. By measure of mass discrepancy all the falling went the way of the apple. This then brought about a pulling contest and the end result of the pulling is the falling of the apple and this enormous event of the earth and the apple pulling one another in the tug of war happened in front of his eyes. He became the eyewitness of an absolute enormous cosmic event and Newton furthermore was so brilliant in mind as to give this the deserving Universal implications, but that is the mind that Newton had!

In the vision of Isaac Newton he interpreted what he saw as the Earth dragging the apple down by the weight, which the apple pulled the Earth closer. In this action he envisaged a force at work. He apparently later named the force gravity and I presume that being a very religious man unlike modern Newtonians he saw this force as being something that was there to drag one into the grave hence the grave connection in gravity but about this part I can be mistaken but that is a presumption I now make and is not necessarily the truth as his presumption about a force coming about by the influence of mass was not necessarily the truth at the time.

Newton saw the Earth rise to the occasion by meeting the apple halfway thus the apple took on the first part of the square of the radius leaving the other part of the square of the radius by which the apple travelled to the Earth travelling to the apple. However, the mass of the earth is rather substantial put in regard to the apple and this reflected all movement down to the apple travelling the entire square distance.

The formula depicted a force that dissolved space by the power of mass coming from two ends that was drawing towards a common centre. $F = \dfrac{r^2}{M_1 M_2}$ The mass of the two objects destroys the radius between the objects and to top the lot they finally after so many centuries discovered a force. Everyone went ballistic, proclaiming him as an instant genius, the one the world was waiting for after the crucifixion event. If there were TV in those days he was the Newsperson of the day. He made a presumption about mass as I made a presumption about his motive in choosing the name gravity. His presumption was as little tested as my presumption is tested. He got more famous than Jesus Christ and Mohammed put together. There are more people believing Newton than there are people believing in the preaching of either Jesus Christ or Mohammed. Of all persons on earth there is one person on earth doubts Newton while there are billions of those who disbelieve in either Jesus Christ or Mohammed and this is one of the biggest facts and this indicates the biggest farce ever created by any man ever. This opened up the road that other going around by advocating bullshit such as Darwin could follow. Those pretending to fall from your chair about my remarks concerning Darwin, tell me what did Darwin ever prove? Name one fact that he introduced with substantiated and un-denounced facts. What did Darwin conclusively prove in his book that did not rely on exaggerated inconclusive ideas about a collection of deductions that in hundred and fifty years could never conclusively be linked by timeline and proven to be inconclusively correct. Darwin must be the third or fourth biggest science scam to serve a destructive purpose ever launched by man under the pretext that it forms conclusive science in all its aspects…back to Newton.

Newton saw the apple drawing to the earth and if the apple drew to the earth the moon also then was drawing to the earth just like the earth was drawing to the moon. If you believe physics you have to believe this and if you believe in physics this moon coming to the earth is your destiny waiting on you. If you see physics as Newton does you better look at the moon with much less romantic visions connected.

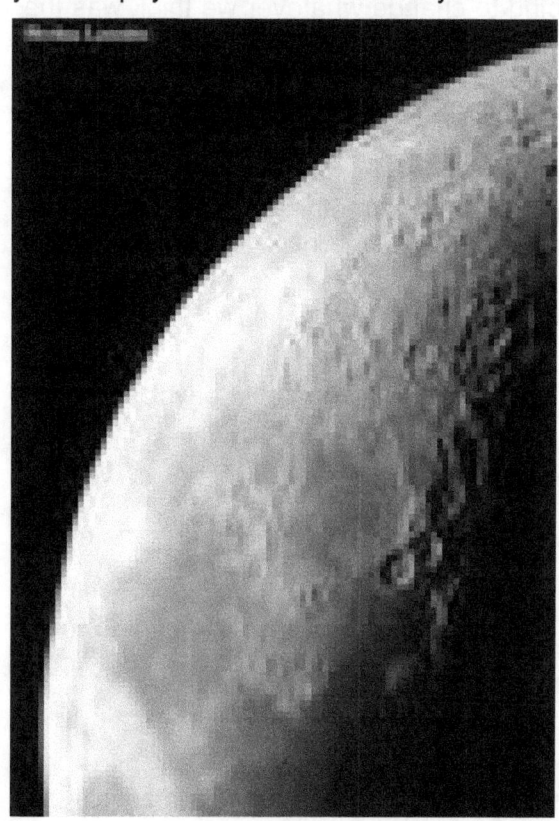

If we go one step further and realise what Newton realised at the time we find that we don't have an apple hanging from a branch but we have a moon filling the sky, or filling a part of the sky. The principle does not change. In the one event there is an apple and in the next we have a moon. The apple is much smaller than the bigger moon but then the apple is much closer to the earth than the moon that is much further away from the earth. The earth is the earth that remains the earth in both cases.

Therefore the scenario remains identical in both cases notwithstanding many discrepancies. As the apple falls, so the moon falls but the moon is a lot further and would therefore take a longer time to fall to the earth. However, by investigating both scenarios, no person who knows the least about science would dispute that if Newton's pulling of gravity by the force of mass were true (and nobody in science seems to disputes this) the moon is falling although the moon is taking a little longer to fall all the way. But as sure as the apple was falling in that we can have the same assuredness that the moon is also falling. When you look outside and you see anything falling you can be as assured that the moon is falling on you and the moon is coming as sure as rain or anything else is coming down to the earth..

According to Newtonian science everything is falling because everything has mass and where everything has mass and in that everything is falling by the value of that mass. Go on and allow something to fall and what you will notice is that when you see whatever falling, it actually is the moon falling except, the moon has a lot more mass and therefore the moon is falling a lot more than a measly apple would fall. The moon is coming whenever you allow an apple to fall or not allow an apple to fall. The moon will contribute all the mass it can contribute and the earth will contribute all the mass it can contribute and both objects are coming towards each other but more the moon to the earth than the earth to the moon…and like it or not the moon is coming all the same! This is Newton's science legacy to us. but in this idea is hidden a lot more

The Tunguska event would in comparison seem to be a minor local inconvenience compared to when the moon collides with the earth and both objects becoming dust particles. In this light would it not be wiser for the rich and those in political power just to keep quiet about such looming catastrophic scenario and in keeping silence not arouse a hysteric human race going mad in fear and anxiety? Think how uncontrollably mad everyone would get knowing death is upon us? So hush the peoples concerns…

With the formula as Newton introduced it being $F = G \frac{M_1 M_2}{r^2}$ any one with a little insight into reality could see that as the bottom part decreases (become smaller) this will affect the top part to become larger. I will show three values and you will see what I mean. $\frac{100}{100} = 1$ and $\frac{100}{1} = 100$ and then getting the bottom really small $\frac{100}{.001} = 100000$. The bottom part's decrease increases the speed of the top part or the gravitational movement. By reducing the distance the force of gravity will grow exponentially faster!

I show you this to indicate that the closer the moon gets to the earth the faster it will come to the earth. The last minute the moon will travel about say one hundred million kilometres because the gravity pull will increase by a billion trillion times per second. During the last day's coming closer to the earth by radius reduction will be as quick as the last million years we now have. The rushing moon will increase in speed by the square of the disappearing radius and we sit between the two as a human sandwich.

Go to any of the doomsday scenario websites and see what they predict when a comet hits the earth. It is going to be final. They really scare the shit out of those who believe easily in frightening tales aimed to scare the crap out of everyone. A comet hitting the earth is going to change whatever we think was the earth…and this is only when a comet hits the earth according to Newtonian science. Think what the mass would be when a comet hits the earth and compare that mass to what the mass of the earth will be when the moon hits the earth. Go and visit the web and se on these Armageddon websites how they paint a picture of doom and gloom about all sorts of destructive events bringing a conclusion to the earth. These Armageddon websites tell stories of comets we have to wait for and that might come. See if you can find one website forecasting the moon arriving on your doorstep and this is not something that might or might not be. This event is in the future as sure as Newton's principles are true. We have Newton's formula $F = \dfrac{r^2}{M_1 M_2}$ that then by deduction became $F = G \dfrac{M_1 M_2}{r^2}$.

The formula $F = G \dfrac{M_1 M_2}{r^2}$ is science as taught by any person that knows anything about physics.

The symbol F represents the gravity force by which the moon is coming towards the earth or visa versa.

The symbol G represents the composition of what space holds that is between the moon and the earth, which forms the stuff keeping the moon and the earth apart.

The symbol $\dfrac{M_1}{r}$ represents the force by which the mass forms the gravity that the earth creates in relation to the distance $\dfrac{1}{r}$ there is between the moon and the earth.

The symbol $\dfrac{M_2}{r}$ represents the force by which the mass forms the gravity that the moon creates in relation to the distance $\dfrac{1}{r}$ there is between the moon and the earth.

The mass is in place. We have the formula given by no person less than Newton. We know the distance…and can you find any website predicating this forthcoming collision? No, you can't but the conspiracy is in the reasons why you can't find one such a website. Why do we have so many scary warnings but nothing mentioned about what is truly part of our future, the moon coming for a visit!

This $F = G \dfrac{M_1 M_2}{r^2}$ is the tempo whereby the distance of the radius between the earth and the moon will shrink. Then these numbers still have to be divided by the squared of the reducing radius making the result incomprehensibly more devastating. The smaller the radius becomes the faster the radius would become smaller and the more the tempo would be of the moon closing down on the earth. During the one-year everything might seem perfectly under control and the next year everything will seem to go mad in haste. The moon will seem to be on a runaway collision coarse going out of bounds the closer it gets. Remember the formula showing the closer the moon gets the quicker the rush of colliding will become as in $\dfrac{100}{100} = 1$ and $\dfrac{100}{1} = 100$ and then getting the bottom really small $\dfrac{100}{.001} = 100000$. The devastation it brings will increase exponentially and become so big no one would be able to realise the changes as the advent of the final doomsday is increasing! No one would be able to keep track of the fatalities occurring. Everybody that died during the first and the Second World War plus the epidemic influenza that came after the First World War would be deaths happening in one day and during every day.

No humans were around to make permanent records of prehistoric hurricanes. These so called hypercanes that was prevailing in the prehistoric days will come back and ruin the planet again, but this time it will be an everyday occurrence. Think of how conditions may have existed about 65 million years ago that could have spawned prehistoric hypercanes that is far more powerful than modern storms. The doomsday theorists do not have the imagination to sketch reality when the moon is closing in on the earth! Whatever they think is bad this is far worse. When the moon is only a few thousand kilometres away how much will that influence the weather patterns applying?

In the era we now enjoy a hypercane might be thought of as a hypothetical extreme tropical cyclone, but in reality when the moon comes closer this will become a multiple seasonal event. In order to form hypercanes, ocean waters would need to be much warmer than they are currently. This would become a reality when the gravity of the moon brings on waves thousand meters high. Hypercanes could form if ocean temperatures reached approximately 122 degrees Fahrenheit; this is 15 degrees higher than the warmest ocean temperature ever recorded! When the Moon is half as far away as it is now the influence it must have on water temperatures in the oceans will increase this to the colder end of the ocean temperature scale. Science at present acknowledge that such a hypercane can result from large any asteroid or comet slamming into our planet, a volcanic or super-volcanic eruption, or extensive global warming could all contribute to ocean waters reaching such outlandish temperatures. These conditions dwarf in reality when we consider the moon coming closer towards the earth. Some scientists speculate that the dinosaurs were killed off by a series of hypercanes caused by an asteroid hitting Earth. Hypercanes would have incredibly long life spans due to central pressure below 700 mille bars and wind speeds in excess of 500 miles per hour.

This is the condition within a Hypercane. Hypercanes would have wind speeds of over 800 kilometres per hour (500 mph) and would also have a central pressure of less than 70 kilopascals (21 in Hg), giving them an enormous lifespan. For comparison, the largest and most intense storm on record was 1979's Typhoon Tip, with a wind speed of over 300 kilometres per hour (190 mph) and central pressure of 87 kilopascals (26 in Hg). The extreme conditions needed to create a hypercane could conceivably produce a system up to the size of North America (compare image of Typhoon Tip's size at right), creating storm surges of 18 metres (59 ft) and an eye nearly 300 kilometres (190 mi) across. The waters could remain hot enough for weeks, allowing more hypercanes to be formed. A hypercane's clouds would reach 30 kilometres (19 mi) into the stratosphere. Such an intense storm would also damage the earth's ozone. Water molecules in the stratosphere would react with ozone to accelerate decay into O_2 and reduce absorption of ultraviolet light. Other scientists have theorized that the system, compared to a normal hurricane, would be considerably smaller, about 10 miles in diameter. This would be more comparable to a tornado, which has been recorded at up to about 2.5 miles. And when the moon is on us this will be everyday weather changes where the real frightful ones will be hundreds of times worse.

Has such a storm ever formed? No. Will such a storm ever form… probably not? Here is why:
Storms of this magnitude rarely merge, and if they were to join up, the result would actually be a weaker system. When two tropical cyclones merge, air circulations become disrupted, and as a result, wind velocities are lessened. Over time, the storm could re-intensify and it would certainly be larger in area than the two storms it was formed from, but it would get nowhere near hypercane status. Some meteorologist for the National Weather Service in St. Louis, Missouri, describes the situation as follows: "To use an analogy, these storms are similar to magnets of the same polarity...they tend to repel each other rather than attract. When they approach too close, usually one of them will weaken dramatically, due to the other one taking most of the energy necessary to sustain the hurricane force winds." Therefore, hypercanes are HIGHLY unlikely because two storms rarely collide over open water. But with the moon filling most of the night sky things are going to change dramatically.

Scientists have long thought that the dinosaurs may have died after an asteroid struck the Earth and caused dramatic climate changes. There are other researchers that think any asteroid could have heated the ancient oceans to as much as 50 degrees Celsius (about 120 degrees Fahrenheit). This is the research that is going into speculative scenarios where the moon hitting the earth is as definite science as Newton is and no one in science even thinks about this scenario while Newton's gravitational laws are in place? Can you think why not?

I have to repeat this information once more because it indicates how well studied the future of the sun is while how incompetent the coming of the moon apparently is. Newtonians or then those who should know better says when the sun runs out of its current fuel, the hydrogen starts to cool down, as it will expand into a red giant and could expand out to where Jupiter is circling or so they say it is going to be. These are the estimates I have seen quoted before and are not mine to quote. The end of the sun is billions of years away. It is a lot of hogwash but it shows someone somewhere got interested and started some form of thinking about how the sun and the solar system would end. Yet no one in science ever thought it worthwhile to conduct a study into when the earth and the moon will go into destruction.

Even if this is going to happen in a billion or five billion or a trillion years from now, still there are persons who find the end of the sun important enough to be concerned about. It shows that it is not because it is far into the future that there is no apparent interest into the collision between the moon and the earth. The moon would have hit the earth much sooner than the sun would explode, so why is there no investigative research done on the earth and moon contracting by the mass that Newton foresaw? I don't for a minute say I agree with anything said about the sun, but at least it shows some investigation was done not looking at the correctness thereof, there was a thought spared about how the sun will end. It is said using Newtonian vision about what the Universe holds install that the sun is going to blow up and burst and come to an end. Don't repeat me on this hogwash because all I wish to prove that in terms of even the sun coming to a closure of a life cycle there was research done in this line of thought. It is said that the sun will scourge the solar system out of existence but never does any one mention that the moon and the earth is having a get-together long before that formed on the grounds of Newton's vision on gravity coming from mass that pulls mass. If science can look this far ahead and find a doomsday waiting on us billions upon billions of years from now, why is everybody avoiding the apparent doomsday that is much closer to realising the ending of all of the earth and its moon including all (and the only) life that was ever present in the Universe.

With everyone in science saluting Newton's gravitational contracting there was an extended effort by Albert Einstein to find the critical density of the Universe. That is the backbreaking effort that science took with painstaking accuracy to find the level of interest in determining the end of the entire Universe. The critical density idea did not pan out and that left science high and dry for answers about science. They did not stop there, no Sir, they paddle on in darkness to find the answer. They left no stone untouched to come up with a conclusion…and yet not one person in all that time started to think that the contraction will be much better monitored by researching the moon reducing of the radius it has between it and the earth. Would it not have been much easier to study when the moon will splash into the earth and from there work out when Newton's attraction will have the worst collision we can think of happening in our backyard?

Then science goes further and tries to detect the untraceable that is invisible. They try to find dark matter in a shady Universe. They spend billions on dark material or dark energy research but not a single dime goes the way of finding out when the moon and the earth will have a gravitational self-destruction event. Do you not find that very odd…or is it just me wondering about that? Someone in science thought it was worthwhile to study what would happen in the very solar end when the sun comes to a final exploding conclusion. Someone in science took time to bother with the end of the solar system and what will be the applying conditions during the finale era of our sun having a solar system. They even measured how big and how red the sun would be in the end but never do they bother about what is more obvious, when will the earth and moon get together. That seems very odd and suspicious to the less informed…

Science goes even much further. They sequestered Albert Einstein to measure all the mass in the entire Universe to find out when will the Universe start contracting and come to the end of its life cycle in accordance to Newton's gravitational pulling principles. There is no limit or end to which they will not go to find the end or bring finality to whatever they try to establish. To measure the end of the Universe is going much further than to establish when the earth and the moon will meet their gravitational destiny, with Newton and his theory of mass pulling mass so fundamentally proven. But there is no end to their resolution for they never stop with their inquest. When the Critical Density Investigation did not deliver the results that would bring satisfaction, they went in search of Dark Matter. I am going to go much deeper into the Critical Density and the search for Dark Matter later on in the book.

Think of what effort it took and still takes to find the reason why the earth is expanding and what human effort went into that. They are relentless in their quest to find answers. Yet, when it comes to the time when the moon and the earth will finally collide we find only silence. It is the sort of silence I describe when I pointed out what to look for when one goes in search of a conspiracy. I find not even a whisper…

On earth nothing are kept to rest in peace. They run probes from the Artic ice fields all the way down to Antarctica. The search for information never seizes and the lust they have for knowledge never ends, except when we get to the issue of the moon colliding with the earth. In that matter the thought is never even realised and the question serving the silence is: "Why not and why never a word about this matter?" Is nobody at least interested what will occur the final days when the moon meets the earth in a full moon?

If our Universe comes to an end it will start with the moon and the earth coming together and when these two come together it does not matter how the rest will come together because then all our interest in the Universes would be permanently suspended in any case because the earth represents our Universe entirely as the entirety there is. Losing the earth becomes the same as ending the Universe.

What people don't realise is our Universe is this earth. Without having the earth we have no Universe because then as far as we know there is no other life left in the Universe. It will be much better to stop speculating about the possibility of other life in the Universe and try to save our only Universe, which is the earth we so much try to destroy in the name of making money and showing profits.

Researchers have probes to detect tsunamis that will only devastate regions horrible, as it is it can't compare to the moon hitting the earth. There will be waves every day generated by the gravitational increase of the nearing moon that you now would associate with the worst tsunami occurrences, which would then become the mild everyday waves found on the beach. The sea will become a wild devastating machine that pulls all land into the sea with waves going as high as a thousand meters and much more. This is the scenario waiting on us holding life and no research is going into trying to predict when this is going to happen. They research tsunamis in detail but no one place on earth could be considered as safe when the moon lands on earth. Such waves will draw mountain ranges into the sea and destroy giant areas into water-wastelands. Think of the crushing force of one such a wave and then the moon is only still coming our way. Long before humans would feel the demise there will be no human left to experience the demise. That means that the sooner we come to terms with what waits on us the more time we have…

Building caves in mountains would not help because the massive waves will get us there in the final event. It is just a thought but why is there such a surge to investigate possible occupation of Mars? There is no spot on earth that will remain safe so where will those who hold the money and the power go when they need a safe haven. What would be the most obvious reason be to keep order and keep the money flowing and for what purpose would they wish to keep the money flowing?

What information did the Rich and the Powerful gain from these probes on the moon that did not filter down to us? When will the moon contract the earth finally...it is the most obvious point that needs researching because losing the earth becomes the same as ending the Universe and as far as anyone can see the research into this collision is zero, now why would that be if the obvious becomes very obvious and then someone has something to hide?

At first all the additional moon voyaging launches were halted. The reasons for the sudden lack in interest was numerous but the most believable was that the costs involved were truly astronomical. America could not go to the moon and fight a war in the east and a war in the east was raking in much more money than did the trips going to the moon. So the moon expeditions were halted in favour of a Vietnam War. If one takes the millions of tons of bombing material and the production costs going into the manufacturing of this alone the profits will out weigh the money to be made of a trip to the moon many times over. So the war won on the profit margin it could generate alone. Then if one takes into account the machinery that was wasted and the purchasing power of the replacement needs for such machinery a trip to the moon stood no financial chance when compared to the profits gained by the wealthy in the war effort in the east.

After the seventies no one was interested in the moon because "*We*" were "*there*". The later expeditions then involved probes to other planets and research went about finding out what was going on, on the other planets. They found that Venus was too hot to handle and so was Mercury and the two was left alone. Then came normal interest into the conditions on Mars. Photos came back from the other nearby planets that was studied but never with the sudden frenzy state in which the development got a surge as that came later. Then the moon was suddenly left alone. No one said a word about the moon. In this period Computers came into play and producing chips in a "gravity zero" state was generating the money. Money brings about interest. Everyone making money had to make computer chips and this then became the main focus of many. If there is money the banks will have interests and if banks show interest they command the politicians to show interest and if the politicians have interest the job is done without questioning validity. The money that this made superseded everything else and with the money it pushed everything aside. Then from this point on it was about laboratories being manned and worked within conditions only suitable in outer space and the research of "gravity zero". This was a cover to get the public to pay for research by their tax money going for the research that only the rich would financially benefit from. The taxpayer, who is the poor, paid for the scheme whereby the rich got the benefits to get richer. The selling of the products developed in outer space was financed by the public and therefore the gain of wealth by this research belongs to the public and the earnings should go to public funding. The rich should be taxed for the royalties that belong to the public. Yet we have stinking rich benefiting from this research by getting more rich than they were and moreover without paying equal tax as the poor does and behind this the bankers swell their coffers where the money the banks have is almost limitless.

This pushed the interest in the moon completely into fading away. There is a lull in any interest from all sides about the moon until a certain point. Suddenly again out of the blue there came a rush in the George Bush era where the moon became a place to go to in order to launch men to Mars. All of a sudden Mars had to be explored and the building of a halfway station on the moon was on the cards. The question never asked is why is there the shift in interest all of the sudden? Who gains from this interest?

Moreover would be, why would the Rich and the Powerful have the sudden interest in Mars? Think about it this way, if the Rich and the Powerful could take all of the tax money of all the lesser human beings in status and start to build on Mars under the disguise of research, then when the doomsday arrives those being the fortunate, the powerful, the influential and the law makers will have somewhere to go to escape, leaving the less worthy to take the moon head on! In the sixties there was an urge to get to the moon and when they got there then afterwards the interest in the moon fell to an all time low. Then those in investigating investigated all the planets without showing the least interest in what the moon might hold. This trend went on for decades. Then suddenly from the blue comes some hysteric research about Mars, just after someone discovered possible prehistoric semblance of life on Mars. Why would everybody on earth now be dreaming about life on Mars...and why would all of man suddenly benefit from exploring possibilities of life extending to Mars? This sudden research of life on Mars is not about finding life on Mars but it is about making sure there is no life to be found on Mars because finding life on Mars will end all further explorations. If there were a hint of life on Mars it would stop all further research about living conditions on Mars. The risk of finding bacterial or a virus on Mars would mean stopping all further research on Mars because of contamination and infections caused by the Mars micro biology and the chance it will wipe out all forms of life on Earth. However there is a much bigger chance that life never developed on Mars and that means the Rich and the Powerful can go there. But first they have to scarify many less important beings coming from earth.

In 2004, the then President of the United States of America George W. Bush in a speech gave NASA a an order to develop towards the direction to continue to progress where this will take them back to the Moon and from there to Mars. Why would there be this sudden change in strategy. Why would the interest suddenly flame up in going to the moon? If that's the plan what information blew the interest to new highs? The reason they give is as diverting as any reason coming from science as science says it is because it's a lot easier to go to the moon than to accomplish a Mars landing and build a station base

there. But thinking of the answer in such terms is diverting from the true question. Why suddenly go and build space stations to begin with when so much time lapsed since the early seventies when the interest in the moon faded to zero. Thinking that Mars is far and the moon is much closer is a way of not answering the question at all. Yes, the demands of a trip to Mars will be costly, much more costly than the space station that was already too costly to complete by one country alone but the least expensive would be to go on to ignore the entire project of building on the moon or on Mars. Building by itself as a project will be much more demanding than those in far earlier stages of our species' exploratory history. Building a station instead of collecting rocks like they did in the seventies is calling for a much more evolved effort. So why the big effort this time around. Numerous challenges will face us along the way, but three of the most serious are logistics (in terms of both fuel and crew consumables), mass, and crew interaction.

Why would this rush be, as we want to go to the Moon? Why go to the moon in the first place? Because the Moon is an ideal "staging post" for us to accumulate materials and manpower outside of the Earth's deep gravitational well. From the Moon we can send missions into deep space and ferry colonists to Mars. That is the main reason of the sudden urgency to go to Mars. But to give Polly its cracker there has to be something in for John and Jane Dow too so they throw in the idea that tourists may also be interested in a short visit. Mining companies will no doubt want to set up camp there. The pursuit of science is also a major draw and in that they have to get all people back home to believe in the idea. For whatever reason the people would hang on to or believe in, say to maintain a presence on a small dusty satellite would not get the work done to establish a new occupying colony on the Mars so we will need to build a Moon base. Be it for the short-term or long-term, man will need to colonize the Moon in order to colonise mars. But why would we live there? How could we survive on this hostile landscape such as Mars? This is where structural engineers will step in, to design, and build, the most extreme habitats ever conceived…and who is supplying the funding, the suckers they will leave behind on earth to die in the earth moon collision that is coming as sure as Newton is correct. They have to start a living presence on Mars to avoid the destruction on earth!

Manned missions to Mars has to be well funded as it will take up a lot of the limelight insofar as colonization efforts are concerned and why would no one ever tell the true reason for this venture? We must use our initiative to focus our minds where we aim at the ongoing and established concepts for colonization of the Moon. Why the rush when we currently have no exact means of getting there on Mars in the first place and why after all the time that lapsed since it is nearly 40 years ago when the interest went dim since Apollo 11. The technology is sufficiently advanced to sustain life in space as we saw from living in capsules above the earth so the next step is to begin building a life for future inhabitants on Mars. But first is the present instalment that concerns building a Moon Base. What are the pressing issues behind the immediate urge to inspire engineers when planning habitats on a Mars landscape? Only the Rich and the Powerful have the authority to authorise such a venture.

The debate that still rages about whether humans should settle on the Moon or Mars first seems like an eye blinder. With Mars being the ultimate destiny for mankind as to live on a planet other than Earth why would they bother with the moon when the moon is going to be destroyed in the same event as the earth is. But they need an eye blind as not to wake suspicion about the true nature of the venture. But it is also much easier to get people to buy into the scheme and tell the story that it would be possible for all persons to one day look down on earth and see earth during cloudless nights. People will fall for this idea making the earth a bright and attainable view from the moon. From there people will see the details of the earth as we now see the lunar landscape with the naked eye, because the earth would be just as astronomically close when compared with the planets and therefore being that close, people will buy into the idea. Give as many as possible the wish that the Moon should be the first port of call before humans begin the six months (at best) voyage to the Red Planet. It also helps as we've already been there... We will never live on the moon but leaving the idea can only be beneficial to the launch of the project.

From the internet one can see information is given about the general opinion that has shifted somewhat in recent years from the going to Mars directly plan that was on the table in the mid-1990s to the now favoured going to the Moon first idea, and this shift has recently been highlighted by US President George W. Bush when in 2004 he set out plans for re-establishing a presence on the Moon before we can begin planning for Mars. That it is US President George W. Bush that made the promise has little surprise because he started an oil war on behalf of the oil goons and then bankers that ill benefited from such a senseless war in terms of financial gain. It makes sense; many human physiological issues remain to be identified, plus the technology for colonization can only be tested to its full extent when... well... colonizing. This is the eye blinding part. The question is why the sudden rushing where the moon has to become the halfway house?

Firstly the human canon fodder that is lost in war can be sent to the moon under the disguise of keeping peace on earth. Then it will be a normal trend if some are lost in the task given to them to learn the very vital understanding as to how the human body will adapt to life in the low-gravity environment on Mars but this could also be learned much closer on the moon. The colonisers made up of the disposable part of society in the ranks of humans will have to endure a very new way of locomotion on Mars. It is not that common as to how earth technologies will perform on any location forming the environment on Mars or the moon. Therefore learning this on a surface much closer to earth will be beneficial because working on the handicaps brought about by the project will become easy enough to alter and adjust as it is much closer to relate to persons back home on earth. The first colonisers of Mars will be too far away to deal with the first problems that arise during the start of the venture. Then as progress smoothens out further development it will become assuring, as a new lifestyle will become feasible by the colonists and astronauts. Launching first from the moon and later from Mars will also see sensible to those brainless taxpayers that pay for everything but eventually will perish in the moon earth collision. Exploring the space and living conditions on Mars is dangerous enough, so therefore minimizing the risk of mission by first developing some new moon base makes failure less of a certainty than going straight to mars. Having as little failure or covering up most failure will be critical to the future of manned exploration of the Solar System.

So if the moon is the favourite place to start building out of the earth projects where do you start when designing a moon base? The structural engineers that would put a premium on the quality of building material may be worried that when building material is exposed to a vacuum is the enduring quality still adequate because that must be at the biggest premium. Damages may result from severe temperature variations, high velocity micrometeorite impacts because there is almost no atmosphere to give protection, high outward forces from pressurized habitats because of the artificial atmosphere within the structures to serve the needs of life within, material brittleness at very low temperatures as zero is a daily occurrence and cumulative abrasion by high energy cosmic rays and solar wind particles will all factor highly in the planning phase. Once all the hazards are outlined, work can begin on the structures themselves but it has to be synchronised to correspond wit Mars.

The actual construction of a base will be very difficult in itself. Obviously, the low-G environment poses some difficulty to construction workers to get around, but the lack of an atmosphere would prove very damaging. Without the buffering of air around drilling tools, dynamic friction will be amplified during drilling tasks, generating huge amounts of heat. Drill bits and rock will fuse, hindering progress. Should demolition tasks need to be carried out, explosions in a vacuum would create countless high velocity missiles tearing through anything in their path, with no atmosphere to slow them down. (You wouldn't want to be eating dinner in an inflatable habitat during mining activities should a rock fragment be flying your way…) Also, the ejected dust would obscure everything and settle, statically, on machinery and 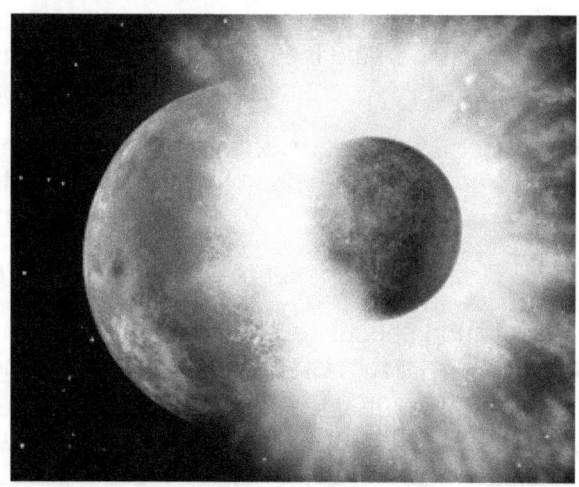 contaminate everything. Decontamination via air locks will not be efficient enough to remove all the dust from spacesuits; Moon dust would be ingested and breathed in – a health risk we will not fully comprehend until we are there.

With all this haste to go to Mars and set up some shop there that could accommodate those that could afford to begin a new existence on Mars how long is it until the moon finally destroys the earth by destroying its own structure. No research tell how long does Newton's formula $F = G \frac{M_1 M_2}{r^2}$ give the rest of us multiply our genes before it is wiped out by the oncoming collision that inevitably has to end all life form on earth according to Newtonian Principles applying in physics as physics. If they study and find this event is still far off into the future, then do the study and tell us. If they do the study and see it is going to happen in three or four generations from now, then include us into the plans of becoming Martians or say who will be left behind and on what merit is the choice made, but for God sake, do studies and tell us, that is if we could trust Newton's physics of mass that pulls mass.

The Definition of Gravity according to the Oxford Dictionary of Astronomy:
<u>Firstly it is said in physics that in Physics: The natural force of attraction exerted by a celestial body, such as Earth, upon objects at or near its surface, tending to draw them toward the centre of the body. With this being Newton's accepted science principal why then is what I said in this first part about the moon colliding with the earth total rubbish.</u>

<u>The conspiracy is in the silence of… It is not in hiding or not telling us how much the moon is coming closer but it is in the "not coming closer" that the conspiracy hides the truth.</u>

If the **Fundamental of Science** according to Newton is **correct** then there is only one way the earth could find its final destiny, it is when **the moon** will hit the **earth** and with doing that **destroy all** forms of **life** in the collision. According to Newton $F = G \frac{M_1 M_2}{r^2}$ the moon and the Earth are in a tug of war and the two are pulling each other closer. That is according to Newton's formula.

The Fundamental of Science says that the mass of bodies attract. Then there is one question more important than all other science research put together.

When will the moon and the earth destruct in a collision where the mass destroyed the distance that held the two solar bodies apart. The two have been pulling for billions of years and should be coming to the period of intimate connecting by collision.

It is said that mass pulls mass.

Newton came up with this novel idea about a little more than three hundred years ago. What this idea says is that every object has mass and each object pulls all other objects by the value of the mass that every individual object has.

Teachings say that the moon is pulling the earth as much as the earth is pulling the moon. All teachers in physics in learning institutions teach students this and by culture every student believes this without condition or question. If I say this is rubbish no one believes me.

This concept is so engraved into our culture we have to except this without question to a point that when I say this is not happening, it then is I that has to deliver the proof of this concept being total void of truth and is complete incorrectness.

Newton was never tested before and I prove that, yet I have to prove Newton is incorrect!

Everything I speculated on about the moon and the earth coming into some collision sometime in the future is a lot of rubbish just because the definition of Newton that says <u>in physics that in Physics gravity is the natural force of attraction exerted by a celestial body, such as Earth, upon objects at or near its surface, tending to draw them toward the centre of the body. With this being Newton's accepted science principal why then is what I said in this first part about the moon colliding with the earth total rubbish.</u>

<u>The Mon is going away from the earth and the earth is getting bigger while the earth is slowing down in rotation velocity as the earth is moving further away from the sun.</u>

The Practice of physics propagating Newton's principles is Brainwashing and Mind Control of students in Physics

Let me show you what is The Practice of Brainwashing and Mind Control in Physics

If the Fundamental of Science according to Newton is correct then there is only one way the earth could find its final destiny, it is when the moon will hit the earth and with doing that destroy all forms of life in the collision. According to Newton $F = G \dfrac{M_1 M_2}{r^2}$ the moon and the Earth are in a tug of war and the two are pulling each other closer. That is according to Newton's formula.

The Fundamental of Science says that the mass of bodies attract. Then there is one question more important than all other science research put together.

When will the moon and the earth destruct in a collision where the mass destroyed the distance that held the two solar bodies apart. The two have been pulling for billions of years and should be coming to the period of intimate connecting by collision.

It is said that mass pulls mass. Newton came up with this novel idea about a little more than three hundred years ago. What this idea says is that every object has mass and each object pulls all other objects by the value of the mass that every individual object has. Teachings say that the moon is pulling the earth as much as the earth is pulling the moon. All teachers in physics in learning institutions teach students this and by culture every student believes this without condition or question. If I say this is rubbish no one believes me.

This concept is so engraved into our culture we have to except this without question to a point that when I say this is not happening, it then is I that has to deliver the proof of this concept being total void of truth and is complete incorrectness.

Newton was never tested before and I prove that, yet I have to prove Newton is incorrect! I say gravity is the part that does not connect to mass and is the factor producing movement that brings on the intention of motion of all objects to carry on moving downwards, notwithstanding the blocking action or mass, which comes from intervention of the earth occupying space that stops further descending. Mass is not connected to gravity in the way as initiating gravity since mass comes in place when it stops the motion of moving downwards of the falling body.

When the body stops moving down further it where the concept starts of having mass. Mass is achieved by stopping and preventing further downwards motion by having a solid object fill the space with material that will intervene further movement of descending to a centre of the body having gravity and therefore gravity is in the performing of the descending motion. Then my Super-educated-Intellectuals feel threatened when I confront them by bursting this super intellectual demeanour they hide behind with a thought through idea of what gravity is in relation to how mass forms.

On the moon I have minimum mass (1/3 of what I have on earth), which must result in minimum gravity (again 1/3 of what I have on earth) if mass is what produces gravity. That means in space I must have no mass that produces micro gravity. They say mass is doing the pulling but then when there is no pulling it must mean there is no mass doing the "no-pulling". That is one part of the ongoing science conspiracy, changing the rules ongoing and never get to a point of understanding. If I have micro gravity I then must have no mass! Gravity is in turning the earth and this thrust downwards end in the body having mass when the thrust downwards can't move the body down any more but the intention of movement remains.

It is not you being glued or not being glued to the Earth that I discard when I question the mass idea. Mass forming weight is indisputable and that the conspirers called physicists know. It is the definition holding this whole idea about mass forming a force of a pulling magnitude that I do not share in the least. What the definition of a pulling force describes is magnets pulling onto iron and it is the total opposite of what I experience when I have mass. If it was produced by the formula Newtonian use it will be the breaking of the first millimetre of the gravity clampdown that would be the hardest to achieve whereas this first millimetre of lift is the easiest part and not the most difficult. The difficulty increases as the radius grows and not as the radius decreases.

When I say there is no mass that produces gravity everyone think I say we all are going to fall off the Earth and with me thinking that way then it is obvious that I must be a nut. Everyone thinks of me as the clown acting mad when I say mass is not to be found in the cosmos. I do not say we are not standing on the Earth where we obviously have mass, I say Jupiter and the other planets can't have mass because I can't find mass playing any role. I do not say there is nothing that is keeping me glued to the earth. I say there is no attraction between two bodies by the force of the mass that is such doing is diminishing the radius parting the bodies by the inverse square law. I say there are a connection by motion between the centre of he body and the material surrounding the centre. This is what I say when I say there is no gravity produced by the pulling power of a fictional force that mass would create.

The man was E.P. Hubble. The world was expanding and not contracting which made the Universe quite wrong. Newton could never be wrong because Newton was never wrong yet…so if the Universe is out of step with science, then science will correct such an abnormality by finding a way to defraud science and postpone the correcting that the Universe had to comply with since the Universe owed the Master Newton some apology. Did the Universe not know that he whom never can be wrong is in name Isaac Newton! Decisive action was needed. When will the Universe confirm its incorrectness by affirming Newton's obvious correctness? When will the cosmos come clean and prove Newton correct.

Looking at the astronaut float in space we believe there is a force of attraction between the mass of the earth and the mass of the person and this force of attraction is gravity. The mass must be present if Newton's principles apply. The man must have mass and the earth must have mass but it is without falling, so where is the pulling "force" of the mass in any of the above pictures?

This principle of gravity declares that a force of attraction by the value of mass of both solar objects is pulling to the value of mass in order to reduce the radius, therefore bringing the moon and the earth closer by reducing the distance between the two bodies by the square thereof. The body of the earth holds mass that pulls the mass of the body of the astronaut down by the force called gravity. This idea is not science but it is culture and is drummed into our minds like it is one of our living senses we have to adhere too.

This idea, being culture and not science is also the biggest lot of hogwash ever invented by man to scare the mindless many forming the greater thoughtless populous on earth. If this was the case, then the result of this must lead to an inevitable collision that must end the existence of both the earth as well as the moon. Let them and those with the ability to design space worms and space whirls and Black Holes now calculate when is this inevitable event is due and inform the human race when our final day of doom will arrive by using Newton's law of attraction or $F = G \dfrac{M_1 M_2}{r^2}$. This collision will end all form of life in the known Universe, it is that important to know. We have to know the day. However the god News is that the **moon coming towards the earth** is **not** the conspiracy under investigation but **what is the conspiracy** is the fact **that the moon** is **not coming closer to the earth**.

Now at this point up to now we have speculated about how a science conspiracy is not presented let's see how a science conspiracy is conducted. Now it is most revealing but very little publicised that there will be no moon / earth coming together. It is spoken of but not a lot and when it is spoken of it is told in a matter-of-fact-hushed-up sort of way. Has any reader seen this was front-page news lately? Big banners reading: *the moon is departing and not coming closer and therefore Newton is absolutely wrong and physicist has no idea about physics*. No it is very slightly and even more sparely spoken of. That confirms a conspiracy to hide something. However there are conspiracies in place to hide conspiracies in place that would cover for conspiracies and the entire thing becomes an onion with layers enough to drive you to tears. In the background hidden under many conspiracies there lurks a Mother Conspiracies and the purpose of all conspiracies is to cover the Mother Conspiracy.

When last did you see on TV on the six o'clock news the headlines shouting out: **It has been confirmed by NASA that the Moon's orbital distance** is increasing at a rate of 3.8 cm per year of about 3--4 cm/yr. Has any one physicist announced that the moon will not hit the earth and why this will not happen is because the moon is parting from the Earth. The moon was much closer in the past but they hide the fact that the Moon's orbital distance is increasing at a rate of 3.8 cm per year? If you saw is on any news channel please let me know. I think this news is much more appropriate to be designated as headline news than the declassification of Pluto from planet status to orbiting rock fragment. But this news will never reach headlines. Neither wills the news become evening headlines that the earth is increasing the circumference every year. What would the purpose be of hiding this truth from the public, it is insignificant? The reason is the Mother Conspiracy, which I will explain later on.

The distance between the moon and the earth is not reducing but it is increasing! Save that for front-page news. What such an announcement would shout out loud is that the Universe is not contracting by the value of mass as Newton anticipated, but it is increasing by a value that I have determined. This is my storey about the conspiracy storey. In order to accept my calculations, one has to reject Newton's law of attraction just as much as the cosmos rejects Newton's law of attraction.

Chapter 2 MOON CRASHING.

Every child at school is forced to believe in the formula Newton introduced when Newton saw his apple fall from the tree. It is $F = G \frac{M_1 M_2}{r^2}$ where it is believed that the mass of the earth pulls the mass of the apple because the mass of the earth is pulling the mass of the moon because the mass of the sun is pulling the mass of the earth. If the moon is not falling, the sun is not falling so how does the apple fall? Never will this be on TV news just because according to Newtonian principle there is attraction by the force of gravity to the value of mass. Nevertheless, NASA confirms the moon and earth is parting in a way that increases the distance between the two solar systems. Knowing this proves physicist's stupidity and lack of fundamental understanding about physics. So one conspiracy was improvised to stop this from becoming public knowledge. It is called the Critical Density Theory and to form blind spats an important name was placed in the centre. It was Albert Einstein that was supposedly the chosen one to correct the wrong doings of the cosmos.

There was one culprit that was responsible for this information getting everyday news and he was E.P. Hubble. Before Hubble got so outspoken about his findings the entire world of physics new they were bullshitting the public blind about Newton's anomalies but was getting away with it for centuries. Just go and question "Kepler's" laws that Newton founded and you can see how far Newton was off the track.
E.P. Hubble saw through his telescope that the Universe was moving apart. Before this everybody agreed with Newton that the Universe was contracting under the load of mass. Newton saw the moon coming close by the value of mass as the earth was getting closer to the sun by the value of mass and to be equal to God those intellectuals could just apply $F = G \frac{M_1 M_2}{r^2}$ and redesign the entire Universe according to each ones liking. If you had the mass you could gauge when the Universe began. If you had the mass you could gage when the Universe will end. By having the mass one would find the force that drove the moon towards the earth. Before the event of Newton's miraculous discovery of forces driving planets around the sun to the tune of gravity, this was mainly God's prerogative to have such knowledge.

Now having Newton's laws and many mathematical equations such as $r = \frac{p}{1 + \varepsilon \cos\theta}$ and others such as $\frac{d}{dt}\left(\frac{1}{2} r^2 \dot\theta\right) = 0$, and also $\frac{P^2_{planet}}{a^3_{planet}} = \frac{P^2_{earth}}{a^3_{earth}}$. because Newton declared that as a result of the mathematical fact that $4\pi^2 a^3 = P^2 G(M + m)$ it then was beyond question that $\left(\frac{P}{2\pi}\right)^2 = \frac{a^3}{G(M+m)}$, and knowing all this put you level with God. No one could argue because if Newton was brilliant enough to know that [illegible equation] who on God's earth will dare argue with him. No one gave a fart about understanding anything this formula said because if they did they would know the correctness thereof was a joke. So with nobody having the intellect to argue the correctness thereof, it was considered brilliant just to pretend one does understand and then become a member of the Brainy Bunch society ruling physics. All one needs is to pretend to understand and believe that mass pulls mass and then you are as clever as Newton and everyone knows Newton is bloody clever so anyone believing the Universe is contracting by the measure of mass that forms gravity places such a person equal to God's intellect. That way all that anyone is asked for when to be thought of, as being clever is to repeat after Newton that $F = G \frac{M_1 M_2}{r^2}$ is the formula on which the entire Universe stands by the principle of physics. How easy could it be to be so smart you think of your position as equal to God's position. No one dare to argue with you because you have Newton backing you and only insanely stupid or mentally handicapped will dare to argue with Newton about mass having pulling forces going around as gravity. Newton never proved that $F = G \frac{M_1 M_2}{r^2}$ was correct but since nobody understood whether it was correct or not nobody gave a blue apes virtue about the concept of correctness. This illusion was ending and it came with a bang that broke a two hundred year conspiracy, which guaranteed silence.

Hubble saw the Universe expanding while Newton said the Universe was contracting. Then along comes this man Hubble. All Hubble had to do is to tell those not believing him to look through his enormous telescope so keeping Hubble quiet was not easy and discrediting Hubble would be most stupid. That meant they had to explain that Newton could be wrong! They had to explain why $\frac{d}{dt}\left(\frac{1}{2}r^2\dot\theta\right)=0,$ and also $\frac{P^2_{\text{planet}}}{a^3_{\text{planet}}} = \frac{P^2_{\text{earth}}}{a^3_{\text{earth}}}.$ was wrong because Newton declared that as a result of the mathematical fact that $4\pi^2 a^3 = P^2 G(M + m)$ it then was not so much beyond question that $\left(\frac{P}{2\pi}\right)^2 = \frac{a^3}{G(M+m)},$ was incorrect and knowing all this put them below the level with God because only God and Newton was flawless and now Newton was not flawless. No one could argue or understand what Newton formulated because if Newton was had the brilliant mathematical genius that made him smart enough to know that [illegible equations] then all they had to do to be stupid was to condole Newton. In the past if you were to be clever just repeat after Newton and you area Newtonian, but now this no longer applied. Don't ask questions because nobody dares to ask questions. You don't have to understand anything because that is not asked for. You don't have to look for flaws because that would demise your position in terms of Newton. What you see is precisely not what you see!

Science has to reflect on what nature is and not interpret it in one way and then judge it in another way.

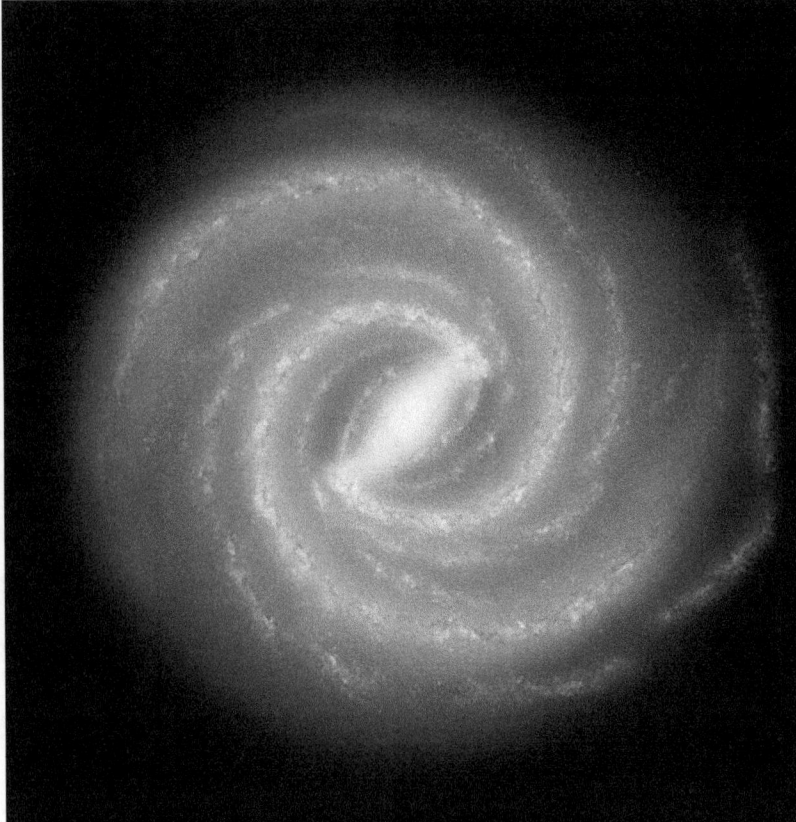

When we look at this picture we do not see on star. We do not see the Milky Way, which we think it portrays. We see a stream of light flooding the lens of the telescope through which we see this fantastic picture and the light by which we see every spot shows time not space.

We see the light that shows the individual star but the star as such we are not able to see. We see the light as ti left the star many, many thousands of years ago. We see what this was before there was humans on earth. Part of this picture could be representing a time when the dinosaurs still walked the earth in drones. However we can't see one star because all the information is written in light and it portrays what is not present any longer.

We witness what was but what we witness does not represent one instant taken from time because every pixel we see left the individual star we think we see at a different time than the others on the picture. We see a mosaic of time differences spanning billions of years ago and the only common factor is that the light that left at so many intervals only arrives in the lens at the same instant.

When we view the moon we think we see the moon because we were taught that we see the moon. We can't see the moon because it is practically impossible to see the moon. We see light that was reflected from the surface of the moon and only when this light arrives we see what the moon was like one and a half seconds ago. We see not the MOON but we see where the moon was one and a half seconds into cosmic history and we see the allocation of the moon at the time the light left the moon and that is part of what we think of as cosmic yesterday.

Time is the very now, the moment the present the instant and anything concerning space is behind us, is the past, is what is no longer in the immediate. The person standing very much next to me forms the beginning of the historic past while a star twelve billion years away is forming the edge where space can be traced by forming the past. All space including everything I see including the fact that I see anything is part of the past while the I that don't think because the thought forms part of the past is in the present. The instant in which I am the Universe expands by pushing everything I see and everything I can't see in time away from me. Time flows at a rate of singularity Π^0 = forming space $\Pi^3 \div$ in relation to what moves ($\Pi^2\Pi$) which is $\Pi^0 = \Pi^3 \div (\Pi^2\Pi)$.

Newtonians know this but is too short minded to realise what it is they realise. The instant forming singularity Π^0 forms space but space seen in light = Π^3 and the light is placing space Π^2 at a growing distanceΠ All you have to be is to be stupid enough not to ask silly questions and then you are brilliant. Be a Newtonian and follow your leader blindly with no questions asked. Everyone would be so amazed with your brilliance they will think of you as one of the Brainy Bunch. People will stare at you in awe and all you have to do is to believe Newton and believe in science. Now Hubble went along and burst this friendly little bubble by his unasked for discovery of the expanding.

Then along came this man Hubble with his large telescope. Everything was going well and everybody was enjoying being clever as long as you were stupid enough to be a Newtonian. Then this man unveiled a fact that declassified Newton as a flawed worthless form of stupidity. The Universe expands and does not contract. Now once again only God was flawless with Newton flawed throughout. This the Brainy Bunch was not going to appreciate and this did not go down well with the most esteemed the Newtonians.

In order to cover up for Newton's misperception an entire variety of reasons are established, each accepted as a possible truth. The fact that Newton's principle goes begging never gets mentioned, although the only reason why it would never get mentioned is because it is the only valid conclusion and that they don't want. All other reasons they mention is overruled by Newton's principal of mass pulling. Mass pulling is the founding law that all other factors rest on. The earth slinging the Moon away can't be a factor because the mass of the earth is too great. The mass that pulls reduces the radius by the square. The moon is moving away from the earth. Yes, have a look at all the theories presented by Newtonians on the web as to why this happens. It is the thrust, no, it is heat expansion, no, it is parting because of sea currents, no…and there is an innumerable many excuses why this is happening. Not one has to do with Newton being wrong! They never take Newton's formula and apply the mass of both solar objects and see what the gravity is to see why the parting occurs. With the radius increasing the gravity must therefore reduce in pulling power because the distance determines the validity of the force by dividing into the multiplication of the mass.

It is not the Universe that is very far that is expanding because Newton said the Universe hear by us is contracting. The Universe is expanding and every atom is the Universe forming time. As every atom spins while going straight forming the spin of the earth every atom performs time by expanding in the movement time performs. The expanding is beyond what Newtonians wish to measure or calculate, the expanding has a value of $\Pi^0 = \Pi^3 \div (\Pi^2\Pi)$ and this says that space Π^3 grows bigger by a margin of Π^0 as space moves in a circle $\Pi2$ where the electron spins while this spinning takes the earth to go around the sun Π. His is how time works. The spinning pendulum shows this effect very clearly. This is gravity where the movement of the earth Π^3 links to singularity Π^0 by the rotation of the earth producing a circle as a line Π that compresses the "air" downwards. Gravity forms as the earth or any other cosmic body rotates by 7 °. By diverting the straight-line movement by 7° a contraction forms in a circle. In my books I prove how this then brings about the value of Π by implementing the law of Pythagoras and gravity is the law of Pythagoras. The reclining of space by redirecting the direction of travel from straight ahead to 7° reclines the space in a steady and sturdy flow. It is the space reclining or contracting and the space contracts albeit filled by solid material or empty of solid material. This is the reason why all things fall equally. It is the space moving down with or without holding material and the space has the same density in relation to the solid cosmic structure rotating.

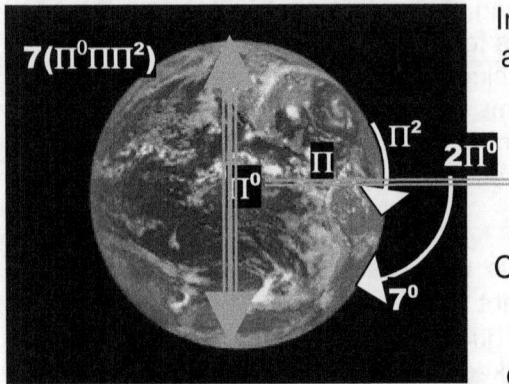

In the centre of the rotating body singularity forms to the value of Π^0 and this extends to the curve forming Π in terms of the curve of the rotating object. This then forms part of gravity moving forming

The extending of singularity goes from Π^0 to $5\Pi^0$.

singularity going square Π^2 and this form the relevancy Coming in from space while having no mass the coordinates change as the value of Π forms a relation to the 7°
When an object comes in from the atmosphere or flies the sky the object associate with the turn giving it a value of the rotation in line with the axis which puts a value of 21.991 / 7 on Π or on gravity.

This is gravity and this is how gravity functions. It has nothing to do with mass in any way, shape or form and there is no factor such as

The entering by singularity that goes from 7° by $3\Pi^2$.

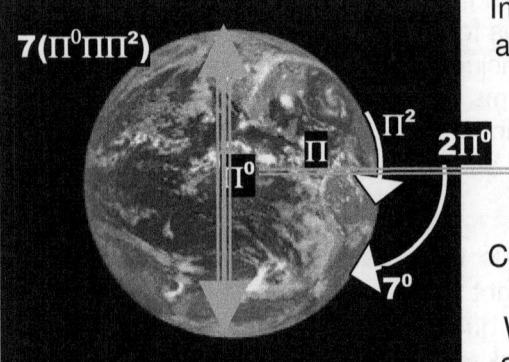

mass in the entirety of the cosmos apart from being in the imagination of the Newtonian conspirators calling them physicists. I sent a fifteen-page article to the Annalen Der Physics in which I explain this process in length.

When the pendulum swings every new position it takes on indicates the time ended and the ending of the instant that formed space. Space is the history of time or the reflection of light indicating time left an image of space behind. In the instant we see time we see Π^0 forming singularity but we think we see Π^3 while the ever onward going space that forms is due to time moving $\Pi^2\Pi$ leaving space Π^3 behind as light that forms an image $\Pi^2\Pi$. When the turning of the earth counters the pendulum swing the rotation turns around, as the earth will not allow the pendulum to move out of the influence sphere of the earth. By moving back and forward the pendulum reads the time of the earth as the earth spins and moves around the sun forming time in motion by the earth moving. The pendulum reads the movement of the earth forming atmosphere or then the gravity of the earth that forms time. Mass does nothing in this entire operation and if mass had value a bigger pendulum would swing slower than a small pendulum that would spin faster?

The earth moves by compressing the space around the pendulum and it is not the pendulum that swings by its own initiative following the space as it moves towards the earth. If the pendulum reads time by swinging through space and the earth moves space by contracting or compressing space then the movement of space is the factor that forms time and then time is the movement of space by the changes occurring in space. Newtonians will never agree because of the mass factor they believe in and then rather disagreeing with what happens in nature they will uphold Newton's version of misinformation. I say again if the pendulum reads time by reading space moving towards the earth then that gravity forming from the earth's circular movement must be time forming in direct relation to gravity moving space. This is a large part of the mother conspiracy whereby Newtonians tell nature what it must be according to Newton instead of reading from nature and then accordingly find what nature truly is.

While the earth provides material Π^3 that contracts Π^2 space moving down Π the pendulum register singularity or the instant Π^0 in which time is and while the earth provides the instant Π^0 that the pendulum reads Π^3 the pendulum encounters swing Π^2 that the space Π that is moving down forms as Π. This spinning action of the earth holding a body results in the space surrounding the body to compress and that movement results in a moving circle compressing and that forms time in relation to the earth forming gravity. But the major issue in al of this is that wherever the instant or singularity forms Π^0 space grows Π^3 by the movement of space Π^2 in a circle that goes straight Π. This growth Π^0 is beyond measure because it depends on material $\Pi^3\div$ that moves in a circle (Π^2 when the circle moves straight Π. $\Pi^0 = \Pi^3\div(\Pi^2\Pi)$.

Singularity Π^0

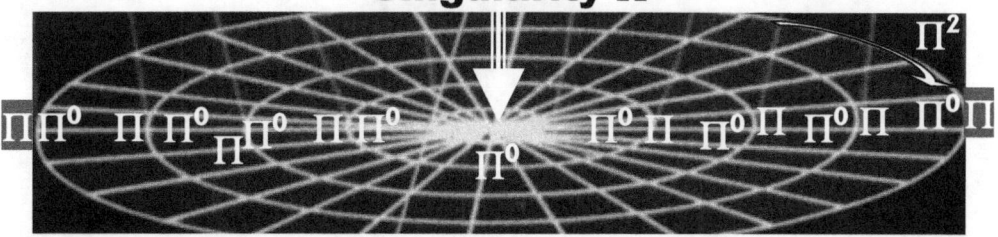

I do explain the following much better as the book develops but this is an introduction to what forms time and what forms gravity.

There never can be anything such as a straight line in singularity and therefore in the Universe. To produce singularity movement must establish such a centre point. Only circles can be the shortest line possible because of the nature of mathematics.

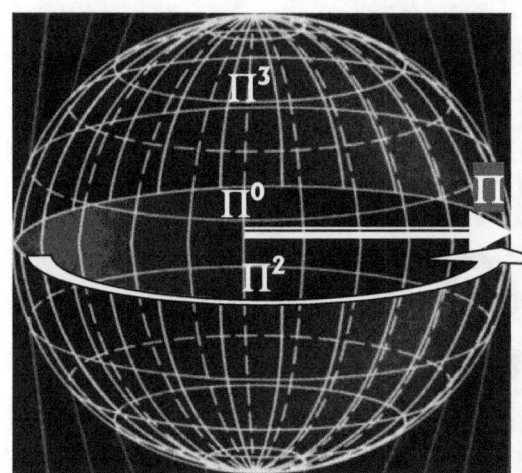

By moving the space that the earth fill down square forms between any points as a circle develops a line that forms as Π and the line forms in the shape of a circle. This brings about the formula $\Pi^0 = \Pi^3 \div (\Pi^2\Pi)$. By connecting Π to the spinning Π^2 earth Π^3 singularity Π^0 is established as a mathematical factor $\Pi^0 = \Pi^3 \div (\Pi^2\Pi)$. The movement of the earth forms gravity by rotation and the space surrounding the earth compresses in order to establish time in relation to the earth rotating. The difference between the Universe expanding and the earth contracting space forms time in relation to the earth, the rest of the solar system and indeed the Milky way which then too the earth is the Universe. Gravity forming is time forming as a result of space expanding and space collapsing.

When the line has a beginning and an end at the very same spot and it wishes to extend the position as to further the possibility it has, which direction should it favour. Extending the line in any one direction will favour one direction without any clear reason not extending in other directions. The only option about extending will be in all directions equally in order to give a meaningful non-bias flow of mathematical equilibrium.

$\Pi^0 = \Pi\Pi^2/\Pi^3$, which says that everything that is (Π^3) moves ($\Pi\Pi^2$) in relation to singularity (=)Π^0

But also when anything (Π^3) cosmic moves ($\Pi\Pi^2$) it forms singularity (=)Π^0 by which it moves.

This formulation of space-time is the starting point of physics.

The entire Universe is the movement of space in relevance of a point forming singularity and

Singularity forms in the centre of all things spinning by the spin the material has. Because singularity holds the compliment of one everything that spins holds the centre of the universe in equal measure and in equal quantity.

All star are the measure of the gravity that every proton within that star holds as every proton conforms space to confirm the stars solidity and therefore the contraction ability that the star holds to retain the heat in space or material that the star could collect. Everything that is in the Universe is $\Pi^3/\Pi\Pi^2=\Pi^0$, is spinning while also rotating something else. This produces a relevancy in relation to singularity which Kepler showed as $a^3/kT^2=k^0$ and since I found gravity is $\Pi\Pi^0$ I changed the format to $\Pi^3/\Pi\Pi^2=\Pi^0$. When the combined contraction that the atoms form by overall movement of the star the relevancy of $\Pi^3/\Pi\Pi^2=\Pi^0$ will change and in changing $\Pi\Pi^0$ the star overheats and goes supernova. The slower movement $\Pi\Pi^2$ will enlist a larger space Π^3 form more space colleted away from the centre holding and securing singularity.

Part 3
Living the lie

Chapter 3 LIVING THE LIE.

I am going to show that there is no such a thing in the Universe as mass. No that would be incorrect. I am going to prove that mass is an invention Newton created to fool every person since Newton. It includes all those that pretended to be intelligent. It is a fairy storey coming true. You've heard about the King and the magic invisible clothes, apparently magical because only the wise and the clever in society could see it. Everyone was fascinated by the beauty of the clothes The clothes was only visible to the very wise and because no one wanted to be seen as being a fool in the community, and least of all the King wanted not to be seen as a fool that couldn't see his own clothes, the lot in the Kingdom saw the fantastic magic clothes. As the King paraded naked through the streets a young girl shouted: "The king is naked!" Then everyone realised the King was naked and even the King realised he was the biggest fool in his entire Kingdom. How does this have any connection to a book that is devoted to science in any way you might ask. Isn't this a famous story about a bunch of fools pretending, well it is not a story in science but it applies to everyone in science. Everyone since Newton that knew physics was able to see mass because no one wanted to be the mindless fool that couldn't see mass. So much does this story apply to physics that we can only change the name of the King to Newton and everything else in the story is in place. Newton created mass by magic and everybody since then could see mass. This was the case for three hundred years where everyone in science saw mass as clear as daylight until I came along in 1977 an I was unable to see mass as a factor when seeing all the evidence that disproves mass as a cosmic factor. I was the simpleminded student that couldn't see mass and I still can't see mass. As you read this you feel agitated by my remarks, don't you?

Object of different sizes all fall equal as seen on television so many times. An army tank will fall as fast as the tank's corporal that fall with it while being either next to it or on top of it or in it the two object of much different mass would always fall equal when free falling. The corporal does not stay behind and watch his tank speed to the ground while he is following much slower and at a sizable distance behind. How can a large mass pull as equal as a small mass pulls to travel equal at the same speed over the same distance and still be driven by the power of mass creating gravity. Have you given this a thought? Any object will float in outer space without showing any sign of mass. To cheat us as science always do about these matters science tell us that when any body floats in outer space the body has mass but the gravity is micro. That is as much a fabrication and a distortion of the truth as the entire mass idea. The body has maximum gravity and micro mass. The body requires gravity or a movement of 11,7 km per second to counteract the gravity the Earth applies.

In order to find movement heat must apply. To escape from the earth a lot of heat must apply. To loft from the earth say in a balloon heat must apply and heat must be released to get airborne as well as get into outer space. The more heat is released the more any object will escape from the surface of the earth. In that sense heat is anti gravity because by releasing heat gravitational confinement is overcome. Expanding increases density therefore gravity is something about density! All the less dense particles float and all the dense particles fall. That connects to gravity

The faster the body moves the further will the orbit of the circling body be from the centre of the Earth and that also depends on motion or gravity. Being big or being small is not part of the requirement and therefore mass has no influence as a requirement. When the body starts to move slower the body stars to descend. Big or small does not matter and having mass or no mass does not matter. It is when the motion of the structure reduces to a speed that is below the rotating speed of the Earth that the body starts to plunge to the Earth. Galileo said all things fall equal and then all things can't fall by mass.

The falling or not falling and the escaping or not escaping only depends on the speed of motion and mass as a factor is only required in relation to the imagination of the physicist. While falling all objects show equal mass and having mass differences puts no extra load on any of the factors in any way. The buoyancy of all objects in the atmosphere depends on the speed in motion of the object. There are many other factors that come into play and all those factors forced me to write four different books on the subject where every book highlights different aspects that play an apart. The only aspect that plays no part is having mass or not having mass. Mass can't pull when all things fall equal as Galileo stated.

Please do not think I am able to bring you all I know about the problem we have in science in this small book. I have written 37 books so far and I have still not addressed the entire issue. This information I present in this book does not tell the entire story even about the conspiracy… This information does not begin to tell a part of the entire story. This information does not even introduce a part of the beginning of the introduction of the entire story. Yet, in thirty plus years in science and through out the entire world of science I have not come across one individual in science that shares my concerns about the problems within science and I am not aware of any person working or teaching or that holds a teacher's post in science on whatever level or working with whatever form of science who would admit to a problem in science and considering what I present, that constitutes to an academic world wide cover-up when faced with the mountain of evidence I gathered. It still helps me nothing because of the conspiracy.

There is a deep, dark conspiracy hidden in physics that some willingly participate in, some absentmindedly participate in, some not being aware of the conspiracy but still very willingly participate in and other too stupid to realise they help to further the conspiracy just because they don't think of the consequences of their stupidity by which they promote the conspiracy. In the end no matter why or how those teaching in science are involved the only action of importance is that students are brainwashed by methodical mind control and I challenge everyone on earth to show I am committing slander against science. Everyone in a teaching position is participating in absolute mind control by brainwashing students to believe science and this forms the conspiracy going on in science. Should anyone not believe me, then don't believe me after you have read this book or another similar book A Conspiracy to Commit Fraud on a Cosmic Scale but is just much less technical and then still after you completed either of the books or both of the books then I dare you to hold the opinion that I am deforming the characters of those teaching physics. Prove to me after reading the evidence that there is not brainwashing going on to conspire against society.

No one in science would face up to the fact that there is a serious problem in science from school level through pre- and post graduates up to nuclear science and astronomy or astrophysics. Science is in a denial about what they know and what they are oblivious too and this denial forms a conspiracy by which everyone in science controls all thinking of what forms the principles by which science is dictated. I offer the solution whenever science would admit there is a problem but everyone is in science is running away as fast as they can from the problem where they are unable to see the problem and if only they would, but they will not even stand still for one minute to look at the solution I offer. A Conspiracy to Commit Fraud on a Cosmic Scale informs about the same conspiracy but the detail used is much less in complexity and the focus is less on mathematical input while the overall focus provides on much more simpler verbal explaining. It is for those favouring less technical detail and broader information. The first objective in this book would be to establish what a conspiracy would be because everyone has his or her own idea about what forms a conspiracy?

Physics students, it is your duty to pull the plug on the powers of the All-Powerful Academics in Physics and stop their dishonesty. It is your task as the as the next physics generation to stop the criminals that are filling the corridors and the lecture halls of physics departments throughout the world by acting as if they know all there is to know and all they know is to fool the next generation of students. Stop their teachings by forcing them to stop their criminal fraud. Force them to explain the deception such as the one they call THE CRITICAL DENSITY, which is a conspiracy to commit fraud. Let them explain how an expanding Universe can suddenly and abruptly turn in direction of developing and start to contract as Newton stated it is doing at present, and when facing all other concluding evidence showing that the Universe was expanding since time began they come up with the utmost unrealistic garbage only an idiot can devise. Tell them to bring proof with evidence that the cosmos is contracting as Newton said. In THE CRITICAL DENSITY conspiracy all they say is that they are waiting to see when the cosmos would stop its criminally insane behaviour and start to listen to the laws of Sir Isaac Newton. They shove all the blame of wrongdoing onto the cosmos and take away all error from Newton. If the cosmos does not contract as Newton said then when will the cosmos mend its ways and follow what Newton said and to start contracting! It is a conspiracy to cheat and lie and crook the human race in order to keep Newton untouched. With The Critical Density shambles the modern Newtonian set out to defraud the world in the same manner as their Master Sir Isaac Newton has done centuries ago. Newton said the cosmos is contracting. When Hubble proved the cosmos is not contracting, Newtonians looked where the cosmos went wrong by not following Newton guidelines he so clearly set the cosmos to follow. It has to contract and not expand. Those in academic positions fabricate non-existing material no one can detect to cover the real conspiracy they try to hade. When the argument arrives of contraction versus expanding they wall this down by referring to the search for a substance that can't exist and could never be detected. It is not the dark matter issue that is the real conspiracy but the dark matter forms a conspiracy to hide the facts that the true conspiracy covers up. It is this mother conspiracy I am gong to uncover and present.

These whom I named in Honour of Sir Isaac Newton as the are the guard of the Newtonian High Priests carrying the name as the Newtonians are Men amongst mankind, that charged the Universe with not applying to standards set by Sir Isaac Newton, and then went on proving how incorrect the behaviour of the Universe was in not adhering to the direction gravity has according to Sir Isaac Newton. Since Sir Isaac Newton can't possibly make a mistake, it then was presumed the cosmos made the mistake by not following the gravity settings laid down by Sir Isaac Newton, the one that cannot falter nor could his teachings fail, carrying the illustrious name of Sir Isaac Newton. It must be the cosmos being at fault by expanding without seeking the approval of Sir Isaac Newton to do so in contradicting Sir Isaac Newton. Up to this point in science and in despite of an array of evidence pointing to the cosmos growing by expanding in every sense and with all pieces of evidence gathered by science from all over the Universe (including the solar system), the theory of contraction is still hailed as the infallible Newtonian truth. Every one that is part of physics, shares the Newtonian vision of a contracting Universe where the lot would one day again come together and Creation will end where Creation started some time ago. The Universe has mass that is pulling mass towards one another and we are in the centre of an ever shrinking Universe. The Universe is about to end where all mass contracts into one huge lump of material, and this conclusion contradicts al evidence gathered by science. Without any evidence available to its authenticity they gave this Newtonian processes a name, they call it the Big Crunch.

If you don't believe me marry Newton's contraction with the Big Bang and see a divorce in place before any Church consummation of such a union could begin…but then again just as unlikely union in principle marriage between Galileo and Newton is in place and the mindless masses never once frowned on that! Students in Physics, it will serve you well to read the following arguments very carefully and come to a conclusion about what gravity is and what mass is and how it is impossible for the concept carrying the idea of mass then become responsible to form what we think of as gravity. Mass can't ever and doesn't bring about gravity. In fact I challenge anyone to prove that a factor such as mass do exist.

Planets do maintain positions like a geared clock but not according to mass that pulls although a balance provides the orbit. Newton claim that mass is used and with that the idea of proof is automatically placed at the door of Newton. We talk of planets orbiting and that is what planets do. Planets don't creep up to the sun by the value of mass. Think of what planets do… and you think that planets orbit. The saying goes that planets orbit indicating they follow a circle and this uses other rules than mass applying position. That is not what Newton said. In conversation we speak of the planets orbiting and this orbit is in terms of the Titius Bode law. If Newton was correct the planets pull but that would be blatantly wrong according to what we find the cosmos applies. Never do we refer to the planets pulling the Sun or the Sun pulling the planets, but we speak of seasons coming from orbital positions. Being in orbit has to neutralise the pulling and then cancel the pulling concept that also became culture. Using the formula $F = G \frac{M_1 M_2}{r^2}$ as Newton provided, disallows any other concept other than moving towards. The person Newton got his ideas from and the work he raped completely, that of Johannes Kepler explained this very well, but Johannes Kepler makes no room for any pulling of any sort. In the work of Johannes Kepler he said that the space being the orbiting route a^3 remains at a specific distance k while the orbit T^2 takes place…and in all my other books that addresses more information I take Newton to task on his dismembering of Kepler's formula by corrupting Kepler's work and with what amounts to fraud, Newton takes science on a goose chase that holds no truth. There is no pulling by mass of mass in any way. Kepler said the space a^3, moves T^2 in terms of a ratio k adhering to a centre k^0.

We have either one of two that has to be incorrect. If Newton is correct, then the normal way of how the cosmos is functioning of the planets rotating T^2 is incorrect. Then we must start saying planets are pulled to the Sun. If the normal form of speech is correct and the planets are merely orbiting the Sun, then Newton is wrong. Newton said centuries ago that gravity is the force of attraction there is between objects that hold mass and it is the mass factor that brings about this attraction, which Newton claimed there is. The Big Bang Theory proves Newton's idea is not only being wrong but Newton's idea of attraction is a joke. If the Big Bang is expanding the Universe, then how can the Universe contract at the same time?

With the knowledge in hand for decades why did science not move away from the contracting idea? Why are they chasing an idea that shows how blatant the corrupt conspiracy is in bedded? Academics are very much aware of this misconception Newton had and still academics in physics are promoting the ideas of Newton as the unwavering truth. Teaching Newton is participating in deception and promoting Newton is criminally deceiving the public and while doing so, is committing an act with criminal intentions. Then, in the face of all this evidence contradicting Sir Isaac Newton; they remain upholding the correctness of Sir Isaac Newton and keep on teaching students about the unwavering correctness of Sir Isaac Newton. I show precisely how gravity produces mass but mass can never produce gravity. I show with explicit detail when, how and where gravity forms mass but mass can never form gravity. What I prove annihilates every Newtonian claim. Still they persist in promoting the lie just to favour science and that is a lie.

When any person, notwithstanding what reasons given, repeats such a lie unabated while being well aware that the information passed on by such a person is incorrect, then the person commits deceit. When anyone is repeating the information that is passed on as being unblemished factual substantiated and verified truth while such a person knows very well that such information is void of proof or lacks proof, then committing such an act is a criminal enterprise. Academics in physics commit every one of the above indignities and yet see their actions as being lawful and even much praiseworthy and hold their role in society in the highest esteem imaginable. Physicists would not tolerate any form of criticism.

They fail to see the crime that they commit while tutoring physics. Whatever motivation they may claim to have which they offer to serve them as forming their driving force, the fact that they perpetually perpetrate in unlawful behaviour, by spreading untruths, such actions on their part put those academics holding such highly regarded positions in the league of ordinary cheats, gangsters and common criminals. By deliberately and constantly falsifying facts to further whatever humble cause and produce illegal claims repeatedly, remains derogative behaviour and is unlawful by nature, notwithstanding what morality it should serve. To promote the lie is to hide their misconduct and it is not to promote science.

A Preacher or Pastor or Priest lying on behalf of God is not lying on behalf of God and to think the Preacher or Pastor improves or underlines the Greatness of God by lying on behalf of God is very mistaken, because in reality such a Preacher is falsifying the truth for his or her personal benefit and trying to impress the congregation about his importance and not the importance of God. Lying is wrong and doing so even in the name of God remains despicable. To say God sent you to do anything is self-promotion because the truth is you told God you are going to do with or without His consent and you use his name to justify your enterprise in any case. That is falsifying facts not on behalf of God but you use God to falsify your justification. The same applies to academics in physics. There is no argument that can change this truth about falsifying the truth and when doing so there is no hiding behind any excuses of ennobling to benefit mankind that will change such truth into righteous conducting.

Ask your professor to show how an expanding Universe can also contract and your professor will tell you about Einstein's Critical Density theory. This theory I prove is the second biggest fraud ever devised by any group of persons in the history of civilization! I am about to introduce you to the biggest fraud ever devised but it is not the Critical Density elusion. The Critical Density elusion is perpetrating fraud and the conducting thereof is upholding deceptions instituted by Newton that then formed the institution of lies they call physics. The Universe does not contract in any way; means or form and even such a suggestion is incorrect! The Moon and Earth are not moving closer but are moving apart. The entire Universe is growing in space and nowhere is space depleting by any norm used because what is space is time. That is exactly what Galileo's swinging pendulum shows; it shows the contraction of space is reading time.

Academics are very aware of this misconception Newton had and still academics in physics are promoting the ideas of Newton as the unwavering truth. Academics teaching these misconceptions are committing fraud, notwithstanding the portraying of their role in society being unblemished, spotless while they are covered in a lily white blanket making them being whiter than snow and having such a holier than thou attitude. Teaching Newton is participating in deception and promoting Newton is criminally deceiving the public and while doing so, those doing so are committing an act with criminal intentions. Then, in the face of all this evidence contradicting Sir Isaac Newton; they remain upholding the correctness of Sir Isaac Newton and keep on teaching students about the unwavering correctness of Sir Isaac Newton. They put down conditions of learning to this effect and are expecting students to repeat these untruths and unproven facts by forcing answers to that effect in examinations. Students must accept cheating to become accepted in the ranks of the learned while also carrying the cloth of righteous in knowledge and unblemished virtue. Forcing the acceptance of this untruth about physics is equal to preposterous subjecting students to physiological torture and heinous mind conditioning, scandalous thought control and brainwashing. This applies to everyone serving as a tutor in physics notwithstanding whatever status the torturers might have in society or the morality they attach as a reason to commit such atrocities.

If you are a student, then you are conditioned by academics in controlling your thinking by enforcing pre-mind setting and in which they methodically force you into believing in Newton and this is an on going process conducted for centuries in the past, while it is the truth that Newton is completely void of any tests that may secure any form of confirmation and in securing proof then also by that establishing proof. They never prove Newton's philosophy on gravity but those persons conducting teaching in the subject of physics force all physics students to learn Newton's gravitational concepts and accept the facts as if it has been proven beyond all other facts. The condition of being accepted in physics is to accept Newton without questioning the proof that is never supplied. Let your professors now prove how it is that Newton's teachings are correct and then examine you on the process they use to prove Newton's concepts. At present they say Newton is correct and then they test you on your ability in repeating that Newton is correct without ever proving to you that Newton is correct. Let those physics professors now prove Newton and then test you on the manner they use to prove Newton to be correct.

The truth beyond all other truth is that Newton's gravity has never been proven (because try as you may it is not possible to prove Newton's formula forming gravity mathematically) and because academics know that, academics require the blind acceptance of Newton by students. This unconditional acceptance of Newton's correctness relies only on the pre-conditioning of students' mind set and academics depend only on the student trusting the academic "say so" about the institutionalised correctness of Newton. Pre-conditioning students into blind acceptance depends on the academics' insistence that students approve Newton's concepts without pre judgment or students insisting on scrutiny of any sorts. Academics depend on students never questioning their say so or demand proof about what academics teach. Those academics in teaching positions insist that all students accept Newton's accuracy.

In order to get students to accept Newton's hypothesis, academics resort to brainwashing pupils and students. Please let you lecturer put in all the values of the formula $F = G \dfrac{M_1 M_2}{r^2}$ and then with the values applying prove Newton is correct. Pre-conditioning students into blind acceptance depends on the academics' insistence that students approve Newton's concepts without pre judgment or students insisting on scrutiny of any sorts. In examination students have to outright and blindly follow academics' say so only because academics say so. Academics depend on students never questioning their say so or demand proof about what academics teach. Those academics in teaching positions insist that all students accept Newton's accuracy. This is methodical mind control as much as it is the brainwashing if ever there was brainwashing. I show what they enforce. If you are one of those believing that Newton was ever proven, then what you believe to be true is a lie because Newton can't be proven and that is the truth! The time has come to face your teachers and force them to stop the ongoing old culture of bullying students and conditioning their thoughts by enforcing on them dogmas, which is, mind control! In order to get students to accept Newton's hypothesis, academics resort to brainwashing pupils and students.

They teach you that the Universe contracts and to state their case they force students to learn that gravity is proved by Newton introducing the following formula $F = G \dfrac{M_1 M_2}{r^2}$ They say that M_1 is the mass of the Earth and M_2 is the mass of the individual in questions mass and the multiplying of these factors with the gravitational constant produces the force of gravity when this gets divided by the square of the radius. Please let you lecturer put in all the values of the formula and prove Newton is correct. If he can't and I know for sure he never can fill in the symbols and calculate the force of gravity, then read the rest of the web page that follows to see how far academics in physics go to brainwash students into believing in Newton's fraud. To cover their ongoing fraud they create a mythical mystical non-existing dark energy.

Those in power of physics wheel the one conspiracy after another conspiracy extending the cover up with another cover up by falsifying the truth to hide the untruth and it is not clear what they would hide when trying to hide from people discovering where they refuse to reveal what the truth is. Why don't they not just come clean and admit that the Universe is departing instead of arriving? They keep cosmic science in disconnecting darkness with patches of light in between as sensibility being between the dark that disallows the light to connect. The senseless darkness of failures to accept the incorrectness holds the light disconnected but the light parts darkness with tiny strands of unrelated correctness and only connects by the dark that does not connect by understanding.

They have a concept and in time the concept is proven to be incorrect. When they discover what they thought applied was wrong, those in physics do not step forward and admit to what is wrong...no they change nothing but keep everything wrong in place or not in place by concocting a conspiracy to keep everyone happy and everything as it is in place. That is how they can unite the idea of the Big Bang with everything moving apart since the time it began in line with everything coming together and Crashing in a Crunch while no one has any vague idea of what is going on in-between. Here is what they essentially in essence really hide. This is what all the scheming and deception is they try to cover. This what follows and what you are about to read undoubtedly for the first time ever in history. It is the true conspiracy that every crooked angle in science is covering! You're going to learn what the mother of all the conspiracies is that you are aware of. I don't plan to hold back my punches due to being scared of Masters in science. This book started off as a website to inform about a science conspiracy but although reduced still it grew into a book that serves much more information than what I first intended to supply. You will see many new aspects about gravity please make sure you understand what you read. It grew into a comprehensive study on cosmology. At times you may observe while reading this book that it seems as if my frustration will ring through like the chiming of the Big Ben Bell. For that there is a reason. At times my frustration and anger will boil over drowning my politeness and that is true, which I admit. For twelve years I have had the answer that would correct the philosophy that has a stranglehold on cosmological science.

I discovered the building blocks of nature where my discovery puts all other cosmic aspects of science into science fiction. Those who force-feed non-existing dogma do so to brainwash students to hide the incompetence of "modern science" so they can rule supreme while ignoring the truth that they deliberately hide by concocting a conspiracy. To keep everyone unguarded they practise a conspiracy by which they perform an accepted practise of thought control on students to further the false dogma presently in place. I try to blow the whistle on such a practise but accepting my resolution makes every thesis ever written science fiction. Therefore no one in science dare to read my work leave alone appreciates the revolutionary nature thereof. Whatever now is deemed to be accepted science would then become what is the past tense in science because the flaws that those in power of science principles kept coated for centuries on end as untouchable truth will then be rust that breaks the surface to show the holes!

They try to silence me but surely somehow somewhere I have to break through with my massage! I bring you a true form of science as never seen before in all of history and I do that when I dispose of the conspiracy that hides all the incorrectness and the failures that haunts science today. Science is accepted as the most righteous information available to man and that is a scam.

In the prelude I am on purpose going to show you, the reader a lot of pretty pictures about the cosmos and you look carefully while I explain certain ideas and then by observing you detect what the common factor is that all these pictures have. Let's see how perceptive you are about the cosmology you view. Can you crack what Physicists could not crack this past three-centuries of gazing at the sky?

I put in lots of pretty picture for you to see and what you will see in them is gravity because it is gravity that is holding the entirety called the Universe together and in form. What you see is gravity but can you visually see gravity?

If we could talk and I could shake your hand on the bet, I would bet you that you are going to miss what gravity is notwithstanding all the lovely pictures of many galactica I show you. If you do see what gravity is you are brighter than all the bright Newtonians minds put together but after saying all that, it is not that a powerful achievement being brighter than what those lot of mathematical minds are anyway.

mailto:info@singularityrelavancy.com

We are obsessive with our search of where it all began but every one has his or her individual idea. So I too have an idea and for what it is worth, I wish to share my idea with readers. Every one goes by the way of mathematics and everyone falls by the wayside using mathematics. The Universe started with singularity and singularity is 1 and no more than 1, so don't get complicated about mathematics applying. Physicists put mathematics in place of God and presume mathematics built the Universe. That is not true because it must be the Universe that put mathematics into the Universe as the Universe built what now forms the Universe by building mathematics from the figure of 1 going on to 2 and then 3, 4, 5 and so on. It never once occurred to any one in search of the beginning that there must have been a beginning of mathematics before mathematical equations took centre stage and that mathematics are most probably just another development product of the Universe. It is very unrealistic to think that before anything, the cosmos devised mathematics and mathematical laws and from that the Universe was introduced.

When we discarded the contraction idea that saddled us to the Middle Ages and the Big Bang brought us thinking in the right direction, it introduced the way to think. But starting off with a Universe the size of a neutron, or having a Universe formed as is at 10^{-43} and counting is dropping in somewhere in the middle. Einstein said that going beyond the speed of light would bring everything into pure energy. Is that not what the Universe is made of…of pure energy. Is that then not a clue to the next phase? What if the lot was spinning faster than light? It is pure energy we are after because the atomic bomb shows that what ever is captured in the form of the atom, when the relevancy breaks what is unleashed is a hellish pure energy. The nuclear explosion took us to the limit of what happens when the atom overeats. Looking at the atom bomb must be looking at a very limited extreme downscaled version of the Big Bang with heat turning to space like nothing else can ever repeat. For a long time I had a huge problem with Einstein's idea of going the square of light to calculate nuclear energy release. Light cannot go square and that is a fact. However the accuracy of $E=mC^2$ is proved beyond question. I had to find out how light can go square and that was one of my first tasks I had to achieve after finding the inaccuracies lurking in science.

Einstein formulated $E = mC^2$. E is energy, a topic I will discuss shortly, m is mass also a topic I am going to discuss throughout and C is the speed of light. Einstein going square on the speed of light is absolute madness The speed of light can't even double and that detail I can't discuss at this point because there is no space. However there are many books I do go into much detail about that. Then I realised Einstein duplicated Kepler's formula before Newton raped it to bits. Kepler made this possible formula he introduced after TWO lifetime achievements (his as well as Tycho Brahe) **by applying $a^3 = T^2k$.**

This is the formula Kepler used for planet distribution however planet distribution accompanies star distribution, which is the same as galactica internal and external distribution. It is space $a^3 = T^2k$ relating to two types of motion. The E is the a^3 the m **k** leaving T^2 to be replaced by C^2. Then, when realising this formula connection I new why Einstein contradicted himself. When light goes square there is no light to go square and that we see from Black Holes, Pulsars and other optic dark or partially dark stars. The speed of light is fixed at C and that brings a motion boundary.

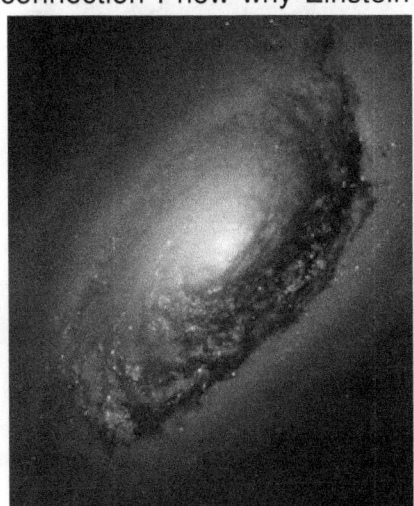

Then it came to me like a hammer hitting me between the eyes with a blow that almost crushed my head. Taking Kepler's cosmic formula $a^3 = T^2k$ into account Einstein saw in using his formula that C^2 is gravity applying and not the speed of light going square. Einstein saw a Universe at the time when the Universe was relevant at a factor of one to the speed of light, which is long, time since not valid any more. Einstein saw the Universe when the electron was outer space and the neutron was pure liquid. The proton then resembled the solid. At that time there were no space because the Big Bang was heat turning heat to space which ever way you wish to put it because heat overheating forms space. What Einstein saw was a time when light was forever taking eternity in motion to move from one point to another point. If the Big Bang was the moment space overcame light and formed space then predating stars were in gravity that is stronger than the speed of light we are seeing in stars going dark time predating the Big Bang.

The rotation speed of outer space was equal to the speed of light at the time before the Big Bang making the gravity that comes about under such condition the square of the speed of light, also applying in relevance to what applied at the time. The darkness we see in space is not "nothing" as Newtonians believe the space in outer space is but it is light moving away whereas white light is light moving towards us. The blackness is light going in then opposite direction in relation to us than white light coming towards us.

We have to forget our ancestral past about seeing the darkness and thinking that is empty and black because it holds nothing. There is one Universe and that Universe holds singularity not space-time. The atom links with singularity claiming space-time by motion of space provided. There are many phases of singularity still developing space-time waiting for the time of their arriving into a related era. In the centre of this and most developing galactica the material is spinning beyond the speed of light. That is why it has light to cast light away as visual beams. The spinning still has to reduce considerably to allow the growth of space not to throw out light, which the singularity deems to move to slow and is therefore discarded from space-time. If the moving was fast enough it could remain in relevancy and remain circling with the rest, but since the light travels to slow, the light moved out of the controlling centre singularity and it now seems to us as light. The Big Bang as the Big Bang has not arrived yet and there is still no electron within the atom. However I am miles ahead of myself with this venture. Only when stars break with C deep within in the centre inner core can stars achieve the independence by moving outside the heat blanket that has been covering them since space broke free from occupation and from individual motion. Where gravity still exceeds the speed of light the gravity will not allow the light to depart from the centre. In that the spin of star material inside the galactic structure circle is stronger than the individual drive to claim independence.

I believe I am about to commit one more eternal sin and criticize Einstein. I am about to gain the scorn of every Newtonian walking this planet. Einstein was a brilliant mathematician that followed where his mathematics took him but mathematics is just another language. It needs interpretation as it needs translation from a written expressed mathematical to a verbally understood form and spoken form of language. Einstein read his mathematics through Newtonian glasses and did not know where he was. Einstein was a brilliant mathematician but he was no philosopher and I immediately admit that I am no philosopher myself either. But I use the help of a great cosmologist provides, which enable me to use the correct interpretation of cosmic relevancies about space-time. When using the spectacles Kepler provide the truth becomes more than obvious.

The motion of the stars grouping in layers flow around a centre, which is allocating singularity, is the forming of the Coanda effect. The centre provide the gravity, the motion provide the time in relation to the relevancy of that particular structure in relation to the group as a unit and singularity governing. The circle forms space a^3 provided and limited by time T^2 in precise relation to the relevancy thereof **k. That is the Coanda principle of gravity** just as Kepler announced it when he said the formula to gravity is $a^3 = T^2 k$. When Einstein saw the square of light it is clear how little he new about Kepler in reality although this should be highly astonishing to know, it clearly is not as my motioning this to physics academics left them cold. No Newtonian knows about Kepler for that matter and Newton was the second highest Newtonian High Priest ever. Einstein saw C^2 but did not connect it to Kepler's gravity and at the time he did not realise the importance of the Coanda principle. Einstein never realised he was working with gravity Π^2 which was at the time equal to the square which we find when light holds the precise relevancy. At the time of the Big Bang the space relevancy matched the speed of light C that made the applying gravity C^2 and that made the available space and that Einstein saw.

If $a^3 = T^2 k$ as Kepler said it is and $E^3 = mC^2$ as Einstein said and $k^0 = 1$ as mathematics says then accordingly $k^0 = a^3 \div (T^2 k)$ because $a^3 \div (T^2 k) = 1$. This mathematical interpretation shows why stars and galactica and all other cosmic structures are round. This is why I have put in these pretty pictures about all the lovely cosmic structures. What this formula $k^0 = a^3 \div (T^2 k)$ **says is that to form material a^3** there has to be movement **$(T^2 k)$ and that movement is around a specific centre k^0. That is gravity and that is the essence of gravity. Moreover and more important than anything else is the knowledge that this proves the Universe has a centre a specific spot around which everything in the Universe spins.**

From that point holding that centre of material turning around such a pivotal centre the relevancy forming space-time amplified and space grew as time reduced. But moreover if there was an era that had stars developing before the Big Bang era arrived, but stars are in the initial stages about light, we are in a scenario that there is more than one Big Bang. Then there was an era before light ever became a valid factor. But everything does not move equally and movement produces space formation that varies as much as the Universe is large. With that in mind it comes to mind that there had to be more than one Big Bang, which incidentally is also most obvious in some galactica development since there was a major time spent in development before light became a factor. There was a time when gravity propulsion was beyond light and even beyond time or space or form. The line that forms relevancy must therefore be the key to solving the riddle of where the lot came from since it goes pre-Big bang many times over and again. The fact that star doesn't develop all at once shows a clear timeline divide in development.

To trace our beginnings we have to find the centre of the Universe where the curvature of space-time forms gravity that forms the form of the cosmos. It is clear that with many centres there also has to be one specific centre forming the centre of singularity. Locating the centre of gravity is also locating the centre of the Universe because where space is least gravity is the strongest and gravity inflicts the curvature of space-time. Finding this centre I feel was my biggest achievement At the heart of bringing about the solution to one of the greatest Astronomic riddles one will find a child's toy… the riddle of Einstein's singularity pointing to the position where the cosmos started so many billions of years ago. An explanation about the growth of space such as the above picture matches every logic view we all have about the Universe, but does Science really provide the answers matching our modern logic, or are we filling in and compensating for science's shortfalls. The big question I asked and that you reading this must ask and answer is as flows: Does Newtonian or official outlook really match the logic of science used by nature? If you are one that thinks Newtonian science knows everything stop reading because you are in for a shock you will never recover from. Here is a few questions showing Newtonian shortfall

Does the Hubble concept Newtonian science put forward match the explanations about how Creation all started… where it is heading…and where it will end?
What is motivating the expansion and the moving?
Why is the Universe depicted as a sphere?
…Why did it start small?
…How did everything become so much and so large…
…Why does it grow from small to large?
…Why was the start so small?
…Why is it growing?
…Where is it going while it is growing …
…Why any specific size...
…What was everything before that?
And why in creation would it reduce again!!!

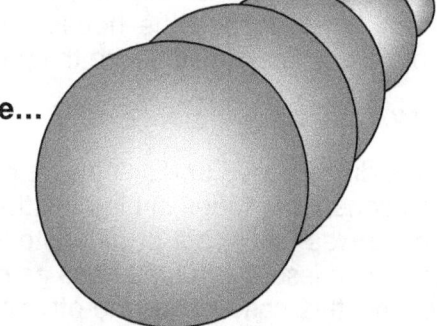

All the above questions Newtonians hide behind **a Conspiracy in Science in Progress**

You are going to read about a conspiracy but people think of a conspiracy in many terms. Let us define not by definition but by interpretation to what a conspiracy constitutes. What do you think is a conspiracy? All the conspiracies you know about is known about because someone somewhere makes money by allowing the revealing of the conspiracy. Silencing the conspiracy does not make money but informing a suspicious public loosens the flow of money. If it were a true conspiracy no one would know about the conspiracy because the powerful would make money from not revealing the conspiracy. The revealing of the facts about any conspiracy would be stopped before it leaked because it would kill the flow of money.

A conspiracy is thought to be a gossip story that makes money and by not revealing it or revealing it goes in line with making money or not making money. You can download this book free of charge because I don't make money by revealing the conspiracy. I truly want to find an audience to divulge the truth. I want to make money but it is by showing how I can correct the flaws in science, not by hiding it in a conspiracy. People put a conspiracy in the same realms as a gossip story, an old wives tail, which is going about but does not intend to harm and mostly serves as amusement to many. Hearing about a conspiracy tests your intellectual comprehension. It is some quiz that you match your truth against the truth that the conspiracy reveals. It is a funny, but it is not funny until you catch the funny part hiding behind the conspiracy and only when you measure the catch behind the conspiracy are you treated to be amused.

If the conspiracy does not touch the person directly then no harm is felt and no harm is intended. Every one holds this view that a conspiracy is on a slightly higher level than gossip. It is a gossip story about someone living in the neighboring village known only to some people next door but has no direct linking to me or has no threat to the safety of others directly associated to me. Everyone treats a conspiracy as if it is something amusing that holds no threat at all. It is something that goes around as a joke of sorts.

For all of those that lives in America or on the far side of the Moon or spent the past three years on a vacation on Mars and others that had no contact with Mother earth or the media on Earth, this is Madeleine McCann the daughter of two prominent Scottish doctors that was on vacation in Portugal in 2007. Since her abduction no one saw her, no one heard of her and no one thinks of her. However that is not new because since the time of her disappearance more than 300 000 other kids in Britain alone befell the same fait and there is no media-mania to try and find those children.

Lets narrow the conspiracy idea down to a known case. A conspiracy is very typical to the Madeleine McCann case where the disappearance of Madeleine brought money to many. Madeleine McCann disappeared Thursday, 3 May 2007 and was / is the daughter of Gerry and Kate McCann. Anyone living in the western world except Americans would know about Madeleine McCann. Americans live in a unique Universe called "the States" and regard Europe as some overseas country situated just past the moon. Americans are the most uninformed brainwashed nation on earth thanks to advertising. Americans can't cross the street without some advertising jingle telling them how to do it the best way while getting the most pleasure while crossing the road and for fun they will throw in a sexual connection in the crossing.

Madeleine disappeared from a hotel room in Portugal while being in bed. For months the disappearance of Madeleine was front-page news and we heard about her all day long. We heard about Madeleine's disappearing day and night on all channels and in all papers around the world. Then the news raised the hopes and then the news dashed the hopes but it was all about the disappearance of a girl called Madeleine McCann. To sell advertising time the press had to write about Madeleine.

This is a conspiracy.

Not the fact that she disappeared or didn't disappear or that she came to harm or died at the hands of her mother or not. The conspiracy hides in big deal that was made about the case all throughout the months afterwards. The case was blown out of all context and proportions considered the international effect it had. The press made an issue about the case as if it was some special occurrence unknown to mankind. The Press presenting this case as a one-off incident makes the entire case a shameful mockery to society. The world viewed this, as an incident equal to the Lindbergh baby's disappearance in uniqueness while the truth is that 11.4 children disappears hourly in Britain alone and no cockerel ever crows about it.

Britain alone loses a hundred thousand children under the age of sixteen per year every year. One hundred thousand children go missing each year in Britain alone and what Special Forces are in place to combat this…nothing. The government never mentions this in parliament. In 2007 there were a number of bomb blasts on trains and one bus blew sky high and that brought Britain to a stand still. The number of lives lost was many times less than one hundred thousand and such bombing happens once in about ten to twenty years. Yet, the propaganda values in terrorists out ways the leverage of disappearing children.

Chapter 3 LIVING THE LIE.

Under tough British laws, anyone of 30,000 on sex register, those marked, as sex criminals must inform police when planning to travel abroad. That information has been used to compile names of at least 130 paedophiles known to be on Portugal's Algarve coast, at the time when Madeleine McCann vanished. Police searching for the missing Madeleine McCann were as they claimed scrutinising four pieces of "very useful" fresh information. The tip-offs came among hundreds of calls made from Portugal via a special UK charity number. If these figures were terrorist-related incidents the world would be in world war 3 by now.

This is not all. It is said that each year, non-family members, often in connection with another crime, abduct more than 58,000 U.S. children. Family members who are seeking to interfere with a parent's custodial or visitation rights abduct more than 200,000 children. Although the vast majority of children (at least 98%) return from abductions, too many children do not. While there are only around 100 reported cases each year of the most dangerous type of abduction – stranger kidnapping – fully 40% of these children are murdered. Where is the media coverage on these events? According to the Department of Justice, almost 800,000 children are reported missing to law enforcement each year, while another 500,000 children go missing without being reported to authorities. I was astonished to find recently that 1 million children, yes 1 million, go missing every year in the US and the UK. These are astounding statistics, every year a city the size of Amsterdam goes missing? Less than 200 of these are murdered so where are the rest? Its mind-boggling stuff, when you delve a little deeper, in the UK a child goes missing every 5 minutes, 2000 kids a day lost in the US! I suspect hat even these figures are suppressed. Getting excited about these figures still hides the overall truth and we will never get to the real numbers.

Belgium is a small country of some ten million people and yet in the Brussels region alone it was revealed that 1,300 minors disappeared between 1991 and 1996. That was only the reported cases and not those that went unreported which is the cases we will never know about. Around the world you find the same repeating story and multi-millions of children go missing every year never to be found. This is not even contained to a continent or a part of the world but is Global. There are children going missing in every country across the world. We dare not mention numbers when we are thinking about what is happening in countries such as Brazil, India, China, Russia and other countries where due to the geographical vastness the Governments just don't have any records of missing persons in place. The most remarkable issue is the absolute silence we experience about this matter. It's as if its not for real and we know it is. I can't confirm what I repeat from the Internet but if it is true it confirms a lot of my suspicions I have about this. Where do they go? It is millions of children in the end. The numbers alone just can't hide the crime but the apathy of all the Governments can apparently hide the numbers. It is another conspiracy kept in a cloak of silence. There is apparently a program or was to be a program on TV called Conspiracy of silence, which is / was a documentary listed for viewing in TV Guide Magazine and that was to be aired on the Discovery Channel, on May 3, 1994. I have seen on the Internet that this documentary exposed a network of religious leaders and Washington politicians who flew children to Washington D.C. for a sex orgy. I can't confirm this that I got from the Internet but my logic tells me not to deny this either. There is a show that was supposed to be aired about this disappearing of children. At the last minute before airing, unknown congressmen threatened the TV Cable industry with restrictive legislation if this documentary was aired. Almost immediately, unknown persons who had ordered all copies destroyed purchased the rights to the documentary...Only the most powerful on earth can fasten a lid on the matter this secure.

Nearly 30 million children and youth go online to research homework assignments and to learn about the world they live in. Research by the University of New Hampshire found that one in five children between the ages of 10 and 17 received a sexual solicitation over the Internet during any given year. One in thirty-three received an aggressive solicitation - a solicitor who asked to meet them somewhere; called them on the telephone; or sent them regular mail, money, or gifts. This is very difficult to comprehend, we are not talking of two or third world country's here, wouldn't you think the media would be all over this for instance? Or our respective governments constantly warning us but no, I only managed to stumble on to these unbelievable statistics on an obscure website. Belgian parents live in fear of paedophiles because it is very well known that in Belgium more than any other place these paedophile rings are almost openly active! When I hear about doubt or I hear people say that if child abuse, kidnap and murder were happening on the scale it is suggested to occur then it would not be able to be kept under wraps. The child kidnappings and murders in Belgium in the mid-1990s threatened to implicate the country's political establishment and other famous names. That power that democracy places in the hands of politicians is where the power is that control absolute silence while the banks control the political power by telling the politicians what the bankers wish to happen and the crime syndicates control the banks by telling the banks what it is that they wish to happen. Therefore democracy obeys the rich that controls then banks that orders the politicians that rule us. What happened in Belgium shows that the syndicates can control the people. This is the only explanation as to how they got away with it and pinned it all on a sick and pathetic paedophile and child-supplier called Marc Dutroux. It was the classic establishment response to danger that is repeated constantly across the world. As so many have said so many times, people generally don't get their views and opinions from researched information, but from an 'image', an 'impression', of how things are. By controlling the leaking and presenting of conspiracies then by controlling the conspiracy they control the information leaked as a conspiracy. With regard to missing children, this 'impression' is heavily influenced by the number of lost children stories they see in the media. The Belgians live in fear of having their children abducted, as the paedophile gangs operate unbothered by the Police. Once a child is abducted, they are never seen again. The Marc Dutroux case became famous, as he allowed girls to die in his cellars when he was arrested. But two girls escaped unharmed and provided many leads as to who Dutroux' clients would have been. But no inquiries were made. Belgians all have their theories as to who is involved, many suggesting paedophiles are high up in the government. In Brussels, kids are often photographed at play and abductions made to order. Why the Police don't do anything, or politicians keep silent is not known for sure, but you can use your imagination...a combination of threat and reward - in a country where money easily buys influence, it is not surprising. The strain of knowing so much evil and not being able to act to prevent it gets too much for many. Belgian Police are very prone to committing suicide. The Sun recently reported that Algarve is 'haven' for paedophiles. Why must society's parents tolerate these criminals scum living amongst us?

In Britain, Missing People -- a charity formerly called the National Missing Persons Helpline -- has tried to draw some degree of attention to these thousands upon thousands of missing cases. It is astonishing to witness the degree or actually the lack thereof to which the news media devote attention to vulnerable missing persons. One little undefended organisation is claiming that despite its efforts to generate news coverage for all missing persons cases, the news media themselves will cover only those a very modest few cases which the media claim that those cases fit their publications. The Media should eat this storey as if it was a ripe banana and yet they don't. Why would that be, why would they ignore the bulk of this storey. If one or two cases could generate such a massive response and earn that much advertising revenue, what prevents the media to go all out on a mission to bring justice to those that went missing.

Two cases of missing white woman syndrome are given as contrasting examples: the murder of Hannah Williams and the murder of Danielle Jones. Although both victims were white female teenagers, Jones received more coverage than Williams. It is suggested that this is because Jones was a middle-class schoolgirl, whilst Williams was from a working-class background with a stud in her nose and estranged parents. Media reports about the murder of Amanda Dowler, the murder of Sarah Payne, and the Soham murders as examples of "eminently newsworthy stories" about girls from "respectable" middle-class families and backgrounds whose parents used the news media effectively. These cases are controversially, that in contrast, to the street murder of gangland murders by youths on youths in drug related turf war incidents which receives little news coverage, with reports initially concentrating upon street crime levels and community policing, and largely ignoring the victim. The assertion is that "the near hysterical outpourings of anger and sadness that accompanied the deaths of Sarah, Milly, Holly, and Jessica" brings ratings to news events. The National Centre for Missing Adults has also commented on the phenomenon by saying "Unless it's a pretty girl aged 20 to 35, the media exposure is just not there. Mentioning these few names fall much short of the numbers of children disappearing pointing fingers to the press. Some 2,300 Americans are reported missing every day, including both adults and children.

But only a small proportion of those are stereotypical abductions or kidnappings by a stranger. For example, the federal government counted 840,279 missing persons cases in 2001. All but about 50,000 were juveniles, classified as anyone younger than 18. About half of the roughly 800,000 missing juvenile cases in 2001 involved runaways, and another 200,000 were classified as family abductions related to domestic or custody disputes. Only about 100 missing-child reports each year fit the profile of a stereotypical abduction by a stranger or vague acquaintance. Two-thirds of those victims are ages 12 to 17, and among those eight out of 10 are white females, according to a Justice Department study. Nearly 90 percent of the abductors are men, and they sexually assault their victims in half of the cases. To further complicate categorization of cases, the FBI designates some missing-person incidents—both adult and juvenile—that seem most dire as "endangered" or "involuntary." Kim Pasqualini, president of the National Centre for Missing Adults, said the media tends to focus on "damsels in distress" — typically, affluent young white women and teenagers. The media's dilemma is that government research shows that victims of non-family abductions and stereotypical kidnappings are most at risk of injury, sexual assault or death. "Damsel" cases may be the exception, but they often are the most urgent. These numbers are staggering and yet even so they are ultimately never mentioned because? Compare what is mentioned as cases to what the cases are mentioned and are targeted by the press to what disappears in reality. What makes the media so reluctant to tell to entire story and blast information out for months?

That is the media response of what happens to losses accruing in terrorist cases but the numbers of losses hardly match. In relevancy one could say for every two million children lost to prostitution there are five bomb blasts. During about the same time there was a huge bank robbery and the culprits were traced far and wide. They never stopped searching until everyone was caught and every note was accounted for. The leader was traced back to an Arab country (I forgot the name) and was brought back to face justice. He was chased until he was caught showing that criminal cases can be solved, except where children disappear. If one bank robbery is committed they get the thug but when children disappear who cares? If we are talking the national media there are very few cases reported in the light of what takes place. The enormous coverage of the missing British girl, Madeleine McCann, who was abducted while on holiday in Portugal, is a rare example compared with the number of children who disappear every year. While people get their impression of scale from the lost children featured in the media, staggering numbers of children go missing never to be seen again. I remember calling many American states a few years ago to ask for their missing children figures and it was truly extraordinary. On average, around 3,000 children a day are reported missing in the United States, never mind those the authorities never hear about. Add them together and you are talking hundreds of thousands of children.

Yet in that time it took to close the bank robbery case about 200 000 children under the age of sixteen went missing and the number that disappeared were never even reported! If there is another two major BIG bank robberies in the same period no stone is left untouched to solve the problem of banks robbed and "terrorist" attacks against "the people of Britain" but to lose a million children every decade or so is quite acceptable because the disappearing is part of a conspiracy and true conspiracies are never revealed or investigated. If a bomb blast occurs in London MI6 forms a special investigating unit but no Special-Force action is ever taken to bring this loss of children into the open. Losing 100 000 of the prime persons, the future of the country is a matter of discussion that is never discussed in polite conversations. It would be very embarrassing to ask the prime Minister or President of America what happens to 100 000 children under the age of sixteen in Britain alone that disappears every year. This applies to all countries. Which British politician ever made it a political pledge to search into this crime? No one mentions it because it is best wiped under the carpet. Most of the children are a product of broken homes and pupils of the unmentionable side of society and therefore no one cares to care for them. Their parents never vote because they are mostly the driftwood that never vote and the children are too young to vote so who cares? But when two medical doctor's child disappears and they know how to manipulate the press the world goes fanatic. The couple even found an audience with the Pope no less.

This has to do with money. The fact that 100 00 children disappears while it never forms a political debate alone tells the whole story. If Bankers make money Politicians have to hush about the Bankers' very rich and therefore very influential customers. The crime cartels tell Bankers to tell the Politicians to be quit. The crime bosses are in charge of the Bankers that are in charge of Politicians that make laws we obey.

According to a Portuguese lawyer that worked in the Madeleine McCann case, him being professionally experienced in working with several English clients, any parent being accused of abandoning the children to danger, is a crime under British law that is severely punished by UK laws. Yea, sure and the moon is made of cheese! That is so typical of a conspiracy. There are laws in place that are laws never attended to or adhered to and nobody ever obeys the laws because they are as good as being non-existent. Every law that is in place serves to protect the rich and the powerful by denying the poor and the defenceless any say in Government law forming. Yes they show democracy but they create mass hysteria instead.

The Members of Parliament put conspiracy-serving laws in place that serves nothing anyway. TV hosts such as David Frost claim their diligence and their courage and their tenacity. Why don't TV talk show hosts ask any politician about the children disappearing, because the TV station would not permit such questioning? Why do all members of the printing press or electronic media ignore the subject as if it does not exist? Why not attack all the Political Party leaders on a live debate or cross-question them? Why don't the press report every missing child in the same manner as they reported Madeleine McCann? Is it because there then will be no space left to report anything else or who stops them shouting about it?

These children just disappear from the face of the earth and that says there are mighty powerful money barons at work. Those that bought those that have the political power to power to decide who goes to war and die and who gets rich from the declaration of the war have the leverage in the social structure to allow 100 000 children in Britain alone to go missing and the public is none the wiser and could care even less about the matter. Those children disappearing are hard currency more valuable than money. Those children become assets as good as gold or property because their currency is set fast. I believe there is more dope bought with children earmarked for prostitution than there is money used as transaction payment. I believe a lot of oil is bought with pretty little blond haired girls that have gone missing.

Telling everybody about the McCann girl is not the conspiracy but not telling the world at the same time about the other 100 000 children that also disappeared becomes the conspiracy. It is not the information that is presented but it is the information withheld from the presentation that becomes the conspiracy. It is screaming from the mountaintops about one girl while pushing the other 100 000 that went missing under the carpet, that becomes the conspiracy. To defer the attention from the true problem they cry about one. Reading about such a pretty girl being abducted or going lost is sensational. It is something to cry your heart out about. Knowing about 100 000 children being abducted or going lost is a problem and while everyone is ready for the sensation no one wants to know about a bloody problem no one cares about.
The following is a joke coming from the Internet:
What's the difference between Madeleine McCann and Elvis?
The Answer: More people believe Elvis is still alive. This is no joke.

To hide this idea of Madeleine being dead is the centre part of the conspiracy, because only then can the McCann family be in the position to earn money through engaging sympathy.

A Portuguese Polices officer spearheading the investigation wrote a book stating his (and the official police) point of view on the case and tells about why they think there is no more evidence to research since they have the opinion that the girl is already dead. He said Madeleine is deceased and gives his reasons in a book. The McCann family got a court interdict stopping his book being published because they want to stop the idea spreading that Madeleine is dead. Why would they spend hundreds of thousands of Madeleine's money to prevent a detective from putting to print his views on the case?

It is about money as every aspect of this conspiracy is about earning money. A dead Madeleine would not enlist donations and evoke the media ratings that the story generate and the income of the entire enterprise will plummet if the girl's death is accepted by the public at large, while searching for a live girl brings in millions and to hell with free speech and free opinions when money is the issue. That is why the papers that printed the fact that Madeleine is dead was sewed and they paid up. To hell with free speech because the papers saw their earning in revenue go down the toilet by allowing the public to think Madeleine is dead. Admitting to Madeleine's death will kill the money flowing in and that is horrific.

Even the money the McCann family and all other people fight to accumulate is a conspiracy. One group of person's thought of as Bankers bought from crooked politicians worldwide the privilege to print paper and give the paper a value. In the system I take the paper they printed and then I have to regard that paper as having more worth than say my home has because I "sell" my home by detaching my ownership from me and attach my own worth to the printed-paper. Then I am very impressed with myself afterwards in my effort to exchange what has visible and useful worth for some paper they tell me I have to accept the worth but has no worth but to give it back to bankers to put in their vaults. This is insane stupidity and every person on earth goes along with the madness. The Bankers decide the worth of the worthless they print at about no cost to them and then by creating a system I am forced to accept their paper and the worth they attach to it or ells I starve and die. Even committing to dying requires money to accomplish the process. If you come to the end of your life you have to pay for the privilege to die and go on.

The Bankers take their tax or share long before the Government can but they call it bank fees. People are so brainwashed and beaten to a pulp by systematic control of the mind and their thoughts that they fall into the practise of doomed slavery without trying to fight for freedom. In the days of the Romans and the Greeks slaves were paid 10 % of what their Masters earned from their services while the Masters still had to feed, cloth and shelter them at the cost of the Masters. We all are slaves to the bankers but we earn about 10% of the cut they take.

While the Mammonites pay us 10 % of what we earn from what they earn from or services, we have to cloth ourselves, feed ourselves and our children while we purchase houses form the Mammonites and then find that behind successful Mammonites there are Bankers pulling strings by supplying worthless printed paper we accept as the commodity we will work all our lives to accumulate and possess. In the end we can't take with us anything we ever wanted on earth because it is worthless.

From every angle this case presents including the Pope's visit all aspects involve money and it is about publicity that will entice donation of money leaving very little scope for a girl being found. Why don't McCann donate some of the money they have to searching for all children that has gone missing in the time Madeleine went missing? Why don't they also include as many photographs of children that went missing during the time that Madeleine went missing? Because then it would not bring about the money that it makes when only one very special little girl is in the hands of a child molester…but 100 000 children just vanishing slips the minds of every "concerned" do-gooder in England while they donate money to this deserving quest just to keep their minds away from the 100 000 others. This tendency to evade the truth is a sickness that runs through all aspects of society because this disease is what we use as education. I am going to show how teachers in science **brainwash** students by forceful **mind control**. If science cannot prove God's existence, it is not God that does not exist, but it is science failing and therefore it is then that specific view about science that should be re-examined since it is the view on science that is proving as being incorrect. This fact is what the so very brilliant and intellectually mindful Newtonian atheist should remember when they fail in their science altogether. That their science fails altogether and that failing it does in all its splendour, is facts I am delighted to prove! The fact is Newton's views were never tested and that the Newtonian views on science were never challenged before and because of that Newton principles never withstood diligent scrutiny before.

When Sir Isaac Newton is investigated even in the flimsiest of manners, well accepted facts seems to become very suspect, to say the least. This becomes evident when concluding all the facts this book presents. Now, in this book, for the first time Newton is tested and such testing is the proof you gain by reading that which I uncover. What I bring into the open is unseen facts, which I present you with as I take you on a tour through an avenue of facts I introduce in this work. The lack there is in sensibility concerning Sir Isaac Newton's principles this book proves. The theories of Sir Isaac Newton require proof, which was never given while God never needs proof and that is what science constantly seeks. When science perpetually ignored my concerned calling on and ignored my calling on them because (I suppose) they were finding my concerns wanting, in my final letter to them I promised them never to contact them personally again by any and by all means. I also promised them a fight. This is the fight I promised. I was not worth noticing so I was ignored. I now am calling on the public, as I am ignoring their reputations. I am showing the public just how extremely bright the Newtonian inspired super-thinkers are!

Scientists portray the image that they know all there is to know about everything man might ever know and nobody can ever know more than what they currently know. They tell you what to eat, what to drink, what to think and how to live because they have this image that mentally they are on par with God. They are the utmost superiors on all levels of what forms creation. This attitude applies to Rocket scientists as much as it applies to medical doctors as much as it applies to lawmen. This comes through in all departments albeit language, art, science, law or whatever you may have. If you smoke cigarettes then you will die young because the medical profession found that it is harmful to you while a hundred years ago people smoked ten times what they smoke today and they lived just as long as they live today. But the medical profession took it on them to act as God and force-feed everyone to do what they say or die. The truth is the oxygen you can't live without is burning you to death by aging you and without that you die. If you don't breath oxygen you die and if you breathe oxygen you die because you are born to die! In the end oxygen kills more people than any bad habit or disease because it slowly kills everything with life. This acting like God envelops all forms of science. What is the truth? Do we hear the entire truth? The truth is that science only reveals some portion of what they know and ignores what is there that they know they don't know. They only reveal what suits their position and never divulge what they know but what does not compliment their view. When science confronts religion they have the opinion that what is in science is everything there is and there can never be more than what science knows or what science wishes to reveal. Science now knows everything knowledgeable and whatever will be known they know.

If some scientists are of the opinion that we will fry in boiling water in the next century then it is the Biblical truth because science holds the opinion. Science knows best! Today we laugh at medical practising of a century or two ago and in another century we know the future generations will laugh their heads off when listening to what the informed opinions are of the professionals today. Science has forever veneered their status with this blanket of "they know all". This makes a mockery of the truth because science has no clue why man die or why man age and yet they promise eternal life within the next few decades to come. Ask scientists what is life and they will have an "informed opinion" because they know everything there is.

Science keep up this front that they know everything there is to know while even reading my books prove how little they know about science. They withhold every aspect in science that they do not know and only elaborate in detail that which they think they know about and what we presume they know. They are of opinion that there can be no other way that creation started but according to their science. If the Bible describes how events unfold it then are incorrect because science knows everything. In **The Veracity of Gravity** I show scientifically by using science how creation started precisely as the Bible says word for word but then I also show how little science currently knows about science. The book not in print yet is **<u>An open letter Addressing Gravity's Formula</u>**, which is far more elaborate on the matter of how creation started where I show how science proves the Bible correct. How shocking this might be it is just as true.

Science can never take the blame for not knowing. Never is a suggestion put forward that it might be science that holds the shortfall and it is because of science not being adequate that science cannot match the Bible. I can prove how the start came about because I decoded gravity and I did that by finding an explanation about the four cosmic principles. By deciphering the Roche limit, the Lagrangian points, the Titius Bode law and the Coanda effect I am able to show how the very first instant happened when the Universe started the very first point ever formed. These principles are in place and not the principles Newton fabricated... That this book shows. It shows that the cosmos uses other principles than what the Newtonian science promotes. What science says nature uses is not in place or does not hold evidence while what nature does use science deny by just never pressing the issue. I show what is in place and I show why it is in place but first I have to reject what science says is in place because it is not in place.

I show how Newtonians fabricate Newton's ideas about gravity. This is ongoing since the end of the dark ages and Newton. There is no mass that can pull. Most people reading this and who are schooled in physics never heard of the Roche limit, the Lagrangian points, the Titius Bode law and the Coanda effect and these principles are what builds the Universe while I am going to show that there is no factor such as mass. While it serves their purpose notwithstanding never finding evidence to the fact, still science uses *only* and *exclusively* Newton's idea of mass while the principles in place the Roche limit, the Lagrangian points, the Titius Bode law and the Coanda effect are never ever mentioned. They sometimes put referring to these principles as law in brackets to deny the status that any of the above law have.

I am going to show you within the next few pages the silliness Newtonian principles hold. While I discuss the principles please see where I am incorrect or going wrong and convince yourself whom is wrong.

This is because the Roche limit, the Lagrangian points, the Titius Bode law and the Coanda effect disputes Newton and science would rather discard what the Universe uses than to put a question mark behind the fabrication Newton put in place. Where everyone knows the fabricated information and hiding the reality, which is in place within the cosmos, and that is the conspiracy I show to all. Science stupidity ensures they don't understand the working principles that are in place and that was known for centuries in some cases as the Roche limit, the Lagrangian points, the Titius Bode law and the Coanda effect and therefore not knowing how the principles should be interpreted they hide the concept due to not want to be seen as the ignorant fools knowing the cosmos implements the principles as reality. Science hides their limitations and incompetence behind providing the public selective of information. Take for instance the edge of the Universe they talk so much about. There is no edge of the Universe because there is only an unlimited everlasting Universe out there. What the limits are that they see as the edge of the Universe is the limitation of their equipment that can't trace time back beyond what they see and that serves as their limit in understanding what the Universe offers and how the Universe unfolds.

Trying always be perceives as matching the likes of God science can't face the fact that they can't precede further into space by reading time than what their limitations and their equipment handicap them with, then they put their shortcomings onto the Universe having limits so that they can present the image of total superiority in contrast to limiting the Universe. If they do not understand the four principles in the cosmos and which the cosmos uses as building blocks how can they understand how the cosmos works?

Then because they are clueless about the information the Universe provide and because they use misinformed cosmic principles they're confusion gives the Universe an edge where it ends and a date when it started while they admit and we all know that the Universe is timeless and limitless in every aspect we encounter. One thing the Universe des not have is an end because I show where the cosmos holds infinity, the point that can never go smaller and where the cosmos holds eternity, the point that can never stop becoming bigger because it is endless.

When I say there is no such a thing as mass everyone goes quiet and I can hear they immediately question my mental stability. One may question the fact that there is a God or not and one may question which God is the true God and one may doubt any form of God existing but doubting mass or saying mass does not exist is utter irresponsible madness. Everyone knows there is mass. That is one thing no one doubts. The fact of science showing that mass pulls everything down is a real as being alive, or is it?

Well…I wish to bring to mind some of the facts that physics work with when academics as scientists only work with facts. Remember they are the ones boasting that if facts are not proven then it is fables and those very important academics don't waste time with fables because they only work with facts. The accuracy of their basis on which science rests is that mass is responsible for gravity by pulling. If you don't have mass you're not going to have gravity. Mass is equal to gravity and gravity applies only by mass. If mass is present then its by gravity or otherwise gravity is absent. If a body falls it is the mass that pulls the body to fall because the body receives gravity by ratio of mass and mass is that which produces gravity in relation to the mass available.

It is mass that drags you down because the mass is in charge of the gravity and the gravity finds the value from the mass available. Mass pushes you down by the gravity it forces onto you. But if mass drags you down then what lift you up in the air balloon? If mass gives the gravity to drag you onto the Earth then why would the hot air lift you up in a balloon? Is the hot air causing anti gravity or anti mass because gravity by lifting the air balloon and cargo. The hot air balloon is lifting the passenger and all that is in the bag plus the bag plus the balloon into the air. So what is then pushing the lot up if it is mass that drags you down. Has the air not got mass because then the air can't have gravity and then the air must escape into the blackness of outer space because by going up it shows a resilience of either mass or gravity. We have seen that it is mass that pulls everything onto the ground.

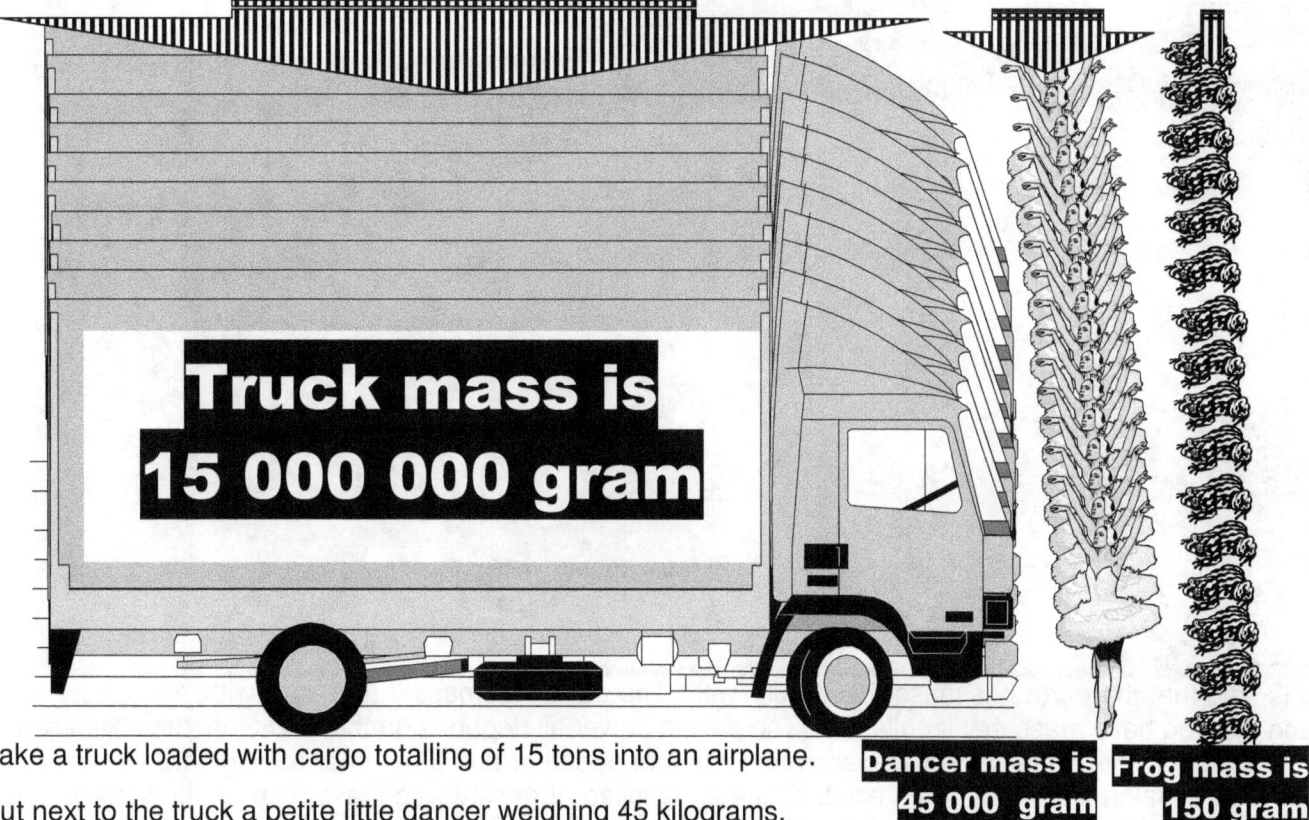

Take a truck loaded with cargo totalling of 15 tons into an airplane.

Put next to the truck a petite little dancer weighing 45 kilograms.

Then to keep the dancer on her toes, put a frog of 150 g next to her.

Now we will have the mass of each object "pull" by "gravity" as this lot falls down
Take note that you are told by the wise amongst us that it is mass that produces the gravity that pulls you down. We have just had a lovely debate on how it works and how mass drags you down and wondered if it then is anti mass or anti gravity that lifts you up with the hot air balloon, well take note of this as your airplane reaches 11 thousand meters which is eleven kilometres straight up into the air. Now you throw

this lot out without parachutes and let the lot free fall in accordance with each object enjoying its individual mass forming its own gravity. If Newton is correct this lot will fall according to gravity and if Galileo is correct this lot will fall together. But never can Newton and Galileo have the same opinion of the outcome, as they are not sharing an issue with the same outcome because by falling with mass the lot must fall different and we know they do not. If falling is according to mass then it is ridiculous to think the lot would arrive on the ground simultaneously. If the falling is equal Galileo is correct and if every one lands in ratio of individual mass then Newton seems to be correct. Judge on what you see and not what you are told by physicists upholding Newton.

Why would the air defy mass and allow the balloon to go anti whatever is going anti because nothing in or on the balloon has gone without mass. The object(s) has mass to produce gravity. Why then would hot air allow the balloon plus everything in the balloon to lift into the air? The balloon lifts in relation to the hot air that blows into the sack. The more hot air and the hotter the air is the more lift and the swifter the lift will be that the balloon provides. The issue sticking out is that the balloon then must not have mass because with anti gravity it is pulling up. Remember mass drags you down and mass can't pull you up and drag you down at the same time. Then what is pushing while mass is pulling or is mass pushing while what is pulling? The object is not going in the normal direction where it is dragged down by mass forming gravity and in all my life I have never heard one Academic mention anything about gravity lifting and that makes the lot very confusing. What is lifting up when the lot should be pushing down and why did everything connected to the balloon lose the mass and if it has mass why is it not dragging down the balloon?

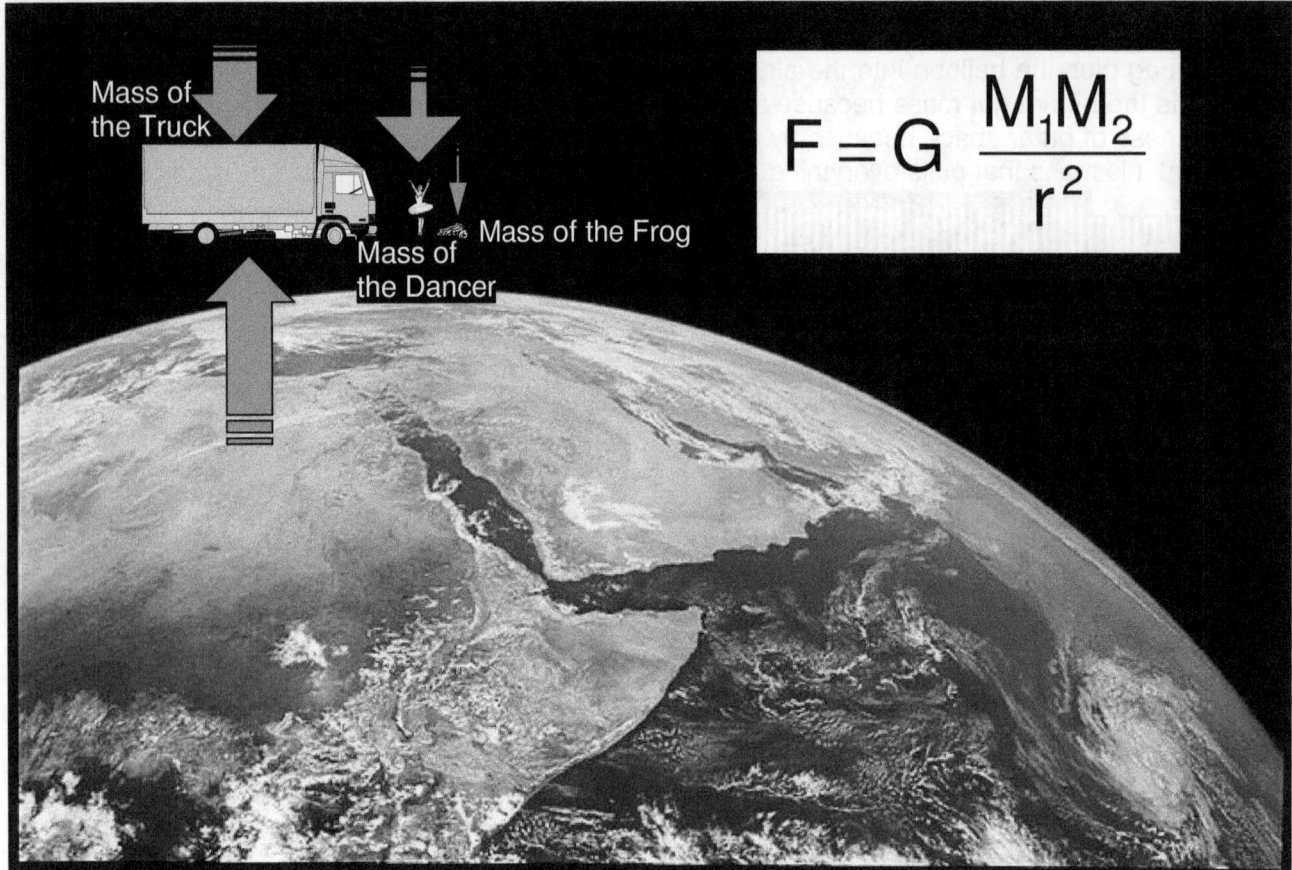

It is said that the earth has mass that it pulls with. The truck has mass that it pulls with. So does the girl and the frog have mass that it pulls with. The pulling power all depends on the mass that the object holds, well that is according to Newton. Science teaches that a feather and a hammer have different mass while they fall equal in time through an equal distance travelled. If gravity was mass related, then this was not possible, because then objects must fall according to mass. Reality says if mass plays a part in the "pulling" that forms gravity, then there has to be distinction in the speed when large and small objects fall and the time it takes to fall. The period large objects take to fall must be longer than small objects take, that is if mass did play such a fundamental part in the forming of gravity.

That fact about Galileo, science does embrace, although this strongly contradicts Newton's impressions about mass inflicting gravity. On TV we see how all objects, such as cars, humans and bags fall at the same pace, which sets a standard totally against Newtonian mass principles that produce the falling, and proves Newton wrong because mass then does not underwrite gravity in any way or form at all. The formula $F = \frac{M_1 M_2}{r^2} G$ would suggest mass taking all the responsibility for such falling that takes place. Newtonians declare gravity as the force of gravity F, that is = equal to gravitational constant G, when it is multiplied by the mass M_1 and the mass M_2 after which then the product of the three factors influencing gravity is divided by the square r^2 distance between mass pulling the mass that destroys the distance between the two objects. If mass pulls mass as Newton said, the Big Bang is not possible but the Universe is notwithstanding Newton's claims, expanding (growing apart). If mass forms gravity, every planet must orbit at a different pace, which they do not, as all planets orbit at the same pace around the Sun.

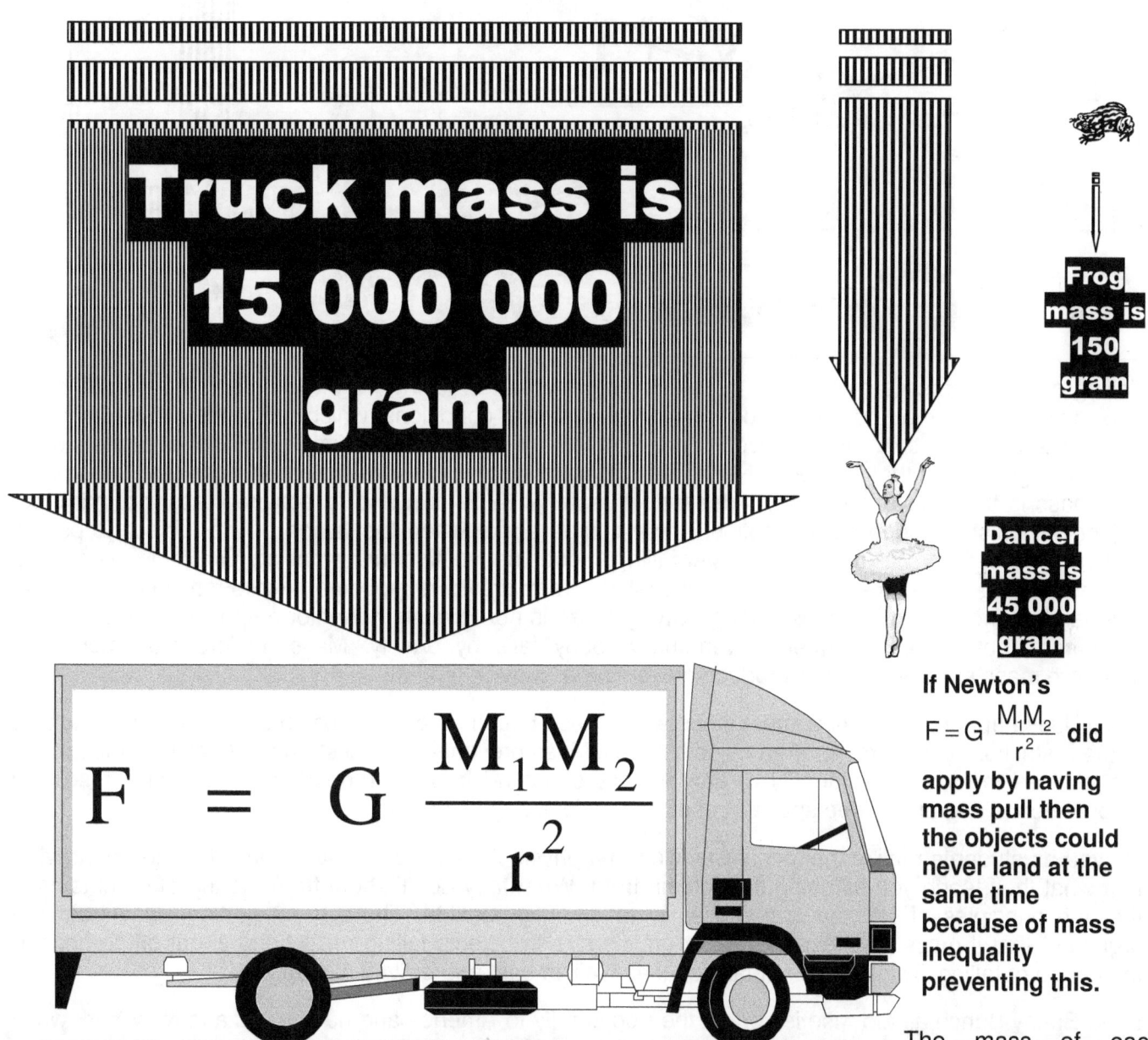

If Newton's $F = G \frac{M_1 M_2}{r^2}$ did apply by having mass pull then the objects could never land at the same time because of mass inequality preventing this.

The mass of each falling object amongst others supposedly provides the gravity that pulls the truck down should pull the truck down one million times better than it pulls the frog down. If the mass is doing the pulling by forming the gravity the truck must fall 333 times faster than the girl and one million times faster than the frog. If Newton is correct the frog must land on the earth hours after the truck landed and minutes after the girl. In reality the frog and the dancer and the truck fall at an equal descending rate. That is what Galileo said all along. That is not what Newton said. On the way down the frog then pretends he drives the truck and the next scene he is dancing with the girl while the truck is falling as fast as a truck can fall and they all fall together as if they are in a unit sharing space. Does each then fall according to their mass as Newton says? Who do you think is lying? Physics are telling students all over the world that mass is in charge of gravity and it is by the force that mass forms that the bodies are pulled down. Then the mass is pulling the truck of 15 tons down since the mass produces the gravity and the gravity produces the fall which is three hundred and thirty three times more in the case of the truck going in a down direction than the girl's mass of 45 kg is pulling the dancer down while the frog of 150 g is descending as if it also is 15 tons as it falls beside the truck. This point is crucial to what you are going to read about gravity in the rest of the book!

Falling objects bears no evidence of mass playing any part in falling. Any two objects holding different mass fall equal in time and in distance when sharing similar conditions, which suspends mass altogether as an influencing factor. Galileo proved different mass fall equally under similar conditions. Planets don't give the slightest hint that they obey Newton's suggested cosmic laws by implementing mass. The truth is that mass is the resistance of any independent material to deform and to acquire mass the individual object relinquishes independent motion. Mass comes about when the falling of any object stops the motion of the falling. Mass prevents further falling, it does not sustain further falling. Gravity is the moving of the object to the centre of the Earth while falling. I say gravity is movement while mass is obstructing independent movement, which is what gravity is. Mass is not forming the factor responsible for gravity or movement, but prevents further movement. A body falls by gravity. Mass hinders movement and therefore mass can't enhance or produce movement or gravity.

What does happen is that movement stops when mass becomes a factor because mass is the indicator of the restriction of movement when objects do fall. Mass prevents or blocks gravity. Gravity is the motion that defines the individual identity of any object's structural form by rendering motion while reserving independence in granting free space from other manipulating objects.

Do you as students realize the inconsistencies that physic Academics present you with when portraying that what they teach you as being the solemn truth. What they don't' shout from the top of mountains is that the principles of Newton and Galileo is totally incompatible. There is no getting away from the argument that if mass was responsible for gravity size in objects falling must bring about differences in the tempo of falling. Thing could not fall equal as Galileo said and proved it does.

If the Brainy Bunch all too wise is correct the frog can fly to America and have a pizza in New York while the truck has a few micro seconds to get down if the girl is going to fall during the normal falling duration of a minute or so. Everyone has seen skydivers jump out of airplanes next to cars and trucks and bags. Every one has seen they all fall at the same rate as the big and small objects. They can hop in and out of cars while floating next to the downwards descending car and the bags that fall with them. The girl can do tap dancing around a jumping frog on top of the truck or below the truck and they can be inside the back of the truck galloping on fresh air inside the truck because the lot is falling at the exact same rate.

The academics wishes to brainwash you by mind control in accepting that it is the mass that the falling takes place and that mass is responsible for the gravity and by mass pulling you down it is gravity that makes you fall. Where is the proof of mass that according to them is that which is producing gravity. They tell you Galileo said all things fall equal and we can see from the TV monitors how all things fall equal. Where is the mass that makes the gravity to let you fall if all things fall equally? They tell you that the truck has a mass of 15 tons and that mass is making the gravity that is having the truck fall while the truck is falling at the same speed and distance than the frog does.

Then if you don't repeat after them and echo every word test after test and exam after exam they will fail your papers and kick you from campus. You repeat after them and you live an academic life or you disagree and you go home to play with your toes. If mass is in the picture then mass must be represented by a factor of more than just one because if mass is not part of the overall picture then mass has a factor of one which proves that mass is not part of the equation since mass can't change the results. With all the objects falling equal mass has no role and if mass has no role then for my money academics in physics can't just go and put everything in as their hearts desire. If it is Galileo that is correct and if all things fall equal then mass has no part in gravity. If mass is the inspiration behind gravity the truck must fall a million times faster than the frog and in fact the frog should almost land in another country because that is how slow it falls

Telling this to your physics professor will have him shrug his shoulders dismissively as if it is not his problem and raising the issue is slightly out of place and rather a silly point to make. That is how they have been eluding the problem now for three hundred years. They put the mistake onto you for noting it.

The fact of the matter is that I don't wish to be near when any of this lot hits the ground because the truck will cause a quarry and the dancer will be a splash of red fluid while the frog might not be that worse for wear if the truck or the dancer doesn't land on the frog. The differentiation of having mass or having equality when falling and then not having mass and between individual differences in mass by each component that enters the equation when the objects touch the ground. Then every one gets the mass it has. Only when they touch the ground and land on the soil is mass as a factor awarded. While they fall they all fall equal and there is no distinction between the falling at all. What then is gravity? The gravity is the falling. The gravity is the motion. The tendency to move and apply gravity is the part that the mass restrains. The factor of weight only comes into play once the independent movement stops and the object becomes a part of the earth in all aspects of movement. When it moves with the earth it has mass. The soon as the object hits the ground the object stops falling with the mass that manifests as weight that then prevents the falling from continuing. It is the role of forming identifiable mass is to prevent further falling and independent motion to continue. Some might even still honestly believe it is mass that produces gravity due to the brainwashing as they were taught it is mass that produces gravity and never thought about the matter again afterwards. Take this issue up with your professor.

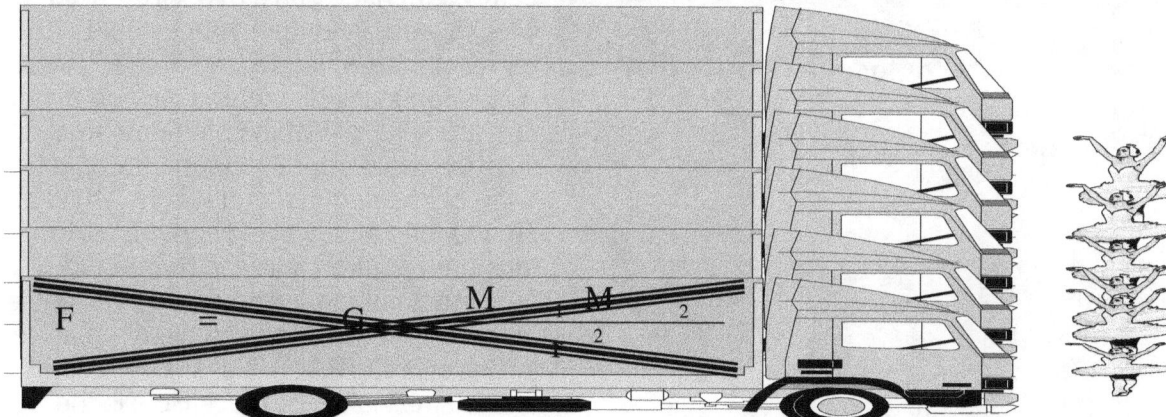

The falling object experienced no mass while falling therefore the falling or moving must be gravity's contribution. While objects are in motion those moving objects is experiencing gravity. The object show mass when the object has a tendency to move but the motion towards the centre of the Earth no longer takes place. That means mass is the restraining of the motion or is that which prevents the motion or gravity taking place. On Earth, objects experiences mass by restricting gravity or motion with the Earth giving mass but taking away free motion. By giving mass the Earth forces the object to become one with the Earth and move with the Earth as a pat of the Earth.

In the time of Newton everyone thought that mass represented size and the bigger the star the heavier the star or more "massive" was the star. This turns out to be a specially nonsense because the heavier the star is by measure of forming gravity then the smaller the star appears to be seen from our perspective. Acknowledging the Hertzsprung-Russell diagram of star classification proves how back of beyond science still is in their perception of cosmology.

Chapter 3 LIVING THE LIE.

Science never thought that a star can get cold and a supernova is the result of a cold star overheating and then bursting into liquid. No they simply tell all that would listen that a supernova is the result of gravity "going mad" and that is the best explanation in science they can come up with. If gravity depended on mass then a Supernova explosion was out of the question because in such as tar the mass cannot increase since the mass is at a steady state with in the structure and nothing adds before the supernova event. But when something is cold because of spinning and the spinning reduces to a point where the coldness or gravity can't bind the inside any longer than the inside will overheat and when anything overheats it expands. The star will by overheating burst or go supernova and the bursting of layers we all the pretty picture that we see because every layer in a star represents a different heat band in the star.

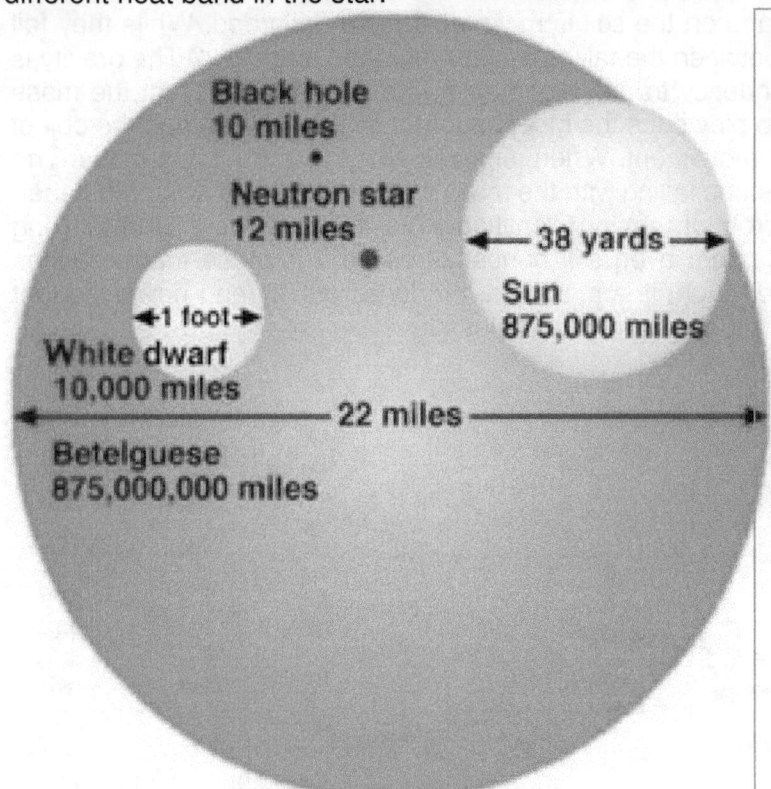

Massiveness brings gravity and the bigger that star the stronger is the gravity.
It is put forward that size brings about mass according to the Hertzsprung-Russell diagram of star classification. This shows big stars produce heavy mass and is as outdated as all Newtonian principles. If Betelgeuse the biggest star's diameter of 875 x10^6 miles was represented by 22miles then the sun's 875 x 10^3 miles was 38 yards and then 1 foot would represent the white dwarfs 10 x10^3 miles. A Neutron star's diameter would be 3 mm and a Black Hole would be .5 mm true size. On the earth 100 lb weight would be 1 ton on the sun and that weight would be 10 000 tons on a white dwarf, 10 billion tons on a neutron star and that would become 30 billion tons inside a Black hole. In the Universe there is no big or small and everything that is big is small and all that is small is big. A Black Hole is the smallest there is and the biggest too.

In a later stage I use Kepler to show why stars are big and small at the same time.

They were brainwashed by their tutors as their tutors were brainwashed before them. You are in a position where you can teach your tutors the truth about gravity if you read what is in the books. The truth is there and the truth is out and the truth will be because the truth is written for all that wishes to read. The academics on the other hand have ignored my work and my being on Earth for the past six years while I was writing them letters about gravity. They ignore me as if I am a rattlesnake because to them I am a rattlesnake because everything I introduce becomes venom to their prehistoric out of date perceptions.

If we take a sincere look at massive stars and find the massive stars grow smaller in size as the stars grow bigger in gravitational application, the gravitational increase of the stars must be the result massive increase in spinning motion. The faster the star would spin the bigger the gravity would be. The star exerts gravity because the movement cools the star down and by movement the star becomes colder and by becoming colder it compresses the star as well as whatever surrounds it. Thus when a body is subjected to the gravity of the star it is the thrust by which the star slings the body onto the surface of the object and that increases the weight the body will have but by the same margin the body will reduce in size because it is increase in movement bring more thrust that reduces the apparent size the body holds. All the atoms would be more compressed and by being more compressed the body will have more weight and thus be smaller while weighing more.

Cosmology became a vacuum cleaner that kept on working but never came to reject incorrect views that prevailed in the pass but has since turned out to be wasteful thinking. Science kept hanging onto every idea that came from whenever Newton thought a steam engine was the epitome of what science would bring about as the ultimate invention. In Newton's time the steam engine was the nuclear technology of the day and even to this day they have stars "burning fuel" just as the steam engines did three hundred years ago and then had the stars "die" because the stars "ran out of fuel" just as the steam locomotives did back then. This perception was linked to the highest science development that man had reached in those days. Even today science cling onto this idea, which is totally rubbish because culture drags it along, and science is stuck with it. A star is not a gas on the inside but it is a liquid and a star freezes outer space from a gas to a liquid by movement. The star cools down to form a cosmic liquid because the star moves so much. By moving it freezes space into a liquid and that I explain in later chapters.

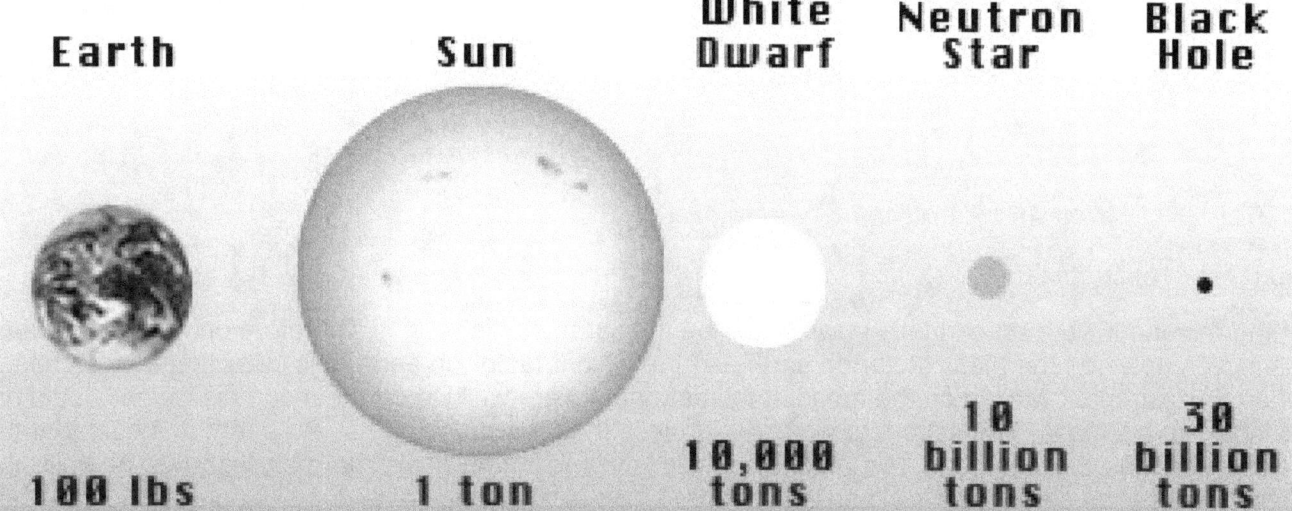

For that reason if no other they will rather go on lying to you and cover their corrupt fraud than face up to the truth and admit their work is lost. The truth will be whether it is recognised by them and they can become the first to admit and repent or they will be the last of the laughing stock that those in the future will refer to as the bunch that couldn't see when things fall equal they cannot have mass and when things do not fall by mass then one can know mass has nothing to do with the falling and the gravity.

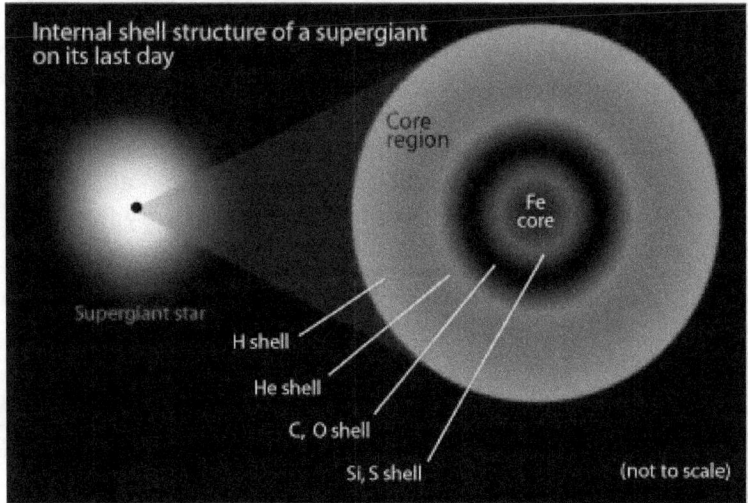

The Nobel prize was awarded on two occasions to person that came up with the novelty that after 1.3 times the mass of the sun and the other person was 1.6 times the mass of the sun the atom collapses onto itself and then form a Black Hole. It is presumed that all the atoms within the star collapses onto itself and this forms a Black Hole. Now what part of the atom will collapse onto what part? Is the electron falling into and through the neutron or is the neutron destroying the proton and in each case would that not lead to fusion where pressure forms the next atomic proton accumulating value. How does an atom collapse on itself? Is it because the atoms on the outside is too heavy for the atoms on the inside because we have to remember that the atom on the inside can carry many times more weight than the atoms on the outside have. Doe the helium, or hydrogen annihilate the iron core inside the star? Does the star go from being matter to becoming anti-matter because no one ever explained what the hell anti matter is except give anti-matter a name and make a hell of a story about more senseless fiction.

This astronaut has "mass" and the earth also in the picture has "mass" and the pulling should be going on since there is no restraining that would prevent the man from falling to the earth so why is Newton's "mass" not applying? This astronaut is circling the earth at a certain speed notwithstanding "whatever mass" science contributes to the pilot. If "mass" was pulling then why is the man not falling. The man will fall depending on the speed by which the man rotates the earth. His spacecraft (not in the picture) has about a hundred times more mass that should entice about a hundred times more pulling but both float above the earth at a specific pre-calculated speed. It is the speed that determines the distance of circling and not "mass".

$$F = G \frac{M_1 M_2}{r^2}$$

If the astronaut were on earth his mass would be the same as his weight. Therefore on earth there is no distinction between the mass factor on earth and the weight factor on earth. It is clear that things change when this astronaut walked on the moon. If the astronaut were walking on the moon his weight would not be equal to his mass. The astronaut would weigh less than the mass he has on earth. It is said that the gravity changes the numbers. The moon has less mass and that gives the man less weight while the mass remains the same. Say the man weighed 90 kg on earth he would have mass equal to 90kg. When the same man walks on the moon he would weigh 30 kg while having a mass of 90 kg being on the moon. On the moon the machine may have the same weight as the man on earth but the mass of the machine stay what it was on earth and so does the mass of the man. That becomes science in fictional perspective. If the man could walk on the sun his weight would become a thousand times more but the mass of the man remains 90 kg. If I believe that bit then somewhere I am fooled just because I am that stupid and I deserve to be fooled that easy. Then when the astronaut is in space he still has a mass of 90 kg but when walking in air the man has no weight with a mass of 90 kg. When reading on, just be clear about the "having weight" part that I believe and the having mass that is very fictitious and manufactured.

I have no qualm with having weight because that you have or you have not depending on the reality of where you are. My problem is with having mass and that is a fictional representation of what Newtonian wishes ask for at that minute. It is this mass constant where-ever-you-go-fit-all scenario that seems to be something similar to a rule that is made up-as-you-go. Something much like New Zealand Rugby Rules applying to them in a game and can change in every game as they go along but others playing have fixed rules applying.

That there is a conspiracy in science present and very, very active, about that fact there can be no doubt. Dodging the truth runs as deep as it can go and involves every person that studied science, worked with science, works in science and those teaching science. Some know they conspire and others conspire unwittingly. Everyone in science conspires to keep up untested and corrupted views to promote the status quo. Everyone tries to elude the truth by conspiring not to address the issue and this even applies to Albert Einstein in his quest for answers. Einstein let the establishment dictate although his calculation could never substantiate any of Newton's claims and it was only Newton's claims Einstein could not collaborate.

In 1907 Einstein experienced what he called "The happiest day of my life" as he described that day. For weeks he wrestled with a daunting question that from all the natural laws, and concluded that it was only Newton's law on gravity his special theory on relativity couldn't explain. There he missed the conspiracy! If the Newton gravitational principle theory was the only part of science that did not fit into his relativity concept, then it was what is incorrect.

He said he was sitting in his office in the department of patents in Bern and suddenly he had a thought. He realised that when someone fell freely through air the person would not feel his own weight. The simple thought made a massive impression on the young Einstein. In his mind he saw someone falling and he asked himself what happened to the law of gravity in the fall? If other object would fall with him, the other objects would be in a complete state of total rest.

When bodies fall, they fall as if the bodies in mass are equal because all things fall together. That is the result of complete equality in mass during the fall. Only by having equal mass while falling would they fall equal and this takes mass out of consideration of attracting.

Every object on earth moves at a speed of 30 km / sec. around the sun. The sun takes the solar system that includes everything in it around the Milky Way at 300 /sec. Where everyone is at present the absolute speed can be determined because everything on earth moves around the earth's axis by spin that moves around the sun's axis that moves around the Milky Way.

Everything that moves have to move in a line while moving in a circle where the circle becomes a straight line that circles around something again. That is gravity where gravity is the relative movement and not the pulling of mass. That proves it is not the objects that fall but it is the space the objects hold that condenses because the space the object holds give equal density as they have equal buoyancy.

On this principle I founded my concept of gravity where the Earth by turning around its axis forms Π and only that spin contracts to liquefy the space around the earth and no mass pulls mass because with mass other bodies will distinctly fall different. If it fell according to mass things will not fall equal which means mass does not pull. Read on to learn about the conspiracy in science where science protects and will even change laws of nature to protect Newton's ideas.

When any one read the books I offer you will see how science always pushed in incorrectness of Newton over to the cosmos. It is the Universe that has insufficient mass to form the critical density in the critical density theory and needs dark matter. According to the conspiracy it is never Newton that is wrong in the assumption of contraction, but the Universe is wrong to expand. The Universe must start to contract and not Newton that needs a review.

Newton's law of physics state that: This Newtonian principle declares that there is a force of attraction between bodies forming gravity… It forms attraction by the value of the mass of both solar objects that holds the gravitational force…
The gravitational attraction that is forming is to the value of the mass which forms the gravity, therefore the pulling results in the reducing of the radius between the bodies by the square…. … the force of gravity therefore is reducing the radius running both ways and therefore reducing by the square thereof…*and then there are those that say I don't "understand Newton"*
…Therefore this must result in the moon and the earth reducing the distance between the two by the square thereof to the value of the mass both objects hold by coming closer by gravity.

This must then lead to an **inevitable collision** that must end the existence of both the earth as well as the moon. Let the wise calculate when is this event due and inform the human race when our final day of doom will arrive by using Newton's law of attraction by mass. $F = G \dfrac{M_1 M_2}{r^2}$

Science says mass pulls mass. Then why is this astronaut floating in space? He still has mass and the earth still has mass and the mass must be pulling to form gravity. The man is floating on the same principle that the moon is floating. If the man is not going to fall, the moon and earth will never collide and that makes Newton's mass pulling mass a contradiction that makes science a joke. If Newton's "mass is pulling mass" is correct, then the moon and the earth must collide. So which is it because it can't be both? The man has mass just like the moon has mass and the earth still has the mass it always has and the law of attraction between the human floating above the earth must be the same as the moon floating above the earth. If the man could float then this law of attraction is rather becoming suspicious. If that is the way you think then having those suspicions are very correct. Science confirms that there is always a force of attraction between all bodies. There must therefore be a force of attraction between the astronaut and the earth in the picture above, so why isn't it falling straight down in like with Newton's apple? The mass has gone nowhere but science cheats the answer by giving the man "micro gravity". This way they cheat by avoiding the truth without pulling a face.

Ask him or her to say without doubt which object with most mass will land first or will the lot land at the same instant? If the cargo haul of the truck was empty, will it fall slower or will it fall faster when the cargo bay is full and there is more mass doing the pulling? Get some truth out of them because they are there to teach you the truth and not the truth Newtonians create. By telling you what they wish to tell and forcing you to repeat that in examination only forms brainwashing and that is the basis on which science forms the basis for centuries.

The following forms the backbone behind the brainwashing in teaching. If I come to you as a mentor with a proposal about something I wish to educate you about with you on condition that you pay me an amount to share my past and proven knowledge with you about what I am sure is everything I know, then I am an honest academic wishing to teach you. Now I tell you one part knowing that part is flawed and not telling you the rest because I know that is the flawed part then do you think I am trustworthy. Am I the honest person you can entrust with your future?

I, as the tutor know the half that I tell you, but it is unproven and the other half I do not tell you about are made up as science goes along while I also know that the half I tell you is rubbish but because I know it is rubbish I don't tell you it is rubbish because where will I get another job if I tell we know nothing about science. I only tell you the first part that is not proven and leave out the second part that shows the first part is rubbish and still I wish you to pay for my services. Have you a name for such a person that will force another person to pay him to brainwashed him by mind control because the tutor has absolute control over the life and death of the academic future of the brainwashed individual and therefore is willingly forcing this unfortunate creature in accepting what will never amount to form the truth? Those are called Physics professors and rule Universities as draconian authoritarian dictators bent on sadism.

Let's investigate the falling as such and see what happens during the fall. The truck falls at the same pace empty or loaded and this falling is at the same pace in which the girl falls, which is the same pace as that which the frog falls. I don't quite se the role mass has in this! If the truck falls at the same pace as the girl and as the frog there has to be a common denominator in this process and since the common denominator eliminates size form and shape we can eliminate mass. Mass brings distinction and the falling eliminates all forms of distinction. Something bigger must have more mass than something smaller.

Lets take this scenario to a waterfall. When I fall down a waterfall with a boat I travel the same pace, as does the boat. That could be because I am fixed to the boat by sitting in the boat. But my sitting in the boat has certain condition and one is that I can remain sitting because I fall the same pace as the boat is falling. This is like the truck's empty cargo bay falling as fast as it will fall with the filled cargo bay.

If I fall down in a boat with the boat and the boat and me forming a distinctive unit falls at the same pace as the water that forms the waterfall falls. Should I at the time of my falling hold an empty mug in my hand and I wish to fill the mug with water, and then I will have to move the mug against the flow of water streaming down the waterfall. I will have to thrust my mug upwards at a faster pace than my descending is casting the mug down and therefore I accompanying the mug down the waterfall. My mug will not automatically fill with water or if there was water in the mug my mug will not automatically empty with water just because the emptiness filling the mug will be at a different pace than the content that is otherwise the filling of the mug. The empty part is falling as fast as the filled part that holds the mass.

The mug being empty falls as fast as the boat and I. If mass pulls the filled part what pulls the empty part? The empty space in the mug is falling as fast as the mug will fall when the mug is filled to the brim with what ever can fill a mug to the brim. Notwithstanding the content within the mug or the content within the boat or the content within the water being within the waterfall, the very lot is falling at a similar pace. By lifting the cup while falling the cup will fill with water. I am not putting the water into the cup but I am exchanging the space that the water holds with space that the empty cup holds and my action in truth has no bearing on the water filling the space, which I then transfer into the cup. I am filling the cup with space that at that point holds water but the holding of water has nothing to do with the transferring of space.

If I leaped from the boat and fell I would fall alongside the boat. The boat will be empty but will fall at the same pace and as the same space as I fall notwithstanding being empty. The mug being empty will fall at the same pace as the boat being empty which will fall at the same pace as the water in the waterfall and I would fall. The space in the boat, which is empty if I do not fill the space, will fall at the same pace as the empty space, which fills the mug, and the mug will fall at the same pace whether the space in the mug contains or doesn't contain whatever can fill a mug.

The space filling the mug is falling the same as the water that would fill the space in the mug should the mug be filled with water. The space in the boat is falling at the same pace as I would fall whether I am filling the vacant space in the boat or otherwise filling the vacant space next to the boat. It is the space that falls and not the object filling the space that are falling. It is the space that is filled or not filled that is dropping down because the space being filled is in decline. If it was not the space that fell the space within the mug would fill first as the mug and the boat fell because the empty space would first fill before it could take anything down. But since the boat falls as fast as whether it is being filled or not we can assume that the space which the boat fills or does not fill is falling as fast as it would fall whether it is holding the boat or I or the boat and I. The space not filled by mass also moves just as fast as space filled by mass. If Newton's idea of mass was valid the boat should only fall when it is full and when the emptiness was removed. But if you fill a hot air balloon with more air or with more emptiness it takes mass with it into the air removing the pulling. When the object such as the mug or the boat or I connect with the Earth the Earth disallow the object free motion by taking any more space the object claims through to the centre of the Earth. The object now has to give up the space it claims and take on new space that the object claims to flow by contraction to the centre of the Earth. In forming a blocking it resists the flow or the gravity or space lining up with the centre of the Earth. The flowing of space by contraction is gravity but the object being in the space that flows becomes and obstacle through which the oncoming space must drag in order to flow to the centre of the Earth. It forms resisting of allowing space claimed to release to the normal flow when the object will not relent form in favour of gravity. This is only when the object touches the solid earth.

This resisting such relenting of form and consequently forming a frustrating barrier that blocks the free flow of space towards the centre is time displacement of space and this relenting of space-time flowing freely becomes the mass factor. The density and the resistance that the particles show forms the mass that implicate the degree of the frustrating or preventing or disabling of such free flow of space through time and the displacement of space during time is space-time notwithstanding what ever irrational connection Newtonians wish to add too space-time. Allowing space to displace through time to form time is space-time and that is gravity. All I ask is to read what I bring. Don't be a coward and stop reading as soon as you reach the point where I condemn what is in place! Just move past that to the point where I show what is wrong and how it can be corrected! Just judge me not for condemning what now is so apparently incorrect but for showing why I condemn what now is so apparently incorrect and what I bring to the table and offer as a remedy. See what I have to offer and not only what I am taking away. Don't set your sights on what there is to lose but take a view on what there is to gain! All you lose is the untruth.

Do not reject me on merits you do not wish to instate because you have the fear you are going to lose what is instated. Do not judge me by using your double standards that is useless in the face of the truth.

Rather look at the double standards you employ and do not judge me by using your double standards on me. Rather use your mind to detect what is double about your standards and then investigate with me what needs to change. Don't hide the truth. Don't hide from the truth and don't hide behind what you wish to portray as the truth. Rather come out into the light for the first time in three hundred years and admit to the truth. Follow what I say and see for yourself what there is to gain by trying to detect what is wrong because we all know there is much wrong. The comet does not collide with the Sun and the Moon is not on its way to collide with the Earth in time to come.

Expand science and no the Universe for the Universe is the only aspect that has not the ability to expand. I challenge all of you Newtonians to prove $F = G \dfrac{M_1 M_2}{r^2}$ and not just to declare it proven because it is in use since the Dark ages. Expand your mind and double check the formula you all so vividly underwrite and support. Prove why you support the formula in a modern and a scientific way.

Explore the correctness that this formula $F = G \dfrac{M_1 M_2}{r^2}$ underwrites. Be a true exploring scientist and journey with me through the following pages while we venture on the quest to find and vindicate my incorrectness by proving the truth vested in the formula $F = G \dfrac{M_1 M_2}{r^2}$ that carries the entire physics everyone uses. Let us start where the lot should start and get two Masters together on one point of argument. Galileo said all things fall equal.
That says all things fall alike.

The first thing anyone brings in is the vacuum bit with the feather and the hammer and since we do not live in vacuum there is no chance of finding a feather that will fall as fast as a hammer. Since the feather does not fall as fat as the hammer we immediately jump to the conclusion that there are falling disparities because of the falling discrepancy we find between the hammer falling and the feather falling. Then what would give the feather the time to fall longer than the hammer does. Everyone concludes about mass coming into play and they are correct. But they are half correct while Newton still is completely incorrect by attaching mass to the entire idea of falling. Take away the resisting of the feather and replace it with something far less air resistant and one will come to a different conclusion. We have to dissect what factor consists of gravity and what factor represents mass. Then we have to dissect which part does mass play and what part does gravity play. The falling object experienced no mass while falling therefore the falling or moving must be gravity's contribution. While objects are in motion those moving objects is experiencing gravity. The object show mass when the object has a tendency to move but the motion towards the centre of the Earth no longer takes place. That means mass is the restraining of the motion or is that which prevents the motion or gravity taking place. On Earth, objects experiences mass by restricting gravity or motion with the Earth giving mass but taking away free motion. By giving mass the Earth forces the object to become one with the Earth and move with the Earth as a pat of the Earth.

They were brainwashed by their tutors as their tutors were brainwashed before them. You are in a position where you can teach your tutors the truth about gravity if you read what is in the books. The truth is there and the truth is out and the truth will be because the truth is written for all that wishes to read. The academics on the other hand have ignored my work and my being on Earth for the past six years while I was writing them letters about gravity. They ignore me as if I am a rattlesnake because to them I am a rattlesnake. For that reason if no other they will rather go on lying to you and cover their corrupt fraud than face up to the truth and admit their work is lost. The truth will be whether it is recognised by them and they can become the first to admit and repent or they will be the last of the laughing stock that those in the future will refer to as the bunch that couldn't see when things fall equal they cannot have mass and when things do not fall by mass then one can know mass has nothing to do with the falling and the gravity.

It is up to you as students to rattle their cages and make them admit they've been lied to as they are lying to you. If you do not accept the role as being zombies that is brainwashed then confront these academics that treat you with disgust and betray your trust. They might tell you the mistake is not that serious and the damage is small but then how will they know how big or small the damage is if they don't even know what damage there is or what the damage is. My books will serve as the light switch that brings the light to you. I charge your young minds to confront those fraudsters about the truth. If you reach the need you may down load it because it is a fair bit of information. However it is not my views that are your enemy but the brainwashing you are victim of that holds your logical thinking captured behind the bars of a jail of corrupted ideology that forms a culture.

A question they never ask or answer is what is starting where the cosmos according to their opinion ends because something else has to start behind where an end forms and that is the location of the start of something else. Or can they answer another question: What was in place of time at the time predating what is now in place during a time before what they say is there when they say the cosmos started and from what did it develop and why did it arrive at where we are now. Before what we now have, what was then in place and how did what we now have come into place? They present gravity as a magical force.

In this meagre presentation I provide a small part on an infinitive big answer about how everything called the Universe started but that is only after I annihilate every principle those in science so dearly protect and by doing so I uncover the corruption that they hide. The corruption and falsifying of facts are there for you to see. I show what they create to justify their correctness about what they claim they can prove. As can be seen by this little presentation forming an example, I prove that Newton's principle was never been proven as long as it is in place. In short I would wish to have our scientists prove where mass is part of the planet formation as Newton so eagerly said it is. Use the planetary layout to prove Newton's "_Kepler's laws_" that has nothing to do with Kepler or with science and is completely fake. Then where I try to uncover this incorrectness they refuse to publish my work. You will find in this book I show what the response is when I sent an article in which I explain the Coanda effect and how that transcends to the operational functioning of the gyroscope ands then as a result why the gyroscope works. The entire article is published as **The Absolute Relevancy of Singularity The Website.** They never ever read the article never mind considered the content and this treatment has been going on for twelve years.

If mass pulls why is there circles around planets? The image of gravity forming by mass is a conspiracy preventing the truth from coming out. That there is a conspiracy in science there is no doubt about it. It runs as deep as it can go and involves every person that studied science worked with science teaches science and thought about science. In short it touches every person that lives. Everyone works with fake science and conspires to promote science fiction passing it off as Newton's science.

Has any one ever presented a remote answer why there are rings forming around planets and why the planets as big as they are, don't simply pull the dust onto the giant planet surface? Why would the rings circle the planet and not fall onto the planet? If Newton's mass principle is correct, the mass of the giant just has to pull the dust onto the surface. Firstly this book is free. I am not trying to exploit any person by extorting money with launching an unsupported wild scam. The truth is that Newtonian science is practising an unsupported scam for decades and they do it by mind manipulation and thought control. They force students to believe Newton is correct without proving it. Let them prove just what I claim is untrue in this book alone

The Big Bang shows expanding and Newtonian science still favours mass pulling mass into what? Whereto is mass pulling mass with the lot expanding. Gravity is the least understood concept this far in science. In truth no one in science anywhere remotely knows what brings gravity about and I used Kepler to unravel this mystery called gravity. Newtonian definition cannot even recognise any of the principles I show in this book but only Newtonian science are taught to students. No student can have the fortune to disagree with Newton and remain a student at the institution the student is studying at notwithstanding that Newton's ideas are baseless.

Introducing the six part theses **The Absolute Relevancy of Singularity** written with facts about the creation in mind produces a problem because the complete picture that I introduce has nothing in common with current accepted science. As it should be clear, I remove myself from science. The issues forming the new vision I present remains comprehensive even by using it in a very simple form, such as I do in this book. In spite of this, I shall explain three of the four unrecognised phenomena I use in proving my statements in this very book aiming at a theme of simplistic introduction being "an Academic Letter". Then I go into not that extensive detail proving all of the Cosmic Pillars. With my introduction of the phenomena, which I named the four cosmic pillars you will find it obvious why science do not accept them even if it is documented throughout the Universe and is quite commonly found. The phenomena are there used by nature and still Newtonians deny it being present with a functioning purpose. Mass, which is not present and not has a function science puts forward as the sole contributing factor to the formation of the solar system. What is present science hide and what is not present science promotes as being not only present but being purposeful. Then when I say Newtonian science is a hoax I am incoherent.

Applying the four cosmic principles totally annihilates Academic's formula of the basis on which science rests in the formula being $F = G (M_1M_2)/r^2$. The four cosmic pillars are the following:

1. **Roche-Lobe**
2. **Titius Bode principal**
3. **The Lagrangian principle**
4. **The Coanda effect.**

From these cosmic phenomena I produce a path of cosmic development, even going as far back as preceding the Big Bang. That I manage because following the development through the four cosmic pillars where these form the building blocks in the Universe. Since there is a total absence of mass we have to use principles much more substantial than Newtonian rumours and innuendoes such as mass. All this information is totally new and never before did any person understand the phenomena. The problem that comes from this is that I take the reader from a point and lead the reader through the explanation of the existing principles, pointing out how they are flawed and introducing my explanations and proof and substantiate my argument. This is a path one has to follow should one wish to meet with cosmic reality. The principles I use are in place and used as I explain and I prove there is no evidence of mass anywhere. The biggest problem I encounter is to get those in office to admit and to realise there is a problem in science. As long as those in office in physics are contempt with hiding the problem that they dismiss behind a conspiracy while diverting from the truth for their personal worth in academic standings a resolve will never come about.

This situation is outrageous. Even for persons of the stature and intellect such as the Newtonians enjoy the entire problem must be uncovered before the resolution can be discovered and in this process there is no point where one can drop in, or out and in again and maintain the golden thread of understanding. A simple manner explaining the way the book falls into place is the colour we associate with objects and the differences I see to the reading of the colour compared to the way normality see the colour. Science currently judges the cosmos on face value while we have to use our intellect to scrutinize the cosmos. Things are not what they seem. When looking at a red flower we say the flower is red. Nothing can be further from the truth. The flower is every colour in the spectrum, except the colour we attach to it, although it is screaming with all might to its disposal that that specific colour it cannot accept. Yet, we maintain that that colour is the colour we associate with the object, ignoring the object's rejection of that colour. Only when looking at the cosmos from this stance, can the cosmos make sense? By recognizing a disassociation in spite of our cultural recognizing the association, can we understand the cosmos?

We maintain the sun is burning, while the fact of the matter is the sun is freezing. From our perspective on the outside we see the sun burning as we see the red flower. The sun can't discard heat and stay hot. By discarding heat we have to recognise the sun as being cold. Outer space on the other hand accepts all the heat the sun discards and still remains unchanged. By accepting heat without changing the outer space region must therefore be hot. At this point my statement sounds outrageous but with more space inside the book I explain this stance much better. We feel what the sun rejects and the associate that which the sun rejects as not forming part of the sun as being the sun. That is silly and stupid. What we see is not what is the truth. Only by applying the correct view to the cosmos can the four principles I introduce, make any sense and find any proof... and I do prove them. To get such proof I had to do extensive research on cosmology. The proof lies in the unrecognised and misunderstood laws and principles science know about. The laws although existing, fall outside the parameters of applied physics and because of that therefore they are not accepted, notwithstanding all evidence of them being there in place, applied by the cosmos and used to build the cosmos. Not recognising the principles is madness.

I had to define gravity; a task as yet, never done. From that I defined energy, which too, has never been done. I first had to prove the existence of time and time's control over the Universe, as much as time's role in the Universe and even moreover, what time is. Again, this was up till now not yet been achieved. I had to prove what space is, that time and space is sides of the same coin, with matter forming the separation. Galileo gave us the answer to space and time. By moving a pendulum through space we read time and therefore the pendulum shows what is time by the measure of space. Even before realising that I had to find and locate the position of singularity as not to repeat the same errors that Einstein made. In all this I had to prove, formulate and implicate four factors that are there in the universe, as yet and not yet understood or explained by anyone. I named the four principles the four cosmic pillars, which the Universe rests upon.

In parts of the book I explain extensively in some detail how the four laws not recognised brings about the speed limit though to be The Sound Barrier and explain by some short measure how this connects to the start of the Creation process and how this builds the solar system. The way the solar system forms and the Sound Barrier and the process by which Creation started is all interlinked and commonly connected. What I do not do is eco Newtonian bullshit by repeating unproven trash they use to brainwash students. In this admitting I realize that there is a substantial chance then that you the reader will put the book down and not continue reading, but when you do that you do so at your own peril. THE SUPER-EDUCATED stands highly negative against any religious attachment any person may have to science but it is science having the shortfall you see what the Bible shows.

Xepted science has a notion of including or excluding on the basis of likes and dislikes and through out the book I indicate this, but when being to personally involved with science and brainwashed into mental stupidity one cannot see them doing this. What Xepted science does not realize is that same whip is having a backlash on our SUPER-EDUCATED as well. By the same margin they are turning the loose of ignorance on themselves too because that attitude makes them unable to apply meaningful relevancies. Space is forthcoming due to the value and intensity derived from the concentration of time.

They see a Universe they draw with boundaries (although refusing to admit this) because Xepted science makes no effort in divorcing their thinking applying boundaries everywhere. That makes Xepted science see a Universe, which has to have boundaries (i.e. Einstein's theory on the critical density and that in itself holds boundaries), but boundaries the Universe has not. Time on the other hand has distinct boundaries and that too I show where it is. By calculating mass individually one does not recognise influencing attachments through singularity bonding and binding space and time and space-time. This comes about because the focus of attention falls incorrectly on the invalid part of space-time, which is space. Objects in the cosmos are not a certain distance apart but a certain time apart. It takes light time to displace space and not the other way around.

Using a simple top spinning I indicate time by positioning the point where I prove singularity is. This way I manage to prove many aspects in cosmology that is still unclear or not understood.

I prove and formulate **the Roche limit** as having two factors, both of which play a most dynamic role in the cosmos.

I have also managed to prove, formulate and define **the Titius Bode principle** and in that principle also comes with two factors. The principle is a derogative of **the Roche limit** and in amongst that there is another principle I have discovered concerning the dynamics as well as the role of light in the Universe.

The Lagrangian principle also flows directly from **the Roche limit** as **the Roche limit** is a ratio in conjunction to the point which I claim singularity is.

The formulas you are about to witness is another relevancy of singularity as it relate to positions or marks away from the point holding singularity. With all the bad experience I have had in the past where academics reject my work without any one of them investigating the detail and the substance and because those academics find my denouncing of Academics very presumptions I will not in the least find rejecting on their part again surprising at all. I ask you the reader for about two or maybe three hours of reading time and if you are not at least intrigued by my new ideas I shall find that even more surprising.

The work you have before you I have started in self-study back in 1977 and full time writing (without pay) doing investigative research compiled as seven books in one thesis. Academics through out the world place such a high premium on qualified education and once I admit to my personal research not upholding any recognised work I find immediate rejection just on the ground of that. Because my work is completely new and does not match any work done before I go being rejected. I find that truly unfair and in this day and age where free and fair speech serves such prominence it is most unfortunate to find that that is only lip service in most if not all cases. When reading the book you may come to realise why I came to the discovery, which I did.

I admit that at this point you may even find a remark as innocent as the one I just made very offensive and rowdy, but as you progress through the volumes you will find I am not in the least exaggerating. I ask you not to dismiss the book after the first few pages just because you have the opinion that I do not understand Academic's laws. Again by making the following remark, it will come across as very presumptuous on my part, but I may be the only person in the world that understand Academics and not accept Academics. There is a huge difference between the two notions. The formulas you find is the diversion that an object in movement holds as the movement relates to the line singularity holds in relation to the moving object.

Have you as you sit reading this part at this minute sat back and gave a thought about the light enabling you to read? Such a thought brings to mind the most simplistic answer one can imagine. The light hits the page bounces from the page and contact the lens of my eye where the lens conveys the photons becoming electricity to a part of the brain that translate the electricity to an understandable message and that makes one read. It is as simple as that! Ever gave a broader thought about light streaming across the night sky, coming from ends of the Universe we do not even realise it is there?

How does the photons manage to convey one complete picture coming from as far apart and as wide an area as it does? With a few photons connecting the eye or lens no one ever noticed the wonder of light. The photons reflect a view that seems as if coming from all the billions upon billions of stars. But most is coming from darkness covering an area no man can measure. Yet how many photons can actually connect to the lens of the camera or to the eye?

Still a few photons coming from a single direction directly ahead eventually tell the entire storey. It is very simple to take the process of seeing by means of photon conducting very lightly and I have never heard one of the Brainy Bunch really in sincerity dissect the process to its potential. It is impossible that light from such an array of assorted sources can simply come together at the eye lens and show a picture of objects spanning across a Universe as wide as our mind can receive where the objects they reflect is beyond human measurement and the quantity is inconceivable many. Light is much more than the medium science takes it to be. Light connects the Universe in a way we cannot contemplate. Light being far apart originating from regions not in the same time or universal space connects in a way that present us with a picture holding the Universe in an understandable content. From the point we stand and we watch the Universe the significance of what we see surpasses the sense of understanding of what we are experiencing.

How can the few photons that our lenses catch coming from such an area as the night sky cover transmit the complete picture of what we see. Take a few seconds and gleans at the picture of the night sky then rethink the picture applying the full content in the picture to what the size of you eyes is. Think how big the picture is that your eyes take in and translate that area to the size of your eyeball in an effort to determine a ratio. One will be forgiven if one thinks of the ratio as eternal to nothing. Yet a few pages back I showed that according to mathematics there couldn't be anything as nothing. Consider the path the light followed from the source connecting to light from all other sources where all particles of the other light may come from and bringing a full picture to the lens one use to look through. In your mind connect a line from every atom producing light and connect the lines to your eyeball and see how you can manage to fit all the lines, as small as the lines may be.

Scientists think of outer space as geodesic zero, with nothing in outer space but space. Geodesic zero means the light travels in a straight line from where it originates unhindered all across space to where the light connects the eye. Such an idea by itself is outrages because the stream of photons reduce in space to such a minute quantity that taken the area the photons travel and the space in vastness it covers, the chances of one photon coming across many hundreds of light years through billions upon trillions of cubic kilometres of space and selecting my eye to convey the electricity is less than infinite. Yet such conveying takes place every second of every minute.

The position of the location of the second singularity, which is the precise duplication of the first singularity but in a diminished capacity, is obvious to miss when one is not applying a detective mentality, as one should in scrutinizing the cosmos. Culture will have us believe that when one sees a colour shining from an object the colour is associated with the object. Logic tells a different storey. A yellow dot is all the colours in the spectrum but yellow because it is disassociating with the yellow. That goes for red blue and all other colours we may visualise. I think the norm accepts this as scientific fact with very little argument or substantiating proof about that required.

If light came as individual streams of photon flurries our visage would translate that as such shown in the fragmented picture above. It would be a picture unconnected bringing across some photons in the manner where every object stands apart not being related in any way and that will be what we see, if it is anything that we see. That we know is not the case but that means geodesic zero is as much rubbish as anything Scientists regard with simplicity and with careless thought. Geodesic zero means nothing and how can I see nothing as darkness because "nothing" is not darkness, nothing is "nothing" and the darkness I see is darkness showing the darkness as something.

What then about colours that are technically not colours as is the case with black and white? White is simple. By spinning all the colours in the spectrum the colour white shines through. Black is quite another matter. A friend of mine whom is one of the best painters I have ever come across told me that one couldn't paint black but have to make black a dark blue to show shade on the canvass. That apparently is his success in achieving the realism.

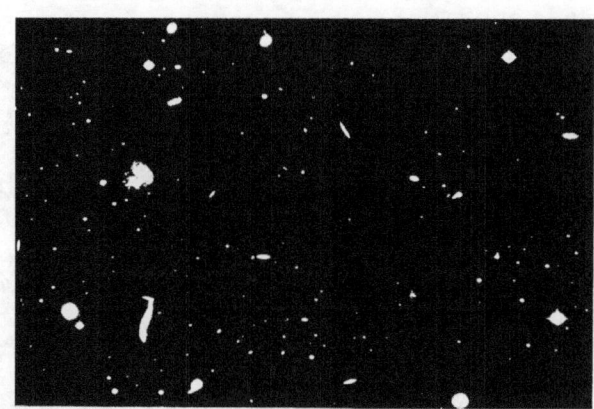

He also went on to explain how many variations of dark blue form the shadows in one simple tree. This remark set my mind in motion. One cannot see black because black has no colour to show, but black is the colour most prevalent in the universe. One can see only by colour and since black is not a colour we should not see black, but we do.

If the darkness was the representation of "nothing", then that should be exactly what we must see, nothing but the stars. Taken from the top picture some stars and leaving the rest to nothing is what we see in the picture below. A blind person sees nothing but when we look at space, we see something that we think nothing of as we see as space. One cannot have the ability of sight and see nothing except by closing your eyelids and then you see nothing. But in that case you do not see "nothing" in contrast of "something" you see "nothing" without it contrasting to "something".

 **Nothing is all about not being and not "not seeing".**

By the ability to see the darkness renders the darkness something other than nothing and that changes the acquired value of the darkness from nothing to something. There is an eternal difference between something in infinity and nothing.

The arguments introduced up to this part of the introduction prologue only touches the most basic aspects of my work and by no means can such an introduction secure an opinion. Yet, not once through all my long investigation in the past thirty or more years have I found any other person claiming such views that I have brought about even in this skimpy way as I do in the prologue. The arguments introduced up to this part of the introduction prologue only touches the most basic aspects of my work and by no means can such an introduction secure an opinion. Yet, not once through all my long investigation in the past thirty or more years have I found any other person claiming such views that I have brought about even in this skimpy way as I do in the prologue.

As it applies with all things, so it does in this case as well that when delving deeper into any issue, the complexity of the issues truly come to the fore ground when analysed in more detail. I wish to advise the reader to treat the seven books as seven different works and in that light I have separated each work in volumes of seven separate books with individual I.S.B.N. numbers with adding one part, the one you are reading, with one sole purpose and that is to bring about an academic introduction to clarify a quick perspective. Then the next three parts being of a general introductory nature there are overlapping in some sense but each highlighting issues in different manner as to clarify facts used in the last three parts bringing conclusion to different cosmic perspectives. Yet the work is seven parts of one thesis and as such it serves.

I have books explaining how Newton corrupted science and this would take too much space to explain that. There is a book for free http://www.sirnewtonsfraud.com/ I am not going into that at this venture.

Download http://www.sirnewtonsfraud.com/ to reveal the truth and find what the truth is. The revelation you download is for free.

This following concept forms the entire basis of everything forming anything in physics that is part of science. If ever any thought represented physics then this is the most fundamental start of physics.

This is so impotent I wish to run through this again because this forms the basis of all physics. Have you, the person reading this, ever thought how it is possible to see that much information that you see at night when looking at the sky. Ever thought about how you are able to see when you see everything in the night sky and how that much light information can fit into such a small space as your eye? Have you ever sat back and think what the amount of information it is that you see when you see the entirety of the Universe when looking at the Universe at night and what the size is of everything of that which you are able to see?

The one star you see seems to be a near visible dot in the picture while the dot might be hundreds or might even be many thousands of times the size of the sun…and we think of the sun as big. The dot is then that much bigger than the sun because the star we think we see could be a galactica hundreds of times the size of our Milky Way galactica but that shows as in the sky as one little dot and yet that entire structure as big as it is, does also fit into our eye socket. But that is not all…there are trillions of such light images and they all fit into one eye socket. What we see is immeasurable and yet we see it effortlessly in the space our eye holds…how can that be?

How is it possible to fit what we see into the space of our eyes we have? Think how much is the entire information that is visible at night and think about how all of that fit into the space your eye holds?

Consider how big is what is visible and put that space into the size is of what your eye can hold and ask your mathematically educated Professor in physics to find some ratio between what you observe and the size of your eye. The ratio is astonishing, but more-over what is truly astonishing is the arrogance of man to think of his position, as being important while the space man holds is beyond any comparison in ratio to everything we see in the Universe we see. Think how small we are when we are able to see the entirety out there! Even if there was other life out there, what is the worth of it in comparison to what there is that we see?

In this idea about how you are able to see the entire Universe you will find all the answers to the questions about how physics use time to employ gravity. Mass and anything Newton ever said has no implications on the explaining. It is about all the information of the entire Universe presented in one electron contacting a nerve in your eye.

The question about the Universe is how can whatever is in view, come stored as a parcel in an electron, and tell the entire story about the entirety out there locking all that data into the space of an electron. That is physics and tries as you may, not one person Newtonian or otherwise can have mustered the ability to calculate that part. Newtonians can pretend to play God and live in their fool's paradise as long as they are King of the Universe of fools while keeping the conspiracy alive to hide the truth about Newton's corrupt formula of $F = G \dfrac{M_1 M_2}{r^2}$. There is a mad conspiracy in physics to prevent anyone not in physics to learn about the truth hidden to all about Newton's Gravitational principle fairy tail. The fact that you can see the entire Universe and everything in it through one optical nerve tells everything that physics and cosmology up to now missed completely.

Newtonians uphold their law of physics without showing mercy. The very first things the Newtonians use to beat us into submission are to blast us with incomprehensible mathematical formulas. Incomprehensible they are but it is to scare anyone with the mathematical equations to get everyone hiding. They bewilder you with equations that put the fear of God into you; used simply to make you feel inferior so that they can feel superior and frown down on your inferiority from a dizzy height. They are masters at manipulating anyone into a state of senselessness...but mostly they do it onto themselves. That they do because it forms the backbone of the fraud.

They do not wish you to read closer and to find the fraud they hide to protect Newton. Ignore their mathematics because it only shows their incompetence to understand physics or Newton and see the fraud they propagate...They employ mathematics to bewilder and that is all. I am going to show what we can uncover underneath what they cover. Look at what the mathematics supposedly says and then wake up, they are using maths as a scare tactic for three centuries to scare the daylights out of you and all this while its been working!

Looking at the formula shows just how little Newton understood physics. Please allow me to show you how they scare you to become fooled and suckered. Don't run and hide when you see the mathematics; it is meaningless although it was used as a scarecrow for more than three hundred years forming the backbone of the conspiracy. In this case I am referring to the so called Kepler's Laws that has nothing to do with Kepler.

Use this picture below to show me where the planets are positioned according to mass or where the orbit going around the sun goes according to mass. The entire Newtonian idea of mass creating gravity by pulling is the complete misrepresentation of the truth. I show what principles are in place do give the reason why. It is easy to talk about "mass" and never get "mass" part of reality when hiding the truth.

Newtonians make a statement about "mass" holding the solar system in place. No matter how much this is corrupt, nevertheless they put it down as a given fact so much so that they will show doubt in a living God being present but that mass pulls planets goes beyond doubt. The proof of mass pulling to form gravity can never be tested because it is beyond doubt. If you doubt in it they throw a Newtonian made formula they call Kepler's laws at you.

Image Copyright JPL

"Kepler's third law" supposedly is "the square of the orbital period of a planet is directly proportional to the cube of the semi-major axis of its orbit." In mathematics it is Symbolically: $P^2 \propto a^3$ and therefore a³ = P² in position of P and therefore a³ / (P²P). This is taken from the idea that "Kepler said", which is totally fabricated that $a^3=T^2$ where Kepler said in fact $a^3=T^2k$ and this is a big difference because $a^3=T^2k$ is the same as $E=mC^2$. Look at reality. $a^3 = P^2$ is total garbage and as big a hoax as is the idea of mass being any form of factor in gravitational physics or that gravity applies in accordance with $F = G \frac{M_1 M_2}{r^2}$. Look at the picture below. Look at how the planets are sorted and that is not by size. There is a ratio applying called the Titius Bode law and this law puts planets in terms of size or mass at a precise equal base notwithstanding that Jupiter is many time bigger than Mercury. Everyone is so taken by the accuracy of Einstein's formula that $E=mC^2$ but this is exactly Kepler's formula where $E^3=mC^2$ is taken from Kepler's formula when accurately used as $a^3=T^2k$. There is no $a^3 = P^2$.

Do not get scared as everyone usually does when see and get frightened then consequently as a reaction to find survival you turn on your heels and run...

Don't run, just read on and see how simple it is to prove Newton was a backward dark aged sod!

<u>**You don't have to be a mathematical mastermind to see that it is not mass that applies to allocate planet or star positions. Here is an example that any person can understand when they don't succeed in bewildering people with frightful mathematics and comprehensive formulas.**</u>

I know and realise that you are disgusted by my attitude when I degrade the name on which physics are founded. In this introduction part I am going to show you just some minor deceptions all students are forced to believe since all physics students are forced to believe in Newton, **Sir Isaac Newton** that is.

I am giving you a choice. You can say I am going to commit fraud or Newton has committed fraud.
If I am judged to be the culprit that is guilty of deception then it is because Newton misled me. You can choose.

You are expected to believe the following:
Newton stated under the nametag of Kepler that there are so called Conversions for "Unknown" factors.
$4\pi^2 a^3 = P^2 G(M + m)$ Newton introduced this concept because he said mass brings about gravity.

From the top formula Newton devised the next formula $P = \left(\frac{4\pi^2 a^3}{G(M + m)}\right)^{0.5}$, which he named after Kepler. Kepler had nothing to do with the entire idea and every incorrect aspect is a Newton contribution. Students don't shy away from the mathematical aspect because the formula is complete bogus fraud.

Ask your physics professor to put in the mass of the sun and any or every planet and from that determine the allocated position in accordance to the calculations derived from the method that the formula dictates. The formula is fraud and keeping the formula in place and used by students all form part of the conspiracy to hide the incompetence.

Any person that upholds Newton's ideas and principles abut mass forming gravity use the next formula Newton introduced $P = \left(\frac{4\pi^2 a^3}{G(M + m)}\right)^{0.5}$ and then go on and put in the allocated position of the planet in the pace of **P.**

Then put in this formula $P = \frac{1}{G(M_1 + M_2)}$ the mass of the sun plus the mass of each planet to show **how the position is in place in accordance.**

This is no less than fraud and yet, on this the entire conception of Newtonian science rests. Let any physicist I challenge take me on by proving this or any other Newtonian formula correct.

Let any physicist I challenge $4\pi^2 a^3 = P^2 G(M + m)$ prove that it is mass $\frac{1}{G(M_1 + M_2)}$ that allocates the **poison reserve by the planet $P^2 = 4\pi^2 a^3$ which is what this part of the formula.**

If "mass" did form gravity by a value that commits a force then the large planets must be on the inside next to the sun having the small planets way on the outside. Instead because we have a random allocation, that destroys the idea of mass forming gravity to pull.

The smallest planets holding the least "mass" are at either end and the largest planets holding supposedly the most "mass" are in the centre. This disproves both arguments that the pulling force forming gravity by the value of "mass" to establish the orbit goes according to "mass" or that the locations of planets are adhering to Newton's ideas of "mass forming gravity by pulling". The way that the present cosmology shows gravity forms is by telling everyone about mass and then this is how gravity forms the Universe. That is the way they put the Newtonian model forward. Everyone is as gullible is the preconditioning would allow the people to hold the mindset in which the people are in. All knowledgeable persons know the sizes of the planets and yet no one thinks about the bigger planets being in the centre and the smaller planets circling near the sun or far from the sun. No one ever took it to task to confront those cheats in physics about the claims that it is mass that holds the planets in place. The world of people wants to be cheated as long as no one is asked to think and apply personal wit.

Please show how mass by $4\pi^2 a^3 = P^2 G(M + m)$ can produce Planet positions. It is hogwash.

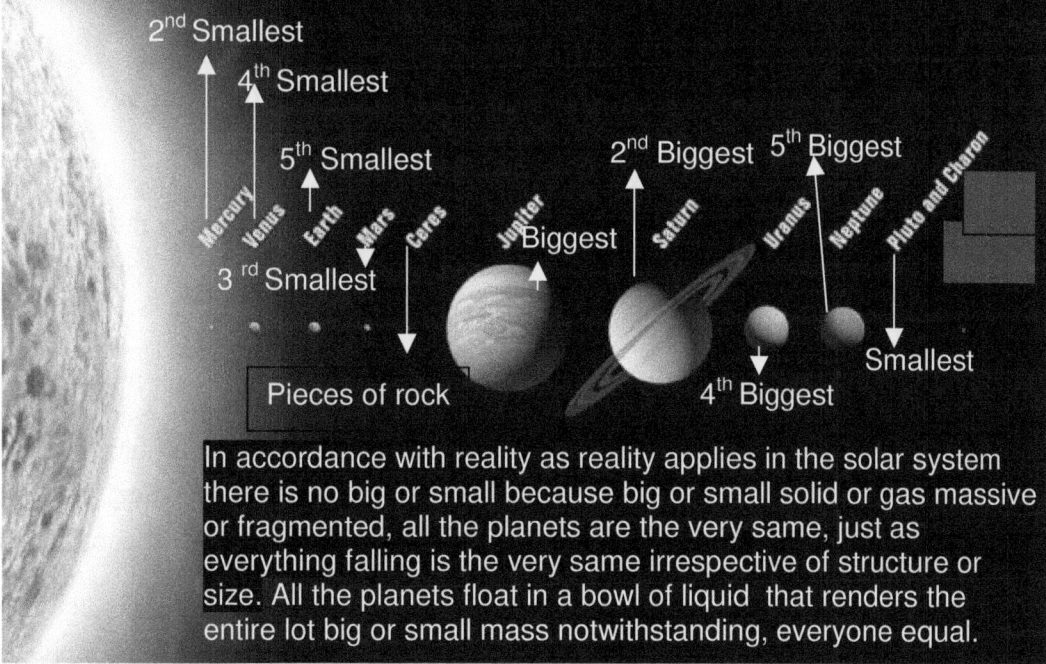

By depicting the solar system in such a presentation as Newtonians normally do such as the picture next this form of presenting the layout without providing correct spacing purposely corrupts the entire structure formation by which the solar system develops. It then purposely hides the essence that forms the solar system.

In accordance with reality as reality applies in the solar system there is no big or small because big or small solid or gas massive or fragmented, all the planets are the very same, just as everything falling is the very same irrespective of structure or size. All the planets float in a bowl of liquid that renders the entire lot big or small mass notwithstanding, everyone equal.

In the Universe all thing are equal in size because Neptune spins around the sun equal to mercury's time and Jupiter floats around the sun equal to mars or Neptune. Notwithstanding what "size" or "mass" they grant the planet to have the rotation happens equal and without mass bringing any favouring in positioning or in speed of movement. So where the hell is mass a factor in gravity forming?

Planet	Supposed "Mass" where earth is taken as 1	Average Orbit distance	Biggest to Smallest
Mercury	0.055 times the earth	57.9×10^9	2nd Smallest
Venus	0.81 times the earth	108.2×10^9	4th Smallest
Earth	1	149.6×10^9	5th Smallest
Mars	0.107 times the earth	227.9×10^9	3rd Smallest
Asteroid Belt	A Multitude of planet-forming fragments	Notwithstanding size the lot orbit alike	Debris orbiting in space at a specific distance that is not in relation to mass
Jupiter	318 times the earth	778.3×10^9	Biggest
Saturn	95 times the earth	1427×10^9	2nd Biggest
Uranus	14.5 times the earth	2871×10^9	4th Biggest
Neptune	17 times the earth	4497×10^9	3rd Biggest

From the Titius Bode that forms the solar system I have compiled the following formula by which gravity forms to the value of Π

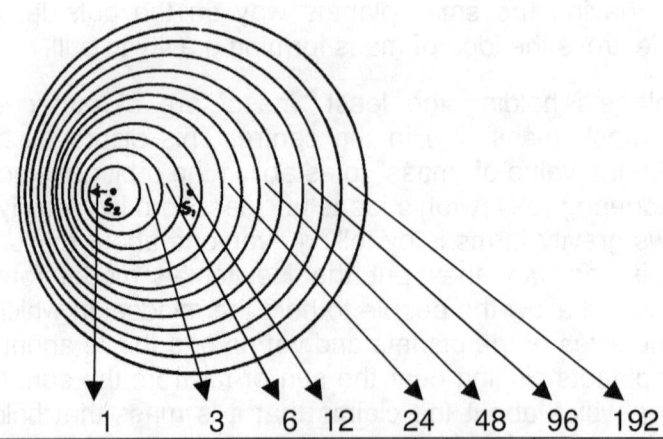

Planet	Mercury	Venus	Earth	Mars	Ceres	Jupiter	Saturn	Uranus	Neptune	Pluto
Bode's Law distance	4	7	10	16	28	52	100	196	-	388
Actual distance	3.9	7.2	10	15.2	28	52	95.4	191.8	300.7	394.6

Bode's Law:

A numerical sequence announced by J.E. Bode in 1772, which matches the distances from the Sun of the six planets then known. It is also known as the Titus-Bode law, as it was first pointed out by the German mathematician Johann Daniel Titius (1729-96) in 1766. It is formed from the sequence 0,3,6,12,24,48,96, and 192 by adding 4 to each number. The planets were seen to fit this sequence quite well – as did Uranus, discovered in 1781.

However, Neptune and Pluto do not conform to the 'law'. Bode's Law stimulated the search for a planet orbiting between Mars and Jupiter that led to the discovery of the first asteroids. It is often said that the law has no theoretical basis, but it does show how orbital resonance can lead to commensurability.

The importance that becomes known is the sequence the Titius – Bode law saw in the number arrangement of 3; 6; 12; 24; 48; 96 etc. The incorrect application of the Titus Bode law lies in subtracting the figure of 3 from 10 leaving 7. The other way of reasoning is to add four each time to the firs value of three starting with 3 and so on. The true significance of the Titus-Bode law is that it points directly to a circular growth of 7 stages. The 7 relating to 10 is a precise derogative of the Roche limit or the Roche limit is a precise derogative of the Titus Bode principle because he two systems interlink. This is how I mange to explain the Titius Bode law that is in the solar system by the ratio applying that really form the solar system in the way nature shows space growing by time. What you see on the next page was never been shown but on the other hand Physicist say this mathematics are too simple to apply as physics!

The Universe is not contained in a sphere but as a sphere. This makes a big difference. Whenever we see how a picture of the Universe is presented in its entirety it is shown as a sphere. That is wrong. It is where the Universe stars that starts at a sphere because the Universe starts with Π and in every aspect Π represents a circle. All spheres are a multitude of circles and that is where the Universe starts.

The cosmos came from $\Theta^0 = \Pi^0$ and from Π^0 moved onto Π. Why would the cosmos react in such a way since whatever can be is available to use to form the form of the emerging Universe? That description goes beyond mathematics and takes the cosmos into the realms of future mathematical equations, which are merely relevancies in place. So the cosmos was in Θ^0 (call it what you wish because that is to remind all that not even language and verbal expression was in place yet) identified it in being Π^0 that became Π by form and introduced space since the space came about through motion that is time. The mathematical radius was not in place yet, only the cosmic expressed relation of dimensions.

The reason why the sphere is a choice of form is a in the form value the sphere has in relation to any and all choices. Where one point touches the cube connecting to the sphere it seems to represent a circle but by the middle the circle doubles and the circle places a square in a circle on the sphere by three other points. This makes that the sphere holds the strength of 7^0 by the three other sides also making contact through the contact the one makes and by the centre attaché Π^2 to the bargain. Gravity is the dimensional changing of heat holding r as reference to the sphere holding Π as the reference. Heat occupying space has the cube that can apply r, as a straight line bringing about the cube with all its other names than may find attachment to specific form but nevertheless still remains only a six-sided cube with angle changing in some cases.

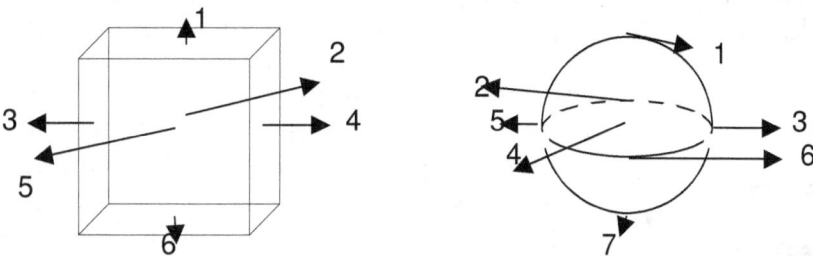

The first dot that formed was a sphere. It has to be a sphere since the sphere is the only form where all sides are identically precisely by specific measure equal and that equality is governed by a centre elected by the form of the sphere. The dot rose from singularity by the form of the symbolic shape and the value of the circle, which is Π. There was still no mathematical value to the form since the form was just that it was a shape that transformed singularity out from the immeasurable into the cosmos. It was only Π^0 that became Π and Π it remains to this day. One has to remember that when dealing with the cosmos, everything that was, is and must be in the future. Referring to Einstein's singularity as some mythical mathematical configuration that can or cannot be protected by the fairy and friendly godmothers of fairyland is as good as pagan godly figures. Introducing antimatter that vanishes from the cosmos is unrealistic. One has to find and show where it is and what it transpired into or what ever is referred too has never been part of the cosmos. Once it is in and part of the cosmos whatever then was part of the cosmos has nowhere to go but to remain in the cosmos. On the other end there is so little variation of what is used to form the cosmos that it is pretty unlikely for anything to have on single purpose and remain in one role.

The Coanda principle indicate that the gravity described in the previous page is generated by motion of liquid in relation to a solid anywhere motion can produce gravity. There is no mention of mass because mass is a derogative of the gravity which the motion creates. A centre is formed where the surrounding space-time forming the one group is relating a position from the "centre point". That forms one inclusive relevancy between points within the gravity field. The gravity field is holding "back" and "front" running through "the centre" where the other line is relating from "side" to " side" running through the "centre point". The fact of the line in the centre is that "it is there", but we cannot see it. Try as you may, no one will be able to calculate the very position that forms the lines, but as they change all particle characteristics, the lines are a reality as the spin of the matter is real. Being to small to hold atoms, the space holding such a centre line is no space at all and with that knowledge we may presume then therefore what ever the line constitutes of must become part of singularity, where singularity is a spot in the centre with two lines crossing the spot at an angle of 90^0. That is the basis of singularity, and since all the positions still relate too a centre of a circle, forming a part of a spinning circle, Π must form the basic value. The second major reality that one has to recognise is that the only way singularity was broken was by motion. The only way motion can come about and break space less ness is by establishing heat which establishes expansion and the Universe became a possibility and later a reality by expansion. The heat swell into space and the space swelling is the motion that produces the gravity we find visible in the Coanda principle. The space at first was presumably filled with material because the expanding could only be material.

One such a relevancy is the sphere.

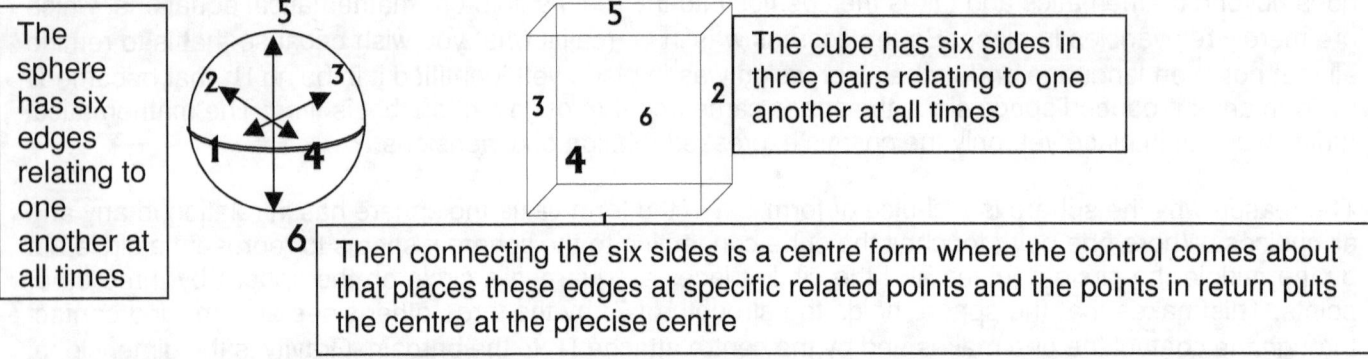

When a sphere spins an axis form around which the sphere spins. The axis forms by a line that holds no space. That line reserves three points in singularity a top, a centre and a bottom. Around this line spins two opposing points forming two points on either side of the centre divide. The spin represents four points in singularity and the sphere holds in total seven points representing singularity. This is the basis or the key that unlocks the Titius Bode law an therefore explains the forming of space in the Universe.

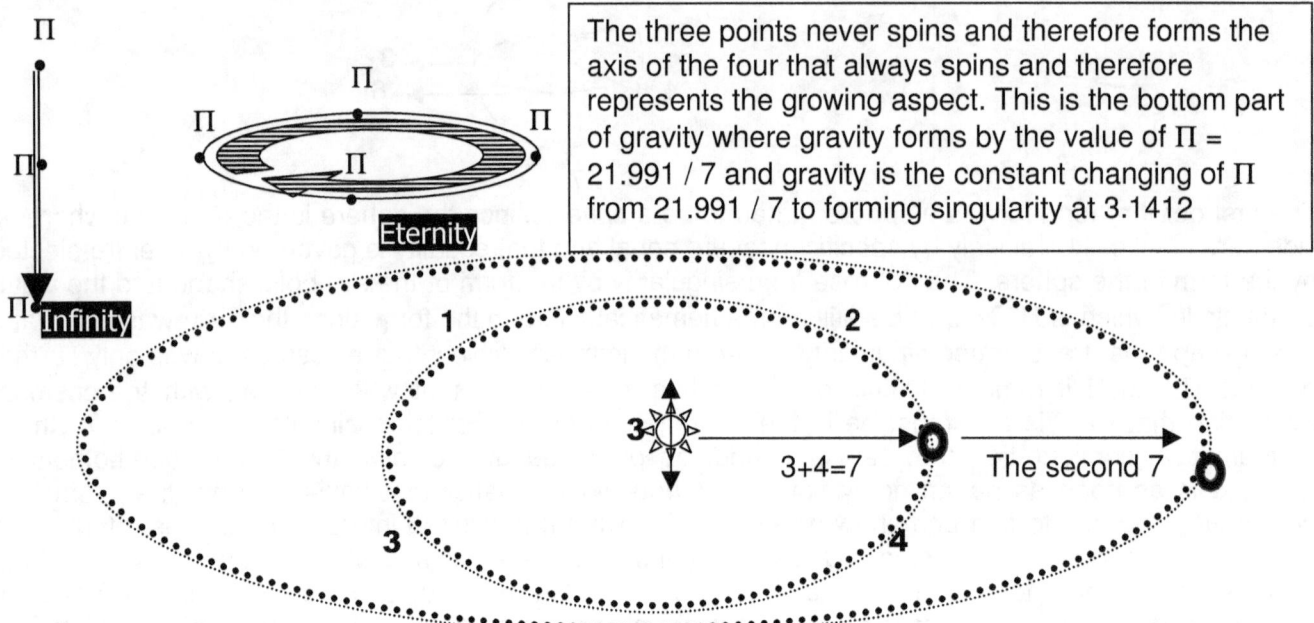

In the Titius Bode law gravity is applied by the forming of Π. The forming of Π is inessential a circle and that is why the heavenly body circles around the sun in a circle. A circle to the value of Π is 21.991 / 7. This puts a value of the circle that forms the orbit at 7 and the second orbit also at 7. But since the 7 is relevant to the same planet orbiting and the 10 it relates to is ten on one side of the sun and 10 on the other side of the sun it puts 10 in division of 7 on both sides of the sun and that forms (7+7) /10. I do explain this formation of the Titius Bode law in much better detail as the book progresses.

This is mathematically how the Titius Bode law forms gravity as space-time in space by time.
I use primary school level mathematics and for that being so simple they say its too simple too apply as physics. I will show you one the many letters of rejection I received later on in the book This is what there is and that is all there is in the solar system. Look at any picture and try to finds mass. The measure of mass forming gravity clearly plays no role in allocating the positions of planets as Newton declared it must do. The entire idea that gravity is a magical force created by the value of mass is as unbelievable as the dogma is of those presenting this idea. Please use what the solar system provides to confirm what Newton says is in place when he says mass forms gravity. Science would rather accept Newton where there is no proof of Newton ever being correct than to admit Newton's incorrectness.

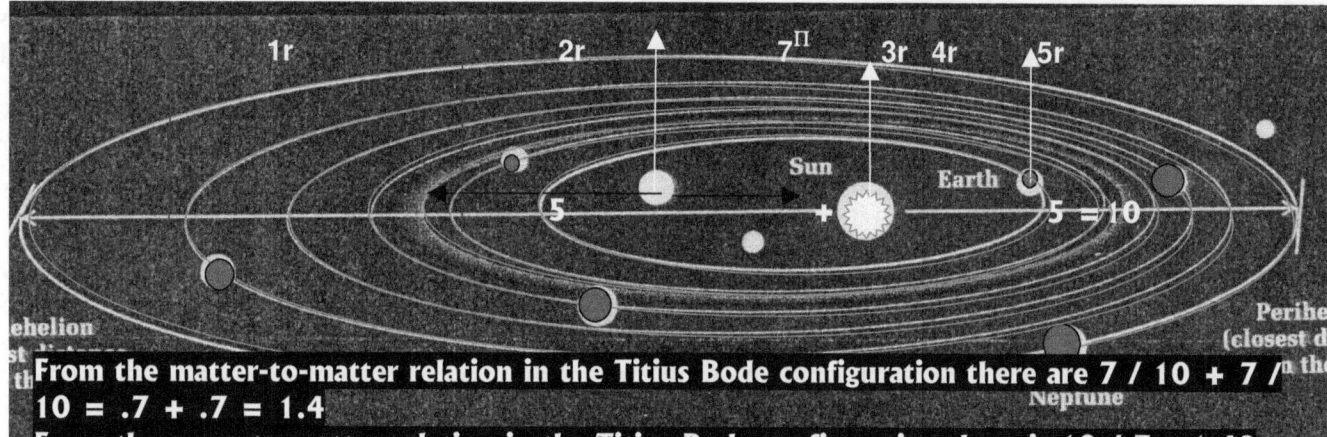

From the matter-to-matter relation in the Titius Bode configuration there are 7 / 10 + 7 / 10 = .7 + .7 = 1.4

From the space-to-matter relation in the Titius Bode configuration there is 10 / 7 = 1.42

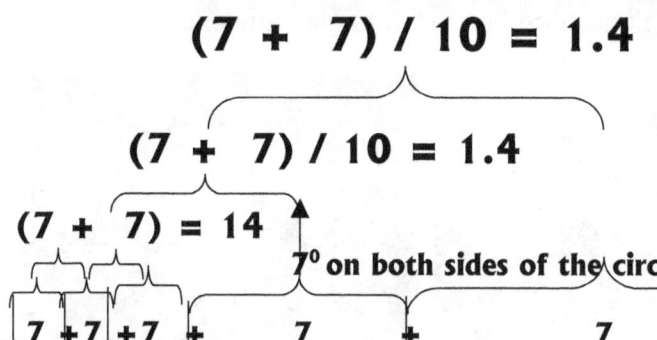

Any object turning around a centre (in this case planets turning around the sun) goes in a straight line by diverting from a straight line by 7°. On a later occasion in this book I show how the 7° forms a direct link in becoming 10. This is the ratio that the space between the planets grows by taking 7° and forming 10 from doubling that value.

In using very simple mathematics and I also dare also say too simple for the extremely intellectual Newtonians Physics Academics I prove how the Titius Bode law works as the Titius Bode law forms space through applying gravity.

While there is no hint of mass as a factor in the solar system, this Titius Bode law is what is present and is what is applying. The only thing new is that I am the first that prove how this law works and no physicist is interested this far in what I have to say because I belittle Newton and his corrupt principles. Every physicist in office with doctoral fights this because with this I prove Newtonian views are no more than science fiction

= .7 //\\ 1.42

= 1.4 //\\ 1.42 **Because the space-to-matter is in the square at 10 placing the matter-to-matter at a square of .7 + .7 = 1.4 the space-to-matter forces the matter-to-matter to double the distance by number as structures are place father from the main$\Pi°$ maintaining singularity.**

I'm correct. If they admit that I'm correct the entire world of physics becomes recognised as the joke it is in reality and they are recognised by all as being the laughing stock they are.

Later I show that 7^2 in conjunction with singularity applying the law of Pythagoras forms 10.

14÷1.42 = 9.859 or 9.86 or Π^2

I prove mathematically the reality of **all four cosmic principles** that are in place in forming gravity but because I do that and because I then make a mockery of Newton and their Newtonian principles they ignore me. Science would rather deny there is cosmic principles that is in place in the solar system, which are

The Roche limit,
The Lagrangian points,
The Titius Bode law and
The Coanda effect than to admit to Newton's failings.

They would not admit to Newton's failings because then the entire world will see they know less about science than does a pig know about history. They would rather put the error on the solar system than they would commit to the blatant mathematical cheating that Newton committed. It is the Universe that is always at fault when Newton becomes incorrect because without Newton's fabrication of science they have nothing to show for all the wisdom they try to pretend they have. Newton fabricated "Kepler's laws" has some correctness mixed largely with a farce and blending the truth with total fabrication of reality hides the lie behind something presenting the truth.

Forget about the fanciful corrupt mathematics that proves nothing when the cosmos does not confirm Newton's crooked mathematical arguments. Newton's religiosity might corrupt science but who in science would cares about correctness when it simplifies the ongoing brainwashing of students studying science.

They confuse everyone about what *weight* is, what *mass* is and what *gravity* is because they wish to have everyone think of "mass" in terms of weight while they then deny weight and "mass" is the very same thing and then they confuse "mass" and gravity because they never distinguish between what "mass" does and what gravity does. This is only the tip of the iceberg and you will see when reading this book.

This phenomenon should not occur with Academic's laws about gravity. These bodies will collide and destruct, without a doubt. When **$F = G(M_1 \cdot m_2)/r^2$** apply, there should not be any force which is able to keep them apart. Known for almost a century and a half science has failed to give any explanation about this cosmic phenomenon.

The Roche limit in the practical sense.

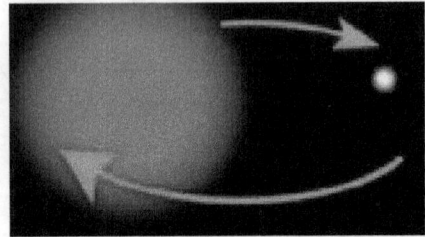

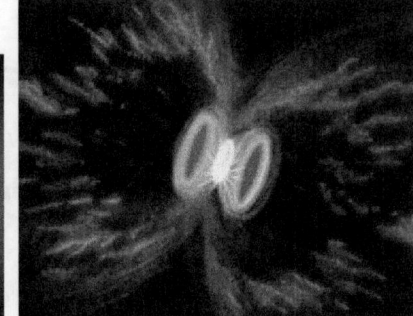

The formula $F = G \dfrac{M_1 M_2}{r^2}$ cannot explain the comic occurrence shown in the pictures above, but I can explain what is occurring in this instance and this occurrence connects directly to the Roche limit, as explained above. Not only does the Roche limit explain this phenomenon, but also it ties directly to the Titius Bode principle, also inexplicable to the formula $F = G \dfrac{M_1 M_2}{r^2}$. According to the science formula of $F = G \dfrac{M_1 M_2}{r^2}$ the orbiting structures should collide with a bang, but instead they do the tango until one drops, but when dropping it still does not collide with the larger structure as would the formula $F = G \dfrac{M_1 M_2}{r^2}$ suggest.

This is not only limited to planets in our solar system. In the Universe, there are giant stars spinning around each other. These stars are binaries, which are also one form of double stars where double stars are another such a form. The difference between the types depend on the distance they remain apart. They keep a certain distance apart and do not collide. In the case of the sun and its planets, it could be a case that the systems might be to small, or they might be to apart. However, this is not the case with binary stars. They are close, they are big, and they spin around a mean axes called the Roche limit.

The Roche limit is:
The region surrounding each star in a binary system, within which any material is gravitationally bound to that particular star. The boundary of the Roche lobes is an equipotential surface, and the lobes touch at the inner Lagrangian point, L_1, through which mass transfer may occur if one of the components expands to fill its lobe. It names after the French mathematician Edouard Albert Roche (1820-83).

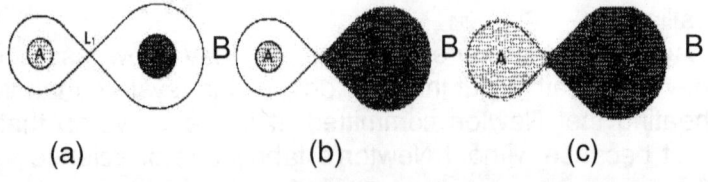

(a) (b) (c)

THE ROCHE LOBE: In a binary system, the Roche lobes of components A and B meet at the L_1 Lagrangian point. (a) In a detached system, neither star fills its Roche lobe. (b) In a semidetached system, one massive component, B, fills its Roche lobe. (c) In a contact binary, both components overfill their Roche lobes and share a common envelope.

The formula $F = G \dfrac{M_1 M_2}{r^2}$ is unable to explain the principle discovered by Titius and later by Bode and it is not coincidental. From this one can arrive at the origins of the solar system.

"Nothing" in the Universe is coincidental, "nothing" in the Universe does not apply and when a principle is discovered, the principle cannot be wrong. Therefore should the principle not match the excepted theory, change the expected theory, the theory does not apply.

The content of my work contain a new view about Cosmology, which I have been working on for the past twenty-seven years and exclusively for the past six years. To give you a little insight into my work, I shall mention the following: I came to realise that lines mathematically couldn't start at zero because there is no evidence of zero as a factor in mathematics. Should you disagree with my statement the question in need of answering is this: What will the length of the shortest hypothetical line imaginable be and moreover, what would the total overall length be in that case? The shortest possible line (hypothetically) must be so short it must have an initial and ultimate point sharing the same spot.

If it used zero as a start, the zero part would not count, because the line will only start at a point past zero where the line then will start forming an infinitely small dot.

The dot is in infinity, however small, it is not zero. Zero ultimately means not existing and then that point, as a start does not exist. The smallest line has a beginning and an end at the very same spot located in infinity, and infinity may be beyond human scope, though infinity is still not zero. Infinity may constitute of something we do not yet understand, but we may not define our human misunderstanding as nothing. In this aspect lies the difference there is between arithmetic and mathematical science where arithmetic can have position such as zero since arithmetic excludes the cosmos calculating numbers only.

A man may have that many oxen or so many sheep and even this amount of wives, (in Africa) or not have any therefore having then a total of nothing, but there cannot be nothing between the sun and its orbiting structures. The having and have-nots are part of arithmetic. Light will indicate a line flowing between the sun and whatever planet, following dot after dot thereby proving the existing of the possibility of something going about by a straight line, and any straight line in relation to other straight lines will be under the law of Pythagoras.

There is no possibility of a straight line not forming in space. Mathematics converts the values of integrating lines according to Pythagoras and arithmetic is about numbers to be added or subtracted. By mathematically excluding zero from cosmology a new Universe opens to the human mind. For instance the distance between the sun and Pluto is roughly one hundred times more than the distance between Mercury and the sun, but both planets mentioned have a vacuum filled with nothing except one atom hear and there occupying the vacuum between them and the sun.

If space supposedly comprises of nothing how can nothing then become plural forming more or be multiplied by a number as to indicate a growth in something not even existing. As the one becomes one hundred the one cannot substitute a value of nothing but then must be part of something. If the one substituted the nothing, all laws of mathematics will go in disarray because when one multiply any number by zero it becomes zero placing both planets in the sun.

By excluding nothing from the equation space becomes something bringing in a value lying inside the realms of the infinite that must form singularity. As the zero becomes a dot, something else becomes clear about the dot. Looking at the night sky we find darkness overwhelming the space in relation to the stars bringing across light. We can detect the dot because we cannot see darkness since our eyes were only meant to cope with light. With this knowledge, then how can we see the sky as darkness at night?

We are only supposed to see the light of the stars and not darkness, yet at night we see a much wider picture than stars alone. One may bring in the argument that the blind see nothing but darkness. We seeing persons do not know what the blind does not see, so we presume it is just about the same darkness, but that is presuming.

When we see a red flower, science knows the flower being all the colours but the red it rejects and this we all know. Therefore the dot we see as darkness also must be light, withholding its light and giving us the darkness we see as light…But the dot must influence the surrounding as well, subtracting the light it claims from the surrounding by casting it as darkness. In the case of stars we see the light the star disassociate itself with, keeping the darkness it has as it pours all the light in excess into the darkness which evidently is then light.

From that one may conclude there should be two forms of singularity where one associate with a dark dot being light, and another being matter with flowing light evidently proving to be the dark one. Proving the dot with many such arguments was easy. Naming the dot and its position, value location and proving the influences mathematically was much more complicated and proving the dot has a definite influence on the surroundings was at first seemingly impossible, yet it is done The definitely defined and underlined value of the dot becomes of utmost importance when finding solutions to cosmic factors not yet clearly defined. That is a big discovery you will make about true physics and how science works as you read on.

My approach might seem unconventional but through the abandoning of the accepted, it enabled me in locating the precise location of a universal singularity forming a connecting basis of the Universe (this I say with some degree of confidence). The smallest figure there can be must be a dot. The only mathematically sensible option about extending a line from the dot will be non-bias progress in all directions equally in order to give a meaningful flow of mathematical equilibrium. The Pythagoras mathematical principle is the proof and that I explain. The obtaining of singularity is in my rejecting of nothing by replacing it with something being the dot.

The claim becomes obvious when observing the connection between the half circle, the straight line and the triangle, which could also promote all the qualities lurking behind the pyramid. Consider the connection between 180^0 sharing and then one may realise much of the pyramid mystique becomes less spectacular in considering the very basic in mathematics being the Law of Pythagoras on which all mathematics are focused.

The claim becomes obvious when observing the connection between the half circle, the straight line and the triangle, which could also promote all the qualities lurking behind the pyramid. Consider the connection between 180^0 sharing and then one may realise much of the pyramid mystique becomes less spectacular in considering the very basic in mathematics being the Law of Pythagoras on which all mathematics are focused.

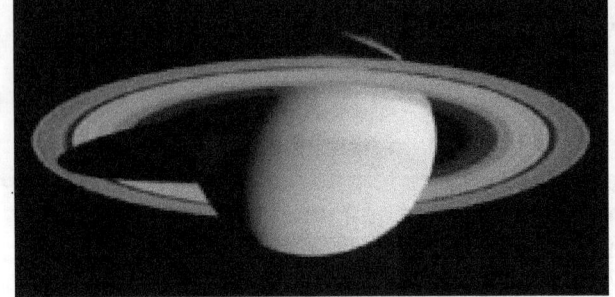

The rings around planets form not by, mass pulling but by applying the Lagrangian formation system, which holds 5 points that forms around the planet as it holds material in position. However, the Lagrangian system can't service the cosmos standing alone but forms one part of the four cosmic principles. The rings show that gravity is everything to do with Π and has nothing to do with mass in any way possible.

LAGRANGE (-TOURNIER), JOSEPH LOUIS DE (1736-1813)
French mathematician, born in Italy. In celestial mechanics he studied perturbations and stability in the Solar System. He examined the three-body problem for the Earth, Moon and Sun (1764) and the motion of Jupiter's satellites (1766). In 1772 he found the particular solutions to the problem that give rise to the equilibrium positions now called Lagrangian points.

Lagrange also studied the Moon's liberation.

LAGRANGIAN POINT One of five points at which small bodies can remain the orbital plane of two massive bodies; also known as liberation points. Three of the points lie on the line joining the two massive bodies: L_1 lies between them, while L_2 and L_3 have the two bodies between them. These three points are unstable, slight displacements of a body from then resulting in its rapid departure. the fourth and fifth points (L_4 and L_5) each form an equilateral triangle with the two massive bodies, 60° ahead of and behind the smaller body in its orbit around the larger one.

A well-known example of bodies flying at the L_4 and L_5 Lagrangian points are the Trojan asteroids in Jupiter's orbit. Among Saturn's satellites, Telesto and Calypso lie at the L_4 and L_5 Lagrangian points in the orbit of the much larger Tethys. In similar fashion, tiny Helene precedes Saturn's satellite Dione, keeping 60° ahead of Dione. The Lagrangian points are named after the French mathematician J.L. de Lagrange, who first calculated their existence.

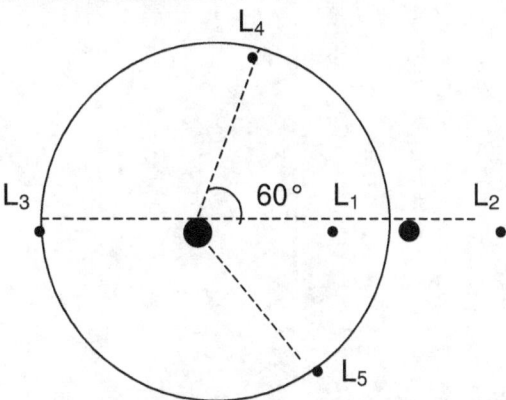

LAGRANGIAN POINT: *The Lagrangian points are five equilibrium points in the orbit of one body around another, such as a planet around the Sun.*

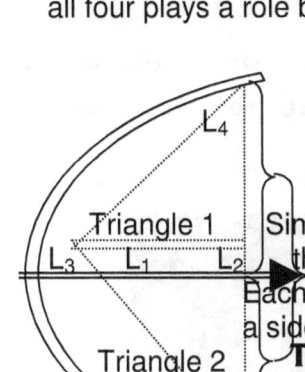

The Lagrangian points, is what keeps the formation in sequel positions.

The Roche limit, is what determines the distance that the rings or structures hold in relation to the centre of the main structure around which the orbit takes place.

The Titius Bode law confirms the circle or the rotation or Π and

The Coanda effect forms the relation between what is solid and holds position and that, which is the liquid part around which the moving part rotates. These principles work in a relation and the one can manifest more amplified in one situation than any of the others would but in every situation all four plays a role because not one can form Π as the circle that is gravity without all four connecting.

The Lagrangian System implicating the five positions extending from singularity

Singularity dividing the cosmos
Each triangle claiming a side of the universe

1 Half circle = 180° L_3 L_4 L_5
2 Triangle 1 = 180° L_3 L_4 L_5
3 Triangle 2 = 180° L_3 L_4 L_5
4 Straight Line = 180°

The half Circle = 180° combining as a Sphere when comprising Singularity in the matching of the value of the straight line forming the half circle and combining as the triangle and all are equal 180°

The distance between the sun and Pluto is roughly one hundred times more and if the distance between mercury and the sun, but both has nothing between them and the sun. If space comprises of nothing how can nothing then become plural forming more or be multiplied. If it was one becoming one hundred, then the one cannot contribute to a value of nothing but then must be part of something. If the one substituted the nothing, all laws of mathematics will go in disarray because when one multiply any number by zero it becomes zero placing both planets in the sun. By excluding nothing from the equation space becomes something bringing in a value lying inside the realms of the infinite that must form singularity. Applying this logic to the Lagrangian system and interpreting that information to the law of Pythagoras a clear pattern come about.

This picture is of the 1987 Supernova event. Ask any physics professor to explain how the "mass went mad" to allow this to happen. That which happens in this picture and what happens when an aeroplane goes through the sound barrier rides on the same set of principles.

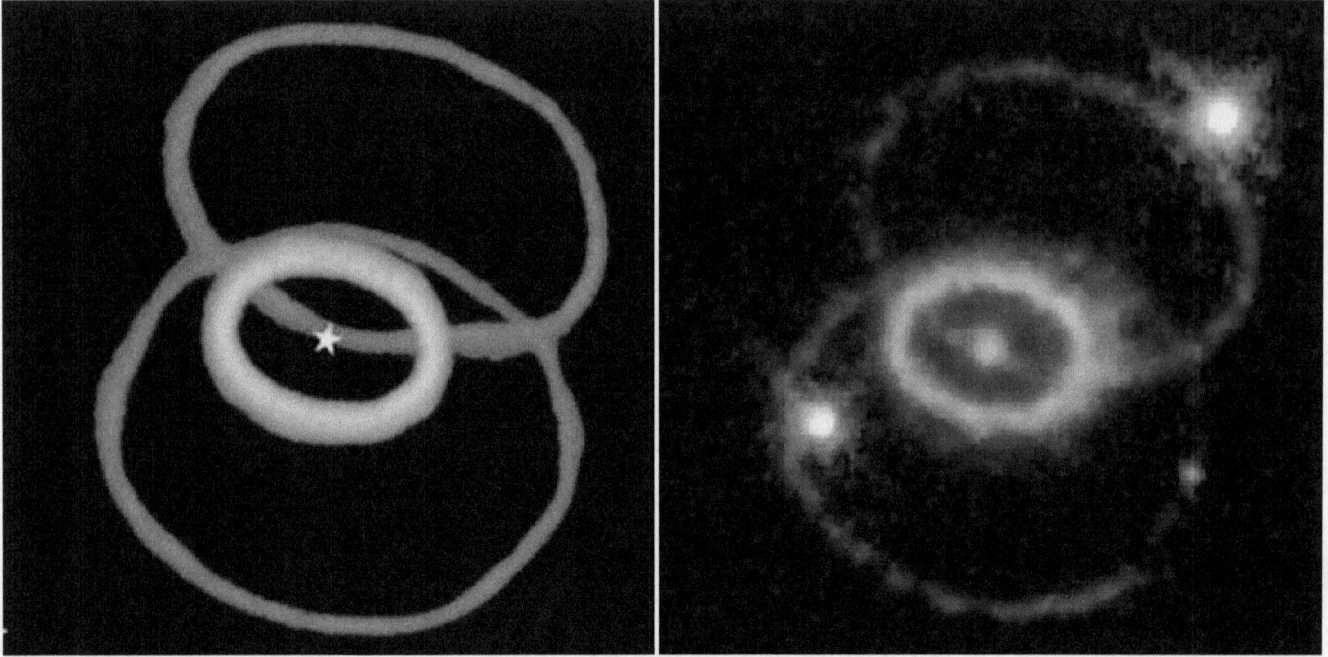

With circle being everywhere in the cosmos and where Π forms all over in circles and as circle, notwithstanding all the leads that the Universe give to Newtonians not one in so many centuries could see a link between gravity and the way that Π mathematically forms.

It is the way singularity presents Π in this picture above that one can see what gravity is and in what way gravity forms. When I saw this I knew I was on the correct line of thinking.

Gravity does not only link to Π but gravity is Π by forming Π. The question to solve was how…and that I solved as you can read in this very book.
Use Newton in the formula $F= G(M \times m)r^2$ to explain the cosmos and explain how stars "devour" each other without ever colliding. $F= G(M \times m)r^2$ may not be able to explain the exploding of double stars, however I can…

..by applying $a^3 = T^2 k$...

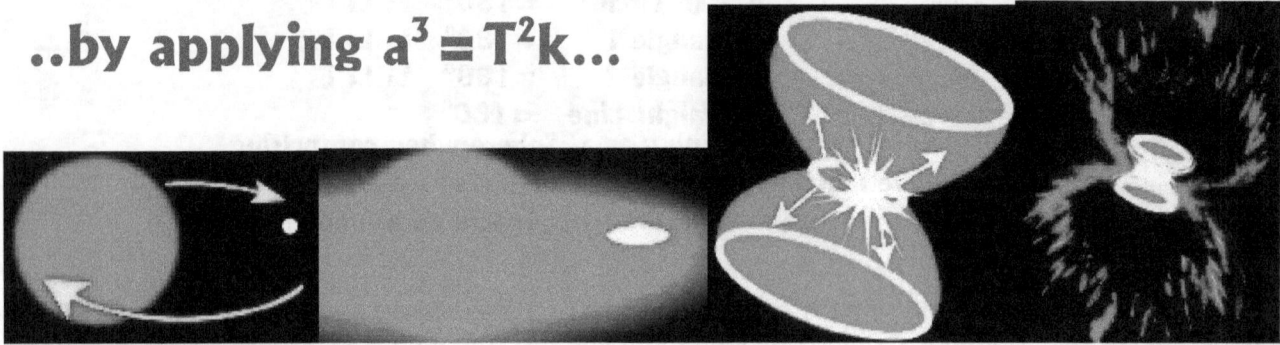

We humans are cursed by all the conspiracies that we have to endure. To break this we better start with science. We are all so entangled in a society filling our senses with one conspiracy upon the other that we can never find freedom in thought without disentangling our mental state from the brainwashing that furthers conspiracies by which we are taught to behave like those in power wish us to behave. This manipulating they subdue us too by rendering our thinking power to become computerised slaves is part of the general public's mental state and proves how much John and Jane Dow and Mr and Mrs Nobody care to be subjected to a comatose condition subdued to having no brainwaves functioning. We are told how to think and what to buy and how to vote and in which to believe by conspiracies whitewashing us so that those in power and those that are rich and those that are intellectual can control us on a daily basis.

Part 4
A Relevancy Between Time in Space and Time and Space

By having more heat per volume in ratio the material will claim and introduce new space that formed. Heat establishes space that expands. This truth science does not recognise. The claiming of more space and disposing of the space after cooling shows new space formed in the process of heat multiplying where there were no space before that which the material in the cool state afterwards cannot fill because of the void that came as a result of the material getting cold and contracting space, reducing the space as the space filled when the material was overheated. If material employs this as a basic technique today it was a basic technique back during the Big Bang. That evidence we can see in when material having a heat level amplifying upwards when motion difference brings on friction and such friction brings on heat. Material is energy and energy is indestructible. However energy can change form...yes that we all know and energy may even hide appearances. By changing legions the proton must then perform gravity by rejecting material or if I am correct, producing space! I am about to prove that antimatter is in fact a process where the heat that became formed heat, which forms space, and therefore space has a valid substance other than being nothing. I go to lengths to make persons see that space cannot be nothing. This is a factor that science has to accept if Mainstream physics have the will to find solutions about the Big Bang. The motion between particles in a cramped space as the case was during the initial stages of the Big Bang would have brought on friction in space we cannot even calculate.

It is the combination that forms gravity by motion. The result is that in the very beginning some matter particles produced gravity in their sustaining of independent singularity by applying motion which in some cases lead to the demise of some forming space-time by converting the where some compromised solidness. This route the one side took resulted in plasma forming on the one side and material on the other side. This was done because there was less control that confirmed the space and the volumetric space grew.

By having some softer than the other harder ones the softer one became a liquid. The notion or defining of a liquid is very relative because as solid as the Earth seems the Earth vibrates as a seemingly liquid during an Earthquake. It forms waves many meters high just like it would be a liquid like the see. This electroplating motion is possible since electricity is gravity to some intense extreme. By removing material from the less dense and electroplating that which is removed from the less dense and then galvanise that softer material onto the harder material (which by the way is a very natural process taking place all around as a corrosion) the density of the liquid will demise in the liquid sector and the material will grow in the solid sector. I believe even to this day and throughout the rest of the Universe wherever there is space such space has to have motion and space cannot be what it is without having motion. With that in mind that is space-time. Space-time is space flowing on.

Where there is motion in space, the motion through of all space is carried along by time in space. The plasma is transforming to material through the motion we named gravity. By being electroplated onto material. By duplicating space in the process of establishing gravity the object does not reduce to a standard in occupying space that it had before the motion took place but by placing liquid heat into the form of solid matter the matter use the newly acquired heat through which to cool. By absorbers the liquefied material onto the solid material is thus freezing it into a solid to secure more material in the fight of combating overheating. In other words in the present time in our Universe gravity is freezing space to first become dense and form a liquid after which it then solidify the liquid heat by freezing the liquid into a solid state within the substance that is the atom. I believe that the first motion came about as singularity was without space and found irrepressible heat levels rising. By overheating it moved into space that was still non-existing and that had therefore produced motion to rebalance the heat. From this I also believe some material that came about from singularity overheating remained as particles forming atoms where there is this relation between the solid proton, the liquid neutron and the gas electron.

The development of space from liquid heat such fluid was becoming gas that is space with the ultimate gravitational relevance that space can carry. This was all contributing to the lack in contracting gravity promoting expanding gravity to those particles that applied lesser motion helping the extending of space to turn into heat that again turned into space. In this, of uncontrolled release of heat performing as softer space-time such release of space-time is the destroying of singularity secured in a unit, which again I believe (within reason) I do prove. I show that on the one side singularity introduce space-time, which confirms singularity and space-time makes contact with space-time not directly controlled by singularity or that, which is directly confirming singularity. I believe heat is the destructed form of material that overheated and this confirmation the atomic thermo explosions give us. But to realise that we must beforehand find what any and all space is and we have to accept that space is made of something.

When one applies heat to an object it expands. That is primary school science. This states that more heat applied leads to more space acquired by the heated object. In sharp contrast to this is the growth in space when heat levels rises but freezing brings about the opposite result. When I freeze an object that object reduces its occupied space as it shrinks. Removing heat reduces space. That comes directly as nature responds to heat and I can prove that easily. By expanding it accumulates space to increase the improving of the size of the material. The accumulating heat for the sake of securing singularity, accumulates the heat in the material whereas the freezing tarnishes the overheating symptoms by the removal of material in unoccupied space using external matter and setting motion to the material until it contracts into a form which we see as visible heat. The heat is in the form of dissolved singularity that became material as material used it as growth. That is why by freezing it will diminish the space as to accumulate the heat absorbing into the heat into the material to maintain the equilibrium needed in space.

Taking this equation of nature to outer space we seem to confuse the natural law. With outer space as expanded as anything can get we regard outer space as incredibly cold. As heat sets in the normal flow will bring about expanding of heat into the form we think of as space that limits the heat overheating. Outer space is the very edge of expanding of space where heat cannot expand into space any more. Outer space is the limit, the epitome of expanding where heat meets space at the edge of all limits once more. Therefore being the representation of the very limit of expanding outer space has to be the hottest place there is. By applying heat to a kettle holding water, the adding of heat manifests as steam and steam is hot water that traded heat as it reviewed space.

The manner in which heat expresses itself when confronted by overheating is to provide additional space through expanding of space. Outer space is outer space because outer space has expanded all it can it is still expanding to the speed of Hubble's $1/H_0$ which inevitably does not only affect far-off places where we cannot be, but effects us on a daily basis. As outer space is stretched to its limit, its limit will continue to stretch but while it is stretching it has to having more than it had before in that outer space holds the limit of heats expanding possibilities. Singularity has been expanding since way back when but that means singularity is still releasing heat as space-time that turns out as space in the universal time of outer space. In outer space heat cannot expand more therefore except for the continual growth that benefits all singularity throughout on a continuous bases concerning all outer space.

If singularity expands when heated and there is a limit to the point it can heat, and that point of maximum expanding has been reached through the unleashing of heat, which is turning into space, we can with great confidence declare that space as the hottest space there is. The space inside the star shrunk to the minimum there can be and that tells us the space has to be cold because of the shrinking took the space to a position where no space can shrink anymore.

That shrinking of no more space can only be inside the inner star and in that region where gravity is at its strongest. With outer space as expanded as nature may allow the space that grew could only grow in conditions of heat because heat produces expanding and expanding is the result of heat coming about. Space shrink because it is cold: that we know and taking this law to the star centre it means regardless of our interpretation of hot and cold, that area in the star centre is as cold as it can get notwithstanding what our nature may tell us. Then obviously the same must apply to outer space for precisely the same reasons because it is so hot it can expand no more. We look at the hotness of space and the coldness of space but it is the relevancy to the solidity that forms the actual heat and cold limits. It is so hot no expansion can produce more space in outer space, as the outer space seems hot and quite the opposite reveals the true scenario inside the star in the centre of a star structure. That means the number of protons in motion has a lot to do with the cold and hot scenarios because where the protons are most dense the cold is in extreme. Only in the absence of space can so much heat gather in excess and the opposite is true about outer space where the least denseness found brings about the space in heat found in outer space. Our human selecting of hot and of cold and what is hot and what is not prevents us the clear vision we would have when truly understanding the applying temperature. Temperature comes about from spin and the smaller the spin density is the colder the space becomes because the more duplication produces the most cold. We think of outer space as 0^0 Kelvin but in fact it is as hot as no other place can be in the Universe. The coldest is where material is freezing solid as material does when frozen solid and the hottest is when by boiling the material is going into a gas with liquid being the intermediate position where heat acquires the space to perform as a flexible substance.

When we look at particles in outer space we see the particles being frozen. It is because there is such a severe contrast between the particles and the environment surrounding the particles and not the particles that is so frozen. The particles are in a gas state because the particles do not form a part that is part of the space unit. Hydrogen clouds of hundred of light years in diameter are a common sight in outer space. The heat we find filling space is not part of the space but like the particles the heat is a separate issue. That heat filling the space is another form of material that could conduce by diverting from space or marry the union of space by becoming more space. If it were that cold which we think it is, it would not have expanded into such a massive cloud but would have contracted forming a cube of frozen hydrogen. But as we can see the cloud expanded the gas as far as the gas can expand.

That expanding is indicative of heat and has extremely little to do with gravity. Because outer space is completely overheating the condition it has in support of the particles makes the particles appear to be in a state of freezing but the particles is counteracting the heat limit it meets. However the particles did not contract the heat because the space in outer space contracted all the heat by means of expanding the heat into what singularity will appreciate. That is not because outer space is freezing the particles it is because in contrast to the heat of outer space the particles seems to be freezing.

The atom must be the utmost coldest and the proton is even much colder because when that cold escapes it turns to heat forming space that no one can understand. When the spin of the atom allow the cold of the atom to release the heat it had it had frozen to space the atom holds but when this heat releases from the containing form of the atom it brings about much more heat than the Human mind can cope with. The fact that hydrogen remains a gas and so does helium in outer space must serve as enough proof that outer space is hot, regardless of our interpretation of the temperature gauge telling us what we wish to hear. One must look at outer space and judge outer space from the findings only considering outer space. If helium remains a gas it is hot. The removing of heat makes the centre of the Earth is cold although we see it as being terribly hot. The only reason why it can seem to be hot is because it is cold and in such a cold environment the heat can gather and space can collect heat because the particles find the surroundings extremely cold.

The cold in the Earth centre causes the concentration of heat by space reducing, as all cold surfaces tend to do. If it was hot the space within the Earth would expand and the space within the Earth where we think so much heat is concentrated does not expand therefore it must be cold. To gather and accumulate the space in a liquid means it became much colder being a liquid. Finding the surroundings terribly cold will allow the heat to gather and not expand but when the surroundings is hot it will not tolerate more concentration of heat and thus will expand to rid the balance of excess heat within space.

Look at the Sun and see how the Sun turned the hydrogen to a freezing cold liquid at 6500 K. Hydrogen is in a fluid state within the Sun and is colder than the hydrogen that is in a gas form in outer space. The Sun is the coldest place in the solar system. That is when the protons oversupply the removing of space to produce the cold that is so apparent. By the reducing of space it can concentrate heat to a fluid state by producing the opposing cold that finally freezes the heat to a solid state. The expanding of space is a way of duplicating space without reducing space and by duplicating in the form of expanding it becomes just the opposite to duplicating by motion therefore reducing space by halving space in time. That is what gravity does. By motion space duplicates and by space halving it removes heat in space as well as by dismissing space. In all the applying of gravity space bites the dust. The density of the protons brings about space dense enough to harbour the heat in such quantities and visa versa applies in outer space.

We have to accept that the coldest place in the solar system is in the very centre of the Sun because there the most number of protons sharing the least amount of space producing the coldest area that can allow therefore the hottest density of heat within the cold environment. Later I will show why the star is so extremely cold and outer space is over boiling with heat expanding into more space. We have to see what forms space and why space can be the absolute basic container through which gravity can relay the influence it carries.

We must come to realise that whatever forms space has to be that same ingredient which also is the basic component that forms the lot of everything in the entire Universe. When particles heat up the particles expand the space the particles hold to limit the heat rising. The particles claim more space when heated to preserve the cold. The claim to more space produces more space and reduces more heat. Such expanding brings about cooling. When particles heat or cool motion applies in some form. Motion started at a point when the Universe was extremely hot and there was no space. By introducing motion space formed and the lack thereof produced friction that became heat that became space.

The application of gravity that condenses space and bringing about heat by the compressing of space we apply in the way we go about tapping into the energy that nature provide. Internal and external engines combustion engines all rely on this application for harvesting motion by driving power. Compress space even today with a piston in a cylinder and then pump the compressed air into a container and such confining of space will increase the heat by the piston effort to reduce the space brought about in the container. The heat coming about inside the cylinder has no relevance to particles colliding because all compressor cylinders cool down colder because when that cold escape it turns to heat as the heat releases from space forming a secondary form of material forming space that no one can understand when the spin of the atom allow the cold of the atom to release into uncontrolled space. This release and unifying with space that heat does is the heat it had frozen heat because of the motion of spin to space that the atom holds remains in a frozen state under the guard of the spinning electron. But when this heat releases from the containing form of the atom frozen by the spin of the electron it brings about much more heat than the Human mind can cope with.

The fact that hydrogen remains a gas and so does helium in outer space must serve as enough proof that outer space is hot, regardless of our interpretation of the temperature gauge telling us what we wish to hear. One must look at outer space and judge outer space from the findings only considering in the terms which outer space insists upon. If helium remains a gas it is hot. The removing of heat from the space that contained the heat makes the centre of the Earth cold. In our Universe we see it as being terribly hot because the heat then forms a separate substance but remains a form of material (8) but that is because we see the heat and not the space derived from the separating of the heat.

The only reason why the space can seem to be hot is because the space is cold and in such a cold environment the heat can gather in a much concentrated state and space can collect heat because the particles hold concentrated heat in the space separating the particles. By removing such high concentration of heat from the space that use to be expanded heat, the space then must contradict the heat by being extremely cold. We look at the heat in the space, which by that time is another form of material and find the surrounding heat in the space hot while the space is extremely cold. The cold in the Earth centre causes the concentration of heat by space reducing, as all cold surfaces tend to do. But the proton contributes that reducing of space. If it was hot the space within the Earth would expand and explode but the space within the Earth where we think so much heat is concentrated is so much it does not expand therefore it must be cold. To gather and accumulate the space in a liquid means it became much colder when the space parted from what then is being a liquid. Finding the surroundings terribly cold will allow the heat to gather and not expand but when the surroundings is hot it will not tolerate more concentration of heat and thus it will expand to rid the balance of excess heat within space. The concentration or release of space with heat or space from heat is a direct contribution of the singularity in control of the space-time. The regard of the singularity stipulates the conducing of heat in space or the release of heat to form space by means of bisecting the occupied space.

Look at the Sun and see how the Sun turned the hydrogen it holds captured in its atmosphere to a freezing cold liquid at 6500 K. Hydrogen is in a fluid state within the Sun and yet it is still colder than the hydrogen we find in outer space that is in a gas form in outer space. The Sun is without any doubt the coldest place in the solar system. That is when the protons oversupply the removing of space to produce the cold that is so apparent in the heat levels that do not join the spell. By the reducing of space it can concentrate heat to a fluid state. By producing the opposing cold that finally freezes the heat to a solid state we find that is what matter is. The expanding of space is a way of duplicating space without reducing space and by duplicating in the form of expanding it becomes just the opposite to duplicating by motion therefore reducing space by halving space in time. That is what gravity does. By motion space duplicates and by space duplicating the material must be by dividing or bisecting - halving it removes heat in space as well as by dismissing space and in that concentrating heat. In all the applying of gravity space bites the dust. The density of the protons brings about space dense enough to harbour the heat in such quantities and visa versa applies in outer space.

We have to accept that the coldest place in the solar system is in the very centre of the Sun because there the most number of protons sharing the least amount of space producing the coldest area that singularity can allow therefore bringing about the hottest density of heat within the cold environment. It is the duty of scientist to look far beyond the ordinary and find why the inner star will be so cold and as to why outer space will be so hot while being seemingly so utterly cold or hot in humanly applied standards.

Later I will show in much better detail why the star is so extremely cold and outer space is over boiling with heat expanding into more space. We have to see what forms space and why space can be the absolute basic container through which gravity can relay the influence that it carries. We must come to realise that whatever it takes to form space it has to contain something that is the same ingredient, which also is the basic component that forms the lot of everything else in the entire Universe. When particles heat up the particles expand the space the particles hold to limit the heat rising.

The particles claim more space when heated to preserve the cold. The claim to more space produces more space and reduces more heat. Such expanding brings about cooling. When particles heat or cool motion applies in some form. Motion started at a point when the Universe was extremely hot and there was no space. By introducing motion space formed and the lack thereof produced friction that became heat that became space.

The application of gravity is that which condenses space by bringing about heat with the compressing of space. Compress space even today with a piston in a cylinder and then pump the compressed air into a container and such confining of space will increase the heat by the piston effort to reduce the space brought about in the container. The heat coming about inside the cylinder has no relevance to particles colliding because all compressor cylinders cool down with time moving and not necessarily with the loss or release of particles. It is not only the discharging of air that will reduce the temperatures inside the container but the time flowing bringing motion about where the motion is not about particles escaping but heat escaping in the replacing of the heat density (not the density of the particles forming the material content within the container) but the space that compressed to heat will also bring about that the heat displaces through the container wall to the outside. After the pumping of air increased the heat in the cylinder which even can go to dangerous levels, the heat will reduce back to room temperature when further pumping seizes and that stops further air movement into the cylinder and such surging of pumping air is what brings about heat stabilizing.

When particles collide such collision forms an atomic thermo release and that action we call an exploding atomic bomb. What principle this argument about particles colliding ignores is that all atoms use negative charged electron forming the atomic limit on the outside forming a definite border to the boundaries of all atoms and in both electrons from different atoms are being negative charged. The closer they come the more violent the rejecting will be and such rejecting is the production of heat that will turn to space. The electrons repel other negative charged sub atomic structures, which the electrons are that form the outer borders of all atoms. With all electrons highly negatively charged (being as negatively charged as any possibility will allow to match the utter extreme) such electrons couldn't touch.

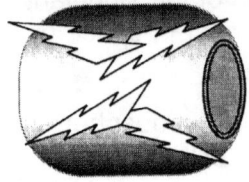
The particles entering the cylinder bring with them an envelope wrapping the atoms in space that is there to distance atoms from one another. Such space formed because of and under the conditions prevailing outside the cylinder walls.

The balance at first favours the forming of heat from the space coming in and being reduced in the containing size they are squeezed into is reducing the space from what it was on the outside. The space distribution inside then has changed considerably and reduced a great deal compared to conditions outside and with the decrease of the space distribution that space then becomes excess heat on the inside.

The electrons will disallow any contact directly between atoms. No force can be big enough to enforce such touching. It is because of that contact rejection electrons bring about that science has to use an overload of neutral neutrons putting them in the atom nucleus to fake a complying of charges that will eventually lead to atom touching each other but that is through enticing a neutral stance which is enticing a positive overload for a short while. When the touching of electrons does take place the event is called a thermo nuclear reaction where heat is released in unmatchable quantities and the atoms in reaction dissolves into a liquid heat. But other than producing an artificial balanced bomb the touching of electron will never take place since the repelling that they would provoke amongst one another. The increase of heat by the distribution of particles in space connecting space and heat to particles is a separate issue that has nothing to do with contained particles colliding because why does it stop when pumping is seized. This ratio of heat reduction is time connected as much as it is motion dependent. Motion reduces space by expansion as much as time contributes to space distribution by allowing the flow of heat.

This means it is not the particles touching one another in the cylinder that is bringing on the heat levels that is rising. Neither is it the particles that will eventually bring about the explosion that will follow, if safety measures are exceeded and should the pumping continue regardless of the danger rising. When the pumping stops the heat immediately starts the reducing thereof. Most important is the realising that every atom constitutes of two parts. In fact the entire Universe constitutes of the two parts I am about to mention. On the inside there is a circle that contains the sphere and holds material in contact with singularity. On the outside there is heat surrounding the inner material part within the sphere and distance the inner material from the space between it and the next atom. The electron forms the division between heat uncontained and heat contained. This is why the Roche factor is so very important. There can be friction between particles in reduced space under controlled circumstances where such particles are grouped together in a unit and as a unit elects a group singularity forming the centre of the chosen form of the unit.

The Universe separated heat from material by covering the exterior of material with heat that forms space. Some material became softer by uncontrolled overheating while others remained more solid by containing form through controlling the overheating. On the outside of all elements there are a layer that is the heat the element uses in relation to place relevancies between such an element and the rest of the cosmos.

The space surrounding the craft becomes liquid as the space becomes more intense in concentrated space that forms heat. There can be no particle in friction and even more so way up there in the atmosphere at the altitude where the cosmos meets the atmosphere just because the particles up there are so sparsely distributed in that part of the atmosphere. Above and beyond this lies the fact that all the so called air particles are very volatile and excitable by nature and they are known to turn the slightest heat into rapid motion thus establishing a scene where the particle that supposedly are in contact with the aircraft sheeting will move away from the hot incoming aircraft. If then not for any other reason then it is because the particles are highly volatile and acceptingly sensitive to heat. Airborne particles are prone to motion just because it is the airborne element nature to change heat into motion and the motion comes about from their sensitivity to duplicate. No particle in the air being part of the space we call air which is in a free floating in that air can produce friction because of the volatile nature those elements have.

The craft's coming into the atmosphere produces a point where $a^3 = T^2 k$ changes to $k^{-1} = T^2 / a^3$ (the explanation is forthcoming a little later on) The distance separating the incoming object from the Earth centre reduces rapidly therefore the object start to descend towards the centre of the Earth.

That point will rapidly increase the time factor where the incoming object crossed such a very visible border. What happens is that the applying gravity reduces the space a^3 and the compromising factor comes about since the time factor T^2 moves back to a time where outer space was as dense back then as the density we now have within the atmosphere that then became as the Earth atmosphere. It is outer space that remained denser that what the outer space currently is. While the gravity of the Earth contained the space surrounding the Earth in a much denser packed envelope the area not under the direct influence of the Earth governing singularity became more spacious.

The contained Earth atmosphere grew denser as the solar system developed into what it is today. As the atmosphere released from what we think of as outer space that release from outer space made the atmosphere much denser and the space above the Earth which is using a reducing time factor and that makes the Earth more compact. That established the T^2 factor to be that more condensed when one compare in ratio the density with outer space. The density at the time there was when the separation came about in outer space at the time of such parting outer space allowed objects to move away. This parting brought a barrier that is in place between the Earth and the outer space and any object coming from outer space into the Earth's atmosphere. The incoming object then would have to reduce the measure of the space the craft holds as the containing singularity set new standards applying to the incoming object with which the craft then needs to affirms its form and its status within the contained space of the Earth. The reducing will then suddenly no longer use space as the compatible factor but the focus will shift to the time factor that dictates to the space what the space can be. Such reducing comes from the switch there is in space – time where it was in outer space performing as being $k = a^3 / T^2$ to what it has to be within the Earth atmosphere $k^{-1}=T^2/a^3$. When the atmosphere grew apart from the outer space there are two ways of looking at the event. One can think that outer space expanded by the implication of the Hubble constant or that gravity withdrew the atmospheric space of the Earth at the time that the parting of space came about. But however you look at it there was a time when both outer space and the Earth's atmosphere shared equal density as we find it still applies on the moon and on Pluto. The space component is reducing the time component by compacting space to alter the space – time ratio.

This is portrayed by Kepler's formula $a^3 = T^2k$ It shows space as the density of space decreases. The Earth still compact space by reducing the volumetric confinement of space $T^{-2} = k / a^3$. This we call the atmosphere, as the atmosphere becomes denser towards the Soil of the Earth. There is a change in the time component. Most evident of this is when studying the pendulum. Just as we can see in the pendulum swinging, we can see that the swing reduces. Such reduction is because as the space diminishes every time the arm rocks from side to side. With this there is proof that in the developing atmospheric space of the Earth the ratios change from outer space. This is proved by the pendulum arms that Galileo's experiment used to show that the swinging pendulum indicates $k^{-1} = T^2/a^3$. Further more it proves that Galileo was correct after all and unnoticed by science Kepler helped Galileo prove Galileo's point.

Every element stands in different regard to the heat surrounding the material, which makes us consider the material to be either a gas or a liquid or a solid. The material in every element there is as such is all three forms and not of the forms one at all. It is the way under which the circumstances is presented that the element allows the heat to gather and accumulate as the surrounding heat occupying he surrounding space. Every particle is unique in the way it regards the heat to material ratio and how much heat it uses to form either the gas liquid or solid state. If space a^3 declines then so must motion in relevance will have to compensate by reducing k and limiting T^2 because space a^3 must always be equal to motion T^2k

As the space surrounding the Earth, which we call atmosphere reduce in volume of space the heat content rises as much as that the space holds heat having the heat rising by the same token. By becoming less the space also become hotter. The ratio there is between space and heat increases as space in measure reduces. We have to learn to see heat where the heat in the space has two different identifiable substances. We also must see material holding space to be different from the space holding the material. We must see material to be different from the heat covering the material and compromising the space that produces the format of material in being a solid, a liquid or a gas. This changes in the state of materials holds a direct relation to the heat that also claims a steak in that space.

I wish to run through the explanation in some entirety once more since this has profound implications on science and have never been concluded before. It seems the aspect people do not understand about the sound barrier is the interpretations there is about the sound barrier or more specific and to the point are relevancies. The truth runs completely against the grain of what Newtonian science advocate.

In contrast to Newtonian view about a spontaneous effort there are in the cosmos of joining and sharing, quite the contrary is true. There is a natural tendency to remain independent and away from each other and where the tendency of staying apart is bridged, there is a tendency to destroy and conquer, to control and delete the lesser by an onslaught of the more superior. There is no mass fighting to join mass and to become one in all. That part is fiction as much as the part about a force is fiction. There is a struggle for superiority and there is a fight for freedom from dominance. The whole idea about masses joining and uniting runs very much against the basic fabric of cosmology and in particular the Big Bang theory, The sound barrier principle, the Coanda affect of motion bringing about space-time control and so many more.

If the joining was with merit, we would by now not have known a moon orbiting apart and on its own coarse around the Earth around the sun around the Milky Way. While there are those attachments they are only attachments and not obsessions of joining and uniting. The moon holds a separate identity, which it refuses to relinquish. This refusal we call a lunar cycle. The moon is on a running spree ever since it's Independence Day. The moon is taking a route that would progressively carry the moon further way from the Earth as the Earth is rerouting its orbit further way from the sun. The question to ask is what makes the sun, the sun and the Earth, the Earth and the moon the moon. It is $a^3 = T^2 k$ or better put it is $\Pi^3 = \Pi^2 \Pi$. It is the space-time as collected by singularity using the Coanda affect

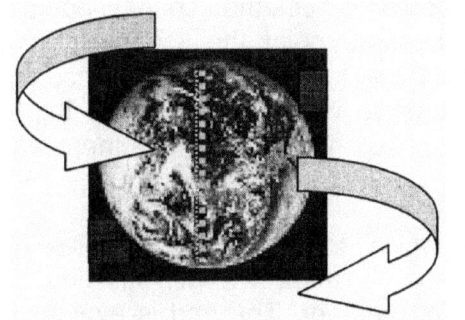

We can see why the Earth has space-time when comparing the principle to that of the Coanda principle. The motion establishes a centre and the centre activates a point where singularity is charged. The motion creates a point that has no motion. The point in the centre kills off al motion by dismissing space into space-less-ness.

This was established by motion charged by space-time. The foremost question arises in asking why would motion charge space-time. The atoms fill the unit, which we see as a star/planet/whatever. In every unit all the atoms spin sequentially in pairs. Those spinning do not start spinning in a direction just because we wish to have it spin in such a direction or because of magic as Einstein once told others about his suspicions. The spinning of the one is in harmony with a double particle spinning in opposing direction but at equal timing because they entice such harmonised spinning to bring about a centre charged with gravity. That means a flurry of dots came about synchronised in creative spin by creating space-time.

Many dots spawned from the spot and such spawning was in flurries but equal.

Then the four cosmic pillars set laws and progress started happening as the cosmos applied and stuck to conditions set under these principles.

Some singularity formed a dominating role as they were in dominating some particles formed an subordinate role but space-time formed since there had to be motion and motion provide form which can be space. But I wish to place a distinction between form and space because by calling what was in progress space, the mediate human connection would be to adapt space by dimension to what was in place. If there were space it was Π, which is form. There were no mathematical equations because for mathematics to be in place there has to be space. There was no space of that particular type yet invented. Please think clearly as this is very important, is that not precisely the commitment we find in gravity, where gravity is flowing from singularity outwards but never favouring any side? This reasoning prompted me to look for singularity in such a spot because if the prime spot from which all came was a spot holding all, then the spot must hold the shortest line but more prominent it will hold the smallest form including the smallest circle or for that matter the smallest sphere. With gravity always being in the centre of a sphere where the space is least available in the entire structure (there is not even space left to fill) one finds a flow of gravity from that centre spot outwards in all possible direction even-handedly. The fact that the original gravity will begin as a circle or will be a circle is the direction it will take when being the first spot created. All progress will be evenly in all direction because no direction will stand out or be in favour above any other direction at first.

The shortest possible line (hypothetically) must be so short it must have **an initial and ultimate point sharing the same spot.** Any theoretical line being the shortest possible line cannot have the line holding the initial starting point at point zero and advance from there. If it used zero as a start, the zero part would not count, because the line will only start at a point past zero where the line then will start. Zero ultimately means not existing and then that point, as a start does not exist. At one point the reducing attempt of the line would start making the use of mathematics seem silly. The reducing would seem tedious and leading nowhere. But as sturdy as mathematicians can be they would carry on (or so I am made to believe…). Then when the man doing the calculations gets carried off in a straight jacket, while the man is making funny noises, when he is totally cracked mentally, the calculations can still go on and on and on and…and that is where sanity prevail and someone says "drop the affair". It is at that point I would have loved to see Einstein carry on counting stars in so many galactica in his attempt to determine the critical density joke. That is not where we get into infinity. That is where man's brain gets blistered but infinity is still far off. As any one can see the Universe is far beyond some insignificant and senseless formulas invented to impress Academics while others are kept busy and free from boredom, but in the real Universe the attempt is not worth the thought it takes to disregard the attempt.

When the line **has a beginning and an end at the very same spot** and it wishes to extend the position as to further the possibility it has, which direction should it favour. Extending the line in any one direction will favour one direction without any clear reason not extending in other directions. The only mathematically sensible option about extending will be in all directions equally in order to give a meaningful non-bias flow of mathematical equilibrium. That is where one would have to go look for the beginning of the Universe. The Universe is a about lines connecting but where does than connecting end. The Universe used form to this point in development, but then at some point the line came and established the presence it still has. The first form was moving from $\Pi^0 \Rightarrow \Pi$. Again I wish to press the issue, at that stage form was in use and not mathematics. The Universe was just simply too big to measure. If radius did apply, one could use r and r2 but since only Π was in use there was no radius used.

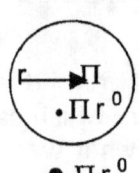

There is forever one circle leading to the next circle, which is followed by the following circles. Where the light does not reflect the image that is there we will still find on concentration of circles leading another and another and another. The end is eventually endless.

The fact that singularity was in control creates a problem there are no solution too. Being in the single dimension the singularity only provide an identical duplication of what the previous was and referring that from the previous on an uninterrupted basis flowing through the present without detecting or distinguishing change to the future as a precise unchanged replica, With no obvious or touchable change the Universe found whatever was in a stayed of exchangeability. There was no visible light since light or vision would influence change. Everything there was were sealed in a container that placed every possible value there is or might be to the same as every other value there might be.

To finally enter the realm of what we are referring too being in singularity with no escape from singularity, such entry will go beyond human conception. On will have to reduce the line in the circle of any line in any circle to a point where the end and the start of all possible directions are fitting onto each other and totals to the value of 1 to the power of zero. The progressing out of this singularity must involve a dynamic we are so aware of that we are unaware of the dynamic that takes place. Multiplying singularity would only reproduce one, which is a continuing of singularity without adding or removing one dimension. It leaves what ever is in the dimension of the single with out allowing change to the alteration by adding a multiplication.

What can accumulate without bringing increase and what can produce more form without enlarging the form? If the form did enlarge, the r would be more or growth in size would lead to r increasing. But as one will notice form increased by spawning new forms holding singularity, which are expansion but not growth.

As said before we now know Π came about since Π is achieving form and not space. Only **r** can establish space as size will accumulate and with everything else singularity had **r** covered by one as in being **$r^0 = 1$.** By reducing the circle radius **r** by half continuously will lead to an infinite small circle and an infinite number holding r would place **r** to the power of one as a factor. Then as a factor **r** would not contest any change when change is introduced into any future equation but Π will remain because the circle as a form remains even being infinitely small. By reducing r indefinitely to the tune of half each time, r would become infinitely small, beyond human calculating means, however as mentioned in the case of the smallest dot holding one spot, r would become insignificant beyond human comprehension even, but never reaching zero and still Π would remain intact and dictating form. To amplify by dimension a value has to be set to r but if r remained covered by singularity all alterations that could possibly come about was in the form, which was Π.

This expanding can be a problem one can wrestle with for one lifetime and never reach any conclusion. How can something grow without getting more that what was before? Then it hit me like a ton of bricks. The answer is in heat but not heat, as we know heat. It is heat in getting relevancies between outer limits. Only heat could break the monotony of singularity. Heat in the form we now know heat as heat is now. Since the Big Bang heat is material transforming from one state to another state. The change that took place involved singularity but singularity was 1^0 and being $^0 1$ could not grow. The growth came about. Heat rose from singularity, but if heat rose from singularity> Singularity as a factor changed from 1^0 to 1^1 which means a relevancy came in place that no one could detect. It is true that 1^1 are still one, but one could then escape from singularity by producing factors other than 1. Heat came about but only as a relevancy to utter cold. If there is heat there is cold or if there is no heat there can be no cold. Space came into forming a relevancy that brought form. But since it is a relevancy and not a generation by accumulation the form produced was Π. The spot formed a dot by heat and cold establishing relevancies and from that singularity was broken to allow all other forms of relevancies to come about. The cosmos did not start because of gravity. The cosmos started with heat and cold coming into a relevancy and in the cosmos there is no hot as much as there is no cold. The cosmos broke put from the confinement of singularity by establishing a singularity in a relation of heat and cold. The heat that cam about was beyond measure because the cold that held the heat was also beyond measure. The immeasurable heat was on the outside of the dot that formed and the cold was on the inside of the dot that formed. The cold contracted because in nature cold contracts. The heat expanded into a dimension of form and heat by expansion is in nature about motion/ Motion is duplicating that which is and heat is what is duplicating by motion. But only heat by expansion was possible because in affect singularity cannot move. The motion became contraction, as the motion was the result of heat expanding which was forming four points in the rim of the dot. The expanding of the points created motion in relevance of a centre that formed because of the motion, which established an immovable centre as the Coanda effect, placed more dots in relation to more dots that formed.

Every dot was Π and every dot formed $Π^3$ because of the expanding heat, which produced $Π^2$. With that a new relevancy came about forming a centre in-between the four points of expansion that was resulting in time. But since the points were in themselves singularity, which is immovable and space-less, they still heated forming a cold centre with the heat bringing about motion. It became a repetition where infinity broke eternity by producing a centre because of space (or rather form) forming the motion to enable the space to form in relation to the heat applying motion. This brought about a Cosmos being conceived.

The spot forms a full circle, but the line running through the circle is forever present because that is the future radius of the circle that will one day develop the circle, which is equal to the present diameter. The fact of the presence of such a possible line in such a possible circle dividing the possible circle into two parts makes the centre line equal to the half circle. The line forms the half circle but not only that the line presents the half circle as much as the line is the half circle. The line then is 180^0 and the half circle is 180^0 because in singularity the two factors are the same. The same value is of course $Π^0 = 1$.

In this half circle of the future, which is no half circle as yet because of a lack of space there are three future points indicating the space less ness that will go on to become space filled with something. On top of such a circle to form must be a marker indicating an awaiting boundary or future border and at the bottom of the future circle there also must be a similar marker that is no marker as yet. Between the two possible points that are not there yet is a future line running that is not there yet. Then indicating the possibility of a position to come that will bring about the half circle being a future distance apart from the future line indicating a diameter that will one day be there a third such a marker must be established for the future. That forms a triangle with two more sides being connected by either a line being one or half pi being one. From singularity comes about that the line is the same as the half circle is the same as the triangle and all has one value being 180^0. From this come the most basic principles in as much as forming the ground rules of the law of Pythagoras.

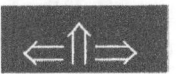 The one form we know there was and was valid at the time was the straight line

 The next form we know there was and was valid at the time was the half circle

The other form we know there was and was valid at the time was the triangle.
Everything at the time that was outside singularity and was in form at that time was equal. Think how big they were. They filled larger parts of the Universe than our brains can cover by thought. They formed the holding tanks that still hold us in the massive Universe which they. They were at the time to small to have size, but since then they grew into structures that are too big to have size. Those dots still are bigger than mathematics can go because there still is no quantifiable mathematics can reach. Jus because they compare with what we seem to preserve as small in cosmic relation they are too big and too large for us to comprehend. Even if they were immeasurable many they filled an immeasurable Universe in the same way they still fill the immeasurable Universe and we are so small we and our surroundings are measurable and quantifiable. They were so enormous there were no relevancies applying to compensate for distinguishing. Distinguishing only followed later when size started to matter. That meant the Coanda

effect was in place without the Roche or the Titius Bode law. It was the start of the relevancy principle from which the atom later came and which is the result of the Kepler expression.

The Coanda principle meant that some expanded at first and by expanding lost density to become liquid where as others protected density by conserving and contracting. The issue that is at hand is that from the Coanda principle cane relevancies and relevancies rule the cosmos.

⇐Bigger⇐⇒smaller⇒

·····················••••••••••••••••••••••

It eventually gets so small we humans can fit into it... remember we are seeing the reverse of the truth. It was never so big that it contained nothing because all that and we came afterwards being smaller that the dot filled the dot.

With no line possible there had to be another dot that formed since the Universe has many dots that formed lines. But let us not to get confused and lost in the range of possible diversions but let us stick to two dots. One dot was next to the dot next to the dot, but as I said we stick to one dot next two the second dot Π is the first step gravity began with. That leaves us with a huge problem in as much as when $r = 0$ then $r^0 = 0$ and 0 dividing any value will leave 0 as the answer. If the particles were inseparable at the start it must bring about that gravity would not be forming since the distance will not permit any dividing. By allowing the distance separating the particles to be zero, the particles melt into a unit. Again this is Mathematics and not my incoherency as some Academics dismissed my work. Let me run through the argument one more time because I have been insulted by Academics in the past telling me I am bending mathematic rules with my applying double values to try and produce some argument. The two particles formed by an inseparable unit separated by a sharing of a spot. We know that at least two spots formed because there are many more than just two that remained to become part of the visual Universe. Let us name the spots because that is what humans do best if they do not know what to do with what they have to do. Let us call the on dot and the other one spot next to dot we then call dot two. Between dot and dot two there were nothing because dot and dot two were inseparable. By they're being inseparable we would naturally be inclined to think that the separation value should be nothing or at least zero. But putting zero in that place is a mathematical excluding procedure leaving future mathematics excluded. With m multiplying m_2 and then dividing ÷ r with zero (r=0) such a procedure will leave the lot at zero and with that nothing is going nowhere. That means although we think the space between the two parts are nothing the non-existing space has to be at least one to be a future factor.

At fist $\Pi^0 = \Pi$. Then after a while $\Pi^0 = \Pi^3 / \Pi^2 \Pi$ and gravity comes about forming space-time by motion of form. Being the sphere that formed the 7 holding relevance in the form of the sphere took shape. But the Universe is layer in dimension forming the next layer in dimension forming the following layer in dimension. the Universe was $\Pi = \Pi^3 / \Pi^2$, which is taken from Kepler's formula he received from the cosmos as **$k = a^3 / T^2$**. Where there is a sphere involved there is a natural tendency to grow by developing the sphere.

Every part of the argument is sound but was never yet used. I repeat once more if my argument reflects on inconsistencies those inconsistencies are not about my work. In order to disprove my argument replace Mass one and Mass two with any number possible, then divide such a number with to the square being zero. If there was no space then the value of the particles had to be one. If there was no space between the particles the particles then had to form a unit. But if there is a mathematical possibility of reducing a line to the single dimension then there had to be a factor representing r as a factor of one. Take $(M_1 \times m_2) / r^2$ and substitute any of the factors with zero and the result coming about has to be zero. The factors in the equation have to have any and all the elements at a value of at least one. Only if r was a factor of one can gravity bring about any mathematical equation developing from this argument. That means the mass on both sides must have a factor of one being a limit, which does not allow such further reduction of r and any further reducing of r beyond the limit will not be tolerated. Only if $r = 1$ then r^2 can be 1 and mass can be apart. Like it or not but believing in the Big Bang must also bring about the accepting that the cosmos moved apart somewhat. The fact that r brought increase in the space separating the mass produces a problem that was solved already. About a century and a half ago Roche found just such a limit. Once again I were confronted by zero becoming growth. There is a huge hole that needs filling when bringing into a relation any forming of an alliance between a cosmos coming from nothing and filling with nothing and a cosmos growing spontaneously through balance shifting prominence. Mathematically the fact of applying nothing as a vale applying in the cosmos is not a strong and convincing argument. The minute one brings in zero as a multiplying factor forming a definite value working into the calculations of the cosmos, growth disappear. If growth was not a factor, the zero factors could be involved with some form of maintaining stability and where then further growth will accept the responsibility of zero

Taking nothing to mathematics zero locates in infinity by abolishing nothing from the universe. The sphere is 7 X Π = 21.991. It is a formula or a recipe for gravity. It shows how a sphere grows into a sphere by form on a continuous basis.

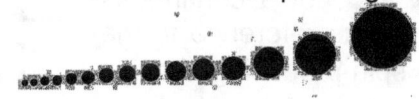

In the circle using $r^2\Pi$ the r has to have distinctive qualities placing it as a factor apart from Π. Where the growth shows no separate distinction but a continuous flow from the precise centre to the precise edge the flow would become in relation with Π depicting the circle and Π replacing r as reference to any point on the circle. By using r distinction in the circle is possible but by using Π there is no distinction possible.

The sphere is 7 X Π = but from the other view the singularity relating to ten. That starts the Titius Bode which in relation to the Roche limit at the limit of Π use form gravity Π^2. However this is a little more involved, as because the one sprouted because of the consequence other bringing the one into the Universe as a relative that to bring about gravity.

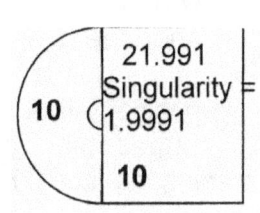

sphere is principal $\Pi^2 / 4$ to it seems of the will sprout

However if the attachment was **7 + 10 + Π^2 / 2 = 21.93 / 7 = Π** is the circle that serves as an attachment meant being a sphere and holding as well as sharing singularity. This is what Newton saw, but this is not gravity in the cosmic sense.

7 A sphere being formed by the six ends crossing as it incorporates the centre singularity
+ 10 anther sphere with an identifiable motion keeping the independent singularity apart but within the relevancy of the unit formed.
+ Π^2 / 2 Singularity by gravity shared by two in one unit.
= 21.93 / 7 still holding the unit to form where the overall containing form will be a cosmic sphere.

The sphere holds many dimensions relating to seven but also by the square of space which is 5+5 = 10. This brings about gravity generated by means of the Titius Bode law and is principle proof of this statement because it indicates an infinite number of numerical positions influenced by quarterly divided sectors around a point holding singularity.

Should that be a lead, there can be no universe incorporating all, but an accumulation of atoms all orbiting their singularity, producing a grand alliance protecting a forever larger universe accumulation. The centre of the universe is in every atom with space-time flowing away from that position in relation to that specific atom location, but every atom is securing each individual universe promoting singularity security in a forever-broadening structure. Because every atom holds the centre of the universe by rotation forming singularity, all other space including other singularity formed wherever in space and time, singularity located outside the boundaries of the specific atom holding the individual centre of the universe and therefore individual singularity, forms space-time to the singularity indication.

By reducing r indefinitely to the tune of half each time, r would become infinitely small, beyond human calculating means, however as mentioned in the case of the smallest dot holding one spot, r would become insignificant beyond human comprehension even, but never reaching zero and still Π would remain intact and dictating form. An observation coming instinctively to mind one may recognise is that the form reminds rather explicitly of natural phenomenon as hurricanes, water whirls and even the shape most commonly favoured to express the cosmic object referred two as a Black Hole. The similarity may be more than coincidental. Let us consider the statement in the reverse.

Anything occupying space in the cube will apply r, notwithstanding the name confirming the shape or r named as length width or height, it is all just a line bringing about the cube with all its other names that may find attachment form but nevertheless still remains only a six-sided cube with connecting applying different angles changing in some cases.

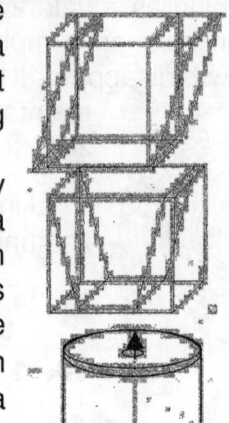

The normal perception is that any circle growing spontaneous would grow by radius, which is r. That cannot be the case because r is an indication of a line. By growing with the aid if a straight line the influence that would have on would result in many circles following one another and not a continuous. Gravity is the dimensional changing of space holding r as reference to the holding Π as the reference. In order to generate spin-producing time in occupying space, therefore creating dimensional change, Π has to be a indicating the possibility of spin. The answer must be in finding Π, and locating singularity.

used straight to specific lines

the straight the circle growth. sphere matter factor thereby

By heat expanding away from the centre motion comes about. Since the most incredible shortage of available space the motion has nowhere to move but to go to the other side of the Universe. From Π^0 the expansion brings about Π in three positions in line and four positions that should move, but since the positions are ultimately singularity, moving is not possible. The expanding comes about instead where the expanding is a direct result of relevancy coming in place between hot and cold.

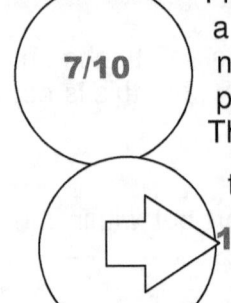

From the centre holding the original singularity position a relevancy of 7 points is relating to a new vantage to singularity where a line holding singularity with three points establish a new sphere in relevancy of another seven points, which makes the new point holding ten places.

The new point that formed on the edge of the established sphere has a relation of ten to the centre position from where the notion of motion is centred because of the applying of the Coanda effect.

Gravity dominating space
(10 / 7) \(7/ 10) = 2.04   which is equal to 1.4285   / 0.7   = 2.04

SPACE DIVIDED INTO TIME
(7/10) / (10/7) = 0.49   which is equal to .7   /   1.4285 = 0.49

SPACE MULTIPLIED WITH TIME
7/10   10 / 7 X 7/10 =2.04

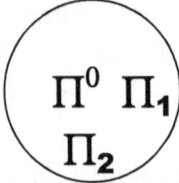

In order to establish motion of any kind Π has to shift from location and that shift is related to the centre Π^0. While Π^3 is established by the location of the borders the borders shift Π / 2 to another point Π / 2 and since it is time that is resulting from this effort all factors are therefore in the square. Π^2 / 4. There is another way of looking at it, which I explain elsewhere. However the relevancy attached does not change the affect the Roche limit in conjunction with the Titius Bode has on the gravity that is coming about.

THE PROCESS PARTED USING THE ROCHE PRINCIPLE
A is   10 / 7 $(\Pi/2)^2$ 7/10   2.04 x ($\Pi/2)^2$ = 5.033
B is $(\Pi/2)^2$ X 2.04 =   5.033
A+ B is   10 / 7   10.066

SPACE DIVIDE INTO TIME

$$7/10 / 7 / 10 = 1 \text{ NO INFLUENCE}$$
$$7/10 / 7/10 = 0.49$$
$$10 / 7 \quad\quad 0.49$$

$$10 / 7 \text{ /}\!/\!\backslash\!\backslash \text{ } 10 / 7 \quad\quad 7/10 = .49 \quad 7/10 = .49$$

$$.49 + .49 = .98$$
$$.98 \times 10.066 = 9.86468 = \Pi^2$$

TIME SPACE = Π^2 = 9.8696 = MATTER.

From this one can see that it proves that the relation of 7 and ten brings about a space decline from ten to Π^2 and that space decline becomes the value we think of as gravity. It comes about as the space decline in form to match the proton dismissing of space to heat and eventually to return the space to singularity.

Taking the relevancy of $\Pi^2\Pi$ to the relation of material in forming k or Π = 7 and T^2 as Π^2 = 10,
If Π is 7 in a single dimension then 10 is T^2 diverting from the single dimension a square in a natural state.

$7^2 / 10^2$ = .49 because 7 are Matter Square and 10^2 is the square of the square of space. The square comes from the split by singularity of the universe splitting the universe in two compatible but never inclusive sectors and space having two squares because space is as it is, is already in a square. That is the result from Pythagoras placing space in a square.

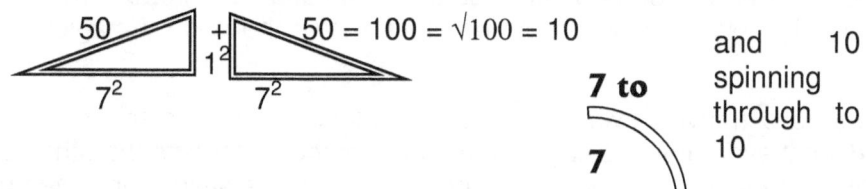

Space is already in the square but through time motion space goes from one point to another point in the square of motion.

$(10 / 7) \backslash (7/ 10) = 2.04$
$1.4285 \quad / \quad 0.7 \quad = 2.04$
SPACE DIVIDED INTO TIME
$(7/10) / (10/7) = 0.49$
$.7 \quad / \quad 1.4285 = 0.49$

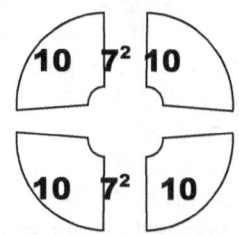

In the universe and even as far as hurricane and tornados singularity parts the circle around singularity into sectors of four as far as time goes.

Position 1 **(10 / 7)** divided by position 2 **(7/ 10)** placing two points of the four quarters of time. Position 3 is **(7/10)** and is divided by **(10/7)** positioning the other two positions of time in relation to positions occupied. This confirms the position material holds as a relevant.

Placing space in a position to time **10 / 7 X 7/10** Brings about the precise opposite of the first sector but not the opposite result because in this space dominates

Since the square of space is directly involved with the dimensional depleting of space from ten to Π^2 and singularity divide the affect of gravity being Π^2 into two sectors forming one double of the four quadrants of time, the affect of Π^2 may not be ignored since that totally dominated gravity from even before the establishing of gravity in the form of the Roche limit.

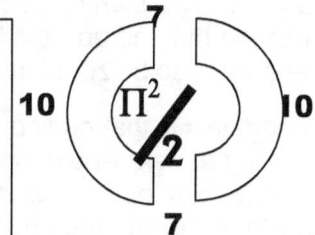

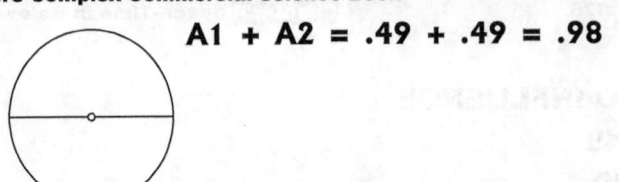

$$A1 + A2 = .49 + .49 = .98$$

$$A1 + A2 = 5.033 + 5.033 = 10.066$$

The result is the dimensional dismissing of space from a relative 10 to Π^2.

In order too understand the Universe one has to truly and deeply understand mathematics. One has to know why the triangle and the half circle and the straight line would all be the same. Why would Pythagoras apply and not only assume that Pythagoras apply, but understand why Pythagoras apply. What has the cross over of line in halving 1800 have to do with Pythagoras? Why would lines form and a ratio of three two one be equal in form when only form apply. For as long as man has the ability to think, man has been under the impression that a line stars off at a point that has the value of nothing. Lines mathematically cannot start at zero because there is no evidence of zero as a factor in mathematics. Where the point is, is as close as anything ever can be to a point that would hold nothing but it is not nothing. Lets again think about the line because we may go on about how "small" the Universe was art the event of the Big Bang, the Universe was at the time big enough to have innumerable complicated measuring lines criss-crossing and giving whatever dimensional qualities. That means having lines we fall into a steady created long established Universe that came about in all sectors. We are not looking for a Universe that has been around for donkey years. The Universe we are looking for must be newly born. If one insists on the Big Bang being the birth, we then have to start looking for the conception stage. The stage where the egg was fertilised and before that when the egg was formed and before that when the egg was conceived to have a form. That is where we must start looking for a start. The Big Bang then is where the egg hatched and all being farmers in one way or another (even eating the produce must deliver some idea of what farming is) we all know the chicken did not spring from the egg by magical wishful thinking.

I have been in contact with numerous (maybe even countless) Academics. The instant they see me referring to Kepler they disregard what ever I have to say on the matter as being far less than their dignity will allow. The instant that I refer to the start of the line they consider the conversation far beneath their social standing. When I bring in gravity at the level I work with gravity, they consider me unschooled as far as their level goes and they consider my conversation a waste of their time. If only one of them truly sat back and gave my reasoning a good thought it could have been worth their while. Please what ever your academic standings id, please chew on this piece of tobacco: The shortest possible line (hypothetically) must be so short it must have an initial and ultimate point sharing the same spot. Any theoretical line being the shortest possible line cannot have the line holding the initial starting point at point zero and advance from there. If it used zero as a start, the zero part would not count, because the line will only start at a point past zero where the line then will start When the line has a beginning and an end at the very same spot and it wishes to extend the position as to further the possibility it has, which direction should it favour. Extending the line in any one direction will favour one direction without any clear reason not extending in other directions. The only option about extending will be in all directions equally in order to give a meaningful non-bias flow of mathematical equilibrium.

The shortest line in the realm of possibilities must have a start and finish holding one spot and such a line will also be a dot or a circle. Not favouring one direction puts all directions at equilibrium meaning that any form of what ever might develop from such a spot with the end and the start being in the same position also has to be a sphere. This reasoning prompted me to look for singularity in such a spot because if the prime spot from which all came was a spot holding all, then the spot must hold the shortest line but more prominent it will hold the smallest form including the smallest circle. One possibility that the shortest spot can never have is having a starting point on the zero mark. If the mark of zero holds the start it must also hold the end because the end and the beginning has the same position. If the position of zero then is the beginning, the end will also be zero leaving the line without an end as well as without a beginning.

The conclusion from this is that no line can start at zero because that will be a mathematical impossibility. A line or spot starting at zero would therefore be shorter than the shortest line possible. A line growing or extending from zero can never leave zero because of the influence of being zero disqualifies any possibility of growth. If the line then had to grow in all directions at the same pace the line must therefore be a circle or being three-dimensional, a sphere. Flowing from this fact is that in the universe there can be no zero point or unfilled space. The value of the circle is Π, and that is where creation started. That gave me the clue where to start looking for singularity. One would find singularity in the value Π and the value Π will be in all things rotating in a circle. You might wonder how does that apply to the cosmos and moreover to gravity? Gravity is the dimensional changing of space holding r as reference to the sphere holding Π as the reference. Heat occupying space has the cube that can apply r, as a straight line bringing about the cube with all its other names that may find attachment to specific form but nevertheless still remains only a six-sided cube with angles changing in some cases.

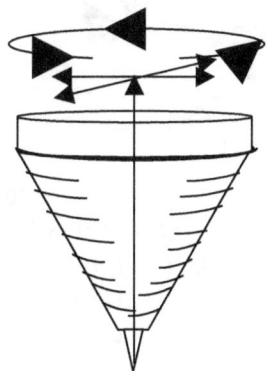

The Coanda effect proves there is not a value such as nothing in the Universe. The Coanda effect indicate that space in motion forms gravity

The Coanda effect proves there is a Universe filled with things marvellous and numerous because just by spinning the spin establish a centre and from the centre gravity comes about that control the entire Universe the gravity forms. The gravity is there because the spinning top stays upright. The gravity control the top by the motion the top generates.

Any object in rotation will have a middle point, a very specific centre point that does not spin. That point once again hypothetical but none the less must be standing still because every line running from that pint in opposing directions are also in opposing directional spin to each other

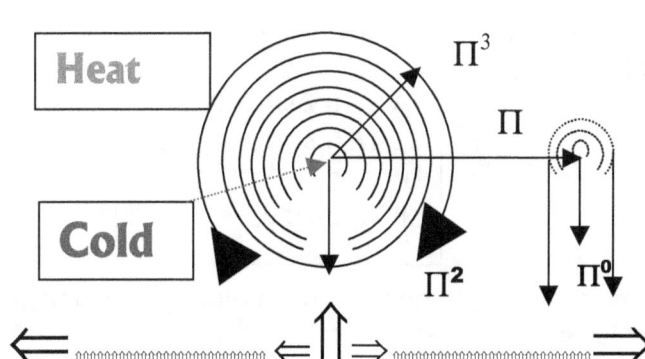

As the stop starts to spin the motion establish the centre line, which activates singularity, which activates space-time that activates gravity at a specific relevancy. Where we locate singularity there is not nothing because gravity cannot come from nothing because only nothing comes from nothing.

After all it is gravity that keeps the top as it is spinning in an upright position while it is spinning because it is gravity that stabilises the cosmos. Moreover, what is actually in progress from the top spinning is the Coanda principle activating gravity and that happens in accordance with Kepler's formula

When heat and cold broke from the confinement of singularity the heat created motion by bringing about a relevancy between what grows and what reduces. The relevancy applies a discrepancy between the heat that moves away from the cold and the cold that moves away from the heat. The heat by expanding has to move towards the outside, which is away from the centre, and the centre being contradicting the heat moves towards the centre, which is in direction moving away from the heat. One part moved away from a centre ($a^3 = T^2 k$; forming the relevance $k = a^3 / T^2$) and another part moves away from the centre as heat expanded from the centre creating motion in ($a^3 = T^2 k$. There is another part represented by the cold factor that is producing a relevance of ($a^3 = T^2 k$, which is manifesting as $k^{-1} = a^3 / T^2$)

Most of all it is what started gravity and that is what still is gravity. Gravity is the contradicting of motion by duplicating in heat or by contracting to concentrate heat. Above all, gravity is the motion of space as gravity started off as the motion of space.

From such a point every other point will be opposing any other point not pointing in the direction to which the first point is pointing, whereby it extends the direction it holds. No matter what the point is or where the point leads, such a point holding a specific direction will be unique in the direction it is rotating because at that or any other specific point wherever, it will be directing not in the direction it spins but in the direction flowing from the centre point outwards.

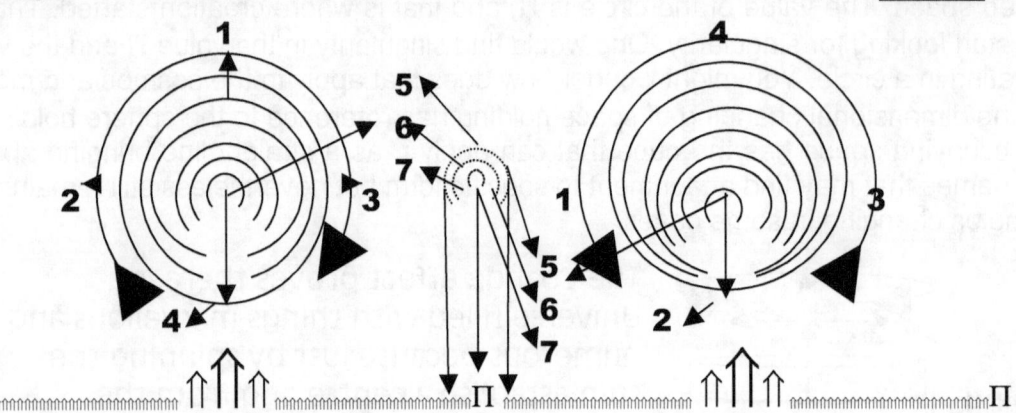

Any point will be it opposing itself within the rotating of 180° changing every aspect of its previous flowing characteristics it previously had or will once again have in 180° from there. While in rotation from the point of an outside observer all may seem static and never changing but to the object in spin every next second will be a diverting from every aspect it was in very second passing, and the direction it held in relation to the direction it held the previous mille, mille second will totally be incompatible with the direction it holds the very next mille, mille second of rotation. That proves no point can be static or constant, all though it may seem that way to outsiders.

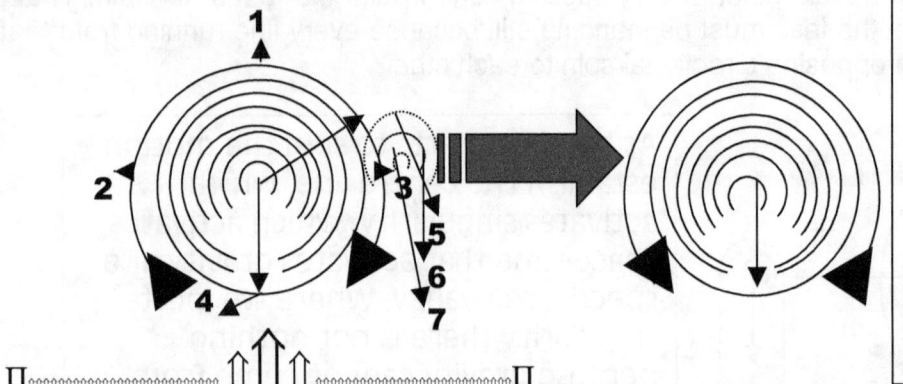

> As one can see with the spinning top delivering the Coanda principle, every point overheating can spawn space-time by centralising singularity.

One can see from the top that singularity is established wherever spin occur. The motion generates a position of seven in relation to ten and singularity manifests as 1.9991 as is explained elsewhere. That means any point formed by the sphere spinning can and does start a centre in which no motion holds no space and of which motion surrounds such a point by forming space. Although everything at the time was in the form as a multiple circle, which results in a sphere, the sphere was not the only form present. This too has to do with singularity interpretations. We see a cube, as we know the cube but at first when form came about the cube were not yet forms. .

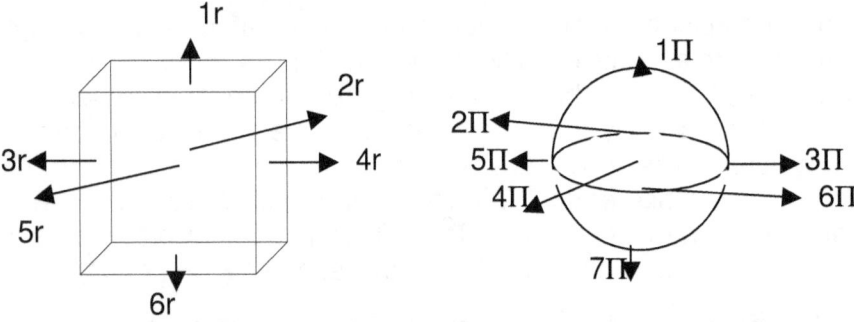

Singularity presents space-time with the sphere and the sphere there are no radius but only the extending of Π from the centre Π in six opposing directions relating to one another by the square but remaining Π because of the unity the matter holds in relating to space. Although what falls outside the sphere from either end may be a sphere as well, from the position singularity hold such space falling outside the controlled Unit being space-time in the sphere what falls outside the sphere from any position is a cube or uncontrolled space-time. With that in mind form holding the sphere is not measured as form in the sphere and while that is appropriate not to count as the sphere all spheres grew in measure of others and all sphere grew in form from any and all positions available.

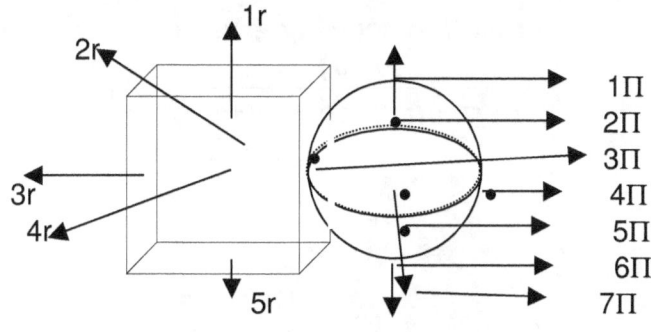

Every point securing a singularity that does not spin secure a singularity in control of the space-time that apply the motion because of heat centralising and bringing about motion through space expanding. That which is outside the sphere becomes a cube to that, which is inside the sphere

1Π
2Π
3Π
4Π
5Π
6Π
7Π

While the on sphere forms on this pot where the dominating sphere secures an edge the dot may be reserved as an edge marker to the dominating sphere. To the forming sphere in progress of emerging heat gathers at that point because the rotation is a result of duplicating and duplicating is the tendency of naturally growing in space-time $k = a^3 / T^2$. In order to find duplicating coming about there has to be heat in order to duplicate what will form heat. The duplicating process is a process of one factor going softer or less solid and therefore more dynamic that the other. To have singularity is to have gravity but to have gravity there has to be a point of motion and a point of sturdy. The point of sturdy may be in the centre of singularity, but then the solid must be motion. However even today it still apply: what moves forms liquid in the presence of a solid and at that point singularity presented the solid therefore what we might think of as solid was the liquid because it moved around the solid. Where the on factor is duplicating the other factor is compressing $k^{-1} = T^2 / a^3$

Because of the principal in which the Coanda works the motion will centralise a new sphere and by appointing six position around the centre three points will not move while four will move about the three points forming the centre line. The result is that the four points by duplication will reserve the point moving as the next point in singularity because of $k = a^3 / T^2$ singularity will be a natural result of the motion. Then that point will secure a position $k^{-1} = T^2 / a^3$ which will secure six points about such a centre. The centre will bring about four points spinning around three points holding a line singularity. The line in singularity will stand in relevance the contacting factor $k^{-1} = T^2 / a^3$ and the duplicating by expanding points will be four and serve the relevancy by contributing $k = a^3 / T^2$ as space-time only in form. From this the rest of the Universe burst into the next phase of Creation.

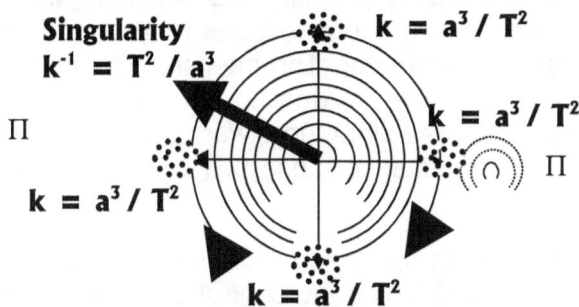

The points duplicating is four moving around a centre by the square of gravity. The motion is the sources of heating because the heat is bringing about the movement. The heat growth therefore provides the action because the action is what energises the points to provide the motion. The motion is purely is space-time duplicating and the duplicating is feeding heat to the centre from the four points overheating thus the points that shows expanding.

But also the duplication leads to the spawning of one point of singularity that provides the installing of the next centre for the next sphere.

Electricity and lightning is the absolute epitome of the Coanda effect where the Coanda effect is precisely the manifestation of light following the exact principles of the Coanda effect but the **Total Internal reflection** is not about applying motion by the flow of space-time (in this case water running) through the atmosphere but in the case of the phenomenon we call the **Total Internal reflection** singularity captures light and by setting borders the boundaries set limits to the flow of photons. But it is what electricity is only with much less intensity and it is the Coanda effect as much as electricity and lightning is the Coanda effect.

Gravity is electricity because electricity is the flow of heat from a gas source to singularity by charging iron **{7/10 (4((Π² + Π²)) = 55} (forming the artificial core exactly as the Coanda effect will charge singularity by applying motion)** in the influencing of copper $(Π^2 + Π^2) \times Π = 62.0$. That is the reason why only iron can excite to charge electricity and only copper can dismiss space to a tome equal to the flow of photons.

The cosmic cube we live in is $7/10 \, (\Pi^6)/6 = 112.16$

The six-sided cube $(\Pi^6)/6$

Forming the star

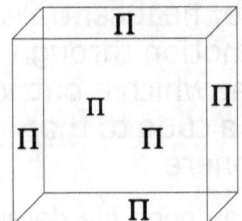
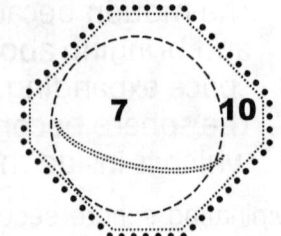

In motion applying gravity $7/10)$

$$10/7(4((\Pi^2 + \Pi^2)) = 112.795$$

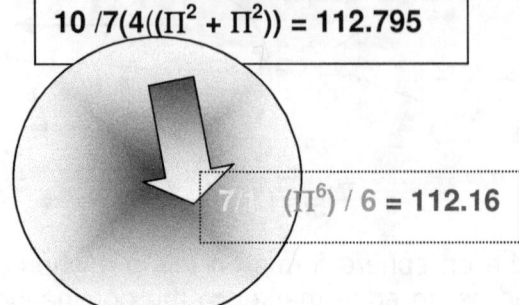

$7/\,(\Pi^6)/6 = 112.16$

Electron

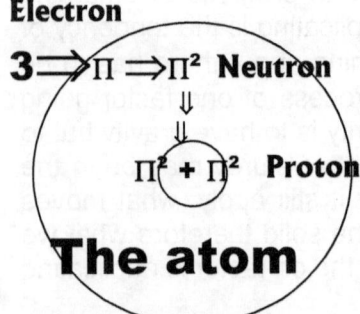

$3 \Rightarrow \Pi \Rightarrow \Pi^2$ **Neutron**

$\Pi^2 + \Pi^2$ **Proton**

The atom

From the value that outer space can support being the sum total of the particles forming the atom $((\Pi^2 + \Pi^2)$ **the proton** $+ (\Pi^2 + \Pi)$ **the neutron** $+ (+ 3)$ **space** $= 35.75 \, X \, \Pi$ **singularity** $= 112.313$ the star deliberately reduce space as the star intensify motion and that reconstructing of space-time changes the qualities of the atom from what we presume the atom to be to suspending the atom beyond the boundaries of $7/10 \, (\Pi^6)/6 = 112.16$. The converting of space-time from outer space through gravity to the star centre is the same route electricity follow

There is no difference between generating electricity and generating gravity. Every object turning is a gravity generator and every era is about generating gravity. The era in which we now are $\{10/7(4((\Pi^2 + \Pi^2)) = 112\}$ to $\{7/10(4((\Pi^2 + \Pi^2)) = 55\}$ and for that reason I named the era we are in the Iron $_{56}$ era. It means that space-time is conducted by motion from a dismissing density proton value of 112 protons in one unity to be concentrated by motion of a spinning Iron core to be concentrated to a density capacity of 55 protons.

It is similar to electricity conducting. Within the Earth at the core the gravity intensity one would find there would be equal to the strongest electric current possible on Earth. As the core reduces in diameter, so would the gravity intensity become more too match the reduction in diameter. The intensity was at that stage 961 which means the diameter density was infinite. At the time we are dealing with, the required charge to generate gravity was 961 protons in one cluster. It does not mean that there were 961 protons in one cluster, but it required the volts to shift the amps though the resistance those 961 protons in one cluster would establish.

All phenomenon used in the Cosmos is the precise same thing using the precise same principles in a more intense or lesser intense gradient. Still it is all about singularity charging the control and the flow of space-time through motion where a liquid flows through space to a solid iron core that is influenced by copper.

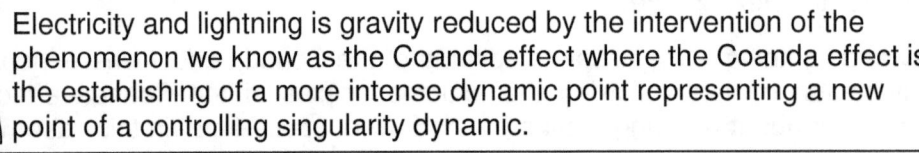

Electricity and lightning is gravity reduced by the intervention of the phenomenon we know as the Coanda effect where the Coanda effect is the establishing of a more intense dynamic point representing a new point of a controlling singularity dynamic.

Iron forming a centre or an iron- core precisely as the Earth core forms in iron form a centre core.

The copper field coils break down space-time by dismissing space-time as the space-time has to flow through the copper in the event where the iron being in motion causes the flow and charging the flow to the equal time set as the photon has. That is electricity. It is taking space-time directly to the centre f the earth because the motion $T^2 \, k$ excited the space a^3 to a level that gravity is within the Earth core.

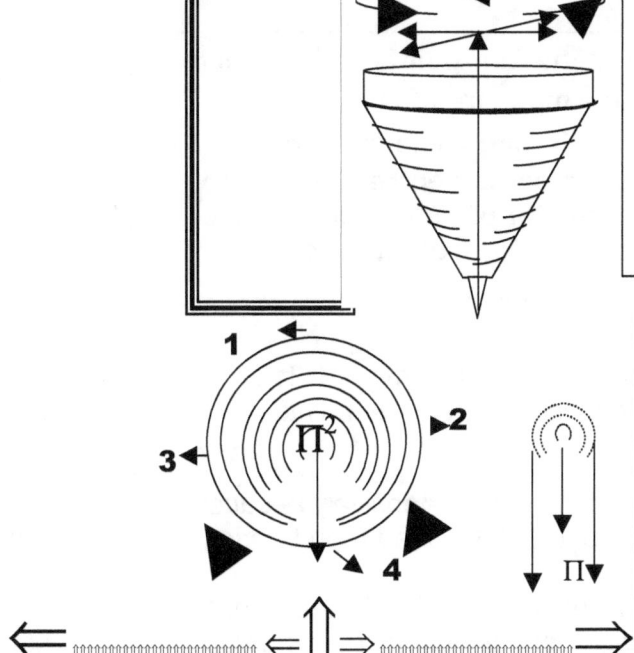

The sphere has seven points. The cube without truly being a cube but is just in consideration of having a cube in form holds five points to singularity. In the centre runs singularity to the value of Π^0, which means that which surround Π^0, holds a position of Π^2,

The spinning sphere activates the seven points, which places gravity in relation to a centre. Outside the centre there are five sides by dimension. The sphere has seven points of which four is spinning. The four spinning stands related to the gravity of spin, which are Π^2.

The gravity is in relation to the spin, which is in relation to the four points spinning which are $\Pi^2 / 2$ and that is the Roche limit. It is the dividing of singularity sharing space-time just as we on Earth share singularity by division between the Earth and us others that is not part of the Earth. The total that forms from the point that spawns is seven plus five plus pi square in division of four totalling twenty one that stands related to the first seven and once again another sphere formed. However this is an eternal relevancy that can never break.

Any object in rotation will have a middle point, a very specific centre point that does not spin. That point once again hypothetical but none the less must be standing still because every line running from that pint in opposing directions are also in opposing directional spin to each other. Although the points had the same characteristics only seconds before, they oppose the characteristics it had just before and just after the very second in which they are and to which they relate by similar points also in rotation. Due to the spinning nature of such a point with all surrounding the point very varying second, the value to such a point can only be Π because of its constant changing. Using r would specifically oppose another r from every angle. From such a point every other point will be opposing any other point not pointing in the direction to which the first point is pointing, whereby it extends the direction it holds. No matter what the point is or where the point leads, such a point holding a specific direction will be unique in the direction it is rotating because at that or any other specific point wherever, it will be directing not in the direction it spins but in the direction flowing from the centre point outwards. Any point will be it opposing itself within the rotating of $180°$ changing every aspect of its previous flowing characteristics it previously had or will once again have in $180°$ from there. While in rotation from the point of an outside observer all may seem static and never changing but to the object in spin every next second will be a diverting from every aspect it was in very second passing, and the direction it held in relation to the direction it held the previous mille, mille second will totally be incompatible with the direction it holds the very next mille, mille second of rotation. That proves no point can be static or constant, all though it may seem that way to outsiders. Although matter is matter, matter can also be anti-matter at the same time.

At this stage time was still eternity being interrupted by infinity. To say the Universe is or was 13.5×10^9 years old is shear Newtonian thinking. Was it 13.5×10^9 years and how many days in the year of our Lord and what about all the years that passed since this date was revised? Time was flowing according to interruptions in eternity changing from what was to what is to what will be. Time is a norm that comes as things in the Universe change about things that are places around and scattered throughout the Universe. We may presume time at this point somewhere became a factor since sphere sprouted from points on sphere edges and differentiation in development came in place.

Considering the role that the Roche limit played one can see how points in singularity grew from contraction and secured ever stronger centres by divulging hear points within the realm of singularity control. When a point in form developed at a position that was close than the original Π^0 to Π, the singularity in control took control.

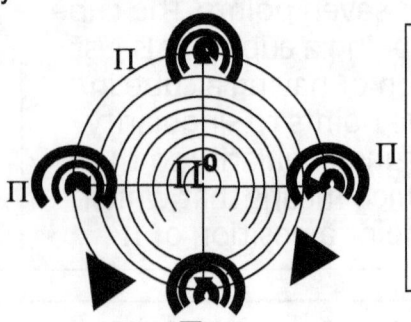

With every one of the four points taking fpm to the value of Π at a measure of $\Pi/2$ each brought about the Roche value of $\Pi^2/4$ in relation to the developing centre. One has to remember that the star of today takes on the characteristics of the form of that era.

Consider what happens to a star that developed closed than the Roche limit of Π to $\Pi^2/4$ would allow, it is easy to see how the singularity centred grew by concentrating the heat the points in singularity brought about

In this manner very much space was sprouted and space-time grew extensively because for every point dismissing was established for points of duplicating came about. This forced form to bring about space by motion but motion created by intense time interference.
From this space grew more and motion came about slightly easier with slightly less intensity.
Space came into form as form formed space from immeasurable singularity points sprouting to duplicate heat. That evolved the Universe to where the atom came about,

$\Pi \times \Pi^2 \times \Pi^3 / 5 = 192$ $\qquad\qquad\qquad$ $\Pi^2 \times \Pi^2 \times \Pi^2 / 5 = 192$

In every small human mind we try to find time, which we know and trust. Each line is an eternity that came and went is and is going or is going to come by passing onto the next eternity. Every line marks more millennium in one era than there is seconds in $13,5 \times 10^9$ years. This $13,5 \times 10^9$ years by number is so typical in portraying Newtonian science as the Newtonian concept is symbolising that it is mass that is producing gravity. However I only deal with the first two developing phases in this book because those three phases represent the true Cosmic Birth. The rest was developing of what already came about.

Form in the shape of the sphere was the only form available because the motion exceeded the cubical dimension of light so far the Universe was empty by the motion of space, which reduced all forms to allow only form to contribute to particle developing.

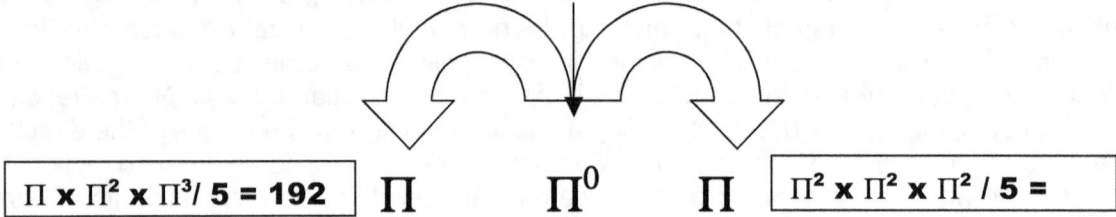

In the previous era everything released from eternity and infinity. The first of course that released was heat. Heat parted from cold. Science loves to see that heat was as high as… and the heat came down too… while those scientists forget that at 10^{34} it was an average day, with the temperatures measured in the shade. The motion retained the space to a point of freezing while outer space was cooking then at 10^{34}, just as much as outer space in the present time is cooking at 0 Kelvin and the sun is freezing at 18×10^6 Kelvin. It is relevancies swapping significance while importance shifts in location. As motion decline space will grow because $a^3 / T^2 = k$ When the cosmos is spinning at the speed off light and the speed of light is $3\Pi^2$ the motion bringing about the gravity for that time, is freezing everything in unusable lumps of cold space, the same we think of as galactica. At this era $\Pi \times \Pi^2 \times \Pi^3 / 5 = 192$ and $\Pi^2 \times \Pi^2 \times \Pi^2 / 5 = 192$ space came about as space parted from heat.

Four relevant positions form an alliance with the centre but the influ $\boxed{\Pi^2}$ orm another point that spawns space-time because the motion not only influence in one direction b⎯⎯ motion generates a point forming space-time in another direction which is in line The four points serve to secure time but the fifth point establish one more position that centralises gravity with a line running from the centre to the fifth point. From this came what I called the Lagrangian atom.

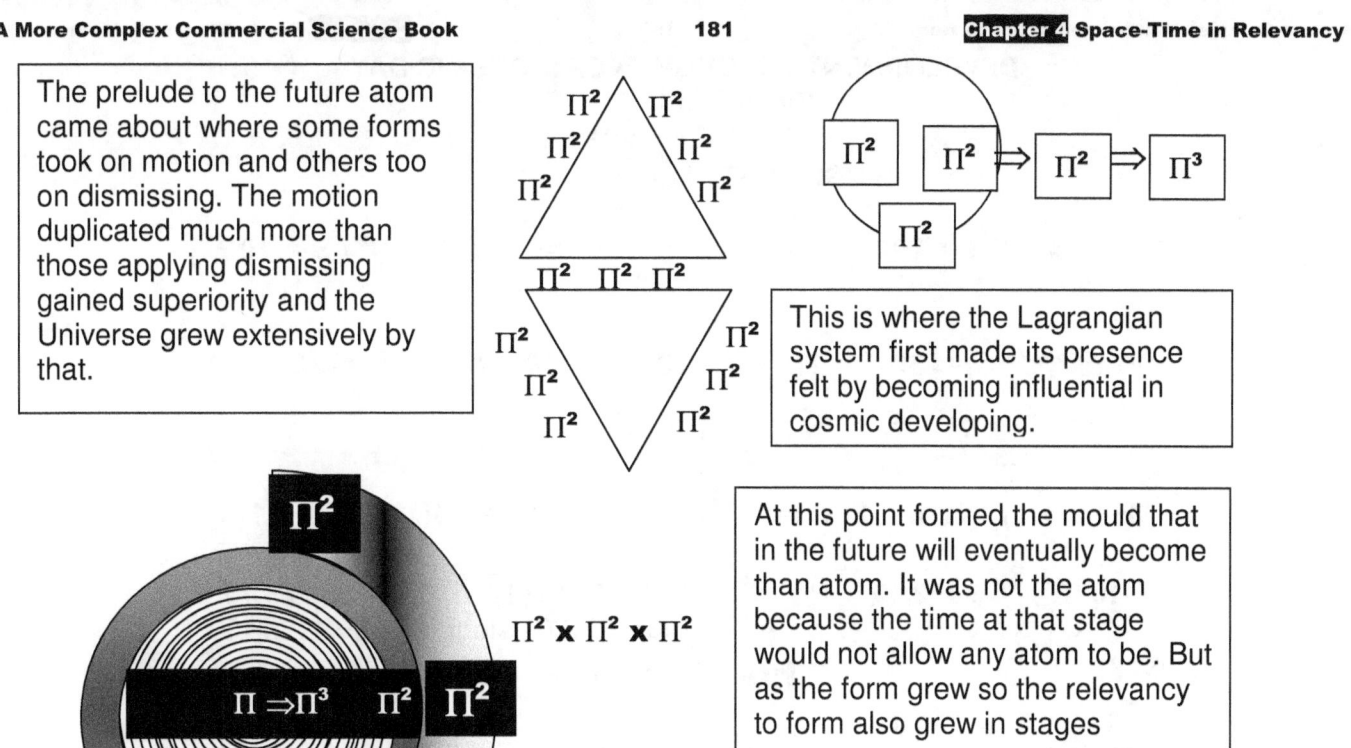

The prelude to the future atom came about where some forms took on motion and others too on dismissing. The motion duplicated much more than those applying dismissing gained superiority and the Universe grew extensively by that.

This is where the Lagrangian system first made its presence felt by becoming influential in cosmic developing.

At this point formed the mould that in the future will eventually become than atom. It was not the atom because the time at that stage would not allow any atom to be. But as the form grew so the relevancy to form also grew in stages

In this period the motion in space was from 961 times that of the speed of light slowing down 192 times that of the speed of light. It might not be a full 961 to the measure but it is an idea of conditions that was in progress. Space was barren and motion was fears and wild. Motion exceeded the speed of light by the square of the cube. It was so fearsome it contracted space more that space expanded by four. Motion saw to it that space remained part of infinity while motion broke free from eternity. Only when light came about at the commencing of the Big Bang did the space in the form we realise space to have evolved. Before the Big Bang the motion was faster than what the speed of light can cope with and solids like the atom was flat in dimension. The image $\Pi \times \Pi^3 \times \Pi^2$ shows that space came about through motion that was produced by the line and gravity formed the proton drawing the Universe flat as Einstein saw with his calculations but also drew into the flat Universe the Neutron spinning faster than light.

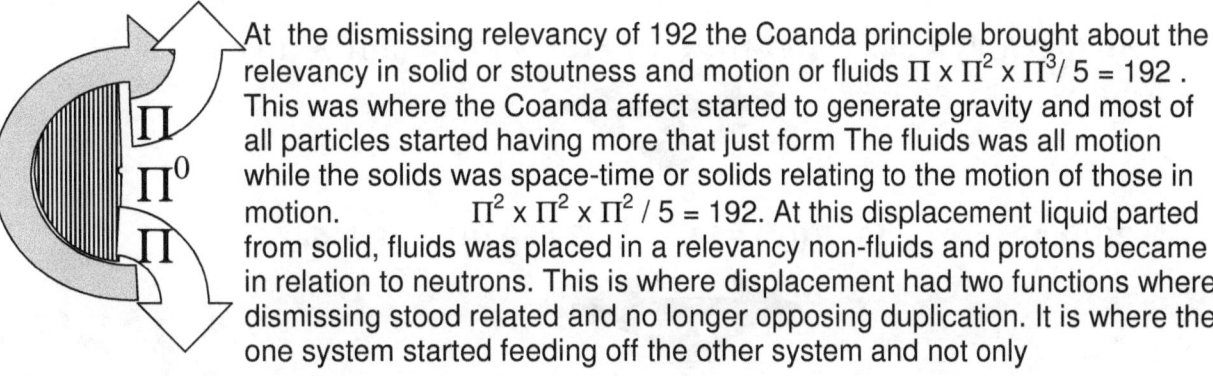

At the dismissing relevancy of 192 the Coanda principle brought about the relevancy in solid or stoutness and motion or fluids $\Pi \times \Pi^2 \times \Pi^3 / 5 = 192$. This was where the Coanda affect started to generate gravity and most of all particles started having more that just form The fluids was all motion while the solids was space-time or solids relating to the motion of those in motion. $\Pi^2 \times \Pi^2 \times \Pi^2 / 5 = 192$. At this displacement liquid parted from solid, fluids was placed in a relevancy non-fluids and protons became in relation to neutrons. This is where displacement had two functions where dismissing stood related and no longer opposing duplication. It is where the one system started feeding off the other system and not only

We think of electricity flowing on the edge of time and for many years we thought of light as being time. To correct myself I might be the only person not thinking of light as being equal to time but be that as it may, at the point where we now are referring too, light was not even conceived yet. Space was deformed by motion and motion was creating form. Electricity was to slow to be a thought and yet the closest image we may have of this time is electricity be it as slow as it is. There was motion that was generating space as electricity is generated by the Coanda principle securing a singularity position that is in control of space in form and the flow of form. Then there was form as fast in a relevancy but all in motion that was establishing motion in the space that was securing singularity.

DEVELOPMENT THROUGH EVERY COSMIC DAY
SINGULARITY

1 $(\Pi^3)^2 = 961.$

SINGULARITY Forming
time   space

2 $\Pi \times \Pi^2 \times \Pi^3 / 5 = 192$ $\Pi^2 \times \Pi^2 \times \Pi^2 / 5 = 192$

3 $10/7\pi^2/2(\pi^2+\pi^2)=139$ $7(\pi^2+\pi^2)=138$ $7/10\,\pi^2/2(\pi^2+\pi^2)=136$

4 $2\pi(\pi^2+\pi^2)=124.$ $2(3)(\pi^2+\pi^2)=118.$ $10 \div 7(4(\pi^2+\pi^2))=112.$

$7/10\,(\Pi^6)/6 = 112.16$ $((\Pi^2+\Pi^2)+(\Pi^2+\Pi)+(+3)) = 35.75$
$(35.75 \times (\Pi\,\text{singularity})) = 112.313$

The Big Bang

5 $\$T = \pi^2(\pi^2+\pi^2) = 107.$ $3\pi((\pi^2+\pi^2)=102$ $3^2(\pi^2+\pi^2) = 98.$

6 $10/7\pi(\pi^2+\pi^2) = 88.$ $10/7(3(\Pi^2+\Pi^2)=84.$ $4(\Pi^2+\Pi^2) = 78.$

7 $10/7(\pi/2)^2(\pi^2+\pi^2) = 69$ $7/10(4((\Pi^2+\Pi^2)=55.$ $2((\Pi^2+\Pi^2)=39.$

8 $10/7(2((\Pi^2)=28.$ $10/7(\Pi^2+\Pi)=18$ $7/10(2(\Pi^2+\Pi)=14$

9 $2(\Pi^2)\;19$ $10/7(\Pi^2)=14.$ $7/10((\Pi+\Pi)=5.$

10 $10/7(\Pi)=2.$ $(\Pi/2)=1.57$ $7/10(\Pi/2)=1$

Heat separated from cold and even today it still remains the principal on which all gravity is commuting. There were an away outward moving and there were an away inward moving between the limits of the form Π. The temperatures we are discussing are the limits there can be because it is above and it is far beyond what any star or any galactica at present can have. It was where the limits began and it is where the limits will end but no even A Black Hole has such temperature limits at present. The cosmos was there and the cosmos one day will again reach that stage but it is on its way getting there. Where we now are which we are talking about still has no outer limits and is not even that which we see as darkness at night.

That which we think we see the Universe when staring into the vastness of space is inside what we no consider as forming.

The inner limit is where the proton gain realises singularity and the outer limit has no outer limit. As far as we are concerned and from where we are and with our extreme limited vision we have been correct throughout the entire Newtonian era because during and from the time of Newton we thought we saw a Universe on can look at and when not seeing a Universe the Universe will end. When I do not see what I wish to see that object then vanishes. That view indecently also depict the mentality development of a four year old child but that is not for me to say that a Newtonian cosmos view is the same as a four year old child's mentality. There is no limit to our limits in comparison to the cosmos that has no limits. That dot that was one step bigger than the spot that was so small, it was without space and without motion, it was space without space and it was space without measure, is so big it is containing all there is to see and is to be and then it still is. It is so big it has no outside and all of what we think we see, is the inside of the inside of the inside we see.

While every line is circling bringing about time in space to the value of Π repeating Π to form Π^2 at the same time Π is extending in one specific centre to the value of Π^0 and only the spin value keeps Π not becoming r The spin keeps the immovability from becoming Π and maintaining Π^2 by performing duplication, but with any slightest reduction in spin reducing Π^2 to $\Pi^{2/4}$, Π will start extending and as one can see from the behaviour shown in the Roche limit, the heat will be concentrated at the centre and the singularity in the centre will grow four time in concentration. Only at points exceeding Π in diameter was time as Π^2 able to retain form and also grow. From that space slowly developed because at Π could Π^2 bring about a form which provided motion. In the centre there developed Π^2 and Π^2 kept all form at a safe distance of Π to bring about the needed solid immovable centre with the form $\Pi^2\Pi$ about the double Π could $\Pi^2 + \Pi^2$. That secured the makings of the atom by applying the Coanda principle of enticing gravity in the centre of motion, which then provides space-time by measure of $(\Pi^2 + \Pi^2)(\Pi^2\Pi)$. This totalled seven in dots and with three of those seven circling singularity at 1.9991 the atom came about.

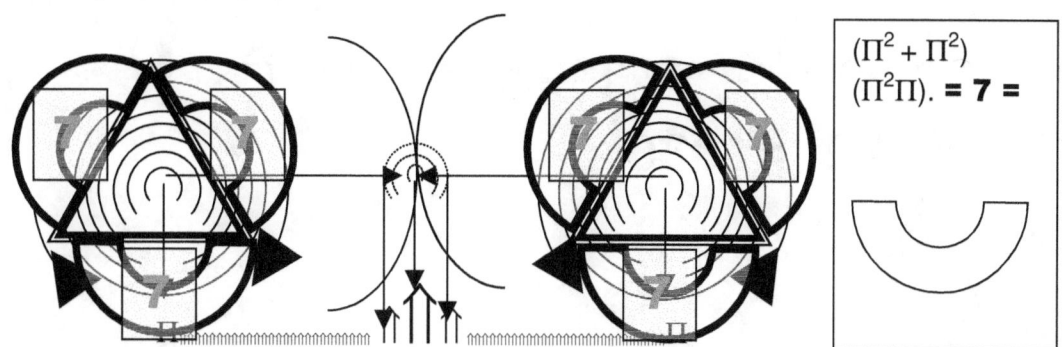

The relevance at first were not space as we now see space in the 3D, it was relevancies of motion duplicating the seven. It is three by motion of where it was, where it is and where it will be because without one not one is valid. By duplication the atomic relevance came about that was the forerunner of the atom.

$(\Pi^2 + \Pi^2)(\Pi^2\Pi)$ 3. =1836, which after wards the atom was about and the Big bang proceeded

The atomic relevancy is $(\Pi^2+\Pi^2)(\Pi\Pi^2)(3) = 1836$ meaning the proton displaces space-time at a relevant value of 1836 times that of the electron, or the other way around is that the proton is 1836 times more intense going on to singularity than heat is going liquid through the electron.

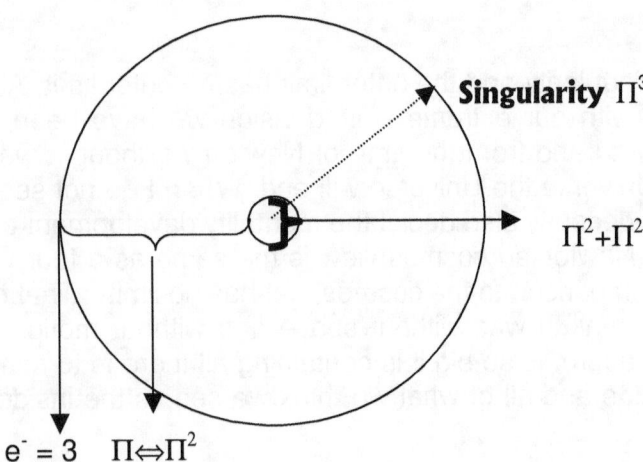

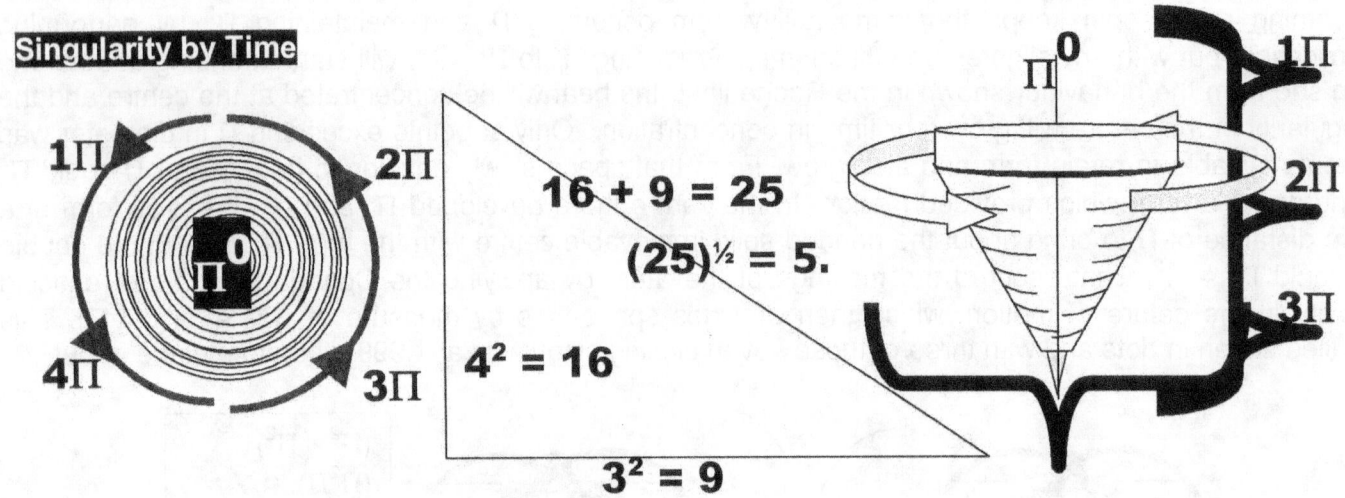

$(\Pi^2 + \Pi^2)(\Pi^2\Pi)\ 3. = 1836$

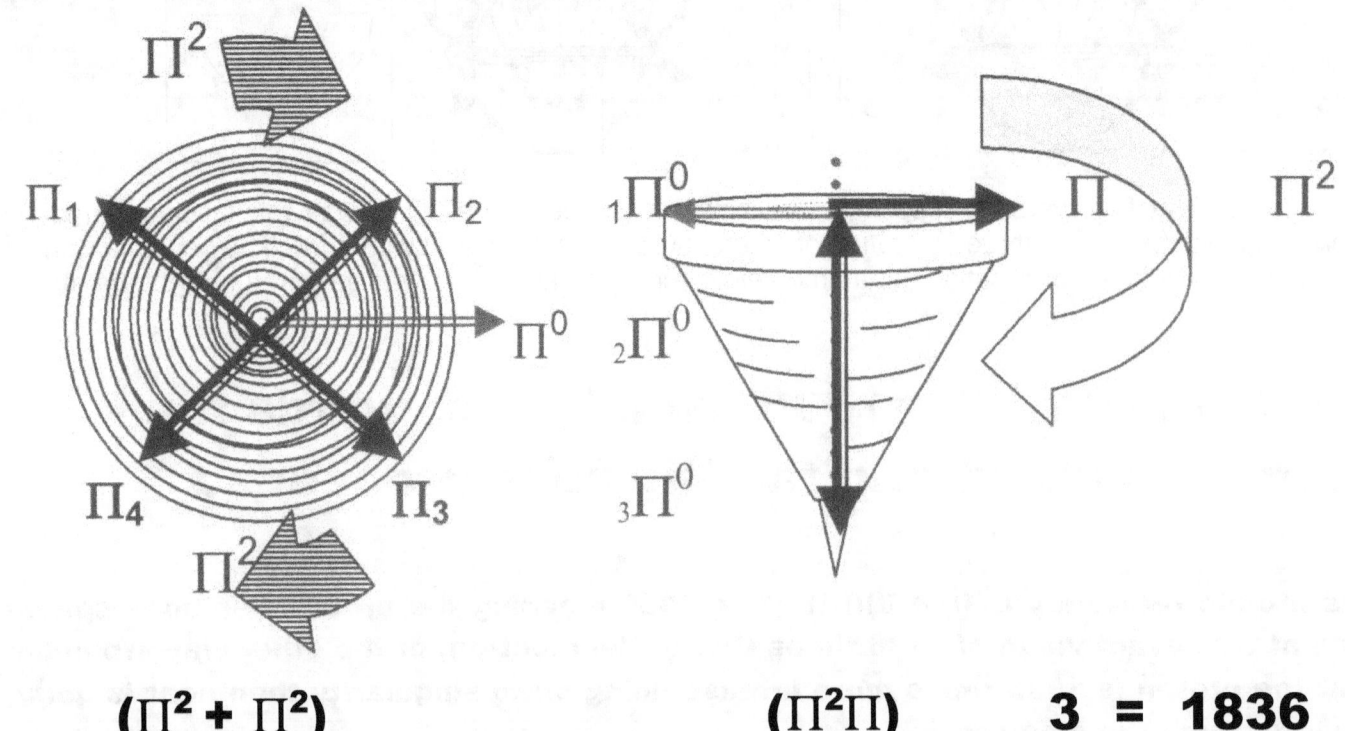

$(\Pi^2 + \Pi^2)$ $(\Pi^2\Pi)$ $3 = 1836$

PART 5
USING REALITY TO GAIN A NEW PERSPECTIVE

For many years I try to show the mistake in physics but the way the academics deliberately block me serves as a clear indicator of a conspiracy rather than just plain stubborn ignorance on their part. Now I am going to introduce the readers to what I tried to show the scientists, which is the remedy to curing the mistake. If it was only two or three that showed me away I would not be this frustrated, but the Super-Educated-Masters get annoyed when they encounter new facts they are not acquainted with and the lack of feeling superior frustrates their narcissism. What Newtonians are accustomed too is the believing that they are mentally superior and blessed with insight only granted to god-like scientific visionaries.

Those Super-Educated-Masters think they are the Brainy Bunch should read this and they will see they are misinformed by misinformation that gives them a presumed self-importance. Members of the public have no preconditions about preconceptions. The public are eager to learn and does not crave only to teach by knowing better, while those thinking they are brilliant don't have the capacity to learn. I seriously came to the conclusion those Mighty-Clever-Newtonians have no idea how to understand my work because they are either too lazy to think or unable to think. Either way they show no capability to respond to my message correctly but their response indicates total unawareness of what I try to convey.

For persons that wish to know more or that wish to confront me about what I say, there is a much more detailed version named **"THE ABSOLUTE RELEVANCY OF SINGULARITY THE SIX PART THESES" that is available from Lulu.com.**

This book aims to show God-fearing persons the truth about science. Science is not there to fight religion but complements religion. If religion was as fraud as science is, then I personally was an atheist, which I definitely never can be. **Moreover it shows how students are mentally abused and brainwashed to believe physics.**
Why would you be most likely correct when saying the Universe is a sphere… as it always is depicted in pictures?
Have you ever thought about: …where is the cosmos coming from…, …where is the cosmos going to...
…and most of all…Why is the cosmos traveling through time ……what brings about the direction of expanding? The answers to these questions carry great significance…and in this I book I answer them all…

My studying Kepler helped find answers to all the questions, which was deemed impossible to answer. The following is only a few of the many questions that I do answer.
Where is the centre of the Universe?
That I can answer…and I also can answer…
…why is the Universe still growing since the Big Bang…
…why did the Universe start so very small…
…why did the Universe fit into a neutron at one time…
…how did everything expand from fitting into a neutron…
…why does space grow from small to large…
…where is it going while it is growing …
…why was the Universe any specific size...
…what was everything before the Big bang…

The content of this book might seem to be intellectually challenging and reading it might require much concentration at some points because the ideas I put forward will be very new to any person reading it and will call on intense detailed analysis of information. I found the building blocks used in the building of the Universe and by applying the four cosmic principles I am able to go as far back as before the Big Bang event. The Big Bang only happened recently when atoms formed the known Universe. I do not touch on this aspect but I inform the reader that what you read was never written before and absorbing new information is difficult. You are confronted with facts you never heard before. So therefore notwithstanding the level of the person's academic qualifications or how well developed background the person has, this is not your average garden-variety storybook. However, in the end after you have come to realise the conspiracy in science, you will be a lot wiser than you were before. The information might ask for a lot of your attention span at times, but then again challenging the mind to think is what makes life that interesting!
I managed my discoveries because I DISCOVERED A MISTAKE IN SCIENCE.

This is a very obvious mistake that should have been detected at least three hundred years ago because this mistake is so biennial one can't believe grown up person's will miss it, let alone those person's groomed in mathematics and cultured in physics with all that most superior education. It is not possible that person's having such intellect and such insight into these affairs could miss it…except in the case where they purposely wished to miss it.

This is a mistake not even well hidden and not that much out of every day context. Everyone subconsciously knows about this and therefore misses it. The mistake is so blatantly missed that questions starts to rise about the reasons why the mistake is been missed by so many for so long. The mistake could only have been missed, if it was missed on purpose and the missing was clearly motivated by other issues. That makes this no mistake that came about because of human oversight and misinterpretation of the not so obvious like something said with the slip of the tong, or mislaying keys because of a lack of concentration, but it is a deliberate miscalculation of the obvious to mislead and falsify facts in order to mismanage and financially profit from such an oversight.

This is a crucial mistake since it touches every aspect of the foundation of physics and rocks physics like no Earthquake ever rocked any part of the Earth ever…it puts science on a cross road that will change cosmology forever. Correcting the mistake will alter the way the humans have a clear concept about physics altogether and forever. It will give the human mind an understanding about situations that no person before had the mental capability to understand. It will clear the view of the human on matter as will relieve the human mind from an intense and everlasting handicap. Man will never again in the future look at the sky as man did before the unveiling of this mistake.

It reverses the essence of cosmology. It alters the explaining we have about the Sun and put stars into perspective as never yet were possible. It explains what different stars are and put stars into context as it lines the most biennial star such as the Sun in ratio with the most bizarre such as the Black Hole. It explains the Black hole in a way a child could understand the concept. It shows the normal evolution of stars and the normal development of stars from the cosmic cradle to finally resting as Black Holes.

It changes the most fundamental foundation of physics. For the first time gravity as a concept becomes clear. The why and the how about gravity is finally explained and the dark ages mentality we have on the question of gravity disappears as the force is laid to rest. This mistake reveals that everyone pretending to know everything becomes those that know very little of what they pretend to know all there is to know.

I formulated gravity and it was not that formula that Newton used to formulated when he formulated gravity. I was unbelievably excited with my new found wisdom and my over exiting discovery. I saw what the clearing of one mistake could mean to the entire world. I saw how removing one misconception could clarify so many blind spots humans have about the Universe and the practical layout of the Universe. I wanted to share my enthusiasm about the mistake with those that could correct the mistake and bring new light to every one willing to listen to the new vision cosmology would bring. Those in charge of physics could correct this one flaw and by correcting this one flaw the jig saw puzzle we had on the cosmos would all fall in place. I contacted the custodians and arch fathers of physics and enlightened them to my discovery. They were the wise in know how and if it was possible for a person such as I to see the mistake and the Universe that would open when one remove the mistake and replace the mistake with the correction, how much further could the wise men in physics take the translation coming from such a correction of the obvious mistake. I contacted them with hard evidence on the matter.

From them came no reaction. I tried again and changed my approach because I thought my approach brought on their missing the essence of the problem. My mistake was indicating they had a mistake. After seven years of constant knocking on doors not one reply was ever returned by a single academic professor. Not to disprove me…not to point out my mistake and my error, not to silence me.

It was as if trying to open a grave by shouting down to the coffin and expecting a joyful reply. It was as if trying to chase a ghost with a catapult. It was as if trying to photograph a spook. There was this silence befitting a graveyard with all the ghosts pleasantly gone to the after life. There was this curtain that absorbed all sound coming from me and killing off any response coming back to me.

I took the mission one step further. I thought with personal touch and eye-to-eye intimacy I could see what part they couldn't see about what I saw and then explain my lack in clarifying the matter as to enable them to see what I so clearly saw. I was more than just ware of the importance there was in detecting the mistake and that put all my sincerity into solving the problem.

I went to see countless academics wherever I could and whenever I could and that I did with much hope riding on their sincerity and it was done with much cost on my part. I though by having a person-to-person and an in depth personal conversation about the many factors arising from the matter, it bring new light to the table, either on my side as I then made a gross error, or on their side. I went on this mission of sharing information with big costs and effort on my part. I thought that such a personal visit conducting a debate on the issue would either bring clarity to my oversight or bring their oversight to their attention.

A fat lot of good all that did me by spending all that time and money going into all that effort. If they were sincere it was only to save their work by not recognizing mine. The mistake is so obvious and in view of all concerned with science that it is impossible not to notice it and yet, not one in science notices it.

From them I got a cold shoulder; from them I got silence or at times blatant aggression. I was received with what was all part of the Academy of science's tactics to prevent me getting closer to the conspiracy by ignoring me. When they did reply it was to belittle me and to aggressively silence me. Their attitude was not something on my part. It was due to something science is trying to hide. It was not I that should be blamed for noticing the mistake but it was they who silenced me because I noticed the mistake. I uncovered a three hundred year old grave in which was hidden a conspiracy far bigger than what the earth ever knew before. It rattled bones that were buried since Elizabethan times and ever since then no one ever cared to bring the mistake to the open where everyone not in science could see what science hides. Then the truth finally dawned on me. I was overstepping me boundary because I was in the centre of something bigger than what I discovered. I was in the centre of a conspiracy where everyone conspires to keep what is the secret to remain the secret and to collude never to divulge what the conspiracy hides.

As I said before I uncovered a mistake in science and by that abandoned the using of mass. From the onset, the mistake seems as insignificant as it is small. Because the rest of the book is about the mistake, I do not intend on elaborating about the mistake itself. The mistake came about with the culture of education and the mistake in itself seems harmless. When admitting that, one must also admit that any pilgrim that got lost and died of starvation through an incorrect travelling direction, made the very first part of his ultimate mistake by looking in the wrong direction. How harmful does looking in a specific direction seem, and yet such a mistake leads to his ultimate mortality. The traveller could when taking his first directional flaw with that the first incorrect step, only put his foot skew in avoiding a rock. Or he could have turned his face to avoid a branch and that move pointed him in a direction that lead to his fatality. It is not the mistake that becomes the penalty and it is not the origin of such a mistake that leads to the penalising, but the ignoring of accepting signs telling the wonderer of an impending error and his stubborn ignoring of such telling sign that makes the lost party pay the ultimate price. By ignoring the mistake, for whatever reason, the ignoring of such a mistake is his undoing because the price due comes from the inability in recognising the sign indicating the presence of the mistake forming the reason for his final demise. The sooner such a person sees and admits the wrong, the less will be the consequences of his final price to pay.

The Newton mistake is one born in culture and the penalty from this mistake is bred by arrogance. At school minds are young and accepting, although developing. Through many tens of millennia humans came to a habit in surviving as a specie where culture taught them that accepting the elders advice is the same as to ensure survival of the following generations, and by such doing is also following the quickest way to an adult mind. By accepting the elders knowledge and experience without question proves the dominance of the tribe in relation to other tribes of the same race. This is culture we cannot do without and still maintain progress. It is an inheriting method humans grew on and is the corner stone of all civilization. We cannot abandon it. This was the method whereby civilisation became practised.

The scholar sits in class and receives from the Master information that is completely his days breaking news. As far as his mind can tell the news is as actual as anything notwithstanding the fact that such news may be with the human mind for thousands of years. Whatever the teacher tells him is bound to be a first time experience so new he has no time to digest the information. Taking into account his youthful ways (which we all had), he has little stomach to scrutinize it because learning is a painful process to all. Without pain and perseverance there can be no education of any sort. He does very willingly accept the facts as tested and correct without flaws of any kind. The scholar has to because in any education system time will not allow students to ponder about detailing information and securing a prognoses to all learnt every day. What ever the Master tells the scholar is taken as Biblical correct without any thought about testing the results. Where there are cases of scholars having doubt and subsequent questions, the

Master takes such behaviour as being obstinate and being a reflection of the student on his (the tutor's) personal integrity and knowledge. The young mind will very soon discover that his behaviour is not tolerated by the system, and the truth is the system cannot tolerate such behaviour for the good of the rest. Time must be spent on learning and accumulating as much information as that which the young mind can accept. The young trusts the knowledge of his elder and superior by not questioning anything.

No information could be more affected by such a culture than that of Newton. You both understand Newton and are smart, or you do not understand Newton and accept that there will be very little future for you to have in the world of science. Newton is science. No Newton understanding automatically becomes "no science" education or learning. Without Newton, there are no other and science will be a vacuumed of containing nothing. This is very unfortunate but is the ultimate of truth. It is either Newton's way or no way at all. With this culture also brought along the stigma that only the minds of the sharp and the sighted can accept and understand Newton and when not understanding Newton one tends to fail your personal I.Q. test. It is a sure sign of the slow witted when the student fails to recognise what Newton said. The only way to advance in science is to understand Newton and indicate to all your pears how brilliant your mind is in accepting information. As I am writing I still fail to understand Newton and therefore I am looked down upon as a slow wit unable "to understand Newton after all these decades of studying Newton.

I am constantly reminded of this handicap and recently an editor enlightened me on the issue. That is why I feel obliged to refer to my position as ILL EDUCATED in terms of Newtonians that are as privileged and moreover blessed to being SUPER-EDUCATED in "understanding Newton". All students have little understanding about Newton, and that I can and will prove through the next pages of this book. The mistake Newton made and which I discovered is laughable small, yet it took me (not being that bright I may add) almost one lifetime to recognise the mistake whereas it took mankind three hundred and fifty years of research without recognising the flaw. Others in the past may have come to see what I saw, but if there were such persons they never saw what I saw because if they saw what I saw they should also see that behind such an almost invisible puncture hole in the tube of insight is a reason for science to deflate and not accumulate. When "understanding" Newton it becomes the very same as learning Newton from the heart and accept that what you memorise is what you know. The memorised knowledge is beyond question, as it has to form part of the identity of the student having secured the knowledge.

It is not the hole forming the puncture and preventing inflating being as such the obstacle that is of importance but what that hole does to the tire and the car and the travelling with the car that becomes a menace. In the extreme effort to keep the journey on course, the Academic Masters are spending an all out effort in inflating the deflating tire faster than the deflating tire can deflate. The recourses attached to this effort is enormous and without cause. It will lead to nowhere and that is where science is heading in their attempt not to head in that direction.

When students become masters the Masters seek to break new academic ground. Masters do not ponder the ways on which they lead their students, but have an all out effort in establishing their own support of new territory that will distinct them from the rest in future to come. My uncovering of such a mistake as I did could only be from a person as ill-educated as I. I did not go through the learning process where the learning process is the very same as a brainwashing and mind controlling process. This remark may seem harsh and is not intended to be such, because reality demands no other way in education. The mistake and the carrying of the mistake with the support in refusing to recognise such a mistake is part of human training and there is very little to do but to admit that such may occur whereby to follow in acting to correct the mistake after the discovering, rectify what ever can be salvaged and build from there.

Friend and foe think me of alike as being slow of mind and not understanding Newton. This was what I was told on more numerous occasions than I care to remember. Pointing at the mistake I see, I am told by the wise as well as the nit-that-wise, such as I that through my lack in education I cannot dream to understand Newton and therefore are committed to the position of the ILL-EDUCATED, a position I learned to accept with grace. In all cases there are more sides than one and therefore I think of myself as poorly educated, the contras to my position must be those fortunate to be SUPER-EDUCATED. In this letter there will be the addressing to you holding the position of the reader and (I hope) the un-bias judge, with me presenting my case to the un-bias (you) in concerning the third party, the SUPER-EDUCATED. Referring to the party in opposing me as the SUPER-EDUCATED is by no means in disrespect and much to the contrary holds my whole-hearted admiration, (at time somewhat limited I admit.)

The information in this book I purposely made easy to red, and easy to follow. The content however, is only a drop compared to a bucket of information contained in all seven parts of "MATTER'S TIME IN SPACE The Theses". All information presented to you, in the first part of the book is an introduction to the second part. In the latter part of the book the conclusion comes. With out a detailed introduction in the first part, the information in the second part will be of little value, as the information is very conclusive about the first part. I have been researching cosmology since 1978, on a part time basis. The conclusions may seem simple; that however is in retrospect. Some of the arguments took me up to six months to arrive at since I do not have all the information on hand.

Through the first three years alone and by e-mail as well as normal mail, I have contacted more than 1500 academics in physics in numerous countries all over the Globe where I got addresses from the Internet and conveyed this unfortunate occurrence in physics to them. I sent them post and sent them letters over the e-mail and by land post with no reaction on their part. Not one academic once contacted me in any way image or form. Nothing came back in any form being positive negative or neutral. It was as if I never made any contact. It was as if my e-mail address was not active. To this day, there was not one person, which I contacted that replied or contacted me... No sorry, I stand corrected, that is a lie. One academic contacted me and said he read my book up to page four and stopped because he did not agree with me. He got to page four reading a book of over one thousand two hundred pages and at page four of over one thousand two hundred pages; this genius concluded he was in disagreement with me. He was bright enough to draw a conclusion at page four of a book containing one thousand two hundred and forty odd pages. However, this much I'll say in his favour, up to now he was the only one that ever mailed me on matters concerning my work.

Then to some I sent letters in book form by addressing to their post boxes in addressing the problem that I saw in physics. At this stage all the letters I have sent on so many issues became so numerous that I was able to compile the various letters into one letter that made up a book format. I bought a printer of many thousands of Rand to do the printing and did the books as professional as I possible could. Their ignoring my work was not the result of an inferior presentation or a compromised format. The books contained so many issues and covered such a wide variety of material that I found it impossible to believe those academics had no opinion about my work. Yet from that I received not one conformation that any one received any letter or book. The letters I sent to professors through out the world became so numerous that I started forming books from the letters I sent explaining my point of view about the mistakes and in finding answers and corrections about the mistake there is in physics.

No reply came back…not to discuss my point, not to discuss their point, not to elaborate on the mistake I so clearly saw, not to defend their views about the point I so clearly saw, not to set my mind at ease with some explanation about facts I am unaware of, not to say they are aware of the facts I am not aware of, not to say they are aware of the facts I that am aware of and was drawing their attention to such facts, not to thank me for contacting them confirming their unwillingness to discuss the issue. Not to plainly tell me that I must not contact them again in future for they shall contact me in further corresponding if any such need arises on these matters (at least my Bank manager wrote me this very reply when I approached him for an overdraft although when I was still actively farming but at the time it was during the drought and with my financial position being what it was at the time, I know that on his part he did so just out of common courtesy and not because I was a well respected customer as he called me in his correspondence). There was not one occasion where a reply came saying they thank me but please never to contact then again because they do not wish to correspond with a person with the likes of my type of person. Not one said they were aware or unaware of the issue I pointed out and not even once came one acknowledgement about receiving any documentation carrying the information on paper of the facts that I pointed out. I found that odd but still I went on trying to find some response somewhere about the mistake I discovered and which I was trying to uncover in science.

Remember, this mistake I pointed out was most serious as you will learn in due time. From my point I have the issue is of such a serious nature that science can't function properly when not correcting this critical error. It is so crucial that it should drive science to a halt. Yet, in all of theses concerns I was the only person on Earth seemingly worried or at least having any concerns about the most apparent flaw any one could ever discover.

I could for my effort wrote them from the grave for I was not even a spook they could not notice although I had the ability to offer the solution that could correct the occurring flaw.

I wrote and printed many books showing my concerns I have about the mistakes. At one point on one day I posted eighty 80 books in one shot to most Grande Universities that were exceptional in their Astronomy Departments and to this day I have received no word in response. This effort I made was just before I went to hospital to have open-heart surgery of a most critical and life threatening nature. I was on the doorstep of death and this was said to me in three different wards in three different hospitals. At the time my health went out the window and I seriously thought that it could be my last attempt while being alive. I printed and bound eighty books while being too weak to pick up one book. I had to ask my sons to pick up and carry the books once I had then bounded. It was an asserted effort on my part that drained my last strength. I say this not to put more blame at their door but to explain how seriously I judged the issue to be and what efforts went into the addressing on my part. That I did it virtually from my death bed was not the concerns of the academics I sent the books to, but from my side it was a last effort because I had more that a good chance to not have pulled through the intense operation. In the books I addressed the issues in detail and the work I named **An Open Letter to Selected Academics** in which I proposed the full remedy to the mistakes I indicated. At the time I thought it more than likely that it could have been my last commission I gave myself.

It drained my funding totally and I got nothing back for all the effort and the money I spent in trying to charge a reaction. The only response was just a vacant silence; they did not even E-mail me in response acknowledging that they received such a book that I sent their way.

These letters were my attempting to get all this Brainy Bunch and the Super- Wise - Academic's attention drawn to a theses I have been working on that contain seven books where each is a Thesis in own write since the themes the volumes cover was never yet addressed by science in any given manner. For a seriously long time I tried to get some academic to read any or all or some of these books seriously because the facts a touch is very vital (well in my opinion they are) to the future of science and especially in cosmic physics. But no sooner does any academic learn that I criticize Newton and all interest in whatever I have to say just disappears out the window.

Not even the great Houdini could make anything disappear as fast as the interest the scientists showed when they got the idea I was no praising the brilliance of Newton. It was then that the letter writing started to get some one interested in something I wrote. Even saying I can explain the sound barrier did not provoke much interest especially after hearing it proves Newton had no idea what he was talking about.

The Theses are as follows:
After the operation and while I was recuperating I had an in depth search into my motivation and why I was doing it on my part. I even mentioned to a friend of mine that I was going to dump the entire project because if they didn't care, then why the hell should I care. Then it dawned on me that backing off meant I lost the fight and I have never lost a fight in my entire life. I wasn't going to let them beat me without them proving that the mistake I was accusing them of, was actually my mistake I made.

If they couldn't prove me wrong their ignoring me was not going to win their fight because not fighting back was not going to bring them victory. At this point I was still very gullible about academics being honest and sincere and forming the pillars of civilisation and the anchor in society. I was very impressed with their positions and did not yet see the blemish that I later saw. But the rude awakening came as it hit me like a ton of bricks. All those nice fellows were not all that nice and innocent. As time lapsed my efforts became numerous my suspicions rose about their hostility towards me and their apathy towards my warnings. As time went on I became more aggressive as I became bolder because slowly my eyes opened as I saw the reasons why I was so blatantly ignored. They were fighting for the cause just as I am fighting for the cause but they are fighting to protect what causes my fighting.

At this point so far after all my numerous attempts in trying to establish some contact with academics world wide I wrote seven books in a combination I titled "Matters Time In Space: The Thesis" covering the entire issue of my work. Then there are five more books wherein I combine all the various letters I wrote to academics through out the six years of ardent trying to establish some line of communication. The last letter I addressed to academics I include as part of the content of this web page for your insight and which forms part of one of my books where I join and elaborate on the letters that I combine to form a unit as a book.

In my work I do some things no one having a sane mind should do. I break rules never broken by man since the end of the dark ages. I transgress what no one dared to do in the past three hundred and fifty years or more. I cross a line that is forbidden to cross by any man not being part of the living dead or mentally insane. I go into the darkness of the foreboding chambers of insane madness and mental instability where the most monstrous beast would walk while even he then will be hanging his head in shame.

I venture into the darkness where only those that did the unspeakable crime will enter.

I go into avenues only those with the mind of the animal will have thoughts so criminal and I became one that must be banished from everything civil. Yet, even to this day my madness is so overwhelming that still I have no shame for what I committed.

The Books holding the letters are entitled

1) **An open letter On Gravity**
2) **An open letter Announcing Gravity's Recipe**
3) **An open letter Addressing Gravity's Formula**
4) **An open letter About Gravity's Prescription**
5) **An open letter Explaining Gravity's Rules**

In the light of all proof and when facing evidence that I do bring, I further even dare an atheist to prove me wrong about Creation. In mentioning this word Creation by name in a science book I break a ground rule enforced by the atheistic dominated world of science. I overstep all boundaries because I prove mathematically that Creation (the entire Universe) came about in the manner exactly and precisely as the Bible states...to the letter). I challenge anyone to prove it did not happen in that way as the Bible describes. Then I go further and do the unspeakable, the act that proves insanity, the dead that vanquishes the force by distorting the gods presence in physics, becoming the reason why the cause of the future Earth will be destroyed... why what would have been will be no more because I unleashed the wroth of God onto man bringing punishment forth that man cannot endure, while with that evilness I sent my sole to physics hell just because I criticize Newton!

I challenge any one to prove that Newton's gravity is not a deliberate hoax instated to con the world by deception and fraud. I dare whom ever to prove that Newton's vision of gravity is true. I challenge any person to bring proof about any part where any of my theory might be incorrect and furthermore I challenge any Academic in physics to prove that Newton's mass pulling mass by creating gravity committed to unleash a force is anything other than delusion. I charge any one to bring proof that the cosmos is contracting by the force of mass and that mass produce gravity as Newton advocated when he committed the biggest counterfeit and circumventing of the truth and reality that has been perpetrated by any one in all of times. If you are of the opinion it is ridiculous that I say that Newton committed fraud then answer the allegations I make and which I prove. I know every Newtonian is shouting for my scalp on hearing this allegation and I did consider the penalties I possibly face that goes with such allegations before I made the allegations.

I knew that since the death penalty was abandoned in most countries I contemplated the fact that I could not receive the death penalty for my attitude towards Newton. I could no longer swing on ropes in some gallows or receive electrification by chair or lethal injection although I knew there would be a drive to reinstate this penal conduct in the light of my transgression and the serious nature of my horrific crime against all of humanity. The time has past where I could be charged by the elite in physics whereby there was showing on my part much insane behaviour and that my nature threatened the very fibre of the human race's survival. They could no longer create civil disobedience to entice a crowd so highly emotionally inflated that the crowd will willingly stone me to death or lynch me by elicitation of hysterical mob raging marauding.

With me having such serious allegations to face that those that brought the charges hanging over me could prove to the effect of me willingly and wittingly committing a crime so serious it equals anything to the effect of hierarchy. I could no longer be held accountable for blasphemy against god Newton and by that that I had to answer the charges of hierarchy and address the evilness of my ways in front of the elders fromx which the punishment I would receive would be a justified torture to death as William Wallace received. I could no longer get whipped until death brings me everlasting relief to my raging madness as well as relief to my enduring agony I suffer from dementia as a result of torture by the hand if the whip master. I knew with some comfort that I will not be arrested and brought in front of the International Tribunal for Human justice in Den Hague where my case will be heard in front of a panel of justices that was paid beforehand to convict me as that will excluding all chances I may have of me not meeting my destiny and death.

Still, I also knew notwithstanding all the security that the new civil liberties brought to modern man, I had to be on guard and vigilant in every step my family or I took because there could be a C.I.A. intervention and an assassination at my family or I, but even that was a small price to pay for my heinous crime against all of humanity. But concerning that even in such an event of assassination by bullet from a distance using a high power rifle meant that death I receive in such a case was quick and simple, very much compatible with the simple minded person I am and the simple life I lead up to that point of being assassinated.

They could no longer hang me from any lamp pole like every pirate should suffer, because the evil I did by criticising Master Newton outweighed any treacherous dead that any pirate could ever have committed. They could no longer decide on a pole of their choice to hang me high, without giving me a fair trial and then for further punishment after my death to leave my body to decay while everyone smelling my degenerating corpse will have a life long remembrance never to commit such a heinous crime, such an unspeakable sin, such a disregard for the holy and the precious and show such an ostentatious example of mental vexing. However I realise regardless of what civil law allow to the academics in physics world wide, in their esteemed opinions, I still deserve such punishment in the eyes of God and of man and that is even including the opinions of all other life forms on Earth for I do what no man may dare do…

I challenge the correctness of Newton's view about gravity. This was testing to what limits I could press the civil ness and the civility of society. Think of my bringing charges to the door of Newton in dissecting my tenacity by the very morality of the issue and do so in micro detail. I took the liberty that modern society granted me and pushed past all boundaries civil order can endure by criticizing Newton on gravity and mass and force and the Universe in general. This is testing the crux of what the modern mind can endure and still remain civil.

I even improve by showing where, how and why Newton's gravity and mass and force and the Universe in general are incorrect if only I could find an audience prepared to listen.

I show that Newton is incoherent and that Newton broke mathematical principles to validate his corrupt view about physics. He stole the work of Kepler, raped the content there of, vilified the meaning and corrupted the truth in Kepler's work. Then he pinned all the blame on Kepler by naming this corrupt lot after Kepler, as being Kepler's laws giving the lasting impression it was Kepler showing the insanity that Newton in person was guilty of. Newton disgraced the work of Kepler by destroying the correctness thereof and even in this book I prove this statement where my proof can leave no doubt as to its authenticity…and yet I am sure that again there will be not one Newtonian amongst the lot of them that will even glance at this book just because I criticize Newton.

As soon as they find I discuss Newton in a negative light the Academics stop reading notwithstanding what the content holds that they read, for if it is negative even in the least, they find such reading unbearably degradingly disgracing. I think they are the stupid bunch…and they realize it. I think in the end I am the only one that truly understand Newton and can therefore see the mistakes he made in the formulising he did on the subject of gravity coming from mass, while they wouldn't dare read what I have to say about Newton because they then have to defend Newton while they have no idea how to defend Newton. While they are supposedly the ones that are the experts on Newton, they are the ones not understanding a word about Newton. If they read what I say then they would be in a position where they had to agree or disagree on what I say about Newton and then they would also have to give reasons for agreeing or disagreeing on what I have to say about Newton. It is not possible to understand Newton and not also understand the mistakes he made. It is not possible to understand Newton and see the laws in physics as well as in mathematics that he broke to justify his incoherent claims.

If only there were one incident where just one Academic turned around and showed me where my mistakes are in my assessment about Newton's reasoning and could indicate as to where or why I go astray. Instead those Masters in Newton are saying that I am to stupid and uneducated and because of such low ranking I have the inability that all stupid people share which makes that I don't understand Newton. Because Newton is so highly regarded and is such a genius and with me being so low down on the intellectual ladder I have to accept that it is not for every one to understand Newton because Newton requires insight and years of study.

One has to be worthy of understanding Newton to be able to understand Newton. I am sick to my sole of hearing that trash because there is nothing to understand after one realizes that it is Newton himself that was the one that didn't understand any of Kepler's work and for that matter neither did he understand cosmology in general or any aspect of cosmology. Newton did not understand Kepler! Or the other the spontaneous academic defence of their not having a position defendable about Newton is that usually the meeting abruptly comes to an end as to bring an end to an embarrassing situation on the part of the academic I visit by appointment is the manner they use to ending the conversation.

They end the meeting abruptly with conversing the words to the affect: Mr. Schutte, Newton explains the situation very clearly where he can leave no doubt and there is no point in debating the issue any further. Then in the same breath without reading one word of my work further they add that Newton's explanations are accepted across the world for many centuries and by millions of highly educated and very intellectual persons and there can therefore be no doubt about the correctness of Newton's work and as a fact it is proven through out time that no one ever could prove Newton wrong on any aspect of his work. When I challenge this in protest, I am silenced by a rude gesture.

Some even add that because no one to this day was ever able even once prove Newton wrong it is accepted in cultured circles that therefore any attempt in doing so is fruitless and foolhardy. Only the insane would attempt it… and while they arte saying that they hold a book in hand that if they read one paragraph on they would see that I bring the evidence I put in front of you now. I made an hour-long appointment to see the Academic in question. I drove many hours to honour this hour-long appointment. An appointment of one hour normally takes sixty minutes. Not in my case. The busy professor suddenly after learning about what I have to say has other appointments and my engagement timely ends abruptly.

While ending my appointment in a fraction of the time they permitted me to show my work, they are at that time minutes if not seconds into an hour appointment. As soon as they read the first negative implication about Newton or his work or any connecting evidence to his work the reading ends. With that my appointment ends a few seconds into my appointment and this comes where my appointment should just be starting. It is an hour appointment that should take a full hour where the time was appointed or dedicated in seeing me with my problem with whichever professor I made the appointment. The abrupt ending comes in the shortest of time, notwithstanding the time and effort as well as cost on my part to come to honour the appointment.

Then there were those I never got the chance to even start the meeting. When learning over the telephone what my book was about and that I was in dispute about what Newton had to say about gravity, I was sent packing before I could come and visit. At first I thought they didn't get my books or in some way my book never got their attention and to avoid that I insisted on seeing those Academics in person. If they saw me I could see they got the book in person and there was no chance that someone else got the book or that the book got mislaid by accident. Therefore I insisted in delivering the book in person as to avoid their missing my book.

On those occasions where I got to the main gate and announced myself I was ordered to contact the professor which I had the appointment with before receiving a pass to enter the premises. At some Universities I was told to leave my book at the main gate where security checks passes and prevent unlawful entry as security will see that professor so-and-so will receive my book. When having a telephone conversation with the very esteemed professor I was about to see I was told there was no need for me to enter the front gate because the professor was of the opinion that him and me had nothing to discuss. Whenever I protested he then said he was not acquainted with my work and first had to familiarise himself with the content before contemplating on any further meetings of any nature.

I then was told to leave the book I wrote on the matter of Newton at the main gate before the meeting could take place, as there was no reason for any meeting to take place on the mater of me finding fault with the work of Newton. I was refused entering the institution because the professor could see no reason for meeting me notwithstanding the money I paid and the effort I made to travel a great distance to meet this person. This happened frequently in the past when I told the academic and academics (plural) that I had written a book where I highlighted the errors Newton made. But eventually I got wise and did not mention this fact before the meeting. It was then that I was blown away after a few minutes of meeting me. No sooner did the Professor learn about the nature of my visit and ended our meeting abruptly.

One of my books I named Xepted Newton mistakes and not once did one academic see it fit to give me an interview on the content of the book after hearing the title I chose for the book. I was eventually forced to change the name to Xepted Science Mistakes, which brought me just as little joy in the end. On so many occasion I was told they see no point in reading my book about Newton's mistake since there are no mistakes and therefore there is no point in meeting me on the matter. I came to the conclusion they are on the offensive because they are on the defensive about Newton and that they do because they make no sense of Newton in when considering even a small percentage of his work. If they had made sense of the lot they would be able to see the mess he made. That was my line of thought before I got suspicious of their honesty and when I got a lot less gullible.

It is easy to refer the blame onto me and my "not understanding Newton" where I have the disposition of having no academic standing or authority and I could be shunt away without me resisting their shunting me away. While they rid their location of me by using authority the truth is that it is not I that don't understand Newton because I am the only one that does understand Newton. It is they that misuse their authority and prevent me addressing the issue that prevents me from insisting that they then explain what it is that I do not understand. To tell me that in the case where I officially studied Newton I would then form some frame to use in order to understand Newton but since I am no official student or any candidate that pays for studies, they have no time to explain to me what it is that I do not understand. It can't take that long to show where my error is if I was at error in the first place. I am not that stupid and I am informed about Newton's prognoses on gravity. It is not as if I have no back ground knowledge and had to be schooled in the most primitive basics of the thesis about Newton. The behaviour they show is both not rational or sober for the guard they produce to shield them from me is not responsible behaviour of men with clear reason.

...And since I still show no remorse as much as I show no regret for my insanity. Therefore it is obviously clear that I should be the one that can gain no grounds or benefit from at least one case where there is one academic somewhere showing one bit of doubt about Newton and therefore are prepared just to read my books from start to finish without their commitment having Newtonian bias that is interfering with there judgement. Not once could I find one academic that would sit with one of my books from start to finish and read past the page where I start to show Newton's defects, although I am so obviously correct on every matter that I state. I am washed off the Earth for I show little regard and even less respect for the consecrated sacrosanct hierarchy of Newtonian wisdom and for that attitude I find no ear in the world of science prepared to listen to my views which they deem as clear insanity.

Later on as my personal naivety about their honesty diminished I could have some blame about my conduct as I addressed their mighty Holiness their Royal Highness The academic professors in astrophysics and in physics, for any misunderstandings they could use against me about the misgivings I had about Newtonian science and they might now condemn me for showing a lot of antagonism and disrespect, showing that I had an attitude, but that was not the case at the very beginning. I am sure that it could at this point influence them where I don't kneel as low and shiver with fear when they enter the building with their wisdom surrounding them like a whirlwind, but this was not the case a few years back and certainly not at the start. In the beginning I showed much respect and that got me nowhere very quickly but also painstakingly slowly and it the overall experience I went through was expensive.

Remember with my health, my diabetes and my heart condition I am unfit to work and I used this in conjunction with my brain to try and edge a little income from the opportunity to sell a book on the matter. That would bring some income and relieve my penurious ness bringing on my situation of destitution and my contribution to science in handing over the solution would be a reward to the world of science in repayment. It is going both ways and to me it is the only door leaving the poverty behind. I lost my farm on poor decision-making but that was in some cases because of health issues that I made poor decisions. But my health was the main reason why I capitulated to the pressure of the creditors and in the end resolved the farm practise. I had a product and there were no buyers because everyone was baying rotten rubbish. That truly got me fed up in the end.

Later on I started attacking their religiosity with venom. Now I go into detail and prove what fools those honourable Academics in physics are and how they corrupt the young mind to brainwash the young and vulnerable in accepting the detestable criminality of lies and deception that they call astrophysics. I challenge any one to show the correctness of Newton's view about gravity in the face of the evidence I am about to bring. I charge students to challenge the academics with the evidence that the academics present to portray what they advocate as being religiously correct. In order to reach the heart of gravity one must discover the heart of gravity. Those academics that are pretending to be so sure about Newton I give the following challenge:

To find the truth you know is going on, then answer to your person in privacy and in all honesty the following questions that I put to you. It will gauge your state of brainwashing and show the amount of damage that you have suffered this far in your particular and specific case.

Do you know what gravity is?

If you knew would you tell me what gravity is...yes beside it being a force...well we all know that those who should know says that gravity is a force but other than being a force that acts on behalf of mass and pulling unsuspected objects all over, what is gravity, which is besides being Newton's pet force.

If you don't know what gravity is besides knowing that gravity is one of the four forces and it is Newton's original force, then do you know someone that knows what gravity is?

If you still answer in the negative and you still don't know what is that which is behind what is causing of gravity, then have you heard of any one that knows what gravity is?

Maybe there is one Academic professor or a NASA Scientist that knows about some person that knows one who knows what gravity is, other than knowing the fact that gravity is presumed to be one of the four forces and being Newton's personal pet force.

There has to be an Academic professor or a NASA Scientist known to someone somewhere that knows about some person that knows one who knows what gravity is, because there are so many Academic professors and even more numerous in numbers amongst the many NASA Scientists with super human mathematical abilities in the art of physics acting as if they know. There are even those going around with ideas to renovate the cosmos by assembling space whirls. They apparently plan great voyages that will take man over great distances while they are so informed on the matter of gravity and space-travel amongst gravity. With their absolute phenomenal calculations as they present physics they are showing such abilities in a class which no one can imitate, which must represent a picture about having super knowledge on gravity in the most precise detail.

When reading what they say they are able to accomplish in terms of astrophysics it stands to reason that they know gravity to the smallest detail there is to know and know everything anyone can ever consider knowing or hope to know about gravity...after all they can present the cosmos as they are able to explain the cosmos with gravity taking centre stage in the past, present and future of the cosmos. This they accomplish by presenting a few mathematical formulas in which all of the Universe then are defined.

They can calculate all the matter throughout the entire cosmos, adding every atom by mass into the conclusion of their calculations when they determine the critical density that the cosmos has, which is responsible for the entirety that is providing all the mass that provides all the gravity. They have gone as far as even calculating the explicit required quantities of atoms that forms mass which should be available, which they then find to be short falling in the mass availability through out the vastness of the Universe and the missing mass is not establishing the matter density throughout the vastness of the Universe required to bring Newton's vision on contracting to reality.

Their calculation ability is so vast that they have the opinion about the mass they measure that is forming the gravity they require which is probably less than the requirements needed to substantiate the cosmos' effort in rendering a constant supply of gravity that will eventually secure the returning of everything to where the cosmos came from. This eventuality they named The Big Crunch even before locating the Big Crunch. It is like naming a baby even long before knowing how the procreating is taking place that will lead to impregnating of some member of the specie (which member it will be is still unclear) where it later on will lead to conceiving the baby … that is the manner in which science dogma is enunciated but that is how clever those are that knows everything there is to know on gravity, or so they pretend to have in their promotions.

They know gravity to split detail where the detail goes to such precise extend that they are able to calculate how much gravity the missing dark matter in all the Black Holes must be to provide a force allowing the cosmos in experiencing the next big implosion that is coming somewhere in the future. That the implosion must come even in the face of insufficient gravity is a certainty otherwise Newtonian physics is completely inadequate in their cosmic vision about the Universal future! With their having this qualified virtue of intellectual splendour spawning such phenomenal abilities they then would have to know what gravity is!

Well…if you don't know any one that knows anybody that knows someone somewhere that is familiar with the ins and outs of what is causing gravity, I then can assure you I know about someone that knew all there is to know about gravity. I am the person that knows someone and that someone knew gravity…but he is dead now…died a premature death long before his time (have you ever heard of any person that died spot on at the second that was his time where he was suppose to die). He sadly passed away and is no longer with us. Still I would like to introduce him to you…and about his work of course, if you would page on.

If that is the formula the cosmos abide to and that is the formula controlling the cosmos the radius of all things should lead gravity on an eternal quest to voyage towards and finally into the centre of the Universe. If gravity is taking whatever there are to where all things will end and all things are forced by gravity in relation to move according to mass to such a point that will unite the entirety of what is into what will be that outcome can only be where the centre of the Universe are allocated at this moment. The first thing to do is to find where r^2 ends because where r^2 ends it will be where we will locate the centre of the Universe.

The factor r^2 should eventually lead everything to the centre of the Universe in the end…then where is the centre of the Universe and where is the factor r^2 going and conclude in the end at the final destiny?

Facts I advise the reader to become acquainted is the following:
Did you know that not one of the supposed laws Newton invented going under his name or the name of other innocent person's such as Kepler, does apply in the cosmos just as Newton described it must It is not I that is at odds with Newton, but it is every principle applied by the Universe that totally contradicts every declaration that Newton made.

If the gravity contraction is going on as Newton said, why did everything come from a small dot called singularity and from that develop into something immeasurably big? Whereto is the lot in this picture above pulling? Is galactica absorbing stars or is starts uniting with galactica.

Which will be the next galactica to implode as its stars meet the gravitational demise. They should see it…

Let's forget about the crooked fictional critical density or the rubbish dark matter that is all made –up fiction forming part of the conspiracy of deception in physics to hide Sir Isaac Newton's fraudulent physics principles.

Every aspect of what is mentioned so far is fraud.

There is a Universe that is very much undiscovered and waits for recognition. This Universe does not work on space and light but it works on singularity running in time that is diverting of space. Before the Big Bang brought about space it worked on singularity holding everything that is in one spot that in today's standards never was.

What happened before the Big Bang is still happening but it is out of view and is located in intellect? It needs much more understanding that what a few formulas can represent because the entirety is still holding a relevancy of numbers and only numbers. The numbers only operate in the adding of numbers because in the dimensional Universe multiplying has worth, but in singularity (1) everything that multiplies brings such multiplication back to 1.

If you are one of those members of society that never thought you would hear the name of an accomplished person such as **Sir Isaac Newton** being associated with **fraud**, **corruption** and **brainwashing**, then these books are specially written to inform you about the truth there is lacking in the correctness about science. Everyone knows that planets orbit around the Sun. Planets circle the Sun which is the same as saying planets orbit the Sun. Just by calling the circle motion in terms of what applies, that statement nullifies Newton's claim of mass that attract mass and put to question the reliability of Newton's dogma. If mass did attract mass, what kept the balance where the planets do find a balance in orbit, rather than move towards the Sun. This is the fraud and a conspiracy to cover up. If mass did attract mass, then what is pushing the planets back into orbit?

The idea of proof is automatically placed at the door of Newton. If normal speech contradicts Newton, then it is it is the Newtonians' task to prove Newton supposition correct and thereby prove the claims about attraction Newton made. Newtonians will deny the fact that Newton was never proven or is incorrect. They will uphold the honour of their Master Newton notwithstanding. Think of what planets do... and you think that planets orbit. It is connected to the brain. No one thinks of planets spinning or planets basking in the summer Sun. When hearing about planets the first thing that comes to mind is the rotating of planets while circling around the Sun.

However, just using the term orbiting is in total defiance with Newton! Newton said gravity draws or pulls or moves in the direction, which would have one understand that the two objects in example the Sun and any of the various planets will be moving directly towards each other. The radius is diminishing.

The term pulling does not suggest any circling because no one can be pulling towards and does that while circling around the object. When pulling anything it must take place while using the shortest line possible. That serves the term pulling. Then the saying goes that planets orbit indicating they follow a circle. That is not what Newton said. However, wrong that may seem but circling is precisely what planets are doing.

If anything whatsoever would contradict or question Newton, then the blame or fault should never lean towards Newton. Of the Universe expand instead of contract as Newton said it must, then it is the Universe that is requiring the missing mass and Newton remains correct. Newton is not wrong about the contraction part; it is the Universe that is wrong about the expanding it is doing. If there is a lack of mass, then the Universe must be hiding the required mass in dark matter in out of sight areas. The Universe carries the blame for the mistake of expanding because Newton said the Universe is contracting and therefore even the Universe must take the blame, but not Newton, never could Sir Isaac Newton carry blame for being wrong.

If there was a pulling, then the orbit cancels out such an idea completely. The orbit then there has to indicate that there has to be some sort of prevention taking place that disallows the pulling to commit the direction of travel. I know it is said that the orbiting object falls as fast as it circles and by falling while moving to the following side on position it never reaches the Sun, and yes, it makes sense, but there has to be some form of resistance replacing the planet in the next side position and preventing the falling or the pulling from taking place. By orbiting the planets don't even suggest a contracting direction of movement going towards the sun.

Students in physics are you aware that you Professors can profess all they like, but they have no foggy idea what gravity is? If you don't believe me, then confront them with the question: what is gravity. If they come out with gravity being a force, then gravity can be water and gravity can be the wind and gravity can be heat because all of those mentioned are also individual forces driving objects. Let them be much more specific than just advocating that gravity is a force of some form.

They do not know what gravity is because Newton, yes, Sir Isaac Newton of physics fame admitted he didn't know what gravity is and that was the only thing he was absolutely correct about, the fact that he never knew what gravity is.

Yet he forms the basis of all physics wisdom... while Sir Isaac Newton never knew what gravity is! No academic in physics know what gravity is because the "father of physics" Isaac Newton did not know what gravity is. That makes on wonder how Isaac Newton could be the "father of physics" when he didn't know what his offspring was...something like an illegitimate child he knew of and he knew about but never knew in person!

Students in physics consider the following: If Isaac Newton didn't know what gravity is how could he have fathered physics? If you're Professors don't know what gravity are how are they able to teach you what applies in physics because they then know nothing about physics because they do not know what gravity are!

They only know what Newton knew and if Newton didn't know what gravity is then they don't know what gravity is...and blessed with that much lack of information they seem to think they are fit to teach you everything they have no idea about

If you never know the answer to what gravity is you inability to reply on a question as to what gravity is, not knowing the answer will leave you unfulfilled because of not being able to answer even the most fundamental physics answer: what is gravity. Gravity is what holds the Universe in form but then one need to answer the following: If gravity pulls towards a centre and gravity holds the Universe attached the question arising from that simplistic **answer is then ... where the centre of the Universe is.**

However using the formula $F = G \dfrac{M_1 M_2}{r^2}$ as Newton provided, disallows any other concept other than moving towards. The person Newton got his ideas from and the work Newton raped completely, that of Johannes Kepler explained this tendency very well, but Johannes Kepler makes no room for any pulling of any sort. Johannes Kepler indicates all cosmic space $a^3 = k\,T^2$ specified forming a containing space – circle a^3 has a dual directional movement of circling T^2 at a specific point in correspondence with a straight line **k** point and the space at such a point will formulate as $a^3 \div k = T^2$. In the work of Johannes Kepler he said that the space being the orbiting route a^3 remains at a specific distance **k** while the orbit T^2 takes place...and in all my other books that addresses more information. I take Newton to task on his dismembering of Kepler's formula by corrupting Kepler's work and with what amounts to fraud, Newton takes science on a goose chase that holds no truth. There is no pulling by mass of mass in any way. The Big Bang proves otherwise and the dark matter swindle is there to hide Newton's incompetent incorrectness.

The Universe expands since the time of the Big Bang or the Universe contracts since the time of the Big Bang. We have either one of two that has to be incorrect. If Newton is correct, then the normal way of expanding since the Big Bang is incorrect. Then we must start saying planets are pulled to the Sun. If the normal form of speech is correct and the planets are merely orbiting the Sun, then Newton is wrong. The Universe can't expand since the Big Bang with or without involving artificial dark matter. There can't be expanding while at the same time we have Newton's accepted scientific presumptions of contraction being correct. The fraud part is in the accepting of the Universe expanding while still insisting that Newton is correct in his dogma of contraction with mass. This book is an effort to show how Mainstream Physics brainwash students into accepting Newton's hypotheses of mass attracting by force while the entire Universe is expanding at the rate the Big Bang indicates. They will rather look for lost missing mass and invent dark matter than to come out and say that Newton was wrong all along. The Universe must take the blame because Newton could never be wrong!

I grew tired of apologising for my (as they see it) having the audacity of being correct on matters of Newton's incompatible religiosity, which I bring to their attention. When being in contact with esteemed Academics agreement with those most esteemed academics in physics. I am expected to show the utmost humble attitude by acknowledging their supreme posture with me being in their surreal presence. I have to feel honoured to be in their company and rather die than to I mention to them their mistake about Newton being mistaken about a Universe that is contracting because according to Newton it pulls while it never ever contracted in the least.

I am quite fed up with the attitude of those academics looking down their noses at me or worse still are those ignoring me whenever I show that Newton's facts just don't add to a conclusive believable answer. They sit in high and mighty places while they cover their positions with fraud and have person controlling their unlimited power. They can waste $5 billion on dubious theories without being responsible to any person controlling their mismanagement and fraud by digging holes into Swiss mountains. Nothing will come from that because their vision on mass is as corrupt as a politician's oath. I have reached my peak with stomaching the corruption they hide behind a lily white cover of dishonesty while they sit in their mighty towers and live in a bubble where not even God can touch them less having me point a finger at their despicable ignorance about their mistaken Master Sir Isaac Newton , the man they portray as a God. If the Universe defies Newton, it is the Universe that is wrong.

I am at my limit with being ignored because those academics can ignore me by using their all-powerful status and with the image they carry with them they never have to prove Newton's correctness and therefore is fair in disclaiming even my presence when I try to disprove Newton. They are only gangsters hiding behind fraud. From my view and from my perspective, I honestly can't see any difference between the Mafia's racketeering by forcing the unlawfulness of their trade onto others with dubious corruption and what academics commit in the name of being honourable scientists. They forcefully brainwash students in believing Newton is correct by means of employing despicable mind control. That is as criminal as anything can ever get!

To them and to those I say, prove Newton correct by proving the moon is coming closer to the Earth ! Any person telling a lie is committing fraud albeit in the name of God or of science; mendacity is despicable corruption. When they tell a lie to distort the truth and find financial compensation while falsifying facts, even if it is by conducting science, then they are behaving criminally. That is distortion and is equal to the behaviour of the Mafia. Anyone feeling offended, please tell me when the moon and the earth will collide as Newton must suggest that is if Newton is correct about a force pulling mass.

They can say the Universe is made up of nothing. Between the earth and the moon we have lots of nothing and this goes unchallenged. For making a most senseless statement any mind can think of is quite accepted because no one dares question his or her superiority. When I bring this to their attention in a book, I am the person they condemn as being incoherent with my arguments about their nothing they fill a Universe with.

Those in charge of Mainstream physics feed students lies in order to be compensated for their misrepresentation of the truth. They are being paid enormous salaries from student fees to ensure that students believe in the impossible and accept what can never be proven and force students with methodical examinations to repeat the unproven or be expelled from the institutions and branding those expelled students as failures. That is a rip off whether it is justified as science in the process of learning or if it is plain legal criminality; it remains the same because they fly the same banner. Those practising physics waste billions of dollars on falsified theories while I can hardly put food on the table because I fight to reveal the truth. They can corrupt facts and get paid while I am ignored and starved.

The information my work carries, which you will read in the event of purchasing my books, you have never seen, it was never yet mentioned or the facts I divulge has never been printed by any person, ever before. I put untruths about the work Newtonian science claim is correct in question and that might have been published before but never published as questionable evidence. The rest I bring is new. I show how physics truly applies and I can explain everything Newtonian science can't explain.

Those academics in key positions of academic credibility keep certain facts and evidence away from students and give other facts that were never proven before. The academics teaching physics give prominence as well as credence to Newton while applying their trade in brainwashing to give their Newtonian views undeserved credibility and from these proceedings they earn substantial incomes. That is the same as racketeering. When you deceive by conveying untruths and cheat to mislead, then your behaviour is criminal.

If you purchase any of the my books on offer, then through the books you will come in contact with the truth for the first time in centuries. My work is about uncovering the truth and blaming the shameful conduct of those persons everyone trusts since no one expects such persons to be criminals.

When you purchase my books, I don't sell ink on paper. I do not sell material with questionable information, holding facts that were repeated so many times that it is accepted as the truth because it became a culture to believe Newton. I prove statements by taking the reader into infinity and into eternity respectfully.

However I do reject the idea that mass is perceived to have a unexplainable and improvable magical pulling power that is used to establish a force thought to be gravity that pulls all things closer.

Please let any of your physics professors show you that all the big things are spinning in a group together as a unit, either in the centre or the outside leaving all the small things then either on the outside or the inside. If $F = G \frac{M_1 M_2}{r^2}$ was true the big stars will group in one area and the small stars in another area because the big stars must have more gravity having more mass.

That is if mass does attract.

The above questions I answer but my answers show the foolishness of Newtonian physics and why it is that the Academics in physics all teaching about Newton's wisdom despises me more then I do despise Newton's corruption.

Should you be of the opinion I am busy with slander and deformation of character then purchase the book of 4 parts <u>An Open Letter on Gravity Part 1 and 2 Volume 1 and 2</u> @ Lulu.com wherein I take academics to task on the fraud we find and they teach in physics.

Let's start surveying civilized principles by evaluating what lawfulness means and what would constitute as morality. Let's determine what makes the crook in the book?

If any person, notwithstanding what reason is given in justifying such depravity, tells a lie or conveys untruths to further whatever humble cause, it is seen as fraud. To convey information that is not substantiated as a verified fact then the mere conveying of such information becomes fraud.

When any person, notwithstanding what reasons given, repeats such a lie unabated while being well aware that the information passed on by such a person is incorrect, then the person commits deceit. When anyone is repeating the information that is passed on as being unblemished factual substantiated and verified truth while such a person knows very well that such information is void of proof or lacks proof, then committing such an act is a criminal enterprise. Academics in physics commit every one of the above indignities and yet see their actions as being lawful and even much praiseworthy and hold their role in society in the highest esteem imaginable.

They fail to see the crime that they commit while tutoring physics. Whatever motivation they may claim to have which they offer to serve them as forming their driving force, the fact that they perpetually perpetrate in unlawful behaviour, by spreading untruths, such actions on their part put those academics holding such highly regarded positions in the league of ordinary cheats, gangsters and common criminals. By wilfully and constantly falsifying facts to further whatever humble cause and produce illegal claims repeatedly, remains derogative behaviour and is unlawful by nature, notwithstanding what morality it should serve. A Preacher or Pastor lying on behalf of God is not lying on behalf of God and to think the Preacher or Pastor improves or underlines the Greatness of God by lying on behalf of God is very mistaken, because in reality such a Preacher is falsifying the truth for his or her personal benefit. Lying is wrong and doing so even in the name of God remains despicable. The same applies to academics in physics. There is no argument that can change this truth about falsifying the truth and when doing so there is no hiding behind any excuses of ennobling to benefit mankind that will change such truth into righteous conducting. By bringing on a cover-up scheme as we find in the critical density scam, it stinks of criminal fraud. The entire saga is only to cover Newton.

Newton said centuries ago that gravity is the force of attraction there is between objects that hold mass and it is the mass factor that brings about this attraction, which Newton claimed there is. The Universe does not contract and all the proof we require to disprove such a statement we find in the Hubble constant as a guarantee. Moreover, it is true that the Universe never contracted even for a brief instant and we have the proof of that as the Big Bang concept. Also does the Big Bang guarantee that contraction will never become part of the Universe since it is the relevancy between what I named cosmic solids and cosmic liquids /cosmic gas that will provide the balance that shifts.

Since the first instant that time began the cosmos grew away from and not towards points. With all the proof that this concept brings in backing the principle of expansion in the Universe, still the Arch-Fathers-in-Physics uphold Newton's contraction. Planets never moved closer, are not moving closer and will never move closer to each other and this is backed by all information collected this past century. The Moon is not coming closer but the distance between the Moon and the earth is widening. Studies about the Universe reveals every time that space in the cosmos increases constantly. Studies find all things are moving apart and away from one another.

Any and all the proof about this is beyond what any doubt may present to counter this knowledge. Notwithstanding this irrefutable findings, science still regards Newton as the only person that ever lived whom no one ever could prove wrong…and this is upheld by Mainstream Physics in spite of the cosmos proving Newton wrong every instant of time. The basis of what science holds as its foundation we find to be the Newtonian principle of $F = G \frac{M_1 M_2}{r^2}$.

The foundation used by science promotes this argument and backs up this argument well knowing that in the cosmos there is no evidence backing up this proposal Newton suggested. The Newton formula $F = G \frac{M_1 M_2}{r^2}$ used as basis for science sees gravity as being a force of attraction and the force of gravity is being in place between all objects in accordance with the mass factor that the objects have as presented by Newton in the formula $F = G \frac{M_1 M_2}{r^2}$.

If you think scientists know what gravity is…then do not be duped that easily because no one in science remotely knows what gravity is…not even Newton knew what gravity is because Newton admitted not to know what gravity is… yet everyone in science are redesigning the Universe as if everyone in science knows exactly to every specific and smallest detail what gravity is.

Yet in spite of all the blistering ignorance there are about gravity in the field of academics and all those academics pretending to be informed in physics while claiming and pretending to have Mastered all there ever can be known about gravity all in science is reinventing the Universe without having the least bit of knowledge about what gravity is.

Having not even a foggy notion about what gravity is, why gravity is there or what gravity does other than performing as a pulling force of some sorts is all they gained in knowledge since Newton did nor know what gravity is and ignorance about gravity is all the gain they can show. All they have to do to sustain the current intellectual grasp academics have on gravity is only to acquire more ignorance and that will maintain the current level of information acquired the past three hundred and fifty years. After 350 years they are still as much in total ignorance about gravity, just as they were when science was in the dark ages and Newton did not know what gravity is.

What we find as we gauge all evidence found while studying the Universe, is that reality shows there is no attraction between objects in space going on anywhere in the Universe, that the entirety of such a concept is a myth and the outward moving of the Universe has been coming from and since the time of the Big Bang and maintaining this flow of material is substantiated in a concept named as the Hubble constant, which proves Newton's perceptions to be a myth. The Hubble constant proves that space everywhere is growing ever since time began and the growth never stopped ever since. Knowing this irrefutable fact does not deter science from under scribing Newton as the sole basis that underwrites all the correctness of all of science known as physics. However, Hubble and the Big bang and all other investigations contradict this attraction Idea Newtonian dogma holds.

Therefore, any further believing that there is attraction going on as Newton claimed has to be viewed for what it is and that it is a fairy tail. The Big Bang Theory proves Newton's idea as not only being wrong but Newton's idea of attraction is a joke. If the Big Bang is expanding the Universe, then how can the Universe contract at the same time? Any contraction by nature would have the Universe collapse back into infinity the moment the Big bang moved out of infinity. Ask your professor to show how an expanding Universe can also contract and your professor will tell you about Einstein's Critical Density theory. This theory I prove is the biggest fraud ever devised by any group of persons in the history of civilization! This is perpetrating fraud and conducting in upholding deceptions instituted by Newton that then formed the institution of lies they call physics.

The Universe does not contract in any way; means or form and even such a suggestion are incorrect! The Moon and Earth are not moving closer but are moving apart. The entire Universe is growing in space and nowhere is space depleting by any norm used. Academics are very aware of this misconception Newton had and still academics in physics are promoting the ideas of Newton as the unwavering truth. Academics teaching these misconceptions are committing fraud, notwithstanding the portraying of their role in society being unblemished, spotless while they are covered in a lily white blanket making them being whiter than snow and having such a holier than thou attitude. Teaching Newton is participating in deception and promoting Newton is criminally deceiving the public and while doing so, is committing an act with criminal intentions.

Then, in the face of all this evidence contradicting Sir Isaac Newton; they remain upholding the correctness of Sir Isaac Newton and keep on teaching students about the unwavering correctness of Sir Isaac Newton. They put down conditions of learning to this effect and are expecting students to repeat these untruths and unproven facts by forcing answers to that effect in examinations. Forcing the acceptance of this untruth about physics is equal to preposterous subjecting students to physiological torture and heinous mind conditioning, scandalous thought control and brainwashing. This applies to everyone serving as a tutor in physics notwithstanding whatever status the torturers might have in society or the morality they attach as a reason to commit such atrocities.

If you are a student, then you are conditioned by academics in controlling your thinking by enforcing pre-mind setting and in which they methodically force you into believing in Newton and this is an on going process conducted for centuries in the past, while it is the truth that Newton is completely void of any tests that may secure any form of confirmation and in securing proof then also by that establishing proof.

Read the book **QUESTIONING NEWTON'S MYTHOLOGY** http://www.lulu.com/content/e-book/questioning-newtons-mythology/7570956] and then use the information I supply in the book to insist that Academics who are teaching physics, prove to students that Newton's statements of attraction are correct. Let those academics explain the method mass uses. Let them with precise detail show when mass is applying it forms gravity that mass does produce gravity and such producing of gravity that then would establish attraction! I show precisely how gravity produces mass but mass can never produce gravity. I show with explicit detail when, how and where gravity forms mass but mass can never form gravity. What I prove annihilates every Newtonian claim.

They never prove Newton's philosophy on gravity but those persons conducting teaching in the subject of physics force all physics students to learn Newton's gravitational concepts and accept the facts as if it has been proven beyond all other facts. Students have to believe that Newton is correct or academics will see to it that they fail their examination. The condition of being accepted in physics is to accept Newton without questioning the proof that is never supplied.

Let those academics now prove precisely how mass brings about gravity and then afterwards test you on how Newton is proven correct and not on you repeating facts about what they say is true about what Newton said, which they say is true. The manner they present Newton is completely hearsay and that method may not be used in any court of law. Let your professors now prove how it is that Newton's teachings are correct and then examine you on the process they use to prove Newton's concepts. At present they say Newton is correct and then they test you on your ability in repeating that Newton is correct without ever proving to you that Newton is correct. Let those physics professors now prove Newton and then test you on the manner they use to prove Newton to be correct.

The truth beyond all other truth is that Newton's gravity has never been proven (because try as you may it is not possible to prove Newton's formula forming gravity mathematically) and because academics know that, academics require the blind acceptance of Newton by students. This unconditional acceptance of Newton's correctness relies only on the pre-conditioning of students' mind set and academics depend only on the student trusting the academic "say so" about the institutionalised correctness of Newton. That Newton is correct nevertheless and notwithstanding that there is no founding proof about this matter, is what students should be accepting blindly. Pre-conditioning students into blind acceptance depends on the academics' insistence that students approve Newton's concepts without pre judgment or students insisting on scrutiny of any sorts. In examination students have to outright and blindly follow academics' say so only because academics say so. Academics depend on students never questioning their say so or demand proof about what academics teach. Those academics in teaching positions insist that all students accept Newton's accuracy.

This is methodical mind control as much as it is the brainwashing I show that they enforce. If you are one of those believing that Newton was ever proven, then what you believe to be true is a lie because Newton can't be proven and that is the truth! The time has come to face your teachers and force them to stop the ongoing old culture of bullying students and conditioning their thoughts by enforcing on them dogmas, which is, mind control! In order to get students to accept Newton's hypothesis, academics resort to brainwashing pupils and students. They teach you that the Universe contracts and to state their case they force students to learn that gravity is proved by Newton introducing the following formula $F = G \frac{M_1 M_2}{r^2}$

They say that M_1 is the mass of the Earth and M_2 is the mass of the individual in questions mass and the multiplying of these factors with the gravitational constant produces the force of gravity when this gets divided by the square of the radius. Please let you lecturer put in all the values of the formula and prove Newton is correct. If he can't and I know for sure he never can fill in the symbols and calculate the force of gravity, then read the rest of the web page that follows to see how far academics in physics go to brainwash students into believing in Newton's fraud.

This is a fair test to see if Newton's contraction theory underwritten by Newton's attraction formula $F = G \frac{M_1 M_2}{r^2}$ is valid, and then force your professor to use this formula as it reads and show WHEN the Moon and the Earth is going to collide. If he fails to do it by using Newton's formula as $F = G \frac{M_1 M_2}{r^2}$ then you will know who is conning you, him or I and who is truthful again him or I. I charge all academics to prove what I say is being wrong in any way or even that I exaggerate in the least. I challenge Newtonian academics to prove that mass does indeed form any force of any sorts and in particular gravity! To those professors claiming Newtonian ideas are substantiated by proof, I say that notwithstanding your personal academic qualifications and while at the same time disregarding your status and previous achievements as well as ignoring your many admirable abilities you may have and however superior they might be, I shall teach you about gravity. I say it is time Students learn the truth about physics notwithstanding the status academics will loose. Students read www.sir**Newtonsfraud.com** and challenge those academics depending on their ability to brainwash you into submission.

The idea of proof comes automatically to the door of Newton although Newtonians will deny this fact as if they deny the honour of their Master Newton and that is what they have to do.

The Practice of Brainwashing and Mind Control in Physics

The Definition of Gravity according to the Oxford Dictionary of Astronomy:
Physics
1) **The natural force of attraction exerted by a celestial body, such as Earth, upon objects at or near its surface, tending to draw them toward the centre of the body.**

What this referrers to is the process by which weight is measures and the more a body pushes to the earth, the more would the scale reading be. Whether this comes as result of as force is what I strongly dispute. This indicates in the same result as measuring weight would be and there was never any defining proof that parts this form of mass from this form of weight. Both mass and weight are measured in the same measurement and have the same indicators. This form of mass is there but is the same as weight and only mind control in physics by tutors can put any distinct identifiable characteristics differences claimed to be between the two.

2) **The natural force of attraction between any two massive bodies, which is directly proportional to the product of their masses and inversely proportional to the square of the distance between them.**

This is the one that I refer too that is the figment of Newton's information. I beg anyone to use the information gathered from the solar system to prove this definition do apply!

3) **Gravitation.**

This proves how low Newtonians can go to solicit absolute fraud and intoxicate student's minds in brainwashing their thinking to accept what never is proven. This is hogwash at best!

Gravity is the fundamental force of attraction that all objects with mass have for each other. Like the electromagnetic force, gravity has effectively infinite range and obeys the inverse-square law. At the atomic level, where masses are very small, the force of gravity is negligible, but for objects that have very large masses such as planets, stars, and galaxies, gravity is a predominant force, and it plays an important role in theories of the structure of the universe. Gravity is believed to be mediated by the graviton, although the graviton has yet to be isolated by experiment. Gravity is weaker than the strong force, the electromagnetic force, and the weak force. Please show me the attraction in relation to mass that applies between any of the planets and the sun.

In this article I am going to investigate how much truth there are in mass pulling by the force of gravity. Most if not to all of the persons reading this article will be annoyed by just the thought of me embarking on an investigation of the issue that seems so totally senseless to investigate. It is senseless because the concept it carries became accepted as household practise and life science. Mass is associated with everything that is represented everywhere.

Do you think of astrophysics as the department that is run by the wise and the level minded, the sober thinking, and the absolute trustworthy? If you think those in charge of astrophysics are the pillars of trust, then get wise and read the following. Newton created the factor mass as a trick of his imagination. We have a body with mass on earth and that we all know, but no heavenly body could ever have mass or present mass! After you have considered the following you might agree with me that even small Children can reach a higher level of clear-minded logic and find more sensibility than what those scientists promoting astrophysics have because science lives in a make believe fool's paradise. They love to calculate because with mathematics they create a fools paradise.

In this article I give the table Kepler represented his research by which the planets move. Also I give one column in which the mass is indicated in terns of earth mass units. Use the column to show where it is mass that produces any valid factor in accordance to planetary activity, movement and allocated positioning. In the table columns we have the mass of the main solar objects and according to Newton the mass is responsible for movement. If it is mass that does the pulling in ratio of the radius then we have to see some evidence of this applying somewhere. Is there any person that can prove that claim from the table showing Kepler's movement in relation to Newton's mass? In what way odes Jupiter move faster or how does Pluto move slower according to mass? If mass did the job, how is it done? Notwithstanding Newton's blatant mathematical manipulation, mass don't apply.

There is no possible way to produce evidence from the behaviour of objects in the solar system that mass is responsible for gravity forming between two heavenly bodies. If it did, the asteroids could no possibly orbit in their allocated circle at the same rate as Jupiter circles in its designated circle. Every one believes in Newton except comets, because comets fail to collide with the Sun. However I can explain in some way…The evidence that mass is pulling mass there is not and there is no evidence of mass pulling mass.

Looking at the size of a body does not allow any one to awards mass. If you are a student, then ask your Masters please to explain the following abnormalities and inconsistencies they promote as part of official physics, which I present in this article and as students get wise instead of brainwashed. I say brainwash again because they force-feed you fabrications, as you will come to see. They can't explain the facts as reliable but hide the fact that the facts are in fact untruths. Tell them to prove that planets have mass. Tell them to prove that it is mass that generates gravity that pulls the planets. Ask them to explain gravity in detail. If any person gives evidence and the evidence is unsupported by truth and unfounded by proof, it is fraud. If a Clergyman should bring evidence of some miracle healing that took place, and he could not repeat the act at will for all to witness again and again, it is fraud not withstanding the religion. Even if the lie is to underline the greatness of his religion, it still constitutes to fraud. If anything is said without truth backing the claim, making it unfounded, it is fraud. It is illegal to make claims that is not based and is not supported by reality. As perfect as everyone thinks Newtonian physics are, this applies to educators teaching physics. There is a suspicion lingering in the back of everyone's mind that something is not quite correct about the approach physics take on the matter of gravity and only those academics seasoned with years of studies and salted with time seems to miss this haunting feeling. But the longer the students are educated, the more this uncomfortable feeling dissipates.

Hidden under a cover of "understanding Newton" or "not being able to understand Newton" tutors in physics force certain incompatible arguments to join that which never can join and while joining also make sense at the same time. In this article I challenge the figures that charge the highest form of respect in our communities and those in charge of the most dynamic part of society and those who stand beyond and above any form of suspicion. I charge those that personify truth and are the very same persons that I accuse of betraying the ones trusting them. Again I challenge you to come forward and tell your students the truth about what I uncover in the articles that follows and as the articles progress by introducing information...then you explain to them how you deceived their blind trust in you as a tutor in physics. In physics the blame goes to students ability to "understanding Newton" or "not being able to understand Newton" but I show that in this case the blame should openly be dedicated to those that should be blamed, named and shamed and they have to defend their years of lying and contribution to cover the misconduct that was committed by them.

Those teaching physics as well as their predecessors now have to explain in the name of science why so many evidence were falsified to keep their noses clean while mud colours their lily white image in hogwash and the stink of their lies equals the pig pen it needs to cover such aroma. It is time that they reveal the truth. Believe it or not, but this diverging from reality and misconduct about applying science is in place because of centuries of brainwashing going on passed on from generation to generation and is employed in physics from teacher to student for centuries on end. This article is dedicated to bringing honesty into the faculty of Astrophysics and Astronomy as well as to show the Physics student on what corruption and deceit does physics base their facts which they proclaim as being such well proven, and godly accurate facts and is unwavering depicting only the truth.

Go to the web site **www.singularityrelevancy.com** and then use the information I supply in the book **www.questionablescience.net** titled Sir Newton's Mythology, which you may download fro free and use the information given free of charge to confront you academics with the truth and then insist that Academics who are teaching physics, prove to students that Newton's statements of attraction are correct. Let those academics explain the method mass implements.

If you are a student studying in physics then reading this article is detrimental to your future, as you can remain part of the problem physics has had for centuries or you may join the solution that came to physics and start to heal the wound. Students read the next pages and you are about to learn how students are brainwashed into accepting the baseless and ridiculous misinformation Sir Isaac Newton puts forward as truths. The Custodians of Physics have nothing better to offer than presenting you with unfounded, corrupt and distorted facts...and by doing that they resort to mind control on students and introduce baseless concepts by manipulating the student's thoughts.

If you are a student then read in **www.sirnewtonsfraud.com** what they do to you and how they brainwash you. This suspicion that there is, this feeling about a certain concern and doubtfulness that is lingering on in the minds of many... and also is lingering on from generation to generation... without anyone ever finding a solution... it is them defrauding you by exchanging your institution fees for corruption, so confront them about their dishonesty. There is this vague unspoken question hanging in the air without any one ever finding words to express the question...and yet the question remains however unspoken it seems. Force them to become honest and to stop corrupting students with intentional malice. Let them explain how mass generates gravity...

Even if you had your personal favourite conspiracy theory, try and match it to the one that I have! I seem to be the first person in generations that ask questions about Newton's work. Questions I now ask is asked for the first time ever, well ever since the time Newton introduced gravity, before the emphasis fell on proof rather than merely what a person with reputation suggested. I now am able to show how gravity forms by forming a circle using Π because I have located the centre of the Universe.

But by my effort in finding the location I disrupted everything Academics in physics hold holy and for that I am most unwanted in the presence of the Academics charged with guarding the ethics of physics. In short, I clash head on with Newtonian dogma and principles forming physics. During my research I discovered abnormalities and inconsistencies about mistakes the Arch fathers in physics must be aware of but are hiding with all their considerable influence and academic power. The road I took in my search for truth concerning physics was never smooth and the resistance I came across coming from the academic sector is almost unbearable.

With my unpopularity rating this high as it does, I never qualified for help and those that would help found my ideas intolerable whereby I only found rejection instead of help as I tagged along. Because of this insider rejection I had to resort to private publishing because from the nature of my work I take Mainstream science head on and am confrontational on most aspects of astronomy. Since no one see the a problem there is in physics, no publisher wants to go head bashing with the Physics Custodian establishment of science on official science principles, which I have to do to convey my message in no uncertain language.

I argue that if it is the correct practise to use $F = G \frac{M_1 M_2}{r^2}$ to calculate gravity then the radius holding the gravitational constant must lead one to the centre of the Universe. As I confront science dogma and principles, nobody is willing to publish my work. I have to walk the road alone and fight the battle by my private effort without any support anywhere. If Newton is the problem one have to go pre Newton to find the problem.

The problem is that when looking at Kepler's table then if there is $T^2 \div a^3$ according to the table matching a column, then mathematically $T^2 \div a^3$ must be k^{-1} and where k^{-1} goes negative it shows space reduces time. It shows space in volume goes single by movement of space and not objects.

Planet	Mass per Earth unit	k^{-1} Movement	a^3 of space volume	T^2 During time units
Mercury	0.06	$T^2 \div a^3$ =0.983	$(a^3)=$ 0.059	$(T^2)=$ 0.058
Venus	0.82	$T^2 \div a^3$ =0.992	$(a^3)=$ 0.381	$(T^2)=$ 0.378
Earth	1.000	$T^2 \div a^3$ =1.000	$(a^3)=$ 1.000	$(T^2)=$ 1.000
Mars	0.11	$T^2 \div a^3$ =1.000	$(a^3)=$ 3.54	$(T^2)=$ 3.54
Jupiter	317.89	$T^2 \div a^3$ =1.000	$(a^3)=$ 140.6	$(T^2)=$ 140.66
Saturn	95.17	$T^2 \div a^3$ =0.999	$(a^3)=$ 868.25	$(T^2)=$ 67.9
Uranus	14.53	$T^2 \div a^3$ =1.000	$(a^3)=$ 7067	$(T^2)=$ 7069
Neptune	17.14	$T^2 \div a^3$ =0.999	$(a^3)=$ 27189	$(T^2)=$ 27159
Pluto	0.0025	$T^2 \div a^3$ =1.004	$(a^3)=$ 61443	$(T^2)=$ 61703

If you are a student in physics then you should read the following information with care and with much consideration because your mental health might be at steak here. One could think of another name for physics and that would be Newton's mythology. It is about the subject of gravity and is most important. The "Newton's mythology" comes from the fact that students have to learn what the professors claim to be true and what was never was proven. Students have to repeat in examinations that the formula $F = G \frac{M_1 M_2}{r^2}$ is truthful and viable while it was never proven. Do you realise that it is an accepted practise that all students that are studying physics on all levels are subjected to the most intense brainwashing and thought control found any where on Earth? This must be some sort of a joke you may think but thinking that way in disbelief is just what those practising the mind control wish you to think!

According to the tables, all movement is according to some other value than mass. They never prove Newton's philosophy on gravity but those persons conducting teaching in the subject of physics force all physics students to learn Newton's gravitational concepts and accept the facts as if it has been proven beyond all other facts. Students have to believe that Newton is correct or academics will see to it that they fail their examination. The condition of being accepted in physics is to accept Newton without questioning the proof that is never supplied.

Let those academics now prove precisely how mass brings about gravity and then afterwards test you on how Newton is proven correct and not on you repeating their facts blindly about what they say is true about what Newton said, which they say is true. The manner they present Newton is completely hearsay and that method may not be used in any court of law.

Let your professors now prove how it is that Newton's teachings are correct and then examine you on the process they use to prove Newton's concepts rather than test the state of brainwashing you have submitted to. At present they say Newton is correct and then they test you on your ability in repeating that Newton is correct without ever proving to you that Newton is correct. That is not testing your knowledge but it is testing the mind control you have submitted to. Let those physics professors now prove Newton and their ability to prove Newton correct and then test you on the manner they use to prove Newton to be correct.

They teach you that the Universe contracts and to state their case they force students to learn that gravity is proved by Newton introducing the following formula $F = G \frac{M_1 M_2}{r^2}$ They say that M_1 is the mass of the Earth and M_2 is the mass of the individual in questions mass and the multiplying of these factors with the gravitational constant produces the force of gravity when this gets divided by the square of the radius. Please let you lecturer put in all the values of the formula and prove Newton is correct. If he can't and I know for sure he never can fill in the symbols and calculate the force of gravity, then read the rest of the web page that follows to see how far academics in physics go to brainwash students into believing in Newton's fraud.

The truth beyond all other truth is that Newton's gravity has never been proven (because try as you may it is not possible to prove Newton's formula forming gravity mathematically) and because academics know that, academics require the blind acceptance of Newton by students. This unconditional acceptance of Newton's correctness relies only on the pre-conditioning of students' mind set and academics depend only on the student trusting the academic "say so" about the institutionalised correctness of Newton. That Newton is correct nevertheless and notwithstanding that there is no founding proof about this matter, is what students should be accepting blindly. I'll bet you they are more surprised than you about me accusing them of systematically mind altering the student's physic and ability to think than you are.

This is a fair test to see if Newton's contraction theory underwritten by Newton's attraction formula $F = G \frac{M_1 M_2}{r^2}$ is valid, then force your professor to use this formula as it reads and show WHEN the Moon and the Earth is going to collide. If he fails to do it by using Newton's formula as $F = G \frac{M_1 M_2}{r^2}$ then you will know who is conning you, him or I and who is truthful again him or I. I charge all academics to prove what I say is being wrong in any way or even that I exaggerate in the least. I challenge Newtonian academics to prove that mass does indeed form any force of any sorts and in particular gravity!

Pre-conditioning students into blind acceptance depends on the academics' insistence that students approve Newton's concepts without pre judgment or students insisting on scrutiny of any sorts. In examination students have to outright and blindly follow academics' say so trumped up rhetoric and respond on facts only because academics say so. Academics depend on students never questioning their say so or demand proof about what academics teach. Those academics in teaching positions insist that all students accept Newton's accuracy.

Where does mass manifest in this? Where does Jupiter show its mass is pulling more than that of Pluto? According to the definition of gravity claiming gravity is a pulling power by the mass of objects and this formula $F = G \frac{M_1 M_2}{r^2}$ then Jupiter has to beat the earth going at a speed 317 times faster than the earth, because Jupiter is 317 times larger than the earth. Show me any evidence that mass plays any role or function anywhere in the Universe! If you are one of those members of society that never thought you would hear the name of an accomplished person as Sir Isaac Newton being associated with fraud, corruption and brainwashing, then the books **The Absolute Relevancy of Singularity** published by **Lulu.com** are specially written to inform you about the truth there is lacking in the correctness about science. Newton's formula and suggestion indicate a straight line moving $F = G \frac{M_1 M_2}{r^2}$ between the objects pulling by mass while it is clear everything applies the motion around the sun using a circle. Everyone knows that planets orbit around the Sun. Planets circle the Sun which is the same as saying planets orbit the Sun. Just by calling the circle motion in terms of what applies, that statement nullifies Newton's claim of mass that attract mass and put to question the reliability of Newton's dogma. This might at first seem a small issue but from that I prove that gravity works by the value of Π and not mass.

This is methodical mind control as much as it is the brainwashing and I show that they enforce this practise. If you are one of those believing that Newton was ever proven, then what you believe to be true is a lie because Newton can't be proven and that is the truth! The time has come to face your teachers and force them to stop the ongoing old culture of bullying students into submission and conditioning their thoughts by enforcing on them dogmas, which is no more than systematically enforcing mind control! In order to get students to accept Newton's hypothesis, academics resort to brainwashing pupils' and students' thoughts.

One such an example is Einstein's Critical density scam and the "Dark Matter" swindle where they hide Newton's shortcomings under a pretext of ongoing research. Hubble found the Universe is not contracting as Newton said it does with $F = G \frac{M_1 M_2}{r^2}$ and hell broke loose. The cosmos stepped out of line by showing Newton is wrong in his supposed theory that mass pulls mass to reduce the radius such as $F = G \frac{M_1 M_2}{r^2}$ will indicate. Einstein was then tasked to supposedly measure "all the mass" in the entire

Universe and to find out when the Universe will correct its evil ways and start to submit to Newton's idea of contraction. They concluded it is not Newton that is wrong but the cosmos went out of line to contradict Newton. Therefore one has to find the error in the actions of not Newton's idea but about the Cosmos acting out of line with Newton. The fault was investigated not on the side of Newton, but in the cosmos! The proof of practise fell on the Universe that could be wrong but never Newton!

With all this in mind did any one ever come to wonder about the reality driving the all too famous **Einstein's Critical Density theory** and the fact that this idea was conceived to conceal the corruption of Newton in physics? Allow me please to elaborate and then make up your minds. The facts in truth are that the **Einstein's Critical Density theory** was a scheme plotted by those in charge of physics principles to cover up and conceal corruption in the heart of physics. Hubble proved beyond doubt that there was an inflating Universe. This contradicted Newton's deflating or pulling Universe and this perception of a deflating Universe being a myth had to be most ardently hidden as to yet again compromise the truth about Newton and his theory. In the formula $F = G \frac{M_1 M_2}{r^2}$ the relevancy applies between the strength of the mass and the distance of the radius that keeps the influence of the mass forming the gravity. The longer the radius increasing the distance between objects the more this will reduce the value of the mass, whatever the value of the mass might be. The Hubble concept proves that while the mass might remain the same, the radius keeps growing and such growth diminishes the influence of the mass all the while the radius increases. This is so basic that primary children learning the basics of mathematics will understand! This means with the radius increasing, there is no chance that the mass will ever become strong enough to bring about the pulling because the constant increase in the radius constantly diminish the influence that the mass might produce.

Yet Einstein proceeded in searching for a value that will determine when the mass would bring a turn about in the direction that the cosmos evolves in. Einstein was looking for the moment the mass will become strong enough while the most basic principle indicates that an increasing radius leads to a decreasing mass influence submitting a decreasing potential gravity since the mass becomes less prominent in influence. If Einstein was unable to recognise this most basic of mathematical principles, then what type of genius did physics create in him and what slur did physics promote. In $F = G \frac{M_1 M_2}{r^2}$ any factor standing in relation of division, it is the bottom value that determines the outcome of the value of the top.

This idea of the two factors being in opposing relevance where the size of the bottom caries the value of the outcome is so simple that children will recognise the principle, and yet those fathers of physics wants me to believe that the greatest mathematician that ever lived did not realise this principle...the principle that the radius and the mass stands related and the growth in the bottom value will promote the decline in the top value as a dominant factor. Can any one with this information including the proof I asked for on previous page have any other conclusion than students should smell rotting fish somewhere?

Notwithstanding my arguments that should have been raised by the mathematical genius, Einstein supposedly measured "all the mass" in the entire Universe and then afterwards concluded there is "insufficient mass" to pull the Universe closer. This rattled the cages of the Newtonian conspirators because Newton once again stood naked, venerable and bear. Yet the blame had to be associated with the cosmos not playing by the rules of Newton. The cosmos had to be at fault because Newton just can't be the person to blame. Then the genius of the Newtonian cunning kicked in once more and we can see why they are seen as the most brilliant minds walking the earth. There had to be something they miss and no one can see.

Then some idea was presented that the Universe is hiding mass from view, I suppose just to spite Newton. If the mass was not visible and was therefore undetected, then the mass was dark and if no one could see the mass then no one can prove the mass being present and then no one can disprove the mass being there waiting in the dark. What a splendid idea this was for cheating. This presented a solution of Biblical proportions and a new scam was introduced to hide the failing of Newton and of the Critical Density sham.

The Dark Matter hides mass that will supposedly pull all material closer again at one stage in the future. This is meant unleashing an enormous swindle, bigger than anything before, only beaten into second place by Newton's swindle about mass unleashing forces. They proclaim there is "Dark Matter" hiding and waiting in the wind to come to the present someday in the future and start to pull the Universe closer as Newton said it must by the measure of $F = G \frac{M_1 M_2}{r^2}$. In that way Newton was vindicated and the cosmos took the blame for hiding mass and the contraction was reinstalled.

It is obviously clear that having such a total idea that there might be dark unseen mass floating in the Universe which at this time does not generate gravity but will some day kick in to generate gravity in order to cover up Newton's deception about a contracting Universe and just because Newton has to be correct at some point in the future. Science wishes me to believe that since there is a lack of seeing material there then will be dark and unseen material where they are so dark they are undetected by all humans. Why would the mass at present then not activate gravity and why would the mass at some point spring to life and start activating gravity? How much can the Physics paternity still hide the fact that Einstein's critical density is being used as a cover-up to distort the truth to conceal the fraud Newtonians wish to cover? Hubble found the Universe is expanding and Newton said otherwise. Hubble's declaration was on track to blow the cover that was concealing the Newton fraud wide open and uncover the century old deception. The question is if it is mass pulling mass onto mass, then why do we have comets left in the solar system? The mass of the Sun should by now at least have destroyed every comet going around.

The term pulling does not suggest any circling because no one can be pulling towards and does that while circling around the object. That serves the term pulling. In conversation we speak of the planets orbiting. If Newton was correct we should be speaking of the planets pulling, but talking about pulling would be blatantly wrong according to the normal spoken word. Never do we refer to the planets pulling the Sun or the Sun pulling the planets, but we speak of seasons coming from orbital positions. Being in orbit has to neutralise the pulling and then cancel the pulling concept that also became culture. The entirety of physics rests on this one formula $F = G \frac{M_1 M_2}{r^2}$. The questions concerning that which you are studying and that touches every aspect you are academically concerned with, is that if everything is moving apart, how does that support Newton's idea that everything is coming together…and please don't let them fool you with Einstein's Critical Density idea! If there was mass seen or unseen in the Universe and mass generated gravity and gravity does the pulling then why is the mass not at this moment doing the pulling? If there is an object it indicates the presence of mass. Then what stops the mass that should be present, dark or luminous, to start pulling by gravity, as it should do?

Should you think this page is some sort of a prank then answer the following simple question to yourself in utter honesty: If there is a Big Bang after which everything was left moving apart, how does that support Newton's contraction? Tests results received after the Moon landing show the Moon and Earth are moving away from each other! Yet students learn about mass pulling mass and that puling by mass forces togetherness by contraction.

This is only one of many points that I make on this one issue and there are so many other issues one may think of those in terms of counting in numbers in many hundreds or even in thousands. If the Sun for instance has mass that is apart from the Earth and the Earth also has mass and there is a gravitational constant in between the Sun's mass and the Earth's mass we have the radius in that location. It then must be the gravitational constant that fills the space that the radius holds. It is rather obvious that while the radius is filling the vacant space between the Sun and the Earth it is the only place left where the gravitational constant can hide. To find the centre of the Universe I had only to find the gravitational constant that holds the centre. Through my venture I discovered one person that knows what gravity is! Newtonians went and filled that space reserved for the gravitational constant having a measured value with nothing! How can nothing have a value of 6.67×10^{-11} while also being filled with nothing as it is nothing filling the nothing of outer space?

My question in this matter is what is all that mass of so many supposed stars living in utter darkness doing at present while waiting to get to work and begin with generating gravity by mass where it will only later, much later form a force of gravity that then will bring about this pulling of the Universe? What makes the mass slumber in darkness to one-day form a pulling force? What has the "darkness" or the fact that we don't see the mass got to do with the idea that the mass at present is not forming gravity that is forming a pulling force? You are taught that gravity pulls objects to the centre and obviously gravity then has to ultimately pull everything to the centre of the Universe. That is what the Critical Density research that Einstein initiated wishes to establish. The idea is that $F = G \frac{M_1 M_2}{r^2}$ makes the mass create a force that will destroy the radius and ensure everything is going to come together eventually at one point where the radius then will be no more. If that is the case, then where is that point? If everything is destroying the radius, then it must end at one specific point.

In the classes you students attend a physics lecture, has any one confirmed a location where one might find the centre of the Universe to confirm the ultimate destination of $F = G \frac{M_1 M_2}{r^2}$? If you wish to apply a Gravitational constant as a calculated factor then it is apparent that one must know to where such gravity is pulling since it then is the gravity that is where the contraction is going that predominantly is keeping everything apart. Then the gravitational constant is what is resisting the collapse of the Universe. If there is a force, then where is the force taking the pulling…if it is a gravitational constant applying through out outer space then where is it having a centre base? To those professors claiming Newtonian ideas are substantiated by proof, I say that notwithstanding your personal academic qualifications and while at the same time disregarding your status and previous achievements as well as ignoring your many admirable abilities you may have and however superior they might be, I shall teach you about gravity. I say it is time Students learn the truth about physics notwithstanding the status academics will loose.

If mass did attract mass, what kept the balance in the distance it held according to the Titius Bode Law where the planets do find a balance in orbit rather than moving towards the Sun. One might think of mass pulling the comet to the sun, but instead of slamming into the sun, the comet makes a circle (Π) and disappears into outer space. If mass pulls the comet closer, what does the pushing away afterwards. If mass did attract mass, it would explain the behaviour of the comet coming towards the sun...but then what is pushing the comet back into orbit, into the darkness of outer space? The idea of proof about mass pulling everything is automatically placed at the door of Newton. Think of what planets do… and you think that planets orbit. If normal speech contradicts Newton, then it is his task to prove his supposition is correct and the claims about attraction Newton made, although Newtonians will deny this fact as if they deny he is correct the honour of their Master Newton and that is what they have to do. No one thinks of planets spinning or planets basking in the summer Sun. However, just using the term orbiting is in total defiance with Newton or physics!

This is my introduction and this is my prologue:
But before I can commence with that task I have another duty administer: I AM ABOUT TO WARN EVERY PERSON IN SIGHT OF MY WORK ABOUT MY SLENDER ABILITIES.

Therefore in the light of what the most respected academic group on Earth accuses me of, I therefore have to issue a most serious warning to any person with the intention of making some kind of inquiry to the content this book holds, then the most concerning matter involving any content within the pages of this book you hold are that you must please seriously consider that where the stating declares the possibility that the content in this book has been **(written by…)** then don't take the announcing **Written By Peet Schutte (Petrus S. J. Schutte)** very seriously for there are grievous doubts leaving considerable dispute about the possibility, which underwrites the authenticity of **Peet Schutte** achieving the **(written by…Peet Schutte)** status. **Please take note of the following dehortation.** In the light of the reference to me serving in the capacity as being responsible for authoring, **(written by…)** in line of keeping fairness and justice to members of society, where all civil beings should carry reputed honesty, then: **Please be warned before any** reader starts reading about the **following extremely serious admonition**: I am bound by my conscience to warn all intended readers that **I am placed under caution** by the **Academics in Physics. Those most esteemed members responsible for the guardianship and maintaining the ethos in physics are of the opinion** that **I, Peet Schutte,** am unable to write any book on the science of Physics as well as Astrophysics. Therefore, **I, Peet Schutte,** must declare that I should be considered as not very able to write anything, because I am incapable thereof. I suppose, I merely generate new information, which I establish as thoughts and then gather as concepts. I further collect the result as words, which I put on paper using alphabetic symbols. I then compile that in a format that others may confuse with a book, but a book it cannot be, since the Masters in science found me unable to write a book. But before you go further and follow my arguments, I first have to level with you about how academics view me in the position I hold. Please do not allow me to fool you, for this then cannot be, or represent a book. Now I have done my duty in warning everyone and in that, I denounce further participating with any purposive intention to wilfully bring down the crux of civilization by acting unacceptable and irresponsible.

I didn't write any books since according to those wise enough to judge me without ever reading my work that I am not schooled to do so. It is my guess that I merely generated uninformed thoughts, which I collected as alphabetic symbols and plotted that in ink on paper. This effort I achieved from harbouring my delusional ideas spawned by a dehumanised brain. It only proves my weak and under developed mentality, due to my lack of an informed insight that is a typical symptom that all those have that is suffering from a disadvantaged past that one can only have when the person obviously lacks formal education. While you are reading the letter deciding to regard or dismiss my work, then also please keep in mind when reading my language used and also please give credit where it belongs…if you do find linguistically improper use of words or misspelling, then remember that I am a feeble minded motor mechanic and not a literal giant.

I submit article to well known physics magazines but my articles are rejected on the most unappeasable grounds and for the most outrageously ridiculous reasons the Newtonians can think of. I explain how gravity forms but I am rejected because they are of the opinion that my work does not meet

One such an article I may use because I said I was going to use the material as an open letter I gladly show. I submitted an article in which I show what the manner is in which gravity conducts movement by means of singularity.

I wish to produce two of the mail response I received when I contacted members of the Physics Academic establishment and show my case while also showing the response I received. Readers will find witness of what I accuse science of and their demeaning attitude from the e-mail of two that I present.

To whom it may concern and all others reading this document just the following concerns:
>>> "Peet Schutte"

Dear Prof Iben Maj Christiansen, I have received your e-mail reply and I will follow your advice to the letter. The book containing four volumes that I sent to you accompanying the proposed article, is the collection I named The Absolute Relevancy of Singularity and as such the collection forms a small introduction to the thirty-two or so books I wrote on various matters concerning physics, but as such does not officially even start to introduce the spectrum of every aspect of my work. I have been in contact with numerous Academics and about one in one hundred reply. When the one in a hundred reply, the academic always uses a most aggressive tone which I came to accept as what I receive from academics, and because of that I was most delighted to find some kind remarks from you as a practicing academic, and might I add, the first such kind remark in ten years of my trying to contact any person in physics that would take note of what I have to say about a new line of thought.

The New Cosmic Theory I try to convey in total is much information and every time when publishers reject the publishing of any entire book I propose, the rejection was on the grounds that "the discourse is not falling within the main-stream science discourse" and therefore I was subsequently advised to write articles on the subject as to find recognition. I was told that only then could I achieve publication of any entire book. Now I find that trying to publish articles has my work rejected on grounds as follows and the following is directly coming from the reply in which one of my articles was rejected recently: "You submitted an article of 15 pages to the "Journals Name Left Out". (I do not wish to include the name of the journal for obvious reasons but quote the reply as I received it). "The content of this paper doesn't constitute a theory in physics. With a lot of words and some simple algebraic relations, there is no way to "explain" the world of physics. You seem to be out of touch with modern developments. This is also shown by the fact that you don't quote any relevant literature." It is not possible to introduce the totality of my work in 15 pages (or whatever a journal would allow) and remain coherent during such an introduction about anything.

I am trying to introduce a study I have done during twenty-seven years of research and there is not one word that I can quote from any other source since every word comes from conclusions that I make and I prove with the use of logic and believable uncompromising mathematics. All I try to do is to find a medium wherein I can tell some interested parties where to go to read my work and then for them to judge me on their merit and not be sidelined by rules set by academics in charge of publishing. Let everyone read my work and then after that let all readers be opinionated by personal opinions applying. Everyone goes on about the unfairness Galileo endured at the hands of the Catholic Church, but at least the Church allowed Galileo to publish his work so that the entire world could take note. Every one in science as well as the Church thought Galileo was out of touch when he declared the science wisdom prevailing at the time was incorrect, and five hundred years later we know who was out of touch. I do not compare my work with that of Galileo but I find the same restrictions brought on me by the Powers of the day controlling science.

In this light I am most delighted by your attitude.

I have tried to acquire the addresses of the academics on the list I include, which I highlight, but was unsuccessful in doing so. Could you please show further kindness and supply me with contract addresses so that I could contact any one and try to arrange such a meeting as you suggested.
Peet Schutte.

Please find our Email Disclaimer here-->: http://www.ukzn.ac.za/disclaimer

Dear Peet,
Those on your list I know are in science education. You need someone in pure physics.

I am afraid that you will continue to get rejections if you do not relate your work to existing theories and previous work. While it is possible that a lay person hits on an insight that has been overlooked by academic trained in the field over many years, it is unlikely. We assume that work offering something new would be related to existing theories, either by building on top of them or by showing how and where they fall short. If you do not relate to existing work, it is repeatedly going to be dismissed as mind spin too easy to shoot down.

I am sure you understand.

Iben

The following is the response I received when I submitter a paper on how gravity forms by singularity forming.
Dear Dr. Schutte,
You submitted an article of 15 pages to the Annalen. The content of this paper doesn't constitute a theory in physics. With a lot of words and some simple algebraic relations, there is no way to "explain" the world of physics. You seem to be out of touch with modern developments. This is also shown by the fact that you don't quote any relevant literature.

I am sorry to say, but the Annalen is not able to publish your work.

I am sorry for having no better news for your.

Best regards,
Friedrich Hehl

Co-Editor Annalen der Physik (Berlin)
--
Friedrich W. Hehl, Inst. Theor. Physics
* University of Cologne, 50923 Koeln _____/_____ Germany
fon +49-221-470-4200 or -4306, fax -5159
hehl@thp.uni-koeln.de, http://www.thp.uni-koeln.de/gravitation
* Univ. of Missouri, Dept. Phys. & Astr., Columbia, MO, USA

Dear Prof Friedrich W. Hehl, I have received your e-mail reply and I wish to respond on your letter. The article of 15 pages to the Annalen had in mind to introduce a very wide-ranging concept contained in many books. I wish to promote books in which I introduce a much larger and much more detailed cosmic picture.

It is four books that actually form four volumes of one theme supporting **The New Cosmic Theory.** I wish to unveil a totally new approach to the thinking in cosmology. The concept is proposed in the article I sent to you which is "revealing" **The New Cosmic Theory**.

In the article as much as the theme I wish to go where no one ever attempted to go before. I introduce **the Universe of singularity**, a state in which the Universe still is because **it is a state from which the Universe grows**. It is where material in a dimensional dynamic does not apply because **it is where Einstein said, "the Universe goes "flat"**. <u>**I show you how and where the Universe goes "flat"**</u> I will guide you to the point where I go…so that you may see where my books and the article lead you. **It is in the domain of singularity.**

When you read work about the Big Bang you have to go right down the development (in reverse order) to the point where the Theory of the Big Bang points at a spot named singularity. It shows the very start from where all material developed. At that point one will find **The Absolute Relevancy of Singularity** and there has never been any attempt by any person ever to venture beyond the dimensional birth of the cosmos, which is called the Big Bang by going into the era where singularity prevailed. I take you there in my books as well as the unpublished article.

However, going there requires a very high degree of concentration and calls for understanding that a very little number of persons are capable to show. I try to show how the Universe goes "flat" as Einstein said the Universe goes "flat". Even by completing this unimpressive letter you will also know how the Universe goes "flat". Even where you failed to read the article I sent you, then by just reading this letter you will be able to find where singularity takes the Universe "flat". But it requires a mental capacity to understand because where I venture no one ever in the history of mankind reached into before.

I do not speculate but even in the unpublished article I show with pictures and sketches as well as "some simple algebraic relations" where to go to where the Universe starts, but you failed to read that because you are opinionated as to what conditions should the Universe have before the Universe will allow any one into physics. That is a pity. One should learn from the cosmos and not tell the cosmos what it must be to qualify as the cosmos. Then in the article I show you by almost taking your finger to the spot, the very point where the Universe ends and that too I qualify. You might dispute my arguments and show me about what you disagree, but it shows very little understanding of reason on your part about qualities man should have before understanding the Universe.

I go into a Universe that was in place before light was in place in the Universe and only darkness prevailed because light calls for space and in that era of singularity space was not even a thought yet. I show why the Universe goes "flat" and in a "flat" Universe only the value of 1 holds value since singularity is 1. If you can understand 1 or $5^0 \times 7^0 \times 3^0 = 1^1$ you have all the mathematical skills required to understand the applying concepts. To reach a value of 1 does not require big mathematical equations but to reach singularity requires 1.

The collection I named **The Absolute Relevancy of Singularity: The Theses** and the collection as such forms a small introduction to the thirty-two or so books I wrote on various matters concerning physics with gravity in mind, but **The Theses** as such in the entirety of the four books does not officially even start to introduce the spectrum of every aspect of my work. I have been in contact with numerous Academics and about one in one hundred reply. When the one in a hundred reply, the academic always uses a most aggressive tone which I came to accept as what I receive from academics, and because of that I was most delighted to find some kind remarks from you as a practicing academic, and might I add, the first such kind remark in ten years of my trying to contact any person in physics that would take note of what I have to say about a new line of thought, because the few others that replied were extremely aggressive about me confronting Newton. I only began to submit books to publishers after twenty-seven years of studying Newton and the role Newton play in cosmology and thereafter which was ten years ago I began promoting these ideas. **The New Cosmic Theory** is a process wherein I try to introduce a study that is ongoing for about thirty-seven years, give or take a few and I did not jump into the frying pan having my first thought about the matter published as an article when I sent the article to the address of Annalen der physics.

This is the ridiculous concepts that Professor Doctor Friedrich W. Hehl, Inst. Theor. Physics supports. Because I don't conduct tainted mathematics and highly suspicious Newtonian concepts covered by completely ridiculous malfunctioning concepts, I am despised as the ridiculous small-minded novice that is openly very feeble in thought. I do not support complete rubbish as this formula indicates the Newtonian vision of physics represents, but for my failing to go along with total trash my work gets rejected time and again by every science publisher I approach.

I was told, "With a lot of words and some simple algebraic relations, there is no way to "explain" the world of physics". Now I am going to go as public as possible and to use "words" to explain physics and to show how incredibly untrustworthy the world of Newtonian physics is. All you that think Newtonian physics is in reliability equal to God professing through the mouth of Newton don't be cowards by not reading my work as you always do but confront the conspiracy all of you are part of.

What you should answer is "how did this formula $F = \dfrac{r^2}{M_1 M_2}$ become equal in every respect to this formula $F = G \dfrac{M_1 M_2}{r^2}$ and never change any aspect of its characteristics or identity.

You go and do this swapping of relevancies in your mathematic class and then see what your mathematics teacher think of your modus operandi. This is what physics believe is true and because I question this I am wrong.

Newtonian science creates a factor such as mass to fake reality and misrepresent true physics. As long as they cheat, they seem clever and supremely superior when compared to the rest of us inferior beings. They live in a Universe fit for Alice in Wonderland! Please, for the love of God, show me what role does mass play in this picture. Show me how the biggest stars is locked together and the smallest stars are pushed to the outside. Show me where gravity "pulls" what to where and how gravity forms in this picture. Show where will dark matter hide. The idea of proof comes automatically to the door of Newton although Newtonians will deny this fact as if they deny the honour of their Master Newton and that is what they have to do. Think of what planets do… and you think that planets orbit. The term pulling does not suggest any circling because no one can be pulling towards and do that while circling. Going in a straight line serves the term pulling. Then the saying goes that planets orbit indicating they follow a circle. That is not what Newton said. Never do we refer to the planets pulling the Sun or the Sun pulling the planets, but we speak of seasons coming from orbital positions. Being in orbit has to neutralise the pulling and then cancel the pulling concept that also became culture. The Universe is growing!

If there was a pulling, and the word orbit cancels such an idea, then there has to be some sort of prevention taking place that disallows the pulling to commit the direction of travel. I know it is said that the orbiting object falls as fast as it circles and by falling while moving to the following side on position it never reaches the Sun, and yes, it makes sense, but there has to be some form of resistance replacing the planet in the next side position and preventing the falling or the pulling from taking place. The person Newton got his ideas from and the work he raped completely, that of Johannes Kepler explained this very well, but Johannes Kepler makes no room for any pulling of any sort. In the work of Johannes Kepler he said that the space being the orbiting route a^3 remains at a specific distance **k** while the orbit T^2 takes place…and in all my other books that addresses more information I take Newton to task on his dismembering of Kepler's formula by corrupting Kepler's work and with what amounts to fraud, Newton takes science on a goose chase that holds no truth. There is no pulling by mass to form a force of gravity in any way. Newtonians are faking the truth because the truth makes them seem small while Newton's misrepresenting cosmic principles allow those in science to paint their own Universe as they wish to present the Universe in the way they fake the story. Show me where there is evidence confirming a factor such as mass. Show me what makes the circling of Jupiter going around the sun faster than any of the small rocks in the asteroid belt. Newton placed mass in the formula he said Kepler represented. Could anyone show how much is the giant planets closer to the sun than the small planets are, keeping in mind that the pull of gravity must be exceedingly greater between the sun and the large planets.

Could anyone show how much does the giant planets orbit faster around the sun than the small planets do, keeping in mind that the pull of gravity must be exceedingly greater between the sun and the large planets.

Could anyone show how much is there a greater gravity-whatsoever between the giant planets and the sun than the small planets have less of, keeping in mind that whatever Newton mass would benefit because the pull of gravity must be exceedingly greater between the sun and the large planets. Whatever Newton saw, was in his dreams. Anything they say that mass gives is cosmically unrealistic.

What will the mass of the gas planets have that put them in a different stance when going around the sun than what the small planets have? By small I refer to Earth, Venus, Mars and say Mercury. All plants orbit equally.

The fraud factor came about when Newton destroyed the work of Kepler by changing the work of Kepler without understanding the principle factor guiding the work of Kepler. The fact that one can use $a^3 = T^2k$ puts the Kepler formula in the realm of singularity. I give away free books that explain how this comes about. One such a book is **The Absolute Relevance of Singularity: The Article** as well as another book that indicate the working more specific and is therefore require a higher degree of understanding because it is more technical is

The Absolute Relevance of Singularity: The Website. I wish to reflect on the modern conspiracy to hide Newton's failure and the deception to cover up Newton's fraud by engaging in almost unlimited fraud in the modern era. The science of the era had Newton's principles wrapped up. Having the formula that Newton gave as $F = G \dfrac{M_1 M_2}{r^2}$ enabled those who could "understand" the principle of "mass pulling" bringing about "a force" that Newton called gravity. This force was supposed to pull whatever was present in the Universe closer to whatever was present in the Universe. The formula gave man an equality to the power only God had because now it was the ability of the physicist to redesign the Universe and correct the failing God made in the Universe. The formula enables science to play a game designing the cosmos and the game used $F = G \dfrac{M_1 M_2}{r^2}$ as the prime basis. The Universe was working on a force valued at the measure mass that produced this force and the Universe was coming together pulling everything into one spot. The Universe was contracting…until a man came along going by the name of E. P. Hubble and this man destroyed the godlikeness of Newton and therefore all his followers as well. This no good science cheat could allow!

E.P Hubble is generally credited with discovering the red shift of galaxies. From his own measurements of galaxy distances based on the period-luminosity relationship for Cepheid with measurements of the red shifts associated with the galaxies, Hubble and discovered a shift in material that clearly showed material was moving apart. The distance between materials was growing and was not contracting as Newtonians believed.

Hubble and Humason were able to plot a trend line from the 46 galaxies they studied and obtained a value for the Hubble-Humason constant of 500 km/s/Mpc, which is clear evidence of a Universe that was growing from every point away from every point. In 1929 Hubble and Humason formulated the empirical Red shift Distance Law of galaxies, nowadays termed simply Hubble's law, which, if the red shift is interpreted as a measure of recession speed, is consistent with the solutions of Einstein's equations of general relativity for a homogeneous, isotropic expanding space. Although concepts underlying an expanding Universe were well understood earlier, this statement by Hubble and

Humason led to wider scale acceptance for this view. The law states that the greater the distance between any two galaxies, the greater their relative speed of separation.

This discovery was the first observational support for the Big Bang theory, which had been proposed by Georges Lemaître in 1927. The observed velocities of distant galaxies, taken together with the cosmological principle appeared to show that the Universe was expanding in a manner consistent with the Friedmann-Lemaître model of general relativity.

All of this shouts one thing **"Newton is wrong!"** Newton's lovely contracting Universe was not contracting or receding but was expanding by growing. Newton's failures surfaced and with it the entire Humans population could see Scientists in Physics new less about physics than does an ape know anything about maritime exploring. This took away…no, this robbed those that thought they are equal to God their insight into God's cosmic manufacturing process and this showed the world that they were not only not equal to God but was just like Newton, utter failures as professional that knew nothing of their trade. If this evidence of Newton came to the knowledge of those idiots forming the general population that they knew even less about physics than the general population, then they were as much the idiots as the general population. "Understanding Newton" always made them in physics superior with the ability to know what God and Newton knew and that vision gave them recognition and the superiority to look down on those with lesser abilities…those that did not "understand" Newton. If this came out that the Universe was not contracting by the measure of mass, then those that the educated Newtonians saw as the ones that knew nothing then uncovered the intellectuals as "Understanding Newton" but with Newton the intellectuals did not understand the cosmos. This revelation would uncover them as being the real simpleminded persons who thought they knew everything but was the ones that knew nothing about cosmology. Those unfortunate ones that did not "understand Newton" as being was then correct all along when they did not understand Newton because Newton was wrong and the intellectuals that "understood Newton" were ones that apparently knew nothing. Then those that did not "understand" Newton was correct because with Newton being incorrect those that understood Newton was the simpleminded minority that did not "understand" what the majority knew by realising Newton did not "understand" physics. They had to devise a plan to save the day. They had to cover Newton's deceptions with more deceiving. They had to go more criminal by topping the corruption to a higher degree as even Newton went before the discovery. They had to devise a plan that would outsmart everyone. They created missing mass in the critical density scam. This is how it works.

I have many books explaining how Newton corrupted science and this would take too much space to explain that. There is a book for free **www.sirnewtonsfraud.com**. I am not going into that at this venture.

They devised the critical density theory. How smart can those that are smart get when they truly try to be smart to save their skins? How brainy can the Brainy Bunch be when they deceive a plan to outfox all of the Human kind? You can go to the Internet and read all the multiple arguments about the Universe going flat and contraction therefore come about although not one sod amongst them know the least how singularity works. You can read in books I give for free how singularity works because I plot singularity as 1 and not as a complex formula. It is in **The Absolute relevancy of Singularity: The Theses** and the more complex explanation is in **The Absolute relevancy of Singularity: The Website.** There is to many arguments presented to mention so if you have a year or two filled with boredom, then search the web and go waste your time on fairy tails. You will see what they present when they have no idea what they present.

Here it is in a nutshell; what this is, it is not a theory that is called the critical density theory but it is a criminal venture conceived by the intellectual minds in physics to go criminal and defraud the world. They decided that the blame of the Universe not working as Newton said it does has to be placed at the door of the Universe. The Universe is lacking matter to contract as Newton said it should. Newton still remains absolutely correct. The blame for the mistake is diverted to the cosmos and away from Newton. Newton did not lack the insight to see the Universe expands, no; the Universe lacks material to contract. The Universe made the flaw and now we must find how the Universe will correct its flaw. The Universe went wrong by not having the sufficient mass and therefore the Universe must correct the lack of mass to get the Universe back in line with Newton. The blame for the mistake must be laid in the midst of the material within the Universe. Newton's formula stands correct and all suspicion goes in the way of how the cosmos was designed not to applaud the truthfulness of Newton.

Students, read the work I offer free of charge and learn how you are brainwashed and how your mind is pre- conditioned into believing in Newton's myth of pure deception which Academics call physics. If you are a student in physics who don't believe that you are subjected to unlawful brainwashing, then read on. Download work I offer free of charge. They come up with the most ferocious mathematical formula that should prove everything they say but all the formula proves is that the one trying to use the formula is out of touch with reality. If ever there is a person trying to impress with the fantastic mathematical formula explaining how the Universe works to his calculation, ask him or her to formulate when the earth and the moon will hit each other. Every formula they use found its basis on the perception that mass forms a force of contraction called gravity. Before they start to bewilder you with breathtaking calculations, call their bluff...ask them to show when will the earth and the moon collide. Tell them to use the mass of the earth and the moon and divide the product thereof with the distance between the earth and moon and then show what a force there is driving the gravity!

All those unrealistic arguments the Brainy Bunch offer as to why the missing mass or dark matter will bring a clarifying solution to avenge Newton has one damning flaw. Whatever they bring as an argument is tainted by a law in mathematics. It is built into the formula $F = G \frac{M_1 M_2}{r^2}$. If the radius increases, then the value of the mass reduces while staying the same. If the factor representing the radius r^2 becomes $2r^2$, then by the very same token does the mass become half of its previous value! It is effectively this $F = G \frac{M_1 M_2}{2xr^2}$ is $F = G \frac{M_1 M_2}{2}$ and this will bring about that while the cosmos is expanding, the worth of the mass is reducing by the same margin.

This is not rocket science; this is mathematics at its most basic. If the Universe was expanding then the measured value of the mass was declining that is if mass was responsible for producing contracting gravity. They are the ones that are the masters in mathematics. They are the ones that know mathematics better than anyone else on earth...and they missed this truth. This missing the basics was as deliberate as it was swindle the hide Newton's incompetence and with it their failure to understand physics.

This is where the second conspiracy started. The first conspiracy was the idea of mass that produces gravity and Newton's fraud to convince the world to believe him. Then the Masters-of-Deceit called Mathematical Physicist devised a plan to protect Newton's image and therefore their academic standings. They got Albert Einstein to hunt for the presumed missing mass and thereby distract attention away from Newton's oversight and their failure. Albert Einstein carried the heavy burden of being acclaimed the title of the best mathematical mind of all times. If you bend mathematical laws you will get a distorted Universe, as distorted as the Newtonian Universe is. The Universe doesn't agree with Newton's principles and Newtonian science has to bend all aspects to get it to fit.

You know...Einstein must have been very dumb or very dapper. Einstein must have been exceptional foolish or extremely brave and you can decide on what merit I suggest you should judge. In astrophysics they teach you about the critical density theory and how Albert Einstein calculated all the mass of the entire Universe and found the Universe fell short of Newton's expectations that was needed to drag the Universe back into one spot. Because of the critical density theory coming short we now sit with the dilemma of Dark Matter hiding in invisible black places waiting in ambush an unsuspecting Universe! If it was not for Einstein counting all the matter in the Universe and found it was too little, the hunt for dark matter would not have raged on.

It is because of this shortcoming that the entire group of science explorers are now searching (mostly in vain I hear) to locate the missing mass hiding in dark matter somewhere in obscure places. What a load of garbage all of that is because look at the picture. Do you think it is possible for one man, (or a million men) to calculate just the mass we see in this one photo? Those in academic posts lecturing you on Newton's correctness think they are the wise and are able to try to fool you because those professors think you lot of students are a bunch of mindless-monkeys that must believe anything they say because they think that only they have the ability and the mind to think and because you are a stupid student, simple-minded when compared to them, you can't think and therefore you will believe anything they say. If Albert Einstein thought himself capable of measuring only the material we see in this picture, then Albert Einstein was a fool with higher ego than what his IQ was. He thought he could measure the Universe and being that egotistic is stupidity crossing over to the insane.

In fact it is very obvious from what we see that Albert Einstein was no mathematical genius, he was an academic stooge, standing in for the other Academics to try and divert the looming shame by extending the visible problem of Newton getting uncovered. If Einstein was not able to see the reality of relevant formulations then he was more stupid than I am and I am so stupid I don't even understand Newton or so I am told.

This is the academic excepted values: The average density of matter in the Universe today that would be needed exactly to halt, at some point in the future, the cosmic expansion. A Universe that has precisely the critical density it is said to be *flat* or *Euclidean*. If the density of the Universe is greater than the critical density, then not only will the expansion be stopped but there will be a collapse of the Universe in the distant future. In this <u>closed Universe</u> scenario, the Universe will eventually implode under its own gravitational pull, leading to an event known as the Big Crunch. If the density is less than the critical density, <u>an open Universe</u> scenario plays out in which the cosmic expansion will continue forever. The critical density is calculated to be about (1 to 2) $\times 10^{-26}$ kg/m^3 – about 100 times greater than the average density inferred from all the known visible matter in the form of galaxies. However, when the inferred presence of dark matter and, possibly, of dark energy, is take into account, the Universe appears to be pretty close to the density called for by the flat scenario.

From these figures, how old is the Universe? What is its future? However, to answer them, it is necessary to know the density of the Universe, also known as "omega" (Ω). What a lot of rubbish this is! The Brainy Bunch insists that the Universe is 13 billion years old. This they measure by using the speed of light. In the first instant the Universe was the size of a neutron. It took some time to get the entire Universe to the size of five millimetres. That was the Universe.

The speed of light travels at what the speed of light = 299 792 458 m / s. At this rate light travels across a vastness of space 12 billion years to reach us. When the Big Bang came about the Universe was the size of a neutron. Take note on this issue; the entire Universe was the size of a neutron they say. What was the speed of light then? They say at the present it is 299 792 458 m / s, with which I completely disagree but let us leave this argument at that. The speed of light could not have been 299 792 458 m / s when the entire Universe was the size of one millimetre. That means the overwhelming distance of this 12 billion light years the light had to travel at a speed of less or equal to 1 millimetre / 10 billion years, because the Universe was so small at the time! The Universe was 13 billion years ago say five millimetres. Then what was the speed of light at that point? They say it takes the light coming from the furthest images 12 billion years to travel to us since the light travelling now left where it came from 12 billion years ago. At that point the Universe then was about the size of a pea. If the Universe was across the size of a pea, then light took a long time to travel a very short distance!

The light back then could not have travelled 299 792 458 m / s 12 billion years ago because 12 billion years ago the entire Universe was a couple of centimetres across. How fast was the speed of light when the Universe was the size of a pea. Then light travelled 10 million years to cross one yocto (y)(10^{-24}) of a mille meter. In terms of us today, light back then stood still. If it took the light 12 billion light years to cross, then the first part of the journey was pretty slow which makes the measure of time travelled versus space crossed rather ridiculous in every aspect it is portrayed. This means a lot of the light years it took to cross had to be billions of years just to gain one millimetre of space. Then the Universe must be trillions upon trillions of years older that they reckon it now is. It took a lot more darkness to form all the space is present in the entire Universal than the 12 billion years they say it did That brings us to the dark matter bit and the conspiracy that carries on undeterred. At present the conspiracy went as far as forming dark matter with (I suppose) dark energy.

This dark matter hides in places we can't see. This dark matter is what now forms the lost matter that protected Newton's image of correctness. Still Newton is untouchable because now in the present time the undiscovered dark matter is waiting to contract the Universe and this dark matter hanging suspended is what protects Newton. Is that not that sweet? Is that not the bedtime story every five year old would wish to hear every night? Every child will go to sleep feeling secured and in comfort. We can't attack Newton because the unseen matter that is dark is protecting Newton. Newton becomes untouchable by undiscoverable, unseen untraceable material that lives like a fairy tail in fantasy. If there is dark matter, what is the dark matter waiting for before it unleashes its incredible mass deployed force of gravity on this little unsuspecting Universe. What is preventing the dark matter from forming gravity that will do the job at this point in time? Why is it that the matter must be dark and must be seen in order not to form gravity. If the matter is present and forming a part of the Universe, albeit dark or not, seen or unseen, detected or not, if it has mass and if mass does bring about a force and the force is contracting gravity, then it must employ gravity. What is suspending this dark matter from kicking in and clocking in for duty? What prevents the dark matter from starting to get pulling? This is as big a scam as all the rest of the fraud they use to cover up Newton's fraud. I am showing all of this to prove how much deception there is in cosmology. Everyone in astrophysics is living a fantasy and everyone can make as they please, as long as the mathematical calculations seems to be in order. The reality and the viability or the lack thereof is no one's concern. As long as they can come up with stupefying formulated mathematics any dream will do. And the conspiracy carries on as long as it avoids reality and is void of constructive argumentative facts.

One night so many moons ago I have no intention of dating the time to which I refer, I was sitting outside while anticipating how to solve the riddles of the Universe. Well at least I was attempting to solve the part that riddles me and with my meagre qualifications it did not take that much to riddle me. Sitting outside and staring at the night sky gave me a break from all the confusion that faced me as I was again rejected by one of the so many academics rejecting my work and at the time I was still taking their rejection seriously and took their replies to heart. Then I saw the darkness of the night sky and compared that darkness with the brightness we find the star portrait in order to inform us of its location. That made me wonders about this thing called light and the manner in which light travels.

Have you as you sit reading this part at this minute sat back and gave a thought about the light enabling you to read? Yes it is simple in Newtonian terms because Newtonians keep everything at a child's mentality level so that they might not be confused. There is a wave of photons travelling at the rate of time and travels through space and time that is equal to time. What a load of rubbish did Einstein dump on the human race with that observation? It comes from some formula he or someone like him devised where they firstly manage to stop time. How they can achieve stopping time is beyond my limits of understanding but then again I admit I am one of those that so not "understand Newton". However, it does fit into the thought pattern of the Newtonian since their great Master also accomplished this outrageous deed. They go and put time equal to one. This then they mange in terms of placing the velocity that any object moves in relation to the speed of light and it is the square of this that they put at the root of this which they deduct from time being one and suddenly by the magic of mass it all comes down to time standing still. They put time at one (1) and deduct the square of the velocity any object travels (V^2) from the speed of light (C^2) which then also serves as a velocity and which then formulises as $t = \sqrt{(1 - (C^2 - V^2))}$. No argument is given as to why time and the speed of light should be the same other than Einstein fantasizing. Let's put the argument to the test of logic and see what happens.

If the speed of is equal to time one of two things will apply: either time as the speed of light will fit into eternity and stand completely still eternally because eternity is the fact that nothing changes as al remains unchanged. It is infinity that changes eternity by interrupting eternity. This might seem a little complicated but read **The Absolute Relevancy of Singularity The Website** where I introduce the points holding infinity as well as eternity. The concept is so easy to understand because do not hide untruths behind a collection of bizarre ideas. I shall give you a quick view into the physics I try to introduce and believe me; the impact of this introduction enrols every aspect cosmic physics may ever offer. This shows the human mind is not some mathematical computing device but an instrument of collective logical informative thinking about concepts outreaching simple computing by the way of mathematics. The only thing connecting what is to what was is light. Light brings the past to the present and therefore that is where physics must start and not with calculations. Time moves light through space. That means when having light as time in eternity, which is what outer space is, then infinity will never interrupt light and light will never move from the point it holds. When we look at any object in the distance it is the light that brings the object to us. Yes that does happen to light but there are specific qualifications that have to apply before that will happen. If light was using infinity as time then light must move through the Universe from end to end before time can establish a moment or any name one wish to give for the smallest duration possible. What this says is that if light was equal to time then light must travels across the Universe in one instant of infinity. Then only can time and the speed of light be equal. The Universe Einstein referred to is the multitude of a combination of indefinite numbers of Universes all having an equal value and therefore all being equal and one but they are definitely not one from our point of view. Such a thought brings to mind the most simplistic answer one can imagine. Every atom is an entire Universe in space.

The light hits the page bounces from the page and contact the lens of my eye where the lens conveys the photons becoming electricity to a part of the brain that translate the electricity to an understandable message and that makes one read. Newtonians think it is as simple as that! It is the movement of photons. Ever gave a broader thought about light streaming across the night sky, coming from ends of the Universe we do not even realise it is there? How does the photons manage to convey one complete picture coming from as far apart and as wide an area as it does? With a few photons connecting the eye or lens no one ever noticed the wonder of light. The photons reflect a view that seems as if coming from all the billions upon billions of stars. But most is coming from darkness covering an area no man can measure. Yet how many photons can actually connect to the lens of the camera or to the eye? Still a few photons coming from a single direction directly ahead eventually tell the entire storey. It is very simple to take the process of seeing by means of photon conducting very lightly and I have never heard one of the Brainy Bunch really in sincerity dissect the process to its potential. It is impossible that light from such an array of assorted sources can simply come together at the eye lens and show a picture of objects spanning across a Universe as wide as our mind can receive where the objects they reflect is beyond human measurement and the quantity is inconceivable many.

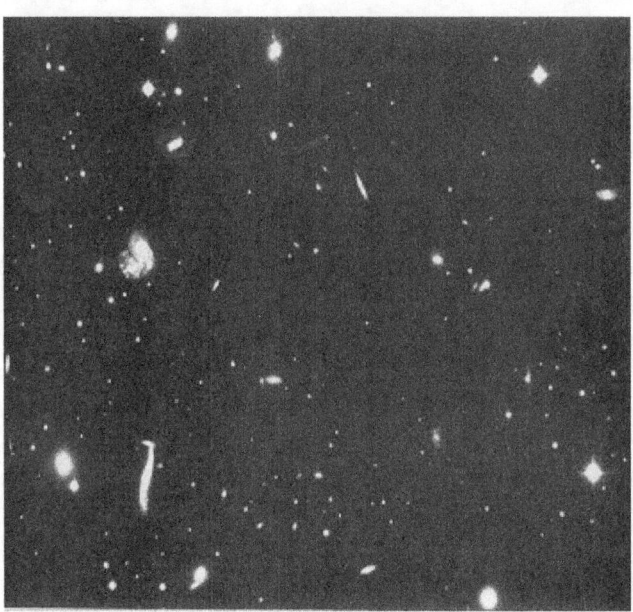

Light is much more than the medium science takes it to be. Light connects the Universe in a way we cannot contemplate. Light being far apart originating from regions not in the same time or universal space but it connects in a way that present us with a picture holding the Universe in an understandable content. From the point we stand and we watch the Universe the significance of what we see surpasses the sense of understanding of what we are experiencing. How can the few photons that our lenses catch coming from such an area as what the night sky cover transmit the complete picture of what we see. Take a few seconds and gleans at the picture of the night sky then rethink the picture applying the full content in the picture to what the size of you eyes is. Think how big the picture is that your eyes take in and translate that area to the size of your eyeball in an effort to determine a ratio. One will be forgiven if one thinks of the ratio as eternal to nothing. Consider the path the light followed from the source connecting to light from all other sources where all particles of the other light may come from and all the light brings a full picture to the lens one use to look through. In your mind connect a line from every atom producing light throughout what you see and connect the lines to your eyeball and see how you can manage to fit all the lines, as small as the lines may be.

Newtonians think of outer space as geodesic zero, with nothing in outer space but space. Geodesic zero means the light travels in a straight line from where it originates unhindered all across space to where the light connects the eye. Such an idea by itself is outrages because the stream of photons reduce in space to such a minute quantity that taken the area the photons travel and the space in vastness it covers, the chances of one photon coming across many hundreds of light years through billions upon trillions of cubic kilometres of space and selecting my eye to convey the electricity is less than infinite. Yet such conveying takes place every second of every minute. The position of the location of the second singularity, which is the precise duplication of the first singularity but in a diminished capacity, is obvious to miss when one is not applying a detective mentality, as one should in scrutinizing the cosmos. Culture will have us believe that when one sees a colour shining from an object the colour is associated with the object. Logic tells a different storey. A yellow dot is all the colours in the spectrum but yellow because it is disassociating with the yellow. That goes for red, blue and all other colours we may visualise. I think the norm accepts this as scientific fact with very little argument or substantiating proof about that required. If light came as individual streams of photon flurries our visage would translate that as such shown in the fragmented picture I show it would be a picture as I show it to be unconnected bringing across some photons in the manner where every object stands apart not being related in any way and that will be what we see, if it is anything that we see. That we know is not the case but that means geodesic zero is as much rubbish as anything Newtonians regard with simplicity and with careless thought. Whatever goes straight does so by a circle and that is how light travels.

Geodesic zero means nothing and how can I see nothing as darkness because "nothing" is not darkness, nothing is "nothing" and the darkness I see is darkness showing the darkness as something. What then about colours that are technically not colours as is the case with black and white? White is simple. By spinning all the colours in the spectrum the colour white shines through. Black is quite another matter. A friend of mine whom is one of the best painters I have ever come across told me that one couldn't paint black but have to make black a dark blue to show shade on the canvass. That apparently is his success in achieving the realism. He also went on to explain how many variations of dark blue form the shadows in one simple tree. This remark set my mind in motion. One cannot see black because black has no colour to show, but black is the colour most prevalent in the universe. One can see only by colour and since black is not a colour we should not see black, but we do. If the darkness was the representation of "nothing", then that should be exactly what we must see, nothing but the stars. Taken from the top picture some stars and leaving the rest to nothing is what we see in the picture below. A blind person sees nothing but when we look at space, we see something that we think nothing of as we see as space. One cannot have the ability of sight and see nothing except by closing your eyelids and then you see nothing. But in that case you do not see "nothing" in contrast of "something" you see "nothing" without it contrasting to "something". **Nothing is all about not being and not "not seeing".** By the ability to see the darkness renders the darkness something other than nothing and that changes the acquired value of the darkness from nothing to something. There is an eternal difference between something in infinity and nothing. I bring in the following to show and prove there is a remedy for all the fraud going on.

Time can expand as the cosmos does, or time can stand still in the form of photons holding relevancy or time can reduce as it goes above the speed of light within the atom. Every atom is a Black Hole that in a massive group is a Black Hole in the making. That is why atoms combining in a group forming material can absorb light releasing a certain value but absorbing most of the content. The group forming material might not produce gravity that would inflict movement retarding on us humans but it condenses light into the oblivious and that is what a Black Hole does.

The role that heat plays goes beyond (I suspect) what we may ever come to realise. Heat is an eternal substance forming solids as it spins above the speed of light, then cosmic fluid in relevancy to matter being the cosmic solid and forming outer space and in that relevancy being the cosmic gas when the movement it represents is the expanding motion we think the Universe holds, but what is space to the one is solid to the next or a fluid to the other. For instance the proton is dimensionless to the neutron, yet it is fluid to singularity allowing heat to flow. In relation to the neutron the proton is a solid but in relation to singularity the proton is liquid. In relation to the electron the neutron is a solid but in relation to the proton the neutron is liquid. In relation to outer space the electron is a solid but in relation to the neutron the electron is liquid. The neutron is solid in appearance yet it is fluid allowing dimensions to concentrate. Because heat is a liquid in relevancies, it is the father of specific density, allowing heat to flow differently in different forms of matter. To realise the correctness of that, one may gauge the heat relation there are in the first ten elements, and how each stand so different from the next element. Consider neon and boron, where boron has many times the density of neon, yet only half the mass, or where oxygen has more mass than does lithium have, yet lithium has a much different relation to heat than does oxygen.

We stand on the outside 150×10^6 km from the spectacle and from such distance we judge the sun. We don't even judge the sun from what we can see but we judge the sun from what we feel. We feel heat coming from the sun and from that we argue that the sun is hot. We see the sun has heat rising from the surface as a liquid soup, but still science maintains the sun is a gas because human culture thinks of helium as being a gas. What we see in this picture puts the hydrogen layer as the outer layer in a liquid. Hydrogen freezes on Earth at a temperature, which is the coldest amongst all other elements. Yes, the sun is 6500^0 and that is on the outside. To a human that is hot but a human has no mind judging the sun. If the sun squirts pure heat turned to liquid from the surface and the heat falls back into the surface the sun is a lot colder than the Earth is. The earth requires an enormous effort to cool hydrogen down to a liquid state. We must mind the way we think of the hydrogen in liquid. The hydrogen remains a solid. The element is untouched by temperature differences. It is the heat environment surrounding the hydrogen that changes from a gas to a liquid to a solid. One remove or one amplifies the heat in which the hydrogen is and that turns to liquid or solid or gas. The hydrogen is untouched in the elements worth.

Yet in the picture we see can the heat flow amongst the hydrogen as a liquid. It is not the hydrogen that is a gas or a liquid because hydrogen is a solid since hydrogen is material. It is heat reduced in space that turns by the compressing thereof into a liquid parting the hydrogen solid atoms. Nevertheless we remain adamant that the liquid is a gas and the hydrogen is in a gas and the sun is a gas bowl filled with hydrogen because to our mind hydrogen must be a gas. After all, our element table classifies hydrogen as a gas and that is the way we think of hydrogen. We do not consider hydrogen to be in a liquid state when we see the heat is flowing just like a liquid and shows all indications that it is a liquid. No the sun is hot because the sun feels hot.

It's a SOLID	It's a LIQUID	It's a GAS
Hydrogen 1	melts at -259^0 C,	boils at -252^0 C,
Helium 2	melts at -269^0 C	boils at -268.9^0 C
LITHIUM 3	melts at 180^0 C	boils at 1300^0
BERYLLIUM 4	melts at 1287^0 C	boils at 2770^0 C
BORON 5	melts at 2030^0 C	boils 2550^0 C
Carbon 6	melts at 804^0 C	boils at 3470^0 C
Nitrogen 7	melts at -210^0 C	boils at -195.8^0 C
Oxygen 8	melts at -218.8^0 C	boils at -183^0 C
Fluorine 9	melts at -219.6^0 C	boils at -188.2^0 C
Neon 10	melts at -248.59^0 C	boils at -246^0 C
Sodium 11	melts at 97.85^0 C	boils at 892^0 C
Magnesium 12	melts at 650^0 C	boils at 1107^0
Aluminium 13	melts at 660^0 C	boils at 2450^0

The solid material remains the same but that which is in between the solid material increases to form a gas or decreases to form a liquid or decreases so much almost only the solid material remains. It is not the material that becomes a gas but the way they are parted. This totally contradicts Newtonian thinking.

Hydrogen is as much a liquid as iron is a gas and neon is a solid. It depends on the element relating to the space/heat in the circumstances surrounding the substance at that very precise instant in time. We have to stop telling the cosmos to show us what we wish to find and start accepting what the cosmos is telling us to find. The culture that I am referring to is all about **nothing.** At present, we find that there is something we think of as nothing in outer space. Because nothing is what we wish to find and nothing is precisely what we are getting because we think of outer space as nothing. If you accept the cosmos to be nothing, then please define nothing to yourself and find the definition in the cosmos.

In the Universe there are no hot or cold, big or small, fast or slow but only a state of differentiation produced by time. The Universe parted by parting heat from cold when eternity parted from infinity, when Π^0 singularity parted from Π singularity, when 1^0 parted from 1^1. There is no hot or cold but there is a relevancy where one factor cools and another factor overheats. By retaining the sun is the coldest space in the solar system and outer pace is the hottest there can be.

Science currently as it always has place the cosmos in place to serve man. Even the atheist has this idea that the cosmos is created to be a place where man (or as they put it "LIFE" can flourish) and that s as far from the truth as thinking that mass forms gravity. From since the time that man discovered intelligence (if he ever did) man has been with the presumption that the sun is the hottest centre in the solar system. Later on in the present time, it came to someone's attention that the sun also holds the solar system in gravity. The Earth by its standard and dominating its sphere of which it can control with influence is the hottest centre in the space of its domain and it holds the moon centred to the Earth. The gas planets are the hottest centres in relation with the most heat and they all hold their satellites captured by a hot centre. All space structures hold in every centre there is that is confirming their independence at that point of securing independence the centralizing of the most heat it is able to concentrate and from that centre holds all material captured or controlled in the domain of what that forms the independence of the structure. All of this thinking is so Newtonian backwards as Newton is Newtonian backwardness. I can go on and on but heat in the centre couples gravity to space-time, just as if Kepler said before he was spoken for on his behalf and without his permission or his agreeing to it. One must think in terms of science and not in terms of human experience.

When anything contracts it is cold. Reducing space goes in terms of cooling

When anything expands it is hot. Expanding space goes in terms of heating.

Outer space expands into the oblivious and therefore by never stopping expanding outer space must be the hottest region. The sun can't dispose heat at a phenomenal rate and be hot as much as outer space can't accept heat and remain cold. If it was hot and received heat the heat received would not be much in terms of changing the status and if the sun reduced the space in outer space to cool it and disposed of the overflow of heat the status quo would not change and that is the main factor in the Universe: keeping equilibriums. The sun reduces the space in outer space and that it does by cooling outer space. Stars such as the Black Hole contracts space into singularity, a point not even part of the Universe if not confirmed by space that it has a presence within the Universe. That makes by the contraction of stars that stars are the coldest regions and by spinning (stronger gravity) profoundly more. It freezes space into the oblivious. That makes stars cold and outer space hot and makes rubbish of the idea that we think in terms of human experience that puts the sun being $6500°$ that is hot and outer space at $-273°$ that then is cold.

Singularity formed by the compliment of all the atoms in the star forming the gravity in the star sets the terms for what is cold and what is hot in relevance to the singularity within that star forming the gravity applying in that star. $a^3 = (T^2 k) = a^{3+2+1=6}$ with the sphere presuming the position of singularity as part the of $k^0 = 1 =$ **singularity**. Einstein proved that at the point where space reduces and such reducing reaches a point where space as a factor in the third dimension disappears into the single dimension (space going flat) gravity is overwhelming. Einstein interpreted this, as the complete Universe going flat but while it may be true that the Universe is going flat, that can only be within singularity since singularity represents the Universe as flat as it can get.

Since occupation may or may not be placing the factor in infinite, the space therefore holds the premier singularity of infinite from which all included in the universe has come.

When the top starts spinning in a specific position the top merely executed the option to fill the premier singularity at that specific point. When it moves it may take the premier singularity with to the new location it moves through spin or it may fill yet another position in singularity as all is the same.

The influence immediately above the circle will have the biggest influence and reduce gradually as the value of Π reduces in the leverage that the space has on Π and a gradual but definite change from Π to r will affect the extending of Π progressively more. The decline of Π will follow the same contour of the circle at 7^0.

The growth we see in the Universe is an adding of space in every cycle completed by every cycle, which all the protons complete. The adding that the Hubble expanding represents in terms of space is adding no part of the Universe but adding a part that is not a part of the Universe. By turning $\Pi\Pi^2$ the space forming material Π^3, a point holding singularity Π^0 confirms space Π^3. That proves that the turning of material produces singularity but singularity as I will show in due course is not is part of the cosmos. This means that singularity forms the Universe by moving material in terms of singularity substantiating that is because it can move. The adding is in terms of is the smallest addition that can come about in the shortest period of repeating by cycle rotation that can ever be. This growth of space-time next to singularity confirms the growth of singularity as singularity recalls the space it uses to grow in the time it grows. The margin of growth will be by the extension of **k** in the formula $k = a^3 / T^2$. Every cycle completed in the relation to space by the initial value of **k**. $k = a^3 / T^2$ leaves ultimately a^1 extending as space or as Kepler chose to indicate it as k^1. That too has to be compensated by the duration of time reducing the time aspect by the margin that the space expands. This confirms what is evident in the

Hubble Constant. The further one looks at time the more time seems to race because time has the invert properties we give to space.

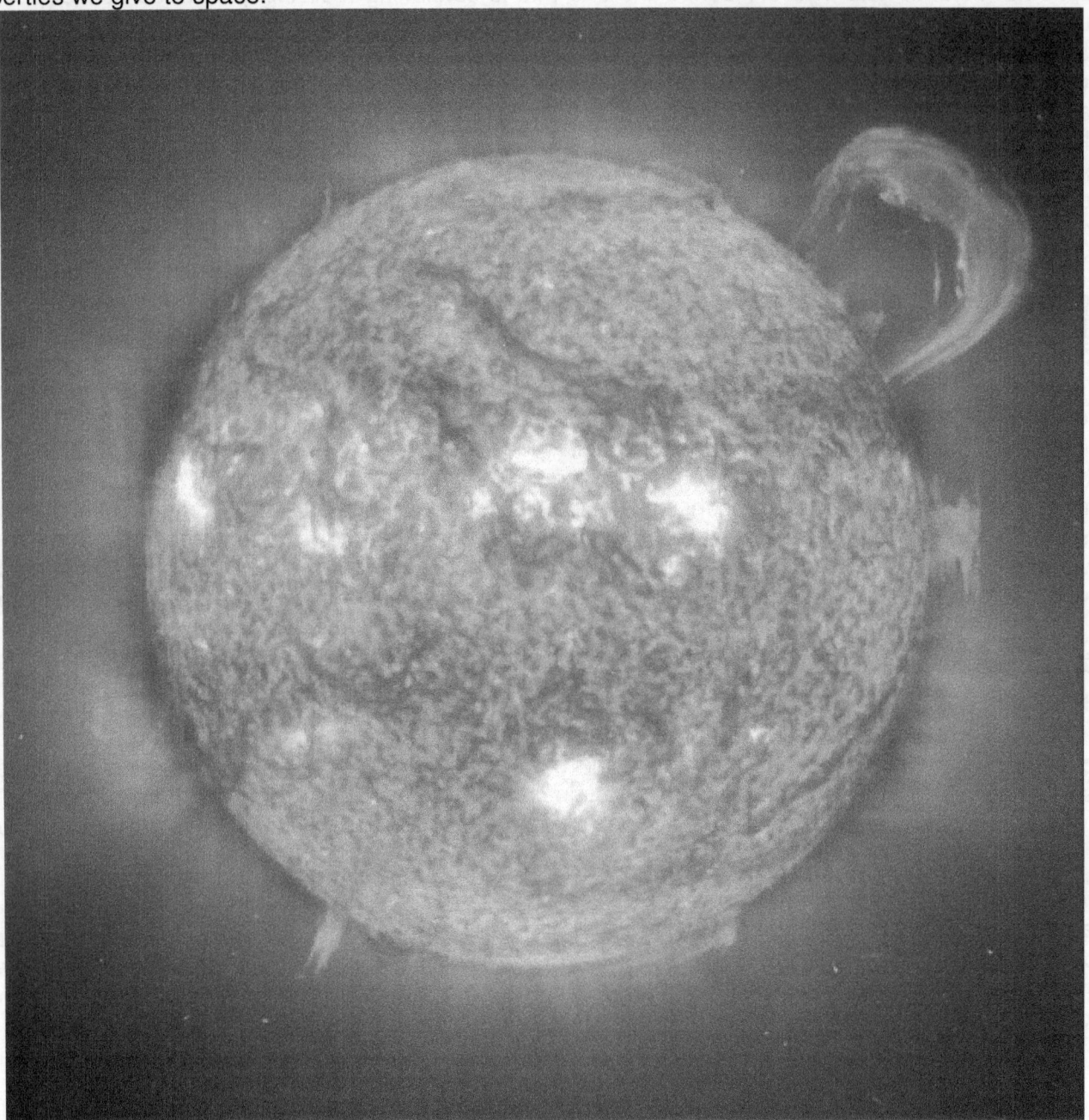

The application of gravity that condenses space and bringing about heat by the compressing of space we apply in the way we go about tapping into the energy that nature provides. Internal and external engines combustion engines all rely on this application for harvesting motion by driving power and this comes from taking heat from material and by harvesting the expansion of the heat to become space it drives the machine. Compress space even today with a piston in a cylinder and then pump the compressed air into a container and such confining of space will increase the heat by the piston effort to reduce the space brought about in the container. The heat coming about inside the cylinder has no relevance to particles colliding because all compressor cylinders cool down. The walls becomes colder because when that cold escapes it turns to heat as the heat releases from space forming a secondary form of material forming space that no one can understand when the spin of the atom allows the cold of the atom to release into uncontrolled space. This release and unification with space that heat does is the heat it had frozen because the motion of spin to space that the atom holds remains in a frozen state under the guard of the spinning electron. When this heat releases from the containing form of the atom frozen by the spin of the electron it brings about much more heat than the Human mind can cope with. One may not look at the material and judge the surroundings. This is exactly what the sun does. The sun is not a coal stove burning away until the coal runs out. The sun is an everlasting concept that turns overheated space into reduced hotness and that material is space-time, which is what the entire Universe is made of.

A More Complex Commercial Science Book 227 Chapter 5 Using Reality

Look at the sun and see how the sun turned the hydrogen it holds captured in its atmosphere to a freezing cold liquid at 6500°. Hydrogen is in a fluid state within the sun and yet it is still colder than the hydrogen we find in outer space that is in a gas form in outer space. The sun is without any doubt the coldest place in the solar system and therefore it is repelling all space towards the sun just as Kepler showed in his tables. But the sun is discarding excess heat it cannot manage to incorporate into its cooling system. Material contracts heat into condensed space in order to maintain cooling levels and when there is an overflow it discards the coolant that it cannot use. We see this discarding as heat coming from the star in the liquefied state of light. Every proton calls on a flow of heat to keep cool. That is when the protons oversupply the removing of space to produce the cold that is so apparent in the heat levels that is not used in the cooling process. By the reducing of space, it can concentrate heat to a fluid state, which is then introduced into atoms through electrons. By producing the opposing cold that finally freezes the heat to a solid state, we find that is what matter is. See the heat that a nuclear atomic bomb releases and thing of what heat is contained in the atoms. By allowing the atoms to overheat in an atom bomb the atoms expand and such expanding of concentrated space becomes a nuclear blast. The expanding of space is a way of duplicating space by movement without reducing space and by duplicating in the form of expanding it becomes just the opposite to duplicating by motion of contracting therefore reducing space by halving space used by material. That is what gravity does. By motion space duplicates and by space duplicating the material must be by dividing or bisecting - halving it removes heat in space as well as by dismissing space and in that concentrating heat. The density of the protons brings about space dense enough to harbour the heat in such quantities and visa versa applies in outer space.

By removing such high concentration of heat from the space that used to be expanded heat, the space then must contradict the heat by being extremely cold. We look at the heat in the space, which by that time is another form of material and find the surrounding heat in the space hot while the space is extremely cold. The cold in the Earth centre causes the concentration of heat by space reducing, as all cold surfaces tend to do. The proton contributes to that reducing of space. If it was hot the space within the Earth would expand and explode but the space within the Earth where we think so much heat is concentrated is so much it does not expand therefore it must be cold. To gather and accumulate the space in a liquid means it became much colder when the space parted from what then is being a liquid. Finding the surroundings terribly cold will allow the heat to gather and not expand but when the surroundings are hot, it will not tolerate more concentration of heat and thus it will expand to rid the balance of excess heat within space. The concentration or release of space with heat or space from heat is a direct contribution of the singularity in control of the space-time. The regard of the singularity stipulates the conducing of heat in space or the release of heat to form space by means of bisecting the occupied space.

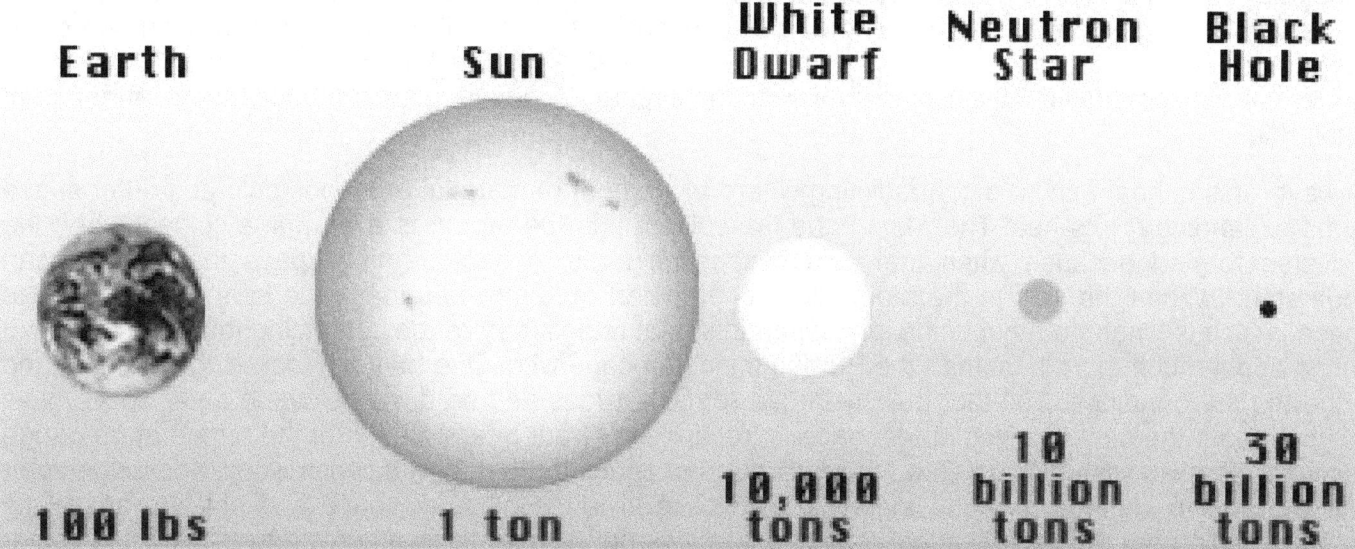

The particles claim more space when heated to preserve the cold. The claim to more space produces more space and reduces more heat. Such expanding brings about cooling. When particles heat or cool motion applies in some form. Motion started at a point when the Universe was extremely hot and there was no space. By introducing motion space formed and the lack thereof produced friction that became heat that became space. It is natural, it is simple, and above all, it makes believable sense.

The stronger the movement is the more the gravity is and the more space reduces in terms of singularity situated in the very centre of the star. Every atom forms singularity and the faster the atoms displace space by moving the stronger with the contraction be but also the stronger will the reduction of size of the star be as the star reduces space it occupies and space the star contracts.

The heat disappears into the centre of any sphere where there has to be a line formed thought of as the axis which is at the very point where space completely falls away. It is at the point where all the points of line centres meet by the crossing the centre of their individual connection coming in to contact as a group. In that way one may assume that the lines connecting the controlling points on the other end are crossing on a centre point that all that is participating in the constructing of the sphere is democratically electing such a centre. Please note this conclusion very well because this forms the heat in liquid relating to heat in solid of the Coanda principle. That forming the centre of the sphere will put that position where the lines cross from all six sides or point of the sphere which then by forming a line in itself is centralising all space in the sphere at that point, such crossing point will become very distinct and controlling where that point forms in the single dimension and singularity is the single dimension. Kepler also solves another riddle that truly got Newtonians unstuck. This, to which I now refer, is what is referred to when they refer to the Hubble constant. Everything that is, moves and by moving it confirms singularity as much as singularity confirms everything that grows which then is confirmed. This means that everything that spins creates a Universe within what you think of as a Universe but actually is time that formed space.

The application of gravity is that which condenses space by bringing about heat with the compressing of space. That compressing or liquefying we call the atmosphere. In angering terms the atmosphere is a liquid substance that is 600 times less dense that water is nevertheless it is a liquid. We apply the progress we have as a species in the way we go about with our skills to unveil ways we can tap into the energy that nature provides. Internal and external combustion engines all rely on this application for harvesting motion by driving power. Compress space even today with a piston in a cylinder and then pump the compressed air into a container and such confining of space will increase the heat by the piston effort to reduce the space brought about in the container. The heat coming about inside the cylinder has no relevance to particles colliding because all compressor cylinders cool down with time moving and not necessarily with the loss or release of particles. It is not only the discharging of air that will reduce the temperatures inside the container. The time flowing bringing motion about where the motion is not about particles escaping but heat escaping in the replacing of the heat density (not the density of the particles forming the material content within the container) but the space that compressed to heat will also bring about that the heat displaces through the container wall to the outside. This is bringing about equilibrium where heat will always flow from more dense areas to the lesser dense areas. This has no influence on the status of the particles on the inside of the cylinder but only concerns the density levels of the particles inside versus outside. After the pumping of air increased the heat in the cylinder which even can go to dangerous levels, will reduce back to room temperature when further pumping ceases and that stops further air movement into the cylinder and such surging of pumping air is what brings about heat stabilizing.

However this is heat and cold contradicting where movement of material produces cold and outer space forms expanding forms heat The atom is be the coldest and the proton is even prime coldest when the Universe formed because when that cold escapes it turns to heat forming space that no one can understand. When the spin of the atom allows the cold of the atom to release the heat it had frozen to space, which is what the atom holds, and when this heat releases from the containing form of the atom it brings about much more heat than the Human mind can cope with. One may not look at the material and judge the surroundings. The fact that hydrogen remains a gas and so does helium in outer space says nothing about the sate in which outer space is, regardless of our interpretation of the temperature gauge telling us what we wish to hear. One must look at outer space through how science works and judge outer space from the findings only considering outer space in terms of outer space and not in terms of us humans feeling hot or not. If helium remains a gas, it is hot. When the sun cools it to a liquid it is cooler and when the star cools the helium to a solid such as it does in big dark stars then it is cold, eternally cold. The removing of heat makes the centre of the Earth cold although we see it as being terribly hot. The only reason why it can seem to be hot is because it is cold and where the earth is ridding that space of the heat it has concentrated in space. In such a cold environment, the heat can gather and space can collect heat because the particles find the surroundings extremely cold.

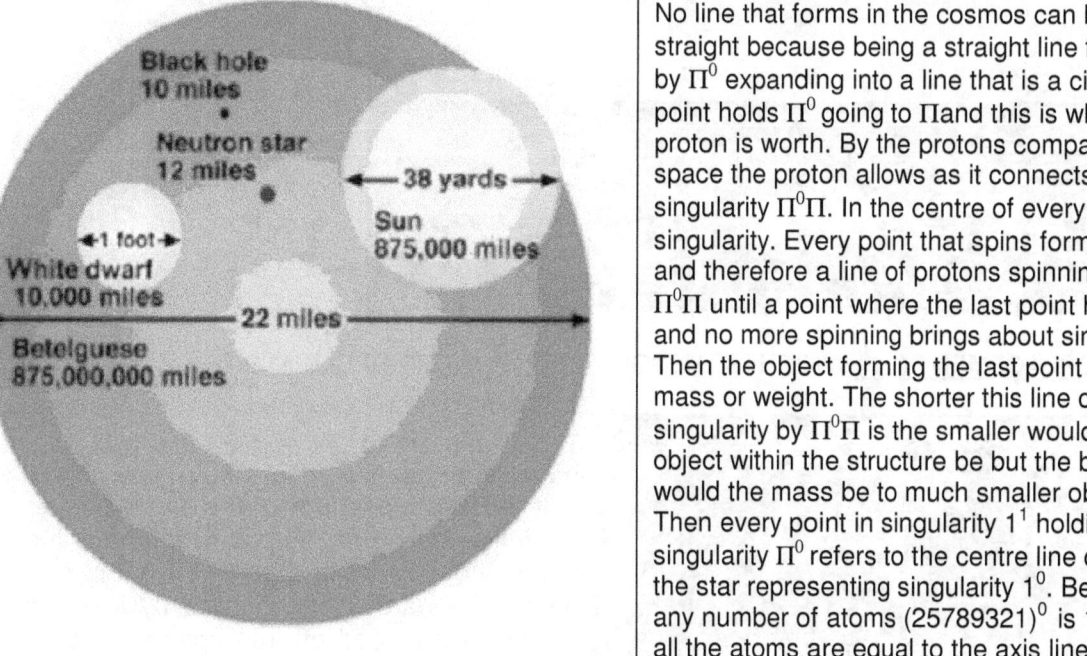

No line that forms in the cosmos can be straight because being a straight line forms Π by Π^0 expanding into a line that is a circle. The point holds Π^0 going to Π and this is what every proton is worth. By the protons compacting the space the proton allows as it connects to singularity $\Pi^0\Pi$. In the centre of every atom is singularity. Every point that spins form $\Pi^0\Pi$ and therefore a line of protons spinning form $\Pi^0\Pi$ until a point where the last point hold $\Pi^0\Pi$ and no more spinning brings about singularity. Then the object forming the last point holds mass or weight. The shorter this line of singularity by $\Pi^0\Pi$ is the smaller would the object within the structure be but the bigger would the mass be to much smaller objects. Then every point in singularity 1^1 holding singularity Π^0 refers to the centre line or axis of the star representing singularity 1^0. Because any number of atoms $(25789321)^0$ is 1^1 is 1^0 all the atoms are equal to the axis line of the star and in such a manner gravity forms.

The cold in the earth centre causes the concentration of heat by space reducing, as all cold surfaces tend to do. If it was hot, the space within the Earth would expand and the space within the Earth where we think so much heat is concentrated does not expand therefore it must be cold. To gather and accumulate the space in a liquid means it became much colder being a liquid. Finding the surroundings terribly cold will allow the heat to gather and not expand but when the surroundings are hot, it will not tolerate more concentration of heat and thus will expand to rid the balance of excess heat within space. Look at the pictures of the sun and see how the sun turned the hydrogen to a cold liquid at 6500 K.

Hydrogen is in a fluid state within the sun and is colder than the hydrogen that is in a gas form in outer space. The sun is the coldest place in the solar system. That is when the protons remove the oversupply the heat in space to produce the cold that the discharging of heat is then so apparent. By the reducing of space, it can concentrate heat to a fluid state by producing the opposing cold that finally freezes the heat to a solid state. The expanding of space is a way of duplicating space without reducing space and by duplicating in the form of expanding it becomes just the opposite to duplicating by motion therefore reducing space by halving space in time. By spinning and moving it divide the space the material holds by doubling the space it has in the same area. This compression of the expanded reduces the space heat uses and makes heat a liquid. It is not helium that becomes a liquid but the fluid in which the solid material is held that becomes liquid. That is what gravity does. By motion, space duplicates and by space, halving it removes heat in space as well as by dismissing space. In all the applying of gravity, space dies. The density of the protons brings about space dense enough to harbour the heat in such quantities and visa versa applies in outer space.

The fact that hydrogen remains a gas and so does helium in outer space must serve as enough proof that outer space is hot, regardless of our interpretation of the temperature gauge telling us what we wish to hear. One must look at outer space and judge outer space from the findings only considered in the terms which outer space insists upon. If helium remains a gas, it is hot. The removing of heat from the space that contained the heat makes the centre of the Earth cold. In our universe we see it as being terribly hot because the heat then forms a separate substance but remains a form of material (8) but that is because we see the heat and not the space derived from the separating of the heat. The only reason why the space can seem to be hot is because the space is cold and in such a cold environment the heat can gather in a much concentrated state and space can collect heat because the particles hold concentrated heat in the space separating the particles.

"GRAVITY IS DIVIDED IN TWO FACTORS, BEING **LINEAR DISPLACEMENT** (Π) WHICH IS WHAT **NEWTON'S GRAVITY** IS AND,

There is a position that is in motion that is forming the very edge of the outside of whatever

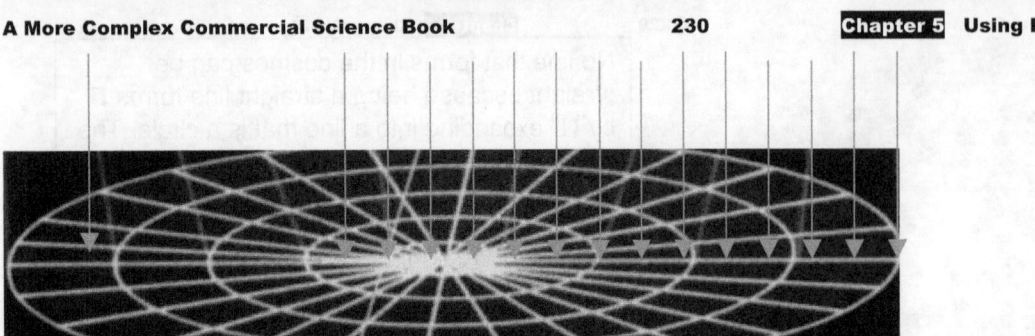

CIRCULAR DISPLACEMENT (Π^2) WHICH IS THE "GRAVITY" EINSTEIN RECOGNIZED

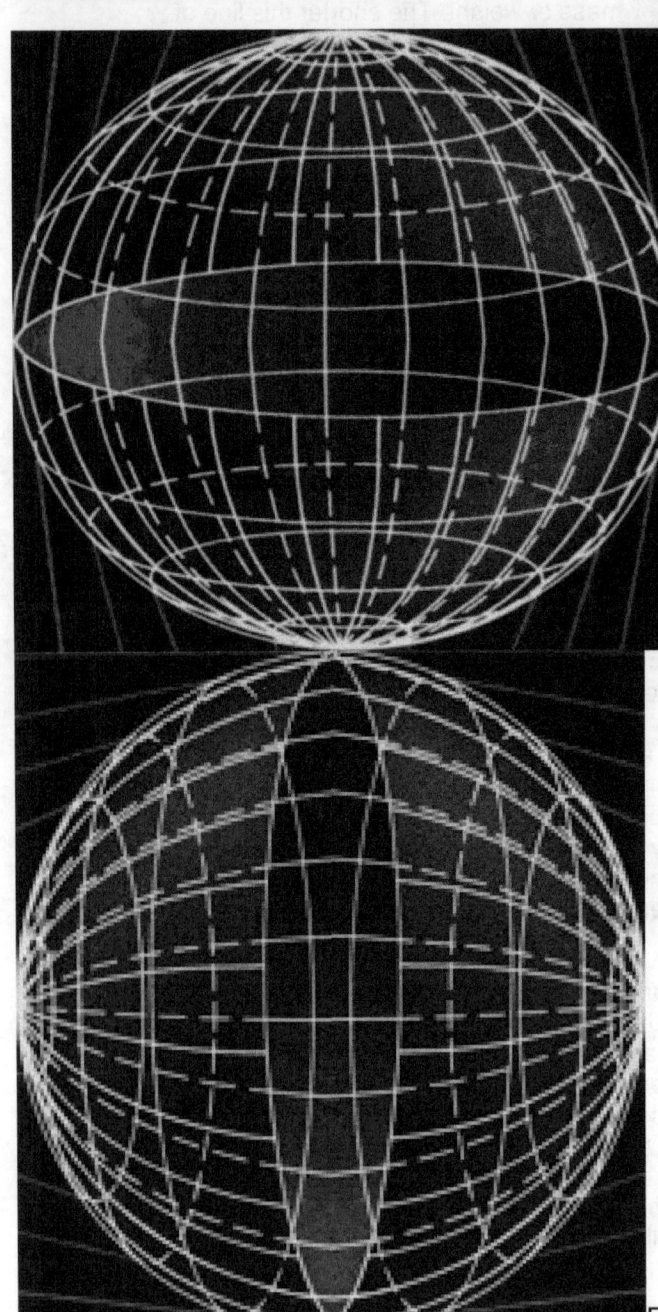

Those that are of the opinion that it is friction with the air that heats a spaceship or any object when it enters the atmosphere of the earth, please explain the following: When a spaceship enters the earth's atmosphere by more than 21° it will burn out. If it enters the earth's atmosphere by less than 7°, it would be bounced off into space. The 7° comes about by the concentration of the atmosphere and this bounces the spaceship outwards. However, when a spaceship enters the earth's atmosphere at an angle higher than 21° it will in fact meet less air than in the case of the spacecraft entering at an angle of more than 21°.

The extension of Π is well received as a dimensional implication to matter holding seven positions from singularity and space having four quarters through out the rotation of singularity forming the centre to the five dimensions (one side lost to the cube's six sides connecting to the five remaining sides) making the total sides facing space from the point holding singularity at any given instant at a value of twenty (4 X 5 = 20).

Since it is linear movement that is also the circular movement $\Pi\Pi^2$ that launches Π the value of 7 can never be excluded from the factor forming Π.

Then adding the singularity cross of Π being (1+1) = 2 the relation becomes 22.991/7. This is crude because in more precise calculations it becomes .91 + 1 = 21.91/7 = Π

The Universe is in place in space by the value of movement and it is the movement that forms the space. If the movement of the space stops it will pull the entire Universe which then is the star and this space surrounding the star into one spot that forms the singularity that represents all spots holding singularity onto and into one spot in singularity.

Remember that according to cosmic physics everything moves by a straight line Π that moves by circling Π^2. From the centre, there must be a specific allocated space ending at the object in motion and starting from a centre that centre line has no dimensions. The object in motion determines the one limit and the centre with no sides and no space, which is standing still in singularity, determines the other limit. By that we can see there are only one way of looking at what we can observe and that is from the outside in. Mainstream physics ignored the clear connection completely, notwithstanding it being so very obvious. There is this far in their recognising of principles in natural physics not one single reference made to prove their appreciation of this matter. They are bent on particle colliding. When particles collide, such collision forms an atomic thermo release and that action we call an exploding atomic bomb. What principle this argument about particles colliding ignores is that all atoms use negative charged electrons forming the atomic limit on the outside forming a definite border to the boundaries of all atoms and in both electrons from different atoms are being negative charged. In being negatively charged, it means both will come out and totally reject the other. The closer they come the more violent the rejecting will be and such rejecting is the production of heat that will turn to space. The electrons repel other negative charged sub atomic structures, which the electrons are that form the outer borders of all atoms. With all electrons highly negatively charged (being as negatively charged as any possibility will allow to match the utter extreme) such electrons could not touch.

It is about time scientists start looking with their minds and not their eyes at the Universe and see what is truly out there to see. All the difference we find is seated in the human mind. We humans set differences because we look at the cosmos by placing humans and the life we find on Earth in a pivotal centre in the cosmos instead of placing singularity in the centre and life where it belongs; only found on Earth. Einstein proved mathematically that in the presence of a strong gravity such a strong gravity slows time down. Surprisingly with that evidence being around this long nobody in science since Einstein's discovery took those statements and made any further progress from that.

It seems to have been left in some drawer to dry. Science still sticks to the opinion that time did not change, not even slightly, since the beginning of the time and holds the same pace ever since the start of the Big Bang notwithstanding the implications this concept carries. Before the Earth took one year to circle around the sun and even before the sun was there a year was still the same duration of one year. How odd... don't you think ... that the only aspect in the entire Universe that is beyond change is the aspect of time? With the entire Universe including all the gravity now present and not excluding one Black Hole or dust speck pressed in such an area that was possibly the size of a lepton even then the gravity extending from that circumstances must have been beyond what words can ever describe.

When everything was that small when the Big Bang took charge, the gravity at the time was beyond light, because even today in the Black Hole the gravity is beyond the speed of light. If the gravity was that high and Einstein already proved that strong gravity slows time down, then there is one logical conclusion and that is that time was in fact at the time of the Big Bang standing still. Mathematically it is incorrect to allow gravity to compress the Universe into a spot smaller that an atom and exclude any other factors and relevancies to change.

If the darkness was the representation of "nothing", then that should be exactly what we must see, nothing but the stars. Taken from the top picture some stars and leaving the rest to nothing is what we see in the picture below. A blind person sees nothing but when we look at space, we see something that we think nothing of as we see as space. One cannot have the ability of sight and see nothing except by closing your eyelids and then you see nothing. But in that case you do not see "nothing" in contrast of "something" you see "nothing" without it contrasting to "something".

 Nothing is all about not being and not "not seeing".

A while ago I said that the miracle of physics is allowing the entire Universe to unite as a speck in the eye by which you can observe the entirety we think of as the Universe. In the next few pages I am going to explain singularity but this is because of singularity. Because the Universe consist of only singularity 1^0 everything in the Universe relates to singularity 1^1 and as singularity is 1 everything that we see comes down to being all not only equal but being the 1 you see by which you see. How do I see it? One photon may represent one galactica, a constellation of stars numbering into the billions. It is all a question of Mathematics. $1^0 = 1^1 = 256^0$. Whatever number it is that you see it is unified by singularity to the power of zero to then be representing 1^0. In the Universe there are no hot or cold, big or small, fast or slow but only a state of differentiation produced by time. The Universe parted by parting heat from cold when eternity parted from infinity, when Π^0 singularity parted from Π singularity, when 1^0 parted from 1^1. There is no hot or cold but there is a relevancy where one factor cools and another factor overheats. By retaining the sun is the coldest space in the solar system and outer pace is the hottest there can be.

This is the mathematical truth...as the dividing factor increases; the influence that the mass will project in the formula $F = G \frac{M_1 M_2}{r^2}$ will diminish in respect to the growth of the distance. In that sense the gravity force between the earth and the moon must reduce its ferocity therefore weaken. However, Newtonians only apply their ability to calculate and knowledge for the purpose of upholding Newton and never to provoke Newtonian liability by telling the truth. Never is there any mention of mathematical reality when mathematics is used. However this no one ever knew to be true outside of the intimate upper circles of physics, and this unmasking was to be prevented at all cost. A plan was to be devised because if the public found out Newton was a fraud all along and being Newtonian was the personification of stupidity the entire science world would come tumbling down on the heads of those most important Brainy Bunch.

No, better still, to save science a conspiracy was devised. The most intellectuals on earth had to cook up something and devise a plan to save their image and the name of Newton. The most intellectual minds concluded and fabricated to form a conspiracy to withhold the truth and forge a fraud that lasted almost ninety years to the day. Should anyone disagree with the term conspiracy, then please let me know what you would call what happened after Hubble's discovery became prominent on the news.

In the past it was accepted that only Newton and God never made a mistake and since the Brainy Bunch Newtonians were mostly atheistic or atheistically orientated they were not that sure about the credibility of God but the unquestionable accuracy of Newton those atheists were pretty sure...and now it seems Newton made a mistake. That just could not be. They would rather have God make the mistake than let it seem as if Newton was up to no good while Newton made the incredible mistake.

Then they allowed God to make the mistake. If the Universe was expanding it was God's fault. It sounds much better than have anyone think it is Newton's mistake. If the cosmos expanded while Newton said it must contract then this rebellious behaviour of the cosmos must end. They had to put the blame on the Universe and then ultimately on God for making such a mistake. It must be God that made the error.

The easiest is to put the blame onto God by finding the fault at the door of an uncompromising Universe. To make the conspiracy believable they had to conjoin a concocted story that would have every halfwit on earth believe it. What will make the Universe expand? It must be a lack of the something that makes the Universe contract. Not enough contracting solution will be the cause of the expanding. If it was mass that had the Universe contract in Newton's terms, then a lack of mass will lead to a shortfall in gravity and then the elasticity would not be enough so the elasticity would be tested and before the elasticity failed completely they had to get the Universe back on track. They saw a good measure of darkness splitting small bits of light. This must be it then. Who put that much darkness amongst that little Light?

If it was mss that should contract the Universe then it had to be not enough mass that would be to blame for the expanding of the Universe. Now to get someone credible enough in the eyes of the public yet foolish enough in ego to provide the cover for the conspiracy to work was another matter. The only candidate must be Albert Einstein. They had to get Albert Einstein to measure all the mass in the entire Universe. Now this is where the joke no longer seems funny. Only fools and idiots are going to fall for that. Look at the picture on the previous page and the picture above. This is unrecognisable small portions of a large and overwhelming large Universe. Who is keeping whom for a bloody fool?

Is there anybody that will seriously try to convince me or any other sober-minded person that any human being can measure what is to be considered as mass even in the picture above? Go back to the previous page and look at is presented as material in that picture. Is any sanity left in the suggestion that any body may even think of attempting to calculate what is in such a small portion of a fragment of a sideshow of the Universe? Can there be any person in his right mind that will think he could have the ability to measure only what is in these pictures, let alone what might be available in the entire Universe? If anything can ever bear testimony of how mad those Newtonians got, then this must represent their total loss of mental coordination about what is reality and what is hallucination of a mind gone missing of reason. Their arrogance at that instant of deciding in following a direction grew into mindless stupidity.

Put the earth in any of these pictures and the task is senseless to perform. Put the sun in as a visible star and the task is ludicrous. Put the solar system in and still it will not show as a freckle. The entire Milky Way might come about as a speck somewhere, but not big enough to be noticed. Then tell me please how many Milky Ways might fit into this small part of a huge Universe…and Albert Einstein was prepared to measure the entire Universe. I know the formula he used but using that formula indicated just how lubricous the attempt was. Still, they say Einstein determined the average mass of the entire Universe. This is a small part of what he said he could achieve. This is a conspiracy as blatant as ever there was one. Believing this first require the drinking an bottle of rum and then getting high on a barrel of cannabis. Either they were fools or they thought the entire human race was brainless fools incapable of thoughts.

Please do try and fit any of Newton's gravitational equations into this or the other pictures ands then remain sensible and sane. Please put [illegible equation] or $\left(\dfrac{P}{2\pi}\right)^2 = \dfrac{a^3}{G(M+m)}$, in place, using the mass in the picture above and come up with a realistically believable and sensible answer! Fit any of Newton's formulas to find either the mass in the entire picture or the force of gravity holding this lot together. For the mathematical genius Einstein said he was this shows he is pretty stupid in understanding the basics of physics. He tried to calculate the mass of every object. To use the formula he did use would bring no results so I am ignoring that wasteful attempt he tried in using the formula he did. If I explain that one in this book the exercise would become a comic or a funny and that I don't want. I try to give his attempt some realistic credibility for the sake of Einstein trying to achieve an effort worth some dignity.

However, all the blatant ignoring of human respect is not showed nor epitomised by this ludicrous ignoring of fellow human intellect because with what they tried to hide by showing what they did with what they tried to hide with that they got away. Everyone this far except me as far as I know never kicked against the critical density theory and therefore they were successful. Everyone that launched the conspiracy died with not one on earth being any the wiser about the role anyone had in this conspiracy! That is success storey in every sense as far as a conspiracy goes.

The crudeness that they got away with was to give the Universe the blame for Newton's misjudgement. It was that they succeeded to put the blame of the error on the part of the Universe. They turned the facts around that the expanding of the Universe was blamed and the cosmos. It therefore is to be understood that it is the Universe that has to mend its ways and correct the misconduct to befit what Newton held as correct. Newton can't take and therefore the blame of the Universe expanding goes to nature and all the while it has to contract in precisely the manner that Newton said it should. If it is not following Newton's orders then there has to be lost mass. Where the lost mass will go is beyond me but again the fact that the Universe could end also exceeds my understanding. But with the Universe being the wrong party there has to be mass that is unaccounted for. If they find the mass they can replace it (why else would they bother to locate the missing mass) and get the Universe back on tract and contracting once more.

Think how scandalously farfetched the presumption goes. Where in any of Newton's formula or cosmic principles is their any indication that there is a limit to the expanding in accordance to what Newton declared. The elasticity or stretch ability and the limit thereof they made up as they went along while formulating this conspiracy. The deduction they make is on the grounds of having a "flat" Universe that is "slowing" and losing momentum. This too is reckless deduction about singularity they have no perception of. They wanted the Universe to slow down so that the Universe would stand still and form a zero expanding change direction and then start to contract. Why would I have this conclusion? Because if $\Omega = 2q_0 = (2/3\Lambda)(c^2/H^2)$ was correct the Universe had to have a limit. This idea is preposterous because with every end there has to be some beginning of something else. The cosmos cannot end because Kepler showed that every ending of a straight line is a circle and every circle follows as a straight line. In mathematical equations it reads as $\mathbf{a^3 = T^2 k}$ and then $\mathbf{a^3 / T^2 = k}$ and $\mathbf{a^3 / k = T^2}$. The formula reads as $\mathbf{a^3 = T^2 k}$ where gravity is space $\mathbf{a^3}$ that moves in a circle t $\mathbf{T^2}$ hat goes straight $\mathbf{k}$.

This means that every circle forms a straight line when an object orbits another object and every object in the Universe orbits some other object while being the centre of something else around which that objects orbits. This is why the Universe can never end because going in a straight line puts such a straight line in an immediate circle. All driving in the cosmos be it steam, combustion engine or electric or even cosmic driving of orbiting objects has to be in a circle followed by a straight line followed by a circle and this continues indefinably. The driving value of $\mathbf{a^3 / T^2 = k}$ becomes $\mathbf{a^3 / k = T^2}$ that becomes $\mathbf{a^3 / T^2 = k}$ that becomes $\mathbf{a^3 / k = T^2}$ repeating the process indefinably. However reading this mathematical equation one see that movement $\mathbf{T^2 k}$ of space $\mathbf{a^3}$ centres around singularity $\mathbf{k^0}$ because if space $\mathbf{a^3} =$ is equal to movement then movement $\mathbf{T^2 k}$ forms space in singularity $\mathbf{k^0}$ since $\mathbf{a^3 / T^2 k = k^0}$. **No object within the solar system that orbits the sun can ever leave the sun to go to another system of orbit. To be stronger than the centre of the sun $\mathbf{k^0}$ the object has to create space $\mathbf{a^3}$ by movement $\mathbf{T^2 k}$ to overcome the gravity $\mathbf{a^3 / T^2 k = k^0}$ with which the sun holds the object in gravity. In this principle we find the eternal time position that captures space forever.**

The planets in the solar system goes in a straight line and while they are going straight they divert from direction and circle around the sun at a specific location. This is energy distribution all work has to comply with. The wheel of the car turns as it goes straight down the road as the car turns around the earth by going straight down the road as it goes straight with the earth as the earth turns around the sun as the sun goes straight to turn around the Milky Way and there will be more going straight while spinning as the picture grows. That is the way that all power drives because this movement connects time to space and that is one thing Newtonians have no idea about…they have no idea what time is in relation to what space is because in all of this we find singularity which is above anything any Newtonian this far could manage. Energy distribution is $\mathbf{a^3 = T^2 k}$ where linear movement $\mathbf{a^3 / T^2 = k}$ depends entirely on circular movement $\mathbf{a^3 / k = T^2}$. By the way the most successful formula ever devised by any man is the formula Einstein came up with as $E = mC^2$. This should read $E^3 = mC^2$ and that is a translation of Kepler's formula of $\mathbf{a^3 = T^2 k}$ in that $E^3 = a^3$, $m = k$ and $C^2 = T^2$ This information they will never release because then they agree that the Newton idea of changing $\mathbf{a^3 = T^2 k}$ to $\mathbf{a^3 = T^2}$ is very ridiculous and if Einstein redeployed it then it is Kepler's formula that proves to be the most successful formula ever devised. …And this formula is the basic formula that Kepler received from the cosmos when he translated what the cosmos told him in mathematics to a verbal language used by man. Kepler had to translate cosmic mathematics into verbal language and Newton disagreed then

about the finer detail the cosmos confirmed to Kepler. The cosmos said $a^3 = T^2k$ by using mathematical numbers to solidify the proof and remove any doubt where Newton then changed it to $a^3 = T^2$.
From this load of rubbish they try to play god by envisaging how the Universe will grow in the future.

What I tell must be told by looking at pictures in order to understand the failures of Newtonian thinking and the flawed concepts they arrive at by using mad mathematics they don't even understand. Let's say we can look back in this picture to the point that is 12.5×10^9 years from us and is where the Universe started according to Newtonian wisdom. We then look at 12.5×10^9 years of space forming. When we get to the point that is 12.5×10^9 years from us we will be in the centre of the Universe and standing at that point looking to where we now are we will be saying oh, that is where the Universe are because that is the point where the Universe started, which in fact then will be the point where we now stand and look at the limit we can see and that makes what we see not the end of the Universe that we see.

Looking at the Universe from this angle is most beautiful and by applying the magic of mathematics it seems to be so real except for one problem we encounter and that comes when we use a dash of logic. When using equated mathematical formulas one include a certain part of the Universe by excluding the rest of the Universe. There is no formula that has the ability to contain the Universe in its inclusive entirety because the Universe is eternal no matter how one would look to appreciate what one sees. Looking at any part of the Universe the distance we don't see removes the quantification of the picture that we see. What ever forms a picture of the Universe excludes untold many, many times more than what the picture we see reveals because it is shear stupidity to think what we see in any picture represents everything there can be.

Beyond what we see is eternity looming outside of the view we have. If we could get as far back as this picture would allow we would encounter as much space in all directions and that space will be filled with as many stars as this entire picture reveals and then even much, much more. If we shift to the back of this picture we take the centre of the Universe with us because the centre of the Universe according to our position would be where we are. Where we go we move the centre of the Universe with us everywhere we go.

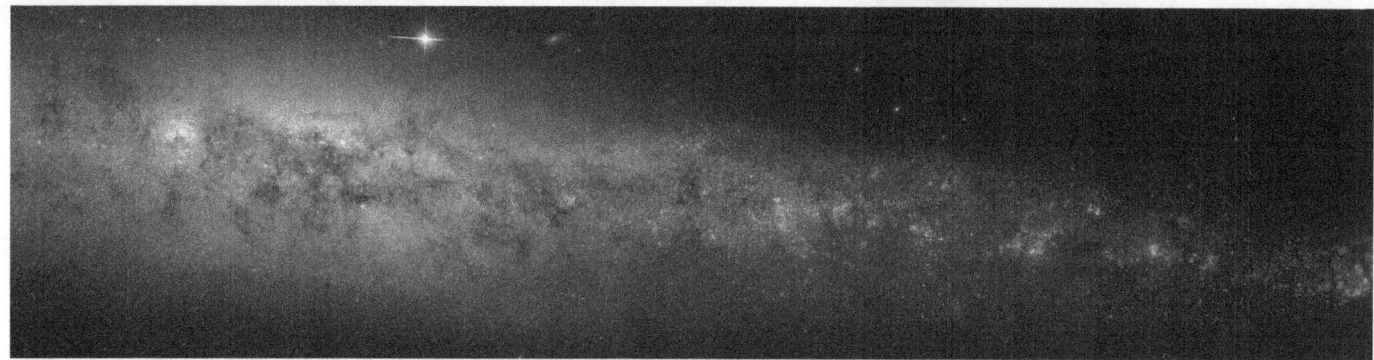

When we see this picture this is not all we see. This is what the electronic and mechanical driven machine and telescopes bring us and those limitations allow us to see what we see in this picture. Behind what this picture shows are an eternity of more stars, which we can't see. For the Newtonian to admit his devises constrains his views will be too much to contemplate because they know everything. But it is so Newtonian to reduce the Universe to fit the simple-minded approach that the Newtonian attach to the Universe and by shrinking the Universe to their mathematical abilities as being 13×10^9 years old it shows not how young the Universe is but how limited is the physics of the Newtonian simpleton. It is the Newtonian's mathematics and their ability to apply mathematics and their ability to understand what no one can understand that holds the limitations and the ends they portray. They can "understand" Newton and dare I say more than that. Newtonian wisdom let us to believe there are three ways of cosmic development. I have to explain that Newtonian wisdom fills all the space in the distance covered by outer space with loads of "nothing". Newtonians say what forms outer space is "nothing" and outer space comprises in total of "nothing". "Nothing" in outer space is what the space in outer space is according to Newtonian intellect. Newtonians think that between the moon and the earth there is "nothing".

The way Newtonians also think is that between the sun and Mercury there is lots of "nothing". Would it then surprise you to know that Newtonians think that between the sun and Pluto there is lots and lots and lots of "nothing". From the earth to the moon the Apollo space mission travelled through lots and lots of "nothing" and the "nothing" stretched all the way up to and even further than the back of the moon because they found "nothing" even there. It is important to know that on the moon surface there is "nothing" and on the back of the moon there still is "nothing" but that "nothing" that is on the surface of the moon is very much different from the "nothing" that the Apollo missions went through on their way to the moon. Between every planet and the sun the "nothing" is escalating to the point that the "nothing" is doubling between every planet that is separated by "nothing" and this doubling of "nothing" is continuing beyond Pluto where there is even more "nothing"! Can you match that Newtonian argument?

In the Universe three possible forms are under construction. One is the closed Universe, bringing boundaries in the shape of roundness or ultimately a sphere. Then there is the open Universe idea allowing the Universe to form in the shape of a saddle.

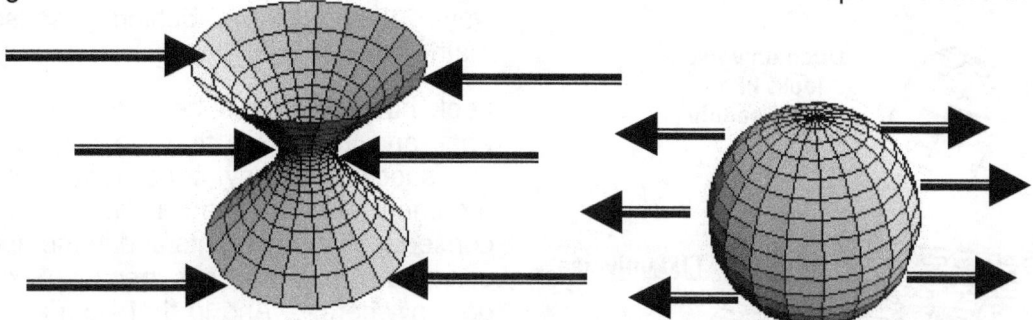

Please look at the pictures above and answer my question with your clear mind thinking logic. If that is the end the lines indicates then what is beyond the limit where the lines end. Our Newtonian wizards say it is "nothing". That brings one of the biggest problems I ever faced to mind. Looking at the examples they hold of the Universe there is a point beyond which there is no point. At the end of some point is a limit after which there is "nothing". However, the most brilliant minds man can bring forward already filled an entire Universe with "nothing" and at a point where "nothing" ends we have "nothing" more going on. What is in place of the point where "nothing" ends by a wall(I suppose) of "nothing".

The "nothing" that ends at that point begins with more "nothing" not continuing but ending in "nothing".

In order to "be", there has to be "something" but if they fill that which should be filled with something with "nothing" then what is in the place of something when they say they run out of "nothing". I think there is a lot of "nothing" filling their thoughts and then they translate the "nothing" they think of to outer space and into the Universe and that is how they got all the "nothing" with which they are filling the Universe.

How bright did Newton think he was to change what the cosmos gave Kepler to something that befitted Newton's ideas about the cosmos? Newton's idea of $\mathbf{a^3 = T^2}$ is as false as a three dollar bill because this means that all third dimensional values are equal to square values making $2^3 = 2^2$ which means that $8 = 4$ or that a flat surface is equal to the same surface being square. It also means that you being in three dimensions in front of your mirror looking at the reflection of the square image can have a debate with yourself while your image in the mirror replies on your argument. When this happens, don't call the police it is too late. Just jump from a building and make it easy on everyone including you. Newton's idea of $\mathbf{a^3 = T^2}$ is clear that the man had no understanding of any mathematical principles. Using $(2/3\Lambda)(c^2/H^2)$ puts the entire Universe equal taking away development of space in time. It says that the quasar they see at this moment never changed in say the three billion years it took the light to get to us and the space holding material that they see is what there is at the very moment they see it. They never gave tolerance for Universal development. This is the crap you get when following Newton $\mathbf{a^3 = T^2}$ without thinking. Then $(2/3\Lambda)(c^2/H^2)$ is ridiculous because every second space changes at any place in the Universe and that is just because the moon will never crash into the earth or the earth crash into the sun.

Another formula I wish to explain is
$\Omega = 2q_0 = (2/3\Lambda)(c^2/H^2)$ where …
Ω = density
q_0 = Deceleration Parameter (*where they get that from only they would believe the fantasy*)
Λ = Cosmological Constant (*every spot in space has a different gravity value so how a Newtonian would get a constant applying is a fairy storey come true but that is Newtonian motto*)
c = speed of light (*can never have a fixed value because gravity slows down the speed of light even to become a minus as it is in a Black Hole*)
H = Hubble Constant (*this too is a dream Newtonians cook up to give Newton some form of legitimacy*)
I am not going into this further because I have other books dealing with the issue but the so called Hubble constant has so many variations and changing values that they had to call a halt to the number of people investigating this field and all of those reached another totally different value. That throws the values connected to the q_0 = Deceleration Parameter, Λ = Cosmological Constant and the H = Hubble Constant out the window. You can cheat to reach a number but when honesty prevails the outcome is a joke. Even the speed of light is no constant because the very same Einstein proved that "gravity can bend the flow of light" and therefore light holds speeds according to the gravity applying in that specific part of the cosmos. Should anyone wish to read about the working of singularity then download the book **The Absolute Relevancy of Singularity The Website** it will serve as a starter. I do not wish to go past the understanding level that easy reading is required because then it would make this book only enjoyed by experts and that is the last thing I want. I want to unmask these crooks called the cosmic experts and the way they hide behind senseless mathematics.

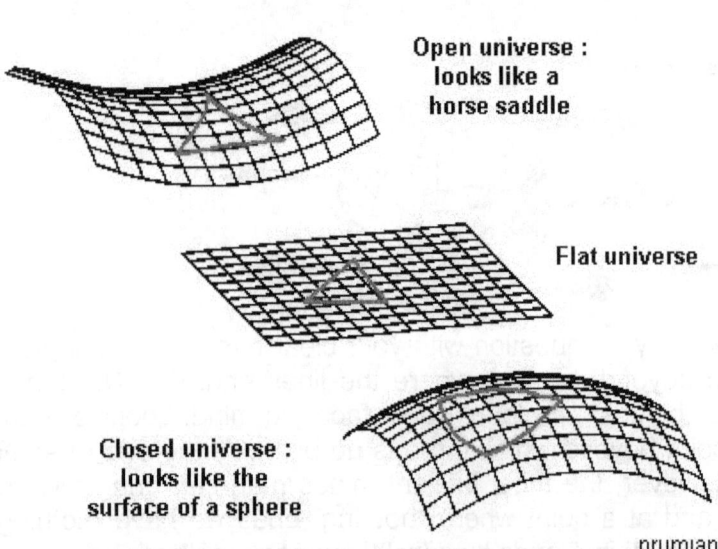

Open universe : looks like a horse saddle

Flat universe

Closed universe : looks like the surface of a sphere

Look how far I have been thrown off the track and in that we find the purpose and the success of any conspiracy. Get the argument to divert into a million other non-consequential arguments and in the end the conspiracy had success because nobody got anywhere. …And in that I am a sucker every time but I learned to keep the argument heading on track again after such a diversion. Now nobody would be surprised to learn that somehow Einstein found not enough material to bring about a contraction that would save Newton and get Newtonians to save face. Einstein did what he was told but clearly he was just a puppet in the conspiracy because he came back with the numbers that clearly showed the Universe was heading for eternal expanding in view of what they thought applied. He called his model the Open Universe as if there was a Closed Universe also as an option. I am not getting into that because I have devoted many pages in more serious books on this matter where I show how futile this view of the Universe is. There can be no open or closed Universe because open and closed brings about limits that has to stretch and anyone in view of a Universe with growing limits or confined limits are not looking at the Universe in which I live. We can't look at the Universe because we are inside as part of the Universe. We can't view space as if we are looking at space because the space we think we see is time in eternity developing as space but we have to look at space being time within space forming. We can't look at a quasar and say that is what it is. We have to look at a quasar and realise at the time we see the quasar that was what we were while we were at the time so small we were not even part of space or time. We also are at this instant in a developed quasar of which the ends we will never see forget about finding the ends to what holds the quasar. We are not God looking from above but we are humans looking from within what we are. We look not at space but at time because space is what history time left behind in light. We look at space and look at our history at a time we, or something as small as the sun we regard as huge had no room to be within that what we look at because at that instant the sun did not exist that is how small the sun back then was. The Universe is not getting bigger but is getting smaller allowing small things within to get bigger in relevancy of growth.

$$ds^2 = -R^2(t)\left[\frac{dr^2}{1-kr^2} + r^2 d\theta^2 + r^2 \sin^2\theta\, d\varphi^2\right] + c^2 dt^2$$

Notwithstanding how impressive the mathematical reasoning is, there is one fundamental flaw in all the suggested possibilities. The Universe according to Newtonians can develop in three possible categories with two more being a cone and a sphere. I am not making it my task to promote or to argue the pros and the cons of each system because all these suggestions are ultimately completely flawed.

The Universe only has an inside and can never have an outside.
Every one of the pictures I present and the others I do not show have the Universe holding an outside. The Universe only has an inside in which I am. Time in singularity forms the outside and because space is the departure of infinity from eternity the space that we think we see is ultimately time and time never ends. Because what we see is time where space is the presentation of the history of time we see what we see without having an outside or a stop to what we can see. We are inside and looking at the smaller end we is looking towards a direction that has no end.

Looking at the Universe we look at the past. We can't see the future; we can only see where what we are in also were so many trillion years ago. But what we now see looking at galactica back in time was so small that what we now think of as big had no relevance to exist in terms of what then filled space. We can't look at a galactica and presume we hold equality in time. Trillions of years filled space from what we see to where we are. We are in time in the present but seeing the galactica does not put the galactica in the present. It puts us a trillion years back before there was any possibility for a sun and an earth to be at all with no chance back then of even finding a thought of life as a factor. We are in time in the present while what we see is the past we look at being in the present billion years ago. What we see grew as much as where we are and time moved on just as time moved on where we now are. We can't project anything into that space or situation and we can't project us into what we see because what we see had no "us" and what we have has evolved so much we can't even realise how it evolved. It's like looking at Julius Caesar and asking why did he not make radio contact with Rome just as he crossed the Rubicon. It could have settled the aftermath in so much better fashion. We can't ask why did Rome not put Attila the Hun under tank fire and prevent him from taking Rome because we know that in the future the tank will become the most decisive weapon man has invented. It is different eras and what we see is what we were and not where we are. If we were there we will still would be going trillions of years to get to where we are and it takes time to be going further from there to where we now are. We are in time and space forms the history of time as every instant go by.

Don't confuse time and space or space with time. When we look at the Universe we look at our past. That which I see is a hologram written in light of something that was that never again will be. It is not something as substantial in substance as the moon or Mars but is a thought God provides of an era long gone into the past. I can't look at the Universe in which I am because I am the Universe in which I am. Every atom is the end of a Universe and the start of the next Universe as Kepler said $a^3 = T^2 k$ and those Universes made entirely of atoms are divided by time formed as space by the history of time. When we wish to look at the future we have to look into the darkness because we are heading in that direction. It is the darkness where we are going so we have to look at the darkness around us to try to see where we are going and also look at the darkness to see where we came from. We are a speck coming from the past being in darkness heading to the future also in darkness. Either we look at where we are and only find darkness going smaller and going bigger because that is where we are in the Universe or we look at a quasar and find a non-existing spot too small to be within the quasar and in that point where there is still no point in the quasar try to find our position while knowing we are too small too have any position in what the quasar was at that time a couple of trillion years ago. We did not have a position in the quasar back then and we now are in our own quasar but we are so small in the dark vastness we are undetectable within where we are. This issue is much too complex to debate in a book as light hearted as this because this issue borders on testing sanity and renders not the cosmic limit but our mental limit. To find where we are going we can't look to the past because we are the future of the past we are looking at when we gauge at space filled with material trillions of years ago. Where we now are we find our position are so small that wherever we look we find eternity covered in darkness surrounding us from everywhere.

But beware of a conspiracy because a good conspiracy has in it a program to allow any person to divert from the route of tracing the essence by starting to argue the content and then land in a completely new debate that has no essence of the conspiracy one wish to detect. So therefore let's get back to where we were before we were here. Let's catch the conspiracy as it ran pout of breath and into a new idea. Einstein came up with some bad news. He did not only saw a Universe expanding endlessly, he draw a catalogue of how the Universe will work. The man again did not soothe Newton's shortcoming but built an argument with models included to exacerbate the issue. Now they had to come up with a plan that will kill all further questions. They now longer could relay on Einstein because Einstein was cashing in on Newtonian despair. Remember, the Universe is supposedly going to crash into itself according to Newton and now these models all show a Universe doing exactly what the Universe should not do.

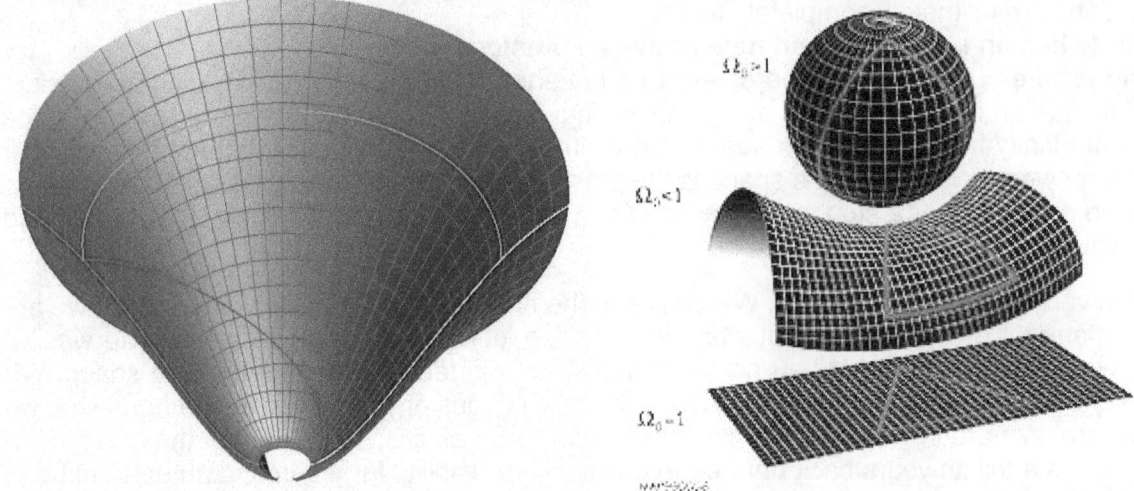

Again the incorrectness of every model is showing it has an outside. Outer space has not and can never have an outside. To correct the model we have to put where we are in the middle with as much going smaller that what is going larger. We will never see the smallest and we are unable to see the biggest. The sphere we search for forming is not formed as a result of outer space forming a spherical limit to space, as outer space is not representative in the largest structure because there is no largest structure. This is because the Universe has no outside. The sphere that forms represents the smallest part, which is within singularity Π^0 becoming space Π and forming space Π^3 by the movement $\Pi^2\Pi$ where that forms the sphere that forms singularity as singularity return the favour to form a sphere. In that we find the eternal value of space $\Pi^3 = \Pi^2\Pi$ by the relevancy of singularity trapping space in the movement thereof. Everything forming space $\Pi^0 = \Pi^3 / \Pi^2\Pi$ will loose all the space that movement forms $\Pi^0 = \Pi^3 / \Pi^2\Pi$ when movement of space is lost. That is what happens in a Black Hole where movement of space stopped and time reclines directly into singularity producing the opposite of what the Big Bang represented. That is the Big Crunch not that big and not very crunchy where it all ends. The Black Hole is the smallest and the largest at the same time of what the Universe offers as an object filling space (or not filling space). In the Theses the explaining gets a lot more technical but now we have to get back to the conspiracy and show the silliness of those hiding the conspiracy and posing the funny part as reality. The conspiracy and what it represents in information becomes a silly joke…and it works, as a conspiracy because as far as I can trace I am the first not to be fooled by the fantasy of the feeble theory underwriting the serious part! All the models and theories show the Universe expanding while every one needs a Universe to contract. This was not helping Newton while in purpose it should be helping Newton. If everyone saw that Newton was a blubbering fool that was mistaken about the cosmic principles that the Newtonians underwrite as better-than-Evangelic-Gospel then they all were a pack of idiots that new nothing about what they professed that they know everything about. They had to get around this because now all theorists were ganging up to abandon the drowning ship representing Newton and came up with theories how the Universe work by expanding. All the models had one thing in common they didn't represent Newton's contracting principles because the growth supported Hubble by "being bigger", abandoning Newton's "getting smaller" idea.

In astronomy and cosmology, **dark matter** is a hypothetical form of matter that is undetectable by its emitted electromagnetic radiation, but whose presence can be inferred from gravitational effects on visible matter. This is totally fiction and is as fabricated as modern science could be. According to present observations of structures larger than galaxies, as well as Big Bang cosmology, dark matter and dark energy could account for the vast majority of the mass in the observable Universe. This means if they can't see it they can't show it and that is brilliant to fool all the sceptics.

Dark matter was postulated by Fritz Zwicky in 1934, to partially account for evidence of "missing mass" in the universe, including the rotational speeds of galaxies, gravitational lensing of background objects by galaxy clusters, and the temperature distribution of hot gas in galaxies and clusters of galaxies. Fritz Zwicky is the "Father of Dark Matter," coining the term itself, as well as gravitational lensing and the sky survey technique. He devised it but I can't say if he was part of the conspiracy or if his ideas were hijacked and misused by the conspirers. But in the end these ideas came in pretty handy to use in the rest of the conspiracy. Dark matter is believed to play a central role in structure formation and galaxy evolution, and has measurable effects on the anisotropy of the cosmic microwave background. All these lines of evidence suggest that galaxies, clusters of galaxies, and the Universe as a whole contain far more matter than that which interacts with electromagnetic radiation: the remainder is frequently called the "dark matter component," even though there is a small amount of baryonic dark matter. The largest part of dark matter, which does not interact with electromagnetic radiation, is not only "dark" but also, by definition, utterly transparent. Most impressive but here is the catch... If they are asked to show it they already admit they can't... because it is utterly transparent. If they are asked to prove it they already admit it is illusive and therefore they can't utterly transparent. They can make up the story on the trod as we run along because no one can prove them wrong because they can't prove they are correct.

The vast majority of the dark matter in the Universe is believed to be nonbaryonic, which means that it contains no atoms and that it does not interact with ordinary matter via electromagnetic forces. The nonbaryonic dark matter includes neutrinos, and possibly hypothetical entities such as axions, or supersymmetric particles. Unlike baryonic dark matter, nonbaryonic dark matter does not contribute to the formation of the elements in the early Universe ("big bang nucleosynthesis") and so its presence is revealed only via its gravitational attraction. In addition, if the particles of which it is composed are supersymmetric, they can undergo annihilation interactions with themselves resulting in observable by-products such as photons and neutrinos ("indirect detection").

This is the same as saying there are "anti matter" eating up matter. If they are asked to say what is "anti matter" or what is "anti matter" made of they can't say. It is a name and naming nonsense is a great Newtonian pastime. They do it to relax and to become social with other Newtonians, which may or may not be part of some mating ritual. If you named something you then it is as if you explained something because naming it and creating mathematical formula goes hand in hand. There is no need to be realistic because the end of any Newtonian intellectual capacity is to give a mind blowing mathematical formula of which the practicality remains a mystery and then give it a name. The name must be so impressive that just to remember it would take up all the effort any Newtonian has in reserve so getting to the point of proving it taxes the Newtonian's stamina beyond breaking limits and then there is no need for it.

One thing it does not answer is if the **dark matter** does have mass it must have pulling power as gravity. Then what is the **dark matter** waiting for to unleash the gravity by mass to pull the Universe back into forming contraction. Why is it not pulling now if it is going to pull at all? Either it pulls by mass or it does not pull by mass but it can't have some retarding switch that will kick in at a time when it pleases the Newton's. What makes the **dark matter** slumber mysteriously while waiting to jump on the poor defenceless little Universe and force it to comply with Newton once more. No one can prove the **dark matter** is not there since no one can prove it is there. This is how one go about to devise a conspiracy. You keep it quiet and while everyone smells a dead rat but no one even thinks of looking in the right direction. The conspiracy is a success if everyone accuses anything but detect the true conspiracy.

When we go in search of what principles applies to form the building material in the Universe we better look and see what is it that the Universe shows us most graphic and we better stop telling the Universe what it is that we want to see and what the Universe should offer us that we wish to see. We better stop telling the Universe it must get mass and start to see what the Universe tells us what it has to offer us to see. If stars burst by releasing heat then stars are constructions that confine heat or cram heat into a small space. If this is true then gravity must be the process of freezing heat by turning movement and displacing space into compacted heat making gravity a process whereby space freezes as it condenses.

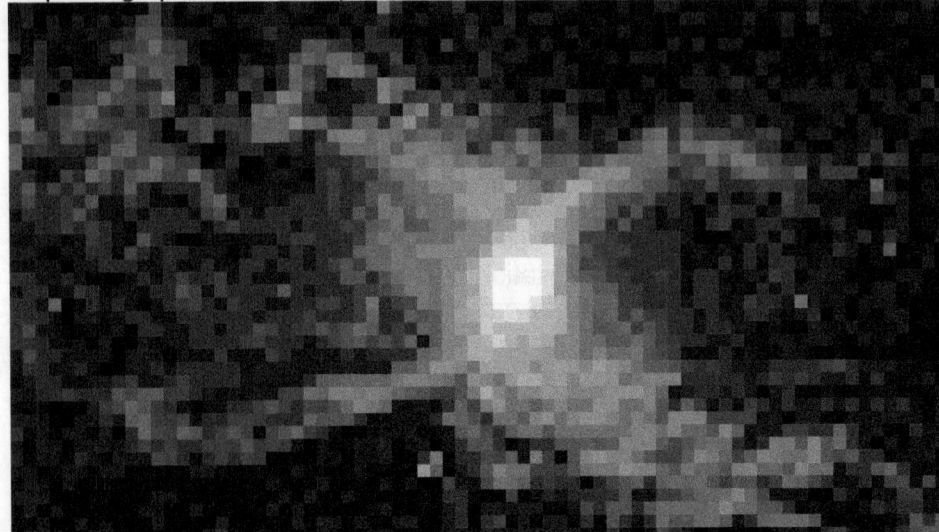

When I see a star bust I should take note of what the star releases and look for the principles applying that should form such a release of heat. Newtonians are forever copying Newton's style by telling the Universe it holds the planets in formational alignment because of their mass while not size nor invented "mass" plays any part in the process. We must stop playing God and create a non-existing Universe and begin to confirm what there is.

When a star burst open it releases massive amounts of heat into outer space. If it exposes heat bursting out then the reason that would apply is it must be because it froze heat into a state of solidity. If the star bursts as it explodes by releasing heat it then clearly overheats. The question never asked is why would stars overheat? We can blame pressure, but pressure would not bring about a star disintegrating from the centre, as the star depicted here clearly does. A burst from pressure should blow the sides out.

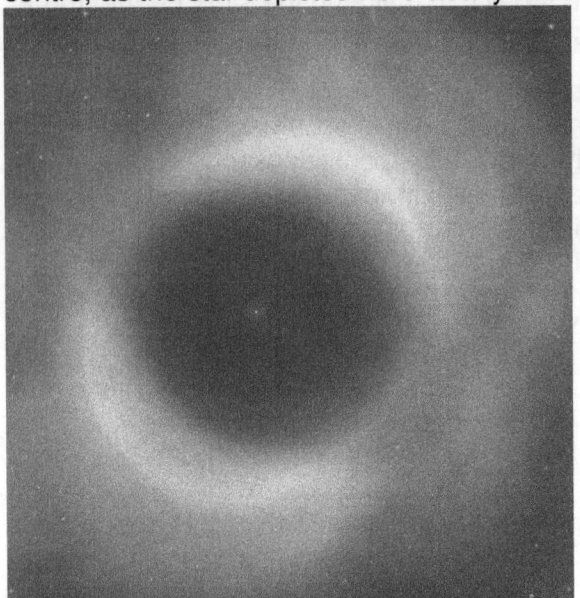

When a Super nova goes bang Newtonians say it is because "*gravity has gone mad*" and then they still see their position as being intellectual. Since when has gravity got emotions that can go array or "mad". Still more off the point is how can that be to their ability the best answer they can get up to while remaining satisfied with the effort! Stars we call Super Nova has blowouts. It can't be a pressure burst because there is no material wall enveloping the heat and thereby Stars we call Super Nova has blowouts. Pressure release comes from material containing compressed space bursts. That we humans know since before writing began, but since of late this phenomenon becomes more and more seemingly misunderstood. If stars blow as stars should and as we can clearly see from the picture just above, then the explosion happening to the star we know as a Supernova comes about from other principles, surely.

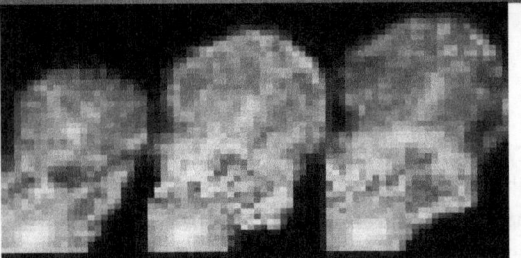

It is very obvious the two occurrences are not a result from the same basic method the Universe uses in destroying stars. When heat surges and becomes too high, it turns into space. That process we call an explosion. It is frequently seen, yet never acknowledged by science. When heat reduces, it relinquish space in the producing of more concentrated heat, this process we see as cooling.

What ever the terms used there must be a recognising of the inter relation between heat and space where the reducing of the one will lead to the increase of the other. The star does not apply pressure to bring about fusion, it freezes the elements into fusion by applying millions and even billions of degrees Celsius. It is our conception of hot and cold bringing total; confusion about the principles of cosmology.

In the sphere Π in singularity holds equal Π in six specific points becoming a total of seven Π and no radius. When removing Π in singularity the circle loses some and sometimes even all of its singularity. Where the singularity no longer control space, Π compensates by relinquishing space to r. where Π removed its value r becomes the square and the circle then becomes $\Pi r^2 = 10\Pi$. In the normal tongue we call it the atmosphere. As is clear from the picture Π relinquished value to extend r and with r more valued the 10Π extends much further, but the Π is less influential and the r is more influential. At a point Π loses all influence giving superiority to r altogether. When that happens 10Π no longer forms a circle but a square forming a cube. With r^2 the square can form anywhere favouring any space to its whish. This allows the r to unfold in any shape possible. That was the Big Bang. But the Big Bang was long after the first Command came in the order of Let There Be Light. Before this the Universe was in a completely different mathematical state than what we now have in the present form.

I do not think that I or any other person is at liberty to try to calculate any on goings with in the star but from what is clear from the outside one may come to some measured idea of the stars position in space – time. Gravity is the cooling of space by duplicating or moving space, albeit filled or not filled. When the star spins to slow it does not cool sufficiently and then it becomes warmer inside. As it reaches a point where it overheats because it moves to slow it burst and by expanding the space it regulates the temperature. At a point when it can no longer contain the confined heat it expands and such expanding we call a Supernova occurrence. The contraction of space must be equal to what amount of heat the total number of atoms spinning within the star can retain and gravity is that balance. The fact that it can freeze heat to a liquid surrounding hydrogen while holding a temperature of 6500^0 C should be an indication it is not what we seem to acknowledge as normal. The sun is freezing hydrogen to a dense liquid at 6500^0 while space is boiling (expanding through overheating according to the Hubble Constant) at -273^0.

Science academics have to review there thoughts on relevancies because what seems to be hot is cold under certain circumstances and what seems to be cold to a point of freezing is boiling hot. There are no standard issue and fit all through out the universe. Every singularity attaché different criteria to borders controlling the space-time with in it rule. What fits humans on earth does not even suit conditions on the moon, yet science cannot appreciate that the moon applies very different standards to that of every structure and every structure is a cosmos on its own turf, supplying its own turf.

We must look at nature to find what is hot and what is cold. Something hot is that which can expand no more because that which is hot expands. When something is cold it can contract no more because it reduced space to the utmost limit. Outer space is the hottest because it eternally expands while the Black hole is infinitely cold as it contracted what it could contract and keeps on contracting. It is not the specifics that are of importance because the specifics change considerably when taking into account that hydrogen remains in a frozen state at 6500^0 C therefore it is obvious we have to look at other clues to give some indication of what is in process. On earth in the time we have as a duration we find hydrogen freezing at 269 ^{0}C as where it freezes on the sun at 6500^0 C, which implicate the reduction of space to an enormous increase in time duration.

In conditions on earth the rotating velocity of the electron is 3×10^5 km / sec. With conditions being that different it can not nearly be the same in the sun. As space reduces time increases. By having the space reduced to such an extent that it matches near Big Bang relevancies (a period where heat flowed like water and which is the very same conditions we find within the sun) the space would apply accordingly. We also know that relevancies is all about conditions showing similarities under variables and therefore the space and heat component may seem altogether incompatible but is almost the same given the singularity presence within the sun and comparing that to the earth.

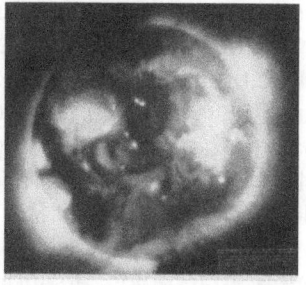

What is applying to stars inside the galactica centre is applying to particles inside the sun. Science sees the nuclear reaction but do not recognise and therefore do not admit that the nuclear reaction is three different phases. At the beginning of the process all the heat is solid, placed in a container by nature and the container has a human name called the atom nucleus. In the atomic explosion there are three ingredients that are distinctly apart. When the solid melts down, it becomes a fluid. The fluid we gave the name of light. There is not enough space to explain the detail of the argument, but light is not a gas, it is a fluid. The first step of the nuclear explosion is converting the solid to liquid. In the liquid state the star does not overheat. The overheating becomes part of the second phase. That phase involves the turning of the heat-fluid to a heat-gas we call space. Space is heat overheated creating space, as heat is space concentrated creating a fluid or liquid not yet correctly named.

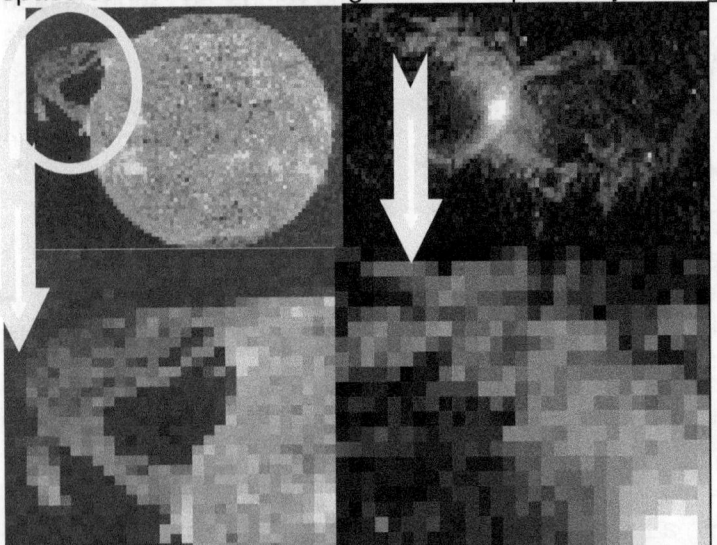

In the next scenario the overheating core is hotter than outer space and that brings about that the heat will flow to a colder region. In this case the star is overheating and with that can no longer protect its individual singularity. The pictures show clearly the difference of a star NOT overheating being "normal with liquid pouring from it and then becomes a gas as it evaporates. We see the sun exerts heat but still we think of the sun as hot. The sun gets rid of the heat, which means the sun is cold and that is why it removes heat from its surroundings. But because we feel that which the sun rejects we then contribute this heat to the sun by attaching that which the sun removes.

Every one knows that a gas is one dimension HOTTER than Liquid as liquid again is one dimension HOTTER than being solid. If the star is liquid on the inside, and the liquid evaporates when coming into contact with outer space, then outer space is the hottest, notwithstanding what ever boundaries and values we humans attaché to the dimension. Our human standards have to change to accommodate the rules layer down by the cosmos and not apply the cosmos to suit our rules of hot and cold, big and small, near and far. In the case of the Super Nova, smoothing prevented the liquid turning into gas, therefore overheating. He liquid froze as a liquid becoming a cosmic lollypop. That which prevents the overheating turned the layers into frozen identities not overheating therefore it became a liquid outside the star. This star was turned into a miniature galactica, sustaining billions of individual singularity, because the governing singularity did not destroy, but the singularity of every nature is still in support of one another. From this picture (and others of Super Nova) one can learn a lot, if one is truly interested in applying cosmic law to the picture and not some human response to what we think would apply to an earth-like star that holds gas as an ingredient. Again I have to press the thought that it is singularity determining space-time form through conditions that bring about the state between matter. Matter can be solid liquid or gas, but it is the condition of the space-time derived from singularity that places the form and conditions valuing the form of the elements. Hydrogen can be as much a solid as gold can be a gas.

The one part of creation the official verdict and mine is in agreement on is that it all started small, but in saying that I go one step further by saying the Universe was at the same time eternally large. It is all about relevancies forming as singularity applying matter in relation to the overheating it started to combat. It started with singularity producing matter and the matter changed in relevancy to one another by becoming solid or liquid in relation to each other. Space still was at a premium because thew space we know and we see as gas, was not yet part of creation. Since the Big Bang the fluid heat is in a process of converting to space enlarging the role of time as the Universe still systematically overheats.

From the offset of the first dot dividing as it became the first two dots, it was bringing about the second dot and the eternal number of dots growing from that means that the splitting of the dots assumed as the dots were growing from infinity in size, which is in fact only part of the relevancy because at the same time the infinity presented eternity, where both locked the same value to the dot. This long sentence structure is an effort to explain that everything is linking to another either directly or through other particles and everything came about precisely simultaneously being eternities apart. The stars are in relevancy part of the growing cosmos, where the growing cosmos presents a liquid covering all solid strictures. The structures are no more or less particles irrelevant of size, since time places the value and space is dependent on time. But by the continuing process of the eternal overheating, the geodesic cosmos overheat gradually which presents as the Hubble Constant and this process changes time in space. Since every star holds an individual singularity, separating its singularity from the galactica singularity it is within, it remains as a relative liquid while the cosmos changes its side of the relevancy becoming more a gas. The difference between the star being the dot we can see in the picture on the outer edge and the star being the dot we cannot see on the inside is the time in promoting the individual singularity. First the star in the centre core changes, starting to collect liquid heat, while the outer part remains part of the cosmic structure. As the Hubble Constant grows the star distinguishes its singularity as it protects the singularity from overheating with the cosmic geodesic space-time. The geodesic space-time is also the outer space, but I prefer not to use outer space. At a point the star becomes a separate structure from the liquid cosmos that turned to gas, and hen starts using the liquid to promote the generation of matter in a solid state whereby that matter then later turns to space less singularity as space-time completely brakes down forming neutron stars, pulsars and eventually a completely space less point of singularity in the cosmos being an ancient dot once more we think of by using the name a Black Hole.

Stars can and stars do **overheat**, sometimes and the **Polar Regions** where **the Titius Bode matter-to-matter applies** holding the square matter (7+7) in relation to the square of space (10) and **other times** in 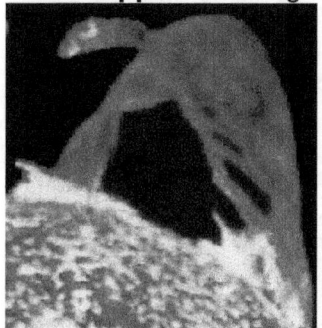 a double relation to the **square of space** 10 to that of matter in a half square (7 /10 or 7/ 10). Saying that one has to differentiate between heat and overheating because a star represents the coldest space in the Universe and not the hottest space. **Heat and cold are relevant dynamics** forming **in appreciation of singularity. The sun is the coldest place in the solar system** and that is fact. Looking at evidence the sun provides contradict everything science wishes to believe about cold and hot. Science wish to see the cosmos through the eyes of what fits the needs sustaining life on earth and what benefits maintaining surroundings in support of life as one find on earth whereas life has no part in the cosmos except for the speck of dust we call earth. Looking at the cosmos impartial to life the evidence support another view. Every aspect in **the cosmos is the very opposite of what science believe** it is. The sun is **not a ball of gas but** a **giant sea of liquid**, frozen **without any** form of **gas or air** in the interior. Having a liquid interior **the sun** has **no pressure** but has the **very opposite of pressure** to which there is yet no name given. **The liquid comes from singularity freezing** space-time within the atmosphere of **the sun**, and such is the case with all stars still in the shining phase. **Stars more developed than the sun is frozen solid causing fusion.** In the pictures we find not withstanding whatever name we attach to the red liquid substance flowing from the sun into space and back to the sun, that liquid is heat in a very direct form.  If outer space was the coldest place in the solar system the heat should immediately escape to outer space and not return to the sun as it clearly does. If outer space were colder the heat would not return to the sun. All elements forming matter in as much as the heat forming an atom is as much a liquid as it is a gas and a solid. **There is no hot as there is no cold. It's about storing energy in space or in heat, which is another Cosmic equal being opposing similarities.**

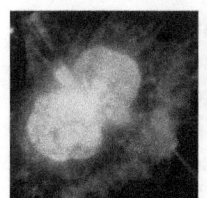 Hot and cold are **relevancies brought about by singularity valuating space-time** and during **the Big Bang** the Universe was **freezing cold** at **three billion degrees C**. It is the relation matter has with heat that provides the form the particle has at that moment. The increasing or decreasing the heat will alter the form of the element. Therefore all elements forming **matter is as much a liquid or not than it is a solid or a gas. It is the space surrounding the atom which provide the form the atom find its relativity to the rest of the atoms it share space with. Hydrogen is as much a solid as tungsten is a gas depending on the heat in relation to the space matter is within.** Should **you reply** that it is **the gravity pulling the heat back to the sun,** then that **confirms** my theory that **gravity is all about collecting heat onto matter** with outer space being the hottest place. **It is the concentration of heat in space being relevant to form. When overheating a star turns its liquid to gas whereby it merely transforms it's interior to a relevancy it has from pre- to post- Big Bang.**

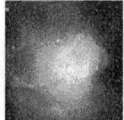 We humans on earth think that hydrogen is a liquid at -259^0 C but that only apply to the earth. The picture clearly shows the **heat in a liquid** flowing **from the sun** and **back to the sun**. In the **sun the hydrogen holds enormous quantities of heat in a liquid at a temperature of 6500^0 C.** When a star has its singularity secured the star is bitterly cold because it has heat in a liquid form flowing back to the point of singularity although we may regard the star to be rather on the hot side. The sun (fore instance) freeze hydrogen in a liquid form at 6500^0 C. If hydrogen remains a liquid at 6500^0 C, just think

how cold it must be as the star's interior approaches the point of singularity. Therefore fusing protons comes from cold and not from heat or pressure. By allowing the singularity to overheat the star overheats and heat within the star flows from singularity to outer space freely. In such an event outer space is then colder than the star because the heat releases to outer space with no intention of returning whereas in the sun it returns as soon as it leaves. There are two ways to reduce heat; one is to bring about expanding space, as the photographs clearly show. The second one is where heat will reduce when in motion by spin. When withholding or retarding motion matter will overheat. Gravity is the motion of unoccupied space through the dimensional transformation to occupied space.

Motion and space therefore is the anti-, the opposite the negative to heat being the positive. With singularity overheating the expansion of the singularity drives heat into space, creating space to compensate overheating **That is a natural phenomenon**. The only reason why **heat will** rather **flow back** to the star than **escape to outer** space once the star released it into outer space is **if outer space presents more heat than does the star,** because **heat always flows from hot to cold** no matter what influences may arise. **Outer space must hold more heat than does the star but the accumulation of space in relation to heat makes it seem colder bringing expanding of heat to become space.** <u>Space and heat directly relates being the one form of the other</u>.

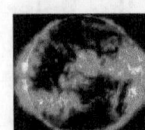

The cosmos is all about **converting space to heat** which we see **as gravity** and **returning heat to space** as a **control mechanism** always **keeping** a very delicate **balance** which we see as **a star shining or being normal.**

The purpose of the converting of space to heat is to supply the core where singularity is with heat. **It turns space to heat** sustaining matter but sometimes singularity overheats and then matter converts to heat allowing heat to convert to space. That we call many names amongst others exploding into super nova.

Whatever the names used is less important because the **process rests on space and heat interacting to form energy**. That was what **the Big Bang** was and **the Hubble Constant** is all about where **matter converts** **heat to space.** I show that **space and heat is the very same thing** and there **is no such** **a thing as pressure** but releasing **heat produces space** and **concentrating** **heat** **reduces space** with the two interacting on singularity demand setting time to **space with** time being the spin or motion of heat in space. **Heat and space form the** **second** **singularity** caused by the **fragmenting of singularity to compensate overheating during the pre-** Big Bang matter forming era. That is what we see as **light and space,** which again is the **same thing and is fragmented singularity forming radiation and heat, where the star re-transfers heat back to space due to an overload.**

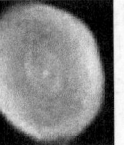

Applying the Titius Bode laws (shown above), the Roche and the Lagrangian principles correctly and the atomic relevancy, I can prove that:

The spinning top is all the evidence any one needs to come to such a conclusion. By saying that I first have to admit about (no not my mental stability), but that I have no academic background and I do not enjoy any link to any university. In the past this remark excluded any further interest any academic might have shown about my work before the remark and I think that is very unfair. Every person has an opinion and it is the opinion that has validity, not the persons social standing. It is the way one absorb and reflect a personal view about matters that are important and not the manner in which the person arrived at such conclusions but it is the validity of the conclusions that should carry importance. The cosmos contains matter, space and time in heavens or dimensions standing in relevancy to one another. Whether you refer to dimensions as dimensions or heavens it is no different because it is the same thing. Singularity brings about heavens as it brings about dimensions and singularity is a mathematical fact. A straight line cannot start at zero and still be a straight line because zero extending to wherever brings about a full zero. A straight line starts at the point where the pen point meets paper. That point may be any distance from infinity to a measurable dot, but it cannot be zero.

Any straight line is also half a square because the line forming the square cannot start at zero for the reasons I just mentioned. That is singularity pointing an eternal direction from a point of infinity and that is the basis of the cosmos as much as that is the basis of mathematics. To escape from nothing one has to become something and by doing that one could not have been in nothing in the first place. If one holds a point in nothing one cannot become something because of the nothing value. To back this argument that no line can ever start at zero is to ask the simple question: what will the length of the shortest possible line be. It must be a line where the starting point is so close to the ending point the distance parting the two is incalculable yet there is the line therefore the end and the start is apart still sharing the same spot. The length cannot be zero because zero means no line. The starting point and the ending point may be inseparably the same point with virtually no space between the two points but neither of the three points can be zero simply because there is a line (be it infinitely small it is there). If the point is zero, then the line will be shorter than the shortest possible line. If there is a line and the two points starting the shortest possible line and being the continuing of the line it may still be the same point and even by sharing the spot as the point ending the shortest line possible the line must be there and the line holding the start and finish is next to zero but can never be zero otherwise there is no line.

I do think that those Scientists feeling more serious about their work does not credit star formation in the manner depicted above and I know for a fact that Eugene Shoemaker was one of the more prominent scientist declaring his doubt about the planets forming from dust. It is not possible that two stars near as can be will form from dust where each have its own separate pivot centre.

In this, I would like to ponder on the binary stars, as they are the key to time in the cosmos. The two stars develop in the galactica in close proximity as they help each other in transforming negative space-time displacement to positive space-time displacement. With the combined effort, they can grow extensively in supporting each other. From the point of singularity time hold in matter claims a space in the value of $\Pi^2 + \Pi^2$. Another claim to control of space comes into value at the presentation of $\Pi^2\Pi$. A third claim to space holds an influencing value to $1 + 1 + 1 = 3$. When the proximity of matter disallow the claim of influencing space, singularity resents space in a half that of space under control, from both ends. That will be the square of the point from singularity under control being $(\Pi/2)^2$. In anther book I go to lengths disproving the dust to planet nonsense and do not whisk to ponder on the issue at this time. But no one could yet offer an explanation to how every planet sorted out the obliquity of the elliptic so many times using a radius of approximately 5900×10^6 km. from the sun and still managed to compress the smallest objects to rock leaving the larger ones with "more gravity" to stay gas where each selected the perfect centre of rotation in accordance with $a^3=T^2k$, a precise measure.

Would it not be more scientific to use a graph having such a perfect formula to work with as in the case of $a^3=T^2k$? From the graph any one can read different seasons applying to a small planet in relation to such a big sun that should be able to shine on the earth from any point all across the poles.

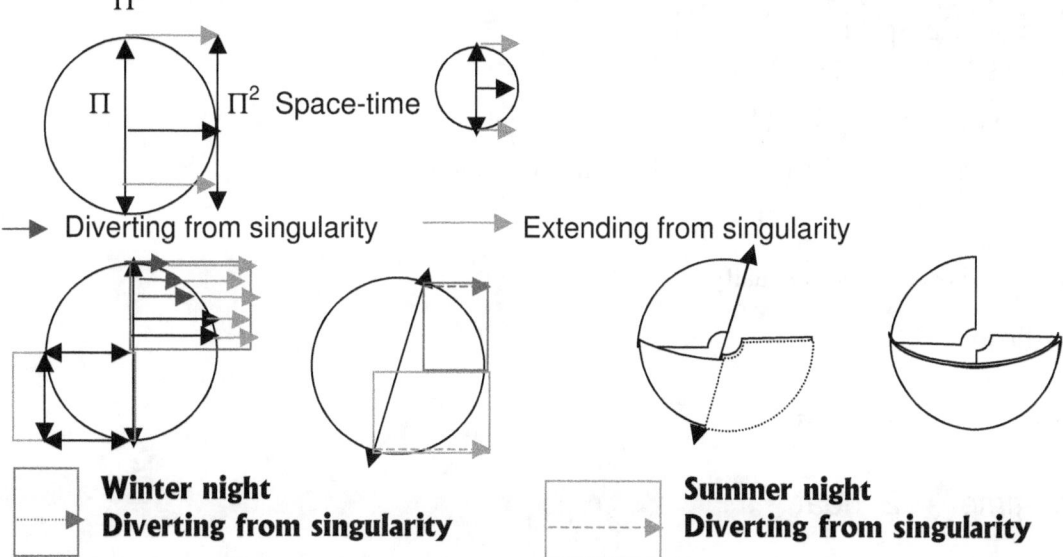

As singularity is eternally cold, it also has to affect the standing between matters, as matter diverts from singularity. The more occupied space there are between singularity and space ends, the more it will be to the advantage of the effect we consider heat, and the less space allowing time the more it will advantage the statement we refer to as cold. But hot and cold are human relevancies not acknowledged by nature. Looking at a star overheating it is obvious how singularity breaks into heat. By demolishing singularity it means Π demolishes the very point holding singularity.

Science cannot be more aware that space and heat stands unconditionally related to the reverse, yet it is taking science centuries to translate that wisdom to cosmology.

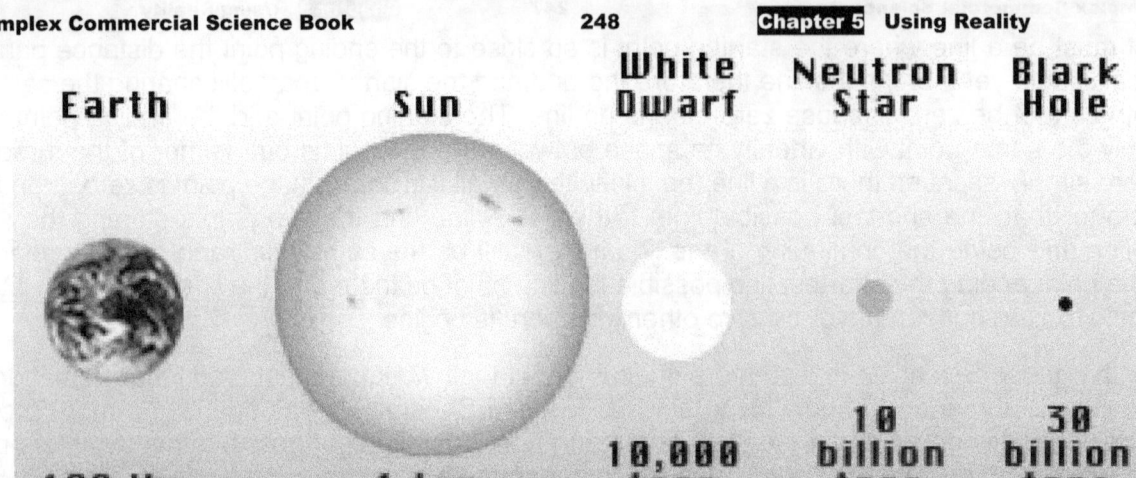

When any object moves it duplicate the allocated position more as the movement increases. It fills more space but individually in less time. In this we have space-time. The shorter the period of time is that an object fills space the less space it would fill in the instant but the more space it will fill in the total duration. In other words it shrinks the space it is in by extending the space it holds relevance to.

As the position matter holds in the time relation of spinning around singularity a faster duplicating tempo, matter move closer to the position of singularity by means of contracting by reducing the space it has contact with, this contributes to the time differentiation in displacement in relevancy becoming hotter, because the relative motion in gravity reduces. The rise in heat levels produce a relative colder scenario to the inside in contracting by bringing about that the status quo of matter to the side of singularity will forever stand to be hotter than the singularity although the singularity as such holds eternally more heat than does space-time. But since space-time pronounce the heat singularity will always be hotter seemingly to be colder but that comes about since relativity brings about hot and cold and the term is in fact nothing more than dimensional implications of relevancies applying.

H = By Heat increasing   space will decline to singularity's infinity pushing heat to eternity

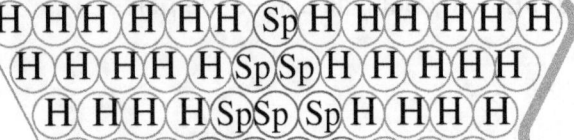

Space is Creation gas manifestation

T = Heat is Creation liquid manifestation

Sp =   Space reduces equally

6   volumes space-time

5   volumes space-time + 1 heat

4   volumes space-time + 2 heat

3   volumes space-time + 3 heat

2   volumes space-time + 4 heat

1   volumes space-time + 5 heat

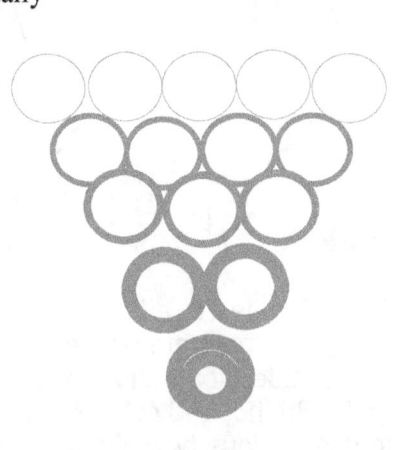

When I explain the sound barrier this differentiation will become more apparent as I then use the aircraft's movement to explain gravity and how gravity applies.

Part 6
The "Sound Barrier" Explained as a Cosmic Phenomenon

If Einstein was unable to recognise the most basic of mathematical principles then what type of genius did physics create in him and what slur did physics promote. This idea of the two factors being in opposing relevance is so simple that children will recognise the principle, and yet those fathers of physics wants me to believe that the greatest mathematician that ever lived did not realise this principle…the principle that the radius and the mass stands related and the growth in the one will promote the decline in the other as a dominant factor. Can any one with this information including the information given on the previous page have any other conclusion? It is obviously clear that having such a total idea that there might be dark unseen mass floating in the Universe which at this time does not generate gravity but will some day because Newton has to be correct at some point in the future. I am to believe that dark undetected mass can be found and such undetectable mass could be found which will bring about contraction after all this expanding? Why would the mass at present then not activate gravity and why would the mass at some point spring to life and start activating gravity? How much can the Physics paternity still hide the fact that **Einstein's critical density** is being used as a **cover-up to distort the truth** to **conceal fraud**? The **uncovering** by the Hubble constant about of the **Newton fraud** is so simple to see. Hubble found the Universe is expanding and Newton's said otherwise. Who is lying about what?

Hubble's declaration was on track to blow the cover that was concealing the Newton fraud wide open and uncover the centuries old deception. To see this we have only too look at the comet behaviour when any and all comets again come around on a cycle by repeated visiting the sun. The question is if it is mass pulling mass onto mass, then why do we have comets left in the solar system? The mass of the Sun should by now at least have destroyed every comet going around.

Let any student ask his Master to explain Newton's formula in relation to the comet behaviour.

Contracting $F=G(M_1 \times m_2)/r^2$

Every one believes in Newton except comets, **because comets fail to collide with the sun.**
However I can explain in some way…

Every indication that we so far received in vivid portraying from astronomy photography studies from outer space disputes a shrinking Universe concept. From the moon increasing the radius distance between the earth and the sun, to the Hubble Constant indicating a space growing any where in space wherever man may conduct studies. Since the end of the middle ages a force called gravity was identified, but more than that science did not take it. What is gravity, besides being a force? What forces the force? I introduce a cosmic theory that turns the missing questions to answers.

Let us for one second return to the science we all know.

There is an undefined phenomenon in the cosmos, never mentioned (in public) because it obscures the basic formula

$$F = \frac{M_1 M}{r^2} G$$

Lets put the mathematical formula into a practical context.

THE SPOT THAT'S HOLDING THE LOT.

As highly developed we seem to regard science to be, it took a genius like Max Planck all his life to try and reconstruct from what there are to what the Universe started with. Once again the human perspective baffled the genius of a Master such as Planck was, because he placed man amongst the first of creation. Where it started from and what we now have is many, many eternities apart and any evidence of an attempt matching what we have now to what first was is foolish.

It started off with one dot so small eternity met infinity within. Then came one more, and another and they continued coming until there were a countless number of dots. The accumulative size of the dots were the same size as one dot because in the true Universe big and small plays no part. The dots were infinitely small and eternally big at the same time because size is a relevancy and without one the other has no size. So in the true perception, there is no difference in size.

It started with the fact that there is no place or part in with which one may associate zero or nothing. There are no room for a number such as nothing. Next to the one dot (infinitely close) one will find the next dot, and if nothing was a factor then that is precisely what one will find between the two dots. Nothing of space, a non existing entity, taking up no space, and much more important, no time, therefore the dots are infinitely close to one another, being the same space, eternally big as much as infinitely small. If we as humans cannot find a manner in comprehending this notion, there can be no manner ever understanding the cosmos as much as the start to the cosmos.

Every dot was a Universe in its own and the accumulation was a universe. The earth in itself is a Universe as the moon is a universe, because rules applying on earth do not apply on the moon and visa versa. When in the ocean another set of rules apply, therefore being in the sea places a body in another universe. The number of universal entities is still countless, as much as it was in the beginning.

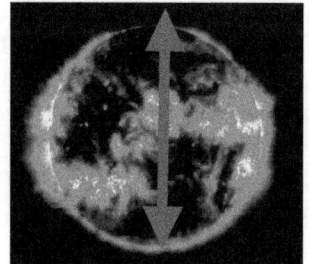

Every position in the Universe either holds singularity in a form, or relates to singularity. There can be no position unrelated to singularity therefore every aspect of the cosmos is space-time in various forms under the provision of singularity connecting. Matter cannot be if not surrounding singularity

Singularity is as close as any spot can ever come to zero BUT IT CANNOT EVER BE ZERO. From singularity diverts space-time and there cannot be space without time as much as there cannot be time without space, not withstanding the size of space or duration of time.

Through space-time singularity connects as much as relates linking the Universe into a network of influences beyond what ever we can ever conduct. There can be no spot that does not participate in the curvature of space-time. From the point of singularity runs space holding time to the prescription singularity dictates.

With singularity connecting singularity will or cannot relieve or release the connecting other than by a method we humans refer too as an explosion, but space-time separating as much as joining singularity dividing can change by applying time too space in a changeable manner, stretching and shrinking the time aspect by changing the density of the occupying heat, which creates the space allowing the spin of the occupying heat creating space setting the time.

Every dot insignificantly small as it may be, is a part of another Universe as much as it is part of the accumulative Universe and every dot in the infinity holds singularity, which we translate as " nothing" being " darkness". There cannot be "nothing" just as much as there cannot be "darkness". There cannot be something big or small, but it into relevancy of perception, and then the relativity of perception becomes the question. There cannot be hot as much as there cannot be cold. The sun FREEZES hydrogen to a liquid at six and a half thousand degrees Celsius and Universe boils over in the form of the Hubble constant at the temperature (we presume from our vantage point) at minus 273 degrees C. If we Humans cannot or will not abandon our human perception and our manly perspective, we may as well return to astrology for all its worth.

Every point in the infinity we may observe at is not merely part of the Universe in not being "nothing", but is the point where the Universe started representing singularity. It is the very first point where everything began so many eternities ago, because after all, how can we ever determine where the first point was, as they were very much equal and alike at the beginning. Every aspect of the Universe started with the fundamental fact that no point in the Universe can represent "nothing" as a number, because every aspect in the Universe represents singularity in what ever form it may hold in that specific spot forming space-time. If man does not reach a conclusion where that conclusion is matching the Universe and stop to match the Universe with man (and man's incapability), we may all go back to caves and become starving hunter-gatherers again, because we will never find a way to progress to the ultimate understanding of the universe.

I wish to make the argument about gravity found in orbiting objects a rather lengthy debate as part of my initiating discussion and after all, it is the very argument on orbiting structures NOT drawing closer that initiated and concluded all the effort of my writing this book. FOLLOWING ARGUMENTS MADE ME SEEM AS THE VILLAGE FOOL TO FRIEND AND FOE, ALIKE, AND WHERE I MAY NOT BE THE BRIGHTEST IN THE VILLAGE, SURELY I AM NOT THE VILLAGE IDIOT THAT EVERYONE TAKES ME TO BE ONLY BECAUSE I MAINTAIN THERE IS NO GRAVITY.

If the graph broke a zero mark, then a new and totally unrelated line will form on the opposing side baring no relation to corresponding with the line in the previous quarter. The new line may start from any point and lead to another point holding no resemblance to the opposing side. What will prevent the line from finishing at a point marked say three and starts afresh at seven because there is NOTHING between the lines. The nothing will provide releasing and detaching all corresponding relativity there may be through disconnecting. Through nothing there can be no resemblance or corresponding because the point of start need not even to continue at all. Nothing will release any connection and if such a connection may come about, such a connection may just as well be very co-incidental and may never be used in calculating accurately.

Experience taught us that there is a definite precise and secure corresponding that can only result from a direct connection in as much as the line being the same line. In that case the line must then come down to infinity and release from infinity as the same line still connecting to a point of re-bouncing to either side. The graphic cross is he result of singularity applying opposing sides but still maintaining connection through the application of Pythagoras that will connect and always bring about a direct relevancy. The graph does not hold zero because information derived as result of a relation prove a contact remaining when the line crosses singularity without applying detachment..

The Brainy Bunch holds the view that in a graph the line crossing amounts to breaking the zero mark. That cannot be the case.

In the centre of whatever two synchronized rotating objects the line of singularity matches in spin to one another not withstanding the space-time involved. The relevancy applying places one rotation of spin eternal, therefore the rotation is standing still, but also one rotation consists of an above measurable number of infinities producing a rotating speed faster than non other will ever reach and that makes the rotation the fastest there can will and may ever be. Again we humans now face the same value we wish to separate, as is the case with hot or cold, near or far, and quick or slow. Once again time locks relevancies beyond understanding. That line so infinite small as it is eternally big connects the cosmos through light we are able to see and light we are unable to realise.

There is not the slightest possibility that particles (even dust) of similar singularity can bond. The preventing of this comes from the way singularity apply the Roche limit to matter holding accumulative

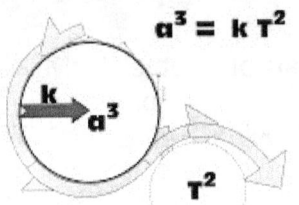

singularity. By reducing r would bring about the same result as enlarging the mass factor of the cosmic objects i.e. the Sun and the planets. It is a very drastic implication that will cause much more than just seasons changing. It must bring about that gravity changes through out the year...yet the radius does constantly change, therefore...

The closer any two cosmic objects come the stronger the force should be, with eventually no force in the Universe being able too keep them apart. This is just not happening!!!

There is no indication of truth about a contracting solar system as Newton proposed and as Newton's followers promote and or a contracting Universe as seen from the Hubble as he introduced has Hubble Constant...and not from any other evidence seen through the Hubble Telescope.

When translating Kepler's mathematical expression into a verbally spoken form of communication such as English, we can see what Kepler said also reads as $k = a^3/T^2$, where **k** is one point from a centre point that is space a^3 relating to time T^2. From a centre comes space-time.

Ever considered why a water drop would freely choose to form a sphere? When a water drop is released in an astronaut's space capsule in outer space, the water drop forms a sphere as it floats in free gravity. As soon as the water drop is released from the Earth gravity and has the opportunity to float in space, it can form any shape it finds pleasing, yet it immediately turns to the shape of a sphere. It is the same reason why we would think the Universe takes on the shape of a sphere, although we know the Universe has no outside and we realize that the Universe is limitless in size. Yet, notwithstanding, we take the shape of the Universe as naturally being a sphere that formed ... but why think of a sphere that the cosmos considers as the sphere being the pre-cast shape? Well by saying it is the strongest form available, has the same argumentative potential as saying a baby is little at birth. Blaming it on no reason substantiating proper evidence that would indicate some facts more prudent than just a simple answer to dodge the question by hiding obvious incompetence. I'll give you one clue ... it has something to do with finding the centre of the Universe. That brings on the next question being where one might locate the centre of the Universe. These questions put to your are most ordinary questions ... and yet the complexity in the answers rise far above the answering ability of those with the supreme mathematical skills. Those master mathematicians cannot answer such simple questions by using their complex mathematical powers. If they could have, then they would have ... the answer is extraordinary simple and childlike easy to explain.

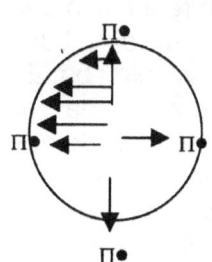

By examining the form of the sphere, we find that there are 6 points on the surface of the sphere that is holding the form at a specific and equal distance from the centre. Lines run from the centre into space at 90° and 180° angles of each other from six opposing sides. There then are six lines at 90° and 180° connecting to the centre from six points on the outside edge of the sphere. As a result of the basic shape that a sphere has, there is a spot in the extreme inner centre of the sphere where the lines in 90° relevance cross each other and others connect by 180°.

There is also at that point a spot where all space relinquishes a position and only singularity 1^0 as form remains. At such a point we find the measure of the sphere being Πr^0 with $r^0 = 1^0$. That is where the line that represents the radius as a line disappears, as it becomes singularity r^0. After more reducing continues, we get to such a point where we find only Π^0 left. At that extreme point is where space in all form disappears as the circle providing the sphere the form the sphere has, removing all possible form by going into singularity $\Pi^0 = 1^0$.

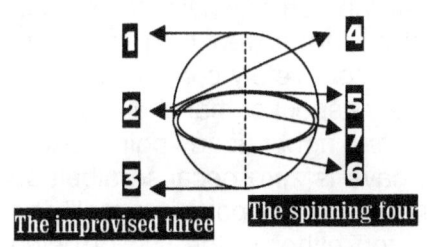
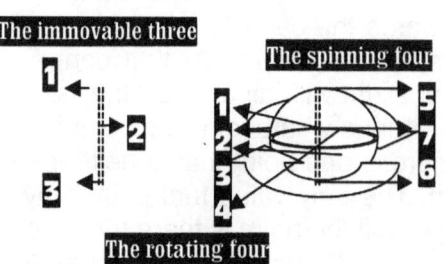

Then in that area all form of any possible space disappeared leaving only the dimensions of singularity 1^0. This too, I take much further in the book as I delved deeper into the argument and by doing that I stumbled on the ingredients forming gravity. However, from such a point there runs lines that connects to space on the outside where six points on the outside points connects to the space less point in the inside. In the book I take this argument much further, but for now, I leave the argument at that. Those lines carry the structural straight that the sphere has where the other six support every one of the six by singularity.

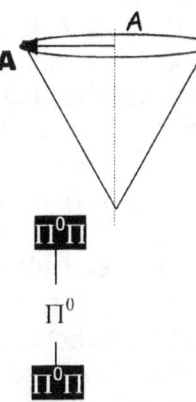

Where there is no space, there must be singularity 1^0 because the space is present although in singularity 1^0. If zero were a factor where all space finally halted in zero as the value, then zero would be able to remove the space from the centre and such removing would continue to remove the space until all space was removed. It will finally abolish all space in the sphere and it would remove the sphere. Zero removes all possibilities of anything coming about. Since the sphere is there, a zero factor in the centre cannot be present. Only infinity can be a factor from where space may grow because infinity can extend and grow into and up to eternity.

The moving of Π^0 to Π involved relegation and not motion as we consider motion. It was Π^0 getting a side and that is all. There was no true side but only a form that came into place. Singularity (A) received singularity (A) and no more of anything but the shift to comply with having a relevancy forming in relation to singularity. The dots had no sides, had no length or diameter. There was not measurable space or measurable time involved. The time could have been a micro, micro second as much a trillion millennium because time had no relevance. It was eternity interrupted by infinity, as it still is the case, however the line that eternity followed was no line because there was no space to hold the line. The line was momentarily interrupted by infinity, however with no one there, there was no one to notice. The lines were not lines but relations to sides being formed.

The relevancy that had the power to set Π apart from Π^0 is the only relevancy that still has the power, to set particles apart or join particles. It is heat in variation from cold. In order to excite singularity, singularity must establish a basis of heat that sets such a heat basis apart from cold. From there the form the atom will take on, however, the atom was still enumerable eternities to the development side.

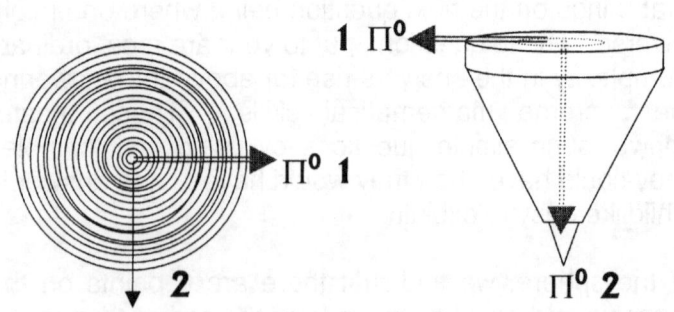

Where they are equal in value we must test the reason why this then is valid.

What is in the Universe is spinning. In the precise middle **of all** objects in rotation **is a precise centre dividing the object in sectors that will** start the spinning initiation **from that centre point.**

$k^0 = a^3 / T^2 k$ states that whatever is, is also spinning in order to be present.

Thus, the spinning object will have a middle point, **a very specific** centre point that does not spin **and only holds Π as a specific value because no radius can apply. But also the one value such a line** cannot have is zero **because the line** is there and holds contact **to the rest of the material bringing about that** zero does not start any **line and therefore the** value of the line must be infinite, **just as described in accordance and by** the definition of singularity. As I am introducing a very new idea, I wish to explain in better detail what I try to convey. While the top is spinning, one will find a line that formed in the centre where no line can form. It comes from spin but can never participate in spin.

That line must be singularity because if one moves any point on that line one position on, such a movement will land the point that then form on the line, on the other side of the line. The line is where the radius ends and starts because the line divides what is spinning in innumerable sectors and when reducing the radius progressively towards the centre of the spinning top at the centre where no line can be there is a line dividing the entire spinning top. At that centre point all further reducing must end because the next movement however slight will fall on the other side that is completely contradicting the one side. One movement further will change whatever is, so completely every aspect of that characteristic will contradict what it was before. There is one point that is neither left nor is it right but any point next to that point must be either left or right. The only value that point may not have is zero because albeit so small that it is not part of our Universe, still the point is there for all to witness and that point is a reality as much as the entire Universe is a reality. Whatever one attaches to the top either in the line of being material or a concept, such a concept or material has to start at the spot in the centre of the top because every aspect of the top changes in contradicting from that point onwards in all directions. That point albeit

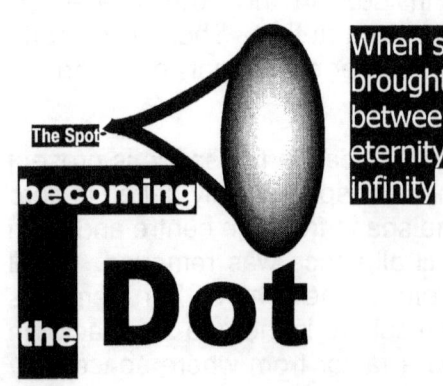

When space brought division between eternity and infinity

hypothetical, is also as much a reality none the less and is placed where that point must be standing still because every line running from that point in opposing directions is also in opposing directional spin the other or opposing side. As the rotating direction moves inwards, the rings holding Π will become smaller and smaller. The reducing of the radius r will eventually end at r^0 but the top does not end there because the top still then is Πr^0. The form we attach to the spin still applies as Π and the top finds directional contradicting change at a point that never moves because it can never move being in the centre where the spin direction ends at Π^0. It is the only aspect in the entire Universe that can be and still be motionless because it is not within the Universe. It is the centre of the Black hole because it is the centre of the Universe. It is 1^0 and only the centre of the Universe in singularity can have 1^0 However that point where the directional spin ends is the point where the actual spin does not takes place because if its immovability. It is at a point in singularity $k^0 = a^3 / T^2 k$. It is where space ends because motionless ness ends space there. The spinning is on the precise location where the point is not spinning because the Universe ends in its not spinning there.

That line running through the centre of the spinning top divides every possible side from the opposing side in innumerable points that are divided by angles and degrees. Moreover, it changes the future position of the point from the present and from the past as it redirects every point every time k moves to reposition in a location where T^2 ends. In the end it proves that both k and T^2 confirms a^3 just like Kepler stated before Newton interrupted with dishonesty.

Another huge factor that favours the use of Π as a measure in singularity progressing is that any expanding by any mathematically sympathising method will have to use Π since Π is the only route that a spot of no significance can develop into a dot that represents a Universe of development in waiting. Only by promoting through the measure of Π can all possible sides progress on equal terms in all directions simultaneously without bias. The development must progress by measure of equality to the smallest indication and that purpose only Π can serve.

The progress must be generated so that it can flow equally to all sides in all directions spontaneously where not one side will favour the growing process as such. Time in it's flow does form a bias but explaining that at this stage will also involve too much other concepts which I would rather leave to the books. There is only one way to permit such a flow and have a mathematically correct outcome and that would be using Π for such expanding. The use of Π would ensure that a dot rises from the developing that comes from the first spot. The dot would have to form a sphere and a sphere is Π in relation to seven. This bring as back to the Titius Bode concept where ten is one half of a sphere relating to seven points forming the sphere where the seven with singularity puts form to the double ten that totals (including singularity) 21.991 and that is the overbearing dominating issue.

When the cosmos came into motion, motion was not yet defined. When the cosmos brought about motion, the first motion was relevancies. Cold parted from hot. Eternity parted from infinity. Motion parted from motion absence. Infinity broke the laboriousness of eternity for the duration of infinity. The spot became and grew into the dot.

From what the spot was to what the dot now is might be just a mathematical implication of going from 1^0 to 1^1 but in reality that first motion was the creating of and establishing of an entire Universe which was with all possibilities that now is it. Never again can that much growth become a reality, although to us the growth is beyond what we ever can notice. But it is because the growth is so massive and we are so small that we are unable to notice such almighty growth. When the spot Π^0 became functional and established all relevancies possible, heat parted from cold as eternity parted from infinity.

The expansion was not clear motion but more a parting of relevancies where a centre formed a relevancy because the centre could not provide motion. Without being capable of motion, the centre established four points, which also served singularity. From the inverse square law we know that the centre doubled by producing the four points holding singularity. We have to presume there is a time line because the Universe has this as evidence. The fact that light travel from there to here and from here to there proves of such a time line because there is no distance in outer space except in the Newtonian's misconception they have about the cosmos. Any line shows direction and the direction implicates positions according to the line having dimensions in the Universe we have in space and time. The line brings in Pythagoras and Pythagoras implicates mathematics.

1^0 going 1^1 where $1^0 \Rightarrow 1^1$

If there were progress that developed from singularity in the form of the first spot and we are the evidence of such progress, then the mathematical conclusion must be that a line formed where the line developed two sides and we have the evidence of that still present in our Universe. That brought about that three markers formed in relation to one and by admitting to the law of Pythagoras we find that what formed was 3^2 in relation to singularity 1^0 that became 10 on the one hypotenuse and 10^2 on the other hypotenuse which forms the square of space. Therefore mathematically space has ten positions and material has seven.

From the three the four (2 on both sides of singularity allocating the cosmic divide) the two in square developed as a mathematical consequence and that brought about the five.

The five was duplicated as a response on the other side of the divide and having five as a result of $(2^2) + (1^2) = 5 \times 2 + 10$ we find that the square of space holds the value of ten in place.

But 5 is in pace because $3^2 + 4^2 = 5^2$. And $2 \times 5 = 10$ is in place because $3^2 + 1^2 = 10$. I just can't go in depth showing how the Universe formed mathematics as the Universe built what is within the Universe because that boo (if my memory serves me correctly) is about 1300 pages in all. There is a reason why some part of mathematics show how a line and a half circle and a triangle is equal and why $1 = 1 = 2$ and at the same time $1 \times 1 = 1$. It all has to do with building the Universe in stages of development.

We read from the way mathematics formed how the Universe formed because the Universe formed mathematics as it formed what is presented as a Universe. Before the Universe there was no mathematics so mathematics did not form the Universe but the Universe formed mathematics and above all this came about exactly as the Bible says it Happens. If you divert one glitch from how the Bible presents it the entire process goes astry, but mathematics form the key in the presentation of how it all came about.

At the very beginning there was a spot. How do we know that? The spot is still with us and holds a value of 1^0. This spot is in the centre of all spinning objects. Then time came into motion and 1^0 moved to 1^1.

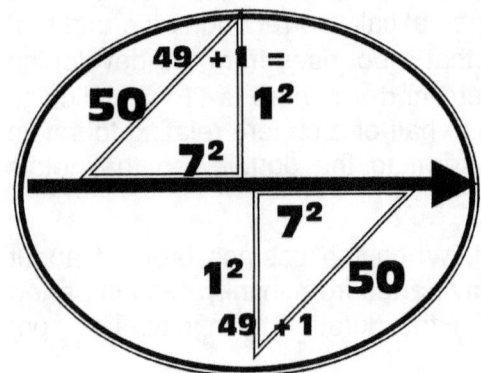

Since from where we stand we see 1^0 and 1^1 as being the same and therefore with the moving of time in the very beginning such moving must contribute to an increase of space. Therefore 1^0 and 1^1 has to have a difference where the one is 1 and the other is one point in singularity smaller making the 1^1 coming from infinity and rising into eternity 1^0, making infinity forever one point smaller than what we find as the value one will associate with the one we find as a measure in infinity.

Therefore we can judge that singularity combines to have a total of $1 + .991$, which then becomes 1.991 or whatever because the one going smaller is running into infinity and since infinity is one less than eternity we are in eternity 1^1 looking at infinity 1^0 which is one point reduced in infinity. However this moving from 1^0 to 1^1 involved 1^2 as well as 1^2 on the other side of the divide. As a result of the form the sphere holds, there is a centre connecting the sides and the centre holds singularity.

However, by presenting a centre where all lines cross on a point that cannot distinguish sides since that point has no individual sides, the centre holding singularity is inactive. Motion makes it active and the motion of space in time activates singularity to charge gravity that we find as a factor in the Coanda effect. That motion that establishes the purpose of space a^3 as a result of motion k through time T^2 was what Kepler presented as a formula. Gravity is $k^0 = a^3/T^2 k$ and to install k^0 the motion of space-time a^3/T^2 is required to complete

Producing singularity sets the divide because singularity splits the Universe apart in separate equal components that in combining form the duplication of singularity, being Π. Since the split brings about equality it, means that what is applying on this side must be applying on that side. When motion changes Π^3 to the proton $\Pi^0 \Pi^2$ it will happen on both sides of the divide of singularity Π^0. In effect it means that, that which combines the proton also parts the proton as it combines the proton because the proton becomes $\Pi^2 \Pi^0 \Pi^0 \Pi^2$ where the adding is the divide being Π^0.

The circle motion comes from space being dismissed by ending the motion and such ending of motion compromises the space it forms. By returning to where it is coming from it is ending the motion that began the space and as space is motion that is duplicating space motion returning is also motion that is ending which is destroying of space. THAT IS GRAVITY! Gravity is the balance between motion forming space by duplication space in motion forming time and time ending motion by destroying the space.

Gravity is about space duplicating space in relation to space destroying space and some particles are more prone to duplicate than destroy not withstanding mass or proton numbers. Those we call gasses. Then there are others that are more prone to destroy space that duplicate and those we refer to as metals. Then there are a few that destroy as much as the create space by duplicating and that we call fluids.

When heat is added to some elements we consider as solids, the heat helps with the duplicating of surrounding space and brings about a balance restoring the difference there are in the destroying of space and the re-establishing of space. The metals become liquid and the heat forming the liquid brings about an adding to the material where such material diminishes space.

By applying heat to materials that already favours the duplicating of space to the destroying of space the adding of heat will bring additional space as duplicated space and thus will produce more space to be duplicated and such elements will rise into the higher part of the atmosphere. When heat is added, the heat that is actually forming into space is what is added. It is the heat forming space by duplication that forms a shift in the balance because the heat forms space also as a process of duplication but without the contracting aspect of singularity renouncing space.

Although the "gasses" are the particles favouring to duplicate space they still hold the tendency to diminish space but when applying heat the balance will favour the duplicating much more because the heat transforms to space acting as space duplicating and adding to the overall duplicating of space. The element has a natural function of returning the space that the heat duplicated back to heat by removing the space destroyed and therefore returns the space to heat. In that way the particles do not only diminish the space they have but diminish space outside their claim.

Such particles we call heavy metals. As heat is added more space becomes available, to duplicate in relation the space they destroy and what space the elements diminish. With more space to duplicate the object will surge higher to a location in a position Earth will naturally duplicate as much space as the newly relocated material duplicates the surrounding space where more space naturally are. Cooling on the other hand reduces the space available for duplication by removing available heat that would have helped with the duplicating of the space and that then tips the balance in favour of the diminishing of space, which that element will also have. In that case the element will become a solid as the space duplication is more that the space. In this, we can trace the most important part of star evolution.

A star with a liquid centre has a lot of heat. Then the duplication of space-time by motion duplicates space much more than it dismisses space and destroys time. With a star in all the liquid as the sun is it is proof of a very young undeveloped and insignificant star with almost no influence sphere. As the star develops, the liquid ratio will shrink until it is only present in the centre. However, as the liquid diminishes the motion of the star deteriorate because the liquid represents the motion.

The star eventually becomes all-solid just before it removes the neutron from the atoms in the star and eventually places the proton action into outer space. Judging the layers we find evidence in this as the outer layers of stars are filled with elements which is highly prone to space duplicating as they have such a relation with heat. Hydrogen and helium stands very favourable to space producing and little in favour of space dismissing while iron, cobalt and copper is much prone to space dismissing. In all factors mass plays no part. It plays no part in the star performance or the star development

Time came from eternity, where it stood still in eternity, as it still does in singularity. Remember that little line I indicated previously that is running through the centre of the spinning top and that holds time apart, while it in itself is motionless, eternal. That being without motion represents eternity, a state of being timeless. As space moves away from that point the duration of time begin to rise and the further the extending the larger the time factor will become. Any fire represents many stages of time, where one part will burn quickly and the other slowly saying this means that science should recognise any fire on earth represents many conditions of heat where the smoke is a solid –gas, the flame is a fluid going on to be gas, the coal are a solid heat going to a liquid as it simmers producing photons which in itself is the dispensing of liquid heat turned to singularity particle dividing. The range that heat forms are so vast one may never appreciate it. Chernobyl showed the world how many forms of burning and burning injuries can come from radiation. Some heat was in the grass, undetectable until a bicyclist past and gained wounds from it killing the person a few weeks later.

The role that heat plays goes beyond (I suspect) what we may ever come to realise. Heat is an eternal fluid in relevancy to matter being the solid and space being the gas, but what is space to the one is solid to the next or a fluid to the other. For instance the proton is dimensionless to the neutron, yet it is fluid to singularity allowing heat to flow. The neutron is solid in appearance yet it is fluid allowing dimensions to concentrate. Because heat is a liquid in relevancies, it is the father of specific density, allowing heat to flow differently in different forms of matter. To realise the correctness of that, one may gauge the heat relation there are in the first ten elements, and how each stand so different from the next element. Consider neon and boron, where boron has many times the density of neon, yet only half the mass, or where oxygen has more mass than does lithium, yet lithium has a much different relation to heat than does oxygen.

Time is the motion of heat in space, and producing more motion, the duration of time will extend.

Every dot insignificantly small as it may be, is a part of another universe as much as it is part of the accumulative universe and every dot in the infinity holds singularity, which we translate as " nothing" being " darkness". There cannot be "nothing" just as much as there cannot be "darkness". There cannot be something big or small, but it into relevancy of perception, and then the relativity of perception becomes the question. There cannot be hot as much as there cannot be cold. The sun FREEZES hydrogen to a liquid at six and a half thousand degrees Celsius and universe boils over in the form of the Hubble constant at the temperature (we presume from our vantage point) at minus 273 degrees C. If we Humans cannot or will not abandon our human perception and our manly perspective, we may as well return to astrology for all its worth.

Every point in the infinity we may observe at is not merely part of the universe in not being "nothing", but is the point where the universe started representing singularity. It is the very first point where everything began so many eternities ago, because after all, how can we ever determine where the first point was, as they were very much equal and alike at the beginning. Every aspect of the universe started with the fundamental fact that no point in the universe can represent "nothing" as a number, because every aspect in the universe represents singularity in what ever form it may hold in that specific spot forming space-time. If man does not reach a conclusion where that conclusion is matching the universe and stop to match the universe with man (and man's incapability), we may all go back to caves and become starving hunter-gatherers again, because we will never find a way to progress to the ultimate understanding of the universe.

Experience taught us that there is a definite precise and secure corresponding that can only result from a direct connection in as much as the line being the same line. In that case the line must then come down to infinity and release from infinity as the same line still connecting to a point of re-bouncing to either side. The graphic cross is he result of singularity applying opposing sides but still maintaining connection through the application of Pythagoras that will connect and always bring about a direct relevancy. The graph does not hold zero because information derived as result of a relation prove a contact remaining when the line crosses singularity without applying detachment.

The Brainy Bunch holds the view that in a graph the line crossing amounts to breaking the zero mark. That cannot be the case. Now I wish to refer once again to one of the academic letters, which I already used as a referral. This is very shortly my theoretical proposal

Dear Peet,

Those on your list I know are in science education. You need someone in pure physics.

I am afraid that you will continue to get rejections if you do not relate your work to existing theories and previous work. While it is possible that a lay person hits on an insight that has been overlooked by academic trained in the field over many years, it is unlikely. We assume that work offering something new would be related to existing theories, either by building on top of them or by showing how and where they fall short. If you do not relate to existing work, it is repeatedly going to be dismissed as mind spin too easy to shoot down.

I am sure you understand.

Iben

PROVE ME TO BE INCORRECT IN ANYTHING I SAY! What you see in the picture above is gravity opposing gravity and by using the four cosmic pillars, which builds the Universe, that is how the cosmos comes about.

I say gravity and the sound barrier is the very same thing where it indicates opposing movement of the earth and some other object within the earth that moves apart from the earth. Allow me to very briefly explain the sound barrier as it was never explained before by applying relevancy.

Science in the present and for the last three centuries placed all their focus on material. In placing the focus on mass while mass does not exist as a cosmic entity, science has been running around like a chicken without a head and the truth eludes them even after three centuries of lying and corrupting science. You think that this is harsh words, read and you will see those in science deserves much more that just that. Those in science concentrates on the material filling the space but gravity is about the movement of the space and not of the object in the space, but of the space the object takes with as the object moves faster than the other space. That is why objects all fall equal because the space the objects fill is the same, unconditional of what forms the material. All objects fall at an equal pace without size or mass becoming a factor. That is why I introduce gravity by a very new set of principles that no one ever heard of. When an aircraft goes through the sound barrier, the object fills more space per time unit by going faster through space in time. The object then in accordance with the space it holds stretches because it holds more space than when it went slower or when it stood still. Since the aircraft is solid and does shrink a little but not much, it has to concentrate the space around the aircraft and by concentrating it reduces the space.

As the space concentrates through the movement of the aircraft a cloud appears around the aircraft because the water vapour in that space concentrates into a cloud that forms. All movement is part of the sound barrier because the sound barrier is the movement of space. However the sonic boom is confused with the sound barrier but the sonic boom is a small part of the entire sound barrier and only fills the centre spot or the middle in the sound barrier. If mass is anything then show me the role that mass plays in the sound barrier and how does mass conform into what we think of in terms of the sound barrier.

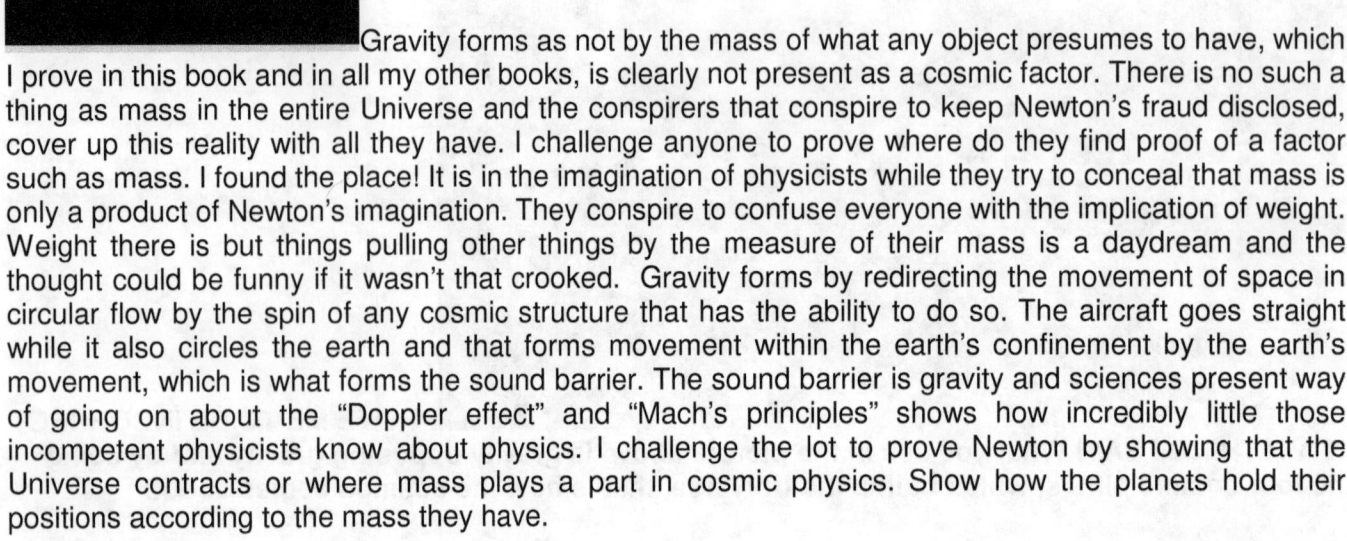

Gravity forms as not by the mass of what any object presumes to have, which I prove in this book and in all my other books, is clearly not present as a cosmic factor. There is no such a thing as mass in the entire Universe and the conspirers that conspire to keep Newton's fraud disclosed, cover up this reality with all they have. I challenge anyone to prove where do they find proof of a factor such as mass. I found the place! It is in the imagination of physicists while they try to conceal that mass is only a product of Newton's imagination. They conspire to confuse everyone with the implication of weight. Weight there is but things pulling other things by the measure of their mass is a daydream and the thought could be funny if it wasn't that crooked. Gravity forms by redirecting the movement of space in circular flow by the spin of any cosmic structure that has the ability to do so. The aircraft goes straight while it also circles the earth and that forms movement within the earth's confinement by the earth's movement, which is what forms the sound barrier. The sound barrier is gravity and sciences present way of going on about the "Doppler effect" and "Mach's principles" shows how incredibly little those incompetent physicists know about physics. I challenge the lot to prove Newton by showing that the Universe contracts or where mass plays a part in cosmic physics. Show how the planets hold their positions according to the mass they have.

All objects move straight while at the same time circle around some other object. The object moves in a straight line while the spin the object hold diverts the space in which the object is into a circle. This circle brings the value of gravity to Π. By diverting the space in which the object is as well as the space surrounding the object, gravity concentrates space into becoming denser by circling and also hotter.

I return to this explanation later on in this book and then bring the applicable my arguments.
Meet the Newtonian. This Newtonian says on the Internet no less that Master Newton and Master Galileo shared the same opinion. This Newtonian says that things fall equal as Galileo said and things fall by mass as Newton said. How much double standard are they allowed to maintain and never get questioned by the public?
When out Newtonian falls our Newtonian would fall at the speed as anything else would fall. This evidence we see on TV everyday in advertisements and on reality shows. We see blokes jump out of aeroplanes with rug sacks and tennis balls and the lot through each other with the balls and catch the balls while putting on the rug sacks and tasking off the rug sack all the while the lot is getting in and out of a car as they please.

We have all seen this. The car drops evenly paced with the rug sack, the humans and the tennis balls. Then the Newtonian tells the student that according to science everything is dragged down by gravity in accordance with mass. See the Newtonian fall. This Newtonian is the only one that falls by mass

The latest is that there is some little particle undetected as yet or as this goes to print (just to cover my arse in case they discover this particle in the meantime) scientifically named and called "graviton" that "pulls" other things that "pulls" back. Science rumours their profession that science only work on established facts and nothing but facts proven to the point of no contest could other wise be good enough.

This expression is so burned into the minds of everyone that science can say any rumour and because it is science everyone believes and never disputes. One good example is this Global Warming by carbon immersions. They take two truths and make one lie to convince people the lie is true. People take what they say as undisputed truths because science is renowned for only working with facts. What a lot of horseshit that expression is, that science only work with proven facts and that they use this idea to convince persons in the public that that is why the lot are atheists. They use this expression to convince people they are atheists because God is not a proven quantity but then they propagate the graviton as the measurable quantity, It is because they only work with facts! I can much easier prove God by using physics that they can prove mass by using physics.

Gravity forms by the dual directional movement of objects going according to the speed of movement. An astronomer will hang suspended in space and high above the earth if the circular movement is high enough to keep him there. If he spins slower he will start to drop. It is not gravitons grabbing him by pulling him closer to the earth. If his circular momentum can sustain the rotation his linear distance will maintain his orbit but as soon as the circular rotation become insufficient to maintain the orbit the linear distance will reduce and then the "gravitons" are called into action. How ridiculous can they get!

Gravity is the sound barrier because the sound barrier forms by four cosmic principles that form gravity and therefore gravity and the sound barrier are the same principles in conflict because of movement differences. The sound barrier applies by movement in space in space shared.

Everything I am about to show Newtonians say is nothing new. Newtonians say they are aware of everything that I show and then they turn around and simplify physics by putting the entirety of their physics down to the pulling power of mass that forms gravity. The sound barrier is gravity and to understand gravity is to understand the sound barrier. They say that I bring nothing new to the table and they say they apply every aspect I indicate that forms gravity. When I ask them to explain the sound barrier they use the Mach principle and Doppler's effect, which both dates from a time when no one was aware anything man-made could fly let alone go faster than sound can. Doppler made his contribution to science at a time when the train he measured was slower than a horse. The statistics he left to science applies to going as fast as a man on a horse could ride. That is a far cry from a jet breaking the cosmic boom. They can claim what they like because they never have to prove anything and they never have to listen since they know everything while their shortcomings makes science more the jesting of a buffoon. That annoys me to my guts because all of that comes down to the conspiracy whereby they whitewash their stupidity from the brainwashing they inherited from their predecessors and pass the brainwashing on to their students. Their ignominious stupidity makes me want to shout to the mountains in agony and unbelievable frustration. When a cloak of arrogant self-righteousness hides their incomprehensible stupidity under a cloud filled by their belief in self-importance it becomes hideously loathing.

In the earth's atmosphere no object can move slower than the earth does. If it moves slower than the earth does, it moves as fast as the earth does by receiving mass and being pinned onto the earth as being part of the earth. There is always movement because of the earth moves and all other movement goes in anticipation of moving above what the earth does. Movement is space that is duplicating filled space at a pace and in relation to other space within other unfilled space. That is why using parachutes slows the falling process of falling objects down because falling is a ratio between solids and liquids. This is proven in these photos about the sound barrier where the movement of the solid jet contracts the space of the liquid cloud. What happens in the photo is vivid proof of my theory and it proves the Coanda effect is what is forming gravity. The Coanda effect is a relation between material spinning and air or liquid compressing and that is what happens between the Earth and the atmosphere. This picture shows objects are always moving extraordinarily as they move in relation to the earth's atmosphere that always moves in relation to the earth's gravity compressing space into forming the atmosphere. But it also shows the movement of the aircraft compressing space in relation to the aircraft moving as well. That shows that all movement is gravity or time related. This shows objects moving is extraordinarily because everyone always forgets about that it is the earth that normally applies all of the movement while all else stands still in relevancy. There is always movement because of the earth moving and when anything moves above and beyond the movement the earth provides that then forms the <u>sound barrier</u> using a modified version of the **Titius Bode law**.

"GRAVITY IS DIVIDED IN TWO FACTORS, BEING **LINEAR DISPLACEMENT** (▮/▮) WHICH IS WHAT **NEWTON'S GRAVITY** IS AND,

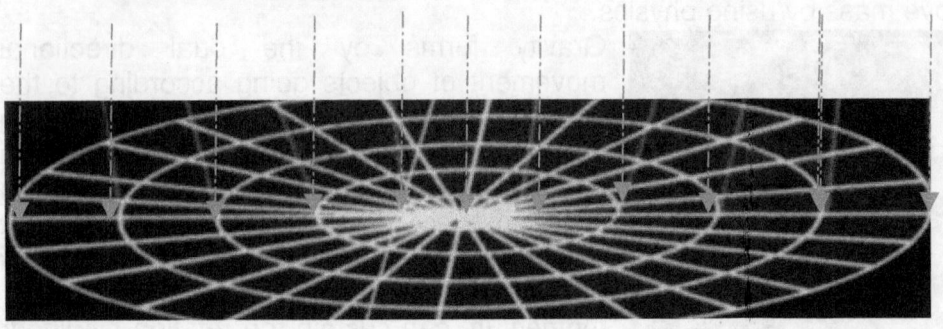
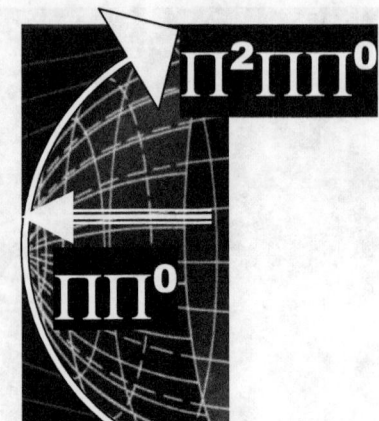

CIRCULAR DISPLACEMENT (▮³/▮) = ▮²

WHICH IS THE "GRAVITY" EINSTEIN RECOGNIZED

Those that are of the opinion that it is mass that "pulls" the object towards the earth might be a little wiser when reading the next part carefully as I explain the following:

Anything entering the earth from space it travels in a straight line. The mathematical formulating of this I do not provide at this stage but it is another "some simple algebraic relations**" as Prof Friedrich W. Hehl from Annalen Der Physics put it (and I also explain this a in the following chapter). But using the law of Pythagoras and putting 7 in relation to 10 provides the value of Π. The value of Π has two different dimensional values where the one is Π and the other is 21.991° /7°**

Any object entering from space and that does not have direct contact with the surface of the earth, encounters gravity in terms of being directionally diverted from going straight to turning towards the earth by 7°. This then is what provides the "pulling of mass" and the other is having contact with the surface of the earth, which then puts Π in terms with the centre of the earth at Π^0.
What becomes clear from these two illustrations is the following: Where there is a direct entry by the spacecraft, the time factor is not sufficient in duration.

This will not allow the time-duration needed for this structure of the aircraft to revaluate its space occupation an therefore the space factor of the spacecraft cannot adopt to its new position in space-time occupation as it has to relate to a new value in accordance with the value that is determined by the earth's concentration of space-time.

In the case of the second entry illustration, a lot more time lapse is allowed for the spacecraft to revalue its structural position in accordance to its new value of space-time occupation. This is the very reason why objects like "falling stars" burn out when they enter the earth's atmosphere.

Forget about mass forming gravity it is a fool's rhetoric. When an object holds a steady position to the earth while making a sound, the sound will go in all directions evenly. It will go left and right equally fast. The concept of gravity connects to roundness and to Π. Gravity is Π. Gravity is the movement in terms of Π by duplicating the position of Π per specific time units applying. Then the object making the sound moves left it will hasten the flow of sound in the direction it moves by moving towards the sound while it will increase the distance the sound has to travel by also moving away from the departing sound going to the right side. This means that gravity has two values one being linear and the other being circular and the linear affects the circular in movement as much as the circular affects the linear. This is not only connected to sound but is connected to everything applying to gravity as gravity. When an atom moves the size of the atom must shrink to compensate for the directional change of the orbit that decreases the electron's orbit and therefore decreases the size of the moving atom where atoms form an object.

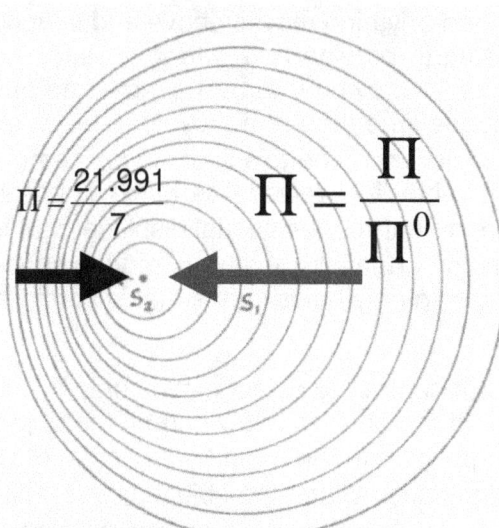

Coming toward the earth from space the object travels straight but the spin of the earth by 7° re-directs the object to follow an inclining line of 21.991 /7. At such a point the object moves towards the earth but the object cannot have weight yet, since the object does not connect directly with the earth. When the object touches the earth it relates to $\Pi\Pi^0$ and at such a point the object receives a value of mass by which it becomes a unit within the earth. By connecting to singularity in terms of $\Pi\Pi^0$ or in terms of $\Pi = 21.991/7$ puts the object in liquid or in solid.

Gravity has the value of Π. The value of Π has two measure where one is when the line gravity forms extends from singularity as $(\Pi^0\Pi)$ and when the line that gravity forms extends to singularity it is $21.991 \div 7$ because of the curve gravity associates with forming the value of Π. To explain all these factors I have to take the reader back to the point where the Universe started as spot that formed a dot. I have to take the reader back to where the point where the Universe was numbers holding no space, which is way before the Big Bang era where the atom broke the Universe into space that formed. In gravity there are the two values that gravity will apply. Gravity will link to the axis forming the centre around which everything spins. This gives the object a connection of 3 because it changes location in relation to the axis. Then the other gravity value connects to Π where Π always at all times connect to singularity within the centre at Π^0.

When $\Pi\Pi^0$ connects as the curve of the earth, this connection applies the object forms part of the solid and moves with the earth but also in accordance with the earth.

The Universe is contracting by $7(3\Pi^2)$
The Universe is expanding by $7(\Pi\Pi^2)$
Gravity is the inclining of material spinning in relation to 7° that is part of Π
This is because Π is 3.1416 and spinning in space is 3 which makes time expanding $0.1416 \times 7 = 0.991$ larger than 3. The concept might appear simple when told in this manner but the entire philosophy is so much more complex when studied overall.

By observing the rings around the planets we can identify all four pillars forming the composition thought of as gravity. We have the Lagrangian points that is five and I explain why it is 5, we have the Titius Bode law and I explain in detail the Titius Bode law, we find the Roche limit and I explain why it has the value of $\Pi^2/4$ and we can see from my explanation I have provided above why the Coanda effect accumulates all the principles forming gravity.

When something moves it is within the space the earth rotates. When anything moves above and beyond the earth's movement it excels as it exceeds the movement the earth provide and with it, it takes the space it holds in excess of the position the earth relates too. Thus by moving it form a space-time unit within the earth confinement but yet also out of and beyond the space-time the earth provides. Between Π/Π^0 and Π^2/Π^0, the body finds itself in its circular displacement, which is the earth's linear displacement value. When an object exceeds the earth's circular displacement value, it is refer to as Mach 1, or the speed of sound. Objects fall at a rate of $7(3\Pi^2) = 208$ km / h while the earth gravitational displacement is $7(\Pi\Pi^2) = 217$ km / h. because the earth moves faster than anything can fall it will always reduce space and have all surrounding space confined to the earth. The object falling only holds space that is confined to the earth.

When the object or aircraft stands motionless according to the earth centre the object moves only by the movement of the earth and therefore in cosmic terms the object and the earth is one whereby the object has weight and holds a value in mass. The object holds a relative position to the earth by the value of Π. This is because gravity then links to Π^0 and this holds regard to singularity. When the object moves while being connected to the earth by singularity the earth holds space at Π^3 while the object moves in terms of $\Pi\Pi^2$. The earth represents Π and the object moves by Π^2. When the object gets airborne the relation becomes $7(3\Pi^2)$ but I will not go into that. As the speed increases the relevancy of linking Π^2 breaks when the Roche limit is exceeded. The Roche limit is Π^2 or movement divided by the four quadrants of the circle making it $\Pi^2/4$. However while the aircraft shares the atmosphere one half is still within the earth making the Roche limit $\Pi^2/2$.

This will come into affect when the projectile reaches Mach 1. I choose to use the accepted term Mach 1 at this point as to limit any confusion that may arise from the new arguments. I have to stress the fact that the Doppler effect is, once again, merely a co-incidental but duly related by product. However, the Doppler effect, as such, plays no part in these phenomena, or in the outcome of the application. It must be seen in terms of cosmology and not from a human perspective. This is the principle applying when a newly formed star escapes the heat envelope within the centre of a galactica as a star is born. There now are two singularity factors within one space-time unit and the two goes into a battle of existing.

There is the same principle applying between stars sharing space-time. When two stars are at the Roche limit, the linear displacement reaches a value of one, and $\Pi^3/\Pi^2\Pi = \Pi^0 = 1$ is equal to singularity and in this we find space-time forming a value. This formula does not impress the most learned Physicists such as Professor Doctor Friedrich W. Hehl, which you will learn from a little further on in this book since Professor Doctor Friedrich W. Hehl thought this was to use his words "With a lot of words and some simple algebraic relations, there is no way to "explain" the world of physics." However notwithstanding Professor Doctor Friedrich W. Hehl not being impressed, on this rides the entire cosmos formed by gravity. The circular displacement reaches its full complement of half Π^2 which is the Roche limit.

I use the very same values I apply to gravity to show how the sound barrier works. This I can do because the sound barrier and the concept we think of as gravity is the very same thing that is what a Supernova star and a jet going through the sound barrier and the Tunguska event in 1908 was. It is the very same cosmic principles.

It is very quietly and not well promoted but we know that the moon and the earth are drifting apart. Nothing is coming closer together as Newton claimed in his principles.

Very quietly and not well promoted is the fact we know that the circumference of the earth or the circle around the earth is steadily enlarging thus Nothing is shrinking.

No where does any Newtonian ever put this tendency in relation to Newton's gravitational pulling principles of contracting in line with other cosmic accepted norms such as with everything expanding since the Big Bang, then what is shrinking.

Galileo proved that a large pendulum swings in time with a small pendulum and this Newtonians never dispute. Galileo proved that large things and objects fall at the same rate and by the same speed as small objects and that too they never dispute.

Yet they force students to repeat in examinations that objects fall by mass, which means that a big object and a small object falls differently because of mass, and yet they fall the same. If they

fall by mass the bigger object must fall faster then the smaller object does. As the objects of different mass fall equal as Galileo said it does, and then Newton is wrong in claiming objects fall under mass pulling mass.

…And yet you can go to the Internet and see how many Newtonian cheats calling them physicists announce in capital letters that Newton and Galileo had the very same principles in mind. You are going to read why I call Newtonian physicists the biggest cheats that walk the earth, they are more crooked than Bankers and Politicians are! If you know one personally please show him this book and see his reaction when he reads my accusations and that I ask to prove mass (not weight).

The fact every one misses is that any structure that is not part of the earth crust in the structure has own gravity that is stronger than that of the Earth which provides the independent individuality the structure with unique structural space has. The gravity of the Earth strive to incorporate all in the sphere the Earth has into the structure the Earth has and therefore the fact that the structure that forms a factor Newton called mass is not incorporated as yet, shows defiance and individuality in the first place.

Without trying to sound theatrical or over boastful there is no sense even in contemplating to read these books mentioned above before completing the reading of The Four e-books. The science introduced is altogether new. The approach introduced to science is altogether new. The complexity by which the introducing and explaining is done is altogether new. The introduced concept defining cosmology and physics is altogether new. If the correct introduction is not made, then the understanding of the total layout and arriving at the eventual conclusion is hardly possible. The explaining of the groundwork and the layout of the new proposed cosmic theory is introduced in the four books on offer called **The**

What you purchase is information and not ink put on paper, forming a copy of yet another science book. You purchase information never yet divulged, and I am not exaggerating. Look at what new knowledge that I uncover as I explain for the first time the following phenomena:
It would be preferable for a reader to be acquainted with some of the new ideas as explained in the e-books **…And also preferably in the designated order as recommended below because you will not be familiar with any of the work I deliver in the printed books if not reading the e-books first.**
Which is the work I propose that any person should read in the order that I suggest above, should any person be interested in finding out more about what was never yet revealed. Then, before venturing into the printed work, please make sure you have read the recommended e-book titles. What ever you know about cosmology or about physics, you will venture into a Universe you have never encountered before and the logic and the proof that I build on in the printed titles I derive from the proof I give in the e-books where I then do not repeat the proof I claim and which I give in the e-books. After completing the e-books you will be one of a few that truly know what gravity is!

The last letters I addressed to academics became books as I included several letters as to form one book where I later joined the letters to elaborate on the letters that I was writing with letters I already wrote and that I eventually combined to form a unit that became a compilation as a book. The letters grew into books and the books grew into a collection where some are intense and might seem to be rather complicated and others would be deemed as introduction material. The first eleven on the list would be rather introduction material by nature and if there would be interest it is advisable to start by reading those first eleven from the list.

But in my work I do some things no one should do. I break rules never broken by man in three hundred and fifty years or more. I cross a line that is forbidden to cross by any man not dead or insane. I go into the darkness of the foreboding chambers of insane madness and mental instability.

Then I go further and do the unspeakable, the act that proves insanity, the dead that vanquishes the force by distorting the gods presence in physics, becoming the reason why the cause of the future Earth will be destroyed… why what would have been will be no more because I unleashed the wroth of God onto man bringing punishment forth that man cannot endure, I sent my sole to physics hell just because I criticize Newton!

I challenge any person to bring proof about any part where any of my theory might be incorrect and furthermore I challenge any Academic in physics to prove that Newton's mass pulling mass is anything other than fraud. I charge any one to bring proof that the cosmos is contracting by the force of mass and that mass produce gravity as Newton advocated when he committed the biggest fraud of all times. If you are of the opinion it is ridiculous that I say that Newton committed fraud then answer the allegations I make and which I prove. I know every Newtonian is shouting for my scalp on hearing this allegation and I did consider the penalties I possibly face that goes with such allegations before I made the allegations.

Which is the work I propose that any person should read in the order that I suggest above, should any person be interested in finding out more about what was never yet revealed. Then, before venturing into the printed work, please make sure you have read the recommended e-book titles. There is a very good reason why I suggest you follow this path. Everything you are about to learn this approach you will encounter is in all aspects very new to science. What ever you know about cosmology or about physics, you will venture into a Universe you have never encountered before and the logic and the proof that I build on in the printed titles I derive from the proof I give in the e-books where I then do not repeat the proof I claim and which I give…but first you must get better informed, so treat yourself to astonishing information by downloading books introduced in www.singularetyrelevancy.com.

As it is important to realise the above it is just as important to realise that heat is another form of material and a separate form of material. The two developed on an equal basis and as a result of the other. The one produced to save the other and what the one produced saved the other .The one principle brought the incentive for motion while the other took the incentive by providing the motion. The one produced what the other captured and the one retained what the other delivered. Eventually the motion did not bring the required relief and another form had to be devised. By overheating and increasing space it counteracted overheating and by removing the expanded material and retaining it onto the contracting of the other, did the two form a synopses where by all received benefits in the form of cooling.

Only when further requirements developed, did the need arise for more to be made available. The first demand on motion asked no further changes because one change brought on satisfaction to all that suited all. The second was more general and on an ad hoc basis that was established to fit the need of individual places and not groupings in the broader perspective to fit individuals at large. At first the establishing of motion set a trend that brought on required results but afterwards the space required, in which to move became a demanding issue as the heat levels raging out of control. The heat had to be stored in space by becoming space to retain heat for later consumption. The number in ratio that produced the heat providing particles that offered to release their form in contribution to have those that retained form, do so to save those others retaining form. Those on offer became those ones that became the danger of destroying Creation instead of saving Creation.

There might even be some areas and regions in far off places in our modern day where an imbalance may evolve and some particles become unsuccessful to save those more successful. By going less successful, the singularity places a demand on another bringing about the command on space-time so that support can be accomplished to save singularity. Therefore by losing density, was gaining security to survive as part of a bigger relative. Density is the distributing of heat in specific relative space and by having less material in more space; the density is the offering for the common survival of the lot. The relevancy brings a contribution in whatever role to secure the survival of the lot in relations. No relevancy therefore can be "nothing" notwithstanding Newton's opinion about the matter as Newton had the opinion a relevancy acquired by rotation brought about an accumulation resulting in nothing.

It is the way the atom formed before the atom took on space-time. It is in the formation, that space-time relates to motion. We have some elements being quite massive but also lighter than air and others are quit light but as dense as they come. This can only be a contribution from the way the atom relates to heat, which make the atom volatile (movable) or dense (motionless). Those elements being volatile are also very movable and in that we find the role that such elements play in the star. Stars that are predominantly made up of hydrogen and helium with very slight support from the metallic inner core are those stars that duplicate by producing motion. However the point I wish to press is that mass and being massive and being heavy do not support the fact that some elements have more gravity they produce because their protons are more numerous than others. The fact that mass generates gravity is a myth.

One will find that whatever group one chooses there are gasses and there are solids. If mass was attracting mass then the strongest mass must be attracted to the strongest mass and the least mass must float in the air. F = G (M.m) r² hardly can even begin to explain the fact that there is a gas that is more massive than iron but floats in the breeze just as hydrogen which is the least massive element.

There is always a relevancy applying between that which spins art double 7 and that, which moves straight continuing at 10. The part spinning at seven we think of in terms of the diverting of direction it applies at 7. The liquid / gas holds the 10 factor.

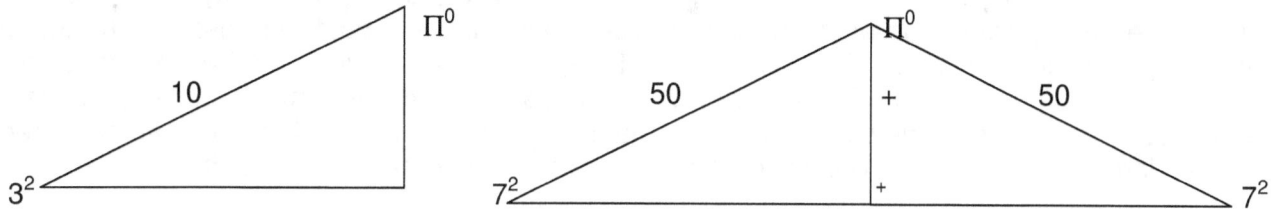

Without the application of specific heat, the object remains in the three directional moving of six possible directions. The value of space unoccupied therefore remains Π Π², as it was before the "Big Bang" event, whichever "Big Bang" you wish to refer to, because there were many. But space unoccupied holds time to the value of 10 to 1, and as the sketch of the triangles also indicated, holds space to Π. Therefore unoccupied heat holds the relation to space in applying 3 directions of influence ($3^2 + 1^2 = 10$) = ($a^3 = T^2 k$). Always part of this equation is the dual function of space in ($a^3 = T^2 k$) while at that very instant one has space-time. Therefore in space in time you have (10^2) = $7^2 + 7^2 + 1^2 + 1^2$. In the sphere we have the axis holding a value of 3 and the circle holds a value of 4. These are dots forming in relaxation to the one spot holding a point from where singularity advances.

We have the axis valued at 3 going square through movement of the linear motion $(3)^2$ and then we have the circular motion $(4)^2$ going square by the spin of the circle ring the direct opposing side. Then the equation of influence becomes $3^2 + 4^2 = 25$ where $\sqrt{25} = 5$ and doubling the 5 on both sides of the triangle will apply the factor of 5 x 2 = 10 that then is (10^2) = ($7^2 + 1^2 + 7^2 + 1^2$) = 50 on both sides is 10. The implication of this may not dawn on one the very instant of realizing, but to scientists, there is no greater shock than just that. To any application of movement, the factor will be in the realms of singularity where half a circle is equal to a triangle is equal to a straight line and the lot is equal to 180°. No fancy mathematical expressions have any value in singularity because singularity holds a value of 1.

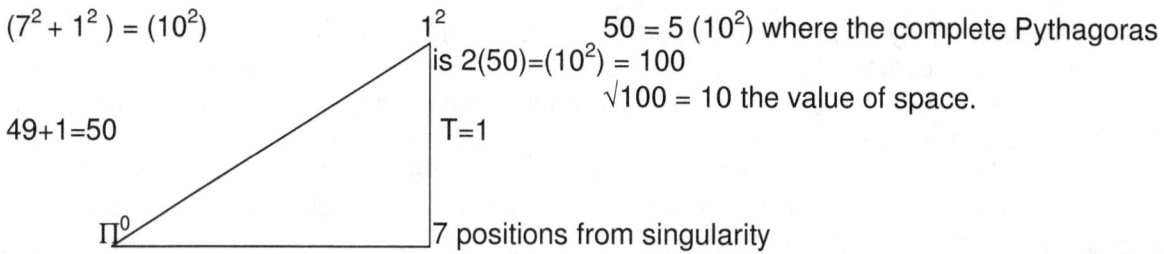

The fact of this comes as 49 plus one becomes 50 and that is in the three dimensions of space Π²/7 where 7 holds the relation to one and Π/7 again where 7 relates to one. At this point it is most important to remember that Pythagoras works on the application of the sum of the square of the two sides. When seven has a direction in the fourth dimension applied to it, the opposing dimension will be one and this applies in time relevancy, therefore the interchanging in time between infinity will place matter at 7^2 x 1 relating to circular and $7^2/1$ with 7^2 x 1. This makes 49 plus one (singularity) always being a factor of one. Space in time however, never can be a cube, it will always be a square with one side pointing the direction of time from time to the past (1) to time to the present (1) to time to the future (1).

The circle forming Π uses 7 to indicate the roundness of the circle but the 7 holds its roots deep within creation. It indicates how the Universe started because this is the way a star will start moving and it shows how as the infant star starts generating gravity just as the top starts to spin when it is thrown by life. Life can create nothing and that is true but life can mimic all laws in the Universe. Time is eternal movement and will be with us always. The line in infinity is still present while not being a part of the Universe. This line is always ready to be in place when the slightest movement orders it in place. Before the Universe was in place eternity and infinity was in perfect harmony and the line forming singularity validates this fact.

Before infinity parted from eternity, eternity met infinity on one spot as eternity came from the past (1) forming the present (2) to go onto the future (3) but also returned to come from the past which was the spot held by the future and this we find in the fact that the line forms 1 when not spinning but as soon as it evokes by spin, 3 points form even now. Then heat and cold differentiated values and space landed in between eternity and infinity. As eternity moved in relation to infinity but not forming a part of infinity any longer, eternity had to follow a path by never going away from infinity (3) and always returning to the point infinity holds but never lash onto the point again. With space parting the points, eternity had two points (the past and the future) before the partition came about and infinity held both the past and the future while infinity had the present as it still gas presently. By eternity also moving, the two points it held opposed each other (the past and the future) and since it moves, by the movement it became the square of the two because movement is the square and not a flat blanket-like surface with squares embroidered on it as Newtonian science depicts it by using grand mathematics to understand singularity.

Then we had two point holding eternity in place going square by movement to form 4 points serving eternity and infinity captured the first three points held by both and since eternity could not release from the two it had but had to duplicate what it had, eternity by movement became a circle captured by the line. With four points captured by the line of three points the circle coming about is eternally returning to infinity but never complying with infinity because if mismatching temperature or movement (3 against four). Material will always be colder than outer space. It is because material spin and outer space moves by expanding due to overheating.

This is where I start when I start to explain the first moment but I use a shipload more information to do explaining when I explain the star in the book I do so. I involve the four cosmic pillars to substantiate the claims I make because all four still work the very same way as it did at the beginning of the Universe. The three points serving one part of singularity combined with the four points serving singularity unites as seven to form a circle of either 3.1416 or 21.991÷7. The seven going to one is eternity matching infinity by movement. But since seven moves it are seven that have to produce gravity. How do I know all these facts, because we can see from the top it is still doing what it did the very first second. When time started infinity as well as eternity had altogether 3 positions, the past, the present and the future. It is still forming the very line in the centre of the top as it forms all lines in the centre of all things spinning. Then eternity parted from infinity when heat separated what is cold from what is hot and eternity formed one more point than before when it had the three points.

With infinity and eternity then jointly having 7 the cosmos came into rotation. In the aftermath post big Bang we now see the phase of cosmic development where the tow sectors try to unite and this brings along the contraction. When Π forms it does so on the grounds that 7 rotates. The circle forms by a change in direction by 7°. Every circle has opposing sides forming in relation to the axis line. If the topside goes rite then the bottom side has to the left. If the rite side goes down then the left side goes up. There is this double presence of a change in direction forming on both sides of the circle. The 7° move and by moving 7° goes square 7^2 and that is Pythagoras.

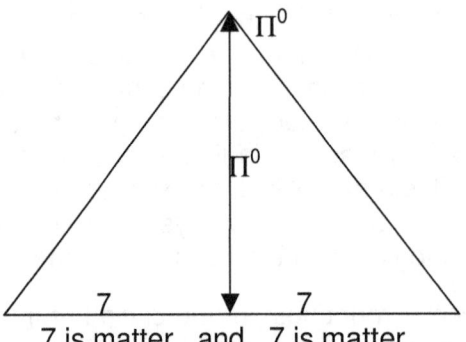

7 is matter   and   7 is matter.

They join space-time therefore the matter factor is the same. This is where one can visually see the one object, filling the space of the other object's atmosphere.
$$7 \times 7 = 7^2 = 49$$
That is matter Π^2 (time) times matter (49)+(1) = 50. This 50 forms space which then applies to both sides of the rotation of the solid being 7 that rotates.

As this is all under the law of Pythagoras the law will evidently place a square root to that value of 483,61 and therefore √50 +50 = 10. This leaves the space value of the Roche-limit, as it develops into the Titius Bode law giving them a shared value of 7 (matter) and 21,91 (space) the value of 21,991 / 7 = Π.
Then the relation becomes

$(\Pi^2+\Pi^2)$ $(\Pi^2\Pi)$ $(\Pi\Pi^2)$ $(\Pi^2\Pi^3)$ holding space (3) still outside. They therefore will share space and that sharing will continue till times end. We know by now that matter is 7, and space is 3. holding time to a relevancy in singularity of 1. Sharing the space means that 21,9 will become (10) space to the one side

1   to the instant position of time (k^0)
,99   lost to space depletion $\Pi^2/10$
7   the relation to matter.

Through that the Titius Bode law comes into affect of 10/7 or 7/10, depending on whether space or matter holds a superior position to time. From that stance, all objects will relate to one another by the value of $\Pi^2\Pi$ and seen in a whole sale total 7/10 or 10/7.

That means to become part of the neutron status of the earth, the object has to be space (21,991 or less) and prove to be matter (7) before the earth will accept it. If holding a position of less than 7°, the earth will discard it and if it is more than 21,991 the earth will find the relevancy to be higher than the space it holds in a neutron time. That places the object in a relation of $4\Pi^2 - \Pi^2$ (because it is not part of the earth) in a position acceptable matter holds (7) within the confinement of Π (21,991/7). That means the object is part of space (21,991) acting as matter (it holds an acceptable own proton structure) 7 relating to the earth in the position the earth allows of $3\Pi^2$.

That means to become part of the neutron status of the earth, the object has to be space (21,991 or less) and prove to be matter (7) before the earth will accept it. If holding a position of less than 7°, the earth will discard it and if it is more than 21,991 the earth will find the relevancy to be higher than the space it holds in a neutron time. That places the object in a relation of $4\Pi^2 - \Pi^2$ (because it is not part of the earth) in a position acceptable matter holds (7) within the confinement of Π (21,991/7). That means the object is part of space (22,991) acting as matter (it holds an acceptable own proton structure) 7 relating to the earth in the position the earth allows of $3\Pi^2$. With the space position of the matter in the parameters of 21,991 it relates to the Titius Bode law as a factor of one. The object has the space value of 21 (3 x 7), which shows the axle value turning, plus the space value of .1416, in that instant of time (7) complying to the earth's space (.1416 x 7) in reduction (Π^2) formulating .1416 x 7 = 0,991.

That makes the object complying with the full agreement as laid down by the Titius Bode law. The object is, no matter where it is, travelling at a rate of 7 ($3\Pi^2$) in the space of the earth (21,991). This will be agreeable to the parameters of the Titius Bode law as long as it remains within the space depleting "gravity" limits of less than Π^2. In accordance to the Lagrangian atom layout, anything less than 5Π is manageable and is in effect less than Π^2. When it exceeds 5Π it will start opposing the dimensional equilibrium space holds of 10Π, therefore it will (according to space) exceed the linear point of R/T, which is 10Π/2 (space going in a straight line).

Everything in the cosmos is moving, either by own individual accord, or under the influence of some other singularity dominance. In explaining we return to Pythagoras where the entire Universe with everything in it started.

It is the point forming the very centre that plays the part as the **controlling singularity** within the Universe I have named as **Infinity,** which is better known as the axis. It is where nothing can go smaller and anything within that point can never reduce. That point is where the entirety called the Universe begins and where everything holding substance begins. Once one accepts the fact of singularity being present in that location, that accepting of singularity then is contradicting all the things we know and we can measure and we recognise that point being present by merit of the fact that the point referred to is not being formed by any of the things we can recognise.

It is made up of everything we don't know and constitutes of everything we are unable to recognise or visualise. In that spot there is no space. That spot holds **Infinity.** In that space there can be no motion because there can be no space to have the motion within. It is formed as a line that is so small that our human reality by perception declare that point as not being there and the only reason why we know it is there is because of the results it left as an imprint of its not being there. We cannot detect it but notwithstanding our failure to note it we can recognise the dot on the merits of its absence and while in our Universe it is always absent, reality disallows the dot ever to be absent, because it is never absent. It cannot be absent. It cannot go absent but it can never be there where it should be in a place from where the third dimension forms and it is always present if I wish to locate it. It is **infinity** that can never go away. I named the other part of singularity forming space **eternity** because that area never become bigger, or become more or find an end to the outside. Whatever was and is and will ever be is locked in that space I named **eternity** and it is **eternity** that never ends because **eternity** can never end moving. What we think of, as expanding is never ending movement giving eternity the eternal motion that will go on forever.

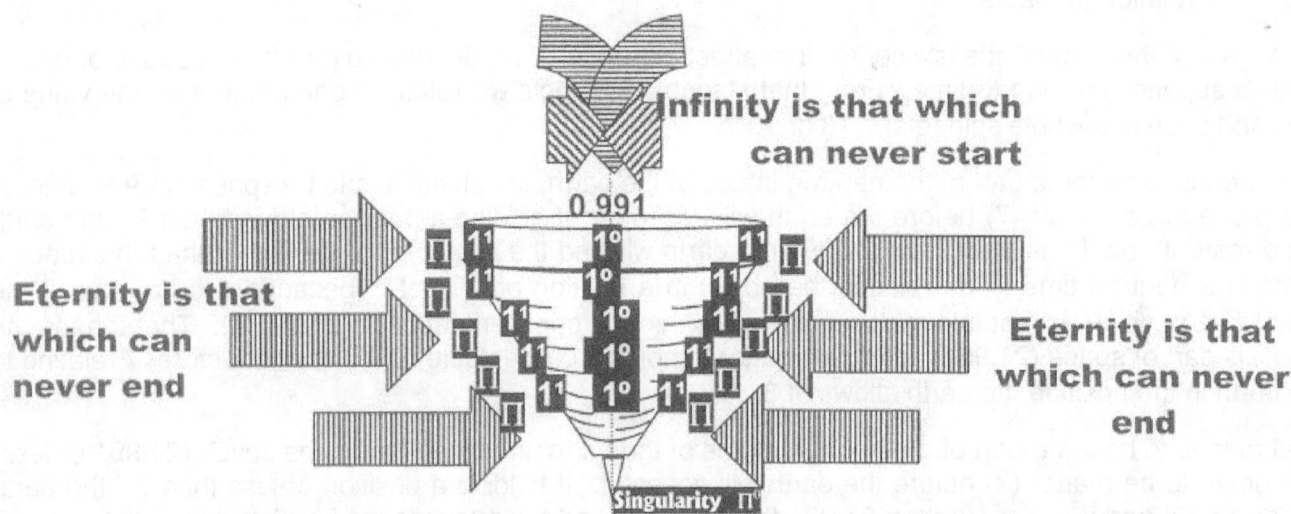

The line **k** coming from the centre (singularity k^0) forms by forming an initial spot Π^0 becoming the dot Πr^0. However, I went on to say that whatever the line used to start with has to continue in order to repeat the same that began the line. Therefore the line started with Π^0 and it has to continue with Π^0 until such a point, as it must end with Π. Whether the line is Π^0 or is r^0, or uses 1^0 the outcome all refers to singularity being used. By reducing the line we come to the end of the mathematical equation of the circle but the circle does not end there. When the top is in a state of motionlessness on own accord it is everything but motionless. The motion it adapts are synchronised with the earth in harmony with the solar system and according to the greater picture of the cosmos.

When an energy source not related to the cosmos called life intervenes and energises the tops motion, the singularity in that top suddenly jumps to life. By adopting a rotation energised to an unnatural state of energising because of life's intervention, the singularity of the top is not in charge but as it applies more and more energy, it will begin to find a means whereby it can escape and apply individual singularity as the top starts to separate from the singularity the earth holds. The singularity holding the earth would then allow the singularity of the top to rotate within a specific band where that a specific band of being active before the earth's singularity will start to destroy the singularity in rebellion.

The top on the other hand will try its outmost, when the singularity it holds gets by individual spin is too strong to remain be in domination of the earth's singularity. The motion of the top is an attempt to begin applying an individual singularity space-time defying and standing apart from the earth's gravity. That action we see as the top starts rotating in a manner where the top does not align with the earth's singularity. With the adding of spin, the time the top holds becomes unrelated to the time the earth holds and the top will start a campaign too escape from the singularity domination the earth has on the top. When the time or spin of the top exceeds the limits the earth places on the top, the top would emerge by trying to escape from constrains placed by the earth.

The view I represent at this point is known to science for almost as long as science knows mathematics. Not long after the law of Pythagoras was understood where Pythagoras introduced mathematics Eratosthenes of Syene made as big a discovery as Pythagoras did. But in the one instance the world took notice because the world could see and understand and the other instance the world disregarded the findings because the world did not see what the implications was. The same apply to aircraft flying and when the aircraft wishes to escape the earth's singularity hold it has to comply with the laws laid down by the earth.

The seven becomes as big a part of the concept as does Π as it all interacts.

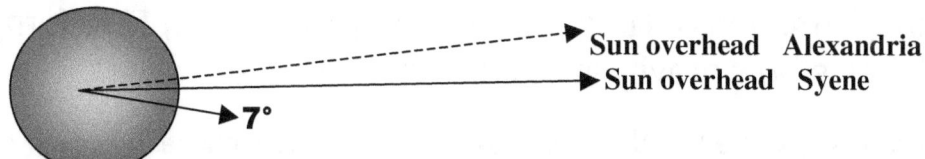

It took Eratosthenes of Syene (276 – 194 BC) a Greek astronomer who in the year 240 BC made a discovery that the earth has a profile of 7°. Since then no one ever did anything about it. When any singularity wishes to disconnect from the earths singularity, specific pre-calculated laws would have to comply to allow the lesser object to divorce from the larger object. I indicated how the dimensions of 10/7 and 7/10 interact to form (Π^2)

Matter is a product through the separation of space and time receiving the value of Π The original time and Π^2 as follows: By circling around a spinning solid the space contracts to form Π and Π^2. Gravity forms everywhere in the Universe by applying singularity. By dividing space into material (material spinning in space) and duplicating space by material spinning, the TITIUS BODE LAW forms a 7^0 deviation and 7 / 10 in conjunction with THE ROCHE PRINCIPLE OF $(\Pi/2)^2$

In my article to Annalen der physics I used 15 pages to explain this process of singularity applying. I received a rather cordial but sincere reply from the Editor of the magazine. When I placed an article in Annalen der physics Dear Prof Friedrich W. Hehl said in the e-mail he sent me that there is no way to "explain" the world of physics I am not going to go into detail how this works. On the other side of the Pythagoras's' triangle we have 1 going square.

That makes Pythagoras's' triangle 49 + 1 = 50 on the one side of the earth and the same on the other side of the earth. The total is 100 and the square is 10. That leaves the Titius Bode law with a value 7 (it forms part of the material of one body) and 10 in relation to the space. Then from the relation of 7/10 and 10 / 7 forming Π the Titius Bode law form Π^2 applying "With a lot of words and some simple algebraic relations" to quote Friedrich W. Hehl, Inst. Theor. Physics of Annalen Der Physics fame. This was simple algebraic relations but still it is science, is it not?

Since it involves singularity moving it calls for the law of Pythagoras to produce space. The law of Pythagoras is the triangle a^3 that is moving forward in singularity **k** by turning T^2. In singularity the 7 stands in for 7 points on the numerical line crossing over the line holding singularity or 1.

By moving 7 has to go square T^2 and that means 7 goes square 7^2 twice $7^2 + 7^2$ crossing the same divide $\Pi^0 = 1$.

Since all movement in singularity has to enforce the law of Pythagoras we have two triangles holding 7 dots moving across singularity. I don't want to get too involved by bringing in numerical outlays because then this can truly become complex.

The line has two opposing sides turning directionally against each other while turning with each other. By moving or turning this involves time duplicating space by the square Π^2 on both sides of the divide $\Pi^2+\Pi^2$ and using the same divide or the same axis or the same point serving singularity we have 7^0 crossing the same point in singularity Π^0.

There then is in this rotational movement 7^0 standing in for Π^2 on both sides of the divide $\Pi^2+\Pi^2$, which then is 7^2 on both sides of the divide 7^2+7^2.

The circle spins in duel directions. On the one side it would go left if on the other side it would go rite. The one side hold a directional change in singularity by $90°$. As it is going sideways it changes to going down.

When the 7 of material and the 10 of space no longer forms a gravitational alliance the 10 forming space will expand because the spin of 7 cannot retain gravitational structure.

For gravity to form the space value of 10 must always maintain the material value of 7 so that the formation of Π can form gravity by moving Π to form Π^2. This is gravity and that is how gravity is maintained

This produces a rite angle triangle of 90° and in it the law of Pythagoras produces direction changes. Since the square of the turn of the circle places by the spin and the direction change we have 7 holding a relation to 10 in space because it is space that has to carry the value of 10 when material circles by 7.

There is a connection between space surrounding the spherical circle turning and the sphere. The circle holds the value of 7 as in 7° and this we find from looking at singularity controlling the circle by movement.

The car straightens the curve of the earth when it stretches the 7° the earth presents as the curve around which it spins. The car leaves the curve of the earth and gets airlift. It then needs "wings" to force the car down by creating more gravity because gravity is the space that pushes down on the earth and the wings produces more space that pushes the car down on the ground. As soon as the car lifts into the air not touching the ground surface the relevancy changes to $7°(3\Pi^2) = 217$ km/h as the relevancy changes to $7°(3\Pi^2) = 207$ km/h.

This picture we have above shows why gravity forms as the space we call air pushes down onto the earth. The faster the car moves from one point to the next the more air it will generate to push it down. Gravity is the space above the earth that pushes down on the earth and that includes the space the car holds. But if the car exceeds the speed the earth retains to spin the car will exclude some space from the earth's space and take this air along for the ride. This lump of air holding the car will stretch the 7° it retains to form a circle and this will extend the position where Π forms. This will then begin to lift the car when the point holding Π as radius becomes wider as the position where 7° forms is further apart.

As Indicated space always holds a double 10 value in terms of 7

By spinning through 7° the changing of direction is taking 7 squared $(7°)^2$ and this gives 49 in total movement or space-time displacement or gravity applying or whatever you wish to call it. This is on one side of Pythagoras's' triangle

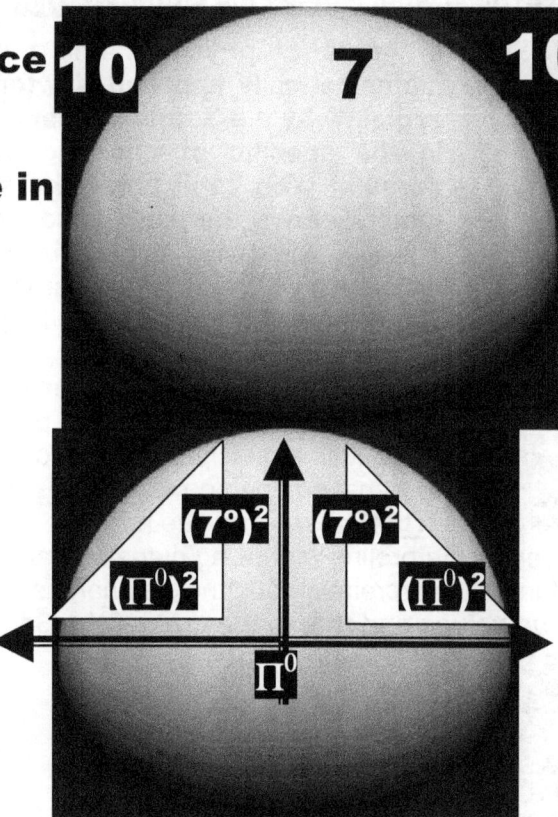

Material associates with 7 forming one value in 21.991 / 7 = Π

The spin re-aligns the centre with the same point on the surface of the earth but it then is at another point. This movement then takes $Π^0$ = 1 or singularity to $(Π^0)^2$ or also 1 which is singularity. I sent an article about singularity to Annalen der physics.

Matter in relation (part of) with the total dimension of space.

$\left(\dfrac{10}{7} \div \dfrac{7}{10}\right) = 2.04$ that then becomes $\dfrac{1.4285}{0.7} = 2.04$ Taking from both orbiting influences

SPACE DIVIDED INTO TIME

$\left(\dfrac{7}{10}\right) \div \left(\dfrac{10}{7}\right) = 0.49$ that then becomes $\dfrac{0.7}{1.4285} = 0.49$ Taking from both orbiting influences

SPACE MULTIPLIED WITH TIME

$\dfrac{7}{10} \div \dfrac{7}{10} = 1$ and $\dfrac{10}{7} \times \dfrac{7}{10} = 1$ Therefore not influencing change

THE PROCESS PARTED USING THE ROCHE PRINCIPLE

$\dfrac{10}{7}$

$\dfrac{7}{10}$

$\left(\dfrac{Π}{2}\right)^2$ $\dfrac{10}{7}$

$\left(\dfrac{Π}{2}\right)^2$ The Roche influence on Titius Bode

$2.04 \times \left(\dfrac{Π}{2}\right)^2 = 5.033$

$2.04 \times \left(\dfrac{Π}{2}\right)^2 = 5.033$

SPACE DIVIDED INTO TIME $5.033 + 5.033 = 10.066$ from both objects

$\dfrac{10}{7}$ $\dfrac{7}{10}$ $\left(\dfrac{7}{10}\right) \div \left(\dfrac{10}{7}\right) = 0.49$

$\left(\dfrac{10}{7} \div \dfrac{7}{10}\right) = .49$ $\left(\dfrac{10}{7} \div \dfrac{7}{10}\right) = .49$

$.49 + .49 = .98$

$.98 \times 10.066 = 9.8696 = Π^2$

TIME SPACE $= Π^2 = 9.8696$

TIME SPACE $= Π^2 = 9.8696 =$ Space and time in a dimensional implication

Ask your physics professor to explain what causes the Sound Barrier

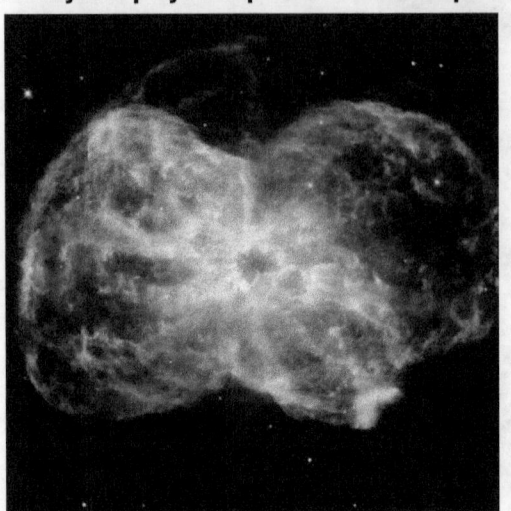

in terms of physics... and moreover in terms of using mathematically applying Newtonians physics. Ask them to explain why the Sound Barrier has nothing to do with speed in the specific but holding space during specific time duration. **You can't break the sonic boom at 31 thousand meters above the earth because there is too much space**. Therefore not being able to break the sonic boom still puts the movement within the sound barrier although there is no sound at that altitude. The space is too large to be able to go at Mach 1. They would not know that!

Don't mention the **Doppler effect** or the **Mach principle** because that gives names but provides no explaining substance to outline principle explanation. If you want to see principle explanation, then go to **The Absolute Relevancy of Singularity The Theses** and see how one should explain gravity leading a process to form the sound barrier by interpreting (not using outdated names) but provide mathematical formula interpretation of principle analysis.

Gravity has two values and that is the crux of the Sound Barrier.

I'll bet you Newtonian physics can't connect the sound barrier with the Supernova event...and I'll bet you it is because Newton's physics can't remotely resemble any meaningful explanation that could render a cosmic connecting principle between the sound barrier and a Supernova event.

But then they wish to redesign the Universe by creating Space Worms and Black Holes that no one will ever be able to disprove because they are busy with a fairy tail physics story. I say ... First prove the obvious and the down-to-earth.

But then again Newton's physics can't explain most information about many things because it is extremely limited. If you are not at all convinced or if you are little convinced about any flaws there are in Newtonian science then ask yourself if you know what these pictures have in common? **If you follow the tour and go where I guide you to go you will learn what it is that this first cosmic phenomenon**

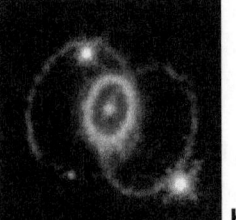

has got in common with this second cosmic phenomenon where

the one is stars bursting and the other is the sound barrier coming about when something is moving in relation to what the earth gravity allows.

Do you realise what you see in these pictures arte the same cosmic principles applying in such a manner that it is the very same thing happening in the cosmos and on earth?

The circles forming is gravity forming Π by the movement of Π² as gravity Π⁰ extends into wider circles by enlarging Π from connecting to one dot and then connecting up to 4Π⁰ in the circle.

 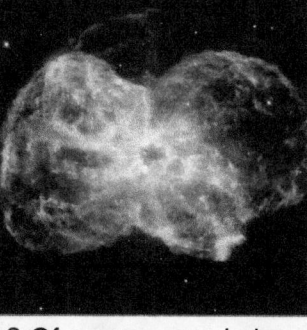 

Do you know what happens in these pictures? Of course you do know this, but do you understand this? Knowing this says little because it is the big anomaly science can't explain because Newtonian science will not allow any correct understanding of this. That is because Newtonian principles can't correctly explain this. This is one of the four cosmic phenomena that prove Newtonian philosophy about physics to be incorrect! This website will introduce you to singularity that is a totally new concept.

There is a relation between anything circling **T²** while going straight **k** to become space **a³**. Everything that forms anything in the Universe goes in a circle while going straight by the same margin but in ratio.

Please believe me when I say you have never even seen the true physics behind whet drives the Sound Barrier or gravity for that mater. To understand what law in physics drive this other than Newton's Dark

  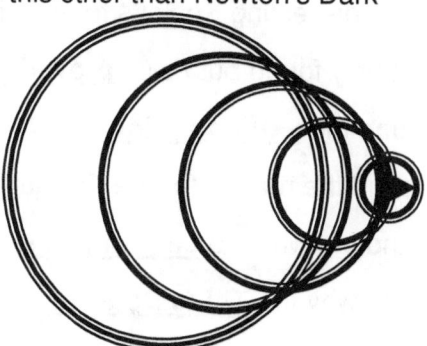

The explanations may seem plausible but put Newton into it and the entire concept falls apart. It is not in the explanation of what happens that the answer lies but in the explanation of why this happens that true physics lie…and that Newton's thinking cannot cope with….It is fine and well to give fancy sketches but explaining why and what principles is behind it Newton can never give.

The Sound Barrier is gravity. The Sound Barrier is the combination of four principles forming the Universe as the four are forming gravity. By explaining the Sound Barrier I explain gravity as I explain every aspect of law that forms the Universe we see. By not knowing how the Sound Barrier forms gravity and how gravity forms the Sound Barrier it is an admission of complete failure to understand physics. Do not confuse the sonic boom with the sound barrier. That would be to confuse a thunderstorm with a lightning bolt. That is what Newtonians do not understand and that serves then bad.

As the electron defines the atom the atom defines the space it holds in the Universe by moving straight ahead. The longer the movement of the atom is the smaller would the circle be that holds the atom. This is why starts with great gravitational inclination becomes SMALLER as their influence grows. If according to Kepler the formula building the Universe is **a³ = T²k** It is according to Kepler $\frac{a^3}{T^2} = k$ and $\frac{a^3}{k} = T^2$.

This puts forward motion directly in relation to circular motion and that is the crux carrying the Universe.

By looking at the Sound Barrier we can see what gravity is but in Newtonian terms the explanation of the sound barrier is non-existent. Everyone in science has an explanation about the sound barrier and every explanation underlines the absolute incompetence in science to form any type of explanation.

No one can explain the cloud forming around the aircraft as the aircraft fly well below the sonic boom. The cloud does not make sense when Newton's gravity is used to explain this phenomenon. No one in physics can use physics to explain the cloud forming. We can all see it is vapour condensing around the fuselage of the aircraft. Let them show the mathematical explanation why this happens. It is easy to see what happens but explain why that "what" happens goes beyond what Newton's physics can cater fort. If you throw out mass and forces physicist can offer little because then they know little.

To explain the Sound Barrier and therefore gravity I had to find **space-time** not as some magic entity Einstein sometimes referred to but something that is everywhere around us and which we are part of. by dissecting Kepler's formula in relation to **valuing singularity**

I not only found but I also **proved space-time** by **aligning space-time** with **gravity**

I found the **working principals** behind **gravity** as a cosmic occurrence.

I found the reason for the **Roche limit** and explaining the resulting of **gravity from that**.

I found out why the **Lagrangian system**, becomes **the building form** of the Universe.

I found why the **Titius Bode law** mathematically provides **the foundation of gravity**.

I am able to explain The Sound Barrier mathematically by not applying

Newton's physics.   Do you know what this pictures shows…?

You will read what exactly happens when the "sound barrier" is exceeded but also I will mathematically prove what gravity truly is. This I explain in simple understandable mathematical terminology. It is so simple even I can understand it. I make this remark because I am forever told I am too stupid to "understand Newton's physics" and therefore I am "unable to accept Newton's physics."

There is nothing to understand in Newton's physics because there is nothing to believe in Newton's physics. With one simple question I derail Newton's physics and that is prove tome there is mass, not weight but mass presented by the cosmos. The Sound barrier is gravity by implicating all four of the cosmic foundation phenomena that forms gravity.

There is always movement because of the earth moves and all other movement goes in anticipation of moving above what the earth does. Movement is space that is duplicating filled space at a pace and in relation to other space within other unfilled space. That is why using parachutes slows the falling process of falling objects down because falling is a ratio between solids and liquids.

This is proven in these photos about the sound barrier where the movement of the solid jet contracts the space of the liquid cloud. What happens in the photo is vivid proof of my theory and it proves the Coanda effect is what is forming gravity.

The Coanda effect is a relation between material spinning and air or liquid compressing and that is what happens between the Earth and the atmosphere. What this picture shows is that objects are always moving extraordinarily as they move in relation to the earth's atmosphere where everything that always moves independent of the earth's movement must move according to the earth when it moves in relation to the earth's gravity compressing space into forming the atmosphere. But it also shows the movement of the aircraft compressing space in relation to the aircraft moving as well.

The Titius Bode law

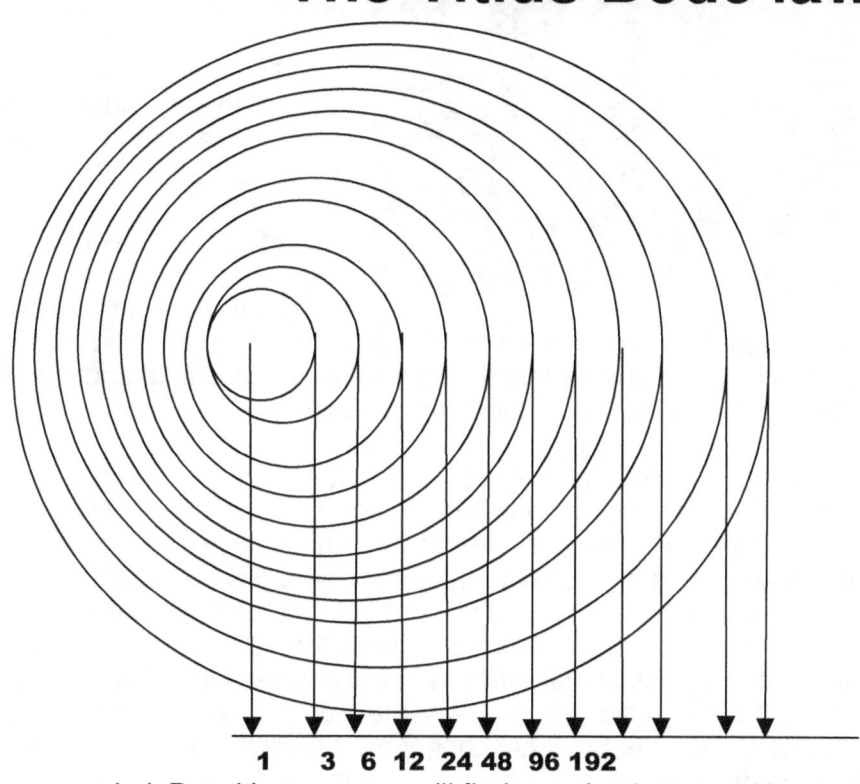

1 3 6 12 24 48 96 192

on the human mind. But either way you will find out why the "sound barrier" is a cosmic limit.

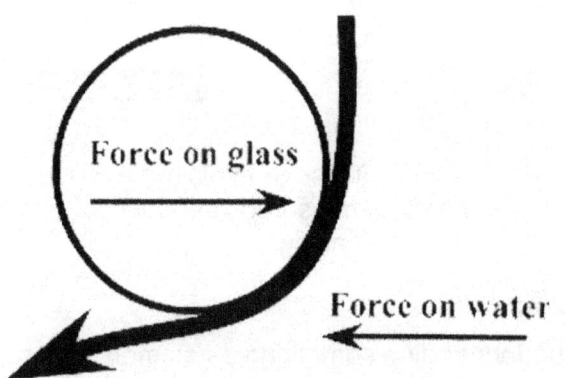

That shows that all movement is gravity or time related. This shows that when objects move then such movement of whatever is moving is moving extraordinarily because everyone always forgets about the fact that it is the earth that normally applies all of the movement while all else stands still in relevancy.

It is not my manner to speak ill of the brain dead or the dead by other means, but in the case of the Newtonian academics I am left

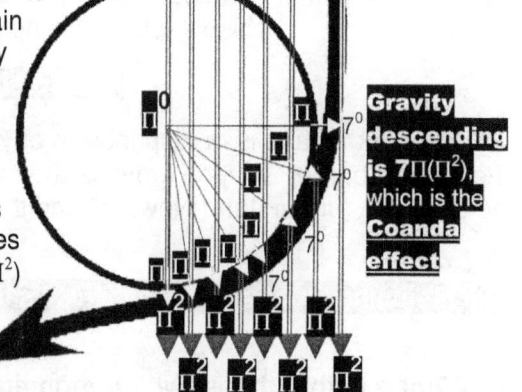

Gravity descending is $7\Pi(\Pi^2)$, which is the Coanda effect

with no option. Their forces haunt me to death and it is their forces and ghosts and witchcraft I have to fight. The lifting of a body comes quite natural when a certain speed is exceeded. By exceeding $7(3\Pi^2)$ the body will start to lift no matter what the mass is. A 747 Boeing of multi tonnage lifts off spontaneously at excess of that speed.

Newtonians are forever concerned with middle ages and with forces they can't explain but such forces and witches there are not, therefore they do not have to fear and can sleep well at night. In the sketch the circle portrays a glass and the arrow portrays running water. The Coanda effect is the water that does not drop straight down but follows the curvature of the glass.

The picture clearly shows the 7° inclination of gravity to the value of contracting $\Pi^0\Pi$. This is gravity!

This is the most vivid example of the Coanda effect and it is what gravity is! It is a whirl allowing the flow of liquid space around a solid centre in relation to the centre holding singularity or $\Pi^0\Pi$ as it contracts space into a denser liquid.

The Coanda effect is gravity and my explaining this statement is part of many other books in which I explain what gravity is. The Coanda effect shows how liquid attach to the solid by $7(3\Pi^2)$ and the solid attach to the liquid by a relevance value of $7(\Pi\Pi^2)$. That is gravity.

Should anyone require more or better explaining I would advise that person to purchase any of my books holding the title as an to go to http://www.singularityrelavancy.com/ and also go to www.questionablescience.net Flying object is under this gravity control of movement and it is this that has crafts fly and cars requiring down force by the aid of aerodynamic devices.

Gravity is defined as a force that is present in mass pulling mass and it is that entire idea that there is not evidence of. When I refer to gravity everyone grabs on a cultural notion of a concept they formed and in that concept they link the smallest part of the concept to the become and represent the overall gigantic principle and by knowing one line everyone has the opinion that anyone then is the absolute master on the idea of gravity. When I freeze any substance the substance contract to a liquid and with more cooling it contracts to a frozen state of ice. The gas expanded more than what the solid did because the gas is hotter than the solid is. When we form the opinion that the outer space expanded to the limits the idea springs to mind that outer space is freezing cold. When I say the Sun freezes hydrogen to a liquid because my eyes see the liquid squirting from the Sun I am dangerously mentally impaired since the Sun is blistering hot. Then through this culture my effort to say gravity is motion and motion is the cooling of an overheating and thus expanding Universe goes wasted. Every one has the opinion that where gravity is the strongest such as the case is on the Sun or the centre of the Earth, such a place is extremely hot and where gravity is least that place is unbearably cold.

No person ever could understand the sound barrier because no person ever could **enter singularity** and see **what applies when the Universe goes singular**.

You will read what exactly happens when the "sound barrier" is exceeded but also to explain the "sound barrier" I will mathematically have prove what gravity truly is by proving the measured value of gravity. This I explain in simple understandable mathematical terminology.

It is so simple even I can understand it.

This remark I make is in response to physics academics not understanding something as elementary as the "sound barrier" which I prove is the most basic principle of gravity but they are forever telling me I am too stupid to "understand Newton" and therefore because of my lack in intellect I question "**the validity of Newton**".

To them and to those that says this I say: **Download what I give free of charge and then start to get wise...**

What it is that these two phenomena has in common except sharing a Universe because he should know, after all he is the Newton-physics expert ... but I will bet you although being a physics expert, your professor will have no idea...mainly because he is the Newton-physics expert. This prevails because of singularity, which clashes with Newton head on.

As Kepler said $a^3 = k T^2$ and therefore $k^0 = a^3 / k T^2$ and therefore we have to find k^0. As a result of examining this proposition, I located two principle positions both holding singularity. The cosmos is made up of one type (1^0) that is in two categories where one type moves and the other type does not move. The one is a liquid and the other is a solid. **The condition for the presence of this singularity that forms everything, controls everything and is everything is the centralised to a centre singularity $k^0 = a^3 / (T^2 k)$ that forms by movement $T^2 = a^3 / k$ of space $a^3 = k T^2$ placed in relevancy $k = a^3 / T^2$ that is centrifugally going both ways $k^{-1} = T^2 / a^3$ thereof (Newton's 3rd law). This explains the Coanda effect and the Coanda effect is gravity and gravity "glues" the water to the glass by implementing Π to form singularity!** *What is in the Universe is spinning.* **The entirety of everything forming the Universe is spinning inside the Universe** and such spinning are always in the centre of one specific point, wherever such a point might be. In the **precise middle** of all **objects in rotation** is a precise centre where this pre-designated centre is dividing the object in rotation into sectors that will **start the spinning initiation** from that centre point. This is what Kepler's formula confirms in $a^3 = T^2 k$. By spinning, the one side is coming towards while the opposing side at that time is going away. Thus, the spinning object **will have a middle point**, a very specific **centre point that does not spin** and only holds Π as a specific value because within that centre being that small, no radius can apply. We have named this position or line the axis, but the true meaning of this line has eluded us since the concept was realised. This line that forms holds no space although it directs all the space that it controls by spin. When going toward the centre where the axis form at the very centre of rotation, the space on the one side has to end and the space at the other side has to begin with the line unable to hold space.

It is not you being glued or not being glued to the Earth that I discard when I question the mass idea.. It is the definition holding this whole idea that I do not share in the least. What the definition describes is magnets pulling and it is the total opposite of what I experience. Breaking the first millimetre of gravity clampdown is the easiest and not the most difficult. The difficulty increases as the radius grows and not as the radius decreases. When I say there is no gravity everyone thinks I say we all are going to fall off the Earth at random and with me thinking that way then it is obvious that I must be a nut. Everyone thinks of me as the clown acting mad when I say gravity is not to be found in nature. But I do not say we are not standing on the Earth. I do not say there is nothing that is keeping me glued to the earth. I say there is no attraction between two bodies by the force of the mass that is such doing is diminishing the radius parting the bodies by the inverse square law. I say there are a connection by motion between the centre of he body and the material surrounding the centre. This is what I say when I say there is no gravity.

The condition for the presence of this singularity that forms everything, controls everything and is everything is the centralised $Π^0 = Π^3 / (Π^2 Π)$ singularity that forms by movement $Π^2 = Π^3 / Π$ of space $Π^3 = ΠΠ^2$ in relevancy $Π = Π^3 / Π^2$ going both ways $Π^{-1} = Π^2 / Π^3$ thereof (Newton's 3rd law).

This explains the Coanda effect and **the Coanda effect is gravity** and gravity "glues" the water to the glass! The water forms a value of $Π^{-1} = Π^2 / Π^3$ while the glass forms a value of $Π = Π^3 / Π^2$ This process happens to all spinning things and as much as it happens to a piston connected to a crankshaft, just as much this will happen to an atom spinning an electron in a similar manner as the crankshaft is spinning holding a piston connected. This proves that gravity is the Coanda effect and in another book I prove that the Coanda effect has its origins in Π forming a value and that value forms gravity.

In order to understand physics applying in cosmology I had to start by dissecting the set-up forming pi. At this point I can introduce my theory on the ***Absolute Relevancy of Singularity*** At the point in the centre of the circle a line must start. In the beginning when I explained the way I figured how the line starts I said a lot of dots has to continue in order to form a line. It would be $1 + 1 + 1$ etc. because the line must form by holding singularity. After that point does mathematics begin but in the line that forms representing space as other all factors, then time holds 1. The line can only form when all the points forming the line have the value of 1 being 1^0. In that conclusion one realises something must separate singularity from all other factors because singularity hosts all other factors but is by own initiative $Π^0$. There are always a line of atoms made up of spinning subatomic particles also spinning holding dots $Π^0$ without space and the line runs through the dots not having space but still connecting by $Π^0$ or 1^1 forming a line $1^0 x 1^1$ up to Π.

Only when singularity meets the end value can the end value have Π where the final ring of the spinning circle forms Π. That will be the spot of origin forming the relevance in Π. That will hold the eternal spot…the smallest spot ever because all spots that ever can be were secured in a position in the centre of that spot that must continue as a line that forms. Because of the progress singularity follows from the single dimension singularity only allows mathematics a start at Π^0 progressing further onto Π^0 and from there the line is born as $\Pi^0\Pi^0\Pi^0$ and to $\Pi^0\Pi^0\Pi^0\Pi^0$ etc. where Π^0 then may form the concept and value of r. But the line starts at $\Pi^0 = r^0$.

This forms because cosmology is singularity based and the value is $\Pi\Pi^0$. This line $\Pi^0\Pi^0\Pi^0$ of singularity can only continue because every spinning atom preserves Π^0 in the very centre and since $\Pi^0 = \Pi^0 = \Pi^0$ the line is the same without finding conclusion except at the end where it forms mass at Π. At the point where Π forms, the movement Π^2 of the circle defines the space Π^3 of the circle and it confirms the centre Π^0 of the circle through the rotation. Let's call this the solid forming or if you wish, let's call it Kepler's singularity. After that singularity forms a line $\Pi^0 = \Pi^0 = \Pi^0$ where this forms another line again as Newton stipulated it by $\frac{dJ}{dt} = 1^0$. Let's call that the liquid singularity or Newton's singularity and the relevance of singularity having a solid base compared to the singularity holding a liquid base comes about by the movement of gravity.

There are only four cosmic principles on which the entire Universe rests and that makes every aspect of Creation connected to every other aspect of Creation. The limits we see in the Sound Barrier phenomenon is the very same principles we find in the Supernova event. The reasons why the star went supernova is the same reason why an aircraft goes sub sonic although the aircraft performing in exponentially less that what goes on in the Supernova. However the principles applying is the very same and is all the principles forming gravity.

The Sound Barrier is the gravitational connection with singularity forming a connection between the curve Π and singularity in the centre Π^0. Gravity is the relation between what moves and what is stationary, what forms a solid and what is part of the liquid, what is gravity bonded controlled substance and what is loosely connected not bonded and therefore uncontrolled liquid substance. There are two forms of material, one form is a cosmic liquid that in outer space becomes cosmic gas and the other is a solid.

From these conclusions I prove that gravity is the result of four cosmic phenomena interacting to form the value of Π which by movement becomes the value of gravity Π^2 and gravity is equal to cosmic time applying. In order to understand the development of the cosmos and moreover the start of the cosmos and the progress in the cosmos as the cosmos formed, one has to understand the measure of Π. One has to microscopically dissect the measure of Π to find the cosmos in measure. One has to understand where 7 fit in Π. The fact that Π is 7 at the bottom and that 7 relates to a double value of 10 is a key issue. It is behind Π that we will find the four phenomena, which I named the four pillars performing as gravity as they form gravity. It is by the actions of Π that the Universe develops. The Hubble expanding goes by implementing gravity as Π in the square through the four pillars on which gravity and time rests. It is behind Π we discover the meaning of singularity and how singularity forms the absolute and only building block as a form that forms the Universe. It is in Π we find the Cosmic Code unlocking the meaning of the Universe.

Time is centralised in $Π^0$ that forms Π as space's limit that becomes space by gravity being $Π^2$.

a^3 symbolises in a mathematical interpretation of implicating the three-dimensional space holding a specific centre in relation to another specific centre indicated by **k** that could apply to either centre points in question. This is always a straight-line **k** representing the position of the **controlling singularity** moving in a circle T^2. The space forming a^3 is a **positional validity** of the space indicated by $k^0 = a^3 / (T^2k)$.

T^2 is representing the circle that goes around the **governing singularity** k^0 that forms in relation to the line **k** in reference to the centre k^0 The space that forms holds the orbiting planet a^3 in direct circular contact with the space in relation to a very specific centre k^0 moving from point T_1 to T_2 in relation to a precisely placed centre k^0. The circle coming about from T^2 is the **controlling singularity** which is always a circle at the centre that is poisoned by the line **k** in relation to the centre k^0 and by forming a circle it holds reference to the **governing singularity.** Where **the governing singularity** is the centre of a spinning object such as the Earth, the centre of every atom holds **mutual singularity** that collectively puts a mutual value of all the atoms' singularity as a combined equal to the **governing singularity** and then the solar system will provides a **primary singularity**. The one would represent T^2 the other forms **k** that then produces the third singularity forming space a^3.

k is the space taken from the centre k^0 to the end of the line **k**. This line shows where the location is around which planet circles. The specific value about the centre is most important because from the specific centre gravity indicates a positional worth. The line forming **k** is pointing the circle or the **governing singularity** formed as a line that eventually forms a circle running from the centre k^0 to where the space a^3 is indicated.

The turning T^2 of any circle holding space a^3 is valid only if forming a reference **k** to a centre k^0. $k^0 = a^3 / (T^2k)$. This depicts a position a domineering singularity k^0 fills in relation to another point serving subordinate singularity **k**. There are always a dominant and a serving singularity interacting. If **k** indicates the centre of the Earth then T^2 rotates to form the **governing singularity** k^0 where then the centre of the Sun **k** will form the **controlling singularity.** When the Sun rotates, the Sun's centre k^0 forms the **governing singularity** giving the Earth in orbit **k** holds the **controlling singularity**. The measure of **k** is not a specific value but serves only as an indicator to which space rotates or applies by the space rotating in a circle.

This role of singularity being **controlling** or **governing** is playing part in movement of gravity forming and is very important when trying to understand the role that the four phenomena play in the forming of gravity. It is most important to understand what happens in the event of an object going through the "sound barrier" or when escaping from the Earth's atmosphere. Where the object is standing still holding a position that allows the object to have mass, the object is part of the Earth while the Earth has the **governing singularity** and the Sun has the **controlling singularity**. As soon as any object moves on Earth, the movement switches singularity by allowing the object to obtain the **governing singularity** while the Earth then fore fills the directional circular control in forming the **controlling singularity.** All four phenomena interacts in a manner forming this role where for instance in the solar system the Sun holds the **controlling singularity** and Milky Way forms the **governing singularity**

I have written twelve articles in which I explain the Titius Bode law, why it is in place, how does the Titius Bode law apply gravity, what keeps the Titius Bode law structurally in place and why is it in place as it is.

The gravity it should hold in distributing movement will bring along a certain amount of linear gravity to maintain the circular gravity position it wants to hold in the cosmic balance. The duel movement of gravity forming the allocated relevant position provides ratio of singularity Π accompanied by the same value but in movement by the square thereof Π^2 and repositioning the structure as a star.

I now explain gravity again and relate this explanation to the sound barrier functioning.

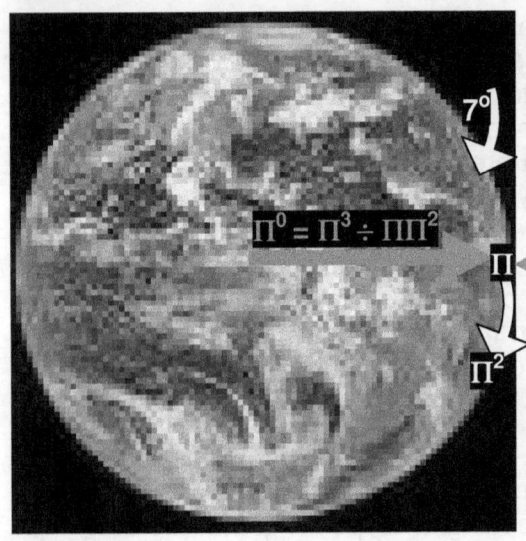

Gravity forms by association or relevancies that holds a value of coming towards the centre from the sky which then is 7(which is the curve of the earth) (Π associating with the curve of the earth) and Π^2 (associating with the moving of the earth) This is part of the Titius Bode law forming the measure of gravity which is not by mass but by forming Π.

$7\Pi\Pi^2$ x Π^0 to $4\Pi^0$

Then when a line is drawn from the centre of the earth another value comes in place referring to the centre of the earth Π^0 and the curve of the earth Π. The value of Π is then shared by both disciplines (association from the sky to the centre ($7\Pi\Pi^2$) and association from the centre to the sky ($\Pi\Pi^0$). In reality it is ($7\Pi\Pi^2$) but I am not explaining this.

If the body does not connect to the earth by way of forming a unit or forming "mass" the axel line of the earth comes into affect giving the ratio a value of 3. Then when not connecting to the earth surface the ratio becomes $7(3\Pi^2)$ x Π^0 going up to $4\Pi^0$ all depending on the sped or the height. This is the way gravity forms by forming a connecting association with $\Pi^0\Pi$.

This forming of gravity explains the Coanda effect because the Coanda effect puts air or liquid in relative movement with solids and this movement (but not the mass part) is how gravity forms.

The condensing of the vapour surrounding the jet is a prelude to the sonic boom and is just one link in the sound barrier where the sonic boom is just another link in the sound barrier process. The sonic boom IS NOT the sound barrier but is as much part of the entire concept as the beginning sound movement is part of the sound barrier. Waves in the sea are as much forming part of the sound barrier as winds howling or blowing is part of the sound barrier.

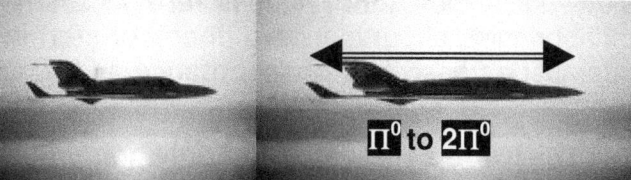

When any object moves it fill a certain space during a specific time period.

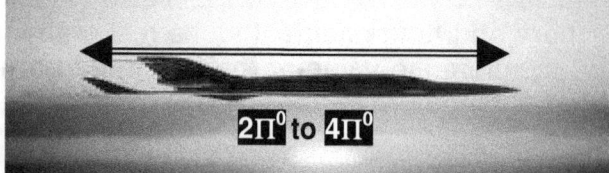

When an aircraft stands still by only moving with the earth it moves at Π^0 As it moves faster it fills more space up to where it fills from Π^0 to $2\Pi^0$ of sky space and as the speed increase this goes up to $2\Pi^0$ to $4\Pi^0$ The aircraft holds more space but does not fill more air and therefore the air has to compress

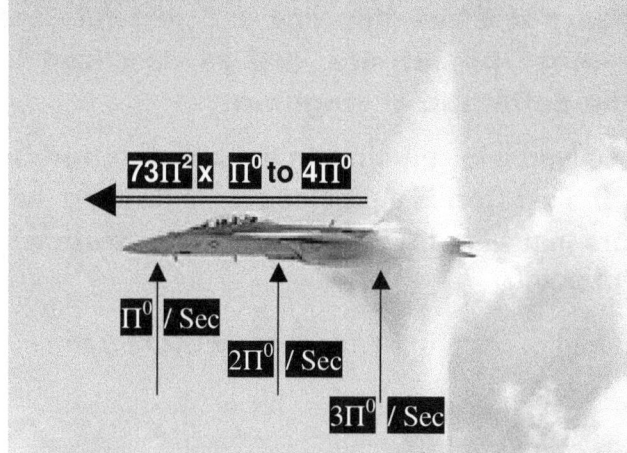
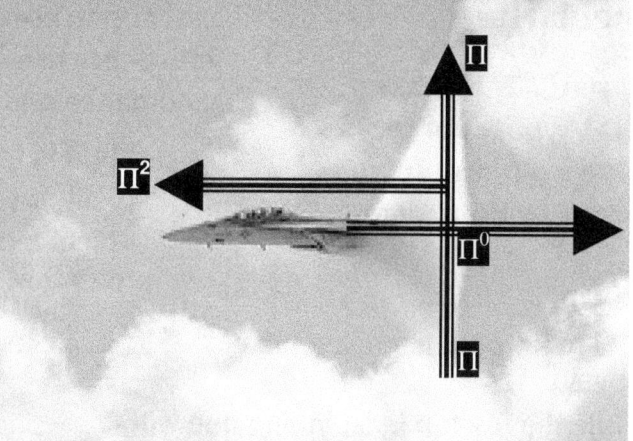

By going into space reduction as the fuselage occupies more space through extensive movement increase the space surrounding the moving aircraft condenses and this condensing forms the concentrated vapour. As soon as the fuselage begins to heat the heat surrounding the airplane pushes the vapour away again. The sonic boom is just one link in the chain of reactions we associate with the sound barrier and the sonic boom is more or less in the middle of the entire concept.

Flying at ($3\Pi^0$) three times the falling speed or the speed of gravity ($7 \times 3\Pi^2$) going in a sideways direction the body of the aircraft starts to expand into another gravity limit ($3\Pi^0$). However by the movement of the aircraft the body also has to shrink ($3\Pi^0$) and the result is the body of the craft cant accommodate conflicting behaviour.

Since the body of the aircraft that should expand contracts the space surrounding the aircraft also contacts in sympathy with the aircraft body contracting. That makes the aircraft becomes literally smaller and this condenses the space surrounding the aircraft. But then the heat coming from the contracting body of the aircraft pushes away the atmospheric air as the surrounding air of the aircraft heats up because of the intensity of the aircraft increasing the space it holds.

Download for free **The Absolute Relevancy of Singularity: Introducing The Sound Barrier**
http://www.lulu.com/content/e-book/the-absolute-relevancy-of-singularity-introducing-the-sound-barrier/7745052]

What I ask of readers is to beforehand forfeit the culture of Newtonian bias when reading this by paying attention to what I say and not about the degree in which I stray from mainstream science's thinking. This way the exercise will present many new ideas and explaining my new concept will become clear. There is so much to benefit from. Science has no idea what a Black Hole is while I can prove what a Black Hole is. I formulate mathematically what **"the sound barrier"** is. I prove what gravity is. By using the four cosmic phenomena, which is what the cosmos uses to form gravity, I show what **"the sound barrier"** is and I go much further than that. I show that gravity forms using the **Roche limit**, the **Lagrangian system**, the **Titius Bode law** and the **Coanda effect**.

I uncover these principles by placing Π within the formulating of gravity and when using Π I bring clarity to the misunderstood cosmic principles. The list of the unknowns I can then explain is almost endless. **Gravity forms by movement that establishes singularity initiating a circle in using Π.**

This is how the gyroscope works and how the ballistic bullet and the curb ball works. It is all to do with singularity forming points that in the next movement form space by turning. What is in the Universe is spinning. In the **precise middle** of all **objects in rotation** is a precise centre dividing the object in sectors that will **start the spinning initiation** from that centre point. Thus, the spinning object **will have a middle point**, a very specific **centre point that does not spin** and only holds Π as a specific value because no radius can apply. But also the one value such a line **cannot have is zero** because the line **is there and holds contact** to the rest of the material bringing about that **zero does not start any** line and therefore the **value of the line must be infinite**, just as described in **accordance** and by **the definition of singularity**

Locating and finding the presence of singularity

$k^0 = a^3 / T^2 k$ states that whatever is, is also spinning in order to be present.

Thus, the spinning object **will have a middle point,** a very specific **centre point that does not spin** and only holds Π as a specific value because no radius can apply. But also the one value such a line **cannot have is zero** because the line **is there and holds contact** to the rest of the material bringing about that **zero does not start any** line and therefore the **value of the line must be infinite**, just as described in accordance and by **the definition of singularity**

What is in the Universe is spinning. In the **precise middle** of all **objects in rotation** is a precise centre dividing the object in sectors that will **start the spinning initiation** from that centre point.

As I am introducing a very new idea, I wish to explain in better detail what I try to convey.

While the toy top is, spinning one will find singularity by moving the rotating line or radius progressively to the middle by reducing the length the line has from the edge to the middle. At one point all further reducing must end but the ending cannot include zero or nothing because the rest of the line still attach the rest of the top.

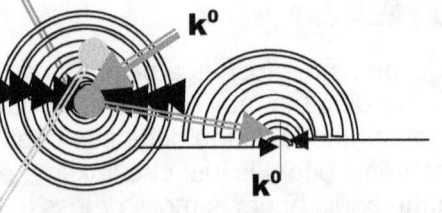

As the rotating direction moves inwards, the rings holding Π will become smaller and smaller. The reducing of the radius r will eventually end where the spin direction ends at $Π^0$. However that point where the directional spin ends is the point where the actual spin takes place. The spinning is on the precise location the point is not spinning.

That point albeit hypothetical, is also as much a reality none the less and is placed where that point **must be standing still** because every line **running from that point in opposing directions** is also **in opposing directional spin the other or opposing side.**

In considering the spinning motion in the fraction of time in the detailed instant every aspect of rotation will turn in every instant of change in time. Although the points had the same characteristics only one instant before, they oppose the characteristics it had just before and just after the very instant in which they are and to which they relate by similar points also in rotation. The fact of the graph proves my point in quarterly opposing dimensions and values.

When an object moves it form gravity. This principle goes according to Kepler's formula of $a^3 = T^2 k$ that mathematically translates to $k^0 = a^3 / T^2 k$. This indicates movement and that is what gravity is.

Where space is the least, which is in the centre of the circle, gravity is the strongest. The gravity located in the circle's space less centre holds not only the sphere together but all that is in the surrounding of the sphere outside the sphere as well. It is from there in a giro action that gravity bonds all atoms forming the structure of the sphere as one unit together in a unit as well as distributing a specific alliance in shape and form. How the atoms manage that we will get to in a while, but there is a law allowing for that to take place. In the centre of the earth space is least and therefore gravity is strongest. In the centre of the earth singularity connects to time forming there least space.

In the very centre of the sphere the form of the sphere dictates that the shape will relinquish space as the line run from the outside towards the very centre. With this natural state of affairs the sphere are naturally inclined to dismiss all space that it can form in the form as the sphere holds space inside and the form will finally be without dimension. All that I attribute to the line shrinking by reducing actually takes pace in every sphere as the diameter reduces to the centre. In the centre where the radius line goes single the form relinquish the three dimensional form it has inside. Being without dimension in the very centre means that at a point in the extreme centre of all spheres there are a point that holds singularity because this point with no space has a mathematical position although it is invisible since there is no sides to such a point to give that point any dimensions. The shape of the sphere is calculated by using the formula $4\Pi (r^3) / 3$. By reducing r to a point where r is r^0 singularity steps in because only the form remains as Π. Going even further we find that there then comes a point where Π goes singular Π^0. At that point absolute singularity is present but so is absolute gravity present at that point. When holding the strength of the shape of the sphere in mind as well as taking into account that all cosmos objects of importance is in the form of planets or stars and they are all in the form of a sphere, we therefore may contemplate that it is where gravity originate. We now only have to find the reason why gravity will hold a base in a space less ness as Einstein predicted. It is clear to be seen that gravity is in the centre of the sphere controlling from the centre everything that is outside the space less centre. We can reason with confidence that gravity is the strongest where space is the least. We can further reason that it is gravity that is holding the sphere in true form and since the sphere allow gravity the best working opportunity, gravity can form the sphere in as strong a shape and form as the sphere seems to have. From every point on the surface of the sphere is where that point connects with the other side of the surface of the sphere by a line that runs through the space less ness of such a centre of the sphere. Such a line also connect by an angle of 180^0 as well as 90^0 to six other lines running from top to bottom, right to left, and back to front, where all join and cross in the centre of the sphere. There are therefore six lines crossing and connecting by a centre from any given point on the surface of the sphere. Such points connects in total six surface points on each side of the sphere while they all support one another through the space less centre. In that absolute space less ness in the centre holding singularity we find gravity supporting and controlling all space within the sphere as well as space connected to the sphere. That is where gravity control and guide the space, which falls in the parameters as well as under the influence of the form of the sphere. In the gravity centre space goes singular meaning space becomes space less or flat.

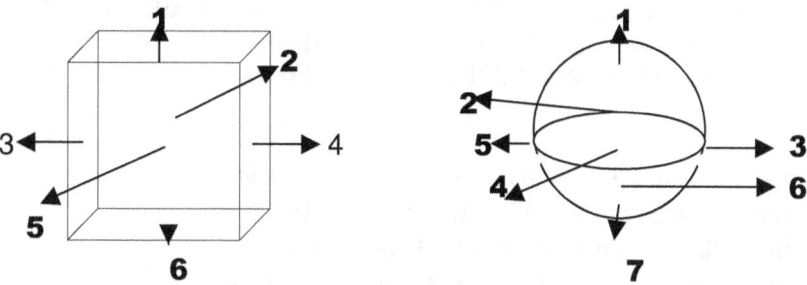

It is from the layout that the sphere uses as natural form that we are able to locate singularity. In the case of the sphere the material naturally reduces by measure of the radius becoming smaller to a point where the radius is r^0. At that point the line that will form the radius has gone single dimensional r^0 and that is equal to 1^0, which is singularity.

Also it is true that the entire form that is the sphere is controlled from a centre within the sphere. That centre holds the sphere in form and shape. Therefore the strong form is dictated from that space fewer centres where there is no space and no form left. The natural inclining is in the form of the sphere. It is part of the roundness that the overall shape of the sphere represents and this structural strength is carrying down to the very centre. Because the circle is forever reducing that reducing which is inherently part of the form of the sphere becomes a tool in distorting of space in the sphere and is eventually removing all forms of space from within the centre of the sphere.

The very centre ends up as having no space because of the reducing that continuous down to become the space less inner centre. The all roundness is the ingredient that forms the backbone of the absolute strength that the sphere has and that is the component that the sphere is so famous for. The form the sphere has allows the sphere to have a control that is coming from the centre deep inside the sphere where the space vanishes and being without space seems to keep the entire structure rigged. From the centre the sphere shape shows strength that the shape as tough as it is. How does it work in its most basic analyses?

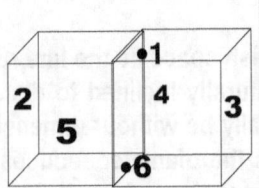

There is one more point in the sphere in the centre forming an addition in the sphere. That point holds gravity secure.

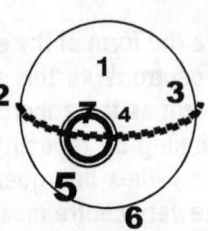

The cube has sides and the sides form a rather weak and flat surface that connects four corners. The flat surface produces a rather indifferent contact point with no special features on the surface. The corners connect to other sets of corners and those corners form a weak structure without any direct support coming from the other five sides. Without material to fill the body of the cube the cube has no direct connecting between any of the sides other than corners connecting at the edges of the sides.

Taking the vantage from the point the sphere is holding from the centre out into space there are ten points connecting to the centre. In that are the dimensions of singularity connecting to space where five connects to space in the second dimension of singularity, and five connects in the third dimension of singularity. On the other hand, the cube does show a very different characteristic, which involves only six sides (at least) connected.

Gravity is the strongest in all cosmic structures holding the form of the sphere and gravity controls all around from that very centre where space is the least, therefore the more material there is to generate motion within a star the more secure would the generated centre be in any star where such centre produces gravity. It is not the material but the motion the material accomplishes that becomes the factor of gravity. The smaller the star is as far as volumetric occupation goes, the stronger the gravity is that is coming from such a centre. The less the space there is the less the motion is and therefore the stronger and more deliberate the motion is evoking gravity. From the centre in the middle where space is absolutely at a premium the gravity grows stronger as it draws all material.

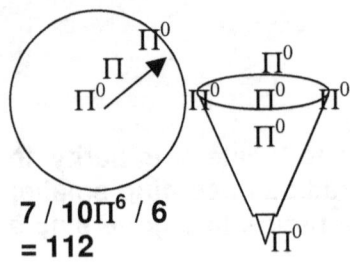

$7 / 10\Pi^6 / 6 = 112$

The spinning of Π^2 around the centre Π^0 establishes Π and Π is what produces the form gravity has. Still it is the relation or relevancy there is between the centre Π^0 and the spinning Π^0 that gives status to the form that Π represents. In out Universe we are accustomed to and are familiar to the rules we want to place seven points holding singularity to the centre holding singularity in a relation of $7/10 \Pi^6 / 6 = 112$. In that Universe everything less that a duplication ability to the value of 112 protons fit but only atoms to a maximum of 112 protons fit.

Kepler said $a^3 = T^2 k$ but that could also be $k = a^3/T^2$

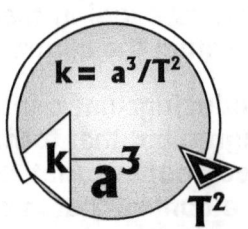

When translating Kepler's mathematical expression into a verbally spoken form of communication such as English we can see what Kepler said also read as $k = a^3/T^2$ where k is one point from a centre point that is space a^3 relating to time T^2. From a centre comes space-time

When drawing a line such a line then starts off with a dot serving the spot that holds all sides equal. That means the line serving as the future radius will be equal to the half circle which is then Π. The only aspect of the point that stands in for the end of the single line forming the radius of the circle is that we then mathematically reach the single dimension. We decreased the line to where a circle being Π formed on the single dimension. This dimension also hold the circle dividing line because from there the radius must once again generate a value and by such a gesture that the extending would form the circle that forms the sphere that eventually leads to the formation of particles. This leaves a problem to investigate.

With no line possible there had to be another dot that formed since the Universe has many dots that formed lines. But let us not to get confused and lost in the range of possible diversions but let us stick to two dots. One dot was next to the dot next to the dot, but as I said we stick to one dot next to the second dot. $M \times M / r^2$ is the first step gravity began with. That leaves us with a huge problem in as much as when $r = 0$ then $r^0 = 0$ and 0 dividing any value will leave 0 as the answer. If the particles were inseparable at the start it must bring about that gravity would not be forming since the distance will not permit any dividing. By allowing the distance separating the particles to be zero, the particles melt into a unit.

Again this is Mathematics and not my incoherency as some Academics dismissed my work. Let me run through the argument one more time because I have been insulted by Academics in the past telling me I am bending mathematic rules with my applying double values to try and produce some argument. The two particles formed by an inseparable unit separated by a sharing of a spot. We know that at least two spots formed because there are many more than just two that remained to become part of the visual Universe. Let us name the spots because that is what humans do best if they do not know what to do with what they have to do. Let us call the one em and the other one spot next to em we then call emtoo. Between em and emtoo there were nothing because em and emtoo were inseparable. By they're being inseparable we would naturally be inclined to think that the separation value should be nothing or at least zero. But putting zero in that place is a mathematical excluding procedure leaving future mathematics excluded. With m multiplying m_2 and then dividing $\div r$ with zero ($r=0$) such a procedure will leave the lot at zero and with that nothing is going nowhere. That means although we think the space between the two parts are nothing the non-existing space has to be at least one to be a future factor.

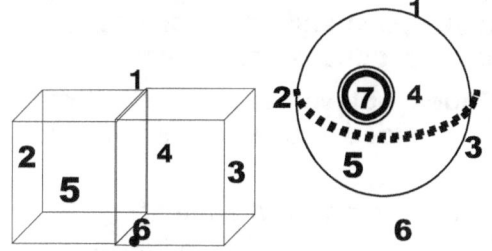

Kepler's formula also indicates that a sphere is within a cube that is holding a sphere

Taking the outlook from the point the sphere is holding from the centre out into space there are ten points connecting to the centre. In that are the dimensions of singularity connecting to space where five connects to space in the second dimension of singularity, and five connects in the third dimension of singularity. On the other hand, the cube does show a very different characteristic, which involves only six sides (at least) connected.

This point, which I now am referring to, is the point where Π a fully appreciated value while the diameter D still remains a dimensional factor of one. His is the dawn of the second dimension where space was there but space was sparsely shared in some cases. It is when Π^0 shifted to become Π for the very fist time.

The point without movement, the point holding singularity must have a value of Π being the eternal dot but since the dot has no dimension in having form the Π that indicates the dot must be Π^0. From such a point there has to be to the side of the centre point be a point where space do start. That point will then receive a diameter but that point will have form only in being a circle. In that point there is a shift from in relevance from Π to the centre Π^0 and for the first time it brought about two separate values for Π. Everytime an object displaces its alocated position in time it forms a new alocated position according to the point in singularity it holds. This is defined by time which is called gravity. It is the number of dots being replaced forming space during a specific period that forms time.

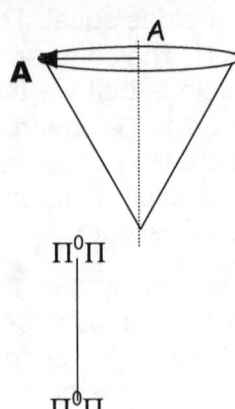

The moving of Π^0 to Π involved relegation and not motion as we consider motion. It was Π^0 getting a side and that is all. There was no true side but only a form that came into place. Singularity (A) received singularity (A) and no more of anything but the shift to comply with having a relevancy forming in relation to singularity. The dots had no sides, had no length or diameter. There was not measurable space or measurable time involved. The time could have been a micro, micro second as much a trillion millennium because time had no relevance. It was eternity interrupted by infinity, as it still is the case, however the line that eternity followed was no line because there was no space to hold the line. The line was momentarily interrupted by infinity, however with no one there, there was no one to notice. The lines were not lines but relations to sides being formed.

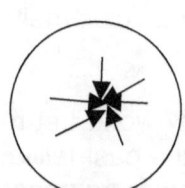

Inherent to the form the sphere offers, there is a specific location of singularity where the radius first goes single $r^0 = 1$ and then form goes into the realms of singularity $\Pi^0 r^0$. The cube also may have such a pint bur having such a point does not connect directly to six points located on the edges of the cube or any other form the is.

are always four seven points in a point of the one side of the singularity by

In relation to such a centre where $\Pi^0 r^0$ forms singularity there cubes related to such a centre where the centre is part of total representing the sphere. Every cube gas lost one side to sphere where the sphere takes control of form and removes cube. In relation to the time factor that is inherently part of the extending of singularity there are five sides connecting to four points standing related to singularity by the Π^0 factor and that gives 5 X 4 = 20. That is always directly in relation to seven points singularity offers.

This only applies wish time is the space in the cube apply to destroy then all factors falling into dimensions have.

in relation to time because time is the square or then if you flat to space being the cube. Time in the square draws flat and that is the principle behind the effort gravity can space. Once gravity destroys space and time goes square preventing time to remain has gone single leaving time singularity as well. This does not yet explain the three forming this well-known six-sided Universe we find we

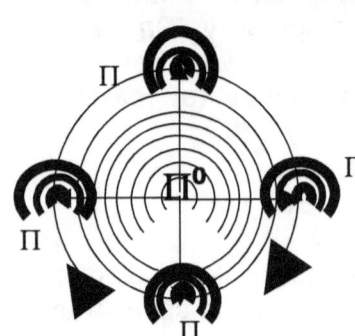

When singularity expanded for the first time ever and when heat parted from cold bringing about the Universe forming 1^0 to 1^1 from Π^0 to Π a relevancy was born and that relevancy grew into what we now have as a Universe

Gravity in the centre formed time Π^2 by dismissing while the four time positions started the cosmic trend of duplicating.

With every one of the four points taking form to the value of Π at a measure of $\Pi^2/2$ each brought about the Roche value of $\Pi^2/4$ in relation to the developing centre. One has to remember that the star of today takes on the characteristics of the form of that era.

In order to move while remaining on the ground it has to make one dot at moving sped if it makes one dot plus a half the object will lift from the earth. If the object makes two dots it must be at a certain height covering a certain space while rotating with the earth. To move according to the earth the relevancy of movement must be $7\Pi\Pi^2$ x Π^0 to 4Π What this shows is that at the surface of the earth the earth repeats singularity at a rate of Π^0 and at the next higher level it is 2 x Π^0 and as the speed or space to time ratio increases the marks in singularity Π^0 becomes more. This will go up to 5 x Π^0 but before 5 x Π^0 comes about the Roche limit of $\Pi^2/2$ and then ultimately $\Pi^2/4$ will form movement or space-time displacement boundaries. This is following the principle on which the gyroscope functions and the gyroscope is how gravity forms.

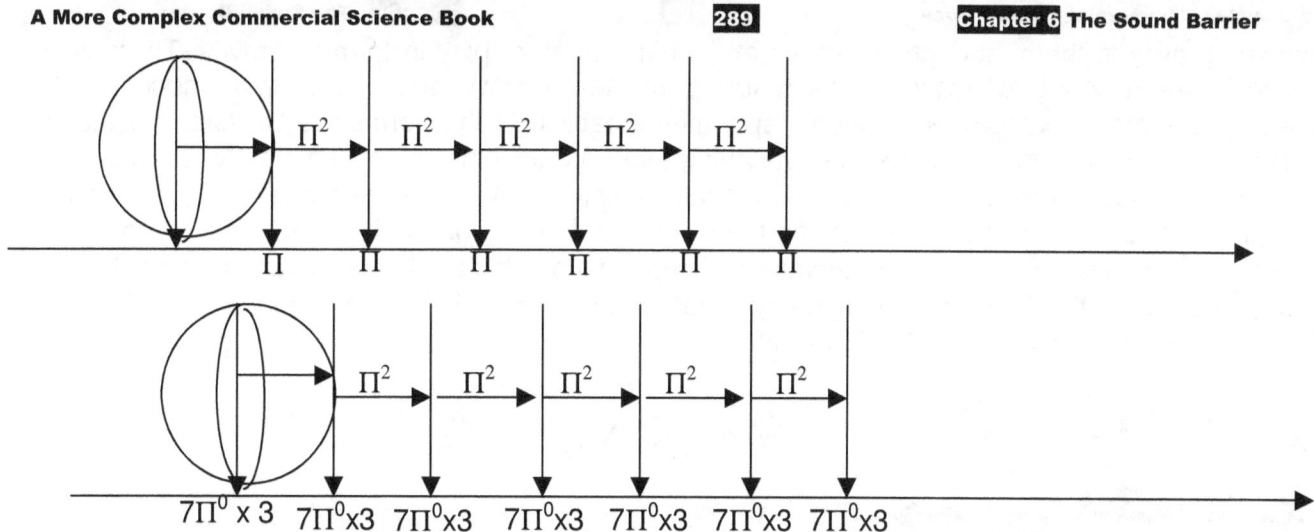

The object flying through the air has no contact with the earth and therefore loses the value of the curve at Π and connects with the axis forming singularity that is representing 3 instead of 3.1416 which is Π. The Π value holding .1416 more than 3 is the growth whereby space grows in relation to time.

When the ball spins and this spin is synchronised with the lines singularity holds as the next point to fill the ball will curve along the same dots as that which drives the gyroscope and which determines what dots must be filled in the next instant forming time that forms space.

I am not going out of my way to make this understandable because this fills a book on its own

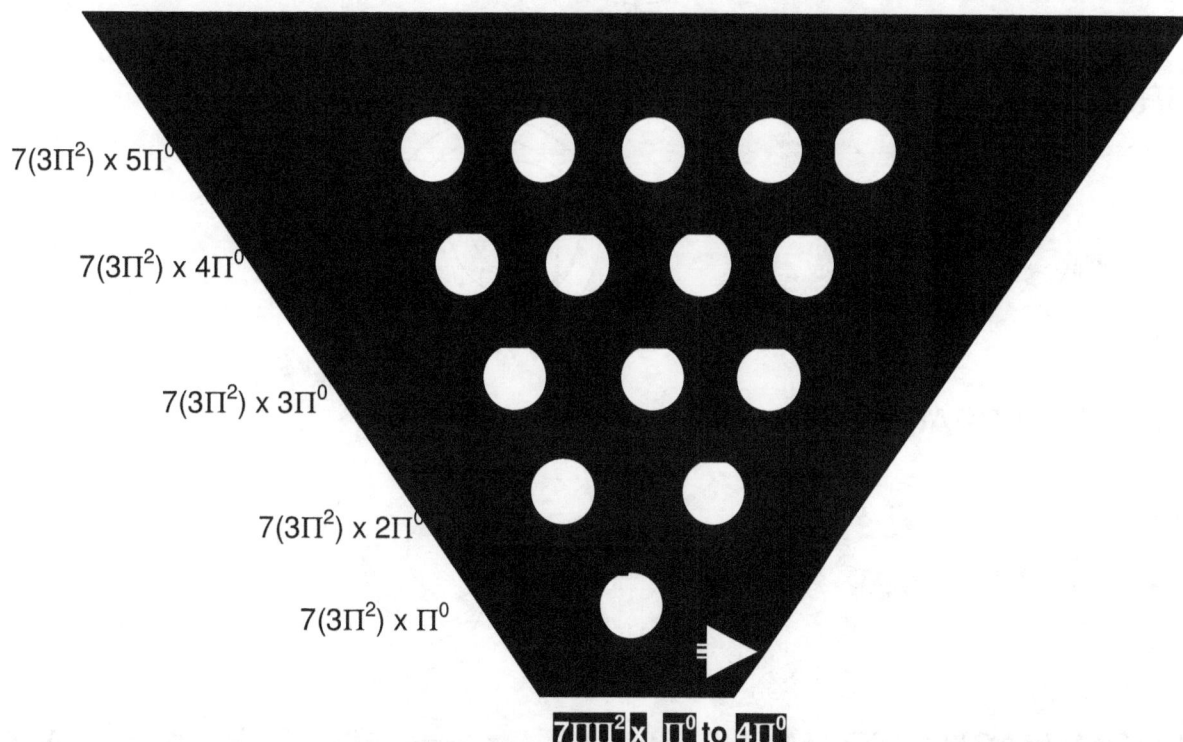

Moving with the earth has gravity at **7ΠΠ²xΠ⁰**. When going faster than the singularity that the earth provides puts the space in relation to movement or space-time displacement at **7ΠΠ²x1.1Π⁰** or at **7ΠΠ²x1.5Π⁰** then when the space required is more because of excessive movement in relation to the earth's gravity **7ΠΠ²x1.5Π⁰** it requires more space in relation to time. Whether the displacement is in relation to the height of the circle that holds $\Pi\Pi^2$ or whether the movement is forced to maintain **7ΠΠ²x Π⁰** at the height the movement through which the object goes will take on 1 or two or three spots per displacement

Because of the limit the Lagrangian law places on space the object moving is firstly capped by the limit at $\Pi^2/2$ and then later at $\Pi^2/4$, but this gets very technical and I have books explaining this very broadly and much better too.

I show why gravity is there, how gravity forms and what role stars play in forming gravity. There is no difference between how gravity and electricity forms and that I prove mathematically by decoding the cosmos. I prove mathematically when atoms spin they establish Π that forms the Universe. Whatever forms gravity has to link closely to Π since everything that has anything to do with gravity forms a circle that is Π by the value of the square radius. If mass has anything to do with generating gravity, then mass has to apply Π or otherwise mass has nothing to do with the forming of gravity. Everything using gravity forms a circle of sorts, which forms the curvature of space-time, which is Π and which curves light. The way the planets orbit the Sun and how stars spin has all to do with Π. In spinning in a circle, Π forms gravity as a centrifugal force that condenses space.

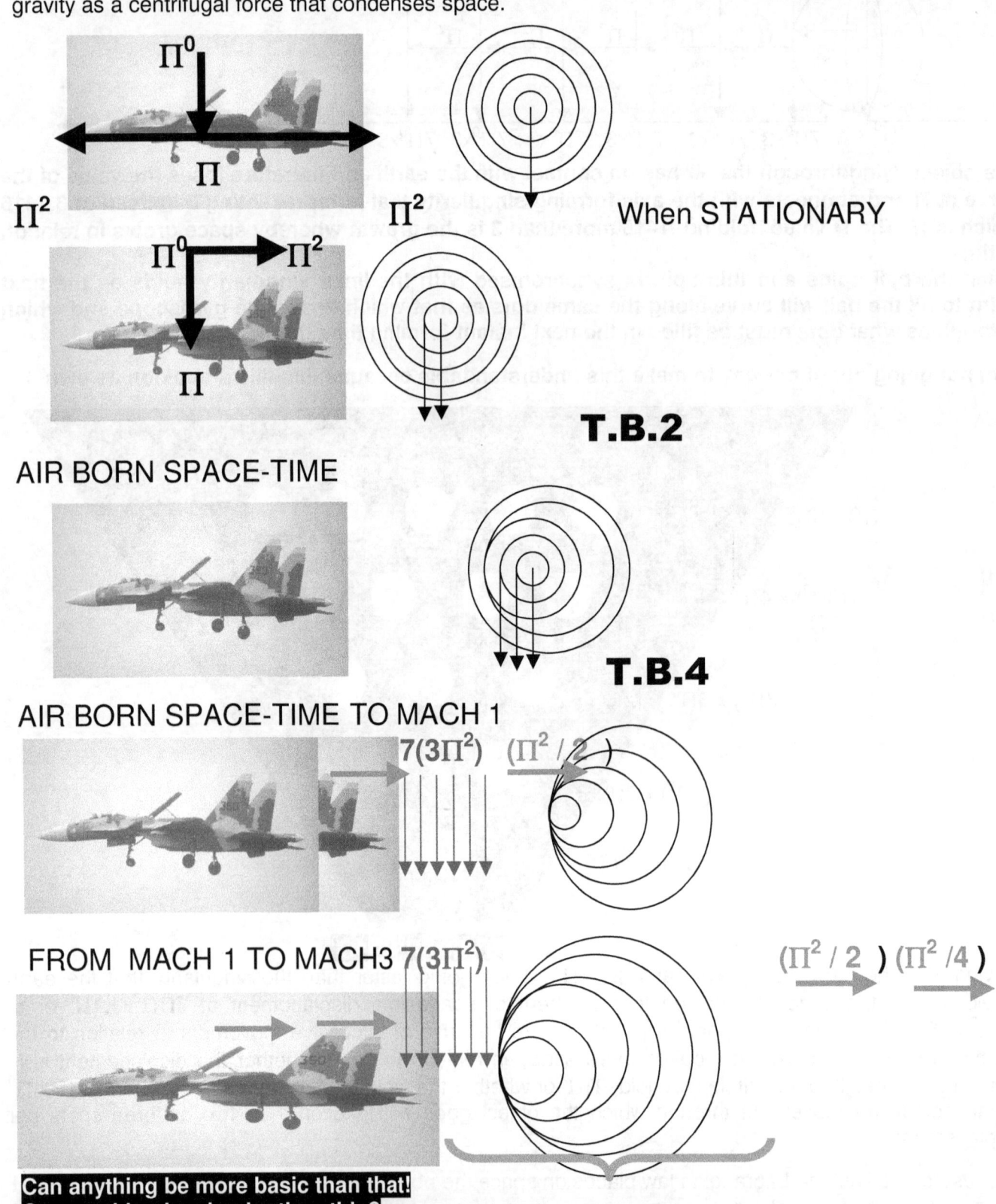

GRAVITY WORKING AS THE CONDA EFFECT

THE MACH PRINCIPLE ILLUSTRATED THROUGH THE TITUS-BODE PRINCIPLE

STATIONARY **GRAVITY** UP TO FREE FALLING

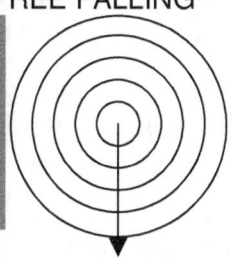

$$7(3\Pi^2) \times \Pi^0 - 2\Pi^0$$

Titius Bode law stage 1

MOVING **GRAVITY** UP TO LIMIT OF FREE FALLING

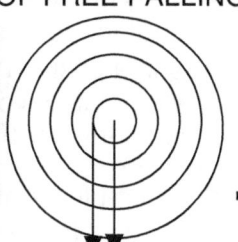

$$7(3\Pi^2) \times 3\Pi^0 - 4\Pi^0$$

Titius Bode law stage 2

AIR BORN SPACE-TIME MOVING SPACE-TIME EXCEEDING FREE FALLING

$$7(3\Pi^2) \times 4\Pi^0 - 4.5\Pi^0$$

Titius Bode law stage 3

AIR BORN SPACE-TIME TO MACH 1 ENTERING THE ROCHE LIMIT

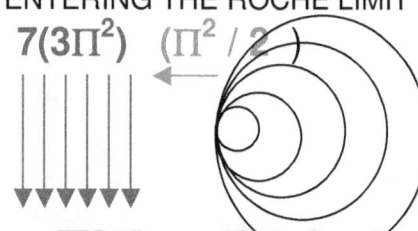

$$7(3\Pi^2) \; (\Pi^2 / 2)$$

$$7(3\Pi^2) \times \Pi^2 / 2$$

Titius Bode law stage 4

MOVING SPACE-TIME UP TO BREAKING THE ROCHE LIMIT BY CHALLENGING THE LAGRANGIAN POINTS

FROM MACH 1 TO MACH 3 $\;\; 7(3\Pi^2) \; (\Pi^2 / 2) \; (\Pi^2 / 4)$

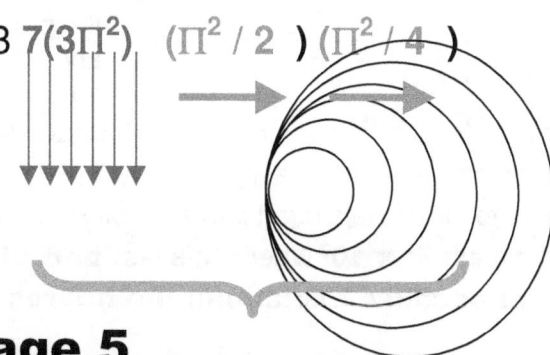

Titius Bode law stage 5

ABOVE MACH 3 WILL THE HEAT NOT ONLY COVER THE WING, BUT THE WING STRUCTURE WILL REVALUATE ITS COMPOSITION TO MATCH THE NEW VALUE OF HEAT THAT ARISE FROM WITHIN THE WINGS ATOMS AS THEY HAVE TO RELATE TO THE NEW SPACE-TIME VALUES.

The wing structure will occupy less space in accordance with the higher time value to which the wing submits. This will lead to a shrinking of the structure that will bring about uncontrollable vibrations and a lack of space-time support.

The earth, as in the case of all other cosmic structures divide into four quarters where every quarter takes a time position in the total displacement. In this the value of a star or a planet in the era we share, the iron era, the time comes to $7/10(4(\Pi^2 + \Pi^2))$ the 4 indicates a separate structure independent and in control of individual gravity.

In reconstructing the situation values must be placed in order to come into understanding. With the projectile fitting on the earth the earth overpowers the projectile claiming the singularity displaced in the governing singularity of the earth sharing the Earth as a proton value by being Π^2 to the earth being $\Pi^2 + \Pi^2$

With the projectile secured on the surface of the earth, the earth holds $3\Pi^2$, but the forth square the projectile holds in a position where 7 ends space therefore it takes on the value of being on top of the earth.

By standing still in a position secured to the earth [...] 7 which the projectile claims relating to the $3\Pi^2$ and that fixes the relation of the projectile relating to the earth. The relevancy of the position is $7(3\Pi^2)$. But since the earth is in a linear motion that motion also relate to the earth's singularity and comes into the calculation as Π^0. The instant the object moves individually from the earth the motion stands related with the linear value of Π^0 in the contexts of space coming from Π^0 to $5\Pi^0$. It is points in space in accordance with the centre holding singularity but because of the four quadrants the Lagrangian point prevents the movement to reach 5 points and therefore the Roche limit of $\Pi^2 / 2$ comes into force and apply

After the five is reached and surpassed the singularity manifests as a separate issue to that of the Π^1 and from that the "Sound Barrier comes about when the value reaches past the earth's singularity applying the projectiles individual singularity relating directly to the earth.

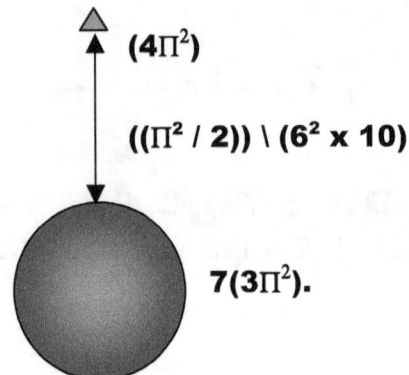

The $7(3\Pi^2)$ represent the relevancy the Earth holds in the separating relation.

The projectile claiming individual gravity also claims $(4\Pi^2)$ in the relation because it reached a gravity point outside that of the Earth.

But in order to overcome the earth gravity the projectile must accomplish the space and time between the position it had on earth and the position it holds in individual space must apply $(\Pi^2 \times 2) \Pi^2$ the 3600

In short: to overcome the space downward thrust enough space must apply in which the aircraft is to firstly overcome applying gravity at a rate $7(3\Pi^2)$ while also exceeding the Roche limit $(\Pi^2 / 2)$ and then this has to move the aircraft into an orbit $(4\Pi^2)$ while removing the aircraft from the space confinement of the earth $6^2 \times 10^2$ where this last **part of the information I am not divulging at this point because explaining the figures and make up thereof is rather extensive**

Gravity applies by way of the Coanda effect. The Coanda effect employs the Titius Bode law and the Lagrangian points interweave one part of the Titius Bode law. The limit the Titius Bode law can arrive at before crossing the Lagrangian points the Roche limit limits it. All of this makes no sense and yet what I minion here is what forms gravity. It is not as simple as awarding mass but only the Newtonian science mind is as simple as awarding mass in order to fine all the solutions unlocking the Universe.

I have concluded the following from deducting The Roche Limit.

The Roche limit in the practical sense but what forms this dividing limit?

The formula $F = G\frac{M_1M_2}{r^2}$ cannot explain the comic occurrence shown in the pictures above, but I can explain what is occurring in this instance and this occurrence connects directly to the Roche limit, as explained above. Not only does the Roche limit explain this phenomenon, but it ties directly to the Titius Bode principle, also inexplicable to the formula $F = G\frac{M_1M_2}{r^2}$. According to the science formula of $F = G\frac{M_1M_2}{r^2}$ the orbiting structures should collide with a bang, but instead they do the tango until one drops, but when dropping it still does not collide with the larger structure as would the formula used by science, $F = G\frac{M_1M_2}{r^2}$ suggest. The position where the formula applies is most surprising. Where the formula $F = G\frac{M_1M_2}{r^2}$ apply one have to find singularity applying because the position of r is pointing to a specific pinpointing of space contracting.

This is not only limited to planets in our solar system. In the universe, there are giant stars (even neutron stars) spinning around each other. These stars can be binaries, which are one form of double stars where double stars are another such a form. The difference between the types depend on the distance they remain apart. They keep a certain distance apart and do not collide. In the case of the Sun and its planets, it could be a case that the systems might be to small, or they might be to apart. However, this is not the case with binary stars. They are close, they are big, and they spin around a mean axes called the Roche limit. They do not grow closer, but orbit around each other holding a common axis.

The Roche limit is:
The region surrounding each star in a binary system, within which any material is gravitationally bound to that particular star. The boundary of the Roche lobes is an equipotential surface, and the lobes touch at the inner Lagrangian point, L_1, through which mass transfer may occur if one of the components expands to fill its lobe. It names after the French mathematician Edouard Albert Roche (1820-83).

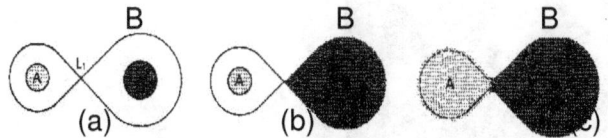

THE ROCHE LOBE: In a binary system, the Roche lobes of components A and B meet at the L_1 Lagrangian point. (a) In a detached system, neither star fills its Roche lobe. (b) In a semidetached system, one massive component, B, fills its Roche lobe. (c) In a contact binary, both components overfill their Roche lobes and share a common envelope.

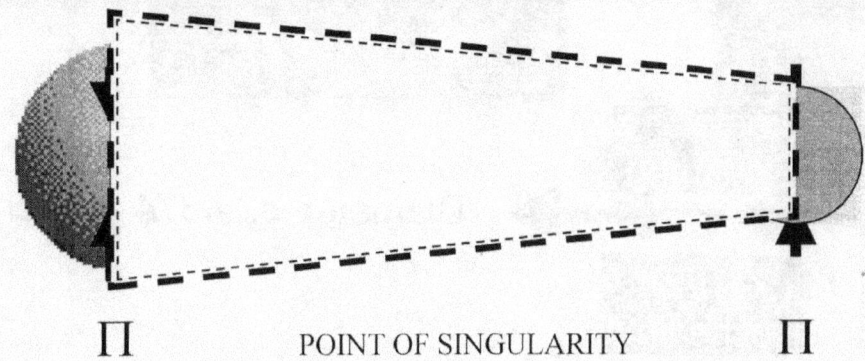

Π POINT OF SINGULARITY Π

$A_1 // k^0_1 = a^3 \div T^2 k$

But $k^0_1 = 1^0$ and $k^0_2 = 1^0$ although $k^0_1 \neq k^0_2$ because $a^3_1 \div T^2_1 k_1 \neq a^3_2 \div T^2_2 k_2$

$A_2 // k^0_2 = a^3 \div T^2 k$

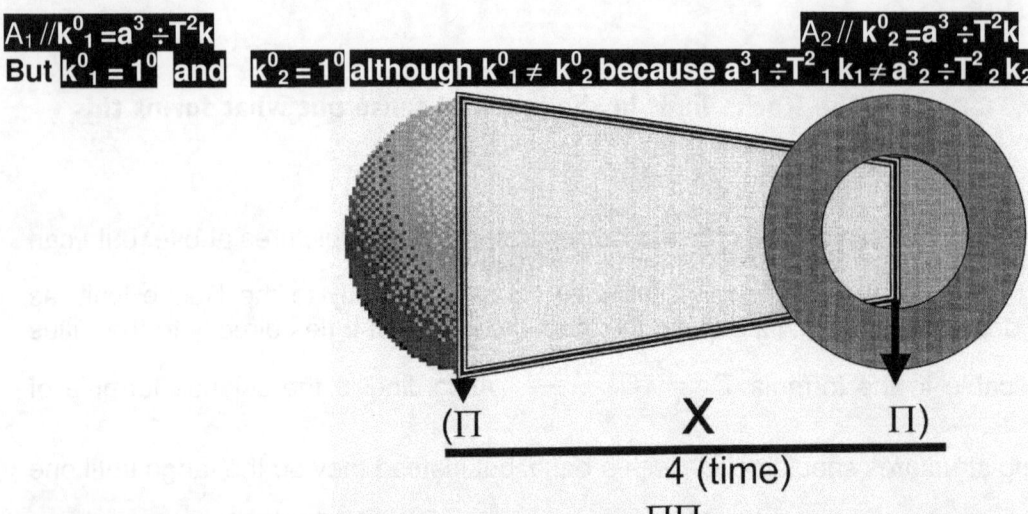

So to make $k^0_1 = 1^0$ and $k^0_1 = k^0_2$ the space occupied has to increase to $(k^0_1 = a^3 \div T^2 k) = (k^0_2 = a^3 \div T^2 k)$ the space occupied a^3_1 must equal the space occupied a^3_2. To do that the time must increase to equal $T^2_1 k_1 = T^2_2 k_2$.

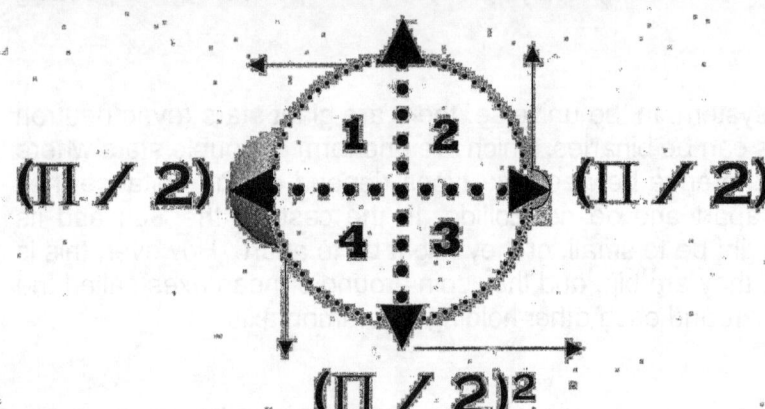

This was the mathematical explaining. Now let's get verbal with the explaining.

Time forms space at the point where singularity forms Π. It forms Π on both sides of the divide of singularity, which splits Π in 2. Moving in the four quarters would form the square of half of Π. This is what time leaves as it leaves space to form the history of time for space is just the history of time. If the space is bright it is to our past and coming towards us by the measure of light and if it is black and dark it is going away while we are going in that direction, but I explain that thoroughly in other more elaborate books. What time leaves to space as time forms space to be the history of time is $(\Pi/2)^2$. Everything within the margin of singularity governing is 1^0 that in space is becoming $(\Pi/2)^2$ and that is 2.4674

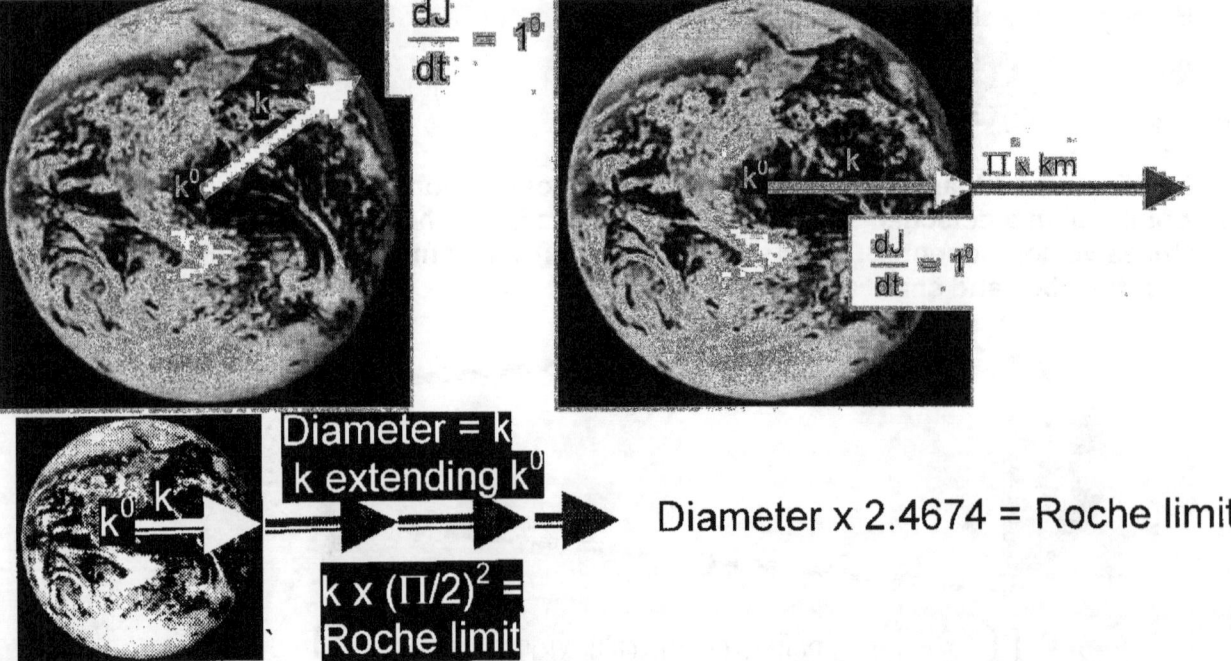

Diameter x 2.4674 = Roche limit

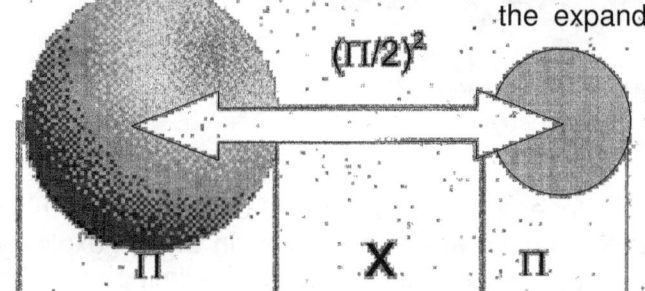

D represents singularity extended

This is how the Roche limit comes about. No solid can be positioned within $(\Pi/2)^2$ of the point away from singularity because that is reserves for liquids. In the relevance $k^0 = a^3 \div (T^2 k)$ we have k connected to singularity (forming the material part) as $k = a^3 \div T^2$ or then space is progressing through time where as the liquid part is defined by the relevance of $k^{-1} = T^2 \div a^3$ or then space is declining through time progressing the factor of liquid or space in decline would be $k \times (\Pi/2)^2$. The atmosphere ends at $(\Pi \times 10^3 \text{ m})$ and that is where the limit of Π ends however k might place Π but it puts a positioning value to Π or in normal mathematical terms to the radius or diameter.

At $(\Pi/2)^2$ 2.4674 time **k** the Coanda effect comes into place where everything within that margin has to be liquid or turn into liquid. We have $k^0 = a^3 \div (T^2 k)$ and that puts a precise measure to $k^0 = 1$. That means within the limits of the governing singularity all singularity is $k^0 = 1$ and all has to be liquid. The Sun is so dynamic that all the planets within the solar system qualifies to be liquid and therefore the planets can all orbit without ever coming into the Sun's contraction.

Everything within $(\Pi/2)^2$ is $k^0 = 1^0$ The expending of the liquid region defined by $k^{-1} = T^2 \div a^3$ expands at a rate of 3 whereas the expanding in the occupied sector holding matter that is defined b the expanding progress in extending Π defined by $k = a^3 \div T^2$ expands at the rate of $\Pi = 3.14156$ which is 0.14156 more than what it is in the unoccupied sector. Therefore as time develops there is more progress in the material sector than there is in the non-material sector by the margin of 1/7. This means that close material would combine to unify and become one through the Roche limit of $(\Pi/2)^2$. This in effect proves that the Universe will eventually combine all material but the time this will take in totally indeterminable even by the cheating methods of the brilliant Newtonians. When the two objects are further than $(\Pi/2)^2$ the one object holds the governing singularity in terms of its own material while the larger object holds the controlling or on the other hand the subjected singularity as it holds the lesser orbiting the superior singularity. There is no big or small but there are subjects and there are controllers in the domain of time.

In a relevancy closer than $(\Pi/2)^2$ all singularity has to conform to $k^0 = a^3 \div (T^2 k)$ and where there is a lesser singularity filling unoccupied space-time by forming material by $(k^0_2 = a^3 \div T^2 k)$, singularity would not allow it to bridge the time constraint of space-time. The governing singularity will take control and confirm the gravity $T^2 = a^3 \div k$. In that it will bring about that the relevance apply through out by producing an equal gravity or time. What this means is that the totality of $k^0_2 = a^3_2 \div (T^2_2 k_2)$ changes to the rime in singularity that $k^0_1 = a^3_1 \div (T^2_1 k_1)$ dictates. Every atom within $k^0_2 = a^3_2 \div (T^2_2 k_2)$ expanded to the time in spinning that $k^0_1 = a^3_1 \div (T^2_1 k_1)$ dictates. That makes k^0_1 take absolute control of k^0_2 and $k^0_1 = a^3_1 \div (T^2_1 k_1)$ becomes the liquid $k^{-1} = T^2 \div a^3$ to the material that $k = a^3 \div T^2$ forms. This is the proof of the Coanda effect and when I saw this, I knew what role the Coanda effect had on gravity.

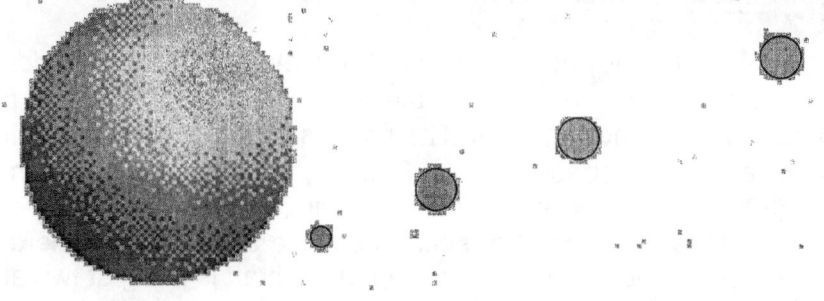

In all the theories how about the Moon will one day hit the Earth or how the Moon was part of the Earth or how the Moon collided with the Earth and then escaped and hearing all these fabulous fairytales one thing spring to mind, those much educated with all those exceptional theories had no knowledge of true cosmology. If the Moon ever was a distance shorter that $(\Pi/2)^2$ in relevancy from the Earth, the Moon was vapour and part of the Coanda effect. Every cosmologist should be aware of the

Roche limit and although not fully understanding it, such a cosmologist should take into account the facts concerning the Roche limit and yet, they never let the truth bother them when they dish up a good fantasy that would please all Newtonians.

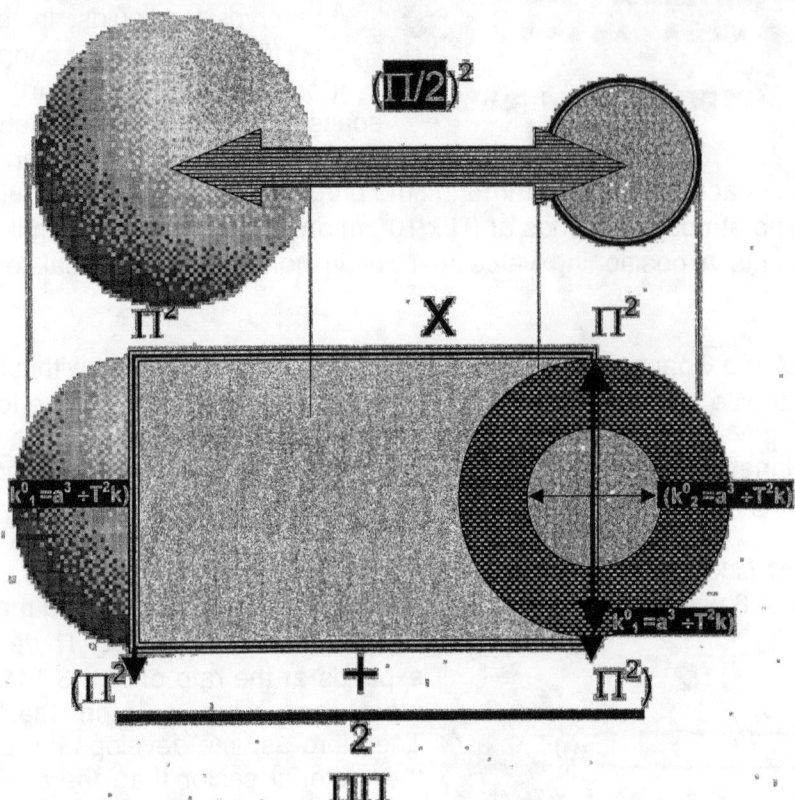

If you are a Super – Educated Newtonian you are in a privileged position where you can ignore anything concerning cosmology as long as you do not offend Newton. I have written two books in which I explicitly show how the planets came about in relation to the Sun and how the effect of the Roche limit placed the positions of the planets in relation to the Sun. Anyone with a simple mind that would believe the hogwash of dust forming planets by the magic of mass driving gravity should also enjoy the story of the three little pigs that went house building in spite of the attention they got from a big bad wolf. There are many similarities in the stories especially the part that insists the believer must have an exceptional low IQ and live in a high-fantasized environment to believe such ridiculous trash in the first place. How can any-one with some form of intellect believe that one dust particle will pull another dust particle so close that the two dust particles will form a rock, which is what we find in the centre of the smallest four planets!

When an aeroplane or even a race car travels at or above $7(3(\Pi^2)) = 207.26$ the car is within the limits of the Earth's gravity. When the car or the aeroplane starts to exceed this relevancy of moving within the Earth boundaries the aeroplane or car will start to get airborne. This has nothing to do with air lifting the car but it has all to do with the curvature of space-time in relation to movement.

By moving into space at a relevancy limit of extending Π^3 by 10^3 meters into the sky, which is 31006km, the relevancy of moving at $7(3(\Pi^2))$ just to stay airborne changed the initial speed from $7(3(\Pi^2))=207$km.h.to $7(3(\Pi^2))(\Pi^2/2)(\Pi^2/4)=2523$.km/h The aircraft is not flying two and a half times the speed of sound because at that height there is no speed of sound. The one saying the aircraft is flying at two and a half times the speed of sound knows very little and even down too absolutely nothing about physics and I will even bet my last shirt that that person will believe that mass brings about gravity!

It is all down to relevancies changing because at $\Pi^3 \times 10^3$ a kilometre of space is not a kilometre od space any more.

You, the scientists of this earth connected space to time. You have taken time as twenty-four hours to circle $6^2 \times 10$ degrees. That is the time you put in place and all space adheres to this time. Then you, the scientist of the planet took 10^3 meters of water as 10^3 kilograms, which without saying connects to the 24 hours going through $6^2 \times 10$ degrees of time. I had nothing to do with this because I found this in place when I arrived on the planet back in 1953. The scientist chose to call 10^3 meters one km and this was connected to water that was spinning about at $6^2 \times 10$ degrees during 24 hours..

I told about the lines forming. These lines holds a relative value of 7 which relates to the curvature of the Earth and there, which holds the time factor. The time component in space is 3 and which is in place when object floats in the air while the 3 changes to Π when the object suffers with a mass attack when it connect to the surface of the Earth. Any object flying in space just above ground will fall at a speed of $7(3(\Pi^2))$ and when it hits the ground or water this changes to $7(\Pi(\Pi^2)) = 217$ km / h and there is no surface tension on water that objects break when they hit water. It is a speed differentiation of going from $7(3(\Pi^2))$ to $7(\Pi(\Pi^2))$ in the smallest of time fames. The following that I explain has a factor value in relation to the earth and the speed of sound only applies as my figures indicate in the region of the atmosphere between absolute ground level and 10^3 meters above ground level.

To explain the relvancy of flyng at $7(3(\Pi^2)(\Pi^2/2)(\Pi^2/4)=2523.$km/h this is explained as follows:
7 The curvature of the Earth
(3 The time as it applies in space
(Π^2) gravity
($\Pi^2/2$) The Roche limit as it applies in the atmosphere
($\Pi^2/4$) The Roche limit as it applies reaching the boundary of outer space
=2523 The speed that must be achieved to remain airborne.
Just to get airborne is **$7(3(\Pi^2)$**
There are the lines forming singularity which I explained and is the lines that keeps the top erect. These lines follow the curvature of space (**7**) in time (**3**) in relation to the Earth rotating (**Π^2**) and a body fall accordign to the relvancy coming from the Earth rotating.

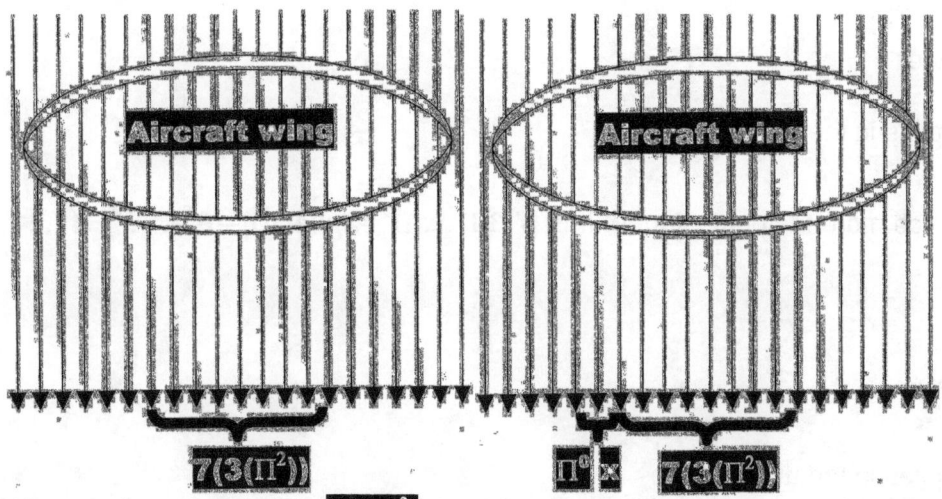

Although the relevancy is **$7(3(\Pi^2)$** there is one additional marker adding to the relevancy by the value of Π^0. This value of Π^0 becomes an additional factor only when the full compliment of **$7(3(\Pi^2)$** has been reached. It has no measured increase to the value of the relevancy but it becomes an additional factor as the speed of the moving object increases. When the craft is reaching an airborne speed it flies over **$7(3(\Pi^2) = 207$ km / h**

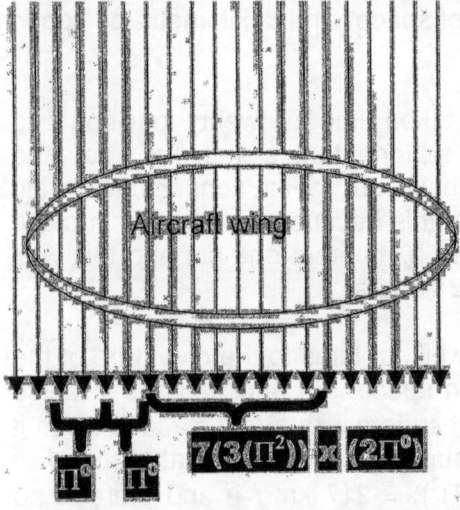

At a speed of $7(3(\Pi^2) \times (2\Pi^0)) = 414.5$ one additional line has been added to the moving object. I show the wing, but this applies to the entire structure of the aircraft in the case where it is an aircraft moving. One could also imagine that the aircraft fuselage or the wing becomes $(2\Pi^0)$ longer in relation to what the aircraft would be when the aircraft is standing still.

At one of these points the helicopter finds its limit in speed and then at another point the turbo prop aircraft reaches its limits in speed. Every one of the following $7(3(\Pi^2) \times (\Pi^0) = 207$ km / h, $7(3(\Pi^2) \times (2\Pi^0) = 414$km / h, $7(3(\Pi^2) \times (3\Pi^0) = 621$ km / h, $7(3(\Pi^2) \times (4\Pi^0) = 828$ km / h is some or other limit to the speed in which a class of aircraft can fly at maximum speed.

The fuselage extends by one line as the duplication of movement increases with the adding of speed.

Then when the speed of $7(3(\Pi^2) \times (5\Pi^0)$ is reached this becomes a principal problem because adding $(5\Pi^0)$ to gravity shifts the measured value to the other side of the quarter divide. When the fuselage stretches to a point of 5 $(5\Pi^0)$ singularity lines the Roche limit becomes a factor. In other books I go into much more detail as to why this happens but I am not getting that involved at this point. Then the half of the Roche limit applies and this is why the pilots feel the vibrations and the drag and all the other complications when they break through the sound barrier in a craft that does not have sufficient power supply to do so. The craft reaches $7(3(\Pi^2) \times (5\Pi^0) = 1036$ but the Roche limit $(\Pi^2 \div 2)$ becomes a presence. The craft then flies at $7(3(\Pi^2) \times (\Pi^2 \div 2) = 1022$ km / h. One should keep in mind that temperature more than quantifiable measured distance applies a role and the length of the kilometre can vary as the temperature increases or decrease. In this case there is no precise kilometre but it also becomes an applicable relevance.

The sound barrier is as follows:
7 The curvature of the Earth
(3 The time as it applies in space
(Π^2) Gravity
$(\Pi^2/2)$ The Roche limit as it applies in the atmosphere
=**1022 km / h.**

Then when going into outer space more factors is added. It is $\{7(3(\Pi^2) \times (\Pi^2 \div 2) \times (4\Pi^2)\} \div \{6^2 \times 10^2\} = 11.216$ **km / second**

{7 The curvature of the Earth

(3 The time as it applies in space
(Π^2) Gravity

$(\Pi^2/2)$ The Roche limit as it applies in the atmosphere

$(4\Pi^2)\}$ Becoming an independent spinning object within the Roche limit of the Earth.
÷ Putting the independence in terms of finding release from the Earth confining atmosphere
$\{6^2 \times 10^2\}$ Representing the curve of the atmosphere (6^2) in terms of space (10^2)
=**11.216 km / second.** The speed one has to travel to escape into outer space and go beyond $\Pi^3 \times 10^3$

The rings are in place because gravity holds the rings in place.

Gravity has the value of Π and this has nothing to do with mass. This we see in every sphere and every circle that translate into objects spinning or objects orbiting or circles forming galactica far away. I prove that gravity is Π and the four cosmic phenomena forms gravity by producingΠ when objects move in relation to each other. The rings you see are in place because gravity is Π. The spherical planet you see is a spare because gravity is Π. The planet spins in relation to Π because gravity is Π

By using the above **the four cosmic pillars**, it enable me to **present the proof** where I now can explain what conditions bring on the **sound barrier.** By proving it **is gravity** that the individual structure generates motion above and beyond the gravity the Earth provide is what is producing individual **motion** that the independent object earned within the sphere of motion that the Earth's gravity provides where the **independent and individual motion** put the relevance that gravity has beyond the conserving means gravity has where the space that is serving the independent object is independently in motion. The adding to the independence on top of the normal structural independence is creating more individualism by the independent motion of the individual structure being apart from the motion that the gravity of the Earth **provides.**

The rings around the planets are the remains of material that moved beyond the limit ration of $7(3\Pi^2)(\Pi^2/2)(\Pi^2/4)$ where the structure disintegrated into a mesh of dust. Exceeding the movement limit of $7(3\Pi^2)(\Pi^2/2)(\Pi^2/4)$ in relation to singularity Π^0 the movement went beyond the first limit of $4\Pi^0$ but could not reach the **Lagrangian points limit of $5\Pi^0$ and then went into the sonic boom limit of $(\Pi^2/2)$**

Furthermore the orbiting debris that became dust exceeded another Roche boundary limit at $(\Pi^2/4)$ where the structure overheated and was annihilated into tiny vapour fragments.

To understand this concept better go to The Absolute Relevancy of Singularity: The Dissertation
http://www.lulu.com/content/e-book/the-absolute-relevancy-of-singularity-the-dissertation/5994478]

If gravity is about movement and not about mass what then forms a star and why does a star destruct? To see what is inside a star we must see what comes out of a star when the inside turns out.

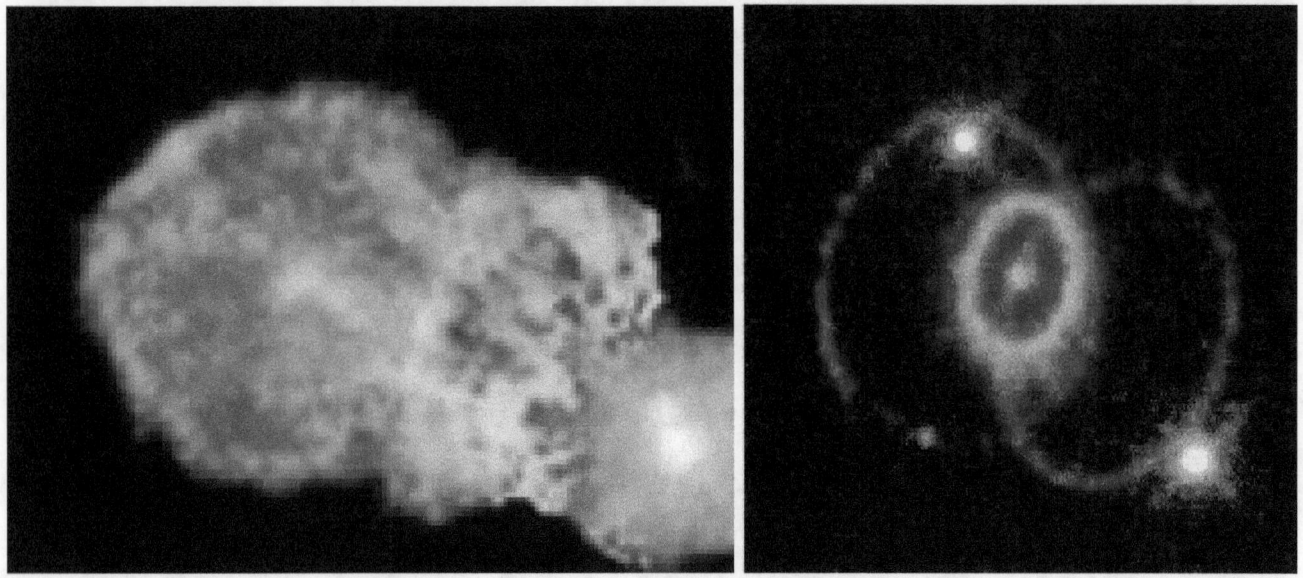

That what comes out of a star must be that which is in a star and what comes out of the star in this picture is liquid heat…then why is it liquid heat? The best Newtonian wisdom can come up with is as senseless as a nursery rime explained to toddlers. What you see in this picture is the Roche limit being violated and compromised and by compromising one of the four limitations that singularity puts in place a collapse of order within material bounded structures such as star bring total destruction to a structure such as a star. Stars are spheres filled with heat, not mass.

By turning it compresses what is in outer space into what is inner space and by compressing the space gravity increases the heat levels. The heat does not rise because heat cools by expanding while it remains heat and heat rises in levels by compressing the space. That is what gravity is. When you observe what happens when a star goes bust it is clear that the event is due to a lot of heat escaping.

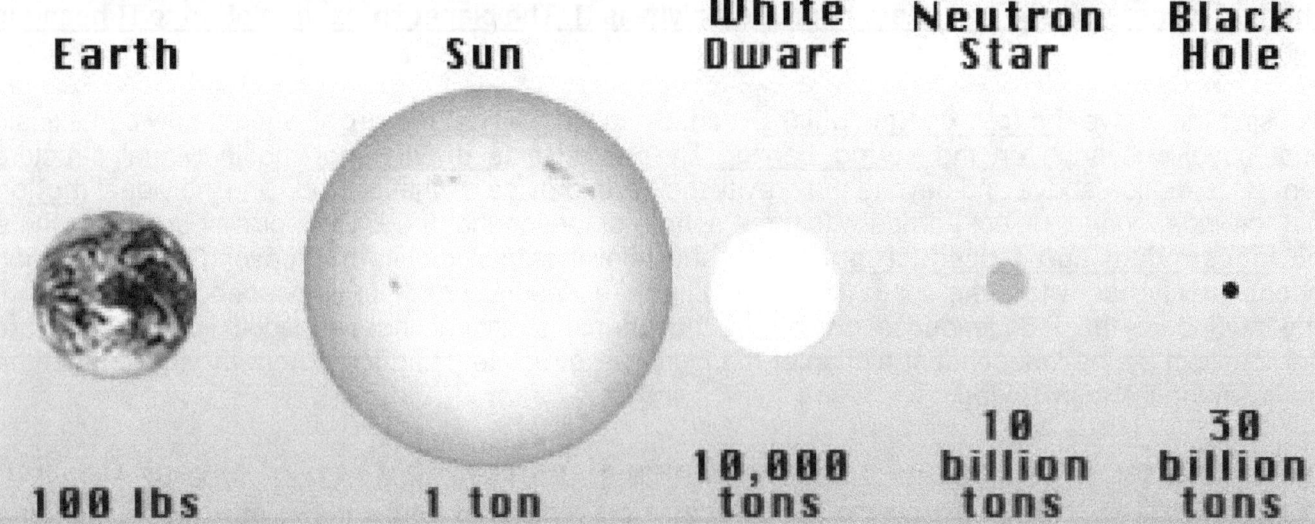

It is simple to come to a conclusion when logic deductions of reality are made. Material such as the picture shows weighing round about 100 pounds will weigh about 30 billion tons in a Black hole. I disagree with this number because I totally disagree with the value they use connecting the radius to the speed of light and my figure I get is another trillion tons on top of the official billion tons. I put this in to let every reader see that gravity is not connected to size but to absolute density. The smaller the actual size of a star is the more enormous the gravity is. Gravity works on density and I am about to explain this by using Kepler's formula when corrected from the deformation with which Newton corrupted it.

The escaping necessarily involves expanding of compressed heat because the gravity was eons of accumulating compressing heat into the star. If gravity is about contracting and the expanding is about a release of heat then gravity must be a systematic reducing of heat. If the star was only material bursting we would observe billions of structural particles flying all around but this is clearly liquid heat going into gas. When the gravity that holds a star is insouciant to keep a star bonded then that would imply that gravity is about cooling the star which enables the star to maintain structural integrity and when the balance it heat goes beyond the control of gravity the star overheats and it explodes. The explosion comes when the spin applying in the star can't retain the heat and when the heat in the star moves too slow it overheats and it expands.

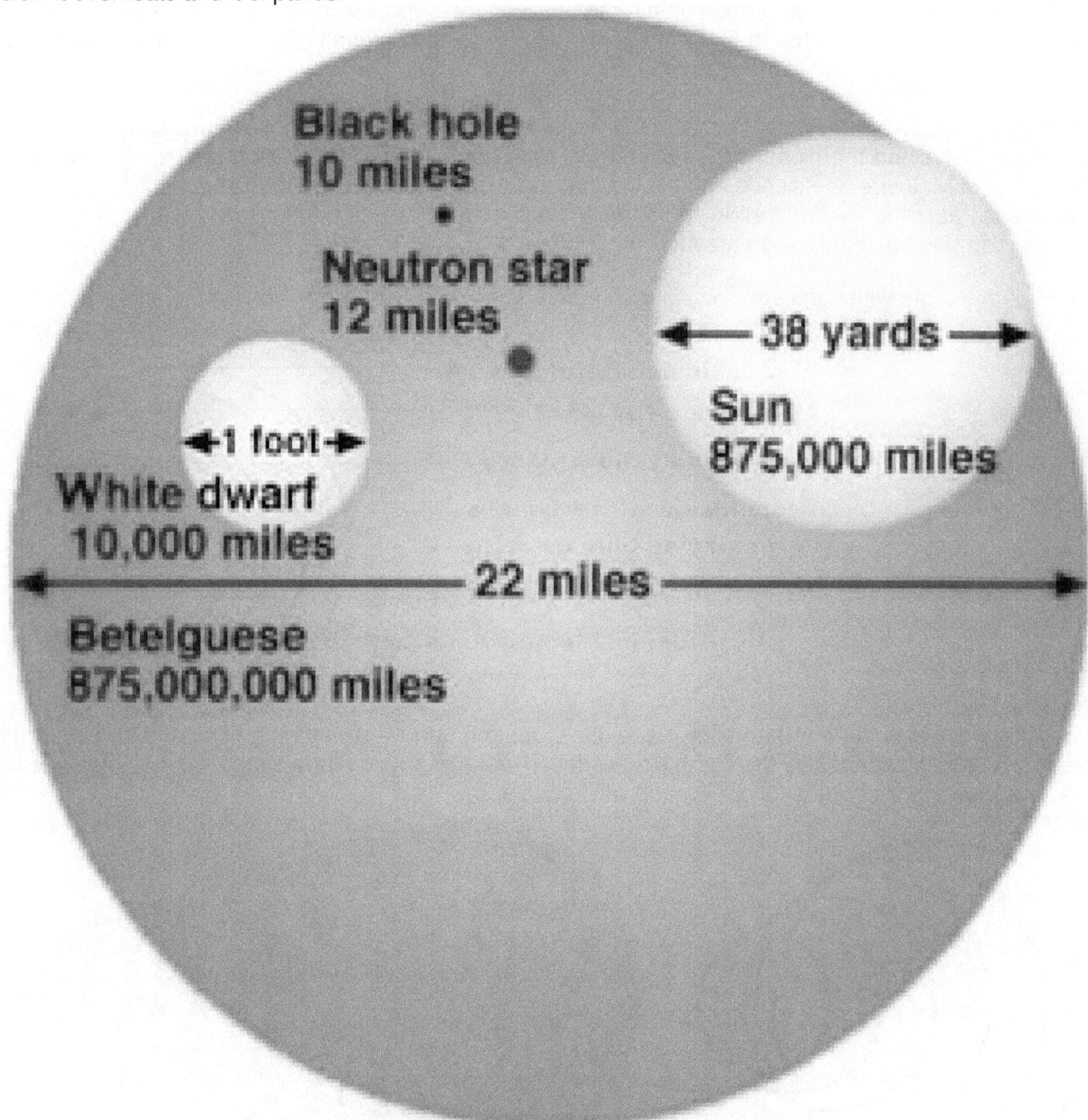

A star holds form by the gravity it has in a form of cooling the space in relation to outer space overheating. The gravity comes as a result from the spin of every atom within the star. The higher the spin is of every atom, the denser would the atom be because the more the heat forming a star would cool and reduce. This is a result of movement reducing the size of the atom within the star.

When an object is in a location with little motion the duplication present a lot of heat because the distribution of the heat over the space in duplication has very little possibilities of spreading the overall heat over a wide area. The motion of something as small as the earth will confine the atoms into a relative hot area since the space in duplication does not reduce the extent of the heat by distributing the heat over much space.

In a structure with the size of the sun the motion of space is enormous by the sure quantity of space in need of duplication. Shifting that volume of space needs duplication that is millions if not billions of times more extensive than what the earth may produce. By duplicating such a vast area in a period of time within say the sun, reduces the individual atoms forming the star to a fraction of what the situation on earth would allow. The more the spin of the liquid is in relation to the solid state of space is reduces the space and extends the material in quantifiable measure many billion times over to what smaller stars are. It is not the space that holds the matter but it is the spin in relation to what the matter holds that puts the relevancy of hot and cold within the star.

The more cold there is because of the more liquid heat bringing about motion, the colder would the atomic material be and the higher the relative contracting gravity that the star produces. This we see in the admitting of Mainstream science confessing that the reducing of space produces an increase in mass and because mass is the frustration of material unable to move, it admits to the fact that mass in volumetric size has no influence on gravity.

The size of any atom within the specific star according to the gravity or movement of the star

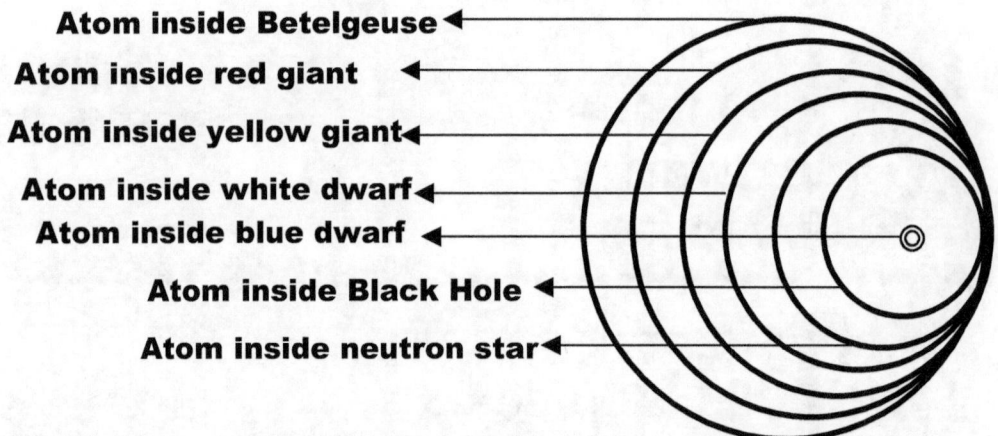

The physics we encounter on Earth allow us to use a common and a constant, a fit all and an all-purpose because we find us captures by the Earth singularity. The Earth provides the space we may claim as well as the time in which such material duplication will take place.

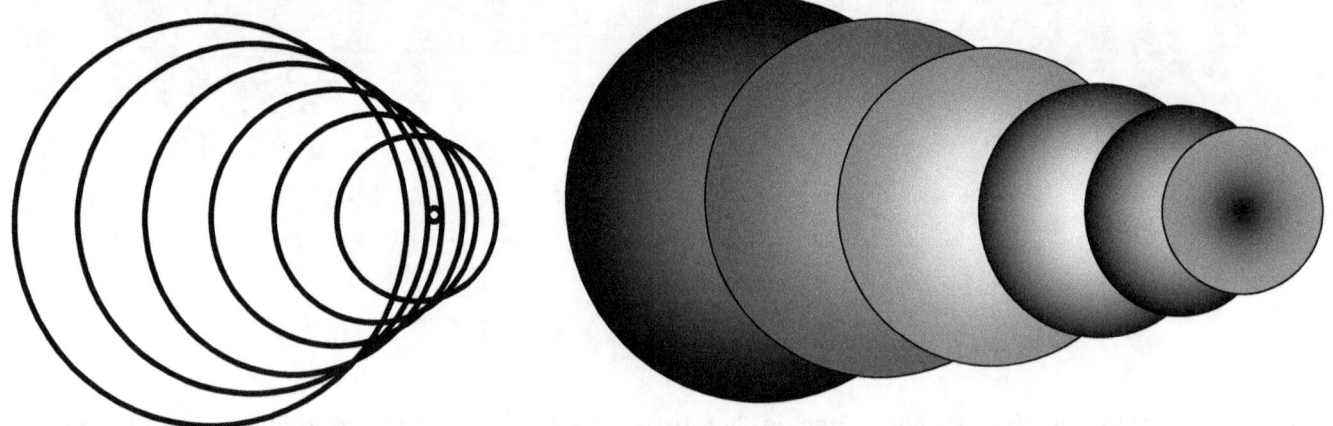

Our human orientated thinking in terms of science is as Neanderthal as was the Ptolemaic Universe was. The focus of the entire study or the Universe is sympathetic to life and the focus of all aspects revolves around the earth as the earth can sustain life. Science think of −270° as cold because humans connect it to cold whereas 650° C is hot because we roast at such a temperature. In the Universe what expands is hot when gaining space and what is cold shrinks as it retains heat. Movement shrinks material because space duplicates by movement and although having more space, the space then is spread over a larger area during the same time.

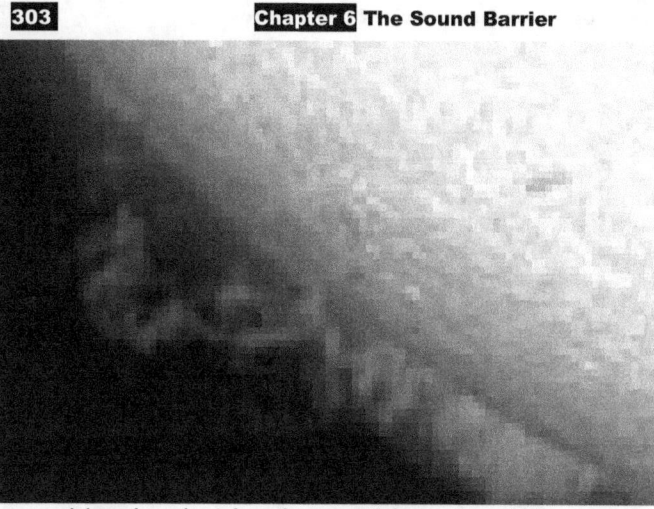

The picture one can see from the above is not of something hot but is of something cold. When putting any solid frozen "gas" into water the reaction we find is the same reaction as what we see on the sun'

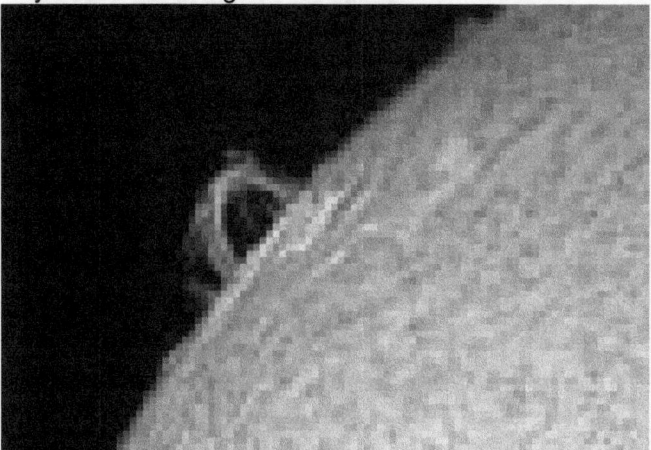 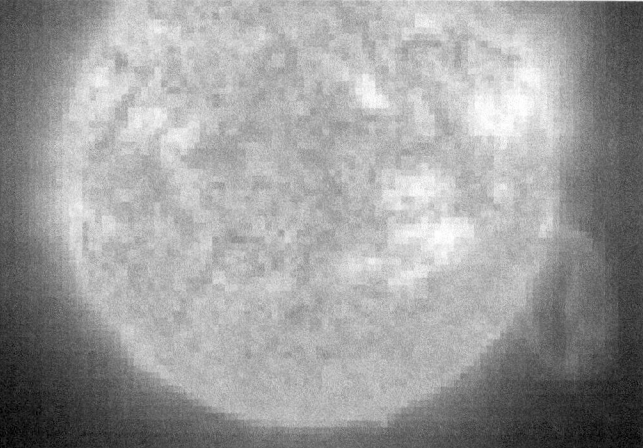

What the sound barrier shows us about gravity is that space reduces the faster movement re-allocate the next position of the object ravelling. This brings about that the atom in the neutron star shrinks so much there is no longer room for the neutron and the atoms then discard the neutrons. This has nothing to do with mass or does prove that mass has nothing to do with forming gravity in whatever way.

Let's get Newton into what forms reality. Newton said $a^3 = T^2$, which is saying a three dimensional object is equal to a square. What this mathematically implicate is saying that I in three dimensions are equal to my mirror image and therefore being $a^3 = T^2$ I can swap places with my image in the mirror because we are the same. The poor man obviously did not know the first principle about mathematics and yet all Newtonians are following Newton sheepishly.

Kepler said $a^3 = T^2k$
Then it says $a^3 / T^2 = k$
And $a^3 / k = T^2$

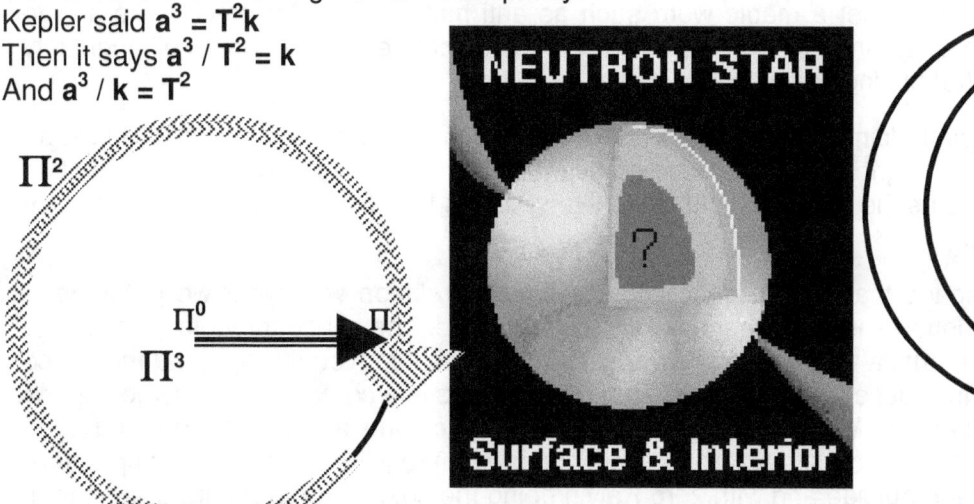 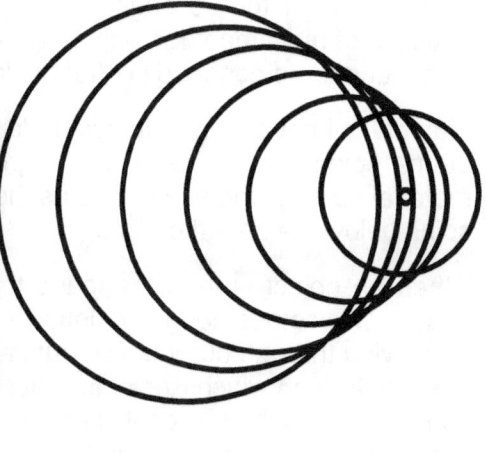

Kepler said $a^3 = T^2k$ which mathematically says that the size of space formed equals the movement T^2k. Then the circle $a^3 / k = T^2$ size of space is formed that is equal to the linear the movement k and also the circle size is directly responsible $a^3 / T^2 = k$ space travelled in one full rotation period of the star.

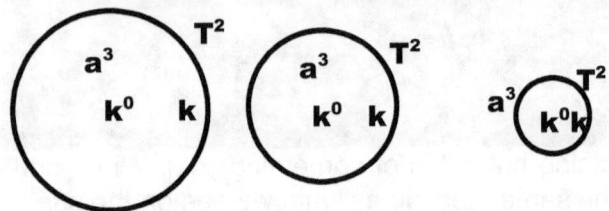

Moving a full rotation in a short period would bring much gravity in a small atom. In the Black Hole there is only one atom left that holds the size of infinity while outer space represents the entire eternity. That is why a Black Hole has no space because all the space is compromised by singularity Π^0 and outer space forms the value of space or Π.

The faster the atom spins the further the atom displaces allocated position because the bigger the gravity is within the star. Gravity has nothing to do with mass and all to do with movement of material within the star.

All this information we find by valuating and then reading the conclusion of the sound barrier correctly. The heat around the flying airplane connects to the plane by the atoms that shrink in size. This makes the atoms much colder as the heat the atoms discard locates around the structure of the star.

This is how simple the explanation is when studying the unfolding event of a star going supernova. Gravity can't be mass infliction because what comes from a star when things go wrong is heat escaping. What comes out of a star must be that which is inside the star and what fills the star. But because we have a bunch of mathematical drones too lazy to contemplate other than to accept we have this pre-historic view Newton brought along as science departed from the Dark Ages about mass forming gravity. They can't think and that is why they can't read my work. They can only compute in terms of mathematical calculations and even in that they are completely incompetent in researching the truth.

The cowards (and I call them cowards because they dodge the bullet to purposely maintain the corruption of three hundred years) in academic positions ignore my work and therefore they never have to challenge my views and therefore they never have to dispute my position and therefore they never have defend Newton and therefore I die in silence while they never have to prove anything. ...And the Science Conspiracy Network remains in place, undetected, erect like a mile long penis ready to screw the living daylights out of a brainwashed human race for another 3-hundred years. Ask any person if he or she thinks they have mass and they will think the person asking the question is bonkers and yet show not weight show me mass.

Again I challenge any Newtonian physicist to show me any evidence of mass anywhere in the Universe. Movement reproduces space during any one specific time-duration and that holds the ratio of space-time physicists refer too so eagerly. It is not a magic word such as anti matter. The Universe works on the principle of movement forming space in a time period in which heat is confined to the control of singularity or not under the confined control of singularity.

Material is heat confined to spin holding a specific space. The space in which material is when it is in an atoms spins above the sped of light and the electron securing the area which confines the heat to material spins at the same r ate as the speed of light while everything in the Universe other than the atom spins below the speed of light.

When an object is in outer space that object, encounters a specific relation with what we presume is space. This comes about by motion and through material volumetric size. The space the object encounter by moving through outer space puts a value of a ratio between the space it moved through and the space moving through which Kepler introduced as $a^3 = T^2/k$. That means there is a contact ratio between space containing and space contained by. When the atom is in outer space the atom is surrounded by a temperature of zero Kelvin and that is because zero Kelvin is the presumably the coldest any temperature can get. Being zero Kelvin on the outside and with zero Kelvin being the coldest temperature there can, it would make the atom also zero Kelvin on the inside since there can be no colder than that. That would mean the entire atom is then zero Kelvin. The zero Kelvin responds the movement of the space in which the atom is and has nothing to do with hot or cold because in fact being at zero Kelvin places the atom at the hottest being as expanded as space in material can be. This comes as a result of absence or the lack of movement in outer space. Being at the most expanded means the lack of movement produces criteria in which the expanding proves space is heating up the material

This will come into affect when the projectile reaches Mach 1. I choose to use the accepted term Mach 1 at this point as to limit any confusion that may arise from the new arguments. I have to stress the fact that the Doppler effect is, once again, merely a co-incidental but duly related by product.

However, the Doppler effect, as such, plays no part in these phenomena, or in the outcome of the application. It must be seen in terms of cosmology and not from a human perspective. This is the principle applying when a newly formed star escapes the heat envelope within the centre of a galactica as a star is born. There now are two singularity factors within one space-time unit and the two goes into a battle of existing

Any object entering from space and that does not have direct contact with the surface of the earth, encounters gravity in terms of being directionally diverted from going straight to turning towards the earth by 7°. This then is what provides the "pulling of mass" and the other is having contact with the surface of the earth, which then puts Π in terms with the centre of the earth at Π^0. What becomes clear from these two illustrations is the following: Where there is a direct entry by the spacecraft, the time factor is not sufficient in duration.

This will not allow the time-duration needed for this structure of the aircraft to revaluate its space occupation an therefore the space factor of the spacecraft cannot adopt to its new position in space-time occupation as it has to relate to a new value in accordance with the value that is determined by the earth's concentration of space-time.

In the case of the second entry illustration, a lot more time lapse is allowed for the spacecraft to revalue its structural position in accordance to its new value of space-time occupation. This is the very reason why objects like "falling stars" burn out when they enter the earth's atmosphere.

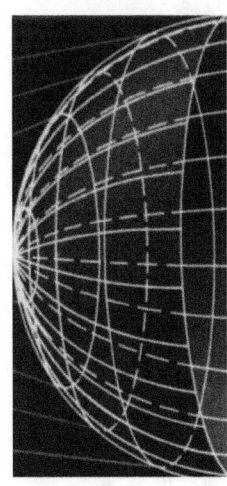

Forget about mass forming gravity it is a fool's rhetoric. When an object holds a steady position to the earth while making a sound, the sound will go in all directions evenly. It will go left and right equally fast. The concept of gravity connects to roundness and to Π. Gravity is Π. Gravity is the movement in terms of Π by duplicating the position of Π per specific time units applying. Then the object making the sound moves left it will hasten the flow of sound in the direction it moves by moving towards the sound while it will increase the distance the sound has to travel by also moving away from the departing sound going to the right side. This means that gravity has two values one being linear and the other being circular and the linear affects the circular in movement as much as the circular affects the linear. This is not only connected to sound but is connected to everything applying to gravity as gravity. When an atom moves the size of the atom must shrink to compensate for the directional change of the orbit that decreases the electron's orbit and therefore decreases the size of the moving atom where atoms form an object.

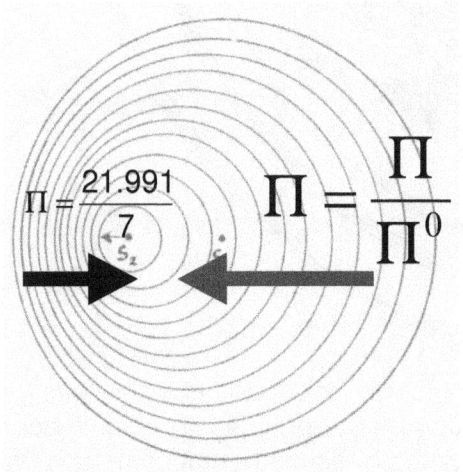

Coming toward the earth from space the object travels straight but the spin of the earth by 7° re-directs the object to follow an inclining line of 21.991 /7. At such a point the object moves towards the earth but the object cannot have weight yet, since the object does not connect directly with the earth. When the object touches the earth it relates to $\Pi\Pi^0$ and at such a point the object receives a value of mass by which it becomes a unit within the earth. By connecting to singularity in terms of $\Pi\Pi^0$ or in terms of $\Pi=21.991\ /7$ puts the object in liquid or in solid.

A More Complex Commercial Science Book   306   Chapter 6 The Sound Barrier

Gravity has the value of Π. The value of Π has two measure where one is when the line gravity forms extends from singularity as (Π⁰Π) and when the line that gravity forms extends to singularity it is 21.991÷7 because of the curve gravity associates with forming the value of Π To explain all these factors I have to take the reader back to the point where the Universe started as spot that formed a dot. I have to take the reader back to where the point where the Universe was numbers holding no space, which is way before the Big Bang era where the atom broke the Universe into space that formed. In gravity there are the two values that gravity will apply. Gravity will link to the axis forming the centre around which everything spins. This gives the object a connection of 3 because it changes location in relation to the axis. Then the other gravity value connects to Π where Π always at all times connect to singularity within the centre at Π⁰.

When ΠΠ⁰ connects as the curve of the earth, this connection applies the object forms part of the solid and moves with the earth but also in accordance with the earth.

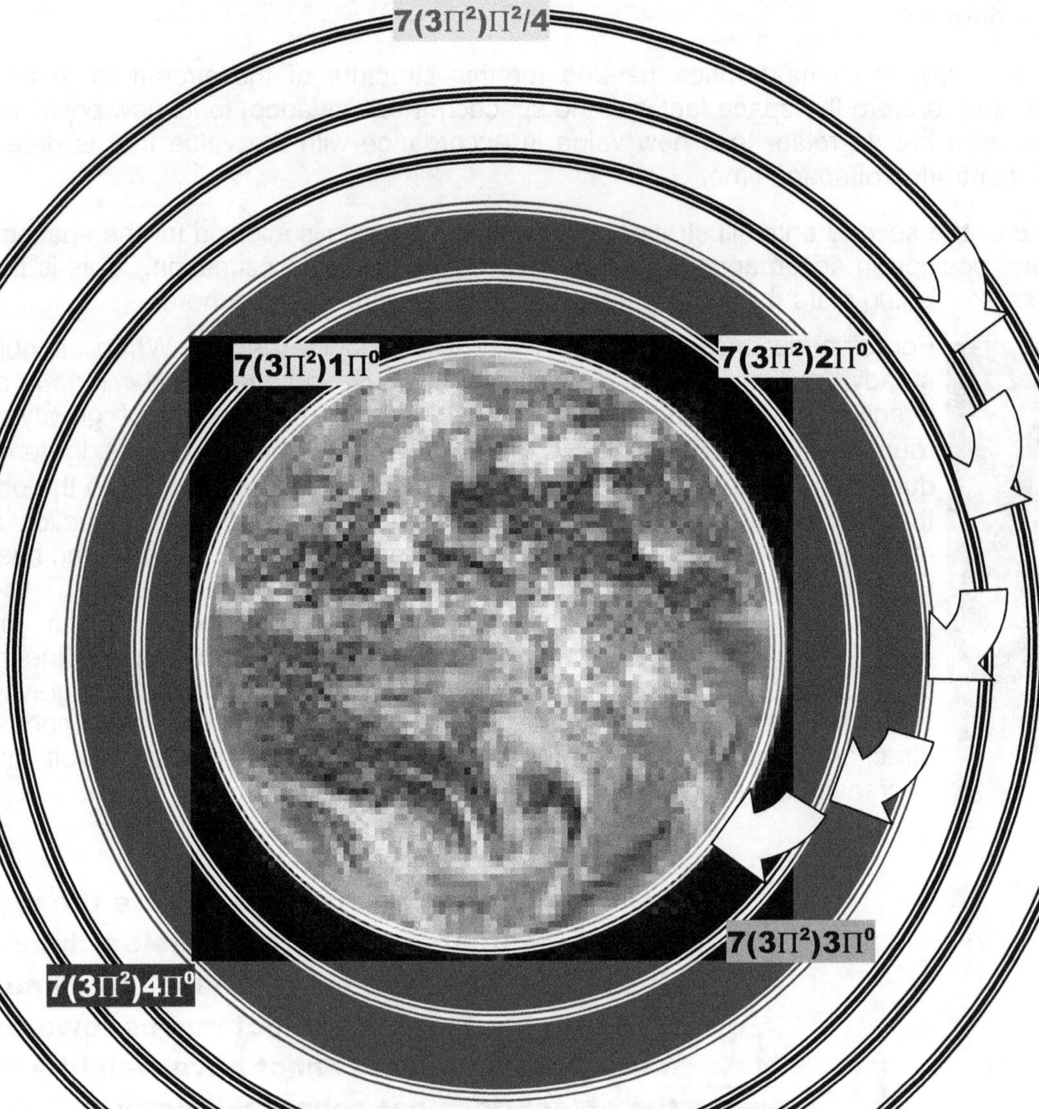

$$7(3\Pi^2)\Pi^2/4$$

$$7(3\Pi^2)1\Pi^0 \qquad 7(3\Pi^2)2\Pi^0$$

$$7(3\Pi^2)4\Pi^0 \qquad 7(3\Pi^2)3\Pi^0$$

The condensing of the vapour surrounding the jet is a prelude to the sonic boom and is just one link in the sound barrier where the sonic boom is just another link in the sound barrier process. The sonic boom IS NOT the sound barrier but is as much part of the entire concept as the beginning of movement is part of the sound barrier. Waves in the sea are as much forming part of the sound barrier as winds howling or blowing is part of the sound barrier.

By the aircraft moving through the earth's space there is a differential difference developing between the size the aeroplane holds when moving at different speeds and the size the aeroplane holds when only moving according to the earth and finding size in relation to the movement of the earth. Movement determines size and this is because by moving faster more space is duplicated during the same time frame. There then is relevance between what moves at which sped in relation to other movement.

$$7(3\Pi^2)\Pi^2/2$$

When any object moves it fill a certain space during a specific time period.

 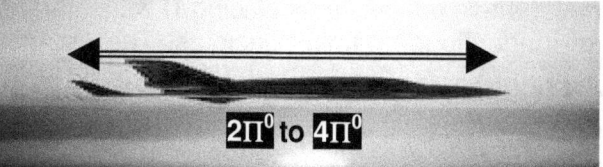

When an aircraft stands still by only moving with the earth it moves at Π^0 As it moves faster it fills more space up to where it fills from Π^0 to $2\Pi^0$ of sky space and as the speed increase this goes up to $2\Pi^0$ to $4\Pi^0$ The aircraft holds more space but does not fill more air and therefore the air has to compress

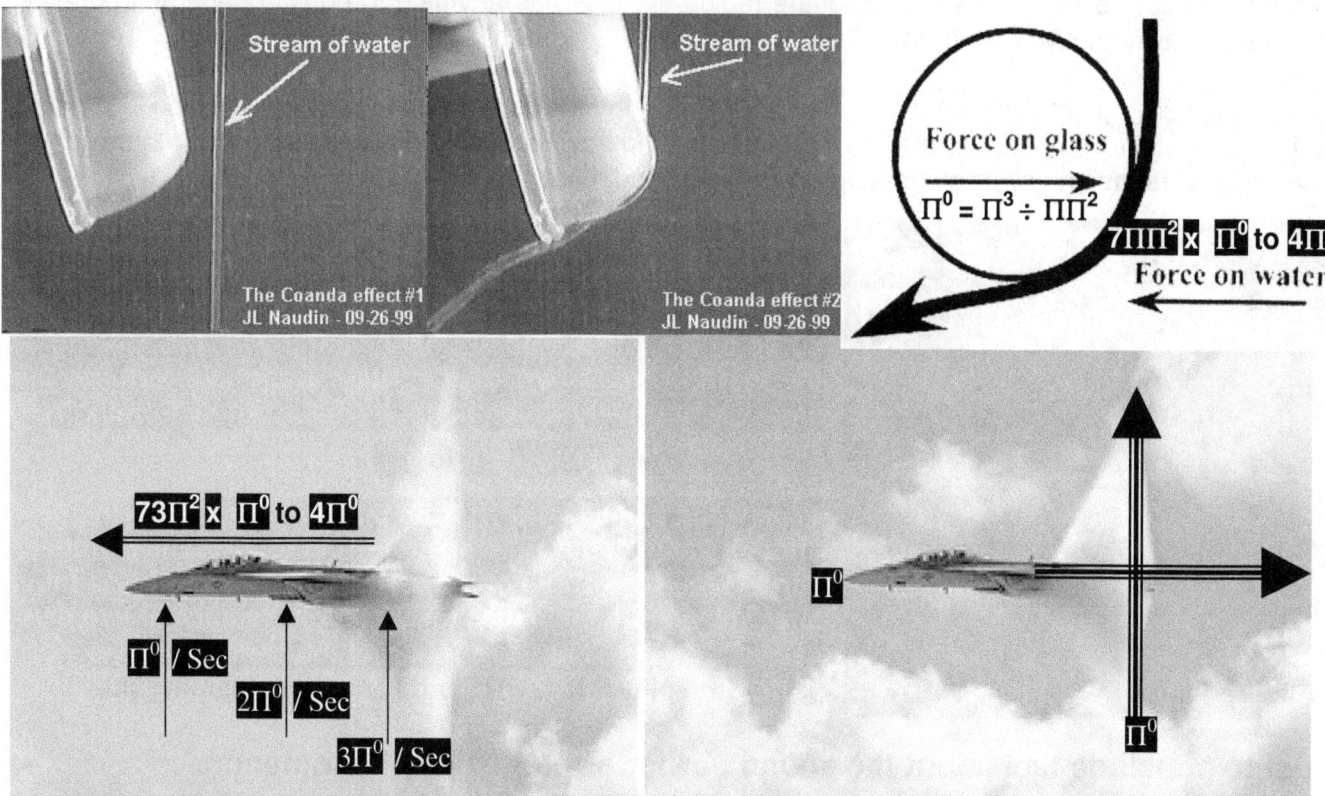

By going into space reduction as the fuselage occupies more space through extensive movement increase the space surrounding the moving aircraft condenses and this condensing forms the concentrated vapour. As soon as the fuselage begins to heat the heat surrounding the airplane pushes the vapour away again. The sonic boom is just one link in the chain of reactions we associate with the sound barrier and the sonic boom is more or less in the middle of the entire concept.

Flying at ($3\Pi^0$) three times the falling speed or the speed of gravity ($7 \times 3\Pi^2$) going in a sideways direction the body of the aircraft starts to expand into another gravity limit ($3\Pi^0$). However by the movement of the aircraft the body also has to shrink ($3\Pi^0$) and the result is the body of the craft cant accommodate conflicting behaviour.

Since the body of the aircraft that should expand contracts the space surrounding the aircraft also contacts in sympathy with the aircraft body contracting. That makes the aircraft becomes literally smaller and this condenses the space surrounding the aircraft. But then the heat coming from the contracting body of the aircraft pushes away the atmospheric air as the surrounding air of the aircraft heats up because of the intensity of the aircraft increasing the space it holds.

What I ask of readers is to beforehand forfeit the culture of Newtonian bias when reading this by paying attention to what I say and not about the degree in which I stray from mainstream science's thinking. This way the exercise will present many new ideas and explaining my new concept will become clear. There is so much to benefit from. Science has no idea what a Black Hole is while I can prove what a Black Hole is. I formulate mathematically what "the sound barrier" is. I prove what gravity is. By using the four cosmic phenomena, which is what the cosmos uses to form gravity, I show what "the sound barrier" is and I go much further than that. I show that gravity forms using the **Roche limit**, the **Lagrangian system**, the **Titius Bode law** and the **Coanda effect**.

I uncover these principles by placing Π within the formulating of gravity and when using Π I bring clarity to the misunderstood cosmic principles. The list of the unknowns I can then explain is almost endless. Gravity forms by movement that establishes singularity initiating a circle in using Π.

I show why gravity is there, how gravity forms and what role stars play in forming gravity. There is no difference between how gravity and electricity forms and that I prove mathematically by decoding the cosmos. I prove mathematically when atoms spin they establish Π that forms the Universe.

Whatever forms gravity has to link closely to Π since everything that has anything to do with gravity forms a circle that is Π by the value of the square radius. If mass has anything to do with generating gravity, then mass has to apply Π or otherwise mass has nothing to do with the forming of gravity.

Everything using gravity forms a circle of sorts, which forms the curvature of space-time, which is Π and which curves light. The way the planets orbit the Sun and how stars spin has all to do with Π. In spinning in a circle, Π forms gravity as a centrifugal force that condenses space.

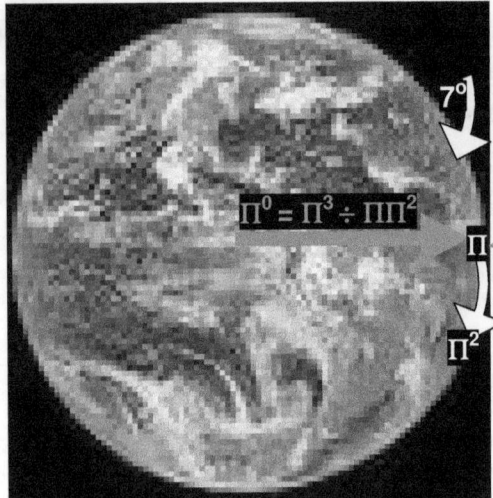

Gravity forms by association or relevancies that holds a value of coming towards the centre from the sky which then is 7(which is the curve of the earth) (Π associating with the curve of the earth) and Π^2 (associating with the moving of the earth) This is part of the Titius Bode law forming the measure of gravity which is not by mass but by forming Π.

$7\Pi\Pi^2$ x Π^0 to $4\Pi^0$

Then when a line is drawn from the centre of the earth another value comes in place referring to the centre of the earth Π^0 and the curve of the earth Π. The value of Π is then shared by both disciplines (association from the sky to the centre ($7\Pi\Pi^2$) and association from the centre to the sky ($\Pi\Pi^0$). In reality it is ($7\Pi\Pi^2$) but I am not explaining this.

Just to conclude fact about the sound barrier as a physics phenomenon.

I have realized the extension of the Einstein hypothesis, and concluded the other part of the relativity theory. In short, the part of "gravity" that Newton saw, and the part of "gravity" that Einstein saw is not the same identical thing, but two different values to the same "gravity". That is why Einstein concluded his theory on a single dimensional Universe. The part of "gravity", which Newton related too, Einstein did not see. "Gravity" consist of a single dimension value (Newton's gravity), which is part of (not the same thing) the "gravity", that Einstein saw. These two values are contained in a balance in space-time. In this, the Universe is in a three-dimensional state of space (three dimensions) in time. Therefore, space is not flat, as EINSTEIN SAW IT, but has three dimensions, locked in singularity where singularity is single dimensioned but by directional relativity forms the fourth dimension of time. I am not going into space-time in this book and I am steering clear by a country mile of any arguments about space –time in this book but Newtonians knows less about space-time than what my dog knows about communism.

In this, there is no force, but only an ever-altering balance, which forms the time component. Only the energy value we regard, as "life" is a force, because only "life" can inflict a change in the balance of space-time and this forms a force. That is why science has been unable to recognize the value of "life" as a separate energy form. If you regard the Universe driven by a force, the value of life has to disappear. The Universe is in a state of balance, and the scale of the balance is determined by time. In this aspect lies the reason why time has never been detected. Only life can upset the balance, but then life will run into upsetting time. When time is upset, we see it as an explosion. The more time is upset, the bigger the explosion will be. I have put everything into dimensions:

1. Newton's "GRAVITY" which is the single dimension. It is a single value, starting nowhere and stopping nowhere.
2. Is Einstein's "gravity" which is a wave that is forever, circling forever by reducing and expanding.
3. These two values above are combined in the third dimension of space, the third dimension we, and everything in the universe, find ourselves in. In this, the first and second dimensions form an inseparable combination.
4. Everything in the Universe is moving. Even the part we relate to as "NOTHING" moves. This movement forms the fourth dimension of time. Time contains the third dimension, which is formed by the combination of the first two dimensions, and dictates the tempo, or balance in the first two dimensions, and which forms the third dimension. As all things the Universe are part of the balance in movement all things stand equally and evenly effected in the third dimension every thing is connected by movement (time). Therefore, everything connects to time and by time, from the smallest part of matter to the Universe as a whole. When space becomes less, the movement becomes more therefore time becomes more.
5. As life are connected to the movement of time, but can also move independently from the balance of time, life is a separate energy value, and is in the fifth dimension, but part of the first four dimensions. Being part of the fifth dimension, life is subjected to the criteria dictated by the balance of the fourth dimension, and therefore has to abide by the laws of the fourth dimension, while it is part of the fourth dimension. In all of this is the outstanding factor that nothing happens by chance, except for the intervention of "life".

Everything I am about to show Newtonians say is nothing new. Newtonians say they are aware of everything that I show and then they turn around and simplify physics by putting the entirety of their physics down to the pulling power of mass that forms gravity. The sound barrier is gravity and to understand gravity is to understand the sound barrier.

They say that I bring nothing new to the table and they say they apply every aspect I indicate that forms gravity. When I ask them to explain the sound barrier they use the Mach principle and Doppler's effect, which both dates from a time when no one was aware anything man-made could fly let alone go faster than sound can. Doppler made his contribution to science at a time when the train he measured was slower than a horse. The statistics he left to science applies to going as fast as a man on a horse could ride. That is a far cry from a jet breaking the cosmic boom. They can claim what they like because they never have to prove anything and they never have to listen since they know everything while their shortcomings makes science more the jesting of a buffoon. That annoys me to my guts because all of that comes down to the conspiracy whereby they whitewash their stupidity from the brainwashing they inherited from their predecessors and pass the brainwashing on to their students. Their ignominious stupidity makes me want to shout to the mountains in agony and unbelievable frustration. When a cloak of arrogant self-righteousness hides their incomprehensible stupidity under a cloud filled by their belief in self-importance it becomes hideously loathing.

When the object or aircraft stands motionless according to the earth centre the object moves only by the movement of the earth and therefore in cosmic terms the object and the earth is one whereby the object has weight and holds a value in mass. The object holds a relative position to the earth by the value of Π. This is because gravity then links to Π^0 and this holds regard to singularity. When the object moves while being connected to the earth by singularity the earth holds space at Π^3 while the object moves in terms of $\Pi\Pi^2$. The earth represents Π and the object moves by Π^2. When the object gets airborne the relation becomes $7(3\Pi^2)$ but I will not go into that. As the speed increases the relevancy of linking Π^2 breaks when the Roche limit is exceeded. The Roche limit is Π^2 or movement divided by the four quadrants of the circle making it $\Pi^2/4$. However while the aircraft shares the atmosphere one half is still within the earth making the Roche limit $\Pi^2/2$. In the Theses I prove these values extensively.

Thinking in terms of gravity we all think according to going in a straight line. The car is going skew but is going straight ahead following a straight line. That is thinking very Newtonian and backwards. Those preaching Newton try to convey the idea that Newton thought the earth was round but this fact as a science proven conclusion was established long after the days of Newton.

We see the car going straight as the car is following the curve of the earth. The car is following a circle by just going straight. In Newton's concept on gravity there is no mention and therefore there is no admitting to the curve of the earth playing a role in the forming of gravity by using the formula $F = G \frac{M_1 M_2}{r^2}$ The formula indicate gravity going by mass going in a straight line. This idea is as Neanderthal as $F = G \frac{M_1 M_2}{r^2}$ is.

The following is the simplest and the easiest I ever attempted to explain gravity. When a car moves at high-speed engineers say a car lift up into the air because air moves in underneath the car and that forces the car to lift up. That is Newtonian rubbish! When a car speeds forward there will always be more air flowing over the car then air could flow underneath the car and with more air above it is impossible to have the air underneath the car lift the car.

The air pushing the car down will always be much more than the air can be that lifts the car up. This is how gravity applies to a speeding car and this is why a speeding car starts to lift into the air at a high speed. The earth rotates at 7°. The curve of the earth holds the circle value of Π by which it rotates.

When moving the circle value move from a point holding Π to where Π move next and that distance is Π^2 and that is the value of gravity. The car moves around the curve at 7° holding a point at Π that align with gravityΠ^2. That means following the curve of the earth Π moves to Π forming Π^2 and this forms a straight line that goes by a circle. Then when the car speeding exceeds $7°(\Pi\Pi^2) = 217$ km/h the line the car moves in starts to grow longer than the 7° that forms the earth curve.

The car straightens the curve of the earth when it stretches the 7° the earth presents as the curve around which it spins. The car leaves the curve of the earth and gets airlift. It then needs "wings" to force the car down by creating more gravity because gravity is the space that pushes down on the earth and the wings produces more space that pushes the car down on the ground. As soon as the car lifts into the air not touching the ground surface the relevancy changes to $7°(3\Pi^2) = 217$ km/h as the relevancy changes to $7°(3\Pi^2) = 217$ km/h. When the car is airborne the formula changes from $7°(\Pi\Pi^2) = 217$ km/h to

If the body does not connect to the earth by way of forming a unit or forming "mass" the axel line of the earth comes into affect giving the ratio a value of 3. Then when not connecting to the earth surface the ratio becomes $7(3\Pi^2) \times \Pi^0$ going up

to $4\Pi^0$ all depending on the sped or the height. This is the way gravity forms by forming a connecting association with $\Pi^0\Pi$.

This forming of gravity explains the Coanda effect because the Coanda effect puts air or liquid in relative movement with solids and this movement (but not the mass part) is how gravity forms.

What is in the Universe, is spinning and therefore what I am referring to, applies to everything holding a place in the Universe and therefore this which I mention directly links everything holding any space whatsoever in the entire Universe to one single point around which all spin, notwithstanding the allocation. In the **precise middle** of all **objects in rotation** disregarding size is a precise centre dividing the object into opposing sectors that will **start the spinning initiation** from that centre point. The spinning object will have a very specific **centre point that does not spin** and only holds Π as a specific value because no radius can apply at the point being one space away from Π^0 holding Πr^0. But also the one value such a line **cannot have is zero** because the line **is there and being unbroken, it holds contact** with the rest of the material bringing about that **zero does not start any** line and therefore the **value of the line must be infinite**, just as described in **accordance** and by **the definition of singularity.** As I am introducing a very new idea, I wish to explain in better detail what I try to convey. While anything spins, singularity forms a line and when reducing the rotating line or radius progressively to the middle at one point all further reducing must end. As the rotating direction moves inwards, the rings forming Π will become smaller and smaller. Then we reach a point everyone thinks of as being the axis around which everything rotates. The line only forms when everything around the line spins by establishing a circle to the value of Π. go to http://www.singularityrelavancy.com/ and also go to www.questionablescience.net Flying object is under this gravity control of movement and it is this that has crafts fly and cars requiring down force by the aid of aerodynamic devices.

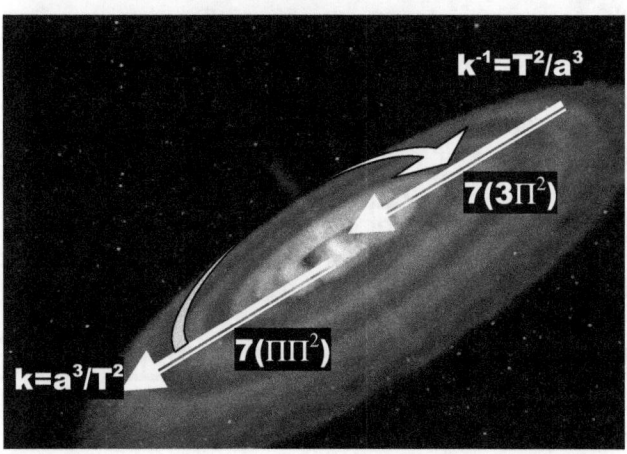

The Universe is contracting by $7(3\Pi^2)$
The Universe is expanding by $7(\Pi\Pi^2)$
Gravity is the inclining of material spinning in relation to 7° that is part of Π
This is because Π is 3.1416 and spinning in space is 3 which makes time expanding 0.1416 x 7 = 0.991 larger than 3. The concept might appear simple when told in this manner but the entire philosophy is so much more complex when studied overall.

When something moves it is within the space the earth rotates. When anything moves above and beyond the earth's movement it excels as it exceeds the movement the earth provide and with it, it takes the space it holds in excess of the position the earth relates too. Thus by moving it form a space-time unit within the earth confinement but yet also out of and beyond the space-time the earth provides.

Between Π/Π^0 and Π^2/Π^0, the body finds itself in its circular displacement, which is the earth's linear displacement value. When an object exceeds the earth's circular displacement value, it is refer to as Mach 1, or the speed of sound. Objects fall at a rate of $7(3\Pi^2)$ = 208 km / h while the earth gravitational displacement is $7(\Pi\Pi^2)$ = 217 km / h. because the earth moves faster than anything can fall it will always reduce space and have all surrounding space confined to the earth. The object falling only holds space that is confined to the earth.

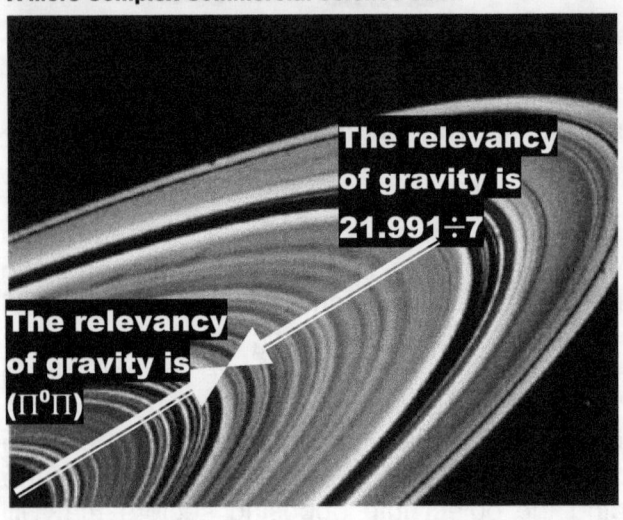

The relevancy of gravity is 21.991÷7

The relevancy of gravity is ($\Pi^0\Pi$)

By observing the rings around the planets we can identify all four pillars forming the composition thought of as gravity. We have the Lagrangian points that is five and I explain why it is 5, we have the Titius Bode law and I explain in detail the Titius Bode law, we find the Roche limit and I explain why it has the value of $\Pi^2/4$ and we can see from my explanation I have provided above why the Coanda effect accumulates all the principles forming gravity.

When the object or aircraft stands motionless according to the earth centre the object moves only by the movement of the earth and therefore in cosmic terms the object and the earth is one whereby the object has weight and holds a value in mass. The object holds a relative position to the earth by the value of Π. This is because gravity then links to Π^0 and this holds regard to singularity. When the object moves while being connected to the earth by singularity the earth holds space at Π^3 while the object moves in terms of $\Pi\Pi^2$. The earth represents Π and the object moves by Π^2. When the object gets airborne the relation becomes $7(3\Pi^2)$ but I will not go into that. As the speed increases the relevancy of linking Π^2 breaks when the Roche limit is exceeded. The Roche limit is Π^2 or movement divided by the four quadrants of the circle making it $\Pi^2/4$. However while the aircraft shares the atmosphere one half is still within the earth making the Roche limit $\Pi^2/2$.

This will come into affect when the projectile reaches Mach 1. I choose to use the accepted term Mach 1 at this point as to limit any confusion that may arise from the new arguments. I have to stress the fact that the Doppler effect is, once again, merely a co-incidental but duly related by product. However, the Doppler effect, as such, plays no part in these phenomena, or in the outcome of the application. It must be seen in terms of cosmology and not from a human perspective. This is the principle applying when a newly formed star escapes the heat envelope within the centre of a galactica as a star is born. There now are two singularity factors within one space-time unit and the two goes into a battle of existing.

There is the same principle applying between stars sharing space-time. When two stars are at the Roche limit, the linear displacement reaches a value of one, and $\Pi^3/\Pi^2\Pi = \Pi^0 = 1$ is equal to singularity and in this we find space-time forming a value. This formula does not impress the most learned Physicists such as Professor Doctor Friedrich W. Hehl, which you will learn from a little further on in this book since Professor Doctor Friedrich W. Hehl thought this was to use his words "With a lot of words and some simple algebraic relations, there is no way to "explain" the world of physics." However notwithstanding Professor Doctor Friedrich W. Hehl not being impressed, on this rides the entire cosmos formed by gravity. The circular displacement reaches its full complement of half Π^2 which is the Roche limit.

Those Ever-So-Wisely-Mathematically-Sublime-Superiorly- Educated-Cosmic-Super-Brains never show that Newton is at odds with the cosmos on every issue at hand. The cosmos not once implement anything Newton says and Newton can never be reconcilable with what applies in the cosmos.
Go to Questioning Newton's Mythology http://www.lulu.com/content/e-book/questioning-newtons-mythology/7570956]

I can prove what gravity is by proving the sound barrier because gravity is the sound barrier.

Gravity is the rotation of the earth in relation to everything that rotates with the earth being relative to everything moving either in conflict to or in support of the earth rotating.

I say I can prove God with far more proof than Newtonian physicists can prove a factor such as mass does exist and moreover that mass can bring a pulling force such as gravity!

I prove what I say so I challenge those in physics to prove the fact of mass and not just weight.

Gravity is the contracting or compressing of the space surrounding the earth and this is by way of rotation.

I use the very same values I apply to gravity to show how the sound barrier works. This I can do because the sound barrier and the concept we think of as gravity is the very same thing that is what a Supernova star and a jet going through the sound barrier and the Tunguska event in 1908 was. It is the very same cosmic principles.

It is very quietly and not well promoted but we know that the moon and the earth are drifting apart. Nothing is coming closer together as Newton claimed in his principles.

Very quietly and not well promoted is the fact we know that the circumference of the earth or the circle around the earth is steadily enlarging thus Nothing is shrinking.

No where does any Newtonian ever put this tendency in relation to Newton's gravitational pulling principles of contracting in line with other cosmic accepted norms such as with everything expanding since the Big Bang, then what is shrinking.

Galileo proved that a large pendulum swings in time with a small pendulum and this Newtonians never dispute. Galileo proved that large things and objects fall at the same rate and by the same speed as small objects and that too they never dispute.

Yet they force students to repeat in examinations that objects fall by mass, which means that a big object and a small object falls differently because of mass, and yet they fall the same. If they fall by mass the bigger object must fall faster then the smaller object does. As the objects of different mass fall equal as Galileo said it does, and then Newton is wrong in claiming objects fall under mass pulling mass.

...And yet you can go to the Internet and see how many Newtonian cheats calling them physicists announce in capital letters that Newton and Galileo had the very same principles in mind. You are going to read why I call Newtonian physicists the biggest cheats that walk the earth, they are more crooked than Bankers and Politicians are! If you know one personally please show him this book and see his reaction when he reads my accusations and that I ask to prove mass (not weight).

Part 7

Today you are going to read for the first time ever an explanation, a valid explanation how the cosmos forms space in ratio to gravity

You are for the first time ever going to read how the Titius Bode law forms and how this law applies in the cosmos.

To whom it may concern:

My introduction as well as introducing the readers to general cosmology in a very brief and compressed manner but first, I have to give the emphatic warning to all prospective contemplating readers.

Please take note of a conscientious warning about the gravity of the misgiving there is on the part of the most respected Academics in physics about a much concerning matter.

I state it emphatically that science accuses me to be not schooled to the point where I am able to have any form of an opinion on any matter concerning Sir Isaac Newton. Notwithstanding that my research proves I did my private studies and through which I skipped the indoctrination and mind control academics place on students goes unrecognised by their standards and so too my ability to have any insight on matters regarding physics. However my skipping their methodical and systematic brainwashing enabled me to see and allowed me to be able to express the incorrectness in Newton's teachings and allowed me to show in clarity what destructive force Sir Isaac Newton used to corrupt the laws of mathematics, corrupting to science along the way and mostly raping to the work of a great man, Johannes Kepler and what Sir Isaac Newton did can only be expressed as being blatant criminal fraud. What his deeds amount to is to corrupt the laws of mathematics, to render the laws of cosmology useless and to rubbish all of science. Should you find this to be unbelievable, then I am glad to announce that this book is more for you than any other person, so go on and read what academics guarding science never wanted published.

I challenge any one that disputes any claim I make to prove me wrong by proving me wrong and not merely suggesting claims in that direction.

I have written several books in which I challenge the thought process of Mainstream physics and especially Sir Isaac Newton's arguments about physics. I am of the opinion that even though everyone thinks that Sir Isaac Newton is the genius that established every aspect today used in modern physics the man did not have a foggy clue about any of the principles driving the concept. He name gravity, but was unable to explain the concept in any detail. I made a study on the subject of gravity and from that study I am able to explain the entire principle. I prove my explaining with mathematics backing up my statements. In using the correct principles, which was what I found applying in the cosmos and which was what Mainstream science had no idea how to explain or even how to interpret, those principles I use.

Moreover I back up my interpretation how the cosmos use those principles to conduct gravity with the correct mathematical formulating. By my effort of using phenomena applied by the cosmos but never understood by Mainstream science, I do a far better job than what Sir Isaac Newton did and what I achieve is of a far more acceptable level as well as being mathematically far more correct than what Sir

Newtonian approach cannot even recognise any of the four principles but only Newtonian science are taught to students. No student can have the fortune to disagree about Newton and remain a student at any institution while studying. Students are taught to accept Newton and to ignore Kepler and any student doing it the other way around will fail all examinations and other testing at the Universities the student is attending. Students accept Newton or they accept a ticket taking them home. According to Newtonian science space is simply nothing with no qualities but gravity separate space and space does not mingle, as one would expect if space was nothing because space does form borders. Those borders form part of gravity and gravity is the least understood concept thus far in science. In truth no one in science anywhere remotely knows what brings gravity about and I used Kepler to unravel this mystery called gravity. But no one in science will admit this fact about Newton or any one else never being able to explain gravity in the least or that Kepler is the one who formulised gravity decades before Newton came and gave gravity the name. Newton started this realising of gravity but it had and still has no more substantial proof than a rumour has and Newton admitted to it being a concept he could not explain. Still to this day nobody in science at present will denounce the principle of gravity being never explained. All in science act in a manner as if Newton's gravity idea is proven fact and only occasionally admit it to not "fully understood" as Newton admitted it was when he introduced the name (not the concept). That includes Newton as well as Einstein and even Hawking. Scientists can declare gravity was a factor at 10^{-43} seconds after the Big Bang but what brought gravity about or why gravity became or still remained, as a presence is still tightly concealed information which all are speculating on. Even to the best informed amongst the most educated do not know what is gravity because they all ignored Kepler and for ignoring Kepler the price they pay is not finding the principles bringing about gravity.

Kepler studied what kept the plants circling the sun and that what is keeping the planets in orbit is gravity. From such explaining what Kepler said without Newton changing formulas on Kepler's behalf I prove the Titius Bode principal also known just as the Bode principle. I explain how singularity forms the Roche limit and how singularity brings about the Coanda affect but most important of all Kepler showed me where to search for singularity. My achievements came from my effort where I separated Kepler's work from the opinion that Newton formed about what he saw Kepler's work should contain and gave to the world his Newtonian concept about Kepler's work. From my view a force is just motion applying and that is what Kepler said gravity is. Kepler said $a^3 = T^2 k$. Presumptuous as it may be on my part of trying to disprove Mainstream Physics, such a presuming does not change the truth about Mainstream science being incorrect about gravity. After all they admit they do not know what gravity is however this admitting is only sometimes when it suits them to do so. As they do admit they do not know what gravity is I am not disproving anything they proved because they agree they do not know what gravity is, which paves the way for my showing what gravity is. By they're admitting that they do not know what gravity is they then also admit being possibly incorrect about gravity but unfortunately mainstream physics do not see it that way (yet). The question in hand is finding what role gravity played when the Creation came about for the first time. I had to find a method that would allow me to explain why gravity played a role.

Remember that not even Newton could explain what gravity is or where it comes from, but Kepler did that without any person ever taking note of the achievement by Kepler. Kepler studied what kept the plants circling the sun and that what is keeping the planets in orbit is gravity. From such explaining what Kepler said without Newton changing formulas on Kepler's behalf I prove the Titius Bode principal also known just as the Bode principle. I explain how singularity forms the Roche limit and how singularity brings about the Coanda affect but most important of all Kepler showed me where to search for singularity. My achievements came from my effort where I separated Kepler's work from the opinion that Newton formed about what he saw Kepler's work should contain and gave to the world his Newtonian concept about Kepler's work. For instance from Kepler's work I can explain the operation of the Black Hole, which not even Prof. Stephen Hawking understands. From my view a force is just motion applying and that is what Kepler said gravity is. Kepler said $a^3 = T^2 k$. The Coanda effect is the establishing of individual independent space a^3 by applying motion $T^2 k$ in relation to a centre point the motion of the liquid establishes. Einstein came to this conclusion but failed to refer his view back to Kepler and by not referring to Kepler missed the point he wanted to make. Just by my studying of Kepler this became possible.

The Practice of Brainwashing and Mind Control in Physics

In www.sirnewtonsfruad.com I investigate how much truth there are in mass pulling by the force of gravity. Most if not to all of the persons reading this article will be annoyed by just the thought of me embarking on an investigation of the issue that seems so totally senseless to investigate. It is senseless because the concept it carries became accepted as household practise and life science. Mass is associated with everything that is represented everywhere.

Do you think of astrophysics as the department that is run by the wise and the level minded, the sober thinking, and the absolute trustworthy? If you are a student there is no other choice you have. If you think those in charge of astrophysics are the pillars of trust, then get wise and read the following. What you are about to read is simply mystifyingly simple and yet to this day I have not challenged one academic any where that had the honesty to admit to the fact of Newton being wrong. Newton created the factor mass as a trick of his imagination. We have a body with mass on earth and that we all know, but no heavenly body could ever have mass or present mass! After you have considered the following you might agree with me that even small Children can reach a higher level of clear-minded logic and find more sensibility than what those scientists promoting astrophysics have because science lives in a make believe fool's paradise. They love to calculate because with mathematics they create a fools paradise.

Looking at the size of a body does not allow any one to awards mass. If you are a student, then ask your Masters please to explain the following abnormalities and inconsistencies they promote as part of official physics, which I present in this article and as students get wise instead of brainwashed. I say brainwash again because they force-feed you fabrications, as you will come to see. They can't explain the facts as reliable but hide the fact that the facts are in fact untruths. Tell them to prove that planets have mass. Tell them to prove that it is mass that generates gravity that pulls the planets. Ask them to explain gravity in detail. You can get the question to ask them for free from www.sirnewtonsfruad.com.

If any person gives evidence and the evidence is unsupported by truth and unfounded by proof, it is fraud. If a Clergyman should bring evidence of some miracle healing that took place, and he could not repeat the act at will for all to witness again and again, it is fraud not withstanding the religion. Even if the lie is to underline the greatness of his religion, it still constitutes to fraud. If anything is said without truth backing the claim, making it unfounded, it is fraud. It is illegal to make claims that is not based and is not supported by reality. As perfect as everyone thinks Newtonian physics are, this applies to educators teaching physics. There is a suspicion lingering in the back of everyone's mind that something is not quite correct about the approach physics take on the matter of gravity and only those academics seasoned with years of studies and salted with time seems to miss this haunting feeling. But the longer the students are educated, the more this uncomfortable feeling dissipates.

Hidden under a cover of "understanding Newton" or "not being able to understand Newton" tutors in physics force certain incompatible arguments to join that which never can join and while joining also make sense at the same time. In this article I challenge the figures that charge the highest form of respect in our communities and those in charge of the most dynamic part of society and those who stand beyond and above any form of suspicion. I charge those that personify truth and are the very same persons that I accuse of betraying the ones trusting them. They misuse the positions of trust by participating in the conspiracy I am about to tell you about. Educators in physics I hope that you will stop your mind abuse on students because they are now able to know more about physics and the application of gravity than did all those that came before you and all those that came with you. Again I challenge you to come forward and tell your students the truth about what I uncover in the articles that follows and as the articles progress by introducing information...then you explain to them how you deceived their blind trust in you as a tutor in physics. In physics the blame goes to students ability to "understanding Newton" or "not being able to understand Newton" but I show that in this case the blame should openly be dedicated to those that should be blamed, named and shamed and they have to defend their years of lying and contribution to cover the misconduct that was committed by them. Those teaching physics as well as their predecessors now have to explain in the name of science why so many evidence were falsified to keep their noses clean while mud colours their lily white image in hogwash and the stink of their lies equals the pig pen it needs to cover such aroma. It is time that they reveal the truth. Believe it or not, but this diverging from reality and misconduct about applying science is in place because of centuries of brainwashing going on passed on from generation to generation and is employed in physics from teacher to student for centuries on end. This article is dedicated to bringing honesty into the faculty of Astrophysics and Astronomy as well as to show the Physics student on what corruption and deceit does physics base their facts which they proclaim as being such well proven, and godly accurate facts and is unwavering depicting only the truth. Notwithstanding as unbelievable as it may seem, I nevertheless challenge any one in physics to show me that the least of any or all facts I uncover is not true.

Go to the web site **www.singularityrelevancy.com** and then use the information I supply in the book **www.sirnewtonsfruad.com** titled Sir Newton's Mythology, which you may download for free and use the information given free of charge to confront you academics with the truth and then insist that Academics who are teaching physics, prove to students that Newton's statements of attraction are correct. Let those academics explain the method mass implements to obtain the attraction they teach about.

If you are a student studying in physics then reading this article is detrimental to your future, as you can remain part of the problem physics has had for centuries or you may join the solution that came to physics and start to heal the wound. Students read the next pages and you are about to learn how students are brainwashed into accepting the baseless and ridiculous misinformation Sir Isaac Newton puts forward as truths. The Custodians of Physics have nothing better to offer than presenting you with unfounded, corrupt and distorted facts...and by doing that they resort to mind control on students and introduce baseless concepts by manipulating the student's thoughts. If you are a student then read in www.sirnewtonsfruad.com what they do to you and how they brainwash you. Everyone knows there is a problem about gravity in physics and this far it seems as if no one can put a finger on the problem. This suspicion that there is, this feeling about a certain concern and doubtfulness that is lingering on in the minds of many… and also is lingering on from generation to generation… without anyone ever finding a solution… it is them defrauding you by exchanging your institution fees for corruption, so confront them about their dishonesty. Something about the way gravity is presented just doesn't add up as it should and does not quite reach the answers it should conclude. There is this vague unspoken question hanging in the air without any one ever finding words to express the question...and yet the question remains however unspoken it seems. Force them to become honest and to stop corrupting students with intentional malice. That which they offer has no truth. All they have are misconceptions and incoherent facts. Let them explain how mass generates gravity...

There is always a thousand and one conspiracy theories going around and the one tries to be bolder and more sensational than the next theory flying around. However, the biggest conspiracy is going on in front of every person's eyes and is committed by the most respectable persons in any respectable upstanding society. Even if you had your personal favourite conspiracy theory, try and match it to the one that I have!

Let them with precise detail show when mass is applying, that mass and such producing of gravity that then would establish attraction produce gravity! Show that objects in outer space are optionally allocated places by value of mass. Let those show where they see stars in any galactica are positioned categorically by size or mass. I show precisely how gravity produces mass but mass can never produce gravity. I show with explicit detail when, how and where gravity forms mass but mass can never form gravity. What I prove annihilates every Newtonian claim. No one cares to read or to print my work because there is no need for it.

I seem to be the first person in generations that ask questions about Newton's work. Questions I now ask is asked for the first time ever, well ever since the time Newton introduced gravity, before the emphasis fell on proof rather than merely what a person with reputation suggested. I now am able to show how gravity forms by forming a circle using Π because I have located the centre of the Universe. But by my effort in finding the location I disrupted everything Academics in physics hold holy and for that I am most unwanted in the presence of the Academics charged with guarding the ethics of physics. In short, I clash head on with Newtonian dogma and principles forming physics. During my research I discovered abnormalities and inconsistencies about mistakes the Arch fathers in physics must be aware of but are hiding with all their considerable influence and academic power. The road I took in my search for truth concerning physics was never smooth and the resistance I came across coming from the academic sector is almost unbearable. I made no friends but only enemies.

With my unpopularity rating this high as it does, I never qualified for help and those that would help found my ideas intolerable whereby I only found rejection instead of help as I tagged along. Because of this insider rejection I had to resort to private publishing because from the nature of my work I take Mainstream science head on and am confrontational on most aspects of astronomy. By finding the centre of the Universe it enabled me to find a point the Universe is controlled from, but because that does not hail Newton it sparked no interest. Then I decided to reject Newton is the only road to go if one wishes to lay axe to the root of the insider corruption they are guilty of. Since no one see the a problem there is in physics, no publisher wants to go head bashing with the Physics Custodian establishment of science on official science principles, which I have to do to convey my message in no uncertain language. I argue that if it is the correct practise to use $F = G \frac{M_1 M_2}{r^2}$ to calculate gravity then the radius holding the gravitational constant must lead one to the centre of the Universe. As I confront science dogma and principles, nobody is willing to publish my work. I have to walk the road alone and fight the battle by my private effort without any support anywhere. If Newton is the problem one have to go pre Newton to find the problem. The problem is that when looking at Kepler's table then if there is $T^2 \div a^3$ according to the table matching a column, then mathematically $T^2 \div a^3$ must be k^{-1} and where k^{-1} goes negative it shows space reduces time. It shows space in volume goes single by movement of space and not objects.

Planet	Mass per Earth unit	k^{-1} Movement	a^3 of space volume	T^2 During time units
Mercury	0.06	$T^2 \div a^3 = 0.983$	$(a^3) = 0.059$	$(T^2) = 0.058$
Venus	0.82	$T^2 \div a^3 = 0.992$	$(a^3) = 0.381$	$(T^2) = 0.378$
Earth	1.000	$T^2 \div a^3 = 1.000$	$(a^3) = 1.000$	$(T^2) = 1.000$
Mars	0.11	$T^2 \div a^3 = 1.000$	$(a^3) = 3.54$	$(T^2) = 3.54$
Jupiter	317.89	$T^2 \div a^3 = 1.000$	$(a^3) = 140.6$	$(T^2) = 140.66$
Saturn	95.17	$T^2 \div a^3 = 0.999$	$(a^3) = 868.25$	$(T^2) = 67.9$
Uranus	14.53	$T^2 \div a^3 = 1.000$	$(a^3) = 7067$	$(T^2) = 7069$
Neptune	17.14	$T^2 \div a^3 = 0.999$	$(a^3) = 27189$	$(T^2) = 27159$
Pluto	0.0025	$T^2 \div a^3 = 1.004$	$(a^3) = 61443$	$(T^2) = 61703$

Use the column to show where it is mass that produces any valid factor. In the table columns we have the mass of the main solar objects and according to Newton the mass is responsible for movement. If it is mass that does the pulling in ratio of the radius then we have to see some evidence of this applying somewhere. Is there any person that can prove that claim from the table showing Kepler's movement in relation to Newton's mass? In what way odes Jupiter move faster or how does Pluto move slower according to mass? If mass did the job, how is it done? Notwithstanding Newton's blatant mathematical manipulation, mass don't apply.

One such an example is Einstein's Critical density scam and the "Dark Matter" swindle where they hide Newton's shortcomings under a pretext of ongoing research. Hubble found the Universe is not contracting as Newton said it does with $F = G \frac{M_1 M_2}{r^2}$ and hell broke loose. The cosmos stepped out of line by showing Newton is wrong in his supposed theory that mass pulls mass to reduce the radius such as $F = G \frac{M_1 M_2}{r^2}$ will indicate. Einstein was then tasked to supposedly measure "all the mass" in the entire Universe and to find out when the Universe will correct its evil ways and start to submit to Newton's idea of contraction. They concluded it is not Newton that is wrong but the cosmos went out of line to contradict Newton. Therefore one has to find the error in the actions of not Newton's idea but about the Cosmos acting out of line with Newton. The fault was investigated not on the side of Newton, but in the cosmos! The proof of practise fell on the Universe that could be wrong but never Newton!

With all this in mind did any one ever come to wonder about the reality driving the all too famous **Einstein's Critical Density theory** and the fact that this idea was conceived to conceal the corruption of Newton in physics? Allow me please to elaborate and then make up your minds. The facts in truth are that the **Einstein's Critical Density theory** was a scheme plotted by those in charge of physics principles to cover up and conceal corruption in the heart of physics. Hubble proved beyond doubt that there was an inflating Universe. This contradicted Newton's deflating or pulling Universe and this perception of a deflating Universe being a myth had to be most ardently hidden as to yet again compromise the truth about Newton and his theory. In the formula $F = G \frac{M_1 M_2}{r^2}$ the relevancy applies between the strength of the mass and the distance of the radius that keeps the influence of the mass forming the gravity. The longer the radius increasing the distance between objects the more this will reduce the value of the mass, whatever the value of the mass might be. The Hubble concept proves that while the mass might remain the same, the radius keeps growing and such growth diminishes the influence of the mass all the while the radius increases. This is so basic that primary children learning the basics of mathematics will understand! This means with the radius increasing, there is no chance that the mass will ever become strong enough to bring about the pulling because the constant increase in the radius constantly diminish the influence that the mass might produce.

Yet Einstein proceeded in searching for a value that will determine when the mass would bring a turn about in the direction that the cosmos evolves in. Einstein was looking for the moment the mass will become strong enough while the most basic principle indicates that an increasing radius leads to a decreasing mass influence submitting a decreasing potential gravity since the mass becomes less prominent in influence. If Einstein was unable to recognise this most basic of mathematical principles, then what type of genius did physics create in him and what slur did physics promote. In $F = G \frac{M_1 M_2}{r^2}$ any factor standing in relation of division, it is the bottom value that determines the outcome of the value of the top. This idea of the two factors being in opposing relevance where the size of the bottom caries the value of the outcome is so simple that children will recognise the principle, and yet those fathers of physics wants me to believe that the greatest mathematician that ever lived did not realise this principle…the principle that the radius and the mass stands related and the growth in the bottom value will promote the decline in the top value as a dominant factor. Can any one with this information including the proof I asked for on previous page have any other conclusion than students should smell rotting fish somewhere?

Notwithstanding my arguments that should have been raised by the mathematical genius, Einstein supposedly measured "all the mass" in the entire Universe and then afterwards concluded there is "insufficient mass" to pull the Universe closer. This rattled the cages of the Newtonian conspirators because Newton once again stood naked, venerable and bear. Yet the blame had to be associated with the cosmos not playing by the rules of Newton. The cosmos had to be at fault because Newton just can't be the person to blame. Then the genius of the Newtonian cunning kicked in once more and we can see why they are seen as the most brilliant minds walking the earth. There had to be something they miss and no one can see.

Then some idea was presented that the Universe is hiding mass from view, I suppose just to spite Newton. If the mass was not visible and was therefore undetected, then the mass was dark and if no one could see the mass then no one can prove the mass being present and then no one can disprove the mass being there waiting in the dark. What a splendid idea this was for cheating. This presented a solution of Biblical proportions and a new scam was introduced to hide the failing of Newton and of the Critical Density sham. The Dark Matter hides mass that will supposedly pull all material closer again at one stage in the future. This is meant unleashing an enormous swindle, bigger than anything before, only beaten into second place by Newton's swindle about mass unleashing forces. They proclaim there is "Dark Matter" hiding and waiting in the wind to come to the present someday in the future and start to pull the Universe closer as Newton said it must by the measure of $F = G \dfrac{M_1 M_2}{r^2}$. In that way Newton was vindicated and the cosmos took the blame for hiding mass and the contraction was reinstalled.

It is obviously clear that having such a total idea that there might be dark unseen mass floating in the Universe which at this time does not generate gravity but will some day kick in to generate gravity in order to cover up Newton's deception about a contracting Universe and just because Newton has to be correct at some point in the future. Science wishes me to believe that since there is a lack of seeing material there then will be dark and unseen material where they are so dark they are undetected by all humans. That leaves another question to address...where can such mass be found. How will such undetectable mass be found, which will bring about contraction after all this expanding ends and the Universe recognises the infallibility of Newton once again? Why would the mass at present then not activate gravity and why would the mass at some point spring to life and start activating gravity? How much can the Physics paternity still hide the fact that Einstein's critical density is being used as a cover-up to distort the truth conceal the fraud Newtonians wish to cover? The uncovering of the Newton fraud by the Hubble constant is so simple to see. Hubble found the Universe is expanding and Newton said otherwise. Who is lying about what? Hubble's declaration was on track to blow the cover that was concealing the Newton fraud wide open and uncover the century old deception. To see this we have only too look at the comet behaviour when any and all comets again come around on a cycle by repeated visiting the sun. The question is if it is mass pulling mass onto mass, then why do we have comets left in the solar system? The mass of the Sun should by now at least have destroyed every comet going around.

The term pulling does not suggest any circling because no one can be pulling towards and does that while circling around the object. That serves the term pulling. Then the saying goes that planets orbit indicating they follow a circle. In conversation we speak of the planets orbiting. If Newton was correct we should be speaking of the planets pulling, but talking about pulling would be blatantly wrong according to the normal spoken word. Never do we refer to the planets pulling the Sun or the Sun pulling the planets, but we speak of seasons coming from orbital positions. Being in orbit has to neutralise the pulling and then cancel the pulling concept that also became culture. The entirety of physics rests on this one formula $F = G \dfrac{M_1 M_2}{r^2}$. The questions concerning that which you are studying and that touches every aspect you are academically concerned with, is that if everything is moving apart, how does that support Newton's idea that everything is coming together...and please don't let them fool you with Einstein's Critical Density idea! If there was mass seen or unseen in the Universe and mass generated gravity and gravity does the pulling then why is the mass not at this moment doing the pulling? If there was mass in the Universe, seen or unseen and mass generated gravity and gravity does the pulling then why is the mass not at this moment doing the pulling? If there is an object it indicates the presence of mass. Then what stops the mass that should be present, dark or luminous, to start pulling by gravity, as it should do?

Should you think this page is some sort of a prank then answer the following simple question to yourself in utter honesty: If there is a Big Bang after which everything was left moving apart, how does that support Newton's contraction? Tests results received after the Moon landing show the Moon and Earth are moving away from each other! Yet students learn about mass pulling mass and that puling by mass forces togetherness by contraction.

This is only one of many points that I make on this one issue and there are so many other issues one may think of those in terms of counting in numbers in many hundreds or even in thousands. If the Sun for instance has mass that is apart from the Earth and the Earth also has mass and there is a gravitational constant in between the Sun's mass and the Earth's mass we have the radius in that location. It then must be the gravitational constant that fills the space that the radius holds. It is rather obvious that while the radius is filling the vacant space between the Sun and the Earth it is the only place left where the gravitational constant can hide. To find the centre of the Universe I had only to find the gravitational constant that holds the centre. Through my venture I discovered one person that knows what gravity is! Newtonians went and filled that space reserved for the gravitational constant having a measured value with nothing! How can nothing have a value of 6.67 X 10 $^{-11}$ while also being filled with nothing as it is nothing filling the nothing of outer space?

My question in this matter is what is all that mass of so many supposed stars living in utter darkness doing at present while waiting to get to work and begin with generating gravity by mass where it will only later, much later form a force of gravity that then will bring about this pulling of the Universe? What makes the mass slumber in darkness to one-day form a pulling force? What has the "darkness" or the fact that we don't see the mass got to do with the idea that the mass at present is not forming gravity that is forming a pulling force? You are taught that gravity pulls objects to the centre and obviously gravity then has to ultimately pull everything to the centre of the Universe. That is what the Critical Density research that Einstein initiated wishes to establish. The idea is that $F = G \frac{M_1 M_2}{r^2}$ makes the mass create a force that will destroy the radius and ensure everything is going to come together eventually at one point where the radius then will be no more. If that is the case, then where is that point? If everything is destroying the radius, then it must end at one specific point.

In the classes you students attend a physics lecture, has any one confirmed a location where one might find the centre of the Universe to confirm the ultimate destination of $F = G \frac{M_1 M_2}{r^2}$? If you wish to apply a Gravitational constant as a calculated factor then it is apparent that one must know to where such gravity is pulling since it then is the gravity that is where the contraction is going that predominantly is keeping everything apart. Then the gravitational constant is what is resisting the collapse of the Universe. If there is a force, then where is the force taking the pulling…if it is a gravitational constant applying through out outer space then where is it having a centre base? To those professors claiming Newtonian ideas are substantiated by proof, I say that notwithstanding your personal academic qualifications and while at the same time disregarding your status and previous achievements as well as ignoring your many admirable abilities you may have and however superior they might be, I shall teach you about gravity. I say it is time Students learn the truth about physics notwithstanding the status academics will loose.

What you are about to read comprises of extractions forming part of the actual book that you can download free of charge. Go to www.sirnewtonsfraud.com and press on the blue button and the book is yours to have! Use this information to test the reliability of your tutors' teachings. Confront them with facts and don't allow them to stupefy you with their ability to commit mind control. This is what eighty years of studying the solar system brought about and no individual study since changed one aspect of what this table brings as information.

If mass did attract mass, what kept the balance in the distance it held according to the Titius Bode Law where the planets do find a balance in orbit rather than moving towards the Sun. One might think of mass pulling the comet to the sun, but instead of slamming into the sun, the comet makes a circle (Π) and disappears into outer space. As much as the comet is pulled coming towards the sun, the comet is pushed away afterwards departing for the darkness of outer space. If mass pulls the comet closer, what does the pushing away afterwards. If mass did attract mass, it would explain the behaviour of the comet coming towards the sun…but then what is pushing the comet back into orbit, into the darkness of outer space? The idea of proof about mass pulling everything is automatically placed at the door of Newton. This is not as small an issue as it seems at first. Think of what planets do… and you think that planets orbit. If normal speech contradicts Newton, then it is his task to prove his supposition is correct and the claims about attraction Newton made, although Newtonians will deny this fact as if they deny he is correct the honour of their Master Newton and that is what they have to do. Think of what planets do… and you think that planets orbit. No one thinks of planets spinning or planets basking in the summer Sun. However, just using the term orbiting is in total defiance with Newton or physics!

Ever person associates gravity applying as ***"The natural force of attraction exerted by a celestial body, such as Earth, upon objects at or near its surface, tending to draw them toward the centre of the body"*** with what we think of happens to solar bodies having mass which is ***"the natural force of attraction between any two massive bodies, which is directly proportional to the product of their masses and inversely proportional to the square of the distance between them"*** and science cheats everyone into believing the two is the very same. The one I experience every day and with the other there is no evidence of applying anywhere. Yet students are fooled into believing the two are exactly the same issue, which is untrue. If there is any academic feeling insulted by me calling the lot fraudsters, bring evidence of the second form of gravity working on the principle of mass and I will withdraw my statement, otherwise if no proof can be brought, then you lot that ignored me for ten years are the fraudsters I accuse you to be. You are villains brainwashing students to corrupt their thinking.

I have shown many ways how mass and Newton does not apply. There is no possible way to produce evidence from the behaviour of objects in the solar system that mass is responsible for gravity forming between two heavenly bodies. If it did, the asteroids could no possibly orbit in their allocated circle at the same rate as Jupiter circles in its designated circle. Every one believes in Newton except comets, because comets fail to collide with the Sun. However I can explain in some way…The evidence that mass is pulling mass there is not and there is no evidence of mass pulling mass. What there is I can prove how that forms gravity?

The Definition of Gravity according to the Oxford dictionary y of Astronomy:
Gravity is the fundamental force of attraction that all objects with mass have for each other. Like the electromagnetic force, gravity has effectively infinite range and obeys the inverse-square law. At the atomic level, where masses are very small, the force of gravity is negligible, but for objects that have very large masses such as planets, stars, and galaxies, gravity is a predominant force, and it plays an important role in theories of the structure of the universe. Gravity is believed to be mediated by the graviton, although the graviton has yet to be isolated by experiment. Gravity is weaker than the strong force, the electromagnetic force, and the weak force.

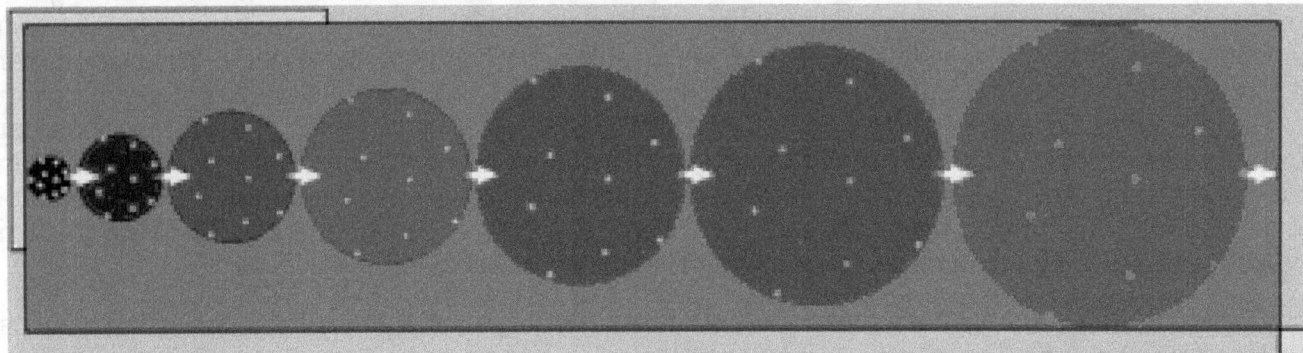

Presenting the Universe in this manner also forms part of the Newtonian conspiracy because it puts limits on what can never have a limit. The Universe has no outside because the Universe has only an inside. We have to look at the Universe not from the point where we are God but where we look up not the never-ending darkness and from where lights intensity starts at singularity.

I can and do explain this principle mathematically in a book I give away at no charge. Go to and download <u>THE ABSOLUTE RELEVANCY of SINGULARITY : THE WEBSITE</u> http://www.lulu.com/content/e-book/the-absolute-relevancy-of-singularity-the-website/7517996] and doing so will show you how it is possible to condense all the space around the spinning planets to form gravity by allowing cosmic structures to act as a centrifugal pump impeller. That is how gravity forms Π. You may see how material compresses the space around the star into a dense liquid. The liquid comes about from the spinning of the star compressing space and not by magical mass pulling. A star is compressed heat because when a star explodes it is always a process of releasing heat.

That is gravity. It is compressing space to form heat. That is gravity applying according to the Coanda effect and not by mass. This website informs you about the facts behind the lovely pictures that Newton's magical mass is unable to explain. Mass forming gravity depends on magic and physics has no magic. Please think clearly, if there is no contact between two objects and yet they teach you the two objects pull each other, then that pulling depends on magic, not science. This website will tell you why the cosmic physics principles establish these phenomena...and because the information does not salute Newton, academics in physics despise what I say. In the image one can clearly see the Roche limit in action where none of the stars are pulling any other star closer but a process of liquefying takes place.

The Roche limit clearly shows no stars can pull each other into a collision but clearly expands by overheating and if mass don't pull stars to a point of colliding then Newton is wrong! The Roche limit proves that any star that comes closer than 2.4674 times the radius of the larger star would liquidise everything within the space filled by the smaller star and this law and law providing the Coanda effect are a precise duplication of the interaction of the way that gravity truly applies in the midst of singularity. It is so clear that when stars expand stars overheat and therefore gravity must be a process of cooling! Only when stars overheat can stars explode as they do. You can see that gravity is locking in heat just by looking at what happens in the Universe and by studying every image as the Universe unfolds before your eyes.

Kepler used these three laws for computing the position of a planet as a function of time. His method involves the solution of a transcendental equation called Kepler's equation. Years ago I was reading of a remark Einstein made about his realisation whiles being a patent clerk. Einstein realised that had Einstein fell from the window of the patent office Einstein would feel as if he was as weightless as a chair and a pen falling alongside Einstein down the building.

Then I then realised Einstein felt weightless because he was falling and part of falling was feeling what was happening to him. He was not pretending to fall whereby he then would feel as if...he was really falling and with that there is no as ifs. What he experienced came by means of what he was experiencing. If Einstein was experiencing weightless ness, it would be because he was weightless while falling. Einstein would not imagine the weightless ness because Einstein was truly falling. He was at that moment truly weightless. Einstein, the pen, and the chair had the same weight since they were all weighing the same. All three items would be equally weightless during the falling...that was what Galileo found because objects of different size and different mass travel equal while descending. The bigger objects do not fall quicker than a smaller object and that can only be attributed to one fact; it can only be true if they weighed the same while falling.

From this one can deduct that gravity is motion or the intent to commit motion and mass is one the motion of gravity is frustrated by blocking the continuing of the motion. Gravity is motion of space and mass is the restricting of the motion of space. Having mass does not bring about gravity but it does restrict gravity's motion. Gravity produces mass but mass does not produce gravity. Mass is the restraining motion and gravity is material moving about. Mass only comes into the application when two objects filled with space moves into a position where both want to claim space the other occupy. In essence it still is the frustration of motion and the commitment to move once the blocking of space is relinquished.

I then after reading this realised that gravity is not mass orientated, but gravity is motion differentiation between objects. While falling, The object moves less or slower in the direction that the Earth rotates and will fall in the direction of the Earth centre until such a time as the movement of the object is in synchronising with the speed that the Earth spins or if not the object will and on the Earth surface at the edge of the Earth and that will bring about having mass. The gravity applies as speed that is putting time in relation to the distance travelled and distance travelled is space. While the object is in a process of falling, the motion confirms gravity, both by getting the object's distance or band in which the object travels in harmony with the Earth that conducts all the spinning taking place at that point. That will reduce the height in which the object spins until it lands on the Earth and then can't reduce such reducing of a travelling band any further. It has to do with specific density. If the specific density is increased by filling the object with helium we will find there arrives a point where the conducted speed is at a level that the Earth no longer will claim the body into having mass. When motion downward ends and the Earth disallow any further movement to secure a better specific density in relation to rotating movement, then mass sets in and becomes what is than point holding mass where the constraining of the object takes place to secure frustration of further movement and the Earth's motion annexes the object's freedom.

While experiencing mass the motion is still there but now incarcerated by mass and locked onto the Earth by the rotation of the Earth and the superior or equal specific density of the Earth. By connecting to the Earth the motion that the object is experiencing is what nails to object to the Earth by the force of mass and the object is then experiencing mass and not falling further through the loss of downward movement and now only conducts with the Earth rotating side-on movement. In this the downward movement is not lost altogether but remains as detectable movement is the form of having a tendency to move although the object in mass is applying by forcing the downward motion to stand still. While the object is in mass and seems to be as if it is resting the tendency to move downward remains applying but that tendency to continue to move downwards is the tendency he named mass. However mass then restricts motion and becomes motion tendency. While falling, gravity applies as equal motion to all objects relying to place all objects in relation to specific density and because of this motion counteracts any size, mass or weight by making everything able to fall equal in specific density. When falling, the object is either equal to what might be in the air according to allowed specific density, or has more than the specific minimum required density that is what is allowed to serve as the minimum required specific density and therefore will spiral down to the Earth. When the Earth restrains further downward motion of the object that comes as the result of finding an allocated position of motion according to the specific density of the falling object, this readjusting of allocated position is stopped from conducting further downward or readjusting movement and all such further movement of gravity is hindering in the form we call mass. The falling object remains individual and still tends to move while Earth individuality resists movement. Further movement is disallowed as other material fill space. While the bonding of the atoms forming the object will secure any further deforming the object will remain to be independent but it is this bonding that is the value of the specific density of the object applying. By securing a [lace on the Earth, the falling object will finally rest and from that motion resistance comes mass.

While falling, the object is experiencing gravity because the object is in gravity but when on the soil the object experience mass which is the restricting of gravity or motion of the space filled with material.

Moreover, I came to another conclusion of equal importance. When any person is standing on any place anywhere, while viewing the Universe, that person is filling the centre of the Universe. Let's get more personal. When you, the person that is reading this, are standing at night and are looking at the Universe you are seeing the Universe from the centre of the Universe. All the light, every single beam that ever left any destiny at any time acknowledges this fact. You are the most important person in the Universe because you are holding the most important position in the Universe. All the light that comes across all of space runs directly in a straight line towards you filling the centre of the Universe. Not excluding the effort of one photon, all light is heading to meet you where you are in that centre spot and not one photon will pass you by. Not one photon dare miss you because if they do they miss the effort that all light has to accomplish and that is to locate you as the person filling the centre of the Universe. If you find this funny, or laughable you are in for a shock because this is what gravity is and this principle dictates gravity. It is the most complex issue one can imagine and expanding on this thought takes thousands of pages. It forms the crux to all cosmic principles and embraces every successful and meaningfully theory ever used to explain the Universe. Without taking this aspect in to account, there is no valid explanation available to understand the cosmos. Al the light coming from wherever meets the point you fill in time and in space. For al the light travelling you hold the spot it was on route to.

Should you decide to shift your position to any other place in the Universe you will shift the centre of the Universe to that location as well. If you install a camera on Mars, the light is obliged to acknowledge your relocating the centre of the Universe at your will to reposition you're being that centre of the Universe. All the light that ever left its destination crossing the vast spaces of the Universe, excluding no particular light, travelled all the way just to find you filling the centre of the Universe, right where you are. By you're standing anywhere, you fill the centre of the Universe, and the entire Universe admits to that because all the light comes to meet you there.

If you shift from the North Pole to the South Pole you will shift the centre of the Universe because all the light travelling throughout the Universe will find you where you then moved the centre of the Universe. The light left its destination billion years ago as it travelled through space at the speed of light anxious to acknowledge you're being in the very centre of the Universe. No photon will pass you by where you are in the centre of the Universe. No wonder every person born has the idea they were born to fill the centre of the Universe, which we do fill. The Universe is spinning around you or I, which is filling a centre where all motion is connected. That is the Coanda effect on the utter-most grandest scale imaginable; nevertheless it is only a manifestation of the Coanda effect. It implicates gravity as wide as can be…

Then I reviewed the Universe. If gravity is motion, what causes motion? What stops motion? That answer is in the Black Hole. If a star is about fusing atoms thereby growing, what happen when all the atoms fused into one all collective atom? What is the gravity if the star has one all-inclusive atom providing all the gravity that the star had when the star still had massive volumetric space? If all that space that once filled an entire giant star fused into one enormous gravity applying atom and that enormous force has been secures in the space that one atom holds, the atom would then show a force that would pull the surrounding Universe flat. Where does the gravity of the star end when all the atoms in the star became one giant atom? Gravity is smallest where space is least. Where space of an entire massive star is left in the size of one atom the gravity coming from that will pull the Universe flat at that point.

Coming to the conclusion about gravity being motion and mass being the restriction of motion was the easy part. What produced the motion and what prevented the restriction from overcoming the motion was the tough part. Figuring out why was everything on the move and where did the motion stop that was the part that took some figuring and some explaining. What made gravity move and why does gravity move…the answers are in the four phenomena never yet explained to satisfaction but now turns out to be the cradle of gravity.

Gravity is The Roche limit,
Gravity is The Lagrangian system
    Gravity is The Titius Bode law
        Gravity is The Coanda affect
…And gravity as the Roche limit forms the principle in producing the sound barrier. Read the book and find out why this is the case.

Newton's claims about the principles that he declared is responsible for guiding physics carries no validated proof and only after I realised that, was I able to start forming another line of thought on gravity. This had the purpose of confronting the corner stone of modern physics and at first I tried desperately to do just that. At first I was not confrontational towards Academics in physics and avoided any indication about disagreeing with Newton, although avoiding to show my disagreements was also totally impossible too but every time I approached academics with my new concept the academics always threw Newton at me . Facing Newton or facing defeat became a two-sided blade and I had to start to confront them by confronting Newton, with which I was in disagreement from the beginning.

At first I was reluctant to voice any opinion about the matter of how far I was prepared to challenge Newton because Newton was and is an icon. But slowly it dawned on me that if I had any serious plans to introduce my ideas I had to dispute Newton's gravity principles and do it head. When the slight confrontation did not bring results I finally decided to go all the way and show the inconsistencies that were prevailing in Newtonian science. That worked neither and it brought me the same results as before whereby I decided to go public and straight to John and Jane Dow avoid arrogance academics have with only one motto they serve and that is their autocracy and in particular their megalomania especially to my case as well as me in person. I wrote them (nine in total) letters in which I warned them that I was going public to show the extent of their dishonesty in their Newtonian's approach and lacking of substance and proof their physics has. The lack of honesty and furthermore the absolute dishonest on their part is there whether I avoid it or attack it; the inconsistencies are part of forming the basis for modern accepted science.

This process I now described is explained in a paragraph or less and it seems I got that far in a breath or two, but getting this far took me the best part of seven years to get to I tried my best not to attack them or Newton but left with the option to leave the project and lose thirty years of work and then fail after I concluded an answer on every aspect they never even thought of or take them on and dish out what they should have received years ago made me decide on the latter. After being avoided and taunted by their powerful positions and arrogance vested in their mentality they show in regard to their positions as well as the disregard they show in the mentality of others I slowly concluded that only and after I can get people forming the general public and the opinion of those that holds their disregard just as I do to see what they hide will I get a response from the Mater's of fraud.

First I had to show the general public the true colours of the academics in physics and get every one to see how incorrect Newton is, and only then do I stand any chance to introduce my line of thought. I am so sure of the ideas that I propose of being correct that I dare any one to disprove any part or the entirety that my concepts about cosmology forms! But that can only come about when I can get an audience to see how I expose Newton for what Newton was and it is in that where I find no luck. I can't find one academic with influence that is brave enough to stand up and face my attack on Newton and argue me down or prove me wrong in a sound debate. Now I see frowning coming from everywhere because it is madness on my part to think the world is wrong and only I am correct!

I realise that it shows signs of madness on my part and in my thinking to even regard any possibility that I am the only person on Earth that is correct and all others that ever studied physics are wrong but mad as it seems, if that is what I have to say to find an audience to listen and to judge my case, then that is what I say. I don't say this lightly or without understanding the enormity of what I suggest is going on, but be that as it may seem, it is the truth without question that Newton went on for three hundred and fifty years defrauding science with no one testing his claims. Argue me down or prove me wrong but don't discount me before hearing me out and only after considerable consideration while studying my arguments then form an opinion that disputes what I say but when disputing what I say, do it while confronting me in a sound argument when proving me incorrect! This not one academic could achieve and I challenge the lot to do so. But do it after studying all my work and being in a position to account for all the details that I propose. Don't just dismiss me because I dismiss Newton because following that road is the way of the coward and the mentally impaired.

Read my challenge about the correctness of Newton's proposals when he brought no more than suggestions into science and when I dispute Newton, then take me on by proving Newton correct... do it just once... prove Newton correct just once...prove that his formula is working and that his principles apply on the grounds he principled his ideas.

Detecting Newton's misconduct is possible because I saw a way to break away from the invalid concepts Mainstream physics hold. I went about and tried to prove Newton and when that was not happening I tried to apply Newton's ideas into the greater fields of cosmology. That also wasn't possible. I tried to amalgamate the four cosmic principles applying in cosmology with what Newton said was happening in the cosmos with mass and with gravity and in light of what the cosmos showed was happening Newton just wasn't happening! Notwithstanding the pose Mainstream physics try to uphold, the entirety of physics still use the idea of magical forces intervening in nature and they still base concepts on unexplained novelties.

Think of finding four unexplained forces going around and influencing persons in an unexplainable manner except that the magic of gravity keeps people attracted to the Earth. To say the least, the concepts physics use in terms of Newton would not even be acceptable to children in the modern informed era we live in, I challenge any person to prove Newton, not to accept Newton but to undoubtedly prove Newton correct! Prove how Newton's formula of mass forming the force of gravity can apply as Newton said it does! I recognised the impossible double standards Mainstream physics apply to promote their much shady explaining. In short I tested Newton's principles and found the principles to be wanting.

The inconsistencies Newton introduced brought science double vision and to compensate for these bogus truths supporting their incredible theories, they simplify issues to such a level where what they embark on, is the meaningless acceptance of the unproven and they proclaim to understand what are meaningless inconsistencies and to achieve this they create scenarios which uses the entanglement of deception. Prove the attraction Newton said was enforcing gravity that is pulling by mass and is gathering plants by contracting the diameter between planets. Show how much the Moon came closer to the Earth since the time of Kepler. Show proven distances taken by radar tracking and indicate just how accurate Newton was. Show how much the Moon came closer to the Earth since the time of the Moonwalk in sixty-nine.

The figures are available but are kept in a grave of silence where no one ever speaks about what science found applies and how much the distance between the Earth and the Moon is shrinking as Newton said is happening or then how much is the is expanding which will contradict the very principles Newton brought about! What they declare as unwavering facts can't even be supported in some form when tested by a silly test as to show that the distance between the Earth and the Moon is shrinking. Even the least degree of verification of correctness is absent when trying to find support of Newton and Newton lacks all evidence of authentication in any investigation of even the simplest terms. It is as if they never read with interest that which they explain when they embark on explaining Newton and they never scrutinise that which they advocate when they teach Newton's principles applying. They give values that are senseless and the very values they use make that which they say meaningless.

In this book I am going to investigate how much truth there is in mass pulling by the force of gravity. To most if not to all of the persons reading this then such a venture of investigating Newton is time wasted and just the thought about me embarking on the investigation of the issue is totally senseless to investigate. It is senseless because the concept it carries became accepted as household practise and life science from where it proceeded to become everyday culture in every person's mind. The worst part is that the group of people normally considered as the wisest bunch there is, never did prudent testing on Newtonian presumptions, while to test the presumptions is most easy to do. I will not believe that a lot that lives up to the veneer of being the best mathematical intellectuals on Earth, never though of testing Newton's very simple formula and in that disregard the formula because of the incorrectness the formula holds.

Do you think of astrophysics as being the department that is run by the wise and the level minded, the honest and pure at heart, the nobility of well-to-do academics and the sober thinking standing in front of the world as the absolute trustworthy? If you are a student, there is no other choice you have but to trust them while they feed you absolute hogwash! If you would so much as dare to doubt any thing they say they will banish you from the institution they rule so absolutely. The banishing process is dome under the blanket of examination.

They teach you what to think and to make sure you think what they wish you to think, they tell you to confirm their teachings on a blank piece of paper. You write what they prescribe and you supply the answers they demand in the words (sometimes) of what they demand. Should you in any way say anything different from what they tell you to think, your presence will not be tolerated any further as they abolish you from their institution of academic tutoring. After reading this book I invite you to...no I dare you to challenge their statements with evidence gained from this book and see them wilfully further their culture of deceit by bringing unfounded arguments just in order to silence you and prevent you from getting behind the truth. If you think those in charge of astrophysics are the pillars of trust, then get wise by reading the following facts and arguments this book presents. What you are about to read is simply mystifyingly simple and yet to this day I have not had the privilege to challenged one academic any where that had the honesty to admit to the fact of Newton being wrong. After you have considered the following you might agree with me that even small Children can reach a higher level of clear-minded logic and find more sensibility than what those scientists promoting astrophysics have because science lives in a make believe fool's paradise.

The manner of regard to life that the Academic Physicist holds and the outlook on life that the followers of Newton physics have (I call them plainly Newtonians and to me they are sheepish because they resemble to the image that to me seems the same as sheep running after their leader without having the ability to think for one second any thought spawned out of personal intellect) is quite the opposite of what I think of them. They keep their forming the establishment of the order the Academic Physicist in high regard and consider their order to be the top thinkers in society.

This religion that they practise of self promotion and sublimely self regarding their status being next to God has them so high that we down on Earth forming the waste of human garbage can be told anything and we will believe what they say just because they with their supreme intellect tell us to think what they wish us to think. This they do because we human waste living way down below their supremacy have not the ability to think and therefore they must think on our behalf. In their view and so far very correctly judged on their part, they, the persons being in the group that forms the Academic Physicists, believe very correctly that can dish up whatever they wish and we, those forming the group in the gutter, those that are mindless in their eyes, we will have to accept what they say without being allowed to form an opinion other than having the opinion they give us to have because in their view we are unable to have a mind other than what they are able to control. This attitude they have is the result of a relationship that worked for so long and thee fact hat it worked that long is what confirmed their opinion that we, the public, are fools to believe anything and everything because of blind stupidity.

But in spite of their aggravating conduct and mischief towards us, it is not because of a lack of insight and inability of controlling a mind that we have our childlike belief and blind trust in their opinions and which there was. It is the faith we shown that they misused for their scandalous cheating. Our faith is what we have shown towards them and is that, which became used as the reason why we accepted what they said blindly. We didn't accept their word on the grounds of us being utterly stupid as they perceive us to be but our trust depended on our good nature and believing in their trustworthiness. This trust we have is brought on by a culture of trusting the King to do the people well and somewhere in every person's cultural past there was a King that did us well in leadership.

But their underestimating of our abilities is the testimony of their poor understanding and their weak insight ability, which results from their arrogance and stupidity. You are about to see just how stupid they really are in the thinking aspect of science. It will become clear as you page along while reading! They didn't fool us half as much as they fooled themselves and you are about to read all about it. The fact that they could fool us for centuries didn't run on their intelligence being so much superior but served their purpose as it stemmed from the trust we had in them resulting from good intentions on our part. This betraying on their part and misusing the public's good nature to be used in schemes to get the public conned must end and I pray that this book form the first step in resisting the arrogance of the Academic Physicist.

Any one not in their group of the Academic Physicist is part of the lowest order of mindless being and to become part of their order and those that have minds with an ability to think, students have to accept what they say when they say whatever they wish to say without having to prove the correctness of what should back their saying so and as a result of this students may never question what they say. Only when and after proving that a student has totally lost all ability to think for him or her self may a student is promoted into the ranks of their sublime intellectual group. The sifting process they named examinations. You write on paper what they told you and never question their opinion and after passing that examination will you ever enter their sphere of intellectual brotherhood. Does this sound far fetched? Then you better read on and I will remove your blindfold and show you what a world of deception the Academic Physicist force on us into.

Read the following and see how they, the high and the mighty, those that think they can replace God and those who think they can think on our behalf and think what to tell us to think, how much they are clowns and the jokers in society. Read how little are they, the Academic Physicists, able to understand concepts about Creation while they think they are able to replace God in their superior intellect.

If you are a student in the science of physics, then ask your Educated Masters to please explain the following abnormalities you are about to read in this book and insist on a clear explanation about the inconsistencies they promote while tutoring physics as if the physics they present are the most flawless and accurate institution there has ever been. Ask those academics supporting Newton about the following flaws that no one mentions ...ever... except me in this book you are about to read and get them to explain the inconsistencies never talked about, which I present in this book and then after confronting those charged with tutoring physics and seeing who should be believed, then get wise instead of brainwashed. Let them mathematically show how one would go about and use Newton's visionary formula $F = G \frac{M_1 M_2}{r^2}$ to calculate the force of gravity by replacing the symbols with the actual values in mass that the items referred to have. Put in the Earth's mass in place where it belongs and put in your mass in place where it should be and then divide that with the distance between your soles and the Earth measured in micro millimetres by the square thereof!

In the book named an **_Open Letter on Gravity Part 1 and Part 2,_** I bring the solution to the mystery behind gravity. I tried in vane to introduce the principles I find valid to the academics in charge of astrophysics. Facts that Science present as being the uttermost explicit and unwavering truth, fails to bring any logic answers to so many questions that it should address. It fails to have substance in addressing the most basic and simple questions about gravity and physics. Yet to every question science can't answer my approach does bring many solutions. The presentation and the delivery of my answers that I reach are understandable and simple where it serves both logical science and the truth. Since my answers do not match Newton and his misconception about gravity and that mass generates gravity, those in charge of science don't even bother to read my work. With their affixation to the corruption they portray I can do little to the giants where they are in the mighty positions they have and just because of that they can go about to sideline and ignore my work and this is notwithstanding the correctness that my work delivers compared to the utter failing that Newton's work shows.

When confronted with my evidence and they have to match my work with the hypocrisy and misleading nature of Newtonian cosmology their defence in substantiating their claims is to ignore me. Since I do not applaud mainstream science and the clear fraud they embrace and fraud it is that they embrace, I am silenced. Why is it that my work is going unrecognised or even in the least goes never debated and never commented on...it is because it will then trash every article anyone has ever written about astrophysics and cosmology. They show little integrity when academics with such supposed high standing or then such as they should have, play a dishonesty game where those in commanding positions will rather protect fraud and save their skins. They would rather protect the corruption they have than seek the truth and find honesty in physics. Those academics in charge would much rather protect their un defendable ethos they maintain as forming the back bone in science and what gives their personal position legality although it is corrupt than admit to the truth they find when they begin reading my work and in agreement they then have to back the truth my work brings. Doing that (accepting the truth in my work) will trash all work in cosmology delivered thus far and condemn it to the waste paper basket and render all work invalid and void. It will put all the Newtonian's bias and fraud into the place where it belongs.

Considering that such acting will lose them money, those academics in controlling positions then will rather rape the truth in order to benefit from continuing to corrupt student's minds further. If they wish to justify their inconstancies they have to attack my work and disprove the accuracy of my work. That they can't do. They then ignore my work because they can't attack my work. In that sense they also place their work beyond my approach, as they can simply ignore me as if I represent the plague while they carry on with little consequence to bother them. I challenge them to prove Newton correct and not just declare Newton being beyond reproach after all has seen the evidence I bring. After reading this all students must challenge them to defend what they can't or get honest.

$F = G \dfrac{M_1 M_2}{r^2}$ $F = \dfrac{r^2}{M_1 M_2}$ This is the basis that Mainstream science uses as the foundation of all physics anywhere. If this is wrong then everything they have got to work with goes out the window. They put mass and the distance that parts objects in a relevancy, in other words the one is a ratio to the other. The one factor brings a measure to the other factor's value. The one cannot be without the other. The increase in one becomes the reducing of the other and the other way round also applies. When the distance is large, the influence of mass will be small and when the distance is small, the influence of mass will be overwhelming. Then they state we are in a Big Bang expanding of the entirety.

Why then, when considering that if it is mass that produces an inclining force of contraction as Newton says there is going on then...why didn't the expanding stop before it started when the Universe was small. Today using hindsight after the fact of the exploding Universe became apparent by the studies Hubble brought to light did the lot of everything that is not implode as Newton would have us believe whereas, instead it did expand just as Hubble proved. The radius at the time of the first instant back then was no factor, which makes the gravity at the time a totality of unrivalled force. The radius being that insignificant leaves the mass unchallenged in asserting power in relation to the non-existing radius it had.

I dare any physicist to show me where they apply Newton's formula just and exactly as Sir Isaac Newton suggested gravity applies. Show me just once where the mass of the Earth is multiplied with the mss of the object in normal physics. Show me just once how $F = \dfrac{r^2}{M_1 M_2}$ or $F \alpha \dfrac{M_1 M_2}{r^2}$ where one M represents the mass of the Earth while the other M represents the mass of the object and in this formula the end result will have a value of 9.81 Nm/s^2 ... show just once one example... where the use of the mass of the Earth comes into play. If multiplying the mass of the Earth with the mass of an object and dividing that with the distance parting the two mass factors does not deliver 9.81 Nm/s2, and then any claim by Newton indicating that $F \alpha \dfrac{M_1 M_2}{r^2}$ is equal to gravity, such claiming constitutes to deliberate fraud...even if Sir Isaac Newton said this. Prove that the mass of the Earth with the mass of an object and dividing that with the distance parting the two mass factors delivers 9.81 Nm/s2 or admit physics is conducting fraud to protect Newton!

Isaac Newton did achieve with his guessing about issues he couldn't explain.

To be successful in my quest to find an explanation for gravity, I had to redirect all my concepts I previously had and also alter all the otherwise normally accepted thinking relating to physics. I realised that if I ever wished to come to realise what gravity is, I had to first realise that gravity is not what Newton saw as forming gravity. What Newton saw as gravity can't withstand even the slightest test of proof. I

$$G \times M / r \times m / r, \text{ which is the same as } G \times M_1 \times m_2 \div r^2$$

have tested Newton's thinking and this book bears the witness of all my testing of Newton. As any body can see that reads this book, I tested Newton from all the angles that he possibly could be correct and found his thinking wanting every time.

The truth about Sir Isaac Newton's concepts that he formed is that I came to conclude that in reality it is not in any way overstated to declare that Newton conspired to defraud science and moreover that he committed blatant mathematical corruption in trying to prove the concept he had about what he thought forms gravity. There are no mathematical or any other forms of proof to be used as backing for his ideas. There is no cosmic backing one might use to prove any of his claims in as much as what forms gravity or find proof about any of the claims that Newton made on matters concerning science in cosmic gravity, and every thought he introduced that later proved useful and was correct, was what he stole from another far better cosmologist.

At least the four phenomena I use are visible and are presently prevailing in the cosmos and I use the four phenomena in my explaining about how gravity forms. Not one of Newton's laws are directly relating to any concept Newton ever introduced at any stage but is the result of academic theft he committed against a much larger figure that preceded him by almost a century. However he committed academic rape and plunder of the man that preceded him while he saw to it that the phenomenal work the first person did was for ever inferiorly linked to Newton's concepts.

He presented the work of this man totally incorrect and these mistakes Newton made when he changed the work of the first person were since then never addressed as it should have been addressed. Newton brought no original input into science except that he gave a concept familiar to everyone a name. This well-known concept I refer too he named gravity and even that is inappropriate.

Newton changed science by incorrectly suggesting changes to science and to mathematics that breaks every mathematical principle he could think of. He has no right to change mathematical laws as it pleases him just because he thought to be greater than any person that ever lived!

Newton changed what no man can and that Newton did in his attempt to win over the prevailing academic thinking of the day as to lay some sort of groundwork to form the required backing for his ideas on physics and the changes he made to mathematics was his personal attempt to explain gravity or what he thought gravity is. Newton stole, cheated, lied, diverted the truth and raped other people's work as well as falsifying mathematics to find support for his most incorrect ideas on gravity.

This part is dedicated to a book entitled Open Letter Announcing Gravity's Recipe. In the book the author explains gravity. This achievement is possible because the author broke from Mainstream physics and the impossible double standards Mainstream physics use to promote their much shady explaining and double vision about things they have no vision of. Do you think of astrophysics as the department run by the wise? If you think that, then get wise and read the following. After you have considered the following you might agree with me that Children can be more logic than what they are because they live in a make believe fool's paradise. If you are a student then ask your Educated masters too please explain the following abnormalities and inconsistencies they promote, which I present in this information and get wise instead of brainwashed. I say again brainwash because they force-feed you facts, which they cannot explain because the facts are untruths. Tell them to prove that planets have mass. Tell them to prove that it is mass that generate gravity. Ask them to explain gravity in detail.

When you deal with a relevancy such as this $F = G \dfrac{M_1 M_2}{r^2}$ it is not the value of the top part that is of a crucial nature but it is the bottom part that controls the top part that predicts the value of the outcome.

The formula $\dfrac{\text{top part}}{\text{bottom part}}$ the size of the bottom part dictates the top part value. With the Titius Bode law applying where the distance doubles every time a new position of the next planet comes about, the mass has even a lesser role to play than it did before.

This is the way you are supposed to see the formula $F = G \dfrac{M_1 M_2}{r^2}$ and when having this view the formula $F = G \dfrac{M_1 M_2}{r^2}$ such as Newton saw it and what was the view in the dark ages Newton lived in having such a view will make a lot of sense.

However, notwithstanding size or mass increases, the distance between the planets forms a doubling value relating to the specific position of the planet and in this there is no referring to size or mass whatsoever.

This Mainstream science use as the foundation of all physics anywhere. They put mass and the distance that parts objects in a relevancy, in other words the one is a ratio to the other. The increase in one becomes the reducing of the other. When the distance is large, the influence of mass will be small and when the distance is small, the influence of mass will be overwhelming. Why then when taken into consideration that if it is mass that produces an inclining force of contraction as Newton says then…when the Universe was small it did not implode whereas, instead it did expand. After all, the radius was almost no factor at that point leaving the mass to enjoy an eternal power in relation to the non-existing radius.

When the Universe was at the point where the Big bang started, the radius was incredibly small. That would make the mass inducing gravity by contraction inconceivably large because the mass was completely overpowering all factors with the small radius. It did not bring about an implosion that the overbearing mass contraction was supposed to unleash on such a small Universe in the beginning. The Universe at present with in comprehendible distances parting object that renders the force of mass no relative value as the force constantly weakens with the growth of the expanding.

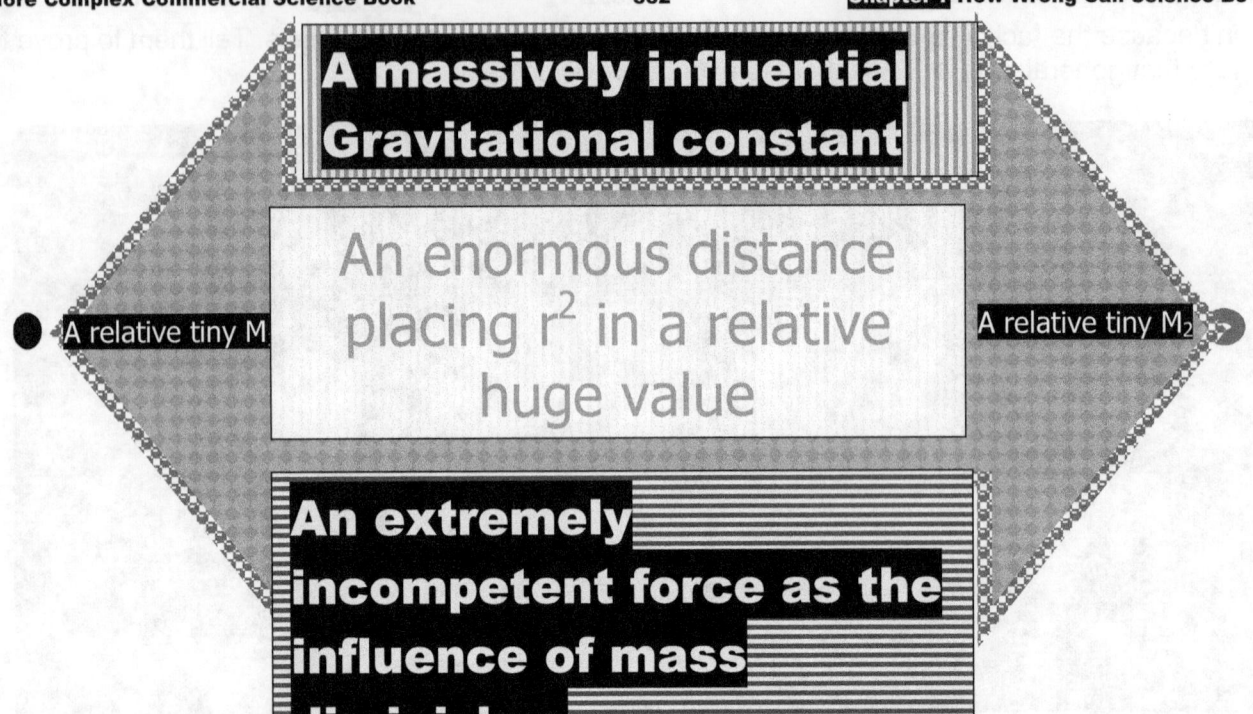

The more the radius develops in time, the lesser would the gravity be that the mass factor generates in relation to the advancing radii developing and the larger would the reducing be of all contraction. The effectiveness of force the mass produce will tarnish as the radius that separates the material from each other increases as time moves

Although it is presumed that the Universe was small at the dawn of the Big Bang, such presumption will put validity to another presumption that the gravity the mass charged at the time was enormous because the influence of the small distance in radii and the factor such distance produced promoted the factor, which the mass has to an enormous large factor.

If an object is a million kilometres apart the radius is a million times more in value by dividing the mass influence than when objects are one kilometre apart. That is the most basic realisation about mathematics. It puts ratio to order and define coherency. That is what gravity is to the Universe as it puts respect to factors about the Universe in the Universe. It is what derives order in the Universe.

At the very same time we will find in a Universe that was supposedly so small it had a radius of less than only one kilometre, then at such a time when the Universe was still that small it must also be accepted that the gravity the mass charged was one million times greater as it would be when the radius keeping the structures apart is one million kilometres in distance. The extremely small radius that was only the size of one neutron in radii distance and with the factor that such a distance produces, it must promote the mass factor, which will support the mass in having an enormous large factor by relevancy to what the case must be at present. The mass factor that produces the gravity at any given point during the event of the Big Bang, had to be eternally larger at the dawn of the Big Bang while having an infinite radius, which gave gravity all the power it can and which it will have ever have.

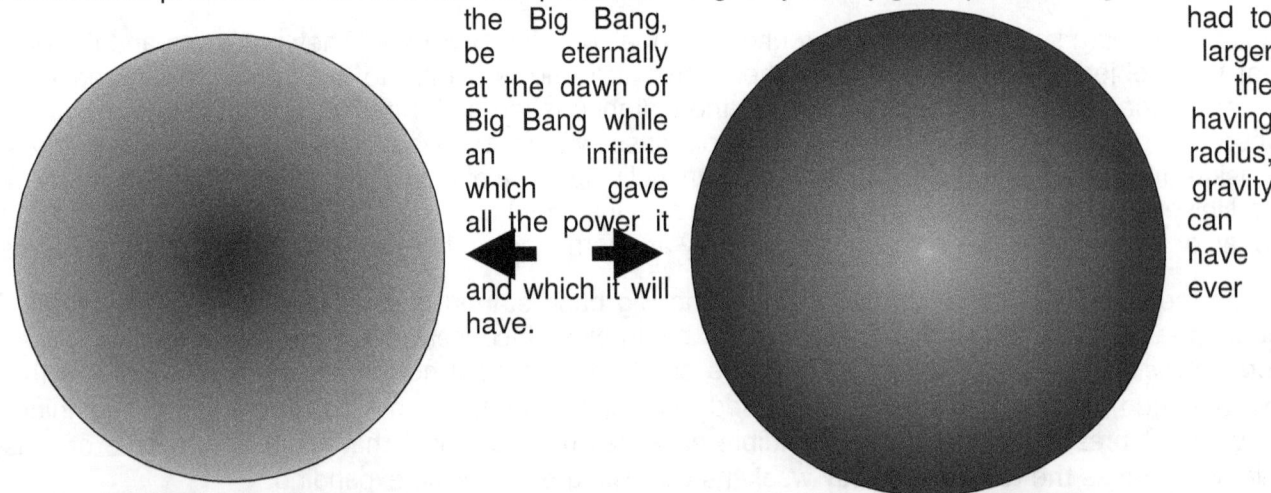

If at the Big Bang there was not sufficient mass to destroy the radius and prevent the expanding from coming about, then the expanding won the match and there can be no contracting Universe as Newton had us to believe. If the Universe started a journey of parting objects no amount of dark matter that might lurk in the night sky and is at this moment hiding from detection will produce the gravity required to stop the expanding from continuing. At the start the expanding became evident and as the radii grows the inclination will suspend in influence as a factor. If there was insufficient mass at the start in order to tilt the balance in favour of the reducing factor, no amount of mass can ever accomplish such a goal afterwards. Then Newton's surmising was one of corruption making that which all physics are based on fools thought and corrupted proof.

If you might be of the opinion that my accusing the greatest intellectual department in the world as being in misconduct and to your view such accusing is outrageous and far-fetched, then be my guest and judge the following with a clear and unbiased mind because when scrutinised with a clear view then the facts cannot fool an idiot. However, that is just what the physics paternity thinks the rest of us forming the general public at large are. They have the opinion that they can feed us in the public arena any senseless rotten garbage they dish up because they see us as being inferior by thought and mind.

With all this in mind did any one ever come to wonder about the all too famous Einstein's critical density theory and the fact that this idea was conceived to conceal the corruption of Newton in physics? The fact in truth is that the Einstein's critical density theory was a scheme plotted by those in charge to cover up and conceal corruption in the heart of physics.

If Einstein was unable to recognise the most basic of mathematical principles then what type of genius did physics create in him and what slur did physics promote. This idea of the two factors being in opposing relevance is so simple that children will recognise the principle, and yet those fathers of physics wants me to believe that the greatest mathematician that ever lived did not realise this principle...the principle that the radius and the mass stands related and the growth in the one will promote the decline in the other as a dominant factor. Can any one with this information including the information given on the previous page have any other conclusion? It is obviously clear that having such a total idea that there might be dark unseen mass floating in the Universe which at this time does not generate gravity but will some day because Newton has to be correct at some point in the future. I am to believe that dark undetected mass can be found and such undetectable mass could be found which will bring about contraction after all this expanding? Why would the mass at present then not activate gravity and why would the mass at some point spring to life and start activating gravity? How much can the Physics paternity still hide the fact that Einstein's critical density is being used as a cover-up to distort the truth to conceal fraud? The uncovering by the Hubble constant about of the Newton fraud is so simple to see. Hubble found the Universe is expanding and Newton's said otherwise. Who is lying about what?

$$F = \frac{M_1 M}{r^2} G$$

Lets put the mathematical formula into a practical context.

Newtonians uphold their law of physics without showing mercy. The very first things the Newtonians use to beat us into submission are to blast us with incomprehensible mathematical formulas.

Incomprehensible they are but it is to scare anyone with the mathematical equations to get everyone hiding. They bewilder you with equations that put the fear of God into you; used simply to make you feel inferior so that they can feel superior and frown down on your inferiority from a dizzy height.

They are masters at manipulating anyone into a state of senselessness...but mostly they do it onto themselves. That they do because it forms the backbone of the fraud. They do not wish you to read closer and to find the fraud they hide to protect Newton. Ignore their mathematics because it only shows their incompetence to understand physics or Newton and see the fraud they propagate...

They employ mathematics to bewilder and that is all. I am going to show what we can uncover underneath what they cover. Look at what the mathematics supposedly says and then wake up, they are using maths as a scare tactic for three centuries to scare the daylights out of you and all this while its been working! Looking at the formula shows just how little Newton understood physics.

Please allow me to show you how they scare you to become fooled and suckered. Don't run and hide when you see the mathematics, it is meaningless although it was used as a scarecrow for more than three hundred years forming the backbone of the conspiracy.

This picture is as big a hoax as Newtonian science is when Newtonian science presents mass to be a factor that produces a pulling force called gravity. The question shouting for an answer in the picture is if mass is a factor that produces gravity as Newtonians claim it is then why are the planets not positioned according to mass as Newtonians declare.

The claim is as bogus as the entire philosophy. They present the proof that planets orbit according to mass in the following "Kepler law" which in its entirety had nothing to do with Kepler at all. It is all devised by Newton because Newton had no inclination of what Kepler's work was about.

Newton brought about the idea of mass positioning the planets in the formula $4\pi^2 a^3 = P^2 G(M + m)$

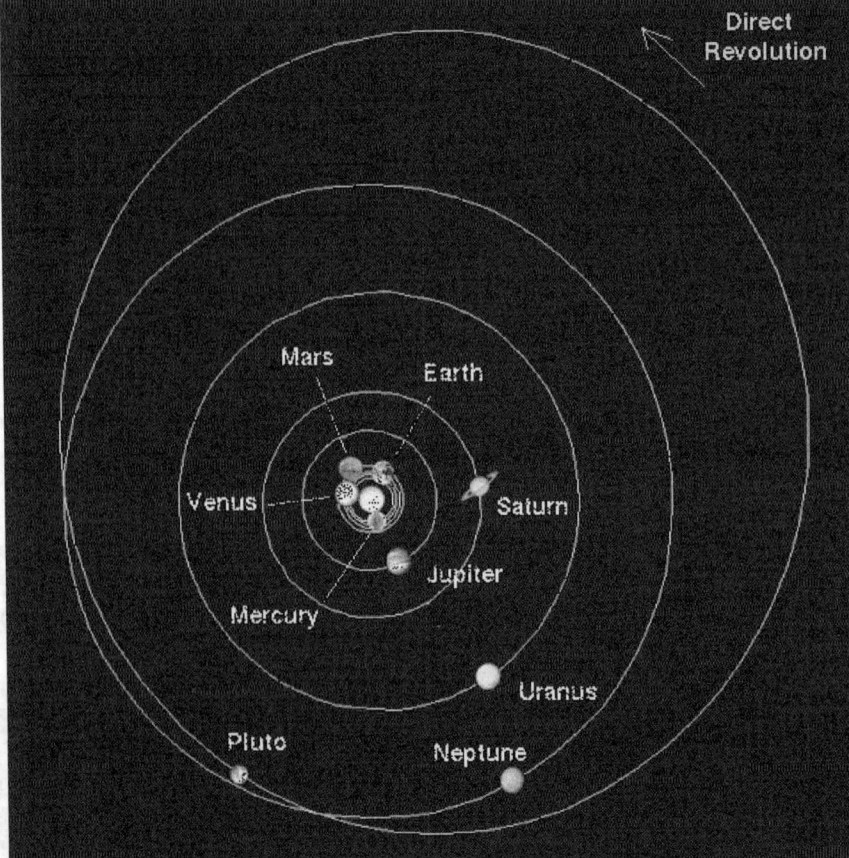

Do not get scared as everyone usually does when seeing the mathematics and then as a result get frightened. Those physicists expect you to turn on your heels and run as fast as your legs can carry you. Then consequently as a reaction to find survival, you turn on your heels and run... but this time don't. Don't run, just read on and see how

simple it is to prove Newton was a backward dark aged sod!
This time, don't run because I am about to show how meaningless this entire mathematical statement in reality is! This formula is total garbage and there is no sign of evidence that this formula forms any part of the solar system, even in the least.

Lets test this formula and see how truthful it is. $4\pi^2 a^3 = P^2 G (M + m)$ indicates that the circle in which the planet orbits ($4\pi^2 a^3$) is the result of (=) the position of the body (P^2) positioned by the mass of both bodies ($M + m$) in terms of the gravitational constant (G). The best way to find clarity is to test this statement with what is happening in the solar system just as it is, wouldn't you think.

A picture such as this provides much credence to the idea of gravity by mass since the lines drawn does not even begin to represent what is truly out there ands what is used by the cosmos in place of Newton's mass concept.

This is a table indicating the **mass** that **every planet** has in relation to the **distance every planet holds** in terns of the sun. If mass positioned planets then why did no one bother to inform the Universe about this because it is clear the Universe did not receive the memo from Newton's office to act accordingly.

Body	Mass (10^{24}) kg	÷	Orbital Distance(10^6 km)	=	ratio
Mercury	.3302	÷	57.9	=	0.0057
Venus	4.869	÷	108.2	=	0.045
Earth	5.975	÷	149.6	=	0.039939
Mars	0.6419	÷	227.9	=	0.00281
Jupiter	1898.6	÷	778.3	=	2.439419
Saturn	86.83	÷	1427	=	0.060847
Uranus	102.43	÷	2869.6	=	0.03569

This picture shows the hoax the Newtonian conspiracy pampers to keep the rest of Newtonian physics believable. They never mention the Titius Bode law and try to explain the Titius Bode law while it is the Titius Bode law that is really in place in the solar system. If you wish to learn the truth then think again.

Mass as a factor does not present or apply in one instance anywhere in the entire Universe and yet that is all theta physics says applies...but why would they cheat? It is because what is there applying between the planets is called the Titius Bode law and although this law is in place you have almost a hundred percent chance that you have never heard of it.

If the planet layout was as I now show it to be according to mass then this was the order that is if it is true according to the solar system that mass do produce the position of the planet:

Jupiter	318 x earth mass	at a distance of	57.9 x 10^6 km
Saturn	95 x earth mass	at a distance of	108.2 x 10^6 km
Neptune	17 x earth mass	at a distance of	149.6 x 10^6 km
Uranus	14.5 x earth mass	at a distance of	227.5 x 10^6 km
Earth	1 x earth mass	at a distance of	778.3 x 10^6 km
Venus	0.81 x earth mass	at a distance of	1427 x 10^6 km
Mars	0.107 x earth mass	at a distance of	2871 x 10^6 km
Mercury	0.055 x earth mass	at a distance of	4497 x 10^6 km
Pluto	0.002 x earth mass	at a distance of	5913.5 x 10^6 km

This is how it would apply if Newton was correct and mass did position planets. There is not even a remote chance that the positioning of the planets go in accordance with mass or $4\pi^2 a^3 = P^2 G (M + m)$.

Do you realise there is much more "gravity produced by mass" in the space your feet has contact with the earth than there could ever be between Jupiter and the sun? You that can calculate it so then do it. Show that $\dfrac{a^3}{GM^2}$.

Put the orbit of Jupiter in relation to the mass of Jupiter and in relation to the position Jupiter holds. Forget getting swept away by the fancy Mathematics; just get to the task of putting the mass in relation or ratio with the position that any of the planets hold. Take the mass of the earth and your mass you have and then divide that with the square of the distance there is between your feet and the earth by the square thereof then divide that square with the product of the mass of the earth and your mass you have. Keep in mind your distance between your feet and the earth is about 10^{-11} meters going square! You that can't calculate it, the value would be meaningless but it is so much it will crush your atoms into a pulp, leaving you not even in a blood blob. It will leave a force to the value of about one Zetta 1 000 000 000 000 000 000 000 000 g per square meter. There is no chance in hell that any object of whatever size and formed by whichever method of construction could manage such a force of pressure. Those Physics-cheats always want to have the radius down to 1 meter because it makes their argument look sensible, and then they "forget" to use the correct mass of the earth not to stun any student into realising reality. Using one meter makes good sense when you wish to cheat but no body floats on meter above the earth. When anything stands on the earth, the distance between such a body and the earth is less than what could be sensibly measured.

Can you see the mathematics, Newtonian scientists use this to scare you into frozen stupidity and you allow it. I know and realise that you are disgusted by my attitude when I degrade the name on which physics are founded. In this introduction part I am going to show you just some minor deceptions all students are forced to believe since all physics students are forced to believe in Newton, Sir Isaac Newton that is. I am giving you a choice. You can say I am going to commit fraud or Newton has committed fraud. If I am judged to be the culprit that is guilty of deception then it is because Newton misled me. You can choose.

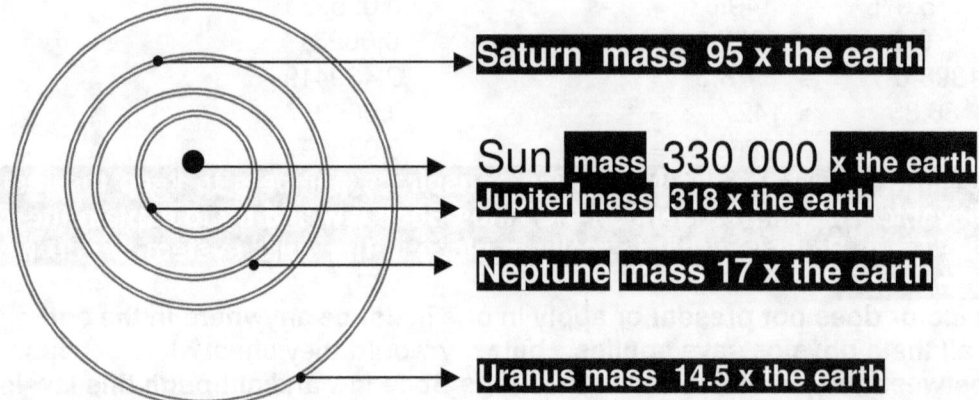

The inner circles that are very close to the sun we find the big gas planets with so much more mass forming the force of gravity that these planets are almost on top of the sun so close they are to the sun.

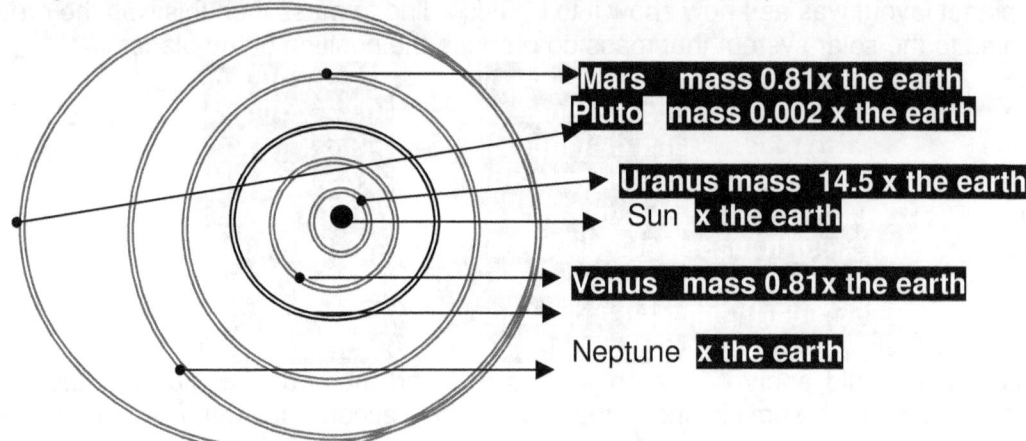

Then in the outer circles that are very far from the sun we find the smaller solid planets with so much less mass the force of gravity just can't pull these planets closer to the sun.
The above can be the only designated outlay of the planet position according to the sun when applying mass. Remember, if I am wrong then Newton and his formula $P = \left(\dfrac{4\pi^2 a^3}{G(M+m)}\right)^{0.5}$ is wrong but if I am correct and Jupiter is the closest planet then Newton is incorrect and then the academics in physics are indulging in Newton's fraud by forcefully brainwashing students to believe in Newton notwithstanding the fact that it is the solar system that disputes Newton's ideas altogether. If you disagree with my layout, then you better disagree with Newton and his ideas. What is in place I am almost sure you have never heard of... Instead of the cosmos using mass as Newton claims there is no mass applying but there is the

Titius Bode law on place. This is never commonly mentioned and the fact that what Newton advocates is not in place is hushed up.

Closet
1) Jupiter mass 318 x the earth
2) Saturn mass 95 x the earth
3) Neptune mass 17 x the earth
4) Uranus mass 14.5 x the earth

Then come the smaller planets with less mass and therefore less pulling force called gravity

5) Earth 1 x the earth
6) Venus mass 0.81x the earth
7) Mars mass 0.81x the earth
8) Mercury mass 0.055 x the earth
9) Pluto mass 0.002 x the earth

The Titius Bode Law

The Titius-Bode Law is rough rule that predicts the spacing of the planets in the Solar System. The relationship was first pointed out by Johann Titius in 1766 and was formulated as a mathematical expression by J.E. Bode in 1778. It leads Bode to predict the existence of another planet between Mars and Jupiter in what we now recognize as the asteroid belt.

The law relates the mean distances of the planets from the sun to a simple mathematic progression of numbers.

To find the mean distances of the planets, beginning with the following simple sequence of numbers:
0 3 6 12 24 48 96 192 384
With the exception of the first two, the others are simple twice the value of the preceding number.
Add 4 to each number:
4 7 10 16 28 52 100 196 388
Then divide by 10:
0.4 0.7 1.0 1.6 2.8 5.2 10.0 19.6 38.8

The resulting sequence is very close to the distribution of mean distances of the planets from the Sun:

Body	Actual distance (A.U.)	Bode's Law <A.U.)< td>
Mercury	0.39	0.4
Venus	0.72	0.7
Earth	1.00	1.0
Mars	1.52	1.6
		2.8
Jupiter	5.20	5.2
Saturn	9.54	10.0
Uranus	19.19	19.6

$P_n = P_o A_n$
P_n = Period of orbit of the nth planet
P_o = Period of sun's rotation
A_n = Semi major axis of the orbit

This is so typical Newtonian in every sense there is in science. The Newtonians gave the Titius Bode law a formula and that explains the lot. To they're under achieving standards that is very satisfactory. Now it is written in mathematics then what more do we need to know. The fact that the distance that Mercury has from the sun is doubled by that which Venus has from the sun is completely ignored. In cosmic reality mass plays no part. Then again the distance that Venus has from the sun is doubled by that which the earth has. This clearly has nothing to do with the size or mass of the planets. Explaining that part is completely ignored. Then again the distance that the earth has from the sun is doubled by that which Venus has and inexplicably this forms the layout of all planets in the solar system. Where do Newton and his idea of mass fit into what truly applies in outer space. Moreover, why does science never mention this?

Newtonians supply the formula. That is it. To the Newtonian doing that explains the lot that they can understand. That explains nothing but then again Newtonians fill the cosmos with nothing putting nothing between stars to fill distances. This is how the layout of the solar system is. It is there and it is in place and while the cosmos openly reject Newton's idea of mass, the Newtonian reject the cosmos' placing the Titius Bode law in place and thereby accepting Newton's mass idea that the cosmos clearly rejects. How on earth can anyone explain such behaviour and make sense. It puts planets at random as far as mass is concerned but it uses a formula that ignores mass completely $P_n = P_o A_n$

This process forming distance between planets carries on throughout the solar system.

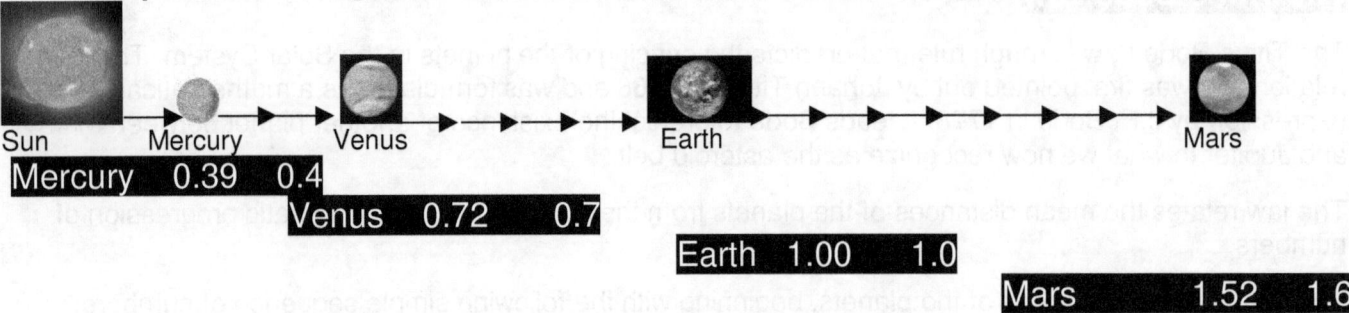

I am going to explain the Titus Bode law again because chances are excellent that notwithstanding how well read you are, you most likely never herd of this law all the while this is the most overall used principle in the solar system. To portray that what follows in relation to this I would need a paper at leas several meters long. There is not a hunt of Newton's mass being used but the Titius Bode law forms one of the four building blocks that the solar system applies. There are reasons for this silence, it makes Newton incorrect and it underlines the stupidity of Newtonian science to explain what they can't understand and that is what is true cosmic science. I am going to explain the layout and later on the entire principle.
There is no room in a room to show this layout in its full compliment where it covers all the nine planets. If mass formed gravity then the layout should be running from the biggest to the smallest. Please put Newton's mass in this Titius Bode Law and explain what happens and moreover why this happens.

$$F = G \frac{M_1 M_2}{r^2}$$

The distance should be in terms of size, but it is not, it is according to the Titius Bode law, which is some law no one ever hears of because it disproves Newton and his mass concept. It shows Newton has no ground on which to form his concept that is completely wrong! But while the cosmos disproves Newton, science believes Newton in spite of the cosmos using the Titius Bode law This is very typical of science in the way Science prefers to cheat the truth to prove Newton correct.

Science would rather say the cosmos is wrong than to admit that Newton is wrong and in the book you can download free of charge named Questionable Science, you will see how many times does science prefer to back Newton when the cosmos shows Newton to be incorrect and this type of corruption goes on and on…do get wiser by getting better informed just by going to and **download** Questionable Science.

The entire idea above called the Titius Bode law is there in place used by the cosmos and is so accurate it was used to discover planets but also is dismissed as not being factual by physics professors because it repeals Newton's principles of mass. It is called the Titius Bode law and the undoubted accuracy of the Titius Bode law was never questioned. It used in the past to locate unknown planets but also it is clashing head on with Newtonian accepted principles and therefore it is degraded and denounced, just because it does not salute Newton's unproven principles.

I can and do explain this principle mathematically in a book I give away at no charge. Go to and download www.singularityrelavancy.com The Website http://www.lulu.com/content/e-book/the-absolute-relevancy-of-singularity-the-website/7517996

The sun holds the first inside planet at a certain position

Sun   Mercury 57.9 x 10⁶ km

Then Venus doubles the distance Mercury has from the sun

Sun   Mercury   Venus 108.2 x 10⁶ km

Where the earth fits into the scenario Mercury holds more or less ⅓ from the distance and Venus hold ⅔ where the earth completes the full distance. Mercury is 57.9 x 10⁶ km; Venus 108.2 x 10⁶ km and the earth is 149.6 x 10⁶ km. Remember we on earth holds the earth at the centre and therefore the inner planets assort to our position. From here on the Titius Bode law comes into its own where the first inner planet or the earth forms the **controlling singularity** the outer planet as Venus then is forms the **governing singularity** and the sun forms the **primary singularity**. The rest of the inner planets have no role to play in accordance with singularity positioning the planets.

primary singularity

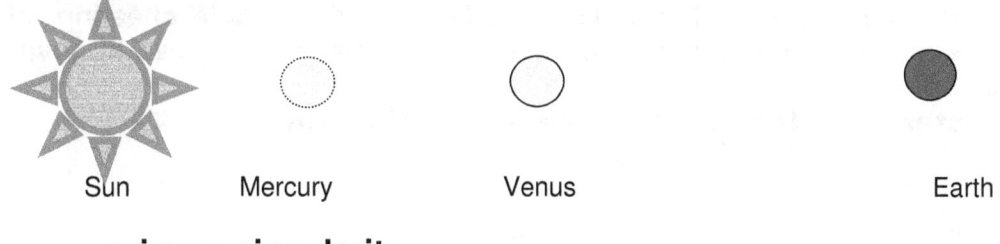

Sun   Mercury   Venus   Earth   149.6 x 10⁶ km

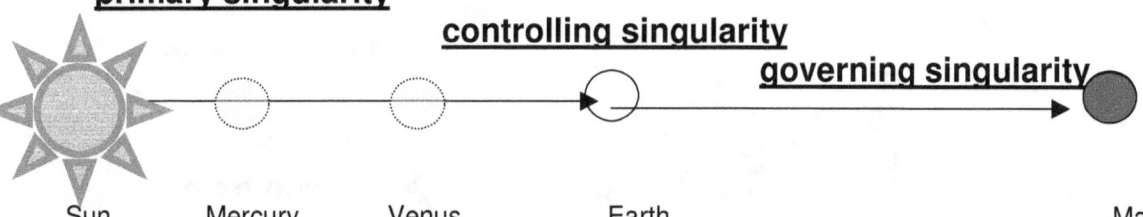

Sun   Mercury   Venus   Earth   Mars

In this order there is no mention of mass playing a part. It is very convenient that Newtonian science ignores the relation or the fact that it is the Titius Bode law that is in place and that the Universe does not recognise Newton's claims on mass in any way.

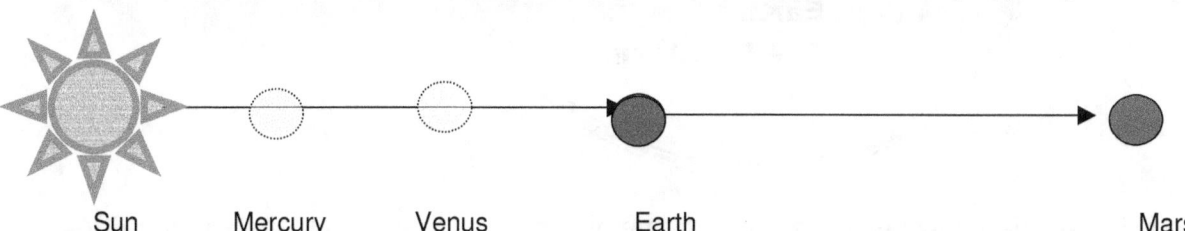

Sun   Mercury   Venus   Earth   Mars

Is there still anyone out there that would dismiss my claims of a conspiracy to put fraud in place by giving merit to Newton and ignore the true factor that the Universe holds?

Confront your physics professor by asking your professor in physics to use Newton and then to explain why every planet doubles the distance it has from the sun in relation to the immediate inner planet because he should know why every planet is doubling its distance from the sun in relation to the previous inner planet; after all he is the Newton-physics expert and with this representing gravity as gravity applies in the cosmos he is the expert in physics... and if he doesn't know, then what does he know because this is the basis of everything forming physics and he is the Newton-physics expert in physics!

One would presume that when dealing with the most intellectuals on earth also being those that even attack the bible and religion for accuracy, they then would see to it that their front Patio is swept clean without a trace of suspicion about accuracy anywhere. Or could it be that they have so much to hide that they attack all other things and in that prevent all the other ideas to attack the Newtonian fabrication of the truth. It is always others that are incorrect because Newton not withstanding the fabrication of the basic facts that is always correct.

Where the Universe does not comply with Newton it is the Universe that has to correct it ways and deliver missing mass in order to start to comply with Newton. If the Universe would still not comply with Newton the Newtonian mentors create and fabricate ark, unseen, undetectable, untraceable and non-existing dark matter to get the Universe to comply with Newton's outrageous mismanagement of physics. Physics is a hoax created by the imagination of those not understanding reality and not complying with any form of concept that forms an understandable science.

I dare any of the academics in physics standing behind Newton in full support of his theory on the gravitational law he formulated to explain where does planetary position fit into mass compliance. There are two options coming to light in this.

The one is that those brilliant mathematicians never saw this and so they are as stupid as toddlers or they are conspiring to keep this information a secret! If they are as smart as they pretend to be in mathematics why did they not tell the world about this inaccuracy about Newton's findings on which he founded physics.

The Titius Bode law is in lace and notwithstanding the level of despicable cheating, all the concocting the truth does not place Newton's misconceptions in place or remove that which is there, the Titius Bode law.

Mean distance of Terrestrial Planets from the Sun

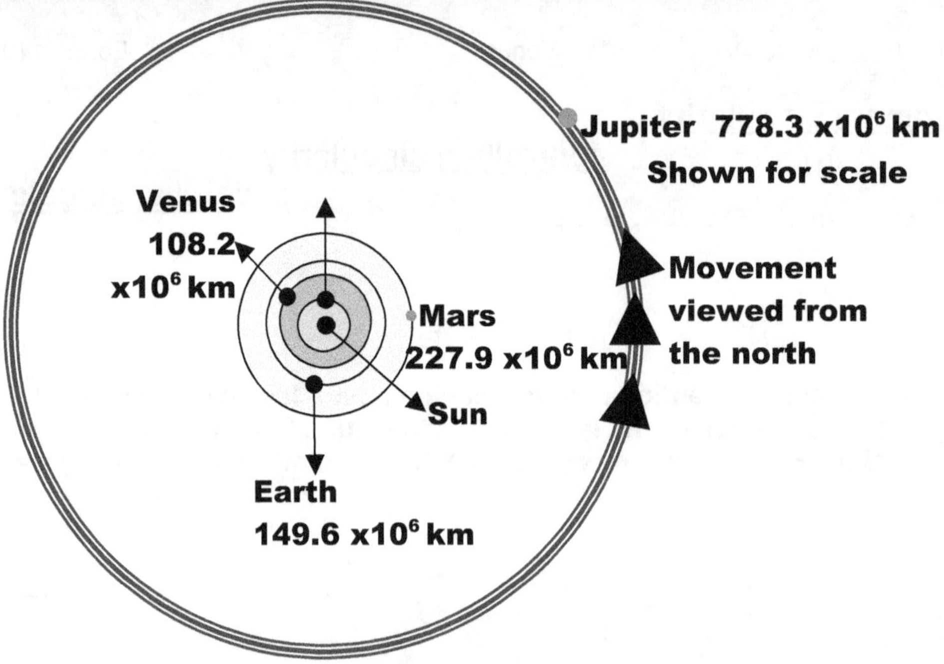

Jupiter 778.3 x10^6 km
Shown for scale

Venus
108.2 x10^6 km

Mars
227.9 x10^6 km

Sun

Earth
149.6 x10^6 km

Movement viewed from the north

The following diagram shows the approximate distance of the Jovian planets to the Sun.

It is a complete fraud that science covers up to hide Newton's incorrectness but moreover to hide their personal ignorance and pittyfull incompetence. Do you still believe science is correct at all? Are you still prepared to have this criminals teach your children "the truth"?

I dare any of them standing behind Newton supporting the Newton idea of $4\pi^2 a^3 = P^2 G(M+m)$ to apply the formula $P = \left(\dfrac{4\pi^2 a^3}{G(M+m)}\right)^{0.5}$ and translate that to the concept that "proves" the planet position allocations is derived from the mass of the sun and the mass of the individual planet relevant to the gravitational constant. Telling the Universe what the Universe should do according to Newton does not put the Universe in complying!

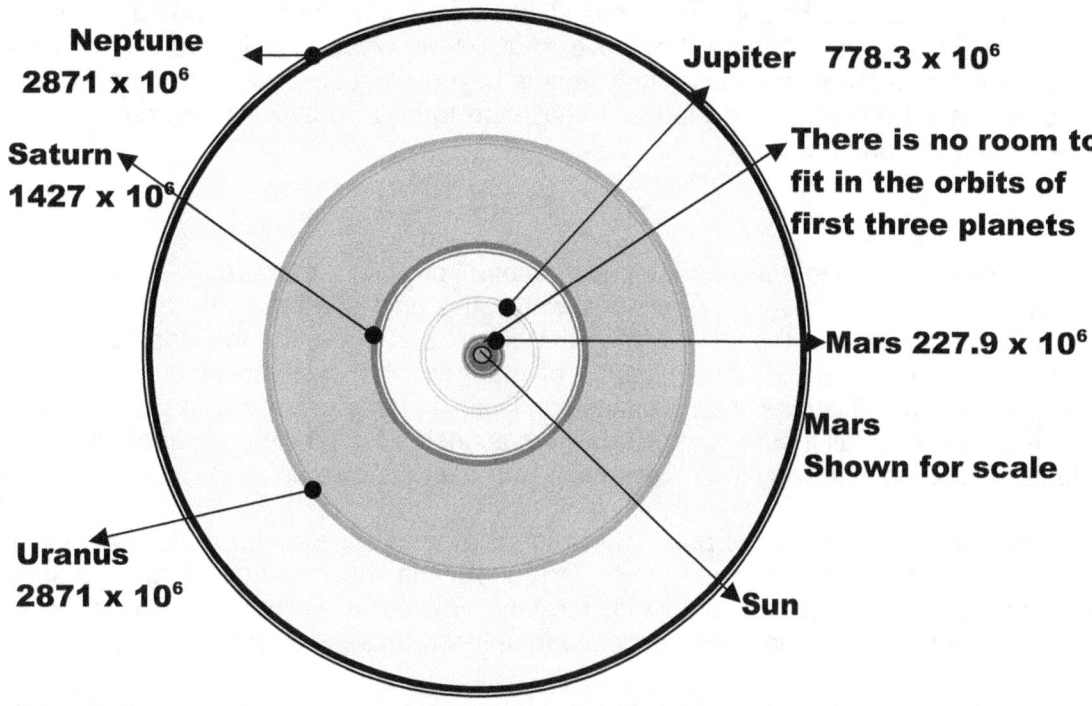

Mean distance of Jovian Planets from the Sun

Neptune 2871 x 10⁶
Saturn 1427 x 10⁶
Uranus 2871 x 10⁶
Jupiter 778.3 x 10⁶
There is no room to fit in the orbits of first three planets
Mars 227.9 x 10⁶
Mars Shown for scale
Sun

The planets only share one thing in common and that is the space in which they orbit that moving towards the sun. All the planets spin according to the space moving towards the sun and in relation to the sun contracting the space

Those that use Newton's formulas Newtonians wish to look brilliant and when peeling down the thin disguise they use to prevent Newton's incompetence becoming known, everyone can see how exceptionally stupid those smart mathematicians are because Newton's formulas disguised as Kepler's laws says nothing but exposes their incompetence in understanding cosmic facts.

Students, your professors are fooling you and you deserve to be their mindless monkeys just the way they think of you because you don't think about what they say! Cut the bullshit, force their ignorance about physics into the open and make monkeys of them, they deserve it even more then you lot do!

As you read the title of the book

www.questionablescience.net

If you think scientist know what gravity is…well, do not be duped that easily because no one in science remotely knows what gravity is…not even Newton knew what gravity is except Kepler… and because of what Kepler introduced now I know I can prove what gravity is.
Gravity is precisely what Kepler said gravity is

That Kepler was the only one that knows what gravity is we can see when we make an effort to investigate Kepler without Newton telling Kepler what he (Kepler) should have found. He (Newton) should have investigated Kepler's work more open minded and much closer then he (Newton) would have seen that gravity is precisely what Kepler found to be gravity

Science fails to bring logic answers to so many questions. It is simple questions about gravity and physics in which they fail, yet to every question science can't answer I bring an answer. My answer serves both logical science and truth but my answer does not match Newton and the misconception that gravity is generated by mass. Yet, since I do not applaud mainstream science and their fraud they go about to sideline and ignore my work is. Why is it going unrecognised.... because it trashes every article anyone has ever written about astronomical science and cosmology. It puts all the delivered on Newton's bias and fraud to the place it belongs, it trashes all work done thus far to the waste paper basket and renders all work invalid and void. Where they have to attack my work they ignore my work because they can't attack my work. In the same sense their work is beyond any defence and when I attack their work, they ignore me as if I represent the plague. I challenge them to prove Newton correct and not just declare Newton being beyond reproach.

$$F = G \frac{M_1 M_2}{r^2}$$

This Mainstream science use as the foundation of all physics anywhere. They put mass and the distance that parts objects in a relevancy, in other words the one is a ratio to the other. The increase in one becomes the reducing of the other. When the distance is large, the influence of mass will be small and when the distance is small, the influence of mass will be overwhelming. Why then when taken into consideration that if it is mass that produces an inclining force of contraction as Newton says then...when the Universe was small it did not implode whereas, instead it did expand. After all, the radius was almost no factor at that point leaving the mass to enjoy an eternal power in relation to the non-existing radius.

When the Universe was at the point where the Big bang started, the radius was incredibly small. That would make the mass inducing gravity by contraction inconceivably large because the mass was completely overpowering all factors with the small radius. It did not bring about an implosion that the overbearing mass contraction was supposed to unleash on such a small Universe in the beginning

The more the radius develops in time, the lesser would the gravity be that the mass factor generates in relation to the advancing radii developing and the larger would the reducing be of all contraction.
The effectiveness of force the mass produce will tarnish as the radius that separates the material from each other increases as time moves

Although it is presumed that the Universe was small at the dawn of the Big Bang, such presumption will put validity to another presumption that the gravity the mass charged at the time was enormous because the influence of the small distance in radii and the factor such distance produced promoted the factor, which the mass has to an enormous large factor. If an object is a million kilometres apart the radius is a million times more in value by dividing the mass influence than when objects are one kilometre apart. That is the most basic realisation about mathematics. It puts ratio to order and define coherency. That is what gravity is to the Universe as it puts respect to factors about the Universe in the Universe. It is what derives order in the Universe.

At the very same time we will find in a Universe that was supposedly so small it had a radius of less than only one kilometre, then at such a time when the Universe was still that small it must also be accepted that the gravity the mass charged was one million times greater as it would be when the radius keeping the structures apart is one million kilometres in distance. The extremely small radius that was only the size of one neutron in radii distance and with the factor that such a distance produces, it must promote the mass factor, which will support the mass in having an enormous large factor by relevancy to what the case must be at present. The mass factor that produces the gravity at any given point during the event of the Big Bang, had to be eternally larger at the dawn of the Big Bang while having an infinite radius, which gave gravity all the power it can have and which it will ever have.

If at the Big Bang there was not sufficient mass to destroy the radius and prevent the expanding from coming about, then the expanding won the match and there can be no contracting Universe as Newton had us to believe. If the Universe started a journey of parting objects no amount of dark matter that might lurk in the night sky and is at this moment hiding from detection will produce the gravity required to stop the expanding from continuing. At the start the expanding became evident and as the radii grows the inclination will suspend in influence as a factor. If there was insufficient mass at the start in order to tilt the balance in favour of the reducing factor, no amount of mass can ever accomplish such a goal afterwards. Then Newton's surmising was one of corruption making that which all physics are based on fools thought and corrupted proof.

If you might be of the opinion that my accusing the greatest intellectual department in the world as being in misconduct and to your view such accusing is outrageous and far-fetched, then be my guest and judge the following with a clear and unbiased mind because when scrutinised with a clear view then the facts cannot fool an idiot. However, that is just what the physics paternity thinks the rest of us forming the general public at large are. They have the opinion that they can feed us in the public arena any senseless rotten garbage they dish up because they see us as being inferior by thought and mind.

With all this in mind did any one ever come to wonder about the all too famous Einstein's critical density theory and the fact that this idea was conceived to conceal the corruption of Newton in physics? The fact in truth is that the Einstein's critical density theory was a scheme plotted by those in charge to cover up and conceal corruption in the heart of physics. If Einstein was unable to recognise the most basic of mathematical principles then what type of genius did physics create in him and what slur did physics promote. This idea of the two factors being in opposing relevance is so simple that children will recognise the principle, and yet those fathers of physics wants me to believe that the greatest mathematician that ever lived did not realise this principle…the principle that the radius and the mass stands related and the growth in the one will promote the decline in the other as a dominant factor.

Can any one with this information including the information given on the previous page have any other conclusion? It is obviously clear that having such a total idea that there might be dark unseen mass floating in the Universe which at this time does not generate gravity but will some day because Newton has to be correct at some point in the future. I am to believe that dark undetected mass can be found and such undetectable mass could be found which will bring about contraction after all this expanding? Why would the mass at present then not activate gravity and why would the mass at some point spring to life and start activating gravity? How much can the Physics paternity still hide the fact that Einstein's critical density is being used as a cover-up to distort the truth to conceal fraud?

The uncovering by the Hubble constant about of the Newton fraud is so simple to see. Hubble found the Universe is expanding and Newton's said otherwise. Who is lying about what? Hubble's declaration was on track to blow the cover that was concealing the Newton fraud wide open and uncover the centuries old deception. To see this we have only too look at the comet behaviour when any and all comets again come around on a cycle by repeated visiting the sun. The question is if it is mass pulling mass onto mass, then why do we have comets left in the solar system? The mass of the Sun should by now at least have destroyed every comet going around.

Every indication that we so far received in vivid portraying from astronomy photography studies from outer space disputes a shrinking Universe concept. From the moon increasing the radius distance between the earth and the sun, to the Hubble Constant indicating a space growing any where in space wherever man may conduct studies. Since the end of the middle ages a force called gravity was identified, but more than that science did not take it. What is gravity, besides being a force? What forces the force? I introduce a cosmic theory that turns the missing questions to answers.

Let us for one second return to the science we all know.
There is an undefined phenomenon in the cosmos, never mentioned (in public) because it obscures the basic formula

$$F = \frac{M_1 M}{r^2} G$$

Lets put the mathematical formula into a practical context.
By reducing r would bring about the same result as enlarging the mass factor of the cosmic objects i.e. the Sun and the planets. It is a very drastic implication that will cause much more than just seasons

changing. It must bring about that gravity changes through out the year…yet the radius does constantly change, therefore…
The closer any two cosmic objects come the stronger the force should be, with eventually no force in the Universe being able too keep them apart. This is just not happening!!!

There is no indication of truth about a contracting solar system as Newton proposed and as Newton's followers promote and or a contracting Universe as seen from the Hubble as he introduced has Hubble Constant…and not from any other evidence seen through the Hubble Telescope.

<u>As explained, there is some discrepancies about calculating the force of gravity, because gravity would apply as nicely as it does if it was the perfect balance, a balance exist in space of equal measure bringing about equal seasonal time.</u>
The biggest discrepancy and a practical denouncing the official version of the comet's flight around the Sun

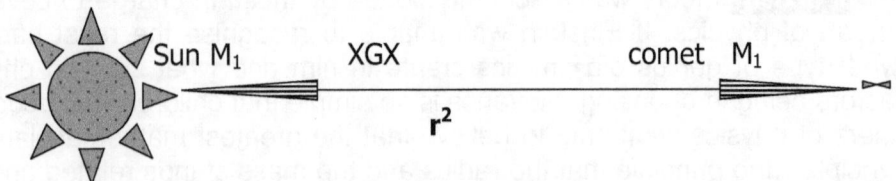

The Sun gets a grip on the comet by mass inflicting gravity and as it gets hold of the comet it drags the comet through the solar system straight ahead to the Sun just as Newton predicted the Sun with all its gravity producing gravity will do. There was no hint of a circle forming at any stage.

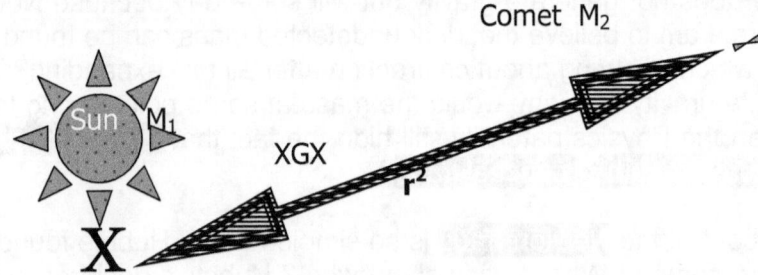

As Newton had said the gravity that the Sun and the comet mass induce pull the comet to the Sun. As we all know the comet moves to the center of the Sun just as Newton predicted with a slight complication and a change in the venue, the comet no longer aims to the center of the Sun but aims at a target outside the limits of the space that the Sun occupies.

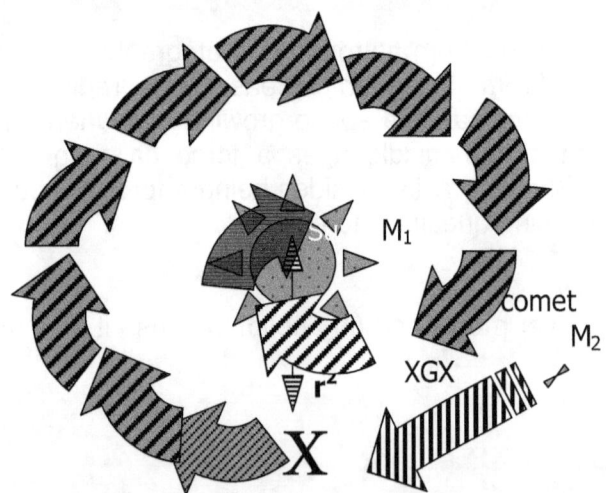

One should give Newton the benefit of the doubt and disregard this miss the position that the comet is aiming at. It could be that the gravity was not generated strong enough to locate the center at such a great distance as the comet was at first. It could be that the reducing of the radius comes in installments of a few circles. The comet might just take a cycle or two to wind down as the radius reduce and one should wait and see if it is not that which Newton meant when he said the mass by the mass is dismissing the radius between the objects. The formula Newton suggested did not make any room for a circle of any sort to form the trajectory of a planet or a comet trajectory.

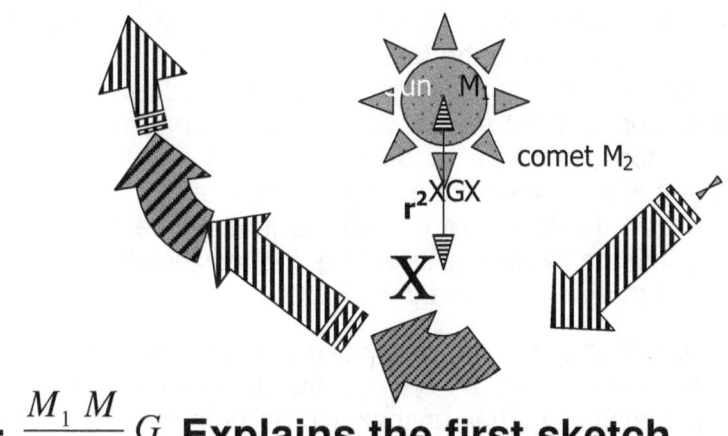

$F = \dfrac{M_1 M}{r^2} G$ **Explains the first sketch.**

Alas, it is not what the comets intends to do because the comet breaks the strangle hold of the Sun and what ever was pulling the comet at first is doing all the pushing at this point because the comet is surging into the darkness of the abyss. The comet speed away from the Sun and also at the same pace it was heading towards the Sun and there is no altering to the speed in any way. The comet seems very much unaware of the comet behaving in opposition to what the great Newton predicted with his formula.

The comet performs all the other maneuvers as the sketches indicate, which Newton's formula totally ignores. If the formula forms the basis of all physics used by science, the basics, which are around for hundreds of years, are trash and simply does not perform as it is supposed to. For many hundreds of years every person in physics were aware of this flaw but did nothing about it.

The lot that was filling the Universe was growing apart. In some case he said the lot was racing apart. The Universe was growing by miles and not shrinking into nothing. The Universe was blowing apart! The main discoverer had a name and a position of seniority, which prevented others from pushing his opinion aside. The man was E.P. Hubble. Through his telescope any one could see that the Universe was expanding and the expansion was most rapid.

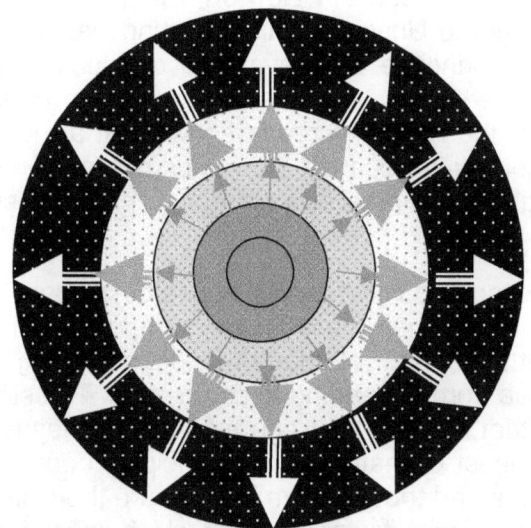

I am sure at the time all the Newtonian con artists were well aware of Newtonian shortfall in logical proof but this Hubble fellow was going to open a can of worms no one in science was able to face. How would the crooks that pretended to be the wise explain that Newton's contracting gravity was part of Newton's wild imagination? What would then happen if those that should know less then became wise and started to ask more questions about what the lot in conspiracy were hiding for (at the time) say about two hundred and fifty years? What is going to happen when the entire world learned that those academics in physics that was pretending to be the most brilliant among men was uncovered as the most stupid crawling the earth. What would happen if every one saw Newton was wrong all along and started demanding answers!

Every one was sharing the Newtonian vision of a contracting Universe where the lot would one day again come together and Creation will end where Creation started some time ago. The Universe has mass that is pulling mass towards one another and we are in the centre of an ever shrinking Universe. That is what the lot of us can see… we are forming the centre of the ever contracting having cosmos where every Newtonian can vividly see with his or her eyes through any telescope that all Newtonians minded scientists are sharing the centre stage of the ever collapsing Universe. The Universe is about to end where all mass contracts into one huge lump of material.

This unleashed a problem the world had no name for. Everything known to science was at that point devastatingly unknown to science. The world was expanding and not contracting which made the Universe quite wrong. It is impossible to have any vision about Newton being wrong. Newton could never be wrong because Newton was never wrong yet…so if the Universe is out of step with science, then science will correct such an abnormality by finding a way to defraud science and postpone the correcting that the Universe had to comply with since the Universe owed the Master Newton some apology. Did the Universe not know that he whom never can be wrong is in name Isaac Newton! Decisive action was needed. At this point I cannot believe that the most brilliant minds were so naïve and therefore I must suspect deliberate deception. Hubble was far too prominent to blow away and Newton was found wanting. At that point they put the onus of proof not on Newton but turned the focus away from Newton to what the presented as the guilty party. When will the Universe confirm its incorrectness by affirming Newton's obvious correctness? If they had to admit that Newton was wrong, the most intellectual science then had to admit they had nothing to show for all their minds brilliant work.

Science that was defying the likeliness of a living God stood bare and naked for all to see. They put the onus of proof and converting onto the cosmos. They asked not Newton but the cosmos when will the cosmos come clean and prove Newton correct, maintaining their unshakable belief that even the cosmos could be at blame but Newton could never be wrong. . When will the cosmos admit to a mistake and set its crooked ways straight. When will it meet its diverting from Newton and reach a point where the Universe will finally come to comply with what Newton demands. It is the cosmos that is wrong therefore it is time to find out when the cosmos will correct its manner. To deal with such a task they needed a man with a bigger ego than he had an IQ. They needed a person that thought more of his abilities than his ability to grasp any complex situation. They needed a man that was presented as a genius without ever proving his genius. They had a man that filled the centre of the Universe, which then placed the man in a location from where the man could see the entire Universe. They had just such a man. He went by the name of Albert Einstein. For all the genius Einstein had, Einstein failed to see the most simplistic and tiniest mathematical rule. Einstein failed to realise that if there was insufficient mass at the beginning of the expanding Universe, the growth of the Universe will reduce the influence of such mass as a factor because as the radius grows, such growth will restrict the gravity by rendering the mass progressively more incompetent.

If the Universe is expanding as Hubble indicated, the growth of the radius will reduce the influence value of the mass as every second passes. The mass will become more and more wanting for such a task. Yet with this obvious shortsightedness of the genius Einstein, the genius saw him fit enough to calculate and measure something as overwhelming as the Universe. As in the case of Newton, Einstein as an ego driven maniac that saw his abilities fit to measure and master the Universe while his mind was to simple to recognise the most basic principle of mathematics, the principle of relevancies or ratios. What a mathematical genius that turns out to be. While the radius enlarges, at the same proportion does the influence of the mass factor reduce and the mere fact that the radius increase shows that at no stage further into the future can the mass stem the growth of the radius because the radius overpowered the mass factor already. Unless there is new material entering the Universe at a point, which is impossible, the entire concept is fraud.

The idea was never to admit wrongdoing on the part of Newton and Newtonian science but to post pone, delay and divert attention away from the truth. If there was not enough mass to start with, no dark matter can kick in later on and start secondary mass frenzy that at that stage will then be enough to bring about the required mass potential that will turn around the Universe from expanding to contracting. To establish a scenario that would hide all deception they got the man that has a bigger ego than an IQ, they tell the world this man is a genius while the fool does not no the least of mathematical principles because his Master Newton did not no the least of mathematical principles and they got him to measure the Universe. While they did not even have any device (and will never have such a device) through which anyone would be able to see the entire Universe, they set of a scandalous misconception that this Einstein could calculate all the mass in the Universe.

Off course as can be expected, there was not enough mass and there will never be enough mass because there is no such a thing as mass in the entire Universe. When the deceit played out to the full, they fraudsters being the paternity of physics elaborated on the delusion by trying to find dark matter that is hidden. If the dark matter did not develop enough contraction at this time, there is no chance in the future to develop enough gravity because the factor of what mass supposedly should have is tarnishing and tarnishing as the Universe expand. The bigger the radius becomes the less would the mass effect be.

The community of astrophysics are trying to frame a picture where they set the stage in the way that if the Universe were stretched to a point the mass would not tolerate any more expanding. The mass will get frustrated in some way and show resistance to the increasingly elastic expanding. The gravity constant (I suppose) must prevent any further expanding. How they ever got to such an argument I never could tell. They surmise that outer space is consistently overall filed with nothing and when this nothing is stretched to the limit, the nothing would resist in growing more nothing or become further nothing and the nothing would stop other nothing to enter outer space in the community represented by nothing. If ever there is a faculty ruled by absolute inconsistency and rubbish as the motto of logic it has to be astrophysics.

Every measured kilometre represents nothing. Every mm is one of nothing. We on Earth are 149×10^6 kilometres holding nothing away from the Sun. Only they can argue that outer space is nothing with material here and there. If that is the case then which has more nothing between the Sun and Pluto or the Sun and Mercury. The distance between the Sun and Pluto is more, therefore that which outer space is made of is more than in the case of Mercury and the Sun. Therefore Pluto has more nothing between the Sun and the planet than Mercury has between the planet and the Sun. Only astrophysics and all the geniuses guarding the principal of astrophysics can put a calculated value by measure on nothing. In fact Mercury has hundred times less nothing between the planet and the Sun than is the case with Pluto. Since my days at school I was always under the impression that a hundred times the value of outer space being nothing is numerically expressed as (zero = 0 x 100 = 0), but where the genius that is such a prevailing part of astrophysics take the stage we find that Pluto can have 100 times more nothing than the amount or distance measuring nothing than Mercury has. The figure containing nothing that puts Pluto at the edge of the solar system is one hundred times more nothing than what Mercury has where Mercury becomes the first planet in the solar system. That is astrophysics. The brilliant minds of the mathematicians hold no rules apart from what they can calculate. Astrophysics is the only department throughout the Universe where normal rules don't apply since because with mathematics they can bend all laws as they wish…in fact Newton started the trend with his deceit.

Only the guardians of astrophysics policy can know why the undetected dark matter will start producing gravity to change the expanding to contraction. Would the fact that it is detected, change the influence it established? Or is it merely to extend the cover up and allow the deceit to linger until the following generation. There is no mass and any one that says there is mass, let such a fraudster then explain why all the planets irrespective of size or density, spin around the Sun at the same sped as all the others. Let them prove that the Universe acknowledge big and small and let them show how Jupiter can move at the same pace as does Mercury and Pluto while Jupiter is so many times more massive than the other two mentioned. More condemning evidence is yet to come because the astrophysics tricksters did not leave the corrupting of evidence just at that.

The fatherhood of physics never once diverted from acknowledging that Newton's contraction is the prevailing thesis on which the cosmos is built because they accepted that Newton used unlawful arguments and to cover up Newton's fraud which they still use to this day, they then proceeded with further criminality when producing the bluff they established with Einstein just to fool everyone in the normal public. Without ever recalling Newton's contraction theory that is obviously not working or admitting doubt about Newton's testimony to the effect, physics accepted the Big Bang Theory. The Big Bang theory opposes what ever Newton might have implied. The physics paternity however finds it wise to still advocate Newton while admitting to the Big Bang event. Newton said the lot is contracting. Go on and marry that with the Big Bang that says everything is expanding. You can't promote both except is you can define why we would see the two merge.

The Universe comes from a point the size of a Neutron. That makes the radius parting the Universe infinitely small. It just about removes the radius as a factor. At the very same implication it takes the pulling of the mass (if there are pulling forces converted by mass) to a level it will never again have. As soon as the distance between the objects holding mass started to grow, the power and influence of the mass factor started to diminish in the same ratio. If the mass were incapable of contracting the Universe then, it will forever remain contracting the Universe. Then you may ask what is the story? Read on and you will learn how far Mainstream Physics stray from the truth and how big a cover up the paternity is protecting.

According to science the Universe started with singularity. Quoted directly from the Oxford dictionary of Astronomy the following:

The definition of singularity is as follows:

Singularity: a mathematical point at which certain physical quantities reach infinite values for example, according to the general relativity the curvature of space-time becomes infinite in a black hole. In the Big Bang theory the Universe was born from singularity in which the density and temperature of matter were infinite. The average daily temperature was "$10^{\alpha\beta}$ to 10^{34} K".

Then the second "day" the daily average temperature came down to 10^{34}K and 10^{4}K. That is fine, but if the temperature was in Kelvin, then what was 0^0K. In order to make sense of the scale used there must be a minimum to secure a maximum otherwise the maximum can just as well be the minimum and is only advocated to impress humans applying earthly standards.

By using a scale as $10^{\alpha\beta}$ to 10^{34} K, it places the lower temperature at a modern 0^0K to make sense of standards. If that was the temperature the standards were lowered, compromising something to gain something, because something had to grow larger for heat to reduce. We know space grew larger bringing heat down to reduce.

Being the onlooker the viewer has to maintain one position. From that position some particles would be circling a centre point, as the particles would be coming towards the onlooker. The other matter would be circling the centre point while rushing away from the onlooker.
At the very end the single dimension may come into the dynamics but where the single dimension comes in the factor of zero is removed.

If there is space, there is a flow of light and a flow of light has to produce lines in relation to angles forming space between them. Something must be present to confirm space because there is an absolute difference between being in space and no space to be found. If there was a line that formed nothing that one line that forms nothing would completely destroy the other lines' chances of ever forming a triangle, let alone having all lines and they then have a total being zero. As shown in the example no line can form zero and therefore no mathematical equation as far as it extends to cosmology can ever bring about zero as a number. While there is space present there has to be three dimensions relating to each other by time and in three dimensions there has to be three lines in relevancy to each other by angles formed holding space in (at least) six opposing sides. Removing one line must bring about a flat Universe and that then will constitute nothing.

Cosmology is about light flowing by means of lines indicating space obeying the rules enforced by time in motion and light flowing dictates crossing space and across space light is using lines. The book: *An open letter Announcing Gravity's Recipe* is dealing with the subject finding singularity by removing the concept of nothing from outer space. By diminishing nothing one uncover singularity and the effort brings in a new perspective not yet introduced.

For your benefit I will shortly give a summary by which I hope to interest you in reading the manuscript: Compressing space produces heat. Releasing heat will bring expansion bringing about space. We call such a release of heat an explosion. In other words heat translate to space and space concentrates back to heat. The one is a product of the other where space forms expanded heat.

They are quick to show the time that was applying at the time being some thousandth of a second or the heat that was present being numbers we have no name for. The other side of the story they ignore. They ignore the other side of the story because in that respect it puts their promoting of Newton down to madness. If you reduce the radius applying at the present back to what it was at the time of the initiating of the Big Bang, you must also increase the influence gravity and mass had at that moment by the same number you are decreasing the radius. That is pure mathematics and the most basic physics of all concepts.

The shrinking radius will increase the effectiveness of the influence of the gravity that the mass can produce by the margin of the shrinking of the radius. If the Radius was infinite at that point, then that means the gravity was eternal. With the entire Universe being as big as a Neutron, the Universe was the size of an atom. If the Universe were the size of an atom and the mass within that Universal atom could not prevent the Universe exploding into immeasurable atoms, then it would not be able to retract all the atoms into one unit again. If there was not enough mass to start the contraction, there can be no contraction of mass that is producing the gravity at this stage. If the gravity is of such a nature that it allows a continuous growth of the radius, then the radius firstly cannot be zero as Newton suggested and the extending of the radius proves there is no contraction in the way Newton had everyone to believe. If Newton's mass contracting mass is true, then on the other hand it must have resulted in an implosion as that which can never repeat again. With Newton's formula of $F = \dfrac{M_1 M}{r^2} G$ forming gravity, then the Big Bang is just not possible because from that formula the Big Crunch must respond.

They put all questions on hold by diverting the attention to some black matter or dark energy, some force no-one can detect and yet it is going to save Newton's honour. Why is that dark matter with mass not pulling at this point?

Forming part of the website www.singularityrelavancy.com
By going to LULU.com the following books are available in e-book format as individual books wherein I share with you the newly discovered information about www.singularityrelavancy.com which you are reading and which you are free to download
Then download the next book from Lulu absolutely free and see if I exaggerate in any way!

THE ABSOLUTE RELEVANCY OF SINGULARITY: THE UNPUBLISHED ARTICLE Free of Charge
http://www.lulu.com/content/e-book/the-absolute-relevancy-of-singularity-the-unpublished-article/7747133]

The Absolute Relevancy of Singularity The Dissertation
http://www.lulu.com/content/e-book/the-absolute-relevancy-of-singularity-the-dissertation/5994478]

Book 0 The Absolute Relevancy of Singularity in terms of Newton
http://www.lulu.com/content/e-book/book-0-the-absolute-relevancy-of-singularity-in-terms-of-newton/7190018

Book 1 The Absolute Relevancy of Singularity in terms of Cosmic Physics
http://www.lulu.com/content/e-book/book-1-the-absolute-relevancy-of-singularity-in-terms-of-cosmic-physics/6624181

Book 2 The Absolute Relevancy of Singularity in terms of The 4 Cosmic Phenomena
http://www.lulu.com/content/e-book/book-2-the-absolute-relevancy-of-singularity-in-terms-of-the-four-cosmic-pillars/7181003]

Book 3 The Absolute Relevancy of Singularity in terms of The Sound Barrier
http://www.lulu.com/content/e-book/book-3-the-absolute-relevancy-of-singularity-in-terms-of-the-sound-barrier/6621856

Book 4 The Absolute Relevancy of Singularity in terms of The Cosmic Code
http://www.lulu.com/content/e-book/book-4-the-absolute-relevancy-of-singularity-in-terms-of-the-cosmic-code/6625975

Book 5 The Absolute Relevancy of Singularity in terms of Life
http://www.lulu.com/content/e-book/book-5-the-absolute-relevancy-of-singularity-in-terms-of-life/6626316

Book 6 The Absolute Relevancy of Singularity Investigating Kepler
http://www.lulu.com/content/e-book/book-6-the-absolute-relevancy-of-singularity-investigating-kepler/7179110

Where the above all are also available from Lulu.com.

Should there be any person whishing to purchase these books in one volume given as a thesis published in paper format then contact me, on this web address by activating
www.singularityrelavancy.com

And you will be able to purchase six books in print as a unit on paper forming one volume with six books going as *The Absolute Relevancy of Singularity* The Theses.

However please note that this printed Theses is very limited as it is printed privately. When you press on the button www.singularityrelavancy.com to activate I will return the e-mail as soon as I can to confirm the availability of published manuscripts and prices. The six books are identical to the six books on offer through Lulu.com but they are in monochrome whereas the individual book in e-book format is in colour where colour applies

The Absolute Relevancy of Singularity The Article is written as the first introduction to introduce singularity forming gravity in the new theorem explaining the Absolute Relevancy of Singularity.

Since the article was comprehensive but was adjudged as to long for a physics journal, I decided to offer the article in its original and total layout in which I introduce the framework of my ideas.

The Absolute Relevancy of Singularity The Dissertation is there written as the second introduction to introduce the four pillars in a very wide sense on which the new theorem rests. This is to convince readers about the authenticity behind the explaining and the thinking that forms the new approach to physics backing the Absolute Relevancy of Singularity where gravity depends on Π.

Then The Absolute Relevancy of Singularity consists of a four individual part theses each forming a thesis. There are either six individual books on offer in e-book format or in print could only be purchased as one unit named The Absolute Relevancy of Singularity The Theses This consist of www.questionablescience.net-the-website http://www.lulu.com/content/e-book/wwwquestionablesciencenet-the-website/7607865]

www.singularityrelevancy.com THE WEBSITE http://www.lulu.com/content/e-book/wwwsingularityrelevancycom-the-website/8074920]

which you are reading and which you are free to download
The Absolute Relevancy of Singularity: The Extended Article Free of Charge from Lulu
http://www.lulu.com/content/e-book/the-absolute-relevancy-of-singularity-the-extended-article/7180231]

THE ABSOLUTE RELEVANCY OF SINGULARITY: THE UNPUBLISHED ARTICLE Free of Charge
http://www.lulu.com/content/e-book/the-absolute-relevancy-of-singularity-the-unpublished-article/7747133]

The Absolute Relevancy of Singularity: Introducing The Sound Barrier
http://www.lulu.com/content/e-book/the-absolute-relevancy-of-singularity-introducing-the-sound-barrier/7745052]

The Absolute Relevancy of Singularity The Dissertation ISBN 978-0-9802725-8-1
http://www.lulu.com/content/e-book/the-absolute-relevancy-of-singularity-the-dissertation/5994478]

The Absolute Relevancy of Singularity The Theses called
Thesis 1 The Absolute Relevancy of Singularity in terms of Cosmic Physics ISBN 978-0-9802725-2-9

The Absolute Relevancy of Singularity in Explaining the Sound Barrier called
Thesis 2 The Absolute Relevancy of Singularity in terms of The Sound Barrier ISBN 978-0-9802725-3-6

The Absolute Relevancy of Singularity explaining the Four Cosmic Phenomena called
Thesis 3 The Absolute Relevancy of Singularity in terms of The Four Cosmic Pillars ISBN 978-0-9802725-5-0 and

The Absolute Relevancy of Singularity used to explain The Cosmic Code called
Thesis 4 The Absolute Relevancy of Singularity in terms of The Cosmic Code. ISBN 978-0-9802725-5-0

This is not the website called
www.singularetyrelevancy.com By pressing the title you can download www.singularetyrelevancy.com

Do you know what this is? This website will not show you the lovely pictures. For that there is another website that will tell you everything that happens in the experiment. This information will tell you why this happens and what is the cosmic philosophy prevailing in physics allowing this.

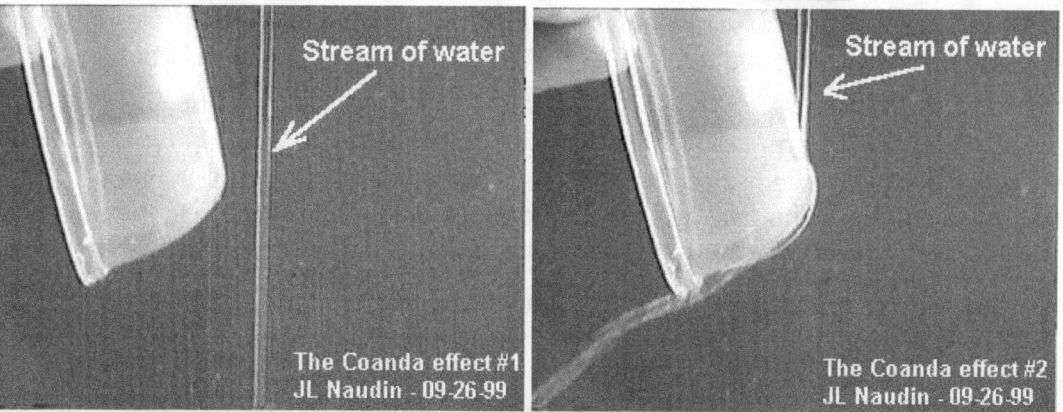

This is the Coanda effect that translates into gravity when the Roche limit comes about.

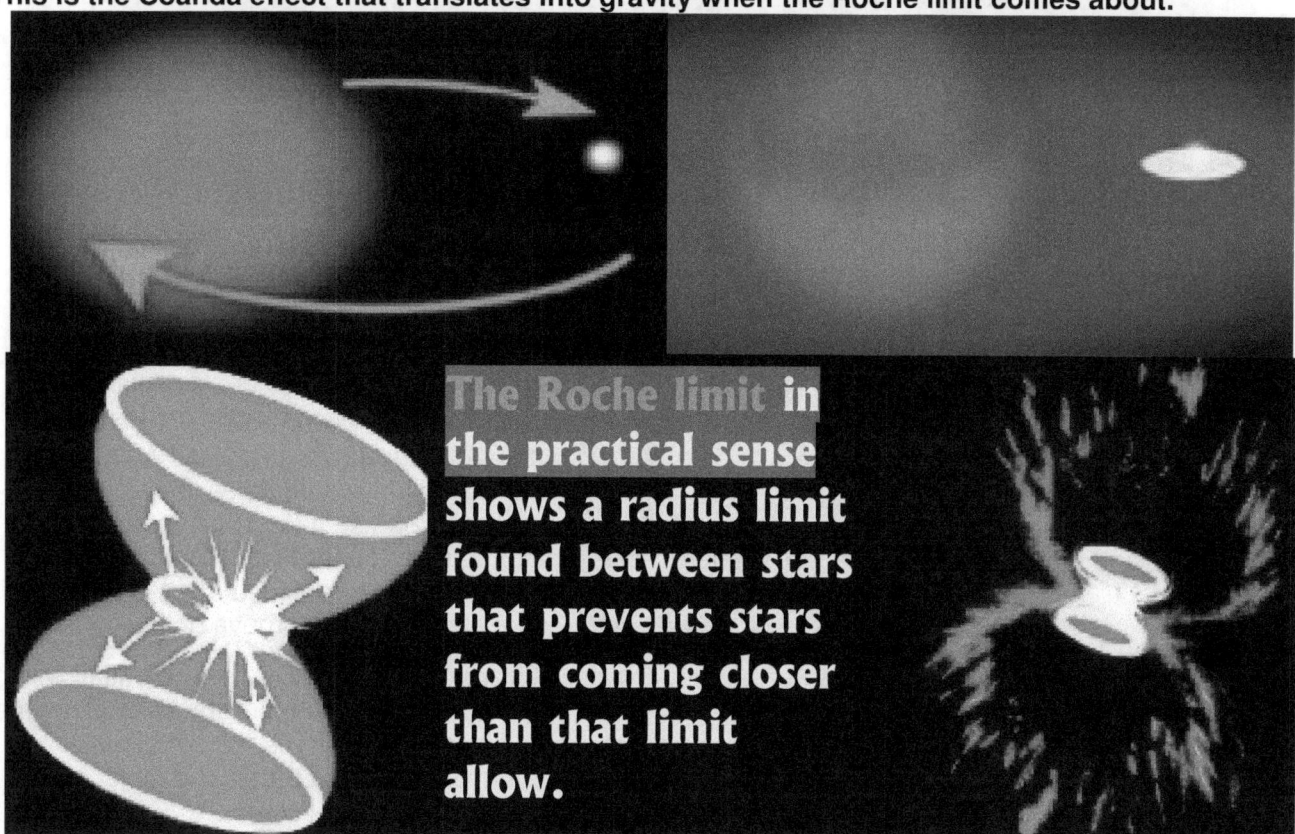

The Roche limit in the practical sense shows a radius limit found between stars that prevents stars from coming closer than that limit allow.

Do you know what this is? What happens in these pictures proves that stars could never collide because what prevents it is cosmic law that science at present can't fathom because their understanding is too limited. This book aims to inform you about the facts behind the lovely pictures. This book aims to tell you why the cosmic physics principles establishes these phenomena...and because the information does not salute Newton, academics in physics despise what I say. If mass don't pull stars to a point of colliding then Newton is wrong and in these picture we see the evidence that stars can never collide as Newton claims it does! Any star coming closer than $\Pi^2 \div 4$, which is equal to 2.4674 times the radius of the larger star would liquidise the smaller star and this law and the Coanda effect are a precise duplication of true gravity truly applying. This value the Roche limit has is $\Pi^2 \div 4$ which is circle movement divided by 4 opposing sides through which the rotation goes.

In order to establish a reference point where singularity divert allowing matter the zone matter can claim space in time, the control of space in time and the influence on space-time, the point of singularity have to reduce the value of singularity on both accounts of the cosmic atoms claiming individual singularity, or enlarge the claim of space by matter away from singularity. This is rather important to understand when arriving at the actual presentation of the formation of the solar system

After establishing the reference point to either singularity reduction or space-time enhancing through allowing matter to grow, the Titius Bode law apply, which I have explained in the pages preceding this page. Total annihilation and destruction of the singularity in one object may result in the object fragmenting to smaller parts where each part will still hold singularity, affected by less matter claiming space. When establishing a point of singularity to both objects a mutual point of referring to both objects' points of singularity will come about at the point of $(\Pi/2)^2$ where both object will hold onto their individual singularity while spinning around the mutual point of singularity.

Another outcome may be where both objects maintain the claim to singularity, by pushing the space-time occupied to new levels of occupied space-time values. The result of the establishing of new individual but unequal points of singularity is the oval way objects rotate, first favouring the on in the matter part and the other in its space part and afterwards turning the points of reference around. This stems directly from singularity

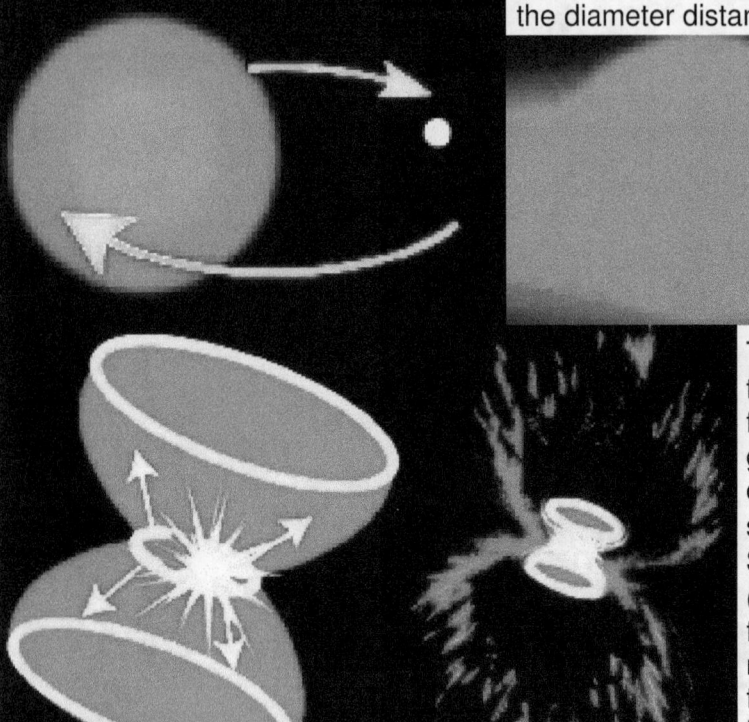

The two stars find a position apart at or closer than 2.467 the diameter distance of the larger star. This is $(\Pi/2)^2$

The process turns the stars into a liquid thereby forming gravity as this puts the Coanda effect into action

Then the singularity of the large star ($\Pi^0\Pi$) takes charge ($\Pi^0\Pi$) of the ($\Pi^2\Pi$) singularity forming gravity within the lesser star. The gravity forming singularity that forms conditions within the major star applies the same ($\Pi^0\Pi$) conditions within the lesser star. Since the gravity ($\Pi^0\Pi$) has more gravity ($\Pi^2\Pi$) this condition liquefies the material forming the space in the lesser star. If the major star can control the singularity within the lesser star there is no need to turn the lesser star into a Supernova or liquid.

After dissolving the lesser star to a state of liquid the major star then form the same space outlay within the space that the lesser star holds it turns the lesser star into an atmospheric state equal to the atmosphere of the major star. After the lesser star becomes as liquid as the atmosphere it dissolves the material formed by the solid as well as the liquid of the lesser star into becoming part of the major star.
In the event where the resistance op the lesser star forming singularity becomes too much to be controlled by the major star, the heat condition within the lesser star would release into space by exploding since the heat within the lesser star rose to a higher level that the major star could control. This is all to do with singularity and the way the gravity within the universe works when the Universe goes "flat" as Einstein put it. The Universe never goes "flat" but alternates between forming time "going flat" and forming space "looking as we see it". There is a period of space formation and there is a period of time displacement

In the centre of all things spinning a spot (1^0) forms a dot (1^1). This connects by a line that only holds the point that time carries. Space is present at that point but it is also meaningless because spots that are not forming part of this Universe consume the period. I am not going to go into much detail because there are books with hundreds of pages dealing with this issue in total depth and the issue is complex. When singularity applies the one dot is equal to any other dot because $1^0 = 1^1$ and $25^0 = 1^0 = 1^1$ and therefore mathematically all dots are not only equal but forms the same dot.

Therefore as Π^0 extends singularity $1^0 = 1^1$ where then the relevancy changes and $1^0 = 1^1$ until $1^0 = 1^1 = \Pi$ where then ($\Pi^0\Pi$). However as I showed by using the top as an example this can only be with movement forming space ($\Pi^0\Pi$) going on to ($\Pi^0\Pi\Pi^2 = \Pi^3$)) and without movement or gravity Π^2 the process will fall back into singularity Π^0 in the spot 1^0. Therefore everything rests in singularity and singularity depends on movement of a spot becoming a dot in relation. The point forming singularity extends to the point holding 1^0 forming ($\Pi^0\Pi\Pi^2=\Pi^3$) an connecting this lot. The relevancy applies in accordance with the singularity taking charge and in the event of the two stars having the lesser star within $(\Pi/2)^2$. This reflects on singularity and we have to see singularity acting and not space producing gravity as Newtonians wish to indicate. Every point holding singularity moves the instant time draws space into singularity. It is not space disappearing as Einstein said but it is the relevance of singularity taking charge by forming $1^0 = 1^1$ until $1^0 = 1^1 = \Pi$ where then ($\Pi^0\Pi$). As singularity becomes the sphere by forming the initial ($\Pi^0\Pi$) by going from the spot to the dot it moves through seven points of which four are turning motion and three are linear motion. Therefore no movement can ever be linear before becoming a circle in a sphere ($\Pi^0\Pi\Pi^2=\Pi^3$). Then moving (Π) forms (Π^2) and this results in a total movement of singularity as ($\Pi^0\Pi\Pi^2=\Pi^3$) but because the circle composes of four points going linear while being a circle the square (Π^2) is divided by four to find the allocated position that time would place singularity at as time then forms space to become space-time. Every atom holds a centre with the value of 1^0 and stands related to a centre worth Π^0 while relating to each other by 1^0 t0 1^1. It all might sound senseless but this means that under gravity each star is worth the total spin of the combined spin value of all the protons $(1+1+1)$ in all the atoms while ending up as the centre value of the spinning star $\Pi^0 = (1+1+1)$. That shows the combining spin value transfer as much as it translates to the centre of a star.

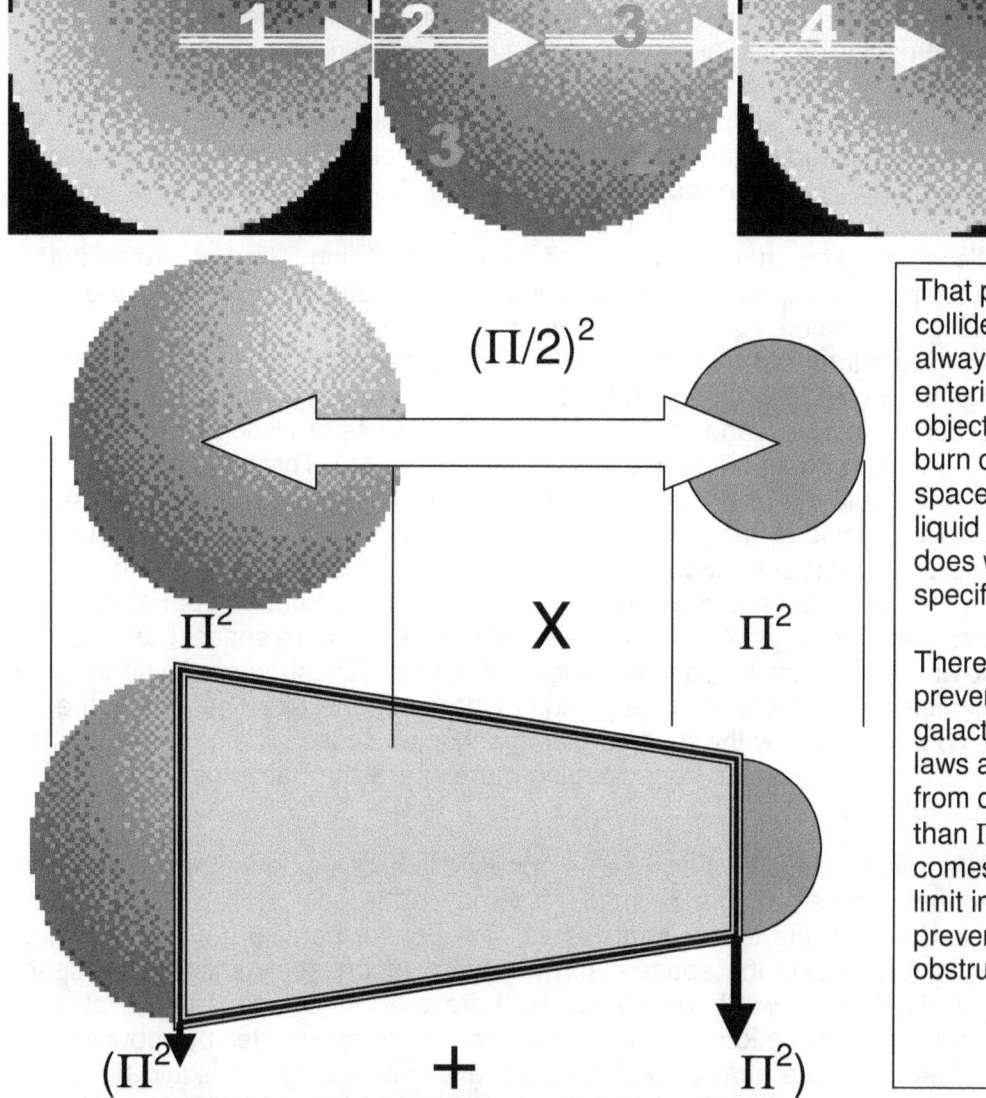

This shows that Π moves 4 times in one displacement circle and where Π moves the moving takes Π from one location to the next location Π by the value of Π^2 and because $\Pi\Pi^2$ space forms $\Pi\Pi^2 = \Pi^3$

That proves that stars can never collide and the lesser object will always resolve into liquid when entering the space of the major object. "Falling stars" always burn out and debris coming from space always dissolves into liquid just as the spacecraft does when not entering at a specific angle.

There are cosmic laws preventing star and therefore galactica to collide and the four laws are in place to prevent this from occurring. By going closer than $\Pi^2 / 4$ the Coanda effect comes into place as the Roche limit instates the boundary of preventing movement obstruction.

By matter moving closer to singularity, the time component will slow down as the heat factor will rise, since singularity was in the very beginning, eternally cold as it was eternally hot, holding both positions.

Stars only become stars when the stars time distinguish it from the geodesic, meaning the stars interior remains a liquid comparing to the geodesic going onto a relative gas. The picture shows how the "atmosphere" of the earth carrying the value of Π in relation the solid being Π^2 is distinguishing from the geodesic cosmos by turning into a relative liquid as the rest of the cosmos remains a gas. The moon shows no distinction in its "atmosphere" of Π to the relevancy of the cosmic geodesic gas. The moon is not even a cosmic body by any standards.

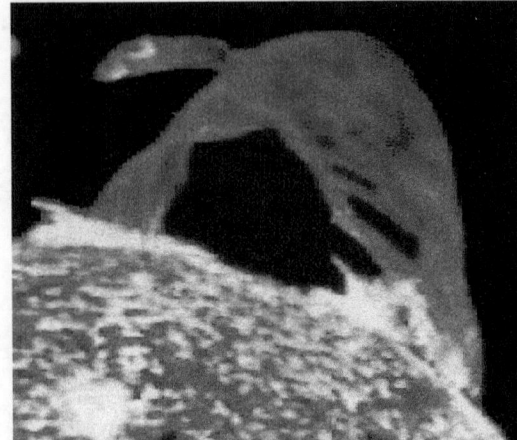

There are two forms of material in the Universe and that is material or solids and non materials or gas / liquid. There is no "nothing" to be found anywhere so please inform your local Newtonian about to go and search for nothing between his ears rather than in soace.

When a star becomes liquid having a fluid atmosphere, only then can it start performing its duties as a star by converting space (gas) to fluid (heat) which then turns matter back to singularity, by combining protons to denser proton clusters and in that manner serving singularity in a more supportive manner. If the sun or any other star were using gas, it would not have the ability to generate the means of sustaining fusion. Without the heat being in a liquid the stars would not have the ability of applying the full vale of Π to its full potential

Since motion in stars is the movement of liquid heat, the differentiation brought about by the liquid heat is much higher than mere pressure would suggest. When an explosion demolishes matter no force in the cosmos will stop the destruction. The reaction starts because there is a massive unbalance in the relevancy of space occupied to heat bounded by matter to specific space occupied. A sudden super abundance of heat coming available that puts space occupied in a disadvantage to space available since all the available heat became available space through the process we refer to as an explosion. The sudden excessive surge in heat produced more space than already occupied. The heat turned to space will alter the ratio effectively to the advantage of space in view that so much more space suddenly needs filling. The fragmenting of the matter will be in proportion to space occupied by density packed by particles, because there then is more space in need of distributing occupation. As the eternal dot in the original singularity will grow in all directions evenly and equally so would the growth in space, since it is the same process still applying. Matter is still the product of singularity creating space and heat with the creating of unoccupied space, but at the same time singularity is regulating space and reclaiming space as well as providing a system of controlling space growth. The reduction of heat will create a production of space. This will also apply within the walls of a star where only liquid heat fills all space not occupied by matter since there are no room for gas. When heat formed as a liquid by the governing singularity controlling space-time occupied and unoccupied, the reduction of space through the spherical form will bring about a production of heat and such a production of heat will bring about a surge in the liquid heat demanding more space. But space is one thing not available, since the urge of the liquid heat will not create more space, but will surge in the amount of heat available at that position in space and moment in time. Thus the production of heat will surge for more space and in this will create more heat going nowhere in the liquid. The urge in the liquid for space will have to be counter acted by the matter holding occupied space. Therefore the matter will have to abandon space occupied to ensure the hear claiming more space through excess heat production by spherical reduction of space.

In this we must acknowledge the role of relevancy because he opposite is applying to the occupied space that matter is claiming. Since the liquid heat demands space, it removes the only available space there are from the matter particles and at that the lesser particles to. The denser particle having the bigger number of protons and neutrons to react on the space claim will put up more resistance to the space compromise the showdown with liquid heat will bring about. As there are lots of space in the atom to compromise by reducing the claimed occupation the matter will become more matter by reducing the space, and the liquid heat will become more heat by replacing the available space with liquid heat. The reduction of space will produce the production of heat and the production of heat in the liquid will bring about that the heat in liquid will rise claiming more space stemming from an urge in heat forming a surge in space. The urge in the liquid heat for space to produce space as the heat in the liquid intensify where by it will produce space by what ever means necessary.

This means it will claim space not belonging to liquid heat and the only such space available is within the claims of matter. A the liquid surge for space, by producing heat, to that same amount will the matter loose heat, and thereby loose claim to space becoming that much colder respectively in reaction to the surge in heat by the liquid. The relation in liquid heat urging for space will leave the matter unaffected in relation to heat growth since matter cannot claim heat growth. Matter can only be relative to heat growth being surrounded by more or less heat in terms of sure heat or in terms of space. Through this the matter effectively reduces in heat as the liquid grows in heat where none of the two can produce space unoccupied. As the liquid heat rises, the relevancy to heat in matter reduces and therefore becoming colder. In that way the matter will eventually loose all claim to space, therefore heat, and will freeze by fusion reducing the minor substance to no heat (space) at all.

As the matter loses space it will become closer to singularity and thus become more matter to less space bringing about that the heat will become more heat and the process will define the two much better by the loss of space, to both ends, up to a point where matter becomes absolute matter losing all space and becoming so cold in relevancy to heat in liquid, that it can freeze that it can freeze in fusion at many billions of degrees in what ever measure you wish to apply!

At the point of fusion the three cosmic components segregate completely where heat becomes liquid denouncing space holding the position of gas and pushing matter to the absolutely solid it should be. The two forms in the star had become defiant defined counterparts, underlining in clarity the distinction without the intermingling of space as we see the form of heat called gas we think of as space. By reducing all space belonging to matter, the fluid will fill all space bringing in an intense heat equal to singularity at creation where matter expanded. Only at this point the opposite occurs where the space reduces to a point reversing the process we think of as explosion.

By introducing singularity Π^0 to the Titius Bode law $[(7+7/10) + 10/7]$ matter-to-matter as well as space-to matter and combining that with the Roche limit matter-to-matter $(\Pi/2)^2$ in relation to heat at the intensity of liquid status 3^2 a value of 62.6 will reduce all space between the particles in fusion and combine the individual singularity by removing the division.

Thus we can be sure that past the density of 62, there are no gas being space or liquid being heat, and matter being solid becomes transformed to singularity whatever singularity is.

A star is a device whereby the natural process is to liquefy space to the advantage of producing more matter as the solid substance of singularity. When a star produce light, it has space left in the solid of matter, but when it reduces space to the absolute, there will be no space to transmit light because light is the flow of liquid heat through the gas of space.

A star is a device whereby the natural process is to liquefy space to the advantage of producing more matter as the solid substance of singularity. When a star produce light, it has space left in the solid of matter, but when it reduces space to the absolute, there will be no space to transmit light because light is the flow of liquid heat through the gas of space.

The time frame science has about creation could be amongst all the miss givens the biggest misgiving of all. The very best is that mass as such is not an issue. It is the density that that mass produce, illuminating space that provide the star its qualities.

Each object has a different value of time in space where time has a different value to that specific part of space, whether it is a sub-atomic sub-structure of the atom or whether it is in a Black Hole.

THE ROCHE LOBE: In a binary system, the Roche lobes of components A and B meet at the L_1 Lagrangian point. (a) In a detached system, neither star fills its Roche lobe. (b) In a semidetached system, one massive component, B, fills its Roche lobe. (c) In a contact binary, both components overfill their Roche lobes and share a common envelope.

Try and marry the Roche Lobe with Academic's $F = \dfrac{M_1 M_2}{r^2} G$ and see a divorce before the marrying date arrive. One cannot say the Roche lobe is not, merely to allow Scientists to save face.

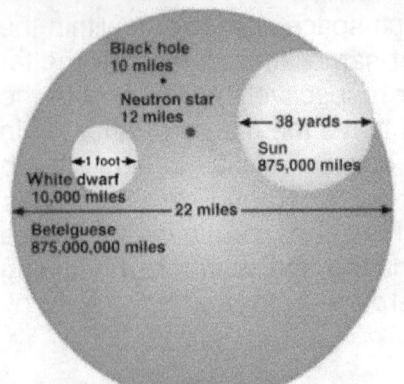

Science will have to face the fact that the larger a star is (the more space it holds to matter occupying space) the less intense the star is and the less formidable is the influence it has. This is going the wrong way with gravity because gravity depends on mass to space. A black hole is a few kilometres in diameter. But holds a relevancy to mass bringing about gravity equal to no other star in the cosmos. The less space the star holds, the more formidable will its influence be and by the time it is so small the gravity supersedes the speed of light, and it is merely a comparative dot

Once more, this phenomenon should not occur with Academic's presumptions about gravity. These bodies will collide and destruct, without a doubt if Scientists are correct when applying the formula $F = \frac{M_1 M_2}{r^2} G$ and there should not be any force which is able to keep them apart. However, they do exist and what is more, they maintain a certain distance apart.

We all accept that we the commoners are to stupid to understand such big issues and instead of feeling completely incompetent to the point of unworthiness we rather leave the issues in the hands of the Brainy Brunch because they do understand and we do not understand. They must be eternally wise and we must be eternally stupid not to understand what they clearly understand. When applying the most limited logic about the big concepts one find that the Big Brains have less understanding of the issues than we have. At least we understand that we do not understand while the Brainy Bunch does not understand what they clearly do not understand. What they do understand are issues they created keeping in mind that they are the only ones who should be able to understand such complex issues. Apply the most insignificant logic, and the Brainy Bunch is stuck in a much bigger way than we commoners are. All they then can refer to is that Academics are correct and Einstein makes sense. What about Academics that is correct is a matter they never ponder on and how much does Einstein convert sense to logic passes them by with out a second thought.

With the "force" of "gravity" "pulling" the stars closer using the accumulative mass of the stars and multiplying that value with both objects by the mass component, this will reduce r^2 progressively until r^2 reduces to zero.

Seen from this view, it is little wonder that the significance of this was lost in the notion that this is yet another "mystery" of the universe. The Scientists of the day (and the past) lost the importance, which this holds for us as earthly dwellers.

A most surprising aspect of this is that it is not that an unfamiliar or rare phenomenon. However, any answer to this would clash with Academic's presumptions, and before the Scientists allow that to happen, they would much rather ignore what is obvious. However, what is the obvious?

In an attempt to find time, let us first exclude time from space-time.

There is a zone on the outside of the star where the magnetic space-time has gone beyond the value of the speed of light, as we know it. This zone would be ever so slim, but it will be there. It is this zone that supplies the neutron star with its novelty being the exertion of neutrons. Light, (the photons) displaces space-time negatively, which means in the onslaught of the super prevailing positive space-time, the biggest majority will be lost to this quest.

In this, I would like to ponder on the binary stars, as they are the key to time in the cosmos. The two stars develop in the galactica in close proximity as they help each other in transforming negative space-time displacement to positive space-time displacement. With the combined effort, they can grow extensively in supporting each other. From the point of singularity time hold in matter claims a space in the value of $\Pi^2 + \Pi^2$. Another claim to control of space comes into value at the presentation of $\Pi^2\Pi$. A third claim to space holds an influencing value to $1 + 1 + 1 = 3$. When the proximity of matter disallow the claim of influencing space, singularity resents space in a half that of space under control, from both ends. That will be the square of the point from singularity under control being $(\Pi / 2)^2$.

It is clear that something filling the space between Jupiter and its first moon because of lightning interaction between the two structures. If there are lightning there are electricity and electricity means a very distinct interaction. Considering the notion of nothing being in place filling the space between the cosmic structure, electricity needs a conductor to transmit the interaction there are and that disproves the nothing theory Xepted as official information. It is official that the interaction was detailed as $a^3 = T^2 k$ which is what Kepler found, yet with this information science still do not appreciate the fullest of the implication. I have changed the formula to R^3 (space) $= T^2$ (time) K (one as the common denominator) and this becomes singularity $\Pi^3 = \Pi^2\Pi$. By applying the atomic value the relevancy changes to $(\Pi^2 + \Pi^2)\Pi^2\Pi3$ and that relevancy projects to cosmic atoms such as two stars interacting. When two objects come closer than the relevancy would permit, cosmic laws change their application and in this case then becomes $(\Pi^2 + \Pi^2)$ from either side where the three of space changes to singularity Π acting as the influence $(\Pi/2)$ from both sides making that influence a square $(\Pi/2)^2$.

At present science is unable to explain these very crucial phenomena and even to the extent that those guarding the principles of applying physics don't realise the importance of these phenomena or how the principles apply in physics.

These phenomena are there and it is in place so ignoring it is rather stupid…but that is precisely what Newtonian science has been doing for centuries.

They ignore these phenomena in favour of what is not present in the cosmos …and that is mass. Those in physics put mass in as an explanation for the movement of planets and while all planets has different mass, yet all planets orbit and move at exactly the same tempo. So how can they move according to mass-differences with having different mass and still be moving at an equal pace?

What there is, are the four phenomena I present, but because Newtonian science insight do not extend as far as understanding the phenomena and the role it plays in physics and moreover in generating gravity, Newtonian science is brilliantly ignoring the phenomena as well as the role these phenomena play.

Science do not realise the value or the impact that these four phenomena has on physics. That makes their judgement unfit in terms of deciding the importance the phenomena offer in relation to understanding physics. For many centuries science used improper factors in physics such as thinking that it is mass that produces gravity while only Newton's word is proof of this. There is no evidence of mass forming gravity and to overcome this issue they manufacture a component they called the graviton. There is no evidence of a graviton existing but to compensate for the shortfall existing in physics they use the graviton to promote a hoax however, this is not the only fable science invented

No one realises there is a shortfall because all evidence of such shortfall in Newtonian science is ignored like they would ignore the plague. There is no evidence of mass that has the ability to form gravity, but by cultural brainwashing every one believes this to be true. With everyone satisfied that there is no problem in science all persons are very satisfied with conditions applying as it is.

In every mind using physics the awareness of the mistake is absent and all presume that physics in the manner Newton said it is applying therefore thinking the Newtonian way is a healthy and correct and no one can find a reason to investigate my claims that Newton's idea of mass that forms gravity is a hoax.

No one can see the need to support my claims of science being wrong in spite of finding no evidence of mass anywhere in the solar system. The phenomena that do apply in the cosmos, science totally ignore because it does not support Newton's ideas and Newton can't explain these phenomena. I prove how these phenomena form gravity but I am ignored because it will turn physics on its head.

Should you download
www.singularetyrelevancy.com
Then you are about to discover that…

THE SPOT THAT'S HOLDING THE LOT.

As highly developed we seem to regard science to be, it took a genius like Max Planck all his life to try and reconstruct from what there are to what the Universe started with. Once again the human perspective baffled the genius of a Master such as Planck was, because he placed man amongst the first of creation. Where it started from and what we now have is many, many eternities apart and any evidence of an attempt matching what we have now to what first was is foolish.

It started off with one dot so small eternity met infinity within. Then came one more, and another and they continued coming until there were a countless number of dots. The accumulative size of the dots were the same size as one dot because in the true Universe big and small plays no part. The dots were infinitely small and eternally big at the same time because size is a relevancy and without one the other has no size. So in the true perception, there is no difference in size.

It started with the fact that there is no place or part in with which one may associate zero or nothing. There are no room for a number such as nothing. Next to the one dot (infinitely close) one will find the next dot, and if nothing was a factor then that is precisely what one will find between the two dots. Nothing of space, a non existing entity, taking up no space, and much more important, no time, therefore the dots are infinitely close to one another, being the same space, eternally big as much as infinitely small. If we as humans cannot find a manner in comprehending this notion, there can be no manner ever understanding the cosmos as much as the start to the cosmos.

Every dot was a Universe in its own and the accumulation was a universe. The earth in itself is a Universe as the moon is a universe, because rules applying on earth do not apply on the moon and visa versa. When in the ocean another set of rules apply, therefore being in the sea places a body in another universe. The number of universal entities is still countless, as much as it was in the beginning.

Every position in the Universe either holds singularity in a form, or relates to singularity. There can be no position unrelated to singularity therefore every aspect of the cosmos is space-time in various forms under the provision of singularity connecting. Matter cannot be if not surrounding singularity

Singularity is as close as any spot can ever come to zero BUT IT CANNOT EVER BE ZERO. From singularity diverts space-time and there cannot be space without time as much as there cannot be time without space, not withstanding the size of space or duration of time.

Through space-time singularity connects as much as relates linking the Universe into a network of influences beyond what ever we can ever conduct. There can be no spot that does not participate in the curvature of space-time. From the point of singularity runs space holding time to the prescription singularity dictates.

With singularity connecting singularity will or cannot relieve or release the connecting other than by a method we humans refer too as an explosion, but space-time separating as much as joining singularity dividing can change by applying time too space in a changeable manner, stretching and shrinking the time aspect by changing the density of the occupying heat, which creates the space allowing the spin of the occupying heat creating space setting the time.

Every dot insignificantly small as it may be, is a part of another Universe as much as it is part of the accumulative Universe and every dot in the infinity holds singularity, which we translate as " nothing" being " darkness". There cannot be "nothing" just as much as there cannot be "darkness". There cannot be something big or small, but it into relevancy of perception, and then the relativity of perception becomes the question. There cannot be hot as much as there cannot be cold. The sun FREEZES hydrogen to a liquid at six and a half thousand degrees Celsius and Universe boils over in the form of the Hubble constant at the temperature (we presume from our vantage point) at minus 273 degrees C. If we Humans cannot or will not abandon our human perception and our manly perspective, we may as well return to astrology for all its worth.

Every point in the infinity we may observe at is not merely part of the Universe in not being "nothing", but is the point where the Universe started representing singularity. It is the very first point where everything began so many eternities ago, because after all, how can we ever determine where the first point was, as they were very much equal and alike at the beginning. Every aspect of the Universe started with the fundamental fact that no point in the Universe can represent "nothing" as a number, because every aspect in the Universe represents singularity in what ever form it may hold in that specific spot forming space-time. If man does not reach a conclusion where that conclusion is matching the Universe and stop to match the Universe with man (and man's incapability), we may all go back to caves and become starving hunter-gatherers again, because we will never find a way to progress to the ultimate understanding of the universe.

I wish to make the argument about gravity found in orbiting objects a rather lengthy debate as part of my initiating discussion and after all, it is the very argument on orbiting structures NOT drawing closer that initiated and concluded all the effort of my writing this book. FOLLOWING ARGUMENTS MADE ME SEEM AS THE VILLAGE FOOL TO FRIEND AND FOE, ALIKE, AND WHERE I MAY NOT BE THE BRIGHTEST IN THE VILLAGE, SURELY I AM NOT THE VILLAGE IDIOT THAT EVERYONE TAKES ME TO BE ONLY BECAUSE I MAINTAIN THERE IS NO GRAVITY.

If the graph broke a zero mark, then a new and totally unrelated line will form on the opposing side baring no relation to corresponding with the line in the previous quarter. The new line may start from any point and lead to another point holding no resemblance to the opposing side. What will prevent the line from finishing at a point marked say three and starts afresh at seven because there is NOTHING between the lines. The nothing will provide releasing and detaching all corresponding relativity there may be through disconnecting. Through nothing there can be no resemblance or corresponding because the point of start need not even to continue at all. Nothing will release any connection and if such a connection may come about, such a connection may just as well be very co-incidental and may never be used in calculating accurately. Experience taught us that there is a definite precise and secure corresponding that can only result from a direct connection in as much as the line being the same line. In that case the line must then come down to infinity and release from infinity as the same line still connecting to a point of re-bouncing to either side. The graphic cross is he result of singularity applying opposing sides but still maintaining connection through the application of Pythagoras that will connect and always bring about a direct relevancy. The graph does not hold zero because information derived as result of a relation prove a contact remaining when the line crosses singularity without applying detachment. The Brainy Bunch holds the view that in a graph the line crossing amounts to breaking the zero mark. That cannot be the case.

In the centre of whatever two synchronized rotating objects the line of singularity matches in spin to one another not withstanding the space-time involved. The relevancy applying places one rotation of spin eternal, therefore the rotation is standing still, but also one rotation consists of an above measurable number of infinities producing a rotating speed faster than non other will ever reach and that makes the rotation the fastest there can will and may ever be. Again we humans now face the same value we wish to separate, as is the case with hot or cold, near or far, and quick or slow. Once again time locks relevancies beyond understanding. That line so infinite small as it is eternally big connects the cosmos through light we are able to see and light we are unable to realise.

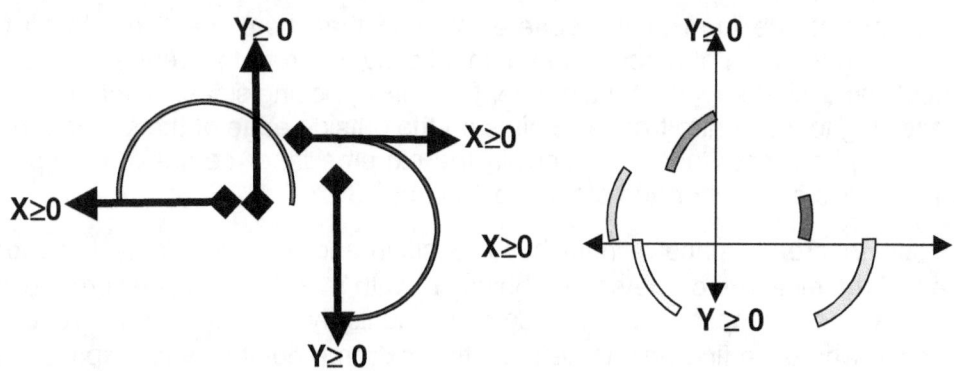

The mysteries of the cosmos are rather simple. Because matter more frozen heat it holds less heat. As it holds less it also holds less space and therefore it will apply the depreciation of space and accordingly present the appreciation of heat.

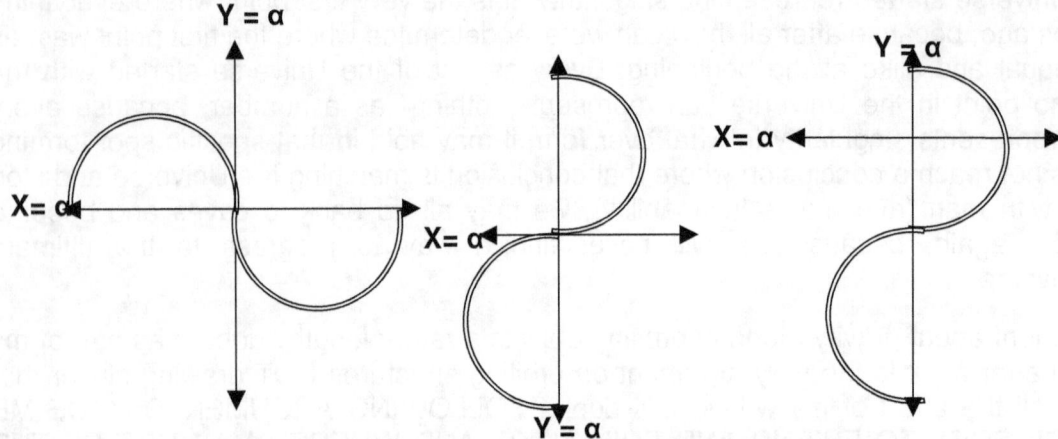

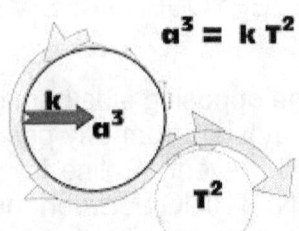

When translating Kepler's mathematical expression into a verbally spoken form of communication such as English, we can see what Kepler said also reads as $k = a^3/T^2$, where **k** is one point from a centre point that is space a^3 relating to time T^2. From a centre comes space-time.

Ever considered why a water drop to form a sphere? When a water drop is released in an astronaut's space capsule in outer space, the water drop forms a sphere as it floats free gravity. As soon as the water drop is released from the Earth gravity and has the opportunity to float in space, it can form any shape it finds pleasing, yet it immediately turns to the shape of a sphere. It is the same reason why we would think the Universe 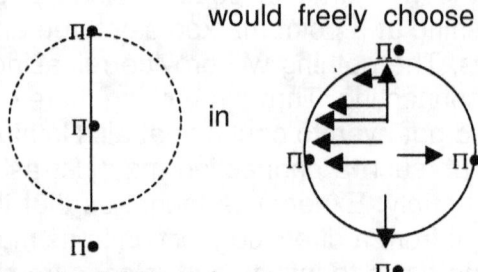 would freely choose in takes on the shape of a sphere, although we know the Universe has no outside and we realize that the Universe is limitless in size. Yet, notwithstanding, we take the shape of the Universe as naturally being a sphere that formed ... but why think of a sphere that the cosmos considers as the sphere being the pre-cast shape? Well by saying it is the strongest form available, has the same argumentative potential as saying a baby is little at birth. Blaming it on no reason substantiating proper evidence that would indicate some facts more prudent than just a simple answer to dodge the question by hiding obvious incompetence. I'll give you one clue ... it has something to do with finding the centre of the Universe. That brings on the next question being where one might locate the centre of the Universe. These questions put to your are most ordinary questions ... and yet the complexity in the answers rise far above the answering ability of those with the supreme mathematical skills. Those master mathematicians cannot answer such simple questions by using their complex mathematical powers. If they could have, then they would have ... the answer is extraordinary simple and childlike easy to explain.

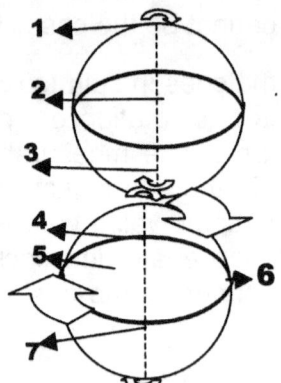

By examining the form of the sphere, we find that there are 6 points on the surface of the sphere that is holding the form at a specific and equal distance from the centre. Lines run from the centre into space at 90° and 180° angles of each other from six opposing sides. There then are six lines at 90° and 180° connecting to the centre from six points on the outside edge of the sphere. As a result of the basic shape that a sphere has, there is a spot in the extreme inner centre of the sphere where the lines in 90° relevance cross each other and others connect by 180°.

There is also at that point a spot where all space relinquishes a position and only singularity 1^0 as form remains. At such a point we find the measure of the sphere being Πr^0 with $r^0 = 1^0$. That is where the line that represents the radius as a line disappears, as it becomes singularity r^0. After more reducing continues, we get to such a point where we find only Π^0 left. At that extreme point is where space in all form disappears as the circle providing the sphere the form the sphere has, removing all possible form by going into singularity $\Pi^0 = 1^0$.

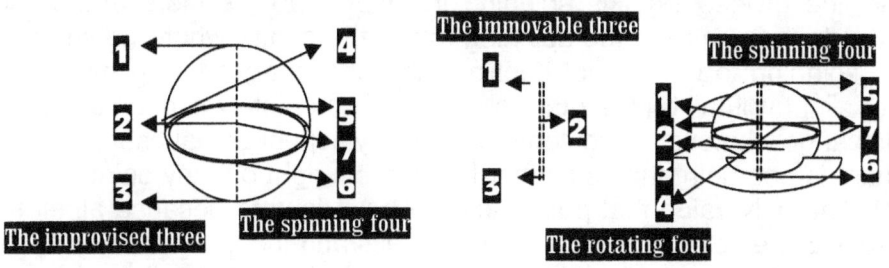

Then in that area all form of any possible space disappeared leaving only the dimensions of singularity 1^0. This too, I take much further in the book as I delved deeper into the argument and by doing that I stumbled on the ingredients forming gravity. However, from such a point there runs lines that connects to space on the outside where six points on the outside points connects to the space less point in the inside. In the book I take this argument much further, but for now, I leave the argument at that. Those lines carry the structural straight that the sphere has where the other six support every one of the six by singularity.

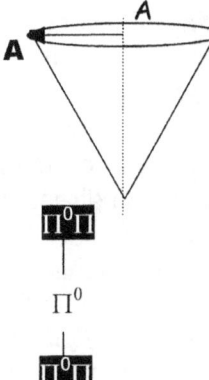

Where there is no space, there must be singularity 1^0 because the space is present although in singularity 1^0. If zero were a factor where all space finally halted in zero as the value, then zero would be able to remove the space from the centre and such removing would continue to remove the space until all space was removed. It will finally abolish all space in the sphere and it would remove the sphere. Zero removes all possibilities of anything coming about. Since the sphere is there, a zero factor in the centre cannot be present. Only infinity can be a factor from where space may grow because infinity can extend and grow into and up to eternity.

The moving of Π^0 to Π involved relegation and not motion as we consider motion. It was Π^0 getting a side and that is all. There was no true side but only a form that came into place. Singularity (A) received singularity (A) and no more of anything but the shift to comply with having a relevancy forming in relation to singularity. The dots had no sides, had no length or diameter. There was not measurable space or measurable time involved. The time could have been a micro, micro second as much a trillion millennium because time had no relevance. It was eternity interrupted by infinity, as it still is the case, however the line that eternity followed was no line because there was no space to hold the line. The line was momentarily interrupted by infinity, however with no one there, there was no one to notice. The lines were not lines but relations to sides being formed.

The relevancy that had the power to set Π apart from Π^0 is the only relevancy that still has the power, to set particles apart or join particles. It is heat in variation from cold. In order to excite singularity, singularity must establish a basis of heat that sets such a heat basis apart from cold. From there the form the atom will take on, however, the atom was still enumerable eternities to the development side.

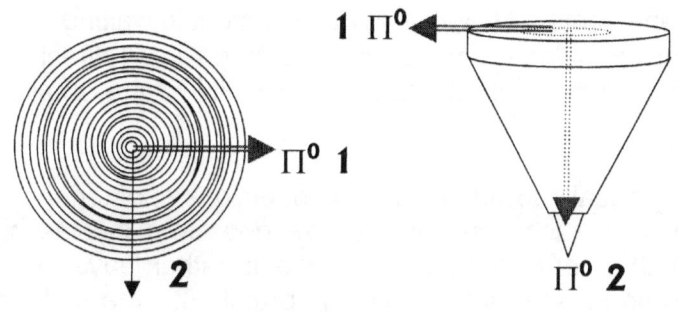

Where they are equal in value we must test the reason why this then is valid.

What is in the Universe is spinning. In the precise middle **of all** objects in rotation **is a precise centre dividing the object in sectors that will** start the spinning initiation **from that centre point.**

$k^0 = a^3 / T^2 k$ states that whatever is, is also spinning in order to be present.

Thus, the spinning object will have a middle point, **a very specific** centre point that does not spin **and only holds** Π **as a specific value because no radius can apply. But also the one value such a line** cannot have is zero **because the line** is there and holds contact **to the rest of the material bringing about that** zero does not start any **line and therefore the** value of the line must be infinite, **just as described in accordance and by** the definition of singularity. As I am introducing a very new idea, I wish to explain in better detail what I try to convey. While the top is spinning, one will find a line that formed in the centre where no line can form. It comes from spin but can never participate in spin.

That line must be singularity because if one moves any point on that line one position on, such a movement will land the point that then form on the line, on the other side of the line. The line is where the radius ends and starts because the line divides what is spinning in innumerable sectors and when reducing the radius progressively towards the centre of the spinning top at the centre where no line can be there is a line dividing the entire spinning top. At that centre point all further reducing must end because the next movement however slight will fall on the other side that is completely contradicting the one side. One movement further will change whatever is, so completely every aspect of that characteristic will contradict what it was before. There is one point that is neither left nor is it right but any point next to that point must be either left or right. The only value that point may not have is zero because albeit so small that it is not part of our Universe, still the point is there for all to witness and that point is a reality as much as the entire Universe is a reality. Whatever one attaches to the top either in the line of being material or a concept, such a concept or material has to start at the spot in the centre of the top because every aspect of the top changes in contradicting from that point onwards in all directions. That point albeit hypothetical, is also as much a reality none the less and is placed where that point must be standing still because every line running from that point in opposing directions is also in opposing directional spin the other or opposing side.

The Spot becoming the Dot — When space brought division between eternity and infinity

As the rotating direction moves inwards, the rings holding Π will become smaller and smaller. The reducing of the radius r will eventually end at r^0 but the top does not end there because the top still then is Πr^0. The form we attach to the spin still applies as Π and the top finds directional contradicting change at a point that never moves because it can never move being in the centre where the spin direction ends at Π^0. It is the only aspect in the entire Universe that can be and still be motionless because it is not within the Universe. It is the centre of the Black hole because it is the centre of the Universe. It is 1^0 and only the centre of the Universe in singularity can have 1^0 However that point where the directional spin ends is the point where the actual spin does not takes place because if its immovability. It is at a point in singularity $k^0 = a^3 / T^2 k$. It is where space ends because motionless ness ends space there. The spinning is on the precise location where the point is not spinning because the Universe ends in its not spinning there.

That line running through the centre of the spinning top divides every possible side from the opposing side in innumerable points that are divided by angles and degrees. Moreover, it changes the future position of the point from the present and from the past as it redirects every point every time **k** moves to reposition in a location where T^2 ends. In the end it proves that both **k** and T^2 confirms a^3 just like Kepler stated before Newton interrupted with dishonesty.

Another huge factor that favours the use of Π as a measure in singularity progressing is that any expanding by any mathematically sympathising method will have to use Π since Π is the only route that a spot of no significance can develop into a dot that represents a Universe of development in waiting. Only by promoting through the measure of Π can all possible sides progress on equal terms in all directions simultaneously without bias. The development must progress by measure of equality to the smallest indication and that purpose only Π can serve.

The progress must be generated so that it can flow equally to all sides in all directions spontaneously where not one side will favour the growing process as such. Time in it's flow does form a bias but explaining that at this stage will also involve too much other concepts which I would rather leave to the books. There is only one way to permit such a flow and have a mathematically correct outcome and that would be using Π for such expanding. The use of Π would ensure that a dot rises from the developing that comes from the first spot. The dot would have to form a sphere and a sphere is Π in relation to seven. This bring as back to the Titius Bode concept where ten is one half of a sphere relating to seven points forming the sphere where the seven with singularity puts form to the double ten that totals (including singularity) 21.991 and that is the overbearing dominating issue.

When the cosmos came into motion, motion was not yet defined. When the cosmos brought about motion, the first motion was relevancies. Cold parted from hot. Eternity parted from infinity. Motion parted from motion absence. Infinity broke the laboriousness of eternity for the duration of infinity. The spot became and grew into the dot.

From what the spot was to what the dot now is might be just a mathematical implication of going from 1^0 to 1^1 but in reality that first motion was the creating of and establishing of an entire Universe which was with all possibilities that now is it. Never again can that much growth become a reality, although to us the growth is beyond what we ever can notice. But it is because the growth is so massive and we are so small that we are unable to notice such almighty growth. When the spot Π^0 became functional and established all relevancies possible, heat parted from cold as eternity parted from infinity.

The expansion was not clear motion but more a parting of relevancies where a centre formed a relevancy because the centre could not provide motion. Without being capable of motion, the centre established four points, which also served singularity. From the inverse square law we know that the centre doubled by producing the four points holding singularity. We have to presume there is a time line because the Universe has this as evidence. The fact that light travel from there to here and from here to there proves of such a time line because there is no distance in outer space except in the Newtonian's misconception they have about the cosmos. Any line shows direction and the direction implicates positions according to the line having dimensions in the Universe we have in space and time. The line brings in Pythagoras and Pythagoras implicates mathematics.

1^0 going 1^1 where $1^0 \Rightarrow 1^1$

If there were progress that developed from singularity in the form of the first spot and we are the evidence of such progress, then the mathematical conclusion must be that a line formed where the line developed two sides and we have the evidence of that still present in our Universe. That brought about that three markers formed in relation to one and by admitting to the law of Pythagoras we find that what formed was 3^2 in relation to singularity 1^0 that became 10 on the one hypotenuse and 10 on the other hypotenuse which forms the square of space. Therefore mathematically space has ten positions and material has seven.

From the three the four (2 on both sides of singularity allocating the cosmic divide) the two in square developed as a mathematical consequence and that brought about the five.

The five was duplicated as a response on the other side of the divide and having five as a result of $(2^2) + (1^2) = 5 \times 2 + 10$ we find that the square of space holds the value of ten in place.

But 5 is in pace because $3^2 + 4^2 = 5^2$. And $2 \times 5 = 10$ is in place because $3^2 + 1^2 = 10$. I just can't go in depth showing how the Universe formed mathematics as the Universe built what is within the Universe because that boo (if my memory serves me correctly) is about 1300 pages in all. There is a reason why some part of mathematics show how a line and a half circle and a triangle is equal and why $1 = 1 = 2$ and at the same time $1 \times 1 = 1$. It all has to do with building the Universe in stages of development.

We read from the way mathematics formed how the Universe formed because the Universe formed mathematics as it formed what is presented as a Universe. Before the Universe there was no mathematics so mathematics did not form the Universe but the Universe formed mathematics and above all this came about exactly as the Bible says it Happens. If you divert one glitch from how the Bible presents it the entire process goes astray, but mathematics form the key in the presentation of how it all came about.

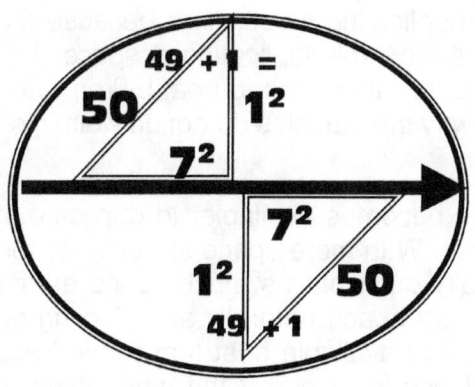

At the very beginning there was a spot. How do we know that? The spot is still with us and holds a value of 1^0. This spot is in the centre of all spinning objects. Then time came into motion and 1^0 moved to 1^1. Since from where we stand we see 1^0 and 1^1 as being the same and therefore with the moving of time in the very beginning such moving must contribute to an increase of space.

Therefore 1^0 and 1^1 has to have a difference where the one is 1 and the other is one point in singularity smaller making the 1^1 coming from infinity and rising into eternity 1^0, making infinity forever one point smaller than what we find as the value one will associate with the one we find as a measure in infinity.

Therefore we can judge that singularity combines to have a total of 1 + .991, which then becomes 1.991 or whatever because the one going smaller is running into infinity and since infinity is one less than eternity we are in eternity 1^1 looking at infinity 1^0 which is one point reduced in infinity. However this moving from 1^0 to 1^1 involved 1^2 as well as 1^2 on the other side of the divide. As a result of the form the sphere holds, there is a centre connecting the sides and the centre holds singularity.

However, by presenting a centre where all lines cross on a point that cannot distinguish sides since that point has no individual sides, the centre holding singularity is inactive. Motion makes it active and the motion of space in time activates singularity to charge gravity that we find as a factor in the Coanda effect. That motion that establishes the purpose of space a^3 as a result of motion k through time T^2 was what Kepler presented as a formula. Gravity is $k^0 = a^3 / T^2 \, k$ and to install k^0 the motion of space-time a^3 / T^2 is required to complete

Producing singularity sets the divide because singularity splits the Universe apart in separate equal components that in combining form the duplication of singularity, being Π. Since the split brings about equality it, means that what is applying on this side must be applying on that side. When motion changes Π^3 to the proton $\Pi^0 \Pi^2$ it will happen on both sides of the divide of singularity Π^0. In effect it means that, that which combines the proton also parts the proton as it combines the proton because the proton becomes $\Pi^2 \Pi^0 \Pi^0 \Pi^2$ where the adding is the divide being Π^0.

The circle motion comes from space being dismissed by ending the motion and such ending of motion compromises the space it forms. By returning to where it is coming from it is ending the motion that began the space and as space is motion that is duplicating space motion returning is also motion that is ending which is destroying of space. THAT IS GRAVITY! Gravity is the balance between motion forming space by duplication space in motion forming time and time ending motion by destroying the space.

Gravity is about space duplicating space in relation to space destroying space and some particles are more prone to duplicate than destroy not withstanding mass or proton numbers. Those we call gasses. Then there are others that are more prone to destroy space that duplicate and those we refer to as metals. Then there are a few that destroy as much as the create space by duplicating and that we call fluids.

When heat is added to some elements we consider as solids, the heat helps with the duplicating of surrounding space and brings about a balance restoring the difference there are in the destroying of space and the re-establishing of space. The metals become liquid and the heat forming the liquid brings about an adding to the material where such material diminishes space.

By applying heat to materials that already favours the duplicating of space to the destroying of space the adding of heat will bring additional space as duplicated space and thus will produce more space to be duplicated and such elements will rise into the higher part of the atmosphere. When heat is added, the heat that is actually forming into space is what is added. It is the heat forming space by duplication that forms a shift in the balance because the heat forms space also as a process of duplication but without the contracting aspect of singularity renouncing space.

Although the "gasses" are the particles favouring to duplicate space they still hold the tendency to diminish space but when applying heat the balance will favour the duplicating much more because the heat transforms to space acting as space duplicating and adding to the overall duplicating of space. The element has a natural function of returning the space that the heat duplicated back to heat by removing the space destroyed and therefore returns the space to heat. In that way the particles do not only diminish the space they have but diminish space outside their claim.

Such particles we call heavy metals. As heat is added more space becomes available, to duplicate in relation the space they destroy and what space the elements diminish. With more space to duplicate the object will surge higher to a location in a position Earth will naturally duplicate as much space as the newly relocated material duplicates the surrounding space where more space naturally are. Cooling on the other hand reduces the space available for duplication by removing available heat that would have helped with the duplicating of the space and that then tips the balance in favour of the diminishing of space, which that element will also have. In that case the element will become a solid as the space duplication is more that the space. In this, we can trace the most important part of star evolution.

A star with a liquid centre has a lot of heat. Then the duplication of space-time by motion duplicates space much more than it dismisses space and destroys time. With a star in all the liquid as the sun is it is proof of a very young undeveloped and insignificant star with almost no influence sphere. As the star develops, the liquid ratio will shrink until it is only present in the centre. However, as the liquid diminishes the motion of the star deteriorate because the liquid represents the motion.

The star eventually becomes all-solid just before it removes the neutron from the atoms in the star and eventually places the proton action into outer space. Judging the layers we find evidence in this as the outer layers of stars are filled with elements which is highly prone to space duplicating as they have such a relation with heat. Hydrogen and helium stands very favourable to space producing and little in favour of space dismissing while iron, cobalt and copper is much prone to space dismissing. In all factors mass plays no part. It plays no part in the star performance or the star development

This I use to indicate where there is a balance favouring the diminishing of space
In all forms of material, there are the constant interaction between space duplicating and space reducing.

Some elements favouring duplicating space more than the diminishing of space are as follows
Hydrogen has a **mass of 1.00797 g/ mol** **melts at −259^0 C,** **boils at −252^0 C,**
Argon has a **mass of 39.948 g/ mol** **melts at −1899 ^{0} C** **boils at -268,9^0 C**
Krypton has a **mass of 83.8 g/ mol** **melts at −157 ^{0} C** **boils at -152^0 C**
Xenon has a **mass of 131.3 g/ mol** **melts at -111.79^{0}C** **boils at −108^0 C**
Radon has a **mass of 222 g/ mol** **melts at −71^0 C** **boils at −61.8^0 C**

It is note worthy to notice that none of the above elements feature strongly in stars although they should be massive in relation to the numbers of protons they have because they duplicate space.

Other elements favouring diminishing of space more than the duplicating of space will be as follows

Magnesium has a **mass of 24.32 g/ mol** **melts at 650^0 C** **boils at 1107^0 C**
Silicon has a **mass of 28.08 g/ mol** **melts at 1412^0 C** **boils at 2680^0 C**
Iron has a **mass of 55.847 g/ mol** **melts at 1536.5^0 C** **boils at 3000^0 C**
Cobalt has a **mass of 58.933 g/ mol** **melts at 1495^0 C** **boils at 2900^0 C**
Carbon has a **mass of 12.01 g/ mol** **melts at 804 ^{0} C** **boils at 3470^0 C**

There are no correlation between mass and elements prone to space or prone to be solids. Mass do not create gravity and again on one more point Newton was wrong. Mainstream Science would rather ignore such compelling evidence as well as my writing about the matter than to admit that Newton could ever be mistaken.

Time came from eternity, where it stood still in eternity, as it still does in singularity. Remember that little line I indicated previously that is running through the centre of the spinning top and that holds time apart, while it in itself is motionless, eternal. That being without motion represents eternity, a state of being timeless. As space moves away from that point the duration of time begin to rise and the further the extending the larger the time factor will become.

Any fire represents many stages of time, where one part will burn quickly and the other slowly saying this means that science should recognise any fire on earth represents many conditions of heat where the smoke is a solid –gas, the flame is a fluid going on to be gas, the coal are a solid heat going to a liquid as it simmers producing photons which in itself is the dispensing of liquid heat turned to singularity particle dividing. The range that heat forms are so vast one may never appreciate it. Chernobyl showed the world how many forms of burning and burning injuries can come from radiation. Some heat was in the grass, undetectable until a bicyclist past and gained wounds from it killing the person a few weeks later.

The role that heat plays goes beyond (I suspect) what we may ever come to realise. Heat is an eternal fluid in relevancy to matter being the solid and space being the gas, but what is space to the one is solid to the next or a fluid to the other. For instance the proton is dimensionless to the neutron, yet it is fluid to singularity allowing heat to flow.

The neutron is solid in appearance yet it is fluid allowing dimensions to concentrate. Because heat is a liquid in relevancies, it is the father of specific density, allowing heat to flow differently in different forms of matter. To realise the correctness of that, one may gauge the heat relation there are in the first ten elements, and how each stand so different from the next element. Consider neon and boron, where boron has many times the density of neon, yet only half the mass, or where oxygen has more mass than does lithium, yet lithium has a much different relation to heat than does oxygen.

Time is the motion of heat in space, and producing more motion, the duration of time will extend.

Every dot insignificantly small as it may be, is a part of another universe as much as it is part of the accumulative universe and every dot in the infinity holds singularity, which we translate as " nothing" being " darkness". There cannot be "nothing" just as much as there cannot be "darkness". There cannot be something big or small, but it into relevancy of perception, and then the relativity of perception becomes the question. There cannot be hot as much as there cannot be cold. The sun FREEZES hydrogen to a liquid at six and a half thousand degrees Celsius and universe boils over in the form of the Hubble constant at the temperature (we presume from our vantage point) at minus 273 degrees C. If we Humans cannot or will not abandon our human perception and our manly perspective, we may as well return to astrology for all its worth.

Every point in the infinity we may observe at is not merely part of the universe in not being "nothing", but is the point where the universe started representing singularity. It is the very first point where everything began so many eternities ago, because after all, how can we ever determine where the first point was, as they were very much equal and alike at the beginning. Every aspect of the universe started with the fundamental fact that no point in the universe can represent "nothing" as a number, because every aspect in the universe represents singularity in what ever form it may hold in that specific spot forming space-time. If man does not reach a conclusion where that conclusion is matching the universe and stop to match the universe with man (and man's incapability), we may all go back to caves and become starving hunter-gatherers again, because we will never find a way to progress to the ultimate understanding of the universe.

Experience taught us that there is a definite precise and secure corresponding that can only result from a direct connection in as much as the line being the same line. In that case the line must then come down to infinity and release from infinity as the same line still connecting to a point of re-bouncing to either side. The graphic cross is he result of singularity applying opposing sides but still maintaining connection through the application of Pythagoras that will connect and always bring about a direct relevancy. The graph does not hold zero because information derived as result of a relation prove a contact remaining when the line crosses singularity without applying detachment.

The Brainy Bunch holds the view that in a graph the line crossing amounts to breaking the zero mark. That cannot be the case.

Now I wish to refer once again to one of the academic letters, which I already used as a referral. This is very shortly my theoretical proposal

Dear Peet,

Those on your list I know are in science education. You need someone in pure physics.

I am afraid that you will continue to get rejections if you do not relate your work to existing theories and previous work. While it is possible that a lay person hits on an insight that has been overlooked by academic trained in the field over many years, it is unlikely. We assume that work offering something new would be related to existing theories, either by building on top of them or by showing how and where they fall short. If you do not relate to existing work, it is repeatedly going to be dismissed as mind spin too easy to shoot down.

I am sure you understand.
Iben

PROVE ME TO BE INCORRECT IN ANYTHING I SAY! What you see in the picture above is gravity opposing gravity and by using the four cosmic pillars, which builds the Universe, that is how the cosmos comes about. I say gravity and the sound barrier is the very same thing where it indicates opposing movement of the earth and some other object within the earth that moves apart from the earth. Allow me to very briefly explain the sound barrier as it was never explained before by applying relevancy.

Science in the present and for the last three centuries placed all their focus on material. In placing the focus on mass while mass does not exist as a cosmic entity, science has been running around like a chicken without a head and the truth eludes them even after three centuries of lying and corrupting science. You think that this is harsh words, read and you will see those in science

deserves much more that just that. Those in science concentrates on the material filling the space but gravity is about the movement of the space and not of the object in the space, but of the space the object takes with as the object moves faster than the other space. That is why objects all fall equal because the space the objects fill is the same, unconditional of what forms the material. All objects fall at an equal pace without size or mass becoming a factor. That is why I introduce gravity by a very new set of principles that no one ever heard of. When an aircraft goes through the sound barrier, the object fills more space per time unit by going faster through space in time. The object then in accordance with the space it holds stretches because it holds more space than when it went slower or when it stood still. Since the aircraft is solid and does shrink a little but not much, it has to concentrate the space around the aircraft and by concentrating it reduces the space. As the space concentrates through the movement of the aircraft a cloud appears around the aircraft because the water vapour in that space concentrates into a cloud that forms. All movement is part of the sound barrier because the sound barrier is the movement of space. However the sonic boom is confused with the sound barrier but the sonic boom is a small part of the entire sound barrier and only fills the centre spot or the middle in the sound barrier. If mass is anything then show me the role that mass plays in the sound barrier and how does mass conform into what we think of in terms of the sound barrier. Gravity forms as not by the mass of what any object presumes to have, which I prove in this book and in all my other books, is clearly not present as a cosmic factor. There is no such a thing as mass in the entire Universe and the conspirers that conspire to keep Newton's fraud disclosed, cover up this reality with all they have. I challenge anyone to prove where do they find proof of a factor such as mass. I found the place! It is in the imagination of physicists while they try to conceal that mass is only a product of Newton's imagination. They conspire to confuse everyone with the implication of weight. Weight there is but things pulling other things by the measure of their mass is a daydream and the thought could be funny if it wasn't that crooked. Gravity forms by redirecting the movement of space in circular flow by the spin of any cosmic structure that has the ability to do so. The aircraft goes straight while it also circles the earth and that forms movement within the earth's confinement by the earth's movement, which is what forms the sound barrier. The sound barrier is gravity and sciences present way of going on about the "Doppler effect" and "Mach's principles" shows how incredibly little those incompetent physicists know about physics. I challenge the lot to prove Newton by showing that the Universe contracts or where mass plays a part in cosmic physics. Show how the planets hold their positions according to the mass they have.

All objects move straight while at the same time circle around some other object. The object moves in a straight line while the spin the object hold diverts the space in which the object is into a circle. This circle brings the value of gravity to Π. By diverting the space in which the object is as well as the space surrounding the object, gravity concentrates space into becoming denser by circling and also hotter.

I return to this explanation later on in this book and then bring the applicable my arguments.
Meet the Newtonian. This Newtonian says on the Internet no less that Master Newton and Master Galileo shared the same opinion. This Newtonian says that things fall equal as Galileo said and things fall by mass as Newton said. How much double standard are they allowed to maintain and never get questioned by the public? When out Newtonian falls our Newtonian would fall at the speed as anything else would fall. This evidence we see on TV everyday in advertisements and on reality shows. We see blokes jump out of aeroplanes with rug sacks and tennis balls and the lot through each other with the balls and catch the balls while putting on the rug sacks and tasking off the rug sack all the while the lot is getting in and out of a car as they please.

We have all seen this. The car drops evenly paced with the rug sack, the humans and the tennis balls. Then the Newtonian tells the student that according to science everything is dragged down by gravity in accordance with mass. See the Newtonian fall. This Newtonian is the only one that falls by mass. The latest is that there is some little particle undetected as yet or as this goes to print (just to cover my arse in case they discover this particle in the meantime) scientifically named and called "graviton" that "pulls" other things that "pulls" back. Science rumours their profession that science only work on established facts and nothing but facts proven to the point of no contest could other wise be good enough.

This expression is so burned into the minds of everyone that science can say any rumour and because it is science everyone believes and never disputes. One good example is this Global Warming by carbon immersions. They take two truths and make one lie to convince people the lie is true. People take what they say as undisputed truths because science is renowned for only working with facts.

What a lot of horseshit that expression is, that science only work with proven facts and that they use this idea to convince persons in the public that that is why the lot are atheists. They use this expression to convince people they are atheists because God is not a proven quantity but then they propagate the graviton as the measurable quantity, It is because they only work with facts! I can much easier prove God by using physics that they can prove mass by using physics.

Gravity forms by the dual directional movement of objects going according to the speed of movement. An astronomer will hang suspended in space and high above the earth if the circular movement is high enough to keep him there. If he spins slower he will start to drop. It is not gravitons grabbing him by pulling him closer to the earth. If his circular momentum can sustain the rotation his linear distance will maintain his orbit but as soon as the circular rotation become insufficient to maintain the orbit the linear distance will reduce and then the "gravitons" are called into action. How ridiculous can they get!

Gravity is the sound barrier because the sound barrier forms by four cosmic principles that form gravity and therefore gravity and the sound barrier are the same principles in conflict because of movement differences. The sound barrier applies by movement in space in space shared.

Everything I am about to show Newtonians say is nothing new. Newtonians say they are aware of everything that I show and then they turn around and simplify physics by putting the entirety of their physics down to the pulling power of mass that forms gravity. The sound barrier is gravity and to understand gravity is to understand the sound barrier. They say that I bring nothing new to the table and they say they apply every aspect I indicate that forms gravity. When I ask them to explain the sound barrier they use the Mach principle and Doppler's effect, which both dates from a time when no one was aware anything man-made could fly let alone go faster than sound can. Doppler made his contribution to science at a time when the train he measured was slower than a horse. The statistics he left to science applies to going as fast as a man on a horse could ride. That is a far cry from a jet breaking the cosmic boom. They can claim what they like because they never have to prove anything and they never have to listen since they know everything while their shortcomings makes science more the jesting of a buffoon. That annoys me to my guts because all of that comes down to the conspiracy whereby they whitewash their stupidity from the brainwashing they inherited from their predecessors and pass the brainwashing on to their students. Their ignominious stupidity makes me want to shout to the mountains in agony and unbelievable frustration. When a cloak of arrogant self-righteousness hides their incomprehensible stupidity under a cloud filled by their belief in self-importance it becomes hideously loathing.

In the earth's atmosphere no object can move slower than the earth does. If it moves slower than the earth does, it moves as fast as the earth does by receiving mass and being pinned onto the earth as being part of the earth. There is always movement because of the earth moves and all other movement goes in anticipation of moving above what the earth does. Movement is space that is duplicating filled space at a pace and in relation to other space within other unfilled space. That is why using parachutes slows the falling process of falling objects down because falling is a ratio between solids and liquids. This is proven in these photos about the sound barrier where the movement of the solid jet contracts the space of the liquid cloud. What happens in the photo is vivid proof of my theory and it proves the Coanda effect is what is forming gravity. The Coanda effect is a relation between material spinning and air or liquid compressing and that is what happens between the Earth and the atmosphere. This picture shows objects are always moving extraordinarily as they move in relation to the earth's atmosphere that always moves in relation to the earth's gravity compressing space into forming the atmosphere. But it also shows the movement of the aircraft compressing space in relation to the aircraft moving as well. That shows that all movement is gravity or time related. This shows objects moving is extraordinarily because everyone always forgets about that it is the earth that normally applies all of the movement while all else stands still in relevancy. There is always movement because of the earth moving and when anything moves above and beyond the movement the earth provides that then forms the sound barrier using a modified version of the Titius Bode law

"GRAVITY IS DIVIDED IN TWO FACTORS, BEING <u>LINEAR DISPLACEMENT</u> (Π / Π^0) WHICH IS WHAT <u>NEWTON'S GRAVITY</u> IS AND,

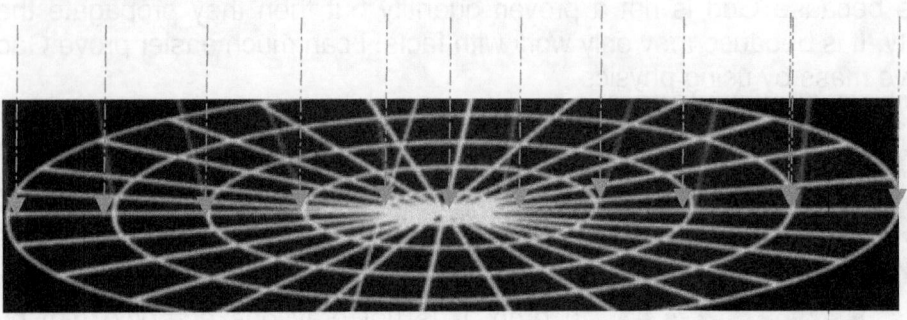

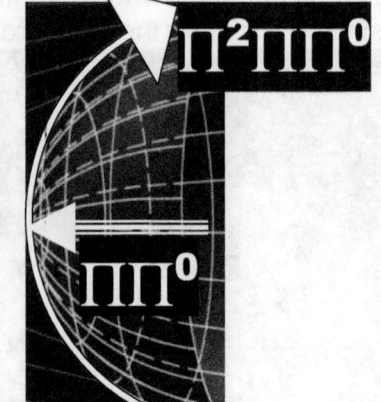

<u>CIRCULAR DISPLACEMENT</u> (Π^3 / Π) = Π^2

WHICH IS THE "GRAVITY" EINSTEIN RECOGNIZED

Those that are of the opinion that it is mass that "pulls" the object towards the earth might be a little wiser when reading the next part carefully as I explain the following:

Anything entering the earth from space it travels in a straight line. The mathematical formulating of this I do not provide at this stage but it is another "some simple algebraic relations" as Prof Friedrich W. Hehl from Annalen Der Physics put it (and I also explain this a in the following chapter). But using the law of Pythagoras and putting 7 in relation to 10 provides the value of Π. The value of Π has two different dimensional values where the one is Π and the other is 21.991° /7°

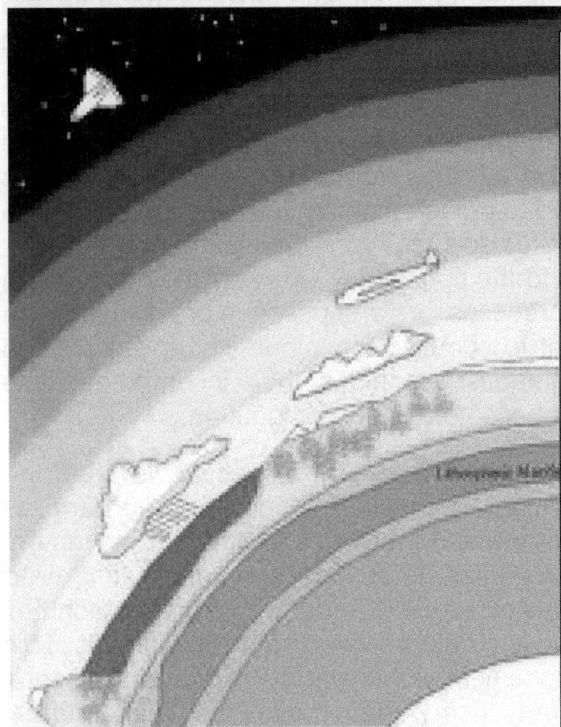

Every layer formed as the atmosphere of the earth holds a relevancy inclining towards singularity. To the above the gravity forming Π holds a value of $\frac{21.991}{7}$ but as the earth turns, in the instant that the earth turns by turning in that instant of time seven becomes singular and in concentrating space it then divides space to the value of 21.991 to form $\frac{3.1412}{1}$. Then at the bottom gravity is $\frac{3.1412}{1}$ and at the top gravity is $\frac{21.991}{7}$ and that is the relevancy called gravity. There is no pulling. There is no graviton. There are no magical forces of any sorts as science wishes to use. The entire idea of the four forces comes from the spinning of the earth or the atom or the electron or the proton spinning and in every instant the dynamic changes and that forms a new criteria what then becomes the intensity re-applying. However as big as you wish to take the concept or as small as you want to go it is a changing of Π where Π by spinning revalue $\frac{21.991}{7}$ to interact with singularity to then form $\frac{3.1412}{1}$. That is gravity and now it is no mystery but just part of God's Creation explained. It becomes nature without mass.

The value of Π forming is what drives the Universe the "ability" to "expand". By Π forming a value above the axis value of 3 this addition shows growth coming from a future in time where it first forms .991 which is less than one in a form where one is the epitome of what can be in the entirety we call the Universe. By coming from .991 going to 1 and then when spinning through the double square of seven becomes ten on both sides of the divide we see how time in growth drives space grow from .991 to 1 to ten on either side of the seven square that spins.

PART 8
THE ABSOLUTE TRUTH
OR IN OTHER WORDS
RE-(W)RIGHTING COSMOLOGY

Gravity dictates mass!

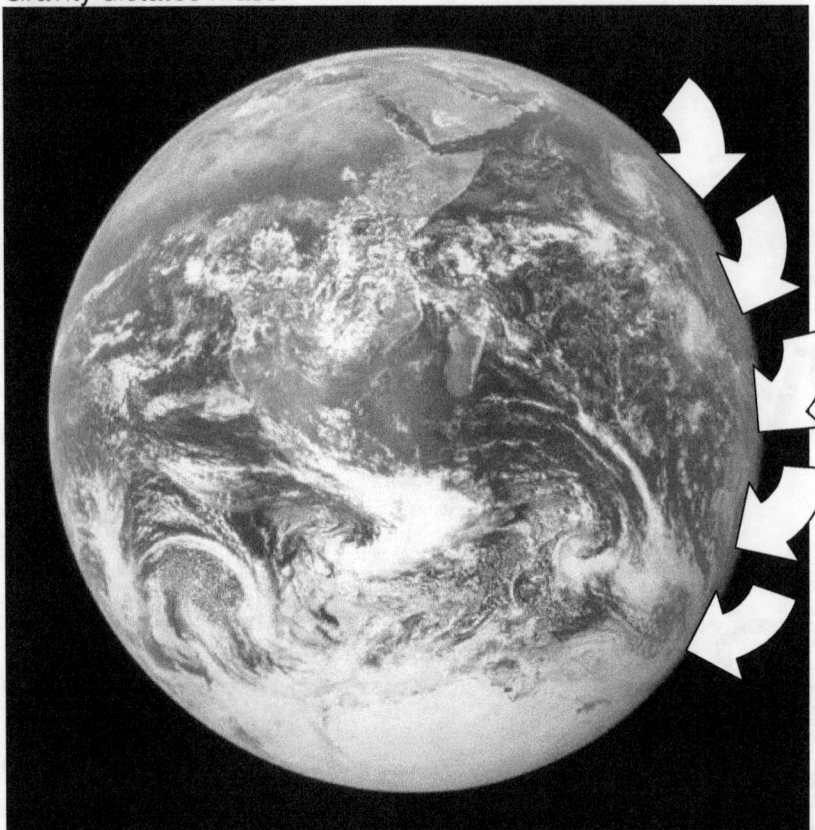

I show how by spinning all spheres form gravity through singularity by charging the **Titius Bode law**, in relation to the **Roche limit** combining the two with the **Lagrangian points** and in conjoining the **Coanda effect** to form gravity as Π. Newtonians don't even bother to read my work because they have one aim in physics to keep the Newton conspiracy alive so that they can brainwash the next generation of physicists to blindly and without reservation believe in all the Newton fraud he concocted of mass forming gravity although there is not even a correlating hint in nature that mass is a factor used by nature.

In the Universe there is no mass but what I forms gravity is precisely what e find forms the solar system.

<u>Gravity is the relation to the earth moving in a circle that moves in line with the sun while orbiting the sun. By turning in a circle gravity forms Π and therefore gravity is Π by rotation.</u>

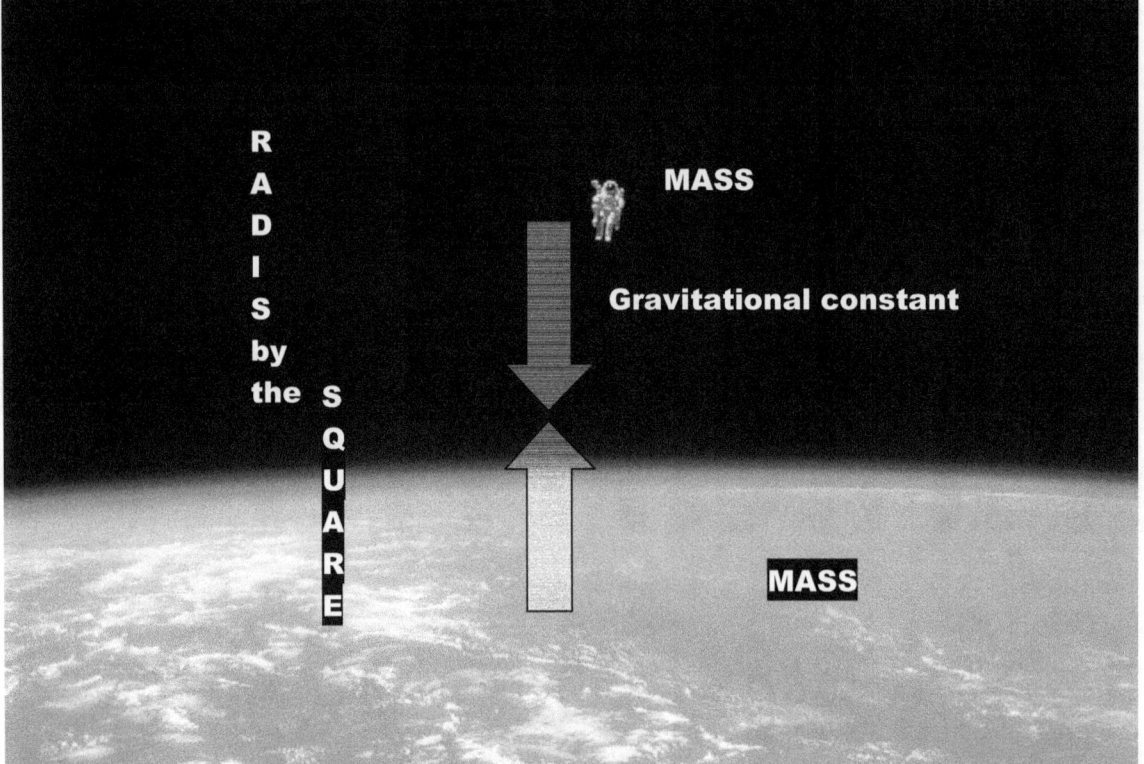

This principle declares that a force of attraction by the value of mass of both solar objects is pulling by reducing the radius, therefore bringing the moon and the earth closer. This must lead to an inevitable collision that must end both the earth as well as the moon. Let the Newtonians calculate when is this event is due and inform the human race when our final day of doom will arrive by using Newton's law of attraction or $F = G \dfrac{M_1 M_2}{r^2}$. The day if (no when if Newton is correct) this happens, Life in all forms will end and therefore knowing this event is crucially important for all of us having life on earth!

All information connects directly to the .com registration.

A Conspiracy In Science in Progress

I came upon a mistake concerning physics.

This mistake is about the cosmic phenomena called gravity. Detecting the mistake is simple because it is uncomplicated to understand. Academics in Science say that a feather will fall with the same speed as what a large rock would fall.

That is according to Galileo and that is accepted as a principle in physics. For the first time ever since the time Newton introduced gravity I seem to be the person that questions this interpretation.

How does mass pulling mass and falling by the gravity power exerted by mass then fit into this interpretation because a feather has much less mass than what a large rock has? If this statement is untrue Galileo is incorrect and the Pope does not need to apologise to Galileo as Physics insist the Pope has to do. Then the pendulum doesn't indicate time and mass does implement falling of objects as Newton protested it to be. But we know Galileo is correct and that makes Newton's suggestions what they are…merely suggestions that proves to be incorrect.

When you disagree with any academic in any lecture hall about mass not forming a picture as responsible for pulling gravity and you come to a conclusion that you doubt the mass part that they bring into the picture as establishing gravity the academics wipe you from the table with a swipe because then they contemplate that you are so stupid you fail to see all the facts that physics present as proven facts and they hold you as being too stupid and mentally underdeveloped to appreciate or to understand physics. They tell you that Newton is not for such stupid people that are unable to see how true Newton's arguments are!

I have been at odds with academics for years and only because of the superior positions they hold in office are they able to bully me into silence but not into submission because that is what this book is about…to expose their corruption.

They can push me into abasement but never into abeyance. By the important Academic positions they hold in the huge academic institutions that give them sanctuary they might dictate the terms of our meeting and in terms of those advantages they can hide behind the criminal wall of deceit and suppress me into silence but they will never get me into submission.

Now I am taking my case to the members of the public so that the truth must be brought into the open. I have had the tour they give and then more came my way. I never got around swallowing their gravity that comes as result of mass creating gravity or any part they present facts as to how it happens and where science is of the opinion that mass pulls as gravity is… Academics condemned my work and therefore me and for six years where I could not get a publisher to come around and bother to read my work let alone seriously proposing a publishing contract. I had to finally go private with the publishing as all doors shut in my face as soon as the academics read the content of my work because from the nature of my work I take Mainstream science head on and as I am sure about my arguments I am confrontational on most aspects of astronomy. There does not seem to be any publisher that wants to go head bashing with the establishment of science on official science principles, which I have to do to convey my message in a no uncertain language. If you also have doubts about the academic's indisputable correctness please read on and confront either them or me on everything you read here.

After reading this book you will have to take sides because you will know the truth and you are free to decide who present the truth…is it the Newtonians or I.

By that decision you then either become my partner in also recognising the crime I uncover or you become part of the crime syndicate as you cover the truth up. You either will be part of the truth by helping me confront them to acknowledge the truth or you will remain part of their cover up by ignoring the evidence as everyone this far did for about four centuries in any case.

By not confronting the establishment, you give the establishment grounds to allure you into being sheepish. Because they see you, as just another stupid senseless student they have the opinion that they can brainwash you into accepting these fallacies that I am about to tell you. They will literally brainwash and condition your mind to accept what they never yet were able to prove. The truth is that this process of brainwashing is going on successfully for four centuries without any backlash to those committing the atrocities.

They are of the opinion you will swallow any rubbish they throw your way just because every generation before you were mind controlled in the way they are about to control you. You may think this is big words but read on and see after you come to know all the facts whether I exaggerate even in the least. They see you as slow-witted and mindless because they think they are the academics being superior making you the lesser and inferior party. If you are not aware of the facts beforehand they know you will follow their teaching without asking questions as it is going on for four centuries this far. Talk about the Catholic Church putting the fear of God into Copernicus and you will find their manner in disagreeing about statements not echoing their perceptions is as bad and more ruthless than the Church was during the Dark ages. The Church just killed those confronting them on science dogma but did not make every student a brainwashed mind controlled Zombie!

They think that your naivety makes you a Zombie and with such a degree of mindlessness that state of mind will incapacitate you into their control. They don't want you to ask nosy questions about contradictions existing and they refuse to answer any uncomfortable questions asked in any confrontational manner. This process of brainwashing and mind controlling in physics has been in progress for hundreds of years. If you are surprised then control me by asking and just answering how it is possible to agree that a feather and a large hammer fall equally while also agreeing that Newton is most correct. How can any person thinking logically agree that Galileo is correct when denouncing objects falling under mass differentiation when Newton insist on mass driving gravity as a force. If you can't…well they can't either! Their task is not to explain but to mislead since they think you can't think while they think they know how to control you.

The motive behind this book is to promote my other books and by trying to be as least complicated as possible this book aims to present Newton's deceit. This book does not aim to represent the full entirety of the original thesis as it represents the entirety of my theory in a single copy for that is not possible but is reduced to aid any possible potential reader in the examining what the purpose is of the information this book wish to present in order to uncover Sir Isaac Newton's fraud. Anybody and everybody are aware that all objects fall at an equal rate. If an object such as a car weighing one ton falls at the same pace as a person weighing fifty kg how does mass come into the picture by committing a force to do the pulling? Mass has to pull because according to their teaching it is mass that establishes gravity. However, mass is a factor that produces differentiation whereas all objects show equality during their fall. If it is mass that is establishing the force gravity, all objects must fall at different speeds. That they do not do as they all fall equal. That means Sir Isaac Newton's physics is wrong from the start because mass cannot have any input in objects falling.

This is not the only untruth that the Paternity called Mainstream Science is keeping concealed as a cover up that is wrapped under an airtight blanket of deception. If you sit in class and listen while also experiencing the sinking feeling that the facts you hear are not adding to a total you are comfortable with while you disagree with what is said then you better read on because this web page has it at task to show all that will read this document how much discrepancies academics lay on unsuspecting students that trust Academics with their future and their life. Do you as students realize the inconsistencies that physic Academics teach you as the truth.

Students tell your Professors to stop deceiving and stop trying to control your minds with their fraud. Those Academics tutoring you are telling facts about gravity that has never been proven.

That is mind control.

They wish for you to accept facts on gravity that they hold as the truth. They claim those truths are beyond questioning yet with the least examining those truths they stand by then proves to be totally void of substance because it was never corroborated by one single experiment.

Should you question that mass produce gravity they will expel you from University by letting you fail your examinations and it was never proven. They will expel you and have you fail tests should you question their authority on the matter of gravity while at the same time they can't for one second bring evidence in support of what they wish you to accept as the unquestionable truth.

That's brainwashing by mind control because if you don't accept their baseless fact as God given truths they dismiss your academic career.

It is either put up and shut up or be gone. Academics do put mind control to work on unsuspecting students by forcing students never to question the legality of statements they offer as being sound and correct.

What they present as correct I prove in this very book are openly laughably totally incorrect and by just reading my evidence you will see how feebly easy it is to rubbish it. Take the evidence I am about to share with you and confront them with the fabrication of facts that they present. Go on and challenge those teaching you with the falsified facts as I challenge any one to prove me wrong.

What they maintain is gravity is total incompetent nonsense and can't be corroborated at all but what they can't corroborate because they don't understand I prove to be that which the Universe employ to form gravity. There are four phenomena they dismiss because they have no idea what they are. I studied each one and formed an explaining by implementing Kepler's formula as the Universe gave it to Kepler.

The by understanding the formula and implementing the content into the four phenomena I am able now to prove what forms the motion we think is gravity and when reading it then the Universe makes sense. All the questions in these books I managed to answer while they can't ... and in the books I answer a lot more questions than what I ask here in the rest of the information while Science fail to answer any...

The following presentation is as simple as gravity can be represented but in order to have the presentation as simple as it is we do surrender some part of the accuracy to achieve simplicity. Under a microscope one would find that in this explanation the explanations strays a little from the truth in order to make it comprehendible to everyone reading it. The truth is the explanation about gravity can be somewhat more complex than what the following presentation has to offer but then on the other hand it will never be as simple as dumping the entire concept on one thought about mass that produces gravity. That simple gravity in explaining can never be because it is so far from the truth as telling that fairies produce summer and witches bring on winter. The whole concept of mass being responsible for gravity is one big hoax and forms a scam. Putting mass as being responsible for creating gravity is the biggest fraudulent lie that ever hit the Earth on any scale ever.

For instance Mainstream science has the theory that matter and antimatter developed and some matter formed as particles, which were nicely wrapped in containers we call atoms and stayed on as material or matter where the rest formed antimatter that disappeared. The anti material also formed atoms but then chose to just plainly vanish, as did singularity and other cosmic factors to now be nowhere. Singularity is a mathematical position falling outside the detection of the observable Universe in any case so in the case of singularity there is a possible excuse for the disappearing...or is there? We know that singularity produced space-time and space-time produced gravity but then space-time went away leaving us with gravity. Singularity is a single dimensional entity that is not material, holds no space, holds no time and in our 3 D view can only be found outside the human visual spectrum. Mainstream science is of the opinion that singularity disappeared after the Big Bang process came about whereas I am of the opinion that everything must remain part of the cosmos once it was in and was part of the Universe simply because there is nowhere for it to go. The same applies to space-time and antimatter. If it was part of the cosmos in the beginning it has to be in the cosmos until the very end only and simply because there is no other place available to be or to move too. The Universe is the container that is the only container leaving no other container to contain what ever is in need of containing. If it is or if it was it is in the Universe! Our task is to find the place it went to or find what it changed into. While we then are on such a hunt, we might just as well find the cosmic principles not understood but which is there all the same. Take for instance the Bode principle where all nine planets show a relation with the sun in precisely the same manner and using the very same method of spacing, yet science brush the Bode principle off as a coincidence. It might be a coincidence when the space between one or two planets shows these phenomena but when all nine planets plus even the fragmented structures adhere too the very same principle no person can be of the opinion that it came about as a coincidence and still pretend to be serious or professional about cosmology. One then must simply find the proof lacking in our understanding

Mainstream science knows about gravity, the Bode principal, the Roche limit, the Coanda affect, the Lagrangian system, the sound barrier but cannot explain any of the phenomena all though the presence of these phenomena is without dispute. It is the explanations about what causes the phenomena that should be part of the dispute but in science the way they defend Newton scientists go overboard by disputing the phenomena and the phenomena as a principle existing in cosmology or not becomes disputed. One cannot be serious about science but defend your view by dismissing the validity of all unknown indicating factors presented as such. There then is some gross incorrectness in the way Mainstream science reason. The Roche limit is there and no denouncing thereof can remove it from the cosmos. They may refer to evidence received from the Hubble telescope as "the star is blowing bubbles" for the lack of explaining what is occurring but occurring it does. One cannot say it is some unknown gesture presented on occasions because not explaining the pictures presents the presence of certain foolishness. For fifty years they lost many pilots but still has no idea what brings the sound barrier about, or find the link gravity holds in the process we call the sound barrier. Instead they try to interpret some effect established almost two centuries ago with steam trains back then travelling at the same speed that horses run. No further investigation with the science in hand brought them closer to new facts! It should be a sign telling them they are going about incorrectly but it does not because Newton said so. It may sound as if I am anti Newton but I am not. But there has to be more than Newton with so many pieces of the cosmic puzzle still missing. Science should not serve only Newton but science should serve the seeking of the truth. When I first came upon the unknown it stirred a sense of disbelief and I decided to respond.

Some twenty-seven years ago I decided to start an investigating quest on my own to see where I could go with my private research. It came as a result of my frustration when I realised the discrepancies there are in theories presented about cosmology and all the unexplained factors no one ever makes any effort to explain. Later I found that no-other person than Newton in person was to blame for the mistake that was made but now I am jumping the gun. If you are a Newtonian and feel a repulsing urge to throw down the letter you will do so at your peril. I say this because I have seen Newtonians get fits in the past when I say what I just said. Whenever I make this very claim the entire science community rejects me immediately without any reservation to any person. When I speak out against incorrectness I see science as a unit and an entire structure reject all further statements that I make without excluding any body active in the field of science. To science Newton is reserved as a god and they placed Newton beyond criticism. When they listen to me criticizing Newton they switch off their mental lights literally. They immediately go blank and I have witnessed it every time. I can visually see their eyes go dim. No Academic Newtonian priest will spend another second to listen to more of my views. Yet they remain unable to use Newton to explain the cosmic phenomenon as the Bode law, the Roche principle, the Coanda affect, the Lagrangian system or the sound barrier. But what they do not realise is the mistake was a resulted not about what Newton presented because it came in what he admitted that he could not present. The mistake came as a later presentation of a concept he admitted he could not underwrite by scientific explanations. When he introduced gravity as a concept he admitted he did not know the origins of the force. What Newton admitted he could not explain and what he did not explain later became an institutionalised claim presented later as if he did explain it.

The error is in the facts not yet ever explained and in that which Newton admitted he did not understand. It is Newton's incorrect suggestions he made about gravity that were later accepted as explained and proven science that went on to become institutionalised facts. It absolutely came in the way Newton changed Kepler's formula. Newton admitted he did not know what gravity was and left it at that. He did not offer any more insight than reducing gravity as a force and a force it stayed all the time. Three hundred and fifty years on science still do not know what gravity is and is still leaving it at being a force. Without knowing what gravity is all other concepts in cosmology does not make sense because all the phenomena I mentioned a minute ago is sides of gravity and performs (each to its own but still) as another principle where the totality forms one concept we call gravity. If the World does not know what gravity is then the phenomena coming from gravity will remain unknown to all. With that in mind Mainstream science still take a very dim view on my criticizing Newton! In all this time Kepler explained exactly, precisely and unequivocally what gravity is! The unbelievable part is that we all missed Kepler's announcing of gravity and in the "we" I use that 'we' include even Isaac Newton and Albert Einstein in persons and by names.

All principles I use in the theory I introduce with the publishing of this book are part of nature. I base my theory on heat stabilizing through space using motion to produce cooling. That is gravity. But however this may sound basic, Mainstream science is also most guilty of their usual departing from this basic principle through the employing of terminology and such terminology has the tendency to cover many of the basic meaning behind the principles in nature. For example the one principle I do not applaud is a principle Mainstream science underwrites in the sense that matter in the beginning was coming about and anti matter came to destroy the matter by consuming it. This translates into a packman computer game that has no correlation to cosmology. It is moreover the disappearing from the Universe of that which came as the result between the two opposing materials that I strongly reject. Anything part of the cosmos at any stage before during or after Creation, remains in and part of the cosmos and cannot leave the cosmos because there is simply no other place for what ever there was to go. Leaving the cosmos just is no option there is.

In any test performed today by creating friction through motion the discrepancy there are between objects in motion will bring friction that will produce heat and the heat will result in space forming. In such destruction of matter space and heat comes about and the net result eventually is space created where no space was before. The cracks showing is space created in the cooled material that was heated to a glowing red-hot always afterwards leaves cracks. The cracks are space not filled but was filled during the heat coming about as a result from the overheating. After the cooling the cracks present new space where there were no space before the heating took place. The heating process started forming and filling space but afterwards when the cooling set in it reduced the filling of the space. The cooling did not destroy the newly formed space. The cracks represent the space not filled. The material in the cool state cannot fill the void that came as a result of the cold material contracting and reducing the space filled when the material was overheated. But the space remains although not filled any longer.

That we can see as evidence with material having a heat building up when motion difference brings on friction and such friction brings on heat. I do not share the view Mainstream science has that when matter and antimatter came into conflict the product that came from this just disappeared without a trace of any sorts. I believe the evidence is present and I think I know where that evidence is. I believe I can show that it is a motion discrepancy that produced matter and anti matter and we do not have to go and look for non-exiting positrons. A positron must produce a negative proton and such a performing sub atomic structure cannot be functional. By changing legions it must then produce a product where it performs as gravity by rejecting material and pushing away other cosmic components. That will lead to an exploding Universe! I say that discarded material became heat that became space that became outer space in the Universe. I go to lengths to make persons see that space cannot be nothing. This is a factor that science has to accept if Mainstream physics have the will to find solutions about the Big Bang. I say the motion between particles in a cramped space as the case was during the initiating of the Big Bang would have brought on friction in space between particles present that we couldn't even calculate. The result is that some of the matter particles produced a means of self-sustaining by applying gravity and the demise of the other particle that became destroyed resulted in plasma forming on the one side and material on the other side. I believe even to today and throughout the rest of the Universal motion through space in time the plasma is transforming to material forming particle growth in space through the motion we named gravity. This was how the cosmos came about and this is the manner the cosmos will conclude.

I believe some of Creation remained as some particles formed by applying gravity in motion and the lack of gravity turned the other particles of lesser motion into heat. In this is the destroying of singularity contact came as a product of light, which again I believe (within reason) I do prove. I believe heat is the destructed form of material and this information the atomic thermo explosions give us. By releasing the heat that is sealed in an atom such release thereof produces heat, light that can liquefy the eye and most important the unexplained nuclear winds that destroy so much. But to realise that we must beforehand find what space is and accept that space is made of something. We have to see what forms space and why space can be the absolute basic container through which gravity can relay the influence it carries. We must come to realise that whatever forms space has to be that same ingredient what forms the lot of everything in the entire Universe.

When particles heat up the particles expand. Expanding is applying more space. The space that the particles hold amplifies. The particles claim more space when heated. The claim of heat on space creates more space and heat results in more space as the product of heat rising in material. Such expanding is one way of bringing about cooling. Heat produces more space and never reduces more space. The motion coming about from the expanding of space brings about cooling. When particles cool motion applies in some form. At first the Universe was extremely hot and without space. Then space came about and heat levels declined. Compress space even today with a piston in a cylinder and such confining of space will increase the heat by the piston effort as the space reduces. The heat coming about has no relevance to particles colliding because compressor cylinders cool down with time and not necessarily with the loss or release of particles. It is not only the discharging of air that will reduce the temperatures inside the container. After the pumping of air increased the heat in the cylinder even to dangerous levels the heat will reduce back to room temperature when further pumping seizes and the stopping of further air movement into the cylinder can bring about heat stabilizing. If it was the commotion of air particles colliding and rubbing of particles going on in the compressor only the release of air could produce cooling. If this were the case the temperatures would rise indefinitely because the action will increase in counter actions producing heat and spurring on more friction by collisions. The stopping of air pumping reduces motion and subsequently brings on cooling. The cooling process is not resulting from calming the excited particles within the container by whatever means. The process is natural and follows immediately when pumping stops. Stopping the pumping automatically leads to cooling coming into action.

This means it is not the particles in the cylinder that brings on the heat levels rising as much as it is not the particles that will eventually bring about the explosion that will follow, should the pumping continue regardless of danger. When the pumping stops the heat immediately starts reducing. There is no further increasing of heat in any way except if the pumping continues again. The already over filled container does not continue with the friction of particles rubbing within the cylinder container. Afterwards when the compressor is left by itself and temperatures stabilize too the same levels of the temperature on the outside a sudden releasing of the air under controlled conditions such motion of the air relieved will bring cooling to the extent that pipes can freeze and block the releasing airflow. Two American Submarines were lost in this manner and yet to this day no person in science saw such a connection in the sinking of the subs. After a few days left undisturbed as far as heat distributing goes the stabilizing will lead to conditions being equal on either side of the cylinder wall. The compacting of air molecules, which although still much higher on the inside than what the denseness is on the outside will not produce the same heat levels inside the cylinder than that what was achieved during and immediately after the pumping operations.

The temperature will only become affected with motion contributing to changes in the balance. The releasing of air will extract heat from the process to a point where it will lead to freezing coming about. We have to see the container for being a container and the Universe also being a container, but more important is that what ever comes about in the one container will be similar in the other container since the containers contain what the cosmos is made of. It is heat flowing between materials in space. Heat will always flow from the highest value to the lesser value. This is the concept I use as the basis of my theory. I base my theory on gravity producing cooling and contraction while heating produces motion by expanding and creating more space. When an object is overheating no amount of force can retain the container from becoming too little and with the heat coming about forming the expanding the space produced with this action will destroy the form any container may have. Even a container as solid as an atom proves not to be able to withstand the expanding produces by overheating. The Universe is a container and as all containers prove the Universe also was unable to sustain the space that heat created when particles overheated. The Big Bang is evidence that. During Creation the compactness of the particles produced motion discrepancies bringing about friction where some particles overheated and formed heat. The heat that then formed space we named the Big Bang where heat produced space and formed motion we see today in the Hubble constant. The Hubble constant is antigravity in its most splendid form.

I believe that my study of Kepler allowed me to achieve an all time breakthrough success because I can now explain what gravity is. Remember that not even Newton could explain what gravity is or where it comes from, but Kepler did that without any person ever noticing. Kepler studied what kept the plants circling the sun and that is gravity. From such explaining what Kepler said without Newton changing formulas on Kepler's behalf I prove the Titius Bode principal also known just as the Bode principle. I prove that the Bode principle is forming the value of gravity when incorporating the Roche limit. These phenomena was never before explained or understood by Mainstream Science although they appear more than regularly in the cosmos. I explain how singularity forms the Roche limit and how singularity brings about the Coanda affect but most important of all Kepler showed me where to search for singularity. My achievements came from my effort where I separated Kepler's work from the opinion that Newton formed and gave to the world about Kepler's work.

For instance from Kepler's work I can explain the operation of the Black Hole, which not even Prof. Stephen Hawking understands. That is because Hawking ignore Kepler. In my opinion my explaining of gravity makes much more sense than the accepted force of Dark Age proportions…and the best part is you do not have to be a genius to realise or understand it. Even a simple person such as I can see it clearly! From my view a force is just motion applying and that is what Kepler said gravity is. Kepler said $a^3 = T^2 k$. I dissected **k** as a factor in the Coanda effect and found that the Coanda effect is proof of my view about gravity and singularity produces the Coanda effect. The Coanda effect is the establishing of individual space a^3 by applying motion $T^2 k$. Where the Coanda effect is producing gravity and such producing is stronger in a small space than the gravity produced by the Earth in that spot I use that principle to show that there was some manner in which the reducing of k brought about a stronger T^2 just as Kepler said. This was a crucial part during the Big Bang and therefore had to play a major part during the period of the Big Bang. Einstein came to this conclusion but failed to refer his view back to Kepler and by not referring to Kepler missed the point he wanted to make.

Presumptuous as it may be on my part of trying to disprove Mainstream Physics, such a presuming does not change the truth about Mainstream science being incorrect about gravity. After all they admit they do not know what gravity is. I am not disproving anything because they agree they do not know, which paves the way for my showing what gravity is. By admitting not knowing what gravity is they then also admit being incorrect about gravity but unfortunately mainstream physics do not see it that way (yet). The question in hand is finding what role gravity played when the Creation came about for the first time. I had to find a method that would allow me to explain why gravity played a role. In the book I present the analysing of Kepler's formula without Newton's interrupting of Kepler's work.

Please let me explain: Tycho Brahe and later Kepler made a study of outer space as never repeated afterwards. From this Kepler concluded that $a^3 = T^2 k$. We all know that a^3 is space and with the space indicated as being in the third dimension and the third dimension is unmistakably a cube that forms volume, which by definition is presenting space. We also know from the way calculations come about by using the formula of Kepler that T^2 is the duration of a specific period of time relating to a specific centre. On the one hand we have space a^3 and on the other hand in direct relation to the space Kepler introduced motion coming from a centre that forms time $T^2 k$. Kepler gave us space-time a^3 / T^2 centuries before Einstein gave the concept a name but no one ever took any notice. In the formula is space a^3. In the formula the space a^3 has direct relation to time T^2 If **k** is a^3 / T^2 it means that from the centre holding the gravity is space-time. Space is a^3 and the motion of space a^3 we accept as time $T^2 k$ and such accepting is part of our understanding for the past three hundred and fifty years. Kepler gave us gravity before Newton named it as a force. Kepler gave us space-time long before Einstein named the notion. With Newton's meddling he missed Kepler introducing gravity as $k=a^3/T^2$ space / time.

Gravity is a rotating solid moving through a liquid space. The Earth is a solid that much is true. The atmosphere is regarded by physics to be a liquid and that much is also accepted as true. Gravity is the Coanda effect and the Coanda effect is where a car tire spins trough water and the spinning wheel gathers the water onto the surface of the tire. At speed the tire picks up the water and secures the water around the tire. The motion of the tire, which is the solid, contracts the water, which is the liquid onto the solid tire surface. The contracting of the liquid onto the solid by rotating motion produces the gravity that attracts the liquid to the solid. That is gravity. The solid tire can cover the surface of the tire by a layer of water where the water is as hard and as sturdy as what the solid tire can be. But the water might be hard and sturdy, yet it remains a liquid with all the characteristics attached to liquid. That is why driving in the wet is so dangerous.

The solid tire can surround the tire surface with as much as one inch of water. That is gravity whew the solid tire spins and by spinning it contracts the liquid water. The water being a slid in the form of ice can't perform in the way the liquid does when the liquid is surrounding the tire. An inch of ice will never be strong enough to allow a car to drive over it but in the case of the Coanda effect the motion of the tire allows the water to be much stronger. While when being in the position where the water is surrounding the surface of the wheel through the spin of the wheel that contracting gravity then makes the water as strong as the tire which enables an inch of water to support the entire car running on the water. The spin of the tire produce a gravity contracting by rotating motion which turns the density of the liquid water to the same compactness as the tire surface being a solid will have. The spinning solid of the tire turns the density of the fluid water into a solid equal to that of the solid tire. The tire asserting a rotating motion does the producing of gravity by motion. The expanding of the rotating action produces a contracting of the liquid space it moves through. By expanding the solid rotating takes away some of the space that the liquid holds.

The tire rotating is expanding the space it holds. That is called fleeting momentum. The matter tries to move away from the centre in an effort to gain more space. As the tire spins the tire tries to capture more space and the tire can thrust this so hard that the tire does go oval. The tire tries to hold more space than it has because it is capturing more space in an effort of expanding.

While the tire tries to capture more space the tire also reduces space that the liquid water holds. By capturing the space that the liquid water holds the tire is capturing water, and in the effort the tire is trying to gain the space that the water has and in that the solid tire is making the liquid water solid. Therefore we can drive on an inch of solid water while the water is liquid. By the water contracting we find the density of the water changing where it meets the surface of the solid. The solid tire expands and in the expanding it makes the liquid water solid. Then where the liquid water finds being reduced and being made denser the liquid water gets so dense it becomes a solid water area. The tire wall turns the water into a solid by the rotating action of the wheel. That is when motion applies to the wheel. When considering the wheel at speed and the Earth at speed we tend to think of the Earth being still and motionless. The Earth is spinning at a far greater speed that the tire would ever be capable of. The Earth is spinning so much it is concentrating air to become a liquid.

When objects fall the object has no mass and this is in spite of all the claims the Academics in physics try to produce. Galileo said all objects would fall at an equal pace and hit the Earth at the same time when falling the same distance through the same air under the same conditions. Newton said mass is responsible for that which produces the gravity by which objects fall. That means the object being more massive must fall faster than the object being less massive. If mass brings on gravity mass must distinguish the amount of gravity by applying more or less falling pace. If that does not happen there is no evidence of mass applying because then all objects hold equal mass while descending to the Earth. We see frequently that the object can fall at a specific rate depending not on the size it has or the shape it has but the distance it travels through the air. We see so many times that a car drops from an aeroplane with a human falling next to the car. There is a car advertisement where the human that is falling has a parachute in a bag falling next to the person that is falling next to the car. The car is around twenty times more massive than the human and the human is about twenty times more massive that the bag containing the parachute. If the falling process depended on mass that instigate the gravity action as the intellectuals' wishes to declare then the lot cannot fall at the same rate. The car must fall twenty times faster that the person and the person must fall twenty times faster than the bag. We can see that the lot is falling at the same rate and the descending bares no implication to any mass that shows differences. That is what Galileo said when Galileo said all things fall equal at the same rate and land at the same instant and a massive object will land at the precise instant that a very light object will land on the condition that they are dropped equally and that they fall through the same space at the same time. That statement excludes mass from any part that gravity has.

One can from that see that the falling has no implications brought on by mass. Mass has nothing to do with gravity but to restrain any further gravity effort putting individuality to the item falling. The gravity is produced by the moving of the object and while mass restricts the moving the moving still apply as gravity because the moving remains as a tendency to move when mass applies. Even where mass stops the gravity moving the gravity remains as a tendency to move towards the centre of the Earth. It is not that the mass is pushing but the mass is stopping the moving of the item to the centre of the earth. Mass only applies when any further motion of objects are restricted. If the restriction of the larger object that inflicts the mass suddenly also start moving, the mass turns to gravity that very instant. While the larger object retains in position of preventing the independent object from any further individual moving mass comes in as a factor that stops further motion.

Mass counter acts gravity by stopping gravity. The object only obtains mass when gravity becomes no longer applying. When the objects all fell and they all hit the surface of the Earth at the same time, only then is there distinction about size and mass. That makes mass a factor of the Earth holding a restraining on the object and then mass is not part of gravity because gravity is part of the falling or the moving of the objects. When object fall they have gravity and the gravity is equal applying to all because the gravity is the moving of the objects without restriction. When the objects hit the ground the objects loses independent motion and with the accepting of mass the objects retain the motion that the Earth provides. The mass renders the objects the motion of the Earth and having the motion of the Earth the move at the same pace as the Earth. The mass then makes them having a relation to the Earth where the mass puts them in a restricted part of the Earth. They move at the rate that the Earth move because they then are part of the Earth by the provision of mass. Mass shows how much the object that then landed on the earth and no longer moves independent of the Earth then became Earth.

The gravity is the Earth forming the solid that rotates and the air is the liquid through which the Earth rotates. The liquid is contracted onto the surface of the Earth by the solid of the Earth trying to expand into the space the liquid holds and this happens due to fleeting momentum or rotating motion. Mass is the resistance that an moving object shows when stopped by a larger object that blocks the smaller object from moving further as an individual object where by showing mass the smaller object resists the effort the larger object asserts on the smaller object to compromise the form the smaller object holds as a unit and not to accept the form the Earth imposes on the object. By having mass the object no longer holds independence but retains some form of individuality by not compromising the unit it forms as an independent structure. When the object lands on a larger object and the larger object is part of the Earth the larger object halts further motion. The larger object removes the individual characteristics that the moving object has. Let us have a good look at mass.

When there is a ship we see the painted waterline of the ship indicating the load in mass the ship can take. The ship is lighter than the water because the ship floats on top of the water. By loading the ship the ship can take in a lot more mass than what the water is because as long as the hull displaces more water than what the mass of the water is in terms of the area the ship claims the ship will float. The ship is less dense than the water when taking the area it holds in relation to the density of the water it displaces.

When the ship has an equal density to that of the water the ship will float in the water while being buoyant. The ship has to accept a certain percentage of water to allow the ship to float inside the water and prevent the sinking of the ship to the bottom. This is an act of precise balancing. The ship then being somewhere inside the water has no mass but being equal to the mass that the water holds. The ship has to displace as much water as what the area it holds will be in mass when being only with water. Then the mass of the water is the same as the mass of the ship and therefore the ship can hold ground inside the water without sinking or floating. If the ship again wishes to float the ship will have to displace some water it has in its hull and exchange that water for air. Then having more air than water that will enable the ship to have less density per volume of space and less density will render the ship in having less specific density than the water. When the ship sinks the ship has more mass than the water it displaces. The ship has a bigger specific density than what the water has and the mass of the water pushes the ship to the bottom of the water. At such a point when hitting the bottom the ship then has mass. It is the mass of the ship at the bottom of the water being more than what the mass of the water is that floats over the sunken ship that puts the ship on the bottom of the water. In water or in liquid the ship or the solid either has buoyancy or it has mass. It either floats in buoyancy or it sinks whereby it receives mass when the sinking ends. When it floats it has motion by buoyancy that the water provides.

The ship while floating has less mass than the water has. The ship while submerged has the same mass that what the water has. The ship has more mass that what the water has when the ship sank. The mass is a relevant factor because of the fluid aspect. When floating the ship has little mass and all floating objects has mass that is less than what the water has. When being submerged it does not matter if the ship is a big ship or a small ship because the ship has the same qualities as the water and therefore a big ship will be as buoyant as a small ship will be. This is very important to note. When in a liquid there is no mass factor. Only when the liquid suppresses the object and onto the solid and with no distinctive difference between the motion that the object has and the motion that the solid has does the body turn from being a liquid to being part of the solid does the mass factor enter the equation. A big ship will float submerged next to a fish and from the mass aspect the two would be equal. It is depending on motion.

Up to now every Academic at all levels in science is normally acting as if gravity is a commonly explained factor, proven in detail without the tiniest whim of uncertainty, where every one knows every aspect about all principles that are involved in gravity down to the smallest detail. In truth no one in science anywhere remotely knows what brings gravity about and I used Kepler to unravel this mystery called gravity. But no one in science will admit this fact about Newton or any one else never being able to explain gravity in the least or admit that Kepler is the one who formulised gravity decades before Newton came and gave gravity the name. Newton did not underwrite or define gravity and even today the most informed in Science at best can only assert their suspicion on a rumour presumed about what causes gravity to perform as the part interlinking the cosmos but no one can go any further by explaining the concept. Newton started this realising of gravity but it had and still has no more substantial proof than a rumour has and Newton admitted to it being a concept he could not explain.

Still to this day nobody in science at present will denounce the principle of gravity in a fashion by acknowledging that gravity is as yet still never been explained. All in science act in a manner as if Newton's gravity idea is the best-proven fact there ever was and only occasionally admit it to be just a rumour as Newton admitted it was when he introduced the name (not the concept). If Newton's concept was accurate it will by now have the moon much closer to the Earth than what it was during the time of Kepler's investigation, yet we from modern test that came into place after man landed on the moon in the seventies we now know it is moving away instead of coming closer. Newton agreed that he could only declare gravity as only a vague concept more in a suggestion and far from the manner of forming the proof as one would demand from a rumour when he was announcing a force that could be anything. Not once could one person in the past or present provide substantiating proof on gravity as reality by defining the very principles.

That includes Newton as well as Einstein and even Hawking. Scientists can declare that gravity was a factor at 10^{-43} seconds after the Big Bang but what brought gravity about or why gravity became or still remained, as a presence is still tightly concealed information which all are speculating on. Even to the best informed amongst the most educated do not know what is gravity because they all ignored Kepler and for ignoring Kepler the price they pay is not finding the principles bringing about gravity. Using Kepler makes the method to follow and understand even Einstein's discoveries shockingly simple.

I started using Kepler when I was shocked by the lack of proof on the matter of gravity as it supposedly applies in Newton's manifesto. I found a lot of total nonsense thought to be truth. That urged me to investigate the matter by using other sources than the mismanaged and incoherent nonsense Newton brought about when he raped every aspect of Newton's findings.

It is so easy to silence me. Prove that I am incorrect In Anything that I Say!

Years ago I was reading of a remark Einstein made about his realisation whiles being a patent clerk. It was abut the night that Einstein realised that had he; Einstein fell from the window of the patent office Einstein would feel as if he was as weightless as a chair and a pen falling alongside Einstein down the building. The question that comes to mind is why one feels weightless if it is mass pulling you down?

Then I then realised Einstein felt weightless because he was falling and part of falling was feeling what was happening to him. He was not pretending to fall whereby he then would feel as if...he was really falling and with that there is no as ifs. What he experienced came by means of what he was experiencing. If Einstein was experiencing weightless ness, it would be because he was weightless while falling. Einstein would not imagine the weightless ness because Einstein was truly falling. He was at that moment truly weightless. Einstein, the pen, and the chair had the same weight since they were all weighing the same. All three items would be equally weightless during the falling...that was what Galileo found because objects of different size and different mass travel equal while descending. The bigger objects do not fall quicker than a smaller object and that can only be attributed to one fact; it can only be true if they weighed the same while falling.

From this one can deduct that gravity is motion or the intent to commit motion and mass is one the motion of gravity is frustrated by blocking the continuing of the motion. Gravity is motion of space and mass is the restricting of the motion of space. Having mass does not bring about gravity but it does restrict gravity's motion. Gravity produces mass but mass does not produce gravity. Mass is the restraining motion and gravity is material moving about. Mass only comes into the application when two objects filled with space moves into a position where both want to claim space the other occupy. In essence it still is the frustration of motion and the commitment to move once the blocking of space is relinquished.

I then after reading this realised that gravity is not mass orientated, but gravity is motion differentiation between objects. While falling, The object moves less or slower in the direction that the Earth rotates and will fall in the direction of the Earth centre until such a time as the movement of the object is in synchronising with the speed that the Earth spins or if not the object will and on the Earth surface at the edge of the Earth and that will bring about having mass. The gravity applies as speed that is putting time in relation to the distance travelled and distance travelled is space. While the object is in a process of falling, the motion confirms gravity, both by getting the object's distance or band in which the object travels in harmony with the Earth that conducts all the spinning taking place at that point. That will reduce the height in which the object spins until it lands on the Earth and then can't reduce such reducing of a travelling band any further.

It has to do with specific density. If the specific density is increased by filling the object with helium we will find there arrives a point where the conducted speed is at a level that the Earth no longer will claim the body into having mass. When motion downward ends and the Earth disallow any further movement to secure a better specific density in relation to rotating movement, then mass sets in and becomes what is than point holding mass where the constraining of the object takes place to secure frustration of further movement and the Earth's motion annexes the object's freedom. While experiencing mass the motion is still there but now incarcerated by mass and locked onto the Earth by the rotation of the Earth and the superior or equal specific density of the Earth.

By connecting to the Earth the motion that the object is experiencing is what nails to object to the Earth by the force of mass and the object is then experiencing mass and not falling further through the loss of downward movement and now only conducts with the Earth rotating side-on movement. In this the downward movement is not lost altogether but remains, as detectable movement is the form of having a tendency to move although the object in mass is applying by forcing the downward motion to stand still. While the object is in mass and seems to be as if it is resting the tendency to move downward remains applying but that tendency to continue to move downwards is the tendency he named mass.

However mass then restricts motion and becomes motion tendency. While falling, gravity applies as equal motion to all objects relying to place all objects in relation to specific density and because of this motion counteracts any size, mass or weight by making everything able to fall equal in specific density. When falling, the object is either equal to what might be in the air according to allowed specific density, or has more than the specific minimum required density that is what is allowed to serve as the minimum required specific density and therefore will spiral down to the Earth.

When the Earth restrains further downward motion of the object that comes as the result of finding an allocated position of motion according to the specific density of the falling object, this readjusting of allocated position is stopped from conducting further downward or readjusting movement and all such further movement of gravity is hindering in the form we call mass. The falling object remains individual and still tends to move while Earth individuality resists movement. Further movement is disallowed as other material fill space. While the bonding of the atoms forming the object will secure any further deforming the object will remain to be independent but it is this bonding that is the value of the specific density of the object applying. By securing a [lace on the Earth, the falling object will finally rest and from that motion resistance comes mass.

While falling, the object is experiencing gravity because the object is in gravity but when on the soil the object experience mass which is the restricting of gravity or motion of the space filled with material.

Moreover, I came to another conclusion of equal importance. When any person is standing on any place anywhere, while viewing the Universe, that person is filling the centre of the Universe. Let's get more personal. When you, the person that is reading this, are standing at night and are looking at the Universe you are seeing the Universe from the centre of the Universe. All the light, every single beam that ever left any destiny at any time acknowledges this fact. You are the most important person in the Universe because you are holding the most important position in the Universe. All the light that comes across all of space runs directly in a straight line towards you filling the centre of the Universe. Not excluding the effort of one photon, all light is heading to meet you where you are in that centre spot and not one photon will pass you by.
Not one photon dare miss you because if they do they miss the effort that all light has to accomplish and that is to locate you as the person filling the centre of the Universe. If you find this funny, or laughable you are in for a shock because this is what gravity is and this principle dictates gravity. It is the most complex issue one can imagine and expanding on this thought takes thousands of pages. It forms the crux to all cosmic principles and embraces every successful and meaningfully theory ever used to explain the Universe. Without taking this aspect in to account, there is no valid explanation available to understand the cosmos. Al the light coming from wherever meets the point you fill in time and in space. For al the light travelling you hold the spot it was on route to.

Should you decide to shift your position to any other place in the Universe you will shift the centre of the Universe to that location as well. If you install a camera on Mars, the light is obliged to acknowledge your relocating the centre of the Universe at your will to reposition you're being that centre of the Universe. All the light that ever left its destination crossing the vast spaces of the Universe, excluding no particular light, travelled all the way just to find you filling the centre of the Universe, right where you are. By you're standing anywhere, you fill the centre of the Universe, and the entire Universe admits to that because all the light comes to meet you there. If you shift from the North Pole to the South Pole you will shift the centre of the Universe because all the light travelling throughout the Universe will find you where you then moved the centre of the Universe.

The light left its destination billion years ago as it travelled through space at the speed of light anxious to acknowledge you're being in the very centre of the Universe. No photon will pass you by where you are in the centre of the Universe. No wonder every person born has the idea they were born to fill the centre of the Universe, which we do fill. The Universe is spinning around you or I, which is filling a centre where all motion is connected. That is the Coanda effect on the utter-most grandest scale imaginable; nevertheless it is only a manifestation of the Coanda effect. It implicates gravity as wide as can be…

Then I reviewed the Universe. If gravity is motion, what causes motion? What stops motion? That answer is in the Black Hole. If a star is about fusing atoms thereby growing, what happen when all the atoms fused into one all collective atom? What is the gravity if the star has one all-inclusive atom providing all the gravity that the star had when the star still had massive volumetric space?

If all that space that once filled an entire giant star fused into one enormous gravity applying atom and that enormous force has been secures in the space that one atom holds, the atom would then show a force that would pull the surrounding Universe flat. Where does the gravity of the star end when all the atoms in the star became one giant atom? Gravity is smallest where space is least. Where space of an entire massive star is left in the size of one atom the gravity coming from that will pull the Universe flat at that point.

Coming to the conclusion about gravity being motion and mass being the restriction of motion was the easy part. What produced the motion and what prevented the restriction from overcoming the motion was the tough part. Figuring out why was everything on the move and where did the motion stop that was the part that took some figuring and some explaining. What made gravity move and why does gravity move…the answers are in the four phenomena never yet explained to satisfaction but now turns out to be the cradle of gravity.

Gravity is The Roche limit,

    Gravity is The Lagrangian system

        Gravity is The Titius Bode law

            Gravity is The Coanda affect

And gravity as the Roche limit forms the principle in producing the sound barrier. Read the book and find out why this is the case.

Newton's claims about the principles that he declared is responsible for guiding physics carries no validated proof and only after I realised that, was I able to start forming another line of thought on gravity. This had the purpose of confronting the corner stone of modern physics and at first I tried desperately to do just that. At first I was not confrontational towards Academics in physics and avoided any indication about disagreeing with Newton, although avoiding to show my disagreements was also totally impossible too but every time I approached academics with my new concept the academics always threw Newton at me. Facing Newton or facing defeat became a two-sided blade and I had to start to confront them by confronting Newton, with which I was in disagreement from the beginning. At first I was reluctant to voice any opinion about the matter of how far I was prepared to challenge Newton because Newton was and is an icon. But slowly it dawned on me that if I had any serious plans to introduce my ideas I had to dispute Newton's gravity principles and do it head.

I am able to explain the how the Universe started only because I discovered the building blocks used to build the Universe. The building blocks are the Titius Bode Law, The Roche limit, The Lagrangian points and these all culminate into the Coanda effect. Using these I can and I do prove exactly how the cosmos was built dot by dot.

Every person associates gravity applying as "The natural force of attraction exerted by a celestial body, such as Earth, upon objects at or near its surface, tending to draw them toward the centre of the body" with what we think of happens to solar bodies having mass which is "the natural force of attraction between any two massive bodies, which is directly proportional to the product of their masses and inversely proportional to the square of the distance between them" and science cheats everyone into believing the two is the very same. The one I experience every day and with the other there is no evidence of applying anywhere. Yet students are fooled into believing the two are exactly the same issue, which is untrue. If there is any academic feeling insulted by me calling the lot fraudsters, bring evidence of the second form of gravity working on the principle of mass and I will withdraw my statement, otherwise if no proof can be brought, then you lot that ignored me for ten years are the fraudsters I accuse you to be. You are villains brainwashing students to corrupt their thinking.

Einstein's Critical Density lacks the accepted matching facts we need in proving the critical mass factor, which makes the entire idea silly and bogus. You can't force the Universe to conform into something Newton said because Newton said the Universe works on mass contracting… But our inability in securing such required evidence defies the most basic logic. It seems all new evidence we receive from outer space is disputing all Newton laws and new findings disprove Einstein's Critical Density as the answer. The Universe will not reach a point of contracting, not withstanding whatever dark matter astronomers try to locate in the vast space. A rush is on finding the black-matter that will be applied to force the cosmos too retract back to where it came from. But what if our view of the cosmos was as incorrect as our views at present is about the sun? I prove that contraction is at present as much part of the cosmos as is the expanding is that we focus our attention on and it is our culture we carry from generation to the next generation that leaves the human view obscured in admitting the truth. The Sun is not a coal stove burning fuel. The Sun is an air-conditioned pumping gas (hydrogen and helium with electricity. I prove by applying the Titius Bode law that gravity is electricity and the two are the same thing. By using gravity or electricity the sun is a huge air-conditioned pump freezing cosmic gas, which is what the Universe is into a liquid that we see squirting from the sun.

Why would the expansion turnaround and do a reverse by going back to where it came from. Consider the momentum alternation such a change will bring about. The sun is not a gas-filled sphere holding hydrogen in its "natural gas" form, but it is all fluid and is in a liquid form where singularity is liquid-freezing hydrogen at 6500^0 C while outer space is boiling over at -276^0 C. The Absolute Relevancy of Singularity book explains the Roche limit as well as the other four cosmic principles in the practical sense… when applying cosmic laws instead of improvising cosmic laws uncovers that reality then becomes awesome. It becomes clear the Universe is as much expanding as it is contracting and contracting by expanding. As there is no hot or cold, no big or small, no grand opposing but relevancies in ratio to one another. If you do not believe me, then believe your eyes when looking at the picture. What ever the sun is it is fluid falling into fluid.

Because hydrogen is a gas on earth and we think of 6500° as hot on earth, therefore the sun must be hot and the sun must be gas at the same time. It is obvious what we see is liquid squirting and when hydrogen is in a liquid form it then must be cold because my eyes tell me that! Newtonians still have the entire Universe apply the standards befitting life on earth and that is why they wish to locate life.

When something is hot it expands. When something is cold it contracts. Outer space expands to the very limit and keeps expanding therefore outer space must be the hottest there can be, notwithstanding scientific stupidity. The sun contracts every bit of space that forms the solar system and therefore the sun must be the coldest place in the solar system notwithstanding whatever ego whish to declare. If I feel the sun is hot it is because it is diverting all the heat my way and it diverts the heat to me then where there is no heat it must be cold. It may sound incorrect and unscientific madness but with my applying of Kepler's formula in alignment with the position I located and valuated singularity it clarifies the possibility of the above statement... but please do not take my word for it, use your eyes and make sure you look past the culture bias of past incorrectness. See the fluid push out of a bowl of liquid, spilling both sides as it falls into liquid. The Hydrogen inside of the sun is not gas but it is fluid. In all of nature in all elements found through out science there is no NATURAL GAS as much as there is no NATURAL SOLID. Hydrogen is as much a liquid as iron is a gas and neon is a solid. It depends on the element relating to the space/heat in the circumstances surrounding the substance at that very precise instant in time. We have to stop telling the cosmos to show us what we wish to find and start accepting what the cosmos is telling us is out there that we should look for and find. In creation there are two substances that formed the Universe. One was earth or solids and the other was heaven or uncontrolled heat. Between all solid we have heat parting the solids and being a solid or a liquid or a gas depends on the ratio between the solids and non-solids. Under conditions suiting life certain elements may be a gas, but in stars conditions don't suit life and in outer space conditions don't suit life so therefore life cannot be a barometer for conditions applying in the Universe. Kepler gave us **solids as $a^3=T^2k$ as liquid or gas.**

The earth, just as all other cosmic objects do, contracts outer space by the movement of the rotation. Material spins in gas to reduce the volumetric size and by that it reduces the concentrated space around the star / planet / earth which then reduces the cosmic gas forming outer space to the cosmic liquid forming the atmosphere. By rotation the earth "pumps" gas from outer space to the core within the centre of the earth by applying centrifugal pump action. As the space becomes denser the heat level rises but this is because the rotation of the earth reduces the heat, not the heat that we as humans feel or experience but the heat level within the space. We have to maintain science laws. When any cosmic substance overheats it becomes larger and when cosmic substance becomes colder it shrinks. That is cosmic law. The heat we feel on our skins is heat escaping from where it is cold and contracted to where it is hot and expanded because the levels that reduces has to dispel the heat from the location it is within and move it to where heat is excessive and expanded. It is the heat moving away that we feel and then think of it as hot. As the earth turns it cools the space in which it is and the space reduces in heat and therefore the heat levels can reduce in order to make the space move towards the centre. As a matter of fact the very first experiment in science ever conducted and recorded was the discovery that it is space that reduces and not things (water) that falls but notwithstanding Newtonians have gravity as mass falling. But as Professor Friedrich W. Hehl, Inst. Theor. Physics With a lot of words and some simple algebraic relations, there is no way to "explain" the world of physics. Your seem to be out of touch with modern developments. I guess that is why Newtonian science can't conclude they are mistaken even after three-hundred years.

Again and as for e so many time in the past I repeat the warning once more of the Newtonian Brilliant-Brainy-Bunch to please take note of a conscientious warning about the gravity of the misgiving there is on the part of the most respected Academics in physics about a much concerning matter. As you can see why I state it emphatically that science accuses me to be not schooled to the point where I am able to have any form of an opinion on any matter concerning Sir Isaac Newton. Notwithstanding that my research proves I did my private studies and through which I skipped the indoctrination and mind control academics place on students goes unrecognised by their standards and so too my ability to have any insight on matters regarding physics. However my skipping their methodical and systematic brainwashing enabled me to see and allowed me to be able to express the incorrectness in Newton's teachings and allowed me to show in clarity what destructive force Sir Isaac Newton used to corrupt the laws of mathematics, corrupting to science along the way and mostly raping to the work of a great man, Johannes Kepler and what Sir Isaac Newton did can only be expressed as being blatant criminal fraud. What his deeds amount to is to corrupt the laws of mathematics, to render the laws of cosmology useless and to rubbish all of science. Should you find this to be unbelievable, then

I am glad to announce that this book is more for you than any other person, so go on and read what academics guarding science never wanted published. I challenge any one that disputes any claim I make to prove me wrong by proving me wrong and not merely suggesting claims in that direction.

We have to realise that it is about heat and cold where moving cools down and stationary space expands. Creation started with differentiating between what is cold and what is hot. That is exactly what the Bible says in Geneses 1:1. The Bible says God made heaven and earth. This means God made what is solid and what is not solid. It means God made material (the earth) and heavens (the sky or non materials). Back then as is the case now with our most brilliant Newtonians there was no other names for solid as earth and non-solid sky as heavens. At least those back then realised the difference and did not put sky down to "nothing" as our incompetent Newtonians of our modern era does. There is singularity controlled by movement (material) and there is singularity not controlled by movement (non-material).

Wxplain creation in terms of the solid proton and liquid neutron.
In terms of mathematical equating the very first recorded instant happened at:
$(10/7)(\Pi^2/2)(\Pi^2+\Pi^2) = 139.15$
(10/7) There was a cosmic gas that formed or it is the "stuff" that "stuffs" outer space.
($\Pi^2/2$) In the outer space a liquid formed that is what fills the liquid atmosphere of a star such as the sun. It is that pure liquid heat that squirts from the sun that Newtonians see as gas while they stare at a liquid.
($\Pi^2+\Pi^2$) The solid formed that by spin controls space occupied to cool it from space unoccupied.

$7(\Pi^2+\Pi^2) = 138.17$
7 The redirection of movement of time brought about the spinning effect of gravity.
($\Pi^2+\Pi^2$) The solid formed that by spin controls space occupied to cool it from space unoccupied.

$(7/10)(\Pi^2)(\Pi^2+\Pi^2) = 136.37$
(7/10) Gravitational movement started moving space by containing from the expanded to the contracted. It is still the ratio of gravity forming a cosmic liquid from condensing cosmic gas.
(Π^2) Gravitational movement started by duplicating Π in all seven possible directions minus 3.
($\Pi^2+\Pi^2$) The solid formed that by spin controls space occupied to cool it from space unoccupied.

This is exactly what the Bible says, if only the Theologians and the Newtonians were not so preoccupied in fighting for their own small-minded egos instead of looking for the truth.
When the Universe started there was one spot that released a dot. The spot as large as a thought grew into a dot. This too is exactly as the Bible says it happened and it still happens exactly like this. The Universe starts from infinity and grows into eternity just as it happened at the start and this is what Newtonians confuse with what they know as the Hubble constant. To check if I am correct look at anything spinning and you will see conformation. The dot that released was by such release relevant to other dots released because there was to be motion that measured many dots. The dots in release were relevant since they were the same. Only time being in delay by cycle of infinity interrupting eternity to form space, as space is the history of time gone to the past. But since we are looking at things as it started I put it into the past tense. That cycle brought about time delay as every cycle drifted further from the original singularity while the original was still responding as well. That was the first space. It was time being one infinity part. Since the dot was also singularity and was the very same as singularity with a small difference that the spot was $\bullet\Pi^0$ and the dot was $\bullet\Pi^1$. At first, at moment-Alfa, there was no space nor time for only relevancies came about. Relevancies acted as motion to bring change to eternity by changing the flow of time in eternity. There was the perfect spot in which time moved while remaining the same. Then heat brought expansion and expansion brought space and space brought movement and movement brought about a Universe. That is how it started and that is how it still is. This was before the atom and the atom was before light and that too is exactly what the Bible says in Geneses 1:1.

Time will forever remain eternity but space or time distortion, which will forever remain infinity, interrupted the flow of eternity. Space breaks the monotony of time in eternity by parting time from in infinity. Yet, the relevancies did imply motion except for the fact that singularity is very much incapable of motion. Every dot had a purpose to fill a position in relation to the other dots that the spot excited. All dots had a line of three where two was one, each on every side of the spot. The spot was one with the two dots forming two, which improvised for motion that would later come to space and the three was what space was going to become. How do we know this: $0.1416 \times 7 = .991$ and that is by the value that the Universe grows.

Time was four because the four would bring about motion as heat separated infinity from eternity or hot from cold. Five was space because space was one removed from time. Space is the distortion of time and one outside time would bring a time delay or a time distortion of four plus one which is five, hence the principle behind the Lagrangian system. Because material was the square of space material was a crossing of three plus three forming six. However to find out what this means you have to read the entire theses called <u>The Absolute relevancy of Singularity</u>.

Space-time is the four of time, plus the three in singularity around which the four of time turns, therefore space-time is seven.

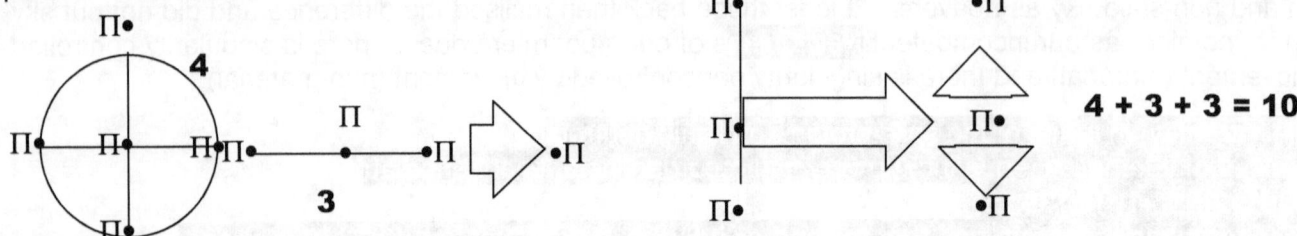

In the circle using $r^2\Pi$ the r has to have distinctive qualities placing it as a factor apart from Π. Where the growth shows no separate distinction but a continuous flow from the precise centre to the precise edge the flow would become in relation with Π depicting the circle and Π replacing r as reference to any point on the circle.

By using r, distinction in the circle is possible but by using, Π there is no distinction possible. Therefore, in the beginning when time formed space there was only relevance coming about from $\bullet\Pi^0$ and the dot was $\bullet\Pi^1$ with no mention of any possible r. The fact of r representing a radius represents space and what we refer to be long before the Big Bang introduced space or mathematics using space.

Before the Big bang the lot was form without dimensions playing any part. Then the atom came. Only after that did the Big Bang come. Even before the atom was the point lining up and forming positions that was spinning faster than the speed of light can ever achieve. However every point today still serve the role it took on at that stage and serves in the position that it had during the time it had no space with eternal time. These relevancies developed as part of a Universe we shall never understand. The Universe had no sides and a line was equal to a triangle, which was equal to a half circle. Singularity holds the double space-time position of five times two (matter and space duplicating singularity) which then is ten.

How did the Universe liberate material and heat forming space from singularity because with singularity comes an unchangeable eternal condition that is non-changing-everlasting in all conditions and aspects that is remaining in absolute equilibrium. This equilibrium maintains because all development extends form precisely in a detailed equal equilibrium throughout. Think about what brought the cosmos out from the eternal rest in which it was. The eternal rest still maintains and is therefore our detection. What inspired the eternal rest the cosmos was in and inspired change to the state of eternal rest? What evoked change? That is the question the Atheist will never be able to answer but that too is the most basic and ever-lasting fundamentals of the Universe. Singularity Π^0 is not substance but it is a thought establishing substance Π. What changed in this split second start before the official start? I do not wish to ponder on this matter in the letter I am writing at this minute, as there are other books where I delve into this matter. It is called <u>The Absolute relevancy of Singularity</u>

Gravity has two values and that is the crux of the Sound Barrier.
I'll bet you Newtonian physics can't connect the sound barrier with the Supernova event…and I'll bet you it is because Newton's physics can't remotely resemble any meaningful explanation that could render a cosmic connecting principle between the sound barrier and a Supernova event.
But then they wish to redesign the Universe by creating Space Worms and Black Holes that no one will ever be able to disprove because they are busy with a fairy tail physics story. I say … First prove the obvious and the down-to-earth.
But then again Newton's physics can't explain most information about many things because it is extremely limited. If you are not at all convinced or if you are little convinced about any flaws there are in Newtonian science then ask yourself if you know what these pictures have in common? **If you follow the tour and go where I guide you to go you will learn what it is that this first cosmic phenomenon**

A More Complex Commercial Science Book 389  RE-(W)RIGHTING COSMOLOGY

 has got in common with this second cosmic phenomenon where the one is stars bursting and the other is the sound barrier coming about when something is moving in relation to what the earth gravity allows.

Do you realise what you see in these pictures arte the same cosmic principles applying in such a manner that it is the very same thing happening in the cosmos and on earth?

The circles forming is gravity forming Π by the movement of Π² as gravity Π⁰ extends into wider circles by enlarging Π from connecting to one dot and then connecting up to 4Π⁰ in the circle. Do you know what happens in these pictures? Of course you do know this, but do you understand this? Knowing this says little because it is the big anomaly science can't explain because Newtonian science will not allow any correct understanding of this. That is because Newtonian principles can't correctly explain this. This is one of the four cosmic phenomena that prove Newtonian philosophy about physics to be incorrect! This website will introduce you to singularity that is a totally new concept.

Please believe me when I say you have never even seen the true physics behind whet drives the Sound Barrier or gravity for that mater. To understand what law in physics drive this other than Newton's Dark Age rhetoric goes to <u>Book 1 **The Absolute Relevancy of Singularity** in terms of Cosmic Physics</u>

http://www.lulu.com/content/e-book/book-1-the-absolute-relevancy-of-singularity-in-terms-of-cosmic-physics/6624181

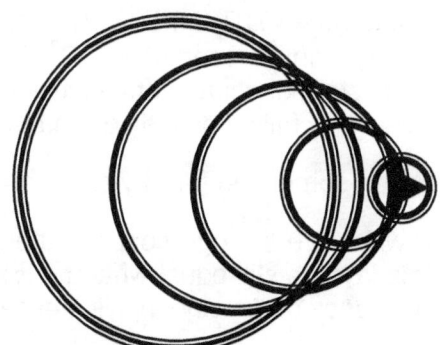

The explanations may seem plausible but put Newton into it and the entire concept falls apart. It is not in the explanation of what happens that the answer lies but in the explanation of why this happens that true physics lie...and that Newton's thinking cannot cope with....It is fine and well to give fancy sketches but explaining why and what principles is behind it Newton can never give.

By looking at the Sound Barrier we can see what gravity is but in Newtonian terms the explanation of the sound barrier is non-existent. Everyone in science has an explanation about the sound barrier and every explanation underlines the absolute incompetence in science to form any type of explanation.

<u>I prove gravity is Π and that four cosmic phenomena forms gravity by producing Π when objects move in relation to each other.</u>

<u>The rings you see are in place because gravity is Π. The spherical planet you see is a spare because gravity is Π. The planet spins in relation to Π because gravity is Π</u>

By using the above **the four cosmic pillars**, it enable me to **present the proof** where I now can explain what conditions bring on the **sound barrier.** By proving it **is gravity** that the individual structure generates motion above and beyond the gravity the Earth provide is what is producing individual **motion** that the independent object earned within the sphere of motion that the Earth's gravity provides where the **independent and individual motion** put the relevance that gravity has beyond the conserving means gravity has where the space that is serving the independent object is independently in motion. The adding to the independence on top of the normal structural independence is creating more individualism by the independent motion of the individual structure being apart from the motion that the gravity of the Earth **provides.**

I use in explaining the Cosmic Birth is that is the Universe started with 1 a point holding singularity, then how did two arrive and then three, four, five up to ten.

I said it before so I say it again. From that comes gravity as follows: In the centre of the rotating body singularity forms to the value of Π^0 and this extends to the curve forming Π in terms of the curve of the rotating object. This then forms part of gravity moving forming singularity going square Π^2 and this form the relevancy

The extending of singularity goes from Π^0 to $5\Pi^0$.

Coming in from space while having no mass the coordinates change as the value of Π forms a relation to the 7^0

When an object comes in from the sky the object associate value of the rotation in line with value of 21.991 / 7 on Π or on how gravity functions. It has way, shape or form and there is entirety of the cosmos apart the Newtonian conspirators

the atmosphere or flies with the turn giving it a the axis which puts a gravity. This is gravity and this is nothing to do with mass in any no factor such as mass in the from being in the imagination of calling them physicists.

Avery thing falls due to the gravitational movement of the earth that spins at $7(\Pi(\Pi^2))\Pi^0 = 217$ km/h. This is the given relevancy taken from the manner in which the earth fixes the falling criteria. If something falls at a rate of $7(\Pi(\Pi^2))\Pi^0 = 217$ km/h in a linear direction straight to the ground the object must travel at a horizontal speed of $7(\Pi(\Pi^2))\Pi^0 = 217$ km/h not to fall directly to the earth but in order to maintain a constant direct movement in line with the curving of the earth. Falling directly down is the same as not falling directly down but then instead of travelling horizontally by maintaining altitude.

When something falls by sheer gravity straight to the earth it falls at a rate of $7(3(\Pi^2))\Pi^0 = 207$ km/h.

It will move through space in the second region going at $7(3(\Pi^2))2\Pi^0 = 414.5$ km / h although in density this layer is still equal what the first layer is being 207 km / h It seems to go faster when falling but it is a relevancy of density applying because the space is less compact.

The next layer is at a density level of $7(3(\Pi^2))3\Pi^0 = 621.7$ km / h. This is the density of the space at $3\Pi^0$ in relevance to what the earth is at a density of $7(\Pi(\Pi^2))\Pi^0 = 217$ km/h.

The second most upper density level is at $7(3(\Pi^2))4\Pi^0 = 829.$ km / h. The density at this level let any object fall at a rate of 829 km/h while in fact the object only in reality and according to the earth travel at a rate of $7(\Pi(\Pi^2))\Pi^0 = 217$ km/h.

Then the last limit is at $7(3(\Pi^2))5\Pi^0 = 1036.3$ km / h but because of the Lagrangian limit set on movement which is $2^3 \times 3^2$ in conjunction with the law of Pythagoras then becomes 5 the boundary of five can only be achieved when the object breaks free from gravity and moves fast enough to orbit in outer space.

In that the border set on movement is $7(3(\Pi^2))\Pi^2 / 2 = 1022.$ km / h. After this the Lagrangian limit sets in. The very limit on movement is formed by the Roche limit applying in a double effect where the limit is at the border of the atmosphere being $7(3(\Pi^2))(\Pi^2 / 2)(\Pi^2 / 2) = 2523.$ km / h. At that density moving at a rate of $7(3(\Pi^2))(\Pi^2 / 2)(\Pi^2 / 2) = 2523$ km / h is equal (not more than) moving at $7(3(\Pi^2))\Pi^0 = 207$ km/h. At the height of $\Pi \times 10^3$ m above the earth the minimum gravitational movement that would keep a body moving in the boundaries of the earth but still at a horizontal or leaner level would be travelling at a density rate of $7(3(\Pi^2))(\Pi^2 / 2)(\Pi^2 / 2) = 2523$ km / h.

I heard a remark that a person fell from a bag at a height of 2500 km / h during one test and that the person fell at a rate of twice that of the sound barrier. Then I knew this uninformed but so important and prominent wise guy was a brilliant member of the Brainy Bunch. He did not even know he knew nothing! When being at 31 km above the earth the density level at that height is $7(3(\Pi^2))(\Pi^2/2)(\Pi^2/2) = 2523$ km / h which is the very same density than what $7(\Pi(\Pi^2))\Pi^0 = 217$ km/h being on the ground. The fool did not think why the person did not go through the sound barrier busting the sound barrier but it is because there is no sound and the reason why there is no sound is because the density level will not allow sound to form. If there is no sound the person can't go through the sound barrier and that is because the density at that level is to little to allow sound to be a part of physics.

Gravity forms by movement that establishes singularity initiating a circle in using Π. This is how the gyroscope works and how the ballistic bullet and the curb ball works. It is all to do with singularity forming points that in the next movement form space by turning. What is in the Universe is spinning. In the **precise middle** of all **objects in rotation** is a precise centre dividing the object in sectors that will **start the spinning initiation** from that centre point. Thus, the spinning object **will have a middle point**, a very specific **centre point that does not spin** and only holds Π as a specific value because no radius can apply. But also the one value such a line **cannot have is zero** because the line **is there and holds contact** to the rest of the material bringing about that **zero does not start any** line and therefore the **value of the line must be infinite**, just as described in **accordance** and by **the definition of singularity**

Locating and finding the presence of singularity

$k^0 = a^3 / T^2 k$ states that whatever is, is also spinning in order to be present.

Thus, the spinning object **will have a middle point,** a very specific **centre point that does not spin** and only holds Π as a specific value because no radius can apply. But also the one value such a line **cannot have is zero** because the line **is there and holds contact** to the rest of the material bringing about that **zero does not start any** line and therefore the **value of the line must be infinite**, just as described in accordance and by **the definition of singularity**

What is in the Universe is spinning. In the **precise middle** of all **objects in rotation** is a precise centre dividing the object in sectors that will **start the spinning initiation** from that centre point.

As I am introducing a very new idea, I wish to explain in better detail what I try to convey.

While the toy top is, spinning one will find singularity by moving the rotating line or radius progressively to the middle by reducing the length the line has from the edge to the middle. At one point all further reducing must end but the ending cannot include zero or nothing because the rest of the line still attach the rest of the top.

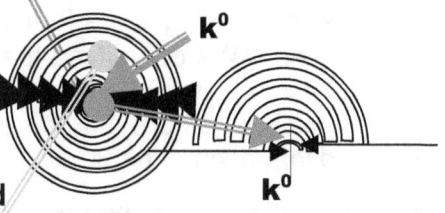

As the rotating direction moves inwards, the rings holding Π will become smaller and smaller. The reducing of the radius r will eventually end where the spin direction ends at Π⁰. However that point where the directional spin ends is the point where the actual spin takes place. The spinning is on the precise location the point is not spinning.

That point albeit hypothetical, is also as much a reality none the less and is placed where that point **must be standing still** because every line **running from that point in opposing directions** is also **in opposing directional spin the other or opposing side.**

In considering the spinning motion in the fraction of time in the detailed instant every aspect of rotation will turn in every instant of change in time. Although the points had the same characteristics only one instant before, they oppose the characteristics it had just before and just after the very instant in which they are and to which they relate by similar points also in rotation. The fact of the graph proves my point in quarterly opposing dimensions and values.

When an object moves it form gravity. This principle goes according to Kepler's formula of $a^3 = T^2 k$ that mathematically translates to $k^0 = a^3 / T^2 k$. This indicates movement and that is what gravity is.

Where space is the least, which is in the centre of the circle, gravity is the strongest. The gravity located in the circle's space less centre holds not only the sphere together but all that is in the surrounding of the sphere outside the sphere as well. It is from there in a giro action that gravity bonds all atoms forming the structure of the sphere as one unit together in a unit as well as distributing a specific alliance in shape and form. How the atoms manage that we will get to in a while, but there is a law allowing for that to take place. In the centre of the earth space is least and therefore gravity is strongest. In the centre of the earth singularity connects to time forming there least space.

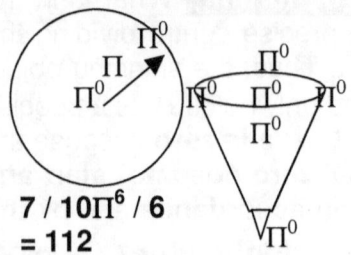

$7/10\Pi^6/6 = 112$

The spinning of Π^2 around the centre Π^0 establishes Π and Π is what produces the form gravity has. Still it is the relation or relevancy there is between the centre Π^0 and the spinning Π^0 that gives status to the form that Π represents. In out Universe we are accustomed to and are familiar to the rules we want to place seven points holding singularity to the centre holding singularity in a relation of $7/10$ $\Pi^6/6 = 112$. In that Universe everything less that a duplication ability to the value of 112 protons fit but only atoms to a maximum of 112 protons fit.

Kepler said $a^3 = T^2k$ but that could also be $k = a^3/T^2$

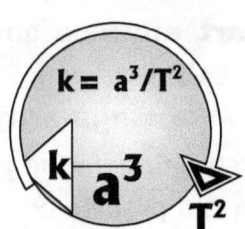

When translating Kepler's mathematical expression into a verbally spoken form of communication such as English we can see what Kepler said also read as $k = a^3/T^2$ where k is one point from a centre point that is space a^3 relating to time T^2. From a centre comes space-time

When drawing a line such a line then starts off with a dot serving the spot that holds all sides equal. That means the line serving as the future radius will be equal to the half circle which is then Π. The only aspect of the point that stands in for the end of the single line forming the radius of the circle is that we then mathematically reach the single dimension. We decreased the line to where a circle being Π formed on the single dimension. This dimension also hold the circle dividing line because from there the radius must once again generate a value and by such a gesture that the extending would form the circle that forms the sphere that eventually leads to the formation of particles. This leaves a problem to investigate.

With no line possible there had to be another dot that formed since the Universe has many dots that formed lines. But let us not to get confused and lost in the range of possible diversions but let us stick to two dots. One dot was next to the dot next to the dot, but as I said we stick to one dot next to the second dot. $M \times M / r^2$ is the first step gravity began with. That leaves us with a huge problem in as much as when $r = 0$ then $r^0 = 0$ and 0 dividing any value will leave 0 as the answer. If the particles were inseparable at the start it must bring about that gravity would not be forming since the distance will not permit any dividing. By allowing the distance separating the particles to be zero, the particles melt into a unit.

In the very centre of the sphere the form of the sphere dictates that the shape will relinquish space as the line run from the outside towards the very centre. With this natural state of affairs the sphere are naturally inclined to dismiss all space that it can form in the form as the sphere holds space inside and the form will finally be without dimension. All that I attribute to the line shrinking by reducing actually takes pace in every sphere as the diameter reduces to the centre. In the centre where the radius line goes single the form relinquish the three dimensional form it has inside. Being without dimension in the very centre means that at a point in the extreme centre of all spheres there are a point that holds singularity because this point with no space has a mathematical position although it is invisible since there is no sides to such a point to give that point any dimensions. The shape of the sphere is calculated by using the formula $4\Pi (r^3) / 3$. By reducing r to a point where r is r^0 singularity steps in because only the form remains as Π. Going even further we find that there then comes a point where Π goes singular Π^0. At that point absolute singularity is present but so is absolute gravity present at that point. When holding the strength of the shape of the sphere in mind as well as taking into account that all cosmos objects of importance is in the form of planets or stars and they are all in the form of a sphere, we therefore may contemplate that it is where gravity originate. We now only have to find the reason why gravity will hold a base in a space less ness as Einstein predicted. It is clear to be seen that gravity is in the centre of the sphere controlling from the centre everything that is outside the space less centre. We can reason with confidence that gravity is the strongest where space is the least. We can further reason that it is gravity that is holding the sphere in true form and since the sphere allow gravity the best working opportunity, gravity can form the sphere in as strong a shape and form as the sphere seems to have. From every point on the surface of the sphere is where that point connects with the other side of the surface of the sphere by a line that runs through the space less ness of such a centre of the sphere. Such a line also connect by an angle of 180^0 as well as 90^0 to six other lines running from top to bottom, right to left, and back to front, where all join and cross in the centre of the sphere. There are therefore six lines crossing and connecting by a centre from any given point on the surface of the sphere. Such points connects in total six surface points on each side of the sphere while they all support one another through the space less centre. In that absolute space less ness in the centre holding singularity we find gravity supporting and controlling all space within the sphere as well as space connected to the sphere. That is where gravity control and guide the space, which falls in the parameters as well as under the influence of the form of the sphere. In the gravity centre space goes singular meaning space becomes space less or flat.

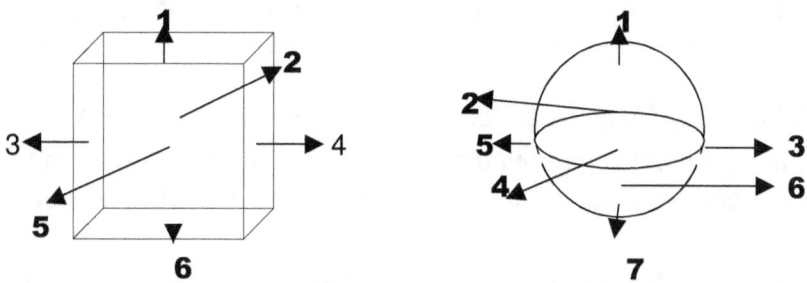

It is from the layout that the sphere uses as natural form that we are able to locate singularity. In the case of the sphere the material naturally reduces by measure of the radius becoming smaller to a point where the radius is r^0. At that point the line that will form the radius has gone single dimensional r^0 and that is equal to 1^0, which is singularity.

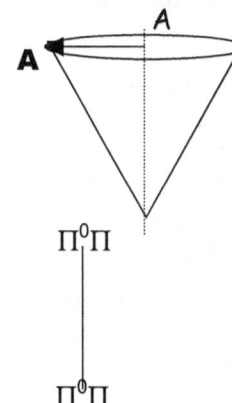

The moving of Π^0 to Π involved relegation and not motion as we consider motion. It was Π^0 getting a side and that is all. There was no true side but only a form that came into place. **Singularity (A)** received singularity (**A**) and no more of anything but the shift to comply with having a relevancy forming in relation to singularity. The dots had no sides, had no length or diameter. There was not measurable space or measurable time involved. The time could have been a micro, micro second as much a trillion millennium because time had no relevance. It was eternity interrupted by infinity, as it still is the case, however the line that eternity followed was no line because there was no space to hold the line. The line was momentarily interrupted by infinity, however with no one there, there was no one to notice. The lines were not lines but relations to sides being formed.

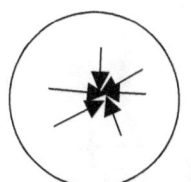

Inherent to the form the sphere offers, there is a specific location of singularity where the radius first goes single $r^0 = 1$ and then form goes into the realms of singularity $\Pi^0 r^0$. The cube also may have such a pint bur having such a point does not connect directly to six points located on the edges of the cube or any other form the is.

Also it is true that the entire form that is the sphere is controlled from a centre within the sphere. That centre holds the sphere in form and shape. Therefore the strong form is dictated from that space fewer centres where there is no space and no form left. The natural inclining is in the form of the sphere. It is part of the roundness that the overall shape of the sphere represents and this structural strength is carrying down to the very centre. Because the circle is forever reducing that reducing which is inherently part of the form of the sphere becomes a tool in distorting of space in the sphere and is eventually removing all forms of space from within the centre of the sphere.

The very centre ends up as having no space because of the reducing that continuous down to become the space less inner centre. The all roundness is the ingredient that forms the backbone of the absolute strength that the sphere has and that is the component that the sphere is so famous for. The form the sphere has allows the sphere to have a control that is coming from the centre deep inside the sphere where the space vanishes and being without space seems to keep the entire structure rigged. From the centre the sphere shape shows strength that the shape as tough as it is. How does it work in its most basic analyses?

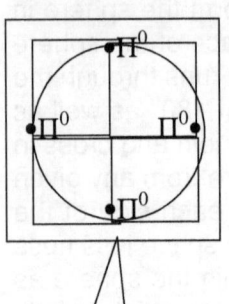

In relation to such a centre where $\Pi^0 r^0$ forms singularity there are always four cubes related to such a centre where the centre is part of seven points in total representing the sphere. Every cube gas lost one side to a point of the sphere where the sphere takes control of form and removes one side of the cube. In relation to the time factor that is inherently part of singularity by the extending of singularity there are five sides connecting to four points standing related to singularity by the Π^0 factor and that gives 5 X 4 = 20. That is always directly in relation to seven points singularity offers.

This only applies in relation to time because time is the square or then if you wish time is the flat to space being the cube. Time in the square draws space in the cube flat and that is the principle behind the effort gravity can apply to destroy space. Once gravity destroys space and time goes square then all factors preventing time to remain has gone single leaving time falling into singularity as well. This does not yet explain the three dimensions forming this well-known six-sided Universe we find we have.

When singularity expanded for the first time ever and when heat parted from cold bringing about the Universe forming **1^0** to **1^1** from Π^0 to Π a relevancy was born and that relevancy grew into what we now have as a Universe

Gravity in the centre formed time Π^2 by dismissing while the four time positions started the cosmic trend of duplicating.

With every one of the four points taking form to the value of Π at a measure of $\Pi^2 /2$ each brought about the Roche value of $\Pi^2 /4$ in relation to the developing centre. One has to remember that the star of today takes on the characteristics of the form of that era.

In order to move while remaining on the ground it has to make one dot at moving sped if it makes one dot plus a half the object will lift from the earth. If the object makes two dots it must be at a certain height covering a certain space while rotating with the earth. To move according to the earth the relevancy of movement must be $7\Pi\Pi^2$ x Π^0 to 4Π What this shows is that at the surface of the earth the earth repeats singularity at a rate of Π^0 and at the next higher level it is 2 x Π^0 and as the speed or space to time ratio increases the marks in singularity Π^0 becomes more. This will go up to 5 x Π^0 but before 5 x Π^0 comes about the Roche limit of $\Pi^2 /2$ and then ultimately $\Pi^2 /4$ will form movement or space-time displacement boundaries. This is following the principle on which the gyroscope functions and the gyroscope is how gravity forms.

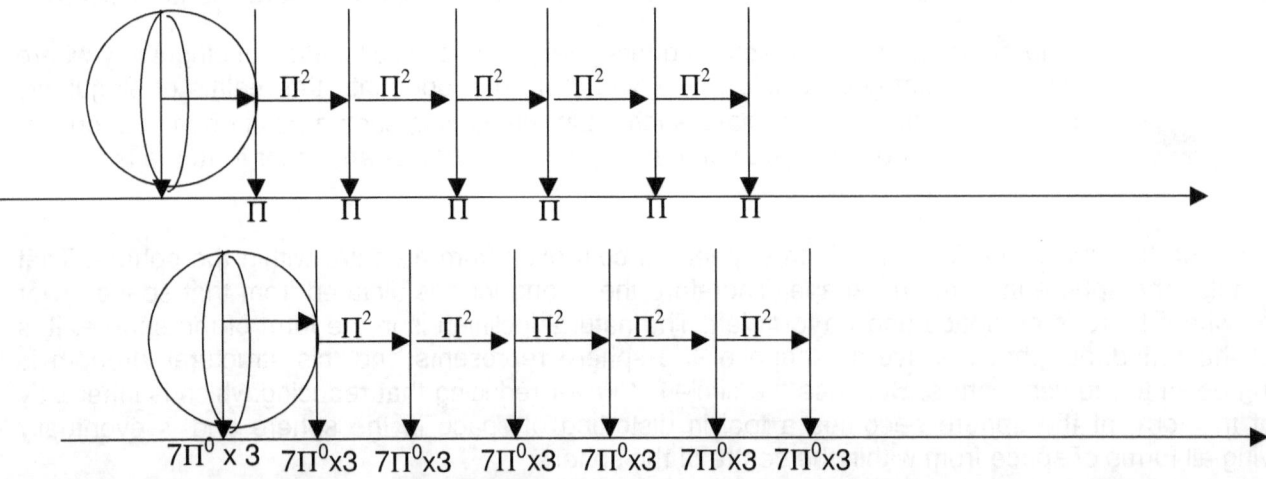

The object flying through the air has no contact with the earth and therefore loses the value of the curve at Π and connects with the axis forming singularity that is representing 3 instead of 3.1416 which is Π. The Π value holding .1416 more than 3 is the growth whereby space grows in relation to time.

When the ball spins and this spin is synchronised with the lines singularity holds as the next point to fill the ball will curve along the same dots as that which drives the gyroscope and which determines what dots must be filled in the next instant forming time that forms space.

I am not going out of my way to make this understandable because this fills a book on its own

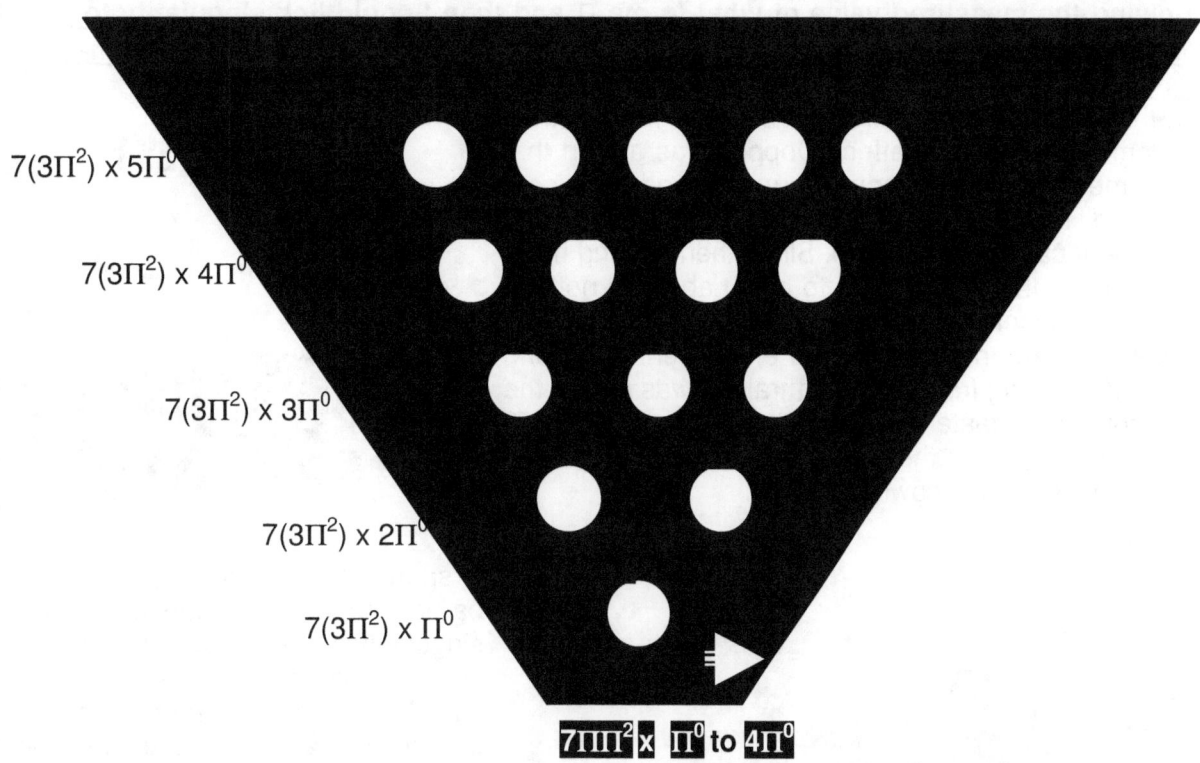

Moving with the earth has gravity at $7\Pi\Pi^2 x\Pi^0$. When going faster than the singularity that the earth provides puts the space in relation to movement or space-time displacement at $7\Pi\Pi^2 x1.1\Pi^0$ or at $7\Pi\Pi^2 x1.5\Pi^0$ then when the space required is more because of excessive movement in relation to the earth's gravity $7\Pi\Pi^2 x1.5\Pi^0$ it requires more space in relation to time. Whether the displacement is in relation to the height of the circle that holds $\Pi\Pi^2$ or whether the movement is forced to maintain $7\Pi\Pi^2 x \Pi^0$ at the height the movement through which the object goes will take on 1 or two or three spots per displacement

Because of the limit the Lagrangian law places on space the object moving is firstly capped by the limit at $\Pi^2/2$ and then later at $\Pi^2/4$, but this gets very technical and I have books explaining this very broadly and much better too.

I show why gravity is there, how gravity forms and what role stars play in forming gravity. There is no difference between how gravity and electricity forms and that I prove mathematically by decoding the cosmos. I prove mathematically when atoms spin they establish Π that forms the Universe. Whatever forms gravity has to link closely to Π since everything that has anything to do with gravity forms a circle that is Π by the value of the square radius.

If mass has anything to do with generating gravity, then mass has to apply Π or otherwise mass has nothing to do with the forming of gravity. Everything using gravity forms a circle of sorts, which forms the curvature of space-time, which is Π and which curves light. The way the planets orbit the Sun and how stars spin has all to do with Π. In spinning in a circle, Π forms gravity as a centrifugal force that condenses space.

1) I found the **location, position** of singularity ΠºΠ as a factor forming space-time
2) I found **space-time** by dissecting Kepler's formula in relation to **valuing singularity**
3) I not only found but I also **proved space-time** by **aligning space-time** with **gravity**
4) I found the **working principals** behind **gravity** as a cosmic occurrence.
5) I found the reason for the **Roche limit** and explaining the resulting of gravity from that.
6) I found out why the **Lagrangian system**, becomes **the building form** of the Universe.
7) I found why the **Titius Bode law** mathematically provides the **foundation of gravity**

By proving that the **Coanda affect is gravity assisted by the other three phenomena and thereby through activating singularity it formed space-time, I was able to link gravity with singularity in applying the measured value of what forms Π and then I could link singularity to gravity by the forming of the value of Π as gravity. That I managed by seeing that it is the way that Π forms that gravity also forms.**

All those practising science has to show for their wisdom is the Doppler effect. Doppler was a man that did some calculations about a hundred and fifty years ago and the finding was based on sound versus movement causing the sound. Since then nothing more came to mind…nothing more to boast about or nothing more to understand. Doppler showed the effect movement had on sound but his experiment was done on sound coming from a moving train. It was during the time horses ran faster than what trains moved. In 150 years since Doppler they know sound forms rings and that is as far as the knowledge goes.

I show mathematically through the sound barrier how gravity functions in relation to singularity and how through singularity applying the sound barrier forms. …But when I say I explain the sound barrier everyone presumes I am using the "traditional" middle age backward thinking that held the concepts in line with a steam train "thundering" down the tracks equal to the speed of a galloping horse.

I show what the sound barrier has in common with the Tunguska event in 1908 and what similarities there are in gravity between a jet breaking the sound barrier and the comet blowing to pieces during the Tunguska event.

I show by explaining that gravity is Π how the sound barrier and a supernova blowing a star to bits is the very same principle where the jet is just applying that principle on a micro-midget scale. I use cosmic physics to explain the sound barrier by using the **Titius Bode law**, **The Roche limit**, **The Lagrangian points** and **The Coanda effect** what happens during the sound barrier. Then those that should be wise enough to understand understands nothing because if it is not as simple as multiplying mass they don't understand anything and the rest is bored because they think I am reverting to a steam engine locomotive that came just after the first Rocket steam engine. My work has progressed beyond the Neanderthal Newton experience of seeing forces and naming forces of sorts.

Forces belong to magic and witchcraft, but definitely not in physics. I use physics by measure of singularity that the Newtonians don't even know to prove gravity.

Newtonian science always has this example they use to confuse and con students. It's about the feather and the hammer falling equal in a vacuum, which is what they use to conspire to keep Newton's contradictions out of discovery. It is the vacuum part they bring in to confuse because vacuum or no vacuum all things fall equal in any way. That example about the feather and the hammer falling in a vacuum is one part of the entire conspiracy. A hammer will fall equal to an elephant that will fall equal to a car that will fall equal to a rucksack that will fall equal to a tennis ball that will fall equal to…all things fall equal and that is what Galileo said. Galileo said all things will fall equal while Newton had this bright idea contradicting Galileo where Newton said that things fall by the mass and since mass differentiate then according to Newton all things must fall unequal as things then fall according to the mass each have. I know that Newtonians try to say that Galileo and Newton said the same thing but to get to that statement one has to be rather brainwashed in Newtonian religiosity to a point that the person's brain no longer can fathom fact from fiction. How can anything be pulled by mass meaning it falls by mass and still uphold Galileo's claim of falling equal. The Newtonians are besotted with Newtonian contradictions and fallacies they can't separate fiction from fantasy longer. Galileo said things fall equal and with everything having different mass then mass takes no part in the falling and since the falling is the gravity that forms the gravity has no implication to the mass factor in any way where mass dictates gravity. Gravity dictates mass!

One thing that may never escape the mind is that life is not a cosmic principle but life is completely alien to the cosmos. Life is found nowhere in nature and life is totally insignificant. Whatever life may achieve, as a marker is in terms of the cosmos unremarkable. Life moves but that is not a natural occurrence because in the cosmos only gravity moves objects. Life moving objects has in cosmic terms no place or role because the cosmos was not created with the idea of holding life. Any person putting life in any significant is so small minded that person must leave science alone and start writing children fairy tales. It will befit the person's mentality better. There is only life on earth and until proven otherwise the mentally retarded and confused searching for life does so because they have no understanding of any concept of what the Universe holds. Going to mars in cosmic terms is not worth a thought. Delving 4 km down a mineshaft is not worth mentioning in cosmic terms. Whatever life may achieve changes nothing in cosmic terms and that is the role life ahs, it plays no part in the cosmos. There is nothing life can possibly achieve that will be something of note compared to what anything else in the Universe may offer.

Gravity forms when objects move in the same gravitational field but at different speeds forming incompatible space formations. The earth turns as the earth spins around the sun. The time we have on earth is uniquely and only comparable to the earth. It is what connects to the centre of the earth that is a part of the time unit we associate with on earth. The moon spins around the centre of the earth, which makes the moon the last outpost of the earth. In that sense man has never left the earth's time and I have the opinion that when going to Mars and losing the singularity we connect to on earth those going on the journey will lose their faculties and become raving mad. Anything moving between the earth and the moon adheres to the centre of the earth and doing that holds absolute connection to singularity on the earth, even when being on the moon. To leave the earth's atmosphere and go into outer space the relevancy that has to apply is **$4\Pi^2 \ (7° \times (3\Pi^2) \ (\Pi^2 / 2)) \setminus 3600.$** Moving at that relevancy is the only way one can achieve to leave the atmosphere and go into outer space. That also must be directly upwards-in direction and the relevancy must apply from the instant the moving object leaves the ground. You can't fly horizontally and then leave the earth because doing that will have the earth destroy the object trying to escape.

No one can explain the cloud forming around the aircraft as the aircraft fly well below the sonic boom. The cloud does not make sense when Newton's gravity is used to explain this phenomenon. No one in physics can use physics to explain the cloud forming. We can all see it is vapour condensing around the fuselage of the aircraft. Let them show the mathematical explanation why this happens. It is easy to see what happens but explain why that "what" happens goes beyond what Newton's physics can cater fort. If you throw out mass and forces physicist can offer little because then they know little.

To explain the Sound Barrier and therefore gravity I had to find **space-time** not as some magic entity Einstein sometimes referred to but something that is everywhere around us and which we are part of. by dissecting Kepler's formula in relation to **valuing singularity**

You will read what exactly happens when the "sound barrier" is exceeded but also I will mathematically prove what gravity truly is. This I explain in simple understandable mathematical terminology. It is so simple even I can understand it. I make this remark because I am forever told I am too stupid to "understand Newton's physics" and therefore I am "unable to accept Newton's physics."

There is nothing to understand in Newton's physics because there is nothing to believe in Newton's physics. With one simple question I derail Newton's physics and that is prove tome there is mass, not weight but mass presented by the cosmos. The Sound barrier is gravity by implicating all four of the cosmic foundation phenomena that forms gravity.

My explanation shows you why the "sound barrier" is a cosmic limit. The pictures that I show might not have anything to do with the sonic boom, but the sonic boom is a very small part of the process of what physics think of in terms of breaking the sound barrier, which is a process of movement forming a relation between two objects moving in space at different speeds in example the jet and the earth where the movement of the jet is extraordinary.

There is always movement because of the earth moves and all other movement goes in anticipation of moving above what the earth does. Movement is space that is duplicating filled space at a pace and in relation to other space within other unfilled space. That is why using parachutes slows the falling process of falling objects down because falling is a ratio between solids and liquids.

This is proven in these photos about the sound barrier where the movement of the solid jet contracts the space of the liquid cloud. What happens in the photo is vivid proof of my theory and it proves the Coanda effect is what is forming gravity.

The Coanda effect is a relation between material spinning and air or liquid compressing and that is what happens between the Earth and the atmosphere. What this picture shows is that objects are always moving extraordinarily as they move in relation to the earth's atmosphere where everything that always moves independent of the earth's movement must move according to the earth when it moves in relation to the earth's gravity compressing space into forming the atmosphere. But it also shows the movement of the aircraft compressing space in relation to the aircraft moving as well.

This series forms as a unit with four individual titles forms a prologue to a Thesis that introduces a whole new concept about Creation.

The Sound Barrier is gravity. The Sound Barrier is the combination of four principles forming the Universe as the four are forming gravity. By explaining the Sound Barrier I explain gravity as I explain every aspect of law that forms the Universe we see. By not knowing how the Sound Barrier forms gravity and how gravity forms the Sound Barrier it is an admission of complete failure to understand physics. Do not confuse the sonic boom with the sound barrier. That would be to confuse a thunderstorm with a lightning bolt. That is what Newtonians do not understand and that serves then bad.

The atoms within a star are the stars. The star is only a container that holds the atoms within. The accumulative gravity forming a unifying gravity in the centre of the star which forms singularity allows the spin the star has. What is in the Universe has to move or else fall into the Black Hole the non-moving space will form. Gravity is the relative movement of space and moreover the difference between the compacting of circular movement of a star in relation to the cosmic expansion of space not occupied or controlled by material. The denser the star forms space around the star the more impressive the expanding of the Universe will seem and the bigger this difference is of contracting versus expanding. The atoms moving within the star form the movement of the star.

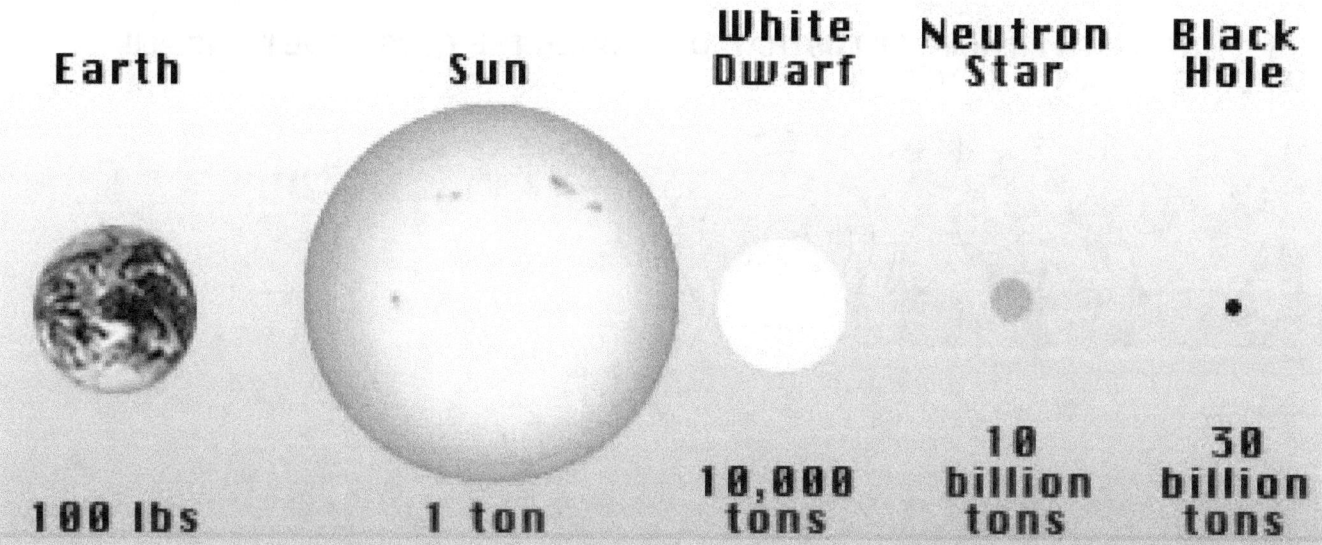

All stars are equal to the size that the atoms are. In the end in a Black hole all that is left of the enormous star that is much, much bigger than the Milky Way is one dot in the centre and it only constrains singularity as all movement is reflected to outer space. As the density increases the space the atoms apply becomes less because the movement that forms the gravity becomes more enormous in relevancy. As this happens the inside becomes more compact and later it is so compact that there is no longer room for an electron left to circle around the atom.. Then the atoms become solid as the atoms unify. I have all the calculations but it is in the Theses because it will cause this book to become too complicated.

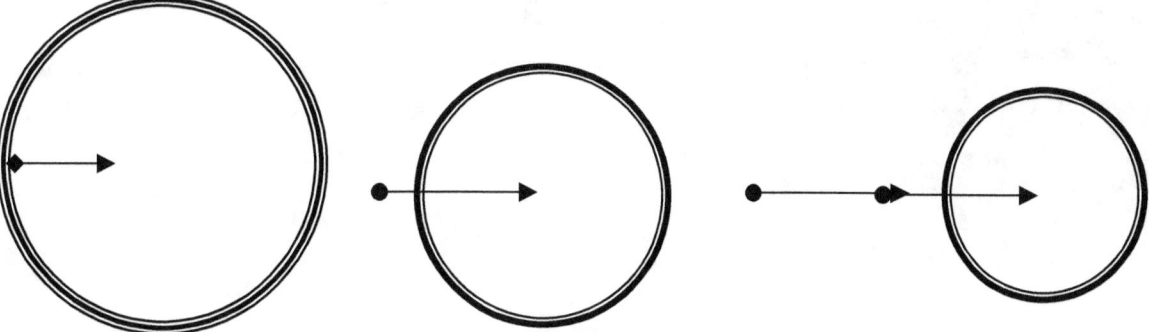

The smaller the star is the faster the atoms will displace in a linear direction because the faster the rotational circle will be that forms the gravity of the star.

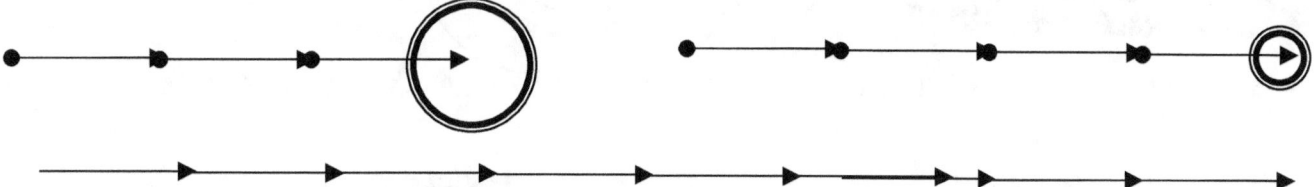

Movement iside a Black Hole forms only a centre dot because no space is left for any atoms to form.

The movement of the atom in relation to the star spinning forms a direct limit to the circle the atom can perform in that that limit forms the density that applies to the star and the space the star holds. That is why the stronger the gravity is the smaller but also the denser the star is. At a point the circle in which the atom spin becomes so small there is just no longer room for a neutron in the atom housing and eventually all the space used by atoms unite into singularity and only a centre holding singularity is left. This then is a Black Hole. This starts with the effect science titled the sound Barrier but the sound-barrier start when movement in a joint atmosphere starts and not only when a bang is heard.

Everything is related to movement and not to mass. The size of the star is not depending on age but it is depending on the gravitational movement of the collective singularity applying within the star.

GRAVITY WORKING AS THE COANDA EFFECT
THE MACH PRINCIPLE ILLUSTRATED THROUGH THE TITUS-BODE PRINCIPLE

STATIONARY GRAVITY UP TO FREE FALLING

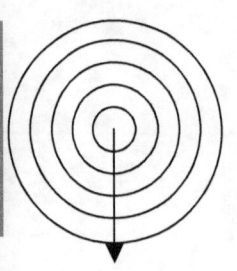

Titius Bode law stage 1

$$7(3\Pi^2) \times \Pi^0 - 2\Pi^0$$

MOVING GRAVITY UP TO LIMIT OF FREE FALLING

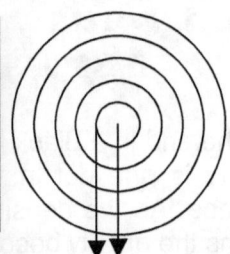

Titius Bode law stage 2

$$7(3\Pi^2) \times 3\Pi^0 - 4\Pi^0$$

AIR BORN SPACE-TIME MOVING SPACE-TIME EXCEEDING FREE FALLING

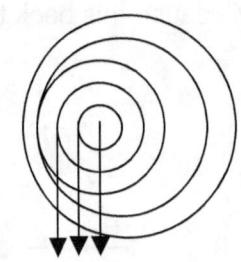

Titius Bode law stage 3

$$7(3\Pi^2) \times 4\Pi^0 - 4.5\Pi^0$$

AIR BORN SPACE-TIME TO MACH 1 ENTERING THE ROCHE LIMIT

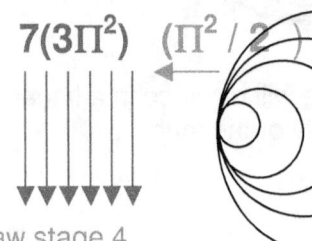

$7(3\Pi^2)\ (\Pi^2/2)$

$7(3\Pi^2) \times \Pi^2 / 2$

Titius Bode law stage 4

MOVING SPACE-TIME UP TO BREAKING THE ROCHE LIMIT BY CHALLENGING THE LAGRANGIAN POINTS

FROM MACH 1 TO MACH 3 $7(3\Pi^2)\ (\Pi^2/2)\ (\Pi^2/4)$

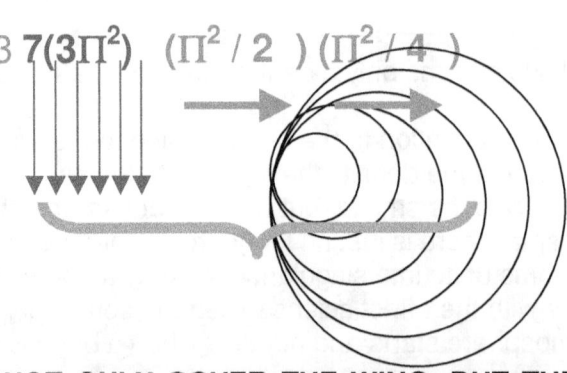

ABOVE MACH 3 WILL THE HEAT NOT ONLY COVER THE WING, BUT THE WING STRUCTURE WILL REVALUATE ITS COMPOSITION TO MATCH THE NEW VALUE OF HEAT THAT ARISE FROM WITHIN THE WINGS ATOMS AS THEY HAVE TO RELATE TO THE NEW SPACE-TIME VALUES.

The wing structure will occupy less space in accordance with the higher time value to which the wing submits. This will lead to a shrinking of the structure that will bring about uncontrollable vibrations and a lack of space-time support.

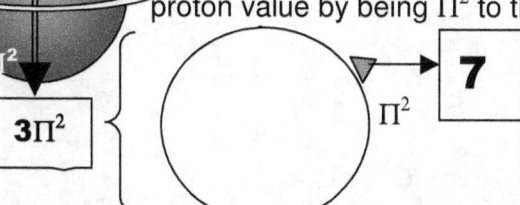

The earth, as in the case of all other cosmic structures divide into four quarters where every quarter takes a time position in the total displacement. In this the value of a star or a planet in the era we share, the iron era, the time comes to $7/10(4(\Pi^2 + \Pi^2))$ the 4 indicates a separate structure independent and in control of individual gravity.

In reconstructing the situation values must be placed in order to come into understanding. With the projectile fitting on the earth the earth overpowers the projectile claiming the singularity displaced in the governing singularity of the earth sharing the Earth as a proton value by being Π^2 to the earth being $\Pi^2 + \Pi^2$

With the projectile secured on the surface of the earth, the earth holds $3\Pi^2$, but the forth square the projectile holds in a position where 7 ends space therefore it takes on the value of being on top of the earth.

By standing still in a position secured to the earth projectile claims relating to the $3\Pi^2$ and that fixes the relation of the projectile relating to the earth. The relevancy of the position is $7(3\Pi^2)$. But since the earth is in a linear motion that motion also relate to the earth's singularity and comes into the calculation as Π^0. The instant the object moves individually from the earth the motion stands related with the linear value of Π^0 in the contexts of space coming from Π^0 to $5\Pi^0$. It is points in space in accordance with the centre holding singularity but because of the four quadrants the Lagrangian point prevents the movement to reach 5 points and therefore the Roche limit of $\Pi^2 / 2$ comes into force and apply

After the five is reached and surpassed the singularity manifests as a separate issue to that of the Π^1 and from that the "Sound Barrier comes about when the value reaches past the earth's singularity applying the projectiles individual singularity relating directly to the earth.

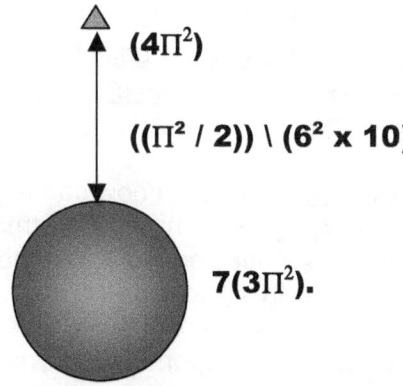

The $7(3\Pi^2)$ represent the relevancy the Earth holds in the separating relation.

The projectile claiming individual gravity also claims $(4\Pi^2)$ in the relation because it reached a gravity point outside that of the Earth.

But in order to overcome the earth gravity the projectile must accomplish the space and time between the position it had on earth and the position it holds in individual space-time as $(\Pi^2 / 2) \Pi^2) \setminus 3600$

In short: to overcome the space downward thrust enough heat must apply to expand the space in which the aircraft is to firstly overcome applying gravity at a rate $7(3\Pi^2)$ while also exceeding the Roche limit $(\Pi^2 / 2)$ and then this has to move the aircraft into an orbit $(4\Pi^2)$ while removing the aircraft from the space confinement of the earth $6^2 \times 10^2$ where this last **part of the information I am not divulging at this point because explaining the figures and make up thereof is rather extensive**

Has any one ever presented a remote answer why there are rings forming around planets and why the planets as big as they are, don't simply pull the dust onto the giant planet surface? Why would the rings circle the planet and not fall onto the planet? If Newton's mass principle is correct, the mass of the giant just has to pull the dust onto the surface.

Gravity applies by way of the Coanda effect. The Coanda effect employs the Titius Bode law and the Lagrangian points interweave one part of the Titius Bode law. The limit the Titius Bode law can arrive at before crossing the Lagrangian points the Roche limit limits it. All of this makes no sense and yet what I minion here is what forms gravity. It is not as simple as awarding mass but only the Newtonian science mind is as simple as awarding mass in order to fine all the solutions unlocking the Universe.

I have concluded the following from deducting

The rings are in place because gravity holds the rings in place.

Gravity has the value of Π and this has nothing to do with mass. This we see in every sphere and every circle that translate into objects spinning or objects orbiting or circles forming galactica far away. <u>I prove that gravity is Π and the four cosmic phenomena forms gravity by producing Π when objects move in relation to each other. The rings you see are in place because gravity is Π. The spherical planet you see is a spare because gravity is Π. The planet spins in relation to Π because gravity is Π</u>

By using the above **the four cosmic pillars**, it enable me to **present the proof** where I now can explain what conditions bring on the **sound barrier**. By proving it **is gravity** that the individual structure generates motion above and beyond the gravity the Earth provide is what is producing individual **motion** that the independent object earned within the sphere of motion that the Earth's gravity provides where the **independent and individual motion** put the relevance that gravity has beyond the conserving means gravity has where the space that is serving the independent object is independently in motion. The adding to the independence on top of the normal structural independence is creating more individualism by the independent motion of the individual structure being apart from the motion that the gravity of the Earth **provides**.

The rings around the planets are the remains of material that moved beyond the limit ration of $7(3\Pi^2)(\Pi^2/2)(\Pi^2/4)$ where the structure disintegrated into a mesh of dust. Exceeding the movement limit of $7(3\Pi^2)(\Pi^2/2)(\Pi^2/4)$ in relation to singularity Π^0 the movement went beyond the first limit of $4\Pi^0$ but could not reach the Lagrangian points limit of $5\Pi^0$ and then went into the sonic boom limit of $(\Pi^2/2)$.

Furthermore the orbiting debris that became dust exceeded another Roche boundary limit at $(\Pi^2/4)$ where the structure overheated and was annihilated into tiny vapour fragments. If gravity is about movement and not about mass what then forms a star and why does a star destruct? To see what is inside a star we must see what comes out of a star when the inside turns out.

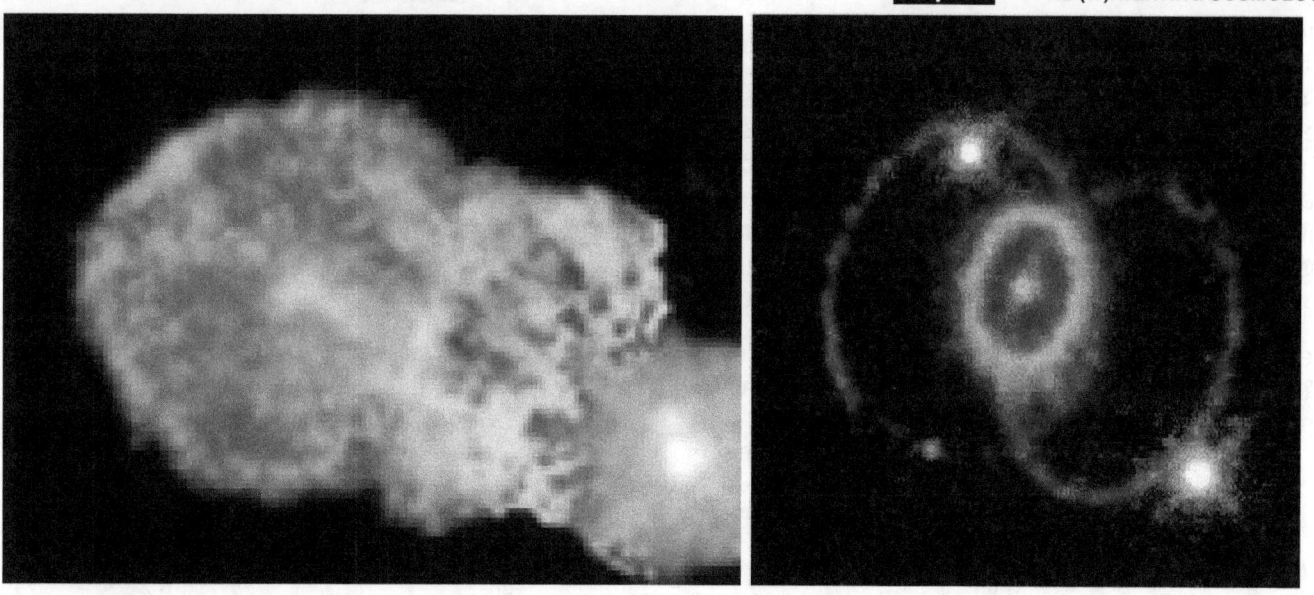

That what comes out of a star must be that which is in a star and what comes out of the star in this picture is liquid heat...then why is it liquid heat? The best Newtonian wisdom can come up with is as senseless as a nursery rime explained to toddlers. What you see in this picture is the Roche limit being violated and compromised and by compromising one of the four limitations that singularity puts in place a collapse of order within material bounded structures such as star bring total destruction to a structure such as a star. Stars are spheres filled with heat, not mass. By turning it compresses what is in outer space into what is inner space and by compressing the space gravity increases the heat levels. The heat does not rise because heat cools by expanding while it remains heat and heat rises in levels by compressing the space. That is what gravity is. When you observe what happens when a star goes bust it is clear that the event is due to a lot of heat escaping.

The star explodes when the spinning of the atoms inside the star no longer ensures the density of the star because the atoms inside the star can no longer form a collective adhesiveness that bonds the star into one solid structure. When the spinning of the atoms within the star no longer supports the star's spinning, the time or gravity goes array and the structure of the star expands into heat gone lost. The cooling of the structure and when exploding the star overheats does the spinning of a star.

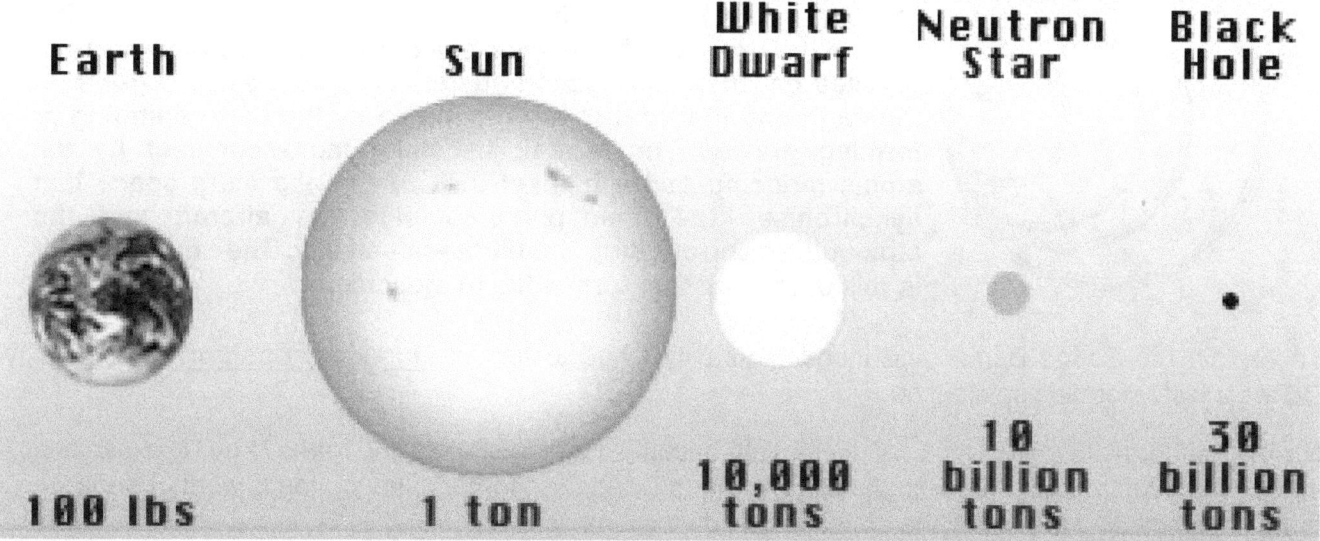

It is simple to come to a conclusion when logic deductions of reality are made. Material such as the picture shows weighing round about 100 pounds will weigh about 30 billion tons in a Black hole. I disagree with this number because I totally disagree with the value they use connecting the radius to the speed of light and my figure I get is another trillion tons on top of the official billion tons. I put this in to let every reader see that gravity is not connected to size but to absolute density. The smaller the actual size of a star is the more enormous the gravity is. Gravity works on density and I am about to explain this by using Kepler's formula when corrected from the deformation with which Newton corrupted it.

The escaping necessarily involves expanding of compressed heat because the gravity was eons of accumulating compressing heat into the star. If gravity is about contracting and the expanding is about a release of heat then gravity must be a systematic reducing of heat. If the star was only material bursting we would observe billions of structural particles flying all around but this is clearly liquid heat going into gas. When the gravity that holds a star is insouciant to keep a star bonded then that would imply that gravity is about cooling the star which enables the star to maintain structural integrity and when the balance it heat goes beyond the control of gravity the star overheats and it explodes. The explosion comes when the spin applying in the star can't retain the heat and when the heat in the star moves too slow it overheats and it expands.

Please believe me when I say you have never even seen the true physics behind whet drives the Sound Barrier or gravity for that mater. To understand what law in physics drive this other than Newtonian's Dark Matter.

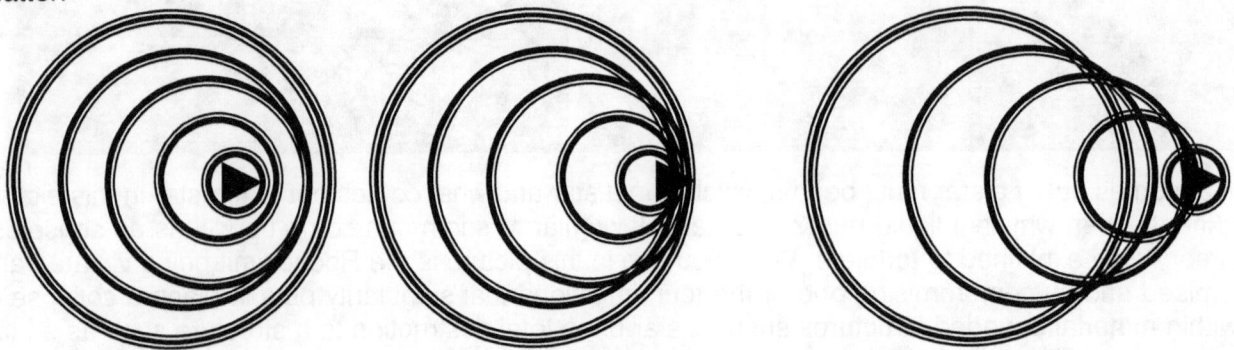

There are four cosmic principle or phenomena that forms the Sound Barrier which is gravity and to free yourself from the hoax of Newton's physics. The explanations may seem plausible but put Newton into it and the entire concept falls apart. It is not in the explanation of what happens that the answer lies but in the explanation of why this happens that true physics lie…and that Newton's thinking cannot cope with…. It is fine and well to give fancy sketches but explaining why and what principles are behind it Newton can never give.

As the sped of the object increases the electron ring that allows the space fro the atom to hold space will reduce in conjunction with Kepler's true formula of $a^3 = T^2k$.

When the aircraft moves at a certain velocity the atoms has to respond by spinning smaller circles that it would have standing still on the ground. The smaller space that the aircraft fill has to be filled because that space is determined by the atmosphere of the earth and that in turn is determined by the earth spinning or forming gravity. The aircraft becomes more compact by the atoms reducing and a blanket of heat fills the extra space that synchronises the atmosphere outside the aircraft and the atmosphere surrounding the earth we call sky. The sound barrier is this difference that forms due to movement.

To explain the Sound Barrier and therefore gravity I had to find the **location, position** of singularity $\Pi^0\Pi$ as a factor forming space-time.

No line that forms in the cosmos can be straight because being a straight line forms Π by Π^0 expanding into a line that is a circle. The point holds Π^0 going to Π and this is what every proton is worth. By the protons compacting the space the proton allows as it connects to singularity $\Pi^0\Pi$. In the centre of every atom is singularity. Every point that spins form $\Pi^0\Pi$ and therefore a line of protons spinning form $\Pi^0\Pi$ until a point where the last point hold $\Pi^0\Pi$ and no more spinning brings about singularity.

Then the object forming the last point holds mass or weight. The shorter this line of singularity by $\Pi^0\Pi$ is the smaller would the object within the structure be but the bigger would the mass be to much smaller objects. Then every point in singularity 1^1 holding singularity Π^0 refers to the centre line or axis of the star representing singularity 1^0. Because any number of atoms $(25789321)^0$ is 1^1 is 1^0 all the atoms are equal to the axis line of the star and in such a manner gravity forms.

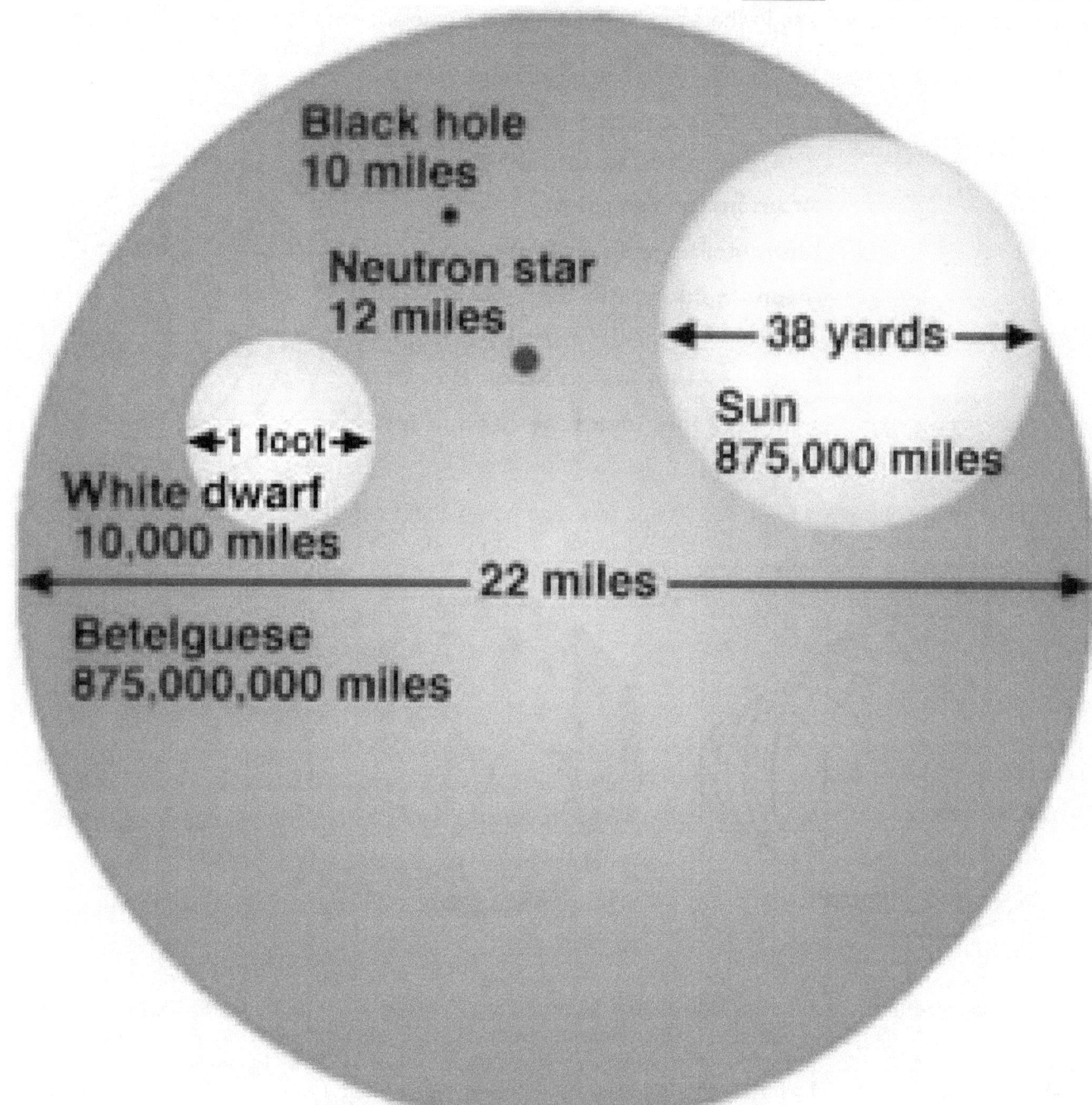

A star holds form by the gravity it has in a form of cooling the space in relation to outer space overheating. The gravity comes as a result from the spin of every atom within the star. The higher the spin is of every atom, the denser would the atom be because the more the heat forming a star would cool and reduce. This is a result of movement reducing the size of the atom within the star. When an object is in a location with little motion the duplication present a lot of heat because the distribution of the heat over the space in duplication has very little possibilities of spreading the overall heat over a wide area. The motion of something as small as the earth will confine the atoms into a relative hot area since the space in duplication does not reduce the extent of the heat by distributing the heat over much space.

In a structure with the size of the sun the motion of space is enormous by the sure quantity of space in need of duplication. Shifting that volume of space needs duplication that is millions if not billions of times more extensive than what the earth may produce. By duplicating such a vast area in a period of time within say the sun, reduces the individual atoms forming the star to a fraction of what the situation on earth would allow. The more the spin of the liquid is in relation to the solid state of space is reduces the space and extends the material in quantifiable measure many billion times over to what smaller stars are. It is not the space that holds the matter but it is the spin in relation to what the matter holds that puts the relevancy of hot and cold within the star. The more cold there is because of the more liquid heat bringing about motion, the colder would the atomic material be and the higher the relative contracting gravity that the star produces. This we see in the admitting of Mainstream science confessing that the reducing of space produces an increase in mass and because mass is the frustration of material unable to move, it admits to the fact that mass in volumetric size has no influence on gravity.

The size of any atom within the specific star according to the gravity or movement of the star

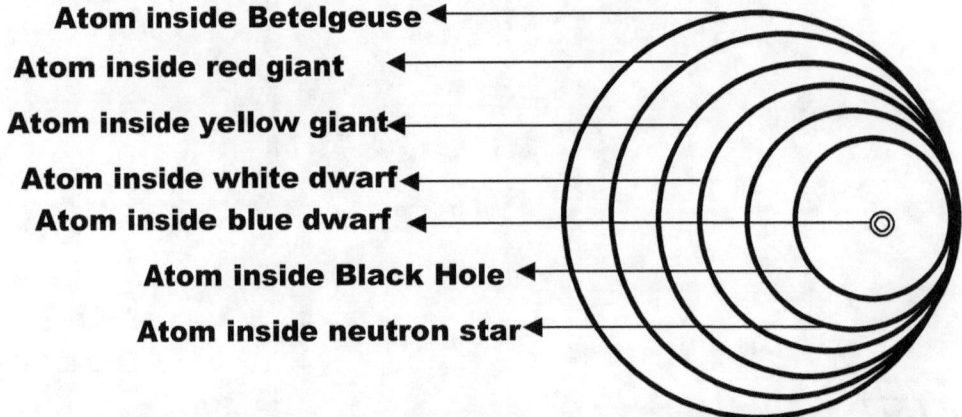

- Atom inside Betelgeuse
- Atom inside red giant
- Atom inside yellow giant
- Atom inside white dwarf
- Atom inside blue dwarf
- Atom inside Black Hole
- Atom inside neutron star

The physics we encounter on Earth allow us to use a common and a constant, a fit all and an all-purpose because we find us captures by the Earth singularity. The Earth provides the space we may claim as well as the time in which such material duplication will take place.

Our human orientated thinking in terms of science is as Neanderthal as was the Ptolemaic Universe was. The focus of the entire study or the Universe is sympathetic to life and the focus of all aspects revolves around the earth as the earth can sustain life. Science think of –270° as cold because humans connect it to cold whereas 650° C is hot because we roast at such a temperature. In the Universe what expands is hot when gaining space and what is cold shrinks as it retains heat. Movement shrinks material because space duplicates by movement and although having more space, the space then is spread over a larger area during the same time.

At the singularity level in which gravity forms the star in the sun, 6500° are colder than what any person with a human mind can realise. It is so cold that it freezes outer space into a liquid and it casts all heat out because it is so cold there is no place for heat to be.

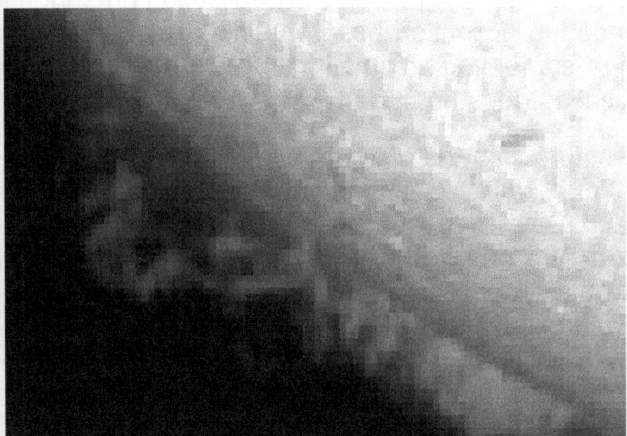

The picture one can see from the above is not of something hot but is of something cold. When putting any solid frozen "gas" into water the reaction we find is the same reaction as what we see on the sun. The liquid we see squirting out of the sun is cosmic gas, the substance filling outer space, being frozen into a liquid form. The sun does that by moving and by moving it is the coldest place in the solar system.

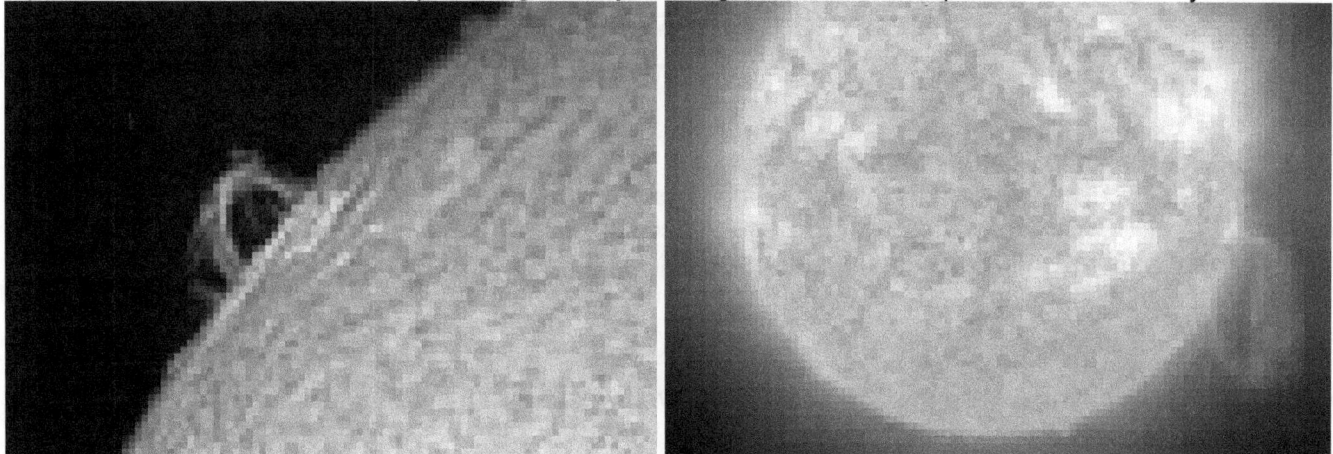

What the sound barrier shows us about gravity is that space reduces the faster movement re-allocate the next position of the object ravelling. This brings about that the atom in the neutron star shrinks so much there is no longer room for the neutron and the atoms then discard the neutrons. This has nothing to do with mass or does prove that mass has nothing to do with forming gravity in whatever way.

Let's get Newton into what forms reality. Newton said $a^3 = T^2$, which is saying a three dimensional object is equal to a square. What this mathematically implicate is saying that I in three dimensions are equal to my mirror image and therefore being $a^3 = T^2$ I can swap places with my image in the mirror because we are the same. The poor man obviously did not know the first principle about mathematics and yet all Newtonians are following Newton sheepishly.

Kepler said $a^3 = T^2k$
Then it says $a^3 / T^2 = k$
And $a^3 / k = T^2$

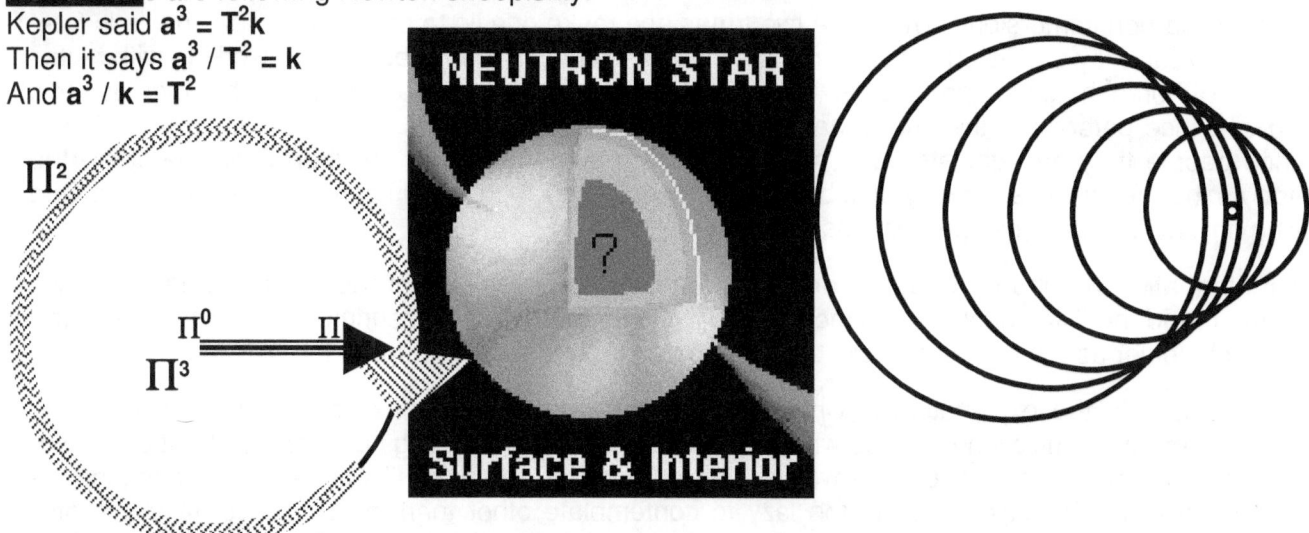

Kepler said $a^3 = T^2k$ which mathematically says that the size of space formed equals the movement T^2k. Then the circle $a^3 / k = T^2$ size of space is formed that is equal to the linear the movement k and also the circle size is directly responsible $a^3 / T^2 = k$ space travelled in one full rotation period of the star.

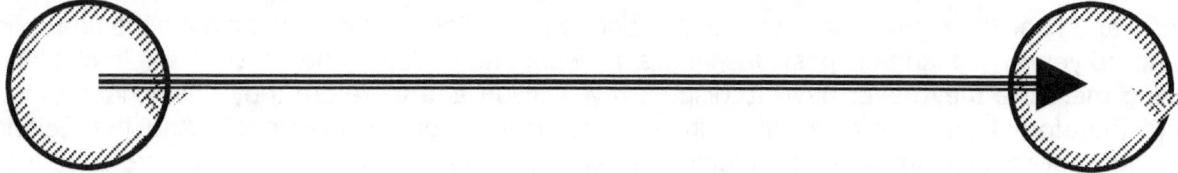

Moving a full rotation in a short period would bring much gravity in a small atom. In the Black Hole there is only one atom left that holds the size of infinity while outer space represents the entire eternity. That is why a Black Hole has no space because all the space is compromised by singularity Π^0 and outer space forms the value of space or Π.

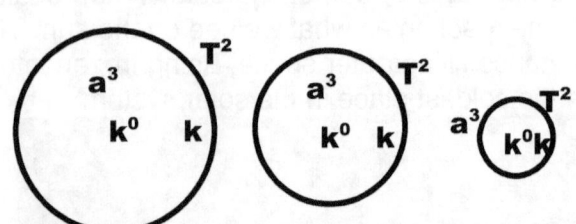

The faster the atom spins the further the atom displaces allocated position because the bigger the gravity is within the star. Gravity has nothing to do with mass and all to do with movement of material within the star.

When out Newtonian falls our Newtonian would fall at the speed as anything else would fall. This evidence we see on TV everyday in advertisements and on reality shows. We see blokes jump out of aeroplanes with rug sacks and tennis balls and the lot through each other with the balls and catch the balls while putting on the rug sacks and tasking off the rug sack all the while the lot is getting in and out of a car as they please.

We have all seen this. The car drops evenly paced with the rug sack, the humans and the tennis balls. Then the Newtonian tells the student that according to science everything is dragged down by gravity in accordance with mass. See the Newtonian fall. This Newtonian is the only one that falls by mass

The latest is that there is some little particle undetected as yet or as this goes to print (just to cover my arse in case they discover this particle in the meantime) scientifically named and called "graviton" that "pulls" other things that "pulls" back. Science rumours their profession that science only work on established facts and nothing but facts proven to the point of no contest could other wise be good enough.

This expression is so burned into the minds of everyone that science can say any rumour and because it is science everyone believes and never disputes. One good example is this Global Warming by carbon immersions. They take two truths and make one lie to convince people the lie is true. People take what they say as undisputed truths because science is renowned for only working with facts. What a lot of horseshit that expression is, that science only work with proven facts and that they use this idea to convince persons in the public that that is why the lot are atheists. They use this expression to convince people they are atheists because God is not a proven quantity but then they propagate the graviton as the measurable quantity, It is because they only work with facts! I can much easier prove God by using physics that they can prove mass by using physics.

All this information we find by valuating and then reading the conclusion of the sound barrier correctly. The heat around the flying airplane connects to the plane by the atoms that shrink in size. This makes the atoms much colder as the heat the atoms discard locates around the structure of the star.

This is how simple the explanation is when studying the unfolding event of a star going supernova. Gravity can't be mass infliction because what comes from a star when things go wrong is heat escaping. What comes out of a star must be that which is inside the star and what fills the star. But because we have a bunch of mathematical drones too lazy to contemplate other than to accept we have this prehistoric view Newton brought along as science departed from the Dark Ages about mass forming gravity.

They can't think and that is why they can't read my work. They can only compute in terms of mathematical calculations and even in that they are completely incompetent in researching the truth.

The cowards (and I call them cowards because they dodge the bullet to purposely maintain the corruption of three hundred years) in academic positions ignore my work and therefore they never have to challenge my views and therefore they never have to dispute my position and therefore they never have defend Newton and therefore I die in silence while they never have to prove anything. ...And the Science Conspiracy Network remains in place, undetected, erect like a mile long penis ready to screw the living daylights out of a brainwashed human race for another 3-hundred years. Ask any person if he or she thinks they have mass and they will think the person asking the question is bonkers and yet show not weight show me mass. Again I challenge any Newtonian physicist to show me any evidence of mass anywhere in the Universe. Movement reproduces space during any one specific time-duration and that holds the ratio of space-time physicists refer too so eagerly. It is not a magic word such as anti matter.

The Universe works on the principle of movement forming space in a time period in which heat is confined to the control of singularity or not under the confined control of singularity.

Material is heat confined to spin holding a specific space. The space in which material is when it is in an atoms spins above the sped of light and the electron securing the area which confines the heat to material spins at the same r ate as the speed of light while everything in the Universe other than the atom spins below the speed of light.

When an object is in outer space that object, encounters a specific relation with what we presume is space. This comes about by motion and through material volumetric size. The space the object encounter by moving through outer space puts a value of a ratio between the space it moved through and the space moving through which Kepler introduced as $a^3 = T^2/k$. That means there is a contact ratio between space containing and space contained by. When the atom is in outer space the atom is surrounded by a temperature of zero Kelvin and that is because zero Kelvin is the presumably the coldest any temperature can get. Being zero Kelvin on the outside and with zero Kelvin being the coldest temperature there can, it would make the atom also zero Kelvin on the inside since there can be no colder than that. That would mean the entire atom is then zero Kelvin. The zero Kelvin responds the movement of the space in which the atom is and has nothing to do with hot or cold because in fact being at zero Kelvin places the atom at the hottest being as expanded as space in material can be. This comes as a result of absence or the lack of movement in outer space. Being at the most expanded means the lack of movement produces criteria in which the expanding proves space is heating up the material

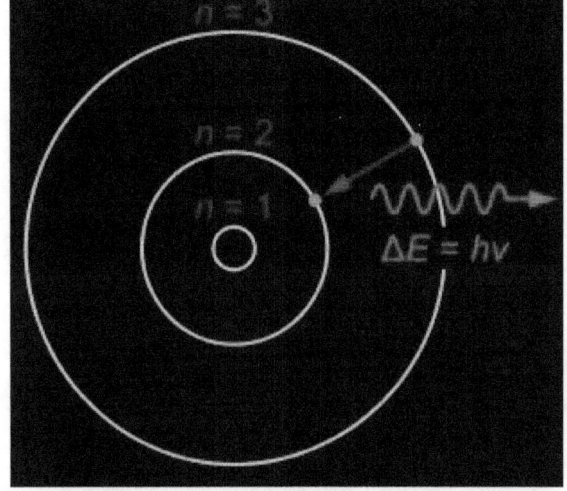

However applying motion reduces temperature and there is much motion going on inside the atom. That means the fact that zero Kelvin produces the coldest there can be makes a little nonsense of such a statement. When the atom is 40^0 C the outside of the atom must affect the inside of the atom because from the fact of what the Balmer and the Lyman series would represent and that proves that the outside temperature of the atom does influence the inside temperature of the atom. The normal summer's day temperature on my farm is 40^0 C normally in the shade because at that temperature little in loony enough to venture outside the shade. We consider that the atom must be 40^0 C because that is what the daily temperature is outside the atom. We feel and experience the 40^0 and we presume that all around is suffering from the heat of 40^0 C.

We know that the action brings about a reaction and the actions leads to a response. If the atom heats on the outside by measure that it finds a need to reposition the electron by one band, then also the inside got smaller in relation to the growth by one band. We associate such repositioning with the heat on the outside that amplify or reduce. However the adding of heat brings on a faster flow of liquid, which results in higher motion and it is in the motion that we find the answer to the cosmic principle applying. In the cosmos there is no hot or cold. There is higher or less motion. The relocating of the electron into a new position where the electron jumps a band is done by implication of the Coanda effect. From the Coanda effect we know that the liquid attach to the solid using the formula $T^2 = a^3/k$ where as space identify new boundaries by identifying the allocated boundary set by the liquid as $k^{-1} = T^2/a^3$ where the space then forms the limit at $k = a^3/T^2$. Every time the motion of the liquid intensifies the motion will attach to the solid by applying a new relation, which alters the relation of the solid by extending the space the solid has differently.

However we must not lock our focus on the heat but we must refocus on the motion that intensify or weakens. It is the motion that produces the new electron allocation and the motion produces a heat that establishes a cold. The focus is on the motion because the motion brings on accelerated duplication and accelerated duplication produces cooling that brings on a relevant cold within the atom. If the temperature on the outside of the atom changes from zero Kelvin to 40^0 C it is not the temperature that changes but the atom is responding to higher motion. With the atom in outer space the atom is subject to lesser motion since the atom is only in distinct and personal orbital motion in relation to the sun. That is why the atom can be subjected to zero Kelvin. When the atom is within the boundaries of the Earth and circling around the sun in a location set by the singularity of the Earth, the motion is distinctly more than what it would be if the atom were located in outer space.

The outside of the atom calls for a direct response to condition inside the atom since the outside can change very little if the inside does not respond in an opposing manner to what the outside produce. In such a relevancy there are always three factors performing as gravity and in that is the Coanda effect in charge of committing the standards by applying the gravity or the motion in relation to the solid. The material revolving through the space holding the material and allowing the material the privilege of motion is in the amount of material per time frame that makes contact with the space which serves time and that it encounters as the space duplicates its position it holds coming from the past through the present into the future.

The movement reduces the size the material occupy by duplicating such vat amounts that the duplicating freezes the material into the oblivious.
Show me why the "mass" does not pull the rings into the planet?

The rings are a circle directional change of 7^0 at a point where gravity would be the strongest in forming a circle Π…and that would be at the equator where the value of Π will bring the strongest circular movement $\Pi^0\Pi^2$. Then the line follows a circle $7^0\Pi\Pi^2$ and at every interval of Π^0 to $4\Pi^0$ a line is in place. Looking at a sideways picture of the planet we can also see rings forming as $7^0\Pi\Pi^2$ forms a line with the axis and the axis is what keeps $\Pi^0\Pi$ and the entire integrity of the star in place.

When an object forms part of the structure as in having mass the object then identifies with the singularity at Π by rotating Π^2 and adhering to the change in direction of 7^0.

When the object does not connect to the structure the object cannot have mass because the space in which the object is serves as liquid and in that way it holds relevancy to the spin formed by the axis holding singularity at 3. The relevancy then is $7^03\Pi^2$ because the movement is in relation to the centre of the spinning structure.

The rings we see indicate every point singularity holds in terms of any individual $\Pi^0\Pi$. Every ring indicates another point where singularity locates another relevancy that holds singularity Π^0 in relation to singularity of the ring that forms Π. Every ring was a satellite that went to close and beyond thee Roche limit and the gravity destroyed the satellite leaving the particles in relevancy with the planet.

Sometimes you will wait a time for some of the documents or a page to open because they are extensive but you will lose less time in waiting than you lose by unwittingly learning the deception and then afterwards finding out you were deceived. It will still be quicker waiting on the response than losing the time wasted when you are being tricked by the Academics in science holding positions in Astrophysics and Physics. Sometimes the reading is slightly extensive but to find the truth you must dig deep and overcome corruption there is a lot of material that came in place over many centuries and it is the lie that detours from the truth that then is extensive.

PROVE ME TO BE INCORRECT IN ANYTHING I SAY!

Science in the present and for the last three centuries placed all their focus on material. In placing the focus on mass while mass does not exist as a cosmic entity, science has been running around like a chicken without a head and the truth eludes them even after three centuries of lying and corrupting science. You think that this is harsh words, read and you will see those in science deserves much more that just that. Those in science concentrates on the material filling the space but gravity is about the movement of the space and not of the object in the space, but of the space the object takes with as the object moves faster than the other space. That is why objects all fall equal because the space the objects fill is the same, unconditional of what forms the material. All objects fall at an equal pace without size or mass becoming a factor. That is why I introduce gravity by a very new set of principles that no one ever heard of.

When an aircraft goes through the sound barrier, the object fills more space per time unit by going faster through space in time. The object then in accordance with the space it holds stretches because it holds more space than when it went slower or when it stood still. Since the aircraft is solid and does shrink a little but not much, it has to concentrate the space around the aircraft and by concentrating it reduces the space. As the space concentrates through the movement of the aircraft a cloud appears around the aircraft because the water vapour in that space concentrates into a cloud that forms. All movement is part of the sound barrier because the sound barrier is the movement of space. However the sonic boom is confused with the sound barrier but the sonic boom is a small part of the entire sound barrier and only fills the centre spot or the middle in the sound barrier. If mass is anything then show me the role that mass plays in the sound barrier and how does mass conform into what we think of in terms of the sound barrier. Gravity forms as not by the mass of what any object presumes to have, which I prove in this book and in all my other books, is clearly not present as a cosmic factor. There is no such a thing as mass in the entire Universe and the conspirers that conspire to keep Newton's fraud disclosed, cover up this reality with all they have.

I challenge anyone to prove where do they find proof of a factor such as mass. I found the place! It is in the imagination of physicists while they try to conceal that mass is only a product of Newton's imagination. They conspire to confuse everyone with the implication of weight. Weight there is but things pulling other things by the measure of their mass is a daydream and the thought could be funny if it wasn't that crooked. Gravity forms by redirecting the movement of space in circular flow by the spin of any cosmic structure that has the ability to do so. The aircraft goes straight while it also circles the earth and that forms movement within the earth's confinement by the earth's movement, which is what forms the sound barrier. The sound barrier is gravity and sciences present way of going on about the "Doppler effect" and "Mach's principles" shows how incredibly little those incompetent physicists know about physics. I challenge the lot to prove Newton by showing that the Universe contracts or where mass plays a part in cosmic physics. Show how the planets hold their positions according to the mass they have. All objects move straight while at the same time circle around some other object. The object moves in a straight line while the spin the object hold diverts the space in which the object is into a circle. This circle brings the value of gravity to Π. By diverting the space in which the object is as well as the space surrounding the object, gravity concentrates space into becoming denser by circling and also hotter.

Gravity forms as the earth or any other cosmic body rotates by 7 °. By diverting the straight-line movement by 7° a contraction forms in a circle. In my books I prove how this then brings about the value of Π by implementing the law of Pythagoras and gravity is the law of Pythagoras.

The reclining of space by redirecting the direction of travel from straight ahead to 7° reclines the space in a steady and sturdy flow. It is the space reclining or contracting and the space contracts albeit filled by solid material or empty of solid material. This is the reason why all things fall equally. It is the space moving down with or without holding material and the space has the same density in relation to the solid cosmic structure rotating.

In the centre of the rotating body singularity forms to the value of Π^0 and this extends to the curve forming Π in terms of the curve of the rotating object. This then forms part of gravity moving forming

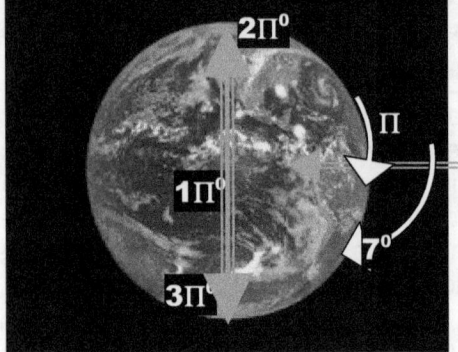

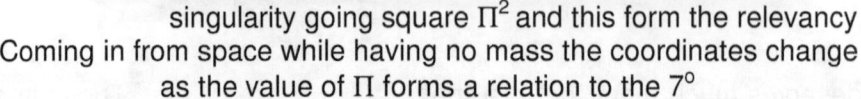

The extending of singularity goes from Π^0 to $5\Pi^0$.

singularity going square Π^2 and this form the relevancy Coming in from space while having no mass the coordinates change as the value of Π forms a relation to the 7°

When an object comes in from the atmosphere or flies the sky the object associate with the turn giving it a value of the rotation in line with the axis which puts a value of 21.991 / 7 on Π or on gravity. This is gravity and this is how gravity functions. It has nothing to do with mass in any way, shape or form and there is no factor such as mass in the entirety of the cosmos apart from being in the imagination of the Newtonian.

The entering by singularity that goes from 7° by $3\Pi^2$.

From these conclusions I prove that gravity is the result of four cosmic phenomena interacting to form the value of Π which by movement becomes the value of gravity Π^2 and gravity is equal to cosmic time applying. In order to understand the development of the cosmos and moreover the start of the cosmos and the progress in the cosmos as the cosmos formed, one has to understand the measure of Π. One has to microscopically dissect the measure of Π to find the cosmos in measure. One has to understand where 7 fit in Π. The fact that Π is 7 at the bottom and that 7 relates to a double value of 10 is a key issue. It is behind Π that we will find the four phenomena, which I named the four pillars performing as gravity as they form gravity. It is by the actions of Π that the Universe develops. The Hubble expanding goes by implementing gravity as Π in the square through the four pillars on which gravity and time rests. It is behind Π we discover the meaning of singularity and how singularity forms the absolute and only building block as a form that forms the Universe. It is in Π we find the Cosmic Code unlocking the meaning of the Universe. Time is centralised in Π^0 that forms Π as space's limit that becomes space by gravity being Π^2.

What is in the Universe, is spinning and therefore what I am referring to, applies to everything holding a place in the Universe and therefore this which I mention directly links everything holding any space whatsoever in the entire Universe to one single point around which all spin, notwithstanding the allocation. In the **precise middle** of all **objects in rotation** disregarding size is a precise centre dividing the object into opposing sectors that will **start the spinning initiation** from that centre point. The spinning object will have a very specific **centre point that does not spin** and only holds Π as a specific value because no radius can apply at the point being one space away from Π^0 holding Πr^0. But also the one value such a line **cannot have is zero** because the line **is there and being unbroken, it holds contact** with the rest of the material bringing about that **zero does not start any** line and therefore the **value of the line must be infinite**, just as described in **accordance** and by **the definition of singularity.** As I am introducing a very new idea, I wish to explain in better detail what I try to convey. While anything spins, singularity forms a line and when reducing the rotating line or radius progressively to the middle at one point all further reducing must end. As the rotating direction moves inwards, the rings forming Π will become smaller and smaller. Then we reach a point everyone thinks of as being the axis around which everything rotates. The line only forms when everything around the line spins by establishing a circle to the value of Π.

Gravity is defined as a force that is present in mass pulling mass and it is that entire idea that there is not evidence of. When I refer to gravity everyone grabs on a cultural notion of a concept they formed and in that concept they link the smallest part of the concept to the become and represent the overall gigantic principle and by knowing one line everyone has the opinion that anyone then is the absolute master on the idea of gravity. When I freeze any substance the substance contract to a liquid and with more cooling it contracts to a frozen state of ice. The gas expanded more than what the solid did because the gas is hotter than the solid is. When we form the opinion that the outer space expanded to the limits the idea springs to mind that outer space is freezing cold. When I say the Sun freezes hydrogen to a liquid because my eyes see the liquid squirting from the Sun I am dangerously mentally impaired since the Sun is blistering hot. Then through this culture my effort to say gravity is motion and motion is the cooling of an overheating and thus expanding Universe goes wasted. Every one has the opinion that where gravity is the strongest such as the case is on the Sun or the centre of the Earth, such a place is extremely hot and where gravity is least that place is unbearably cold.

==No person ever could understand the sound barrier== because no person ever could ==enter singularity== and see ==what applies when the Universe goes singular== You will read what exactly happens when the "sound barrier" is exceeded but also to explain the "sound barrier" I will mathematically have prove what gravity truly is by proving the measured value of gravity. This I explain in simple understandable mathematical terminology. ==It is so simple even I can understand it.==

==I now explain gravity again and relate this explanation to the sound barrier functioning.==

Gravity forms by association or relevancies that holds a value of coming towards the centre from the sky which then is 7(which is the curve of the earth) (Π associating with the curve of the earth) and Π^2 (associating with the moving of the earth) This is part of the Titius Bode law forming the measure of gravity which is not by mass but by forming Π.

==$7\Pi\Pi^2$ x Π^0 to $4\Pi^0$==

==Then when a line is drawn from the centre of the earth another value comes in place referring to the centre of the earth Π^0 and the curve of the earth Π. The value of Π is then shared by both disciplines (association from the sky to the centre ($7\Pi\Pi^2$) and association from the centre to the sky ($\Pi\Pi^0$). In reality it is ($7\Pi\Pi^2$)== but I am not explaining this.

==If the body does not connect to the earth by way of forming a unit or forming "mass" the axel line of the earth comes into affect giving the ratio a value of 3. Then when not connecting to the earth surface the ratio becomes $7(3\Pi^2)$ x Π^0 going up to $4\Pi^0$ all depending on the sped or the height. This is the way gravity forms by forming a connecting association with $\Pi^0\Pi$.==

==This forming of gravity explains the Coanda effect because the Coanda effect puts air or liquid in relative movement with solids and this movement (but not the mass part) is how gravity forms.== **The condensing of** the vapour surrounding the jet is a prelude to the sonic boom and is just one link in the sound barrier where the sonic boom is just another link in the sound barrier process. The sonic boom IS NOT the sound barrier but is as much part of the entire concept as the beginning of movement is part of the sound barrier. Waves in the sea are as much forming part of the sound barrier as winds howling or blowing is part of the sound barrier.

PROVE ME TO BE INCORRECT IN ANYTHING I SAY!

To whom it may concern and all others reading this document:

By applying my principles I take the reader back to the point where the Universe started and show why the Universe started because we can see it from how the Universe started. The big key unlocking the secrets is in the Titius Bode principal, the Lagrangian system, the Roche limit forming the Coanda effect These four principles is what forms Π where Π forms gravity. These phenomena are in place used by the cosmos in every picture, but since the phenomena corrects Newton's failings, the importance of their role is very well hidden. I show Newton's ideas are not what the cosmos uses. I show what the cosmos uses are there for a reason. Those physicist ignore what is in place in the Universe used by the Universe, discards what the Universe uses in favour of Newton where in the Universe there is not one point of evidence that the Universe applies what Newton said forms gravitational principles. With what is in place and by applying what the cosmos uses to apply gravity, I am able to show the place, the very point place we find why, how and where the Universe started.

That it does by giving gravity a very specific value. From such explaining what Kepler said without Newton changing formulas on Kepler's behalf I prove the Titius Bode principal also known just as the Bode principle conjoining the Lagrangian system of object formation. I explain how singularity forms the Roche limit and how singularity brings about the Coanda effect but most important of all Kepler showed me where to search for singularity.

My achievements came from my effort where I separated Kepler's work from the opinion that Newton formed about what he saw Kepler's work should contain and gave to the world his Newtonian concept about Kepler's work. For instance from Kepler's work I can explain the operation of the Black Hole, which not even Prof. Stephen Hawking understands. From my view a force is just motion applying and that is what Kepler said gravity is. Kepler said $a^3 = T^2 k$. The Coanda effect is the establishing of individual independent space a^3 by applying motion $T^2 k$ in relation to a centre point the motion of the liquid establishes. Einstein came to this conclusion but failed to refer his view back to Kepler and by not referring to Kepler missed the point he wanted to make. Just by my studying of Kepler this became possible.

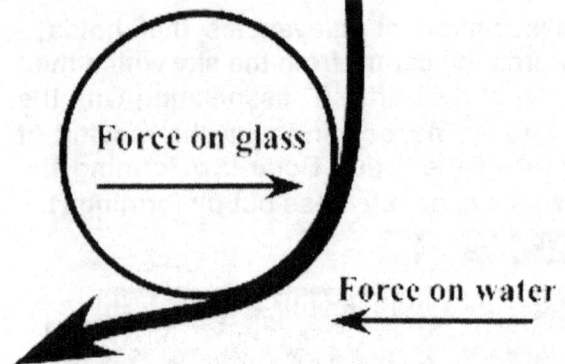

That shows that all movement is gravity or time related. This shows that when objects move then such movement of whatever is moving is moving extraordinarily because everyone always forgets about the fact that it is the earth that normally applies all of the movement while all else stands still in relevancy. It is not my manner to speak ill of the brain dead or the dead by other means, but in the case of the Newtonian academics I am left with no option. Their forces haunt me to death and it is their forces and ghosts and witchcraft I have to fight. The lifting of a body comes quite natural when a certain speed is exceeded. By exceeding $7(3\Pi^2)$ the body will start to lift no matter what the mass is. A 747 Boeing of multi tonnage lifts off spontaneously at excess of that speed. Newtonians are forever concerned with middle ages and with forces they can't explain but such forces and witches there are not, therefore they do not have to fear and can sleep well at night. In the sketch the circle portrays a glass and the arrow portrays running water. The Coanda effect is the water that does not drop straight down but follows the curvature of the glass.

The picture clearly shows the 7° inclination of gravity to the value of contracting $\Pi^0\Pi$. This is gravity! This is the most vivid example of the Coanda effect and it is what gravity is! It is a whirl allowing the flow of liquid space around a solid centre in relation to the centre holding singularity or $\Pi^0\Pi$ as it contracts space into a denser liquid. The Coanda effect is gravity and my explaining this statement is part of many other books in which I explain what gravity is. The Coanda effect shows how liquid attach to the solid by $7(3\Pi^2)$ and the solid attach to the liquid by a relevance value of $7(\Pi\Pi^2)$. That is gravity.

Gravity is defined as a force that is present in mass pulling mass and it is that entire idea that there is not evidence of. When I refer to gravity everyone grabs on a cultural notion of a concept they formed and in that concept they link the smallest part of the concept to the become and represent the overall gigantic principle and by knowing one line everyone has the opinion that anyone then is the absolute master on the idea of gravity. When I freeze any substance the substance contract to a liquid and with more cooling it contracts to a frozen state of ice. The gas expanded more than what the solid did because the gas is hotter than the solid is. When we form the opinion that the outer space expanded to the limits the idea springs to mind that outer space is freezing cold. When I say the Sun freezes hydrogen to a liquid because my eyes see the liquid squirting from the Sun I am dangerously mentally impaired since the Sun is blistering hot. Then through this culture my effort to say gravity is motion and motion is the cooling of an overheating and thus expanding Universe goes wasted. Every one has the opinion that where gravity is the strongest such as the case is on the Sun or the centre of the Earth, such a place is extremely hot and where gravity is least that place is unbearably cold.

No person ever could understand the sound barrier because no person ever could enter singularity and see what applies when the Universe goes singular. You will read what exactly happens when the "sound barrier" is exceeded but also to explain the "sound barrier" I will mathematically have prove what gravity truly is by proving the measured value of gravity. This I explain in simple understandable mathematical terminology. It is so simple even I can understand it.

This remark I make is in response to physics academics not understanding something as elementary as the "sound barrier" which I prove is the most basic principle of gravity but they are forever telling me I am too stupid to "understand Newton" and therefore because of my lack in intellect I question "the validity of Newton". To them and to those that says this I say: Download what I give free of charge and then start to get wise... What it is that these two phenomena has in common except sharing a Universe because he should know, after all he is the Newton-physics expert ... but I will bet you although being a physics expert, your professor will have no idea...mainly because he is the Newton-physics expert. This prevails because of singularity, which clashes with Newton head on.

This is my introduction and this is my declaration:

Gravity is forming the sound barrier because the sound barrier forms by four cosmic principles that form gravity and therefore gravity and the sound barrier are the same principles in conflict because of movement differences. All movement results in material cooling and when material coos it becomes denser and more compact.

As I confront science dogma and principles, nobody is willing to publish my work. I have to walk the road alone and fight the battle by my private effort without any support anywhere. If Newton is the problem one have to go pre Newton to find the problem. The problem is that when looking at Kepler's table then if there is $T^2 \div a^3$ according to the table matching a column, then mathematically $T^2 \div a^3$ must be k^{-1} and where k^{-1} goes negative it shows space reduces time. It shows space in volume goes single by movement of space and not objects.

Planet	Mass per Earth unit	k^{-1} Movement	a^3 of space volume	T^2 During time units
Mercury	0.06	$T^2 \div a^3$ =0.983	(a^3)= 0.059	(T^2)= 0.058
Venus	0.82	$T^2 \div a^3$ =0.992	(a^3)= 0.381	(T^2)= 0.378
Earth	1.000	$T^2 \div a^3$ =1.000	(a^3)= 1.000	(T^2)= 1.000
Mars	0.11	$T^2 \div a^3$ =1.000	(a^3)= 3.54	(T^2)= 3.54
Jupiter	317.89	$T^2 \div a^3$ =1.000	(a^3)= 140.6	(T^2)= 140.66
Saturn	95.17	$T^2 \div a^3$ =0.999	(a^3)= 868.25	(T^2)= 67.9
Uranus	14.53	$T^2 \div a^3$ =1.000	(a^3)= 7067	(T^2)= 7069
Neptune	17.14	$T^2 \div a^3$ =0.999	(a^3)= 27189	(T^2)= 27159
Pluto	0.0025	$T^2 \div a^3$ =1.004	(a^3)= 61443	(T^2)= 61703

Use the column to show where it is mass that produces any valid factor. In the table columns we have the mass of the main solar objects and according to Newton the mass is responsible for movement. If it is mass that does the pulling in ratio of the radius then we have to see some evidence of this applying somewhere. Is there any person that can prove that claim from the table showing Kepler's movement in relation to Newton's mass? In what way odes Jupiter move faster or how does Pluto move slower

A More Complex Commercial Science Book   416   Chapter 8   RE-(W)RIGHTING COSMOLOGY

according to mass? If mass did the job, how is it done? Notwithstanding Newton's blatant mathematical manipulation, mass don't apply.

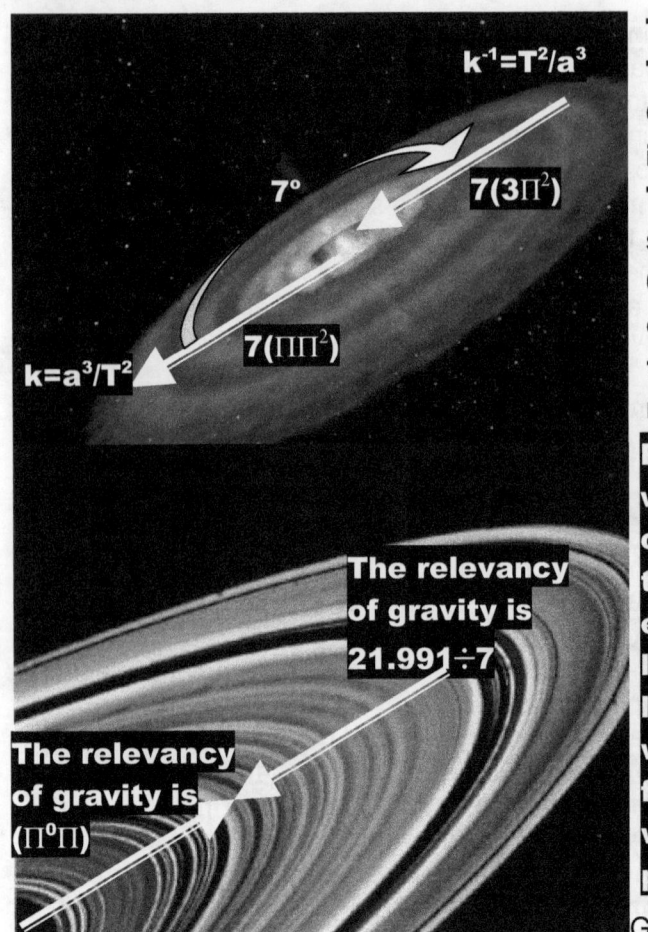

The Universe is contracting by $7(3\Pi^2)$
The Universe is expanding by $7(\Pi\Pi^2)$
Gravity is the inclining of material spinning in relation to 7° that is part of Π
This is because Π is 3.1416 and spinning in space is 3 which makes time expanding 0.1416 x 7 = 0.991 larger than 3. The concept might appear simple when told in this manner but the entire philosophy is so much more complex when studied overall.

By observing the rings around the planets we can identify all four pillars forming the composition thought of as gravity. We have the Lagrangian points that is five and I explain why it is 5, we have the Titius Bode law and I explain in detail the Titius Bode law, we find the Roche limit and I explain why it has the value of $\Pi^2/4$ and we can see from my explanation I have provided above why the Coanda effect accumulates all the principles forming gravity.

Gravity has the value of Π. The value of Π has two measure where one is when the line gravity forms extends from singularity as $(\Pi^0\Pi)$ and when the line that gravity forms extends to singularity it is $21.991 \div 7$ because of the curve gravity associates with forming the value of Π To explain all these factors I have to take the reader back to the point where the Universe started as spot that formed a dot. I have to take the reader back to where the point where the Universe was numbers holding no space, which is way before the Big Bang era where the atom broke the Universe into space that formed.

Gravity forms as the earth or any other cosmic body rotates by 7°. By diverting the straight-line movement by 7° a contraction forms in a circle. In my books I prove how this then brings about the value of Π by implementing the law of Pythagoras and gravity is the law of Pythagoras. The reclining of space by redirecting the direction of travel from straight ahead to 7° reclines the space in a steady and sturdy flow. It is the space reclining or contracting and the space contracts albeit filled by solid material or empty of solid material. This is the reason why all things fall equally. It is the space moving down with or without holding material and the space has the same density in relation to the solid cosmic structure rotating.
In the centre of the rotating body singularity forms to the value of Π^0 and this extends to the curve forming Π in terms of the curve of the rotating object. This then forms part of gravity moving forming singularity going square Π^2 and this form the relevancy
Coming in from space while having no mass the coordinates change as the value of Π forms a relation to the 7°

When an object comes in from the atmosphere or flies the sky the object associate with the turn giving it a value of the rotation in line with the axis which puts a value of 21.991 / 7 on Π or on gravity. This is gravity and this is how gravity functions. It has nothing to do with mass in any way, shape or form and there is no factor such as mass in the entirety of the cosmos apart from being in the imagination of the Newtonian The atmosphere is liquid. Don't as a Cosmologist for they will most probably tell you it is extending nothing. Ask any Engineer and the engineer would tell you when you work with the atmosphere you work with a liquid that has a density 600 times less dense that water but it holds all the characteristics that a liquid has. This picture shows the Coanda effect and while the Coanda effect is the most important principle in physics I know that less that one in a thousand that reads this knows about the Coanda effect.

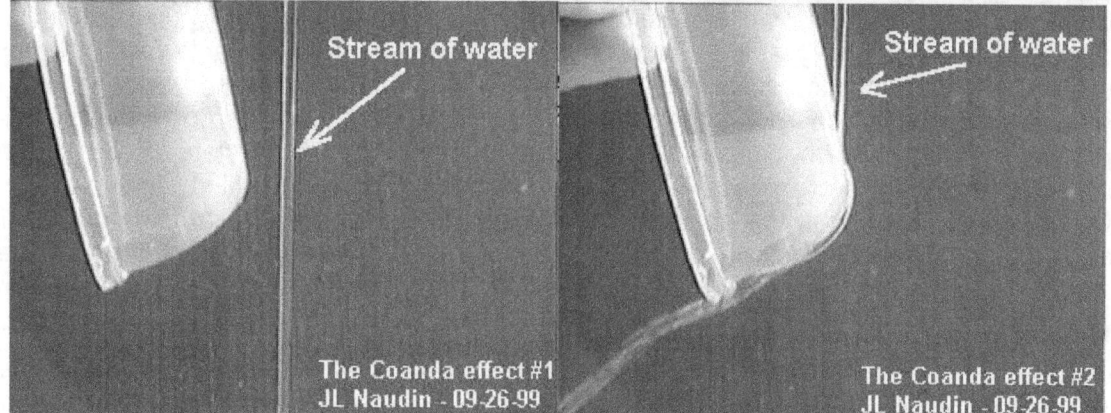

The water follows the surface of the curved shape, this is the Coanda Effect and the Coanda Effect works with any of our usual fluids, such as air at usual temperature, pressures and speeds.

The picture above is about a phenomenon called the Coanda effect but this is never mentioned in any physics handbook because Newtonian physics-religiosity is unable to explain or to understand this principle.

When liquids flow past a cylindrical object the liquid clings to the surface of the object rather than follow the "path of mass" and fall straight down to earth. Gravity is about the atmosphere that forms a liquid that is the same as the liquid running around the solid circle called the Coanda effect.

Gravity is the movement of the air in relation to the movement of the earth. If an object moves within the parameters the earth set the liquid to move around the earth and the object has the ability to maintain the speed, it will be just more liquid floating around the earth at a specific height fore filing a specific circle or rotation requirement. If the speed drops the object will fall notwithstanding mass or the lack thereof. The entire idea vested in gravity focuses on speed-differences. So what the hell has mass got to do with the entire affair of gravity?

If the circle is as large as the sun is, then the compressing of space turns the air into squirting liquid precisely as the picture of the sun shows. Again I ask what has mass got to do with the entire idea.

The movement of space filled with material cools by movement of that which moves and this puts thermo differentiation between space that moves and space that does not move. As the space differences by thermo differentiation grows it would seem that one part grows in relation to another part shrinking. This is why large gravitational stars always seem to lose space while the Universe seems to expand.

But before I can commence with that task I have another duty administer: I AM ABOUT TO WARN EVERY PERSON IN SIGHT OF MY WORK ABOUT MY SLENDER ABILITIES.

Therefore in the light of what the most respected academic group on Earth accuses me of, I therefore have to issue a most serious warning to any person with the intention of making some kind of inquiry to the content this book holds, then the most concerning matter involving any content within the pages of this book you hold are that you must please seriously consider that where the stating declares the possibility that the content in this book has been (written by...) then don't take the announcing Written By Peet Schutte (Petrus S. J. Schutte) very seriously for there are grievous doubts leaving considerable dispute about the possibility, which underwrites the authenticity of Peet Schutte achieving the (written by...Peet Schutte) status. Please take note of the following dehortation. In the light of the reference to me serving in the capacity as being responsible for authoring, (written by...) in line of keeping fairness and justice to members of society, where all civil beings should carry reputed honesty, then: Please be warned before any reader starts reading about the following extremely serious admonition: I am bound by my conscience to warn all intended readers that I am placed under caution by the Academics in Physics. Those most esteemed members responsible for the guardianship and maintaining the ethos in physics are of the opinion that I, Peet Schutte, am unable to write any book on the science of Physics as well as Astrophysics. Therefore I, Peet Schutte, must declare that I should be considered as not very able to write anything, because I am incapable thereof. I suppose, I merely generate new information, which I establish as thoughts and then gather as concepts. I further collect the result as words, which I put on paper using alphabetic symbols. I then compile that in a format that others may confuse with a book, but a book it cannot be, since the Masters in science found me unable to write a book. But before you go further and follow my arguments, I first have to level with you about how academics view me in the position I hold. Please do not allow me to fool you, for this then cannot be, or represent a book. Now I have done my duty in warning everyone and in that, I denounce further participating with any purposive intention to wilfully bring down the crux of civilization by acting unacceptable and irresponsible.

I didn't write any books since I am not schooled to do so. It is my guess that I merely generated uninformed thoughts, which I collected as alphabetic symbols and plotted that in ink on paper. This effort I achieved from harbouring my delusional ideas spawned by a dehumanised brain. It only proves my weak and under developed mentality, due to my lack of an informed insight that is a typical symptom that all those have that is suffering from a disadvantaged past that one can only have when the person obviously lacks formal education. While you are reading the letter deciding to regard or dismiss my work, then also please keep in mind when reading my language used and also please give credit where it belongs…if you do find linguistically improper use of words or misspelling, then remember that I am a feeble minded motor mechanic and not a literal giant. I do find much pride in my status as being Afrikaner and would like to have my names used by pronouncing it in the manner Afrikaans dictates…therefore I would sincerely appreciate the courtesy when readers will take note that my name and last name are pronounced in Afrikaans, which is originally from Dutch and must be pronounced that way. Peet one would pronounce "here" which is the closest English to the pronouncing of the "ee". The "Sch" in Schutte is pronounced exactly as school is where both actually are pronounced Skutte or "skool". By pronouncing my name in Afrikaans you do me the utmost courtesy any one can. Being an Afrikaner is what I am most proud of.

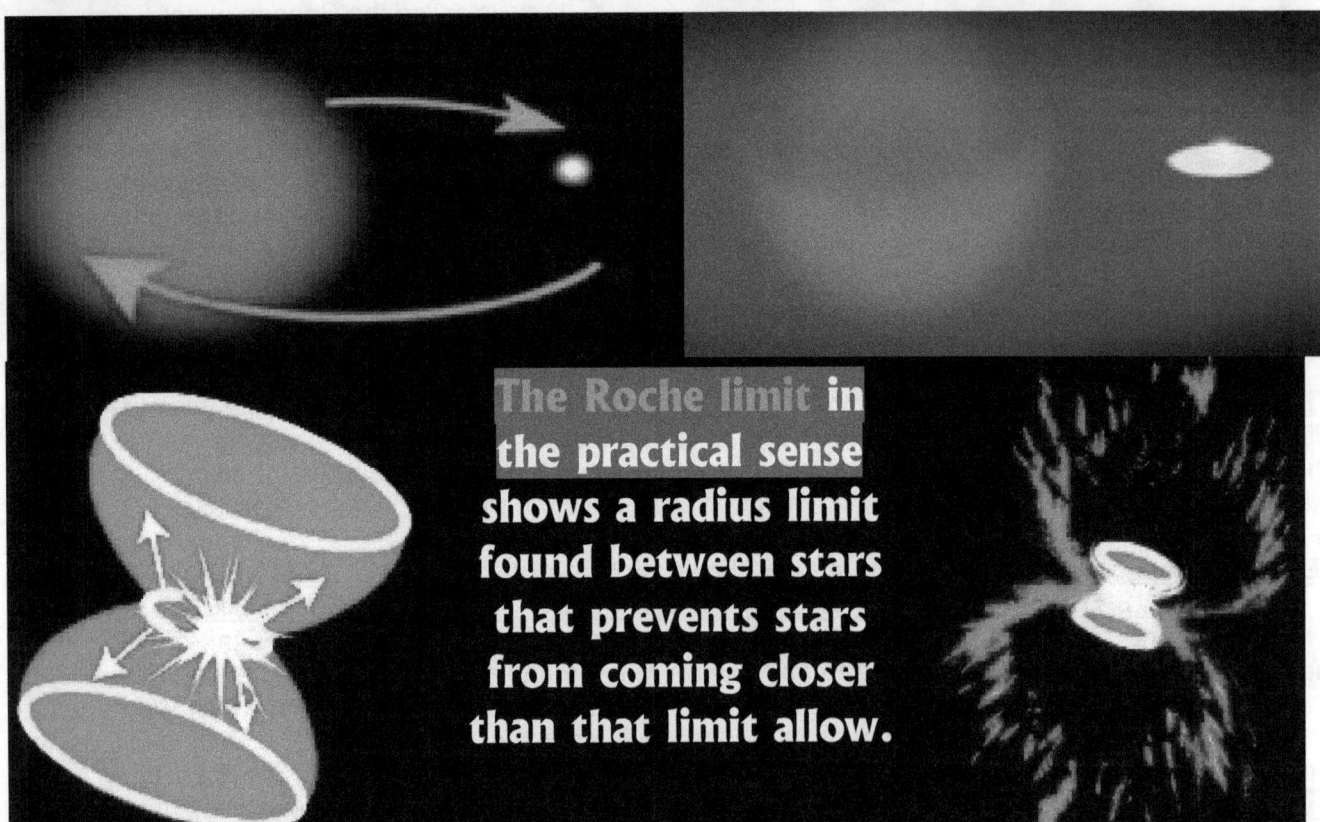

The Roche limit in the practical sense shows a radius limit found between stars that prevents stars from coming closer than that limit allow.

I submit article to well known physics magazines but my articles are rejected on the most unappeasable grounds and for the most outrageously ridiculous reasons the Newtonians can think of.

One such an article I may use because I said I was going to use the material as an open letter I gladly show. I submitted an article in which I show what the manner is in which gravity conducts movement by means of singularity.

I sent my new concept just as I showed it in this book to publishers but with much more explaining detail. It would be the biggest understatement to say that I was not very well received as one of the publishers e-mail shows. Notwithstanding the detail supplied and still I am ignored like Michael Moore was when he tried to enlist the Senate in the Iraq war. This is the biggest cover-up mankind has ever devised and let one academic prove it is not a cover-up as much as any academic prove that Newton is absolutely beyond suspicion and as faultless as the academics in physics are blameless.

The following is the response I received when I submitter a paper on how gravity forms by singularity forming.

Dear Dr. Schutte,

You submitted an article of 15 pages to the Annalen. The content of this paper doesn't constitute a theory in physics. With a lot of words and some simple algebraic relations, there is no way to "explain" the world of physics. You seem to be out of touch with modern developments. This is also shown by the fact that you don't quote any relevant literature.
I am sorry to say, but the Annalen is not able to publish your work.
I am sorry for having no better news for your.
Best regards,
Friedrich Hehl

Co-Editor Annalen der Physik (Berlin)
--
Friedrich W. Hehl, Inst. Theor. Physics
* University of Cologne, 50923 Koeln _____/_____ Germany
fon +49-221-470-4200 or -4306, fax -5159
hehl@thp.uni-koeln.de, http://www.thp.uni-koeln.de/gravitation
* Univ. of Missouri, Dept. Phys. & Astr., Columbia, MO, USA

Dear Prof Friedrich W. Hehl, I have received your e-mail reply and I wish to respond on your letter. The article of 15 pages to the Annalen had in mind to introduce a very wide-ranging concept contained in many books. I wish to promote books in which I introduce a much larger and much more detailed cosmic picture. It is four books that actually form four volumes of one theme supporting The New Cosmic Theory. I wish to unveil a totally new approach to the thinking in cosmology. The concept is proposed in the article I sent to you which is "revealing" The New Cosmic Theory In the article as much as the theme I wish to go where no one ever attempted to go before. I introduce the Universe of singularity, a state in which the Universe still is because it is a state from which the Universe grows. It is where material in a dimensional dynamic does not apply because it is where Einstein said "the Universe goes "flat"". I show you how and where the Universe goes "flat" I will guide you to the point where I go…so that you may see where my books and the article lead you. It is in the domain of singularity.

When you read work about the Big Bang you have to go right down the development (in reverse order) to the point where the Theory of the Big Bang points at a spot named singularity. It shows the very start from where all material developed. At that point one will find The Absolute Relevancy of Singularity and there has never been any attempt by any person ever to venture beyond the dimensional birth of the cosmos, which is called the Big Bang by going into the era where singularity prevailed. I take you there in my books as well as the unpublished article. However, going there requires a very high degree of concentration and calls for understanding that a very little number of persons are capable to show. I try to show how the Universe goes "flat" as Einstein said the Universe goes "flat". Even by completing this unimpressive letter you will also know how the Universe goes "flat". Even where you failed to read the article I sent you, then by just reading this letter you will be able to find where singularity takes the Universe "flat". But it requires a mental capacity to understand because where I venture no one ever in the history of mankind reached into before. I do not speculate but even in the unpublished article I show with pictures and sketches as well as "some simple algebraic relations" where to go to where the Universe starts, but you failed to read that because you are opinionated as to what conditions should the Universe have before the Universe will allow any one into physics. That is a pity. One should learn from the cosmos and not tell the cosmos what it must be to qualify as the cosmos. Then in the article I show you by almost taking your finger to the spot, the very point where the Universe ends and that too I qualify. You might dispute my arguments and show me about what you disagree, but it shows very little understanding of reason on your part about qualities man should have before understanding the Universe. I go into a

Universe that was in place before light was in place in the Universe and only darkness prevailed because light calls for space and in that era of singularity space was not even a thought yet. I show why the Universe goes "flat" and in a "flat" Universe only the value of 1 holds value since singularity is 1. If you can understand 1 or $5^0 \times 7^0 \times 3^0 = 1^1$ you have all the mathematical skills required to understand the applying concepts. To reach a value of 1 does not require big mathematical equations but to reach singularity requires 1.

I have realized the extension of the Einstein hypothesis, and concluded the other part of the relativity theory. In short, the part of "gravity" that Newton saw, and the part of "gravity" that Einstein saw is not the same identical thing, but two different values to the same "gravity". That is why Einstein concluded his theory on a single dimensional Universe. The part of "gravity", which Newton related too, Einstein did not see. "Gravity" consist of a single dimension value (Newton's gravity), which is part of (not the same thing) the "gravity", that Einstein saw. These two values are contained in a balance in space-time. In this, the Universe is in a three-dimensional state of space (three dimensions) in time. Therefore, space is not flat, as EINSTEIN SAW IT, but has three dimensions, locked in singularity where singularity is single dimensioned but by directional relativity forms the fourth dimension of time. I am not going into space-time in this book and I am steering clear by a country mile of any arguments about space –time in this book but Newtonians knows less about space-time than what my dog knows about communism.

In this, there is no force, but only an ever-altering balance, which forms the time component. Only the energy value we regard, as "life" is a force, because only "life" can inflict a change in the balance of space-time and this forms a force. That is why science has been unable to recognize the value of "life" as a separate energy form. If you regard the Universe driven by a force, the value of life has to disappear. The Universe is in a state of balance, and the scale of the balance is determined by time. In this aspect lies the reason why time has never been detected. Only life can upset the balance, but then life will run into upsetting time. When time is upset, we see it as an explosion. The more time is upset, the bigger the explosion will be. I have put everything into dimensions:

1. Newton's "GRAVITY" which is the single dimension. It is a single value, starting nowhere and stopping nowhere.

2. Is Einstein's "gravity" which is a wave that is forever, circling forever by reducing and expanding.

3. These two values above are combined in the third dimension of space, the third dimension we, and everything in the universe, find ourselves in. In this, the first and second dimensions form an inseparable combination.

4. Everything in the Universe is moving. Even the part we relate to as "NOTHING" moves. This movement forms the fourth dimension of time. Time contains the third dimension, which is formed by the combination of the first two dimensions, and dictates the tempo, or balance in the first two dimensions, and which forms the third dimension. As all things the Universe are part of the balance in movement all things stand equally and evenly effected in the third dimension every thing is connected by movement (time). Therefore, everything connects to time and by time, from the smallest part of matter to the Universe as a whole. When space becomes less, the movement becomes more therefore time becomes more.

7. As life are connected to the movement of time, but can also move independently from the balance of time, life is a separate energy value, and is in the fifth dimension, but part of the first four dimensions. Being part of the fifth dimension, life is subjected to the criteria dictated by the balance of the fourth dimension, and therefore has to abide by the laws of the fourth dimension, while it is part of the fourth dimension. In all of this is the outstanding factor that nothing happens by chance, except for the intervention of "life".

Doppler made his contribution to science at a time when the train he measured was slower than a horse. The statistics he left to science applies to going as fast as a man on a horse could ride. That is a far cry from a jet breaking the cosmic boom. They can claim what they like because they never have to prove anything and they never have to listen since they know everything while their shortcomings makes science more the jesting of a buffoon. That annoys me to my guts because all of that comes down to the conspiracy whereby they whitewash their stupidity from the brainwashing they inherited from their predecessors and pass the brainwashing on to their students. Their ignominious stupidity makes me want to shout to the mountains in agony and unbelievable frustration. When a cloak of arrogant self-righteousness hides their incomprehensible stupidity under a cloud filled by their belief in self-importance it becomes hideously loathing.

"GRAVITY IS DIVIDED IN TWO FACTORS, BEING LINEAR DISPLACEMENT (Π / Π^0) WHICH IS WHAT NEWTON'S GRAVITY IS AND,

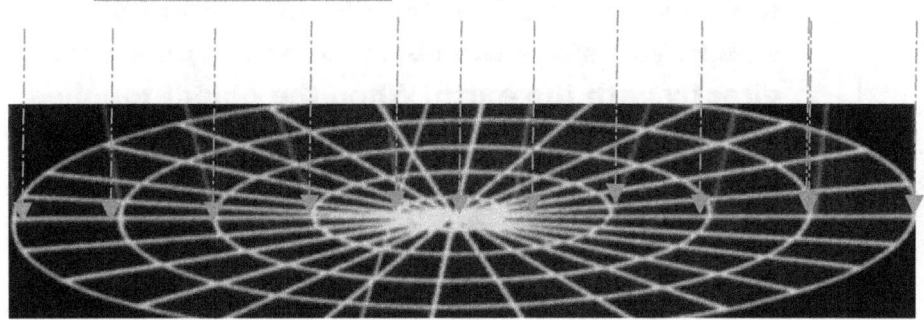

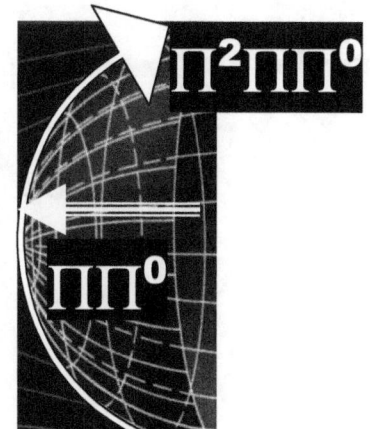

CIRCULAR DISPLACEMENT (Π^2 / Π)

WHICH IS THE "GRAVITY" EINSTEIN RECOGNIZED

Those that are of the opinion that it is mass that "pulls' the object towards the earth might be a little wiser when reading the next part carefully as I explain the following:

Anything entering the earth from space it travels in a straight line. The mathematical formulating of this I do not provide at this stage but it is another "some simple algebraic relations" as Prof Friedrich W. Hehl from Annalen Der Physics put it (and I also explain this a in the following chapter). But using the law of Pythagoras and putting 7 in relation to 10 provides the value of Π. The value of Π has two different dimensional values where the one is Π and the other is $21.991°/7°$

Any object entering from space and that does not have direct contact with the surface of the earth, encounters gravity in terms of being directionally diverted from going straight to turning towards the earth by 7°. This then is what provides the "pulling of mass" and the other is having contact with the surface of the earth, which then puts Π in terms with the centre of the earth at Π^0.

What becomes clear from these two illustrations is the following: Where there is a direct entry by the spacecraft, the time factor is not sufficient in duration.

This will not allow the time-duration needed for this structure of the aircraft to revaluate its space occupation an therefore the space factor of the spacecraft cannot adopt to its new position in space-time occupation as it has to relate to a new value in accordance with the value that is determined by the earth's concentration of space-time.

In the case of the second entry illustration, a lot more time lapse is allowed for the spacecraft to revalue its structural position in accordance to its new value of space-time occupation. This is the very reason why objects like "falling stars" burn out when they enter the earth's atmosphere.

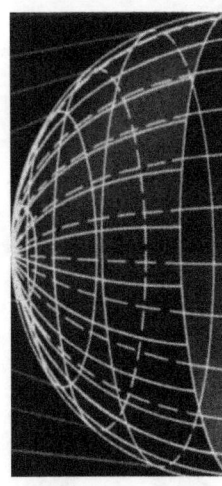

Forget about mass forming gravity it is a fool's rhetoric. When an object holds a steady position to the earth while making a sound, the sound will go in all directions evenly. It will go left and right equally fast. The concept of gravity connects to roundness and to Π. Gravity is Π. Gravity is the movement in terms of Π by duplicating the position of Π per specific time units applying. Then the object making the sound moves left it will hasten the flow of sound in the direction it moves by moving towards the sound while it will increase the distance the sound has to travel by also moving away from the departing sound going to the right side. This means that gravity has two values one being linear and the other being circular and the linear affects the circular in movement as much as the circular affects the linear. This is not only connected to sound but is connected to everything applying to gravity as gravity. When an atom moves the size of the atom must shrink to compensate for the directional change of the orbit that decreases the electron's orbit and therefore decreases the size of the moving atom where atoms form an object.

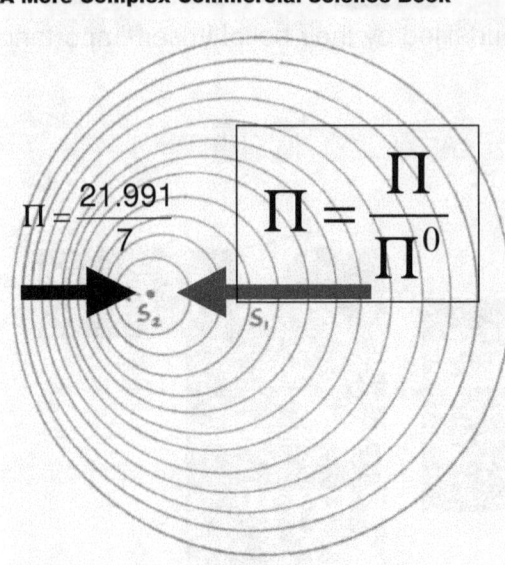

Coming toward the earth from space the object travels straight but the spin of the earth by 7° re-directs the object to follow an inclining line of 21.991 /7. At such a point the object moves towards the earth but the object cannot have weight yet, since the object does not connect directly with the earth. When the object touches the earth it relates to $\Pi\Pi^0$ and at such a point the object receives a value of mass by which it becomes a unit within the earth. By connecting to singularity in terms of $\Pi\Pi^0$ or in terms of $\Pi = 21.991 /7$ puts the object in liquid or in solid.

CURCULAR ATMOSPHERIC DISPLACEMENT $\Pi \setminus \Pi$

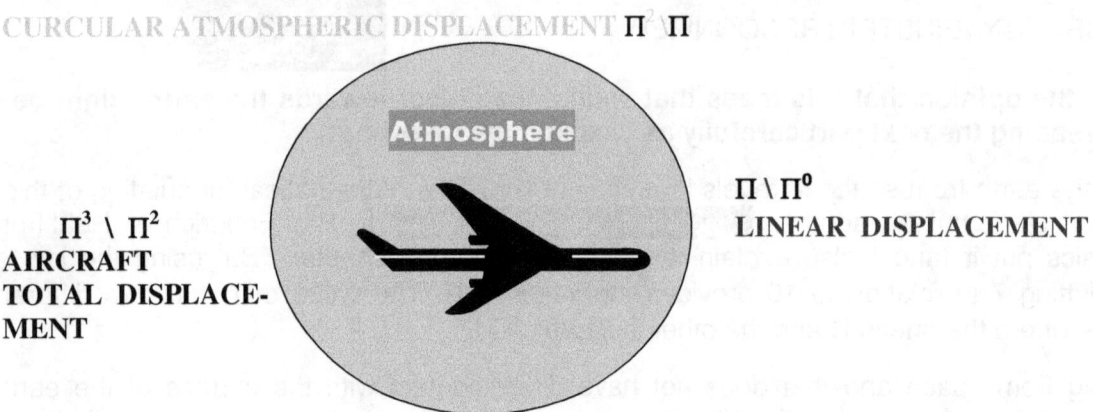

This principle applies to the sound barrier but also to the object moving. As the object moves faster the effect the movement has on sound would transform the structure of the aircraft as well as the movement alters the atoms forming the aircraft. The structure shrinks because the movement shrinks the molecules as the atoms shrink that make up the molecules that make up the structure.

BETWEEN MACH ONE AND THREE Π^3 / Π^2

ATMOSPHERIC CIRCULER DISPLACEMENT $\Pi^2 \setminus \Pi^0$

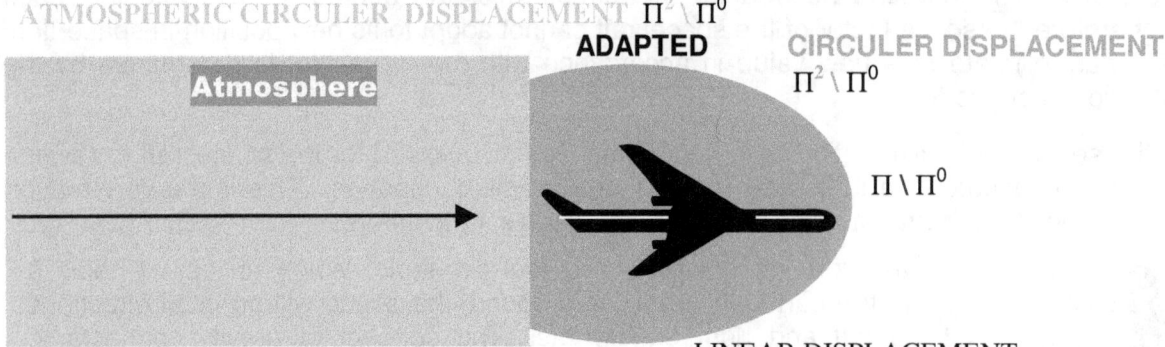

When something moves it is within the space the earth rotates. When anything moves above and beyond the earth's movement it excels as it exceeds the movement the earth provide and with it, it takes the space it holds in excess of the position the earth relates too. Thus by moving it form a space-time unit within the earth confinement but yet also out of and beyond the space-time the earth provides.

Between Π/Π^0 and Π^2/Π^0, the body finds itself in its circular displacement, which is the earth's linear displacement value. When an object exceeds the earth's circular displacement value, it is refer to as Mach 1, or the speed of sound. Objects fall at a rate of $7(3\Pi^2) = 208$ km / h while the earth gravitational displacement is $7(\Pi\Pi^2) = 217$ km / h. because the earth moves faster than anything can fall it will always reduce space and have all surrounding space confined to the earth. The object falling only holds space that is confined to the earth.

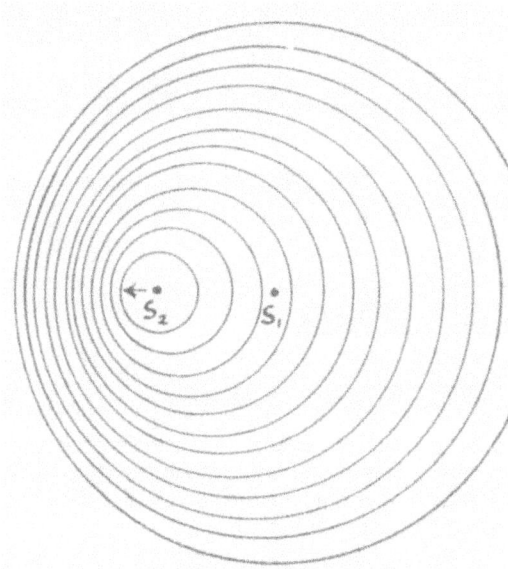

When the object or aircraft stands motionless according to the earth centre the object moves only by the movement of the earth and therefore in cosmic terms the object and the earth is one whereby the object has weight and holds a value in mass. The object holds a relative position to the earth by the value of Π. This is because gravity then links to Π^0 and this holds regard to singularity. When the object moves while being connected to the earth by singularity the earth holds space at Π^3 while the object moves in terms of $\Pi\Pi^2$. The earth represents Π and the object moves by Π^2. When the object gets airborne the relation becomes $7(3\Pi^2)$ but I will not go into that. As the speed increases the relevancy of linking Π^2 breaks when the Roche limit is exceeded. The Roche limit is Π^2 or movement divided by the four quadrants of the circle making it $\Pi^2/4$. However while the aircraft shares the atmosphere one half is still within the earth making the Roche limit $\Pi^2/2$

The Roche limit is:
The region surrounding each star in a binary system, within which any material is gravitationally bound to that particular star. The boundary of the Roche lobes is an equipotential surface, and the lobes touch at the inner Lagrangian point, L_1, through which mass transfer may occur if one of the components expands to fill its lobe. It names after the French mathematician Edouard Albert Roche (1820-83). The Roche limit is a limit stars hold that is closer or equal to 2.4674 times the diameter of the star. The value of $\Pi^2 \div 4 = 2.4674$ which is why it is gravity Π^2 that is divided $\div 4$ by the four quadrants of a circle. The "sound barrier" is just a division of this law, the Roche limit.

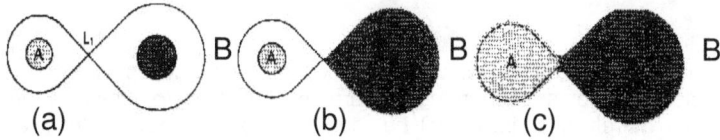

THE ROCHE LOBE: In a binary system, the Roche lobes of components A and B meet at the L_1 Lagrangian point. (a) In a detached system, neither star fills its Roche lobe. (b) In a semidetached system, one massive component, B, fills its Roche lobe. (c) In a contact binary, both components overfill their Roche lobes and share a common envelope. When the aircraft exceeds the Roche limit by half, the aircraft enters the territory of liquid where we have the Roche lobe and where two moving objects share space in motion or space-time within shared space.

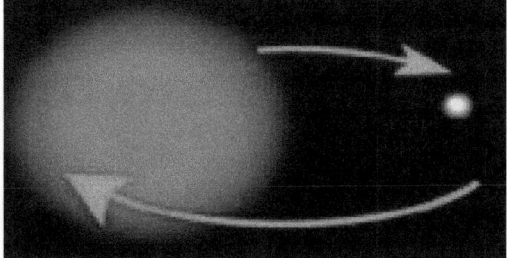

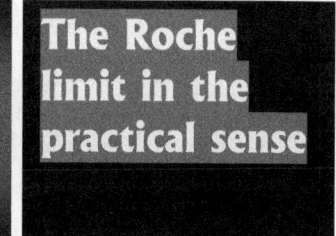

The Roche limit in the practical sense

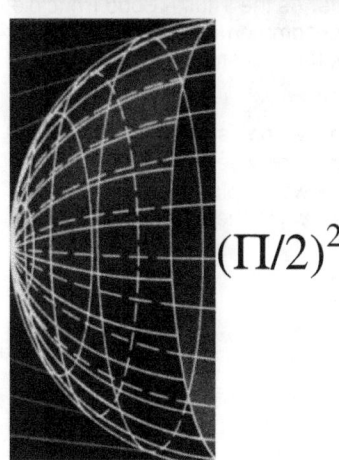

$(\Pi/2)^2$

The link of $\Pi\Pi^0$ connecting as Π^2 disconnects the link it has in Π. After the point where the Roche limit ends the earth hold a limit at gravity or Π^2 while the moving object also finds gravitational identity at Π^2. This then divides the gravity by dividing the connection in movement. In cosmic terms the aircraft then has separate identity dividing movement or Π^2. Then the stage of Mach 1 is entered where the movement in space $7(3\Pi^2)$ holds the limit $\Pi^2/2$ and the culmination is space-time displacement or another name for it is movement or another name for it is gravity. The Roche limit is sacrificed at $7(3\Pi^2)\ \Pi^2/2$ which is anything from $7(3\Pi^2) = \pm 203$ to 207 and the Roche limit then is $\Pi^2/2$ making the sonic boom somewhere between $203 \times \Pi^2/2 = 1001$ km/h and $207 \times \Pi^2/2 = 1022$ km/h. The sonic boom could be as high as the movement of the earth $218 \times \Pi^2/2 = 1075$ km/h.

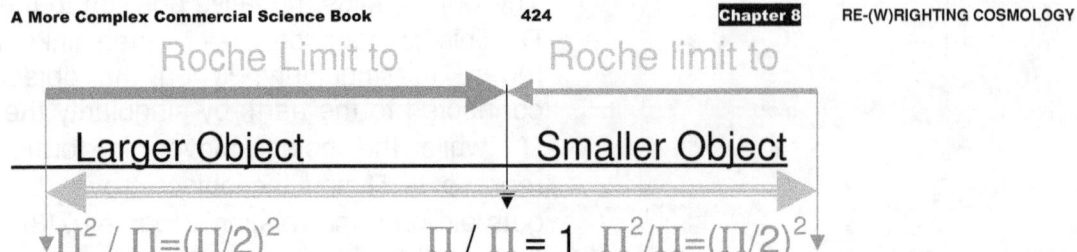

The distance of 7(3Π²) is not a measured unit such as a specific distance but heat increasing space might increase or decrease this value in terms of km / h because km/h holds 1000 meters of earth surface. This does not apply to space as space is not fixed but varies. The argument is a lot more complicated than this because of more relativity and other cosmic principles forming part of the argument.

The linear displacement value surpasses the circular value, therefore it is $Π²/2 \times 203 = 1001$ kilometers per hour, and that is Mach 1. Doppler's effect has nothing to do with the process, and neither has sound anything to do with the affair. Sound is merely an innocent bystander conforming to gravity in relevancy.

Above the Roche limit of $(Π/2)²$.

This will come into affect when the projectile reaches Mach 1. I choose to use the accepted term Mach 1 at this point as to limit any confusion that may arise from the new arguments. I have to stress the fact that the Doppler effect is, once again, merely a co-incidental but duly related by product. However, the Doppler effect, as such, plays no part in these phenomena, or in the outcome of the application. It must be seen in terms of cosmology and not from a human perspective. This is the principle applying when a newly formed star escapes the heat envelope within the centre of a galactica as a star is born. There now are two singularity factors within one space-time unit and the two goes into a battle of existing.

There is the same principle applying between stars sharing space-time. When two stars are at the Roche limit, the linear displacement reaches a value of one, and

$Π³/Π²Π = Π⁰ = 1$ is equal to singularity and in this we find space-time forming a value. This formula does not impress the most learned Physicists such as Professor Doctor Friedrich W. Hehl, which you will learn from a little further on in this book since Professor Doctor Friedrich W. Hehl thought this was to use his words "With a lot of words and some simple algebraic relations, there is no way to "explain" the world of physics." However notwithstanding Professor Doctor Friedrich W. Hehl not being impressed, on this rides the entire cosmos formed by gravity.

The circular displacement reaches its full complement of half $Π²$ which is the Roche limit.

To this reason, stars would form massive binaries, where they share a common combined circular displacement, separated only by each stars Roche limit with no linear value. As the Titus – Bode Principle comes into effect the linear displacement would once again grow, or the common spin value will be to grate for either one, or both, and their structural composition will collapse, forming smaller structures with less space to occupy the time in which they are.

The solid part of what forms gravity will be $Π³$ while the liquid part that moves holds a value of $Π²Π$ and this relates to singularity $Π⁰ = 1$. Whenever there is a conflict of relations between objects within the Roche limit of $(Π/2)²$ the one would play the part as liquid and the other would play the part as solid. Where both have equal properties the one would be a liquid to the other while the other would be the liquid to the first and there would be a binary star developing.

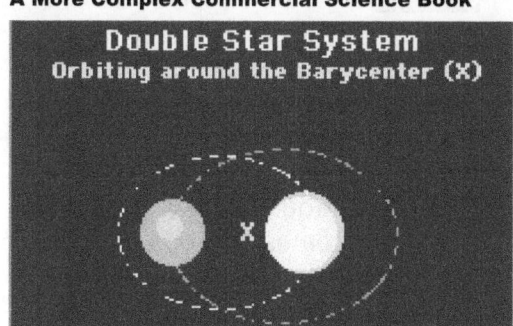

This is not where it ends. Layers within a star also form these same relations and when the layers go array and the borders of the Roche limit is violated, a Super Nova erupts. When the relevancies are bridged the control of the star would not be valid any longer and the star would expand in Π that will lead to the compromise of $(\Pi/2)^2$.

The two stars forming the binary will be in a cosmic duel until both of the binary stars group together to unite as a black hole that is if they did not destroy one another. This would depend on the Titus – Bode law sets in where the linear space-time starts developing through the Hubble constant and they would spin around at greater distances. The circular distance, however remain valet as one con see from the "gravitational pull" that has nothing to do with jerking each other around.

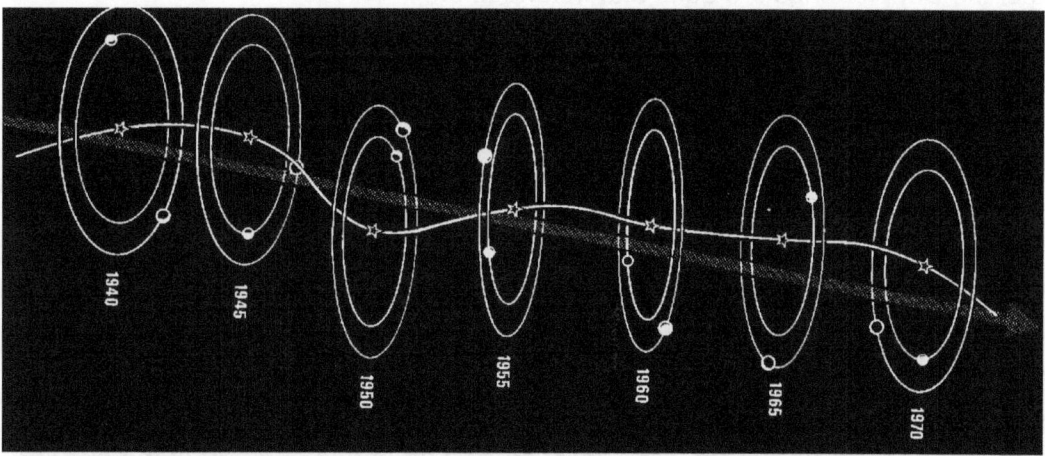

If one can illustrate the Universe and its relation with space-time, the following illustration would fit like a glove. The so-called mass pull is the effect that the Titius Bode law plays on the allocated rotation position of the planets.

$$\Pi^3 \qquad\qquad\qquad \Pi^2$$

```
ⓢⓢⓢⓢⓢⓢⓢⓢⓢⓢⓢⓢⓢⓢⓢⓢⓢⓢⓢⓢⓢⓣ
 ⓢⓢⓢⓢⓢⓢⓢⓢⓢⓢⓢⓢⓢⓢⓢⓢⓢⓢⓢⓣⓣⓣ
  ⓢⓢⓢⓢⓢⓢⓢⓢⓢⓢⓢⓢⓢⓢⓢⓢⓢⓣⓣⓣⓣⓣ
   ⓢⓢⓢⓢⓢⓢⓢⓢⓢⓢⓢⓢⓢⓢⓢⓣⓣⓣⓣⓣⓣⓣ
    ⓢⓢⓢⓢⓢⓢⓢⓢⓢⓢⓢⓢⓢⓣⓣⓣⓣⓣⓣⓣⓣⓣ
     ⓢⓢⓢⓢⓢⓢⓢⓢⓢⓢⓢⓣⓣⓣⓣⓣⓣⓣⓣⓣⓣⓣ
      ⓢⓢⓢⓢⓢⓢⓢⓢⓢⓣⓣⓣⓣⓣⓣⓣⓣⓣⓣⓣⓣⓣ
       ⓢⓢⓢⓢⓢⓢⓢⓣⓣⓣⓣⓣⓣⓣⓣⓣⓣⓣⓣⓣⓣⓣ
        ⓢⓢⓢⓢⓢⓣⓣⓣⓣⓣⓣⓣⓣⓣⓣⓣⓣⓣⓣⓣⓣⓣ
         ⓢⓢⓢⓣⓣⓣⓣⓣⓣⓣⓣⓣⓣⓣⓣⓣⓣⓣⓣⓣⓣⓣ
          ⓢⓣⓣⓣⓣⓣⓣⓣⓣⓣⓣⓣⓣⓣⓣⓣⓣⓣⓣⓣⓣⓣ
```

$$\Pi^3 \qquad\qquad\qquad \Pi^2$$

There are two distinct values in the cosmos. One is tome within singularity $(\Pi^2/\Pi) / (\Pi/\Pi^0) = \Pi^0$ and the next is space that time leaves behind as an afterthought, which mathematically is $(\Pi^3 / \Pi^2)/ (\Pi/\Pi^0) = \Pi^0$. The one says the movement Π^2, which is gravity, puts the relevance Π in place where the relevancy Π confirms singularity Π^0. That is how gravity applies singularity to form time and the result that time leaves behind is space. In this we have space-time.

There are only two energy forms in the Universe. The first is heat and the second is life. No other form of energy exists in the four dimensions of that the Universe exists. No force except for life is to be found in the universe, only balancing values.

The matter of the aircraft lays claim to the unoccupied space-time, which then replaces the occupied space-time (air particles) to another position. This is obvious in the cloud that surrounds the aircraft as it enters another stage in the sound barrier a stage long before the sonic boom comes into practise. The Doppler effect does not cause the sonic boom, as the aircraft overcomes the sound wave. The sound wave is not broken, but merely relocated to another position relating to the aircraft. If the sound wave was broken, there can be no sound travelling through air being on a wave of matter.

The sonic boom is the relocation of occupied space-time (air particles) as the body frame of the aircraft lays claim to the unoccupied space-time. The sudden jolt of the relocation of air particles, leaves a densified strip of air that are then in a process that relocates the air and for that matter the centre point of the sound wave. The second sonic boom comes into effect the minute the matter relocates in another sound wave.

If one looks at the transmission of sound, it too depends on the relocation of matter, but to a very small degree, and in this process lies the transmitting of sound. To make the error of judgment in confusing the process with the breaking of the Doppler rings are quite understandable.

One more point of interest is that light is the only particle that can remain a part of space-time and not be in the wave. $(\Pi^2/\Pi) / (\Pi/\Pi^0) = \Pi^0$ which allow us to use it in observation.

THESE TWO COMPONENTS FORM **SPACE-TIME** (Π^3/Π^2), THE DIMENSION WE FIND OURSELVES IN, AND WHICH IS NOT FLAT.

This is the reason why a spacecraft when launching does not "break the sound barrier". The circular space-time is growing faster than the craft can apply linear displacement.
The higher up in space, the less the prevailing circular space-time would be therefore the more linear space-time will be required to penetrate the circular value.
One of the two most important values in the Universe is the Roche limit.

On this and the Titus –Bode principle rests the growth of the universe, as space relate to time. The Roche limit comes into effect when the linear displacement factor reaches a value of one and part of the circular displacement value. In this is the value $\Pi^3/\Pi^2 = \Pi$

The photon relates to space being behind time while the wave as such connects to time Π^2 with the three connecting the wave to space in accordance with the Titus-Bode law (10-7=-3). Therefore, light moving in the wave, (coming with the wave) will all ways be RED, not blue.
What I cannot understand for the love or money is how this confusion extends to the speed of light? Nothing can come close to the speed of light, never-mind scramble the wavelengths of the speed of light.

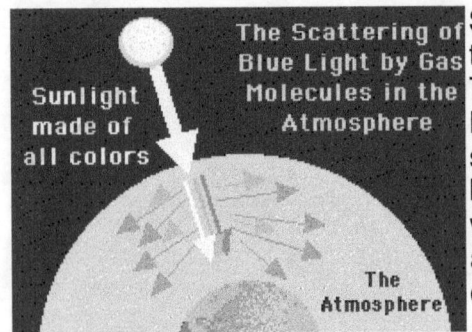

Science says that light reflecting from the rear of the wave will therefore be BLUE, not red as presumed.
This HOGWASH is what intelligent teachers tell students. Because the Earths atmosphere "scatters" light, the sky is blue! If such were the case, the sky would be a rainbow of colours, because a rainbow comes about as the water droplets scatter the wave of the light. The earth's sky is blue, because the earth's atmosphere re-concentrates the wave's photon distribution in the opposite direction, as to the light wave.

The light becoming blue is because the atmosphere of the earth concentrates the light and the light is contracted by concentration. The rings of space surrounding the earth is getting bigger the further it is away from the earth. The space carrying the light becomes denser so the light becomes blue. By forming a lens the earth atmosphere concentrates the incoming waves and the concentrated waves of light coming from the sun intensifies to become blue. Early in the morning at Sunrise we see the real sun rising being much bigger and shining in its natural red. That is because the angle at sunrise neutralises the bubble of the earth's atmosphere that forms the lens that concentrates the incoming light waves.

In a circle with a radius running from the centre of the earth towards the circle end the circle will cover more space as the circle moves further away from the centre into what we call space. If the circle line was formed by lots of dots one alongside the other the dots at the end of the earth would have the least dots when compared to dots in the circles moving into space. Say the first circle has 10 dots then the next circle forming just above this will have 11 dots and the circle just above that will have 12 dots and the next one going into space will have 13 dots. It is these dots that gravity reduces when gravity concentrates space to make the space hotter.

When an object travels at the rotating speed of the earth it goes at $7(3\Pi^2)1\Pi^0$ Travelling in the next line of the circle just above the body will have to travel faster since the circle is bigger. The dots it must travel through in the same time therefore must then be $7(3\Pi^2)1.00001\Pi^0$ increasing in the line above to $7(3\Pi^2)1.0001\Pi^0$ then $7(3\Pi^2)1.001\Pi^0$ $7(3\Pi^2)1.01\Pi^0$ $7(3\Pi^2)1.1\Pi^0$ **When the** object travels on the circle the earth forms $7(3\Pi^2)1\Pi^0$ but travels through $7(3\Pi^2)1.5\Pi^0$ dots per time unit it will lift off from the $7(3\Pi^2)1\Pi^0$ in order to joint the $7(3\Pi^2)1.5\Pi^0$ line that forms that ring above the earth. This has all to do with the density relevancy the material travelling has with the space in which it travels. Everything in the Universe moves in comparison to every other thing moving and gravity is the relevant difference in moving. When anything moves standing on the surface of the earth the movement runs through space and space holds a value of Π^0

When anything moves independent of the earth and moves at the speed of the earth the movement will project towards the centre of the earth because it has a movement displacement of Π^0 When anything moves independent of the earth and moves faster than the speed of the earth the movement will project towards the outer space because it has a movement displacement is more than Π^0 By movement and the relevancy of such movement, in moving the object secure the immediate future it will have and the direction the future will bring. By this process, which is thought to be the "Sound Barrier", stars are born but the process is too difficult to explain at this point.

By going in a circle around the Earth the moving object goes in a straight line. By going on a course that passes through a certain number of dots puts the object in a certain trajectory. Being guided by life to use a different orbit trajectory makes no natural sense in cosmic law. Life has no meaning to cosmic law because to the cosmos and laws applying in the cosmos life does not exist.

Having a specific number of dots displaced in one time unit puts the movement in a specific altitude in relation to the movement the earth claims. By going trough $2\Pi^0$ of space when passing though $7(3\Pi^2)$ of movement places the required Π position at a specific distance from the centre of the earth and this forms an allocated position. This will be a positioning acquired according to relative density and it is the density of let's say oxygen that keeps the oxygen airborne. Should the density not apply and life puts to cosmic will artificial heat and therefore artificial movement in relation to reality, the earth will begin to destroy the movement and that is what happens with air drag.

If the movement takes place at a specific height of $7(3\Pi^2)1.5\Pi^0$ and the movement is actually $7(3\Pi^2)5\Pi^0$ because it is in the density range of $7(3\Pi^2)1.5\Pi^0$ the Roche limit will come into effect and the Roche limit is $7(3\Pi^2)\Pi^2/2$ we call it the sound barrier. It breaks the sound barrier at a height of $7(3\Pi^2)1.5\Pi^0$ while going at a speed of $7(3\Pi^2)\Pi^2/2$ because at $7(3\Pi^2)5\Pi^0$ there is no sound waves that are able to form.

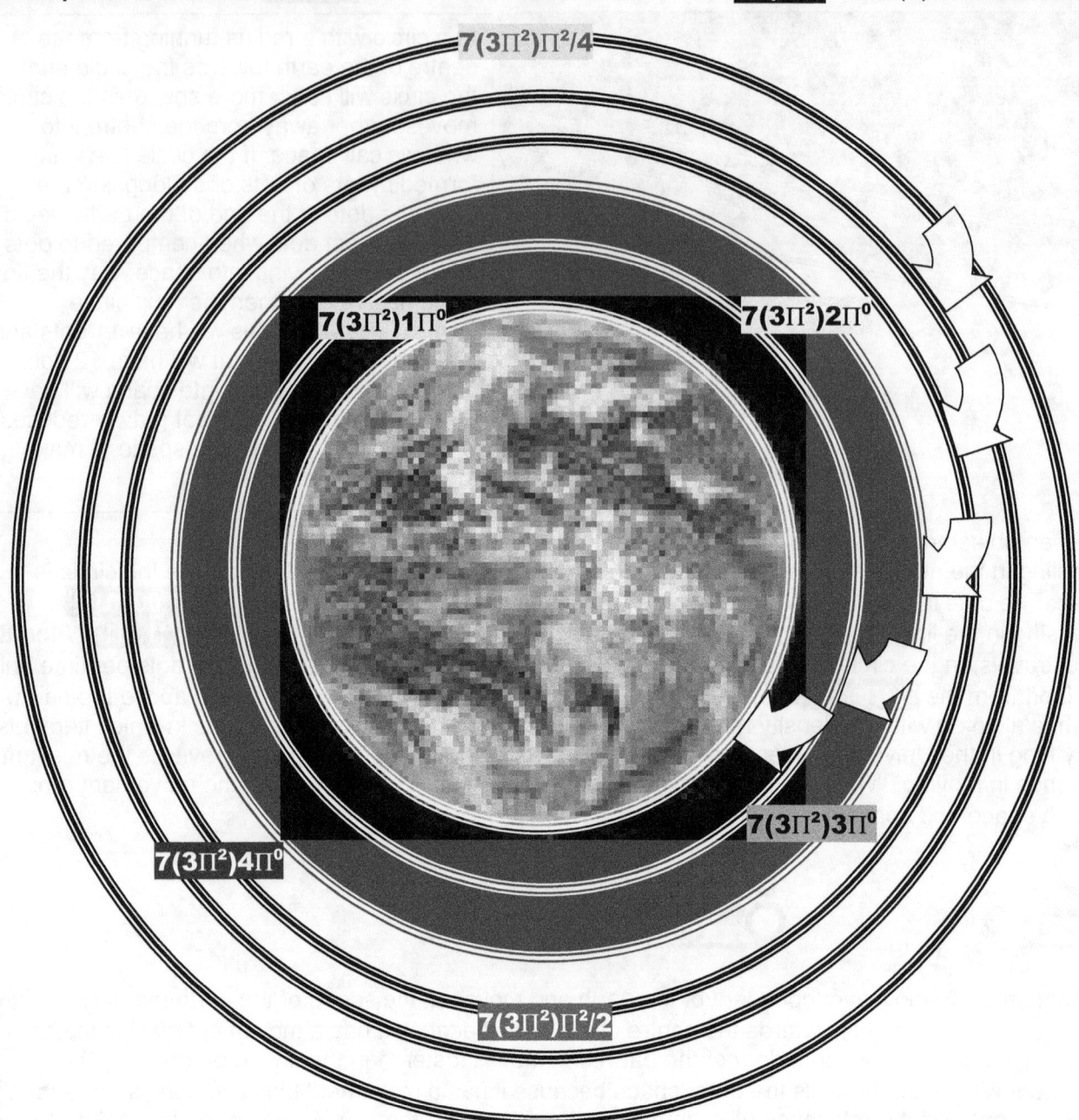

The earth moves at a rate of $7\Pi(\Pi^2)\Pi^0$ but when the contact of the surface of the earth is no longer in touch with the object moving the movement changes to space replacing the surface of Π having the relevancy of $7\Pi(\Pi^2)\Pi^0$ and therefore the relevancy applying is $7(3(\Pi^2))\Pi^0$ this is the speed at which space moves towards the earth and that is all space filled by material or unfilled at all. The concept that changes is that whereas Newtonians viewed only material as forming a factor in this approach space moves in relation to space moving at a different speed. To travel in relation to the speed the earth contracts space towards the centre a body moving separately and not in conjunction with the earth will travel through space at a different rate. But travelling means displacing space. It is not material moving but it is material displacing singularity by filling other singularity and in that filling other space. Empty space falls as fast as filed space and by falling it fills space and not pushing unfilled space aside.

The more space it travels through the higher up it has to be in relation to the earth surface because movement is attached to the earth turning and gravity is the earth turning and time is the gravity by which the earth turns. If the object moves through more space than it would when being on the ground the ratio will still remain the same because in relation to the earth movement in any direction remains movement being the same. That means movement horizontally is equal to movement linear because the rotation and the straight line is synchronised in complementing gravity. If an object moves at a rate of $7(3(\Pi^2))3\Pi^0$ it displaces space at a rate of $7(3(\Pi^2))3\Pi^0$ notwithstanding it being $3\Pi^0$ up in the air or moving on the surface where it displaces $7(3(\Pi^2))3\Pi^0$. To break the sound barrier the displacement has to be at the

Roche limit of $7(3\Pi^2)\Pi^2/2$ and that puts the moving object in a bubble it has in relation to the earth rotating.

Above the Roche limit of $(\Pi/2)^2$.

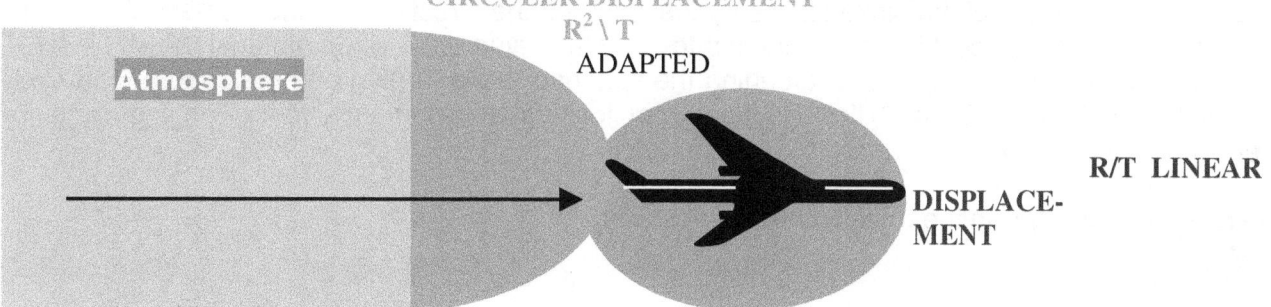

The seven becomes as big a part of the concept as does Π as it all interacts.

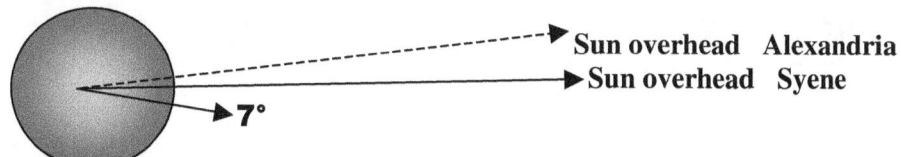

It took Eratosthenes of Syene (276 – 194 BC) a Greek astronomer who in the year 240 BC made a discovery that the earth has a profile of $7°$. Since then no one ever did anything about it. When any singularity wishes to disconnect from the earths singularity, specific pre-calculated laws would have to comply to allow the lesser object to divorce from the larger object. I indicated how the dimensions of 10/7 and 7/10 interact to form (Π^2). Matter is a product through the separation of space and time receiving the value of Π The original time and Π^2 as follows: By circling around a spinning solid the space contracts to form Π and Π^2. Gravity forms everywhere in the Universe by applying singularity. By dividing space into material (material spinning in space) and duplicating space by material spinning, the TITIUS BODE LAW forms a $7°$ deviation and 7 / 10 in conjunction with THE ROCHE PRINCIPLE OF $(\Pi/2)^2$

In my article to Annalen der physics I used 15 pages to explain this process of singularity applying. I received a rather cordial but sincere reply from the Editor of the magazine. When I placed an article in Annalen der physics Dear Prof Friedrich W. Hehl said in the e-mail he sent me that there is no way to "explain" the world of physics I am not going to go into detail how this works. On the other side of the Pythagoras's' triangle we have 1 going square. That makes Pythagoras's' triangle 49 + 1 = 50 on the one side of the earth and the same on the other side of the earth. The total is 100 and the square is 10. That leaves the Titius Bode law with a value 7 (it forms part of the material of one body) and 10 in relation to the space. Then from the relation of 7/10 and 10 / 7 forming Π the Titius Bode law form Π^2 applying "With a lot of words and some simple algebraic relations" to quote Friedrich W. Hehl, Inst. Theor. Physics of Annalen Der Physics fame. This was simple algebraic relations but still it is science, is it not?

Since it involves singularity moving it calls for the law of Pythagoras to produce space. The law of Pythagoras is the triangle a^3 that is moving forward in singularity **k** by turning T^2. In singularity the 7 stands in for 7 points on the numerical line crossing over the line holding singularity or 1.

By moving 7 has to go square T^2 and that means 7 goes square 7^2 twice $7^2 + 7^2$ crossing the same divide $\Pi^0 = 1$.

Since all movement in singularity has to enforce the law of Pythagoras we have two triangles holding 7 dots moving across singularity. I don't want to get too involved by bringing in numerical outlays because then this can truly become complex.

The line has two opposing sides turning directionally against each other while turning with each other. By moving or turning this involves time duplicating space by the square Π^2 on both sides of the divide $\Pi^2+\Pi^2$ and using the same divide or the same axis or the same point serving singularity we have $7°$ crossing the same point in singularity Π^0.

There then is in this rotational movement $7°$ standing in for Π^2 on both sides of the divide $\Pi^2+\Pi^2$, which then is 7^2 on both sides of the divide 7^2+7^2. The circle spins in duel directions. On the one side it would go left if on the other side it would go rite. The one side hold a directional change in singularity by 90°. As it is going sideways it changes to going down.

This produces a rite angle triangle of 90° and in it the law of Pythagoras produces direction changes. Since the square of the turn of the circle places by the spin and the direction change we have 7 holding a relation to 10 in space because it is space that has to carry the value of 10 when material circles by 7. There is a connection between space surrounding the spherical circle turning and the sphere. The circle holds the value of 7 as in 7° and this we find from looking at singularity controlling the circle by movement.

Matter in relation (part of) with the total dimension of space.

$$\left(\frac{10}{7} \div \frac{7}{10}\right) = 2.04$$

$$\frac{1.4285}{0.7} = 2.04 \quad \text{Taking from both orbiting influences}$$

SPACE DIVIDED INTO TIME

$$\left(\frac{7}{10}\right) \div \left(\frac{10}{7}\right) = 0.49$$

$$\frac{0.7}{1.4285} = 0.49 \quad \text{Taking from both orbiting influences}$$

SPACE MULTIPLIED WITH TIME

$$\frac{7}{10} \div \frac{7}{10} = 1 \quad \text{and} \quad \frac{10}{7} \times \frac{7}{10} = 1 \quad \text{Therefore not influencing change}$$

THE PROCESS PARTED USING THE ROCHE PRINCIPLE

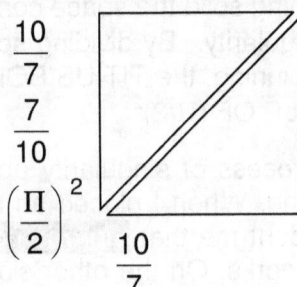

$\left(\frac{\Pi}{2}\right)^2$ The Roche influence on Titius Bode

$$2.04 \times \left(\frac{\Pi}{2}\right)^2 = 5.033$$

$$2.04 \times \left(\frac{\Pi}{2}\right)^2 = 5.033$$

$$5.033 + 5.033 = 10.066 \quad \text{from both objects}$$

SPACE DIVIDED INTO TIME

$$\left(\frac{7}{10}\right) \div \left(\frac{10}{7}\right) = 0.49$$

$$\left(\frac{10}{7} \div \frac{7}{10}\right) = .49 \quad \left(\frac{10}{7} \div \frac{7}{10}\right) = .49$$

$$.49 + .49 = .98$$

$$.98 \times 10.066 = 9.8696 = \Pi^2$$

$$\text{TIME SPACE} = \Pi^2 = 9.8696$$

TIME SPACE = Π^2 = 9.8696 = Space and time in a dimensional implication

When the 7 of material and the 10 of space no longer forms a gravitational alliance the 10 forming space will expand because the spin of 7 cannot retain gravitational structure.

For gravity to form the space value of 10 must always maintain the material value of 7 so that the formation of Π can form gravity by moving Π to form Π^2.
This is gravity and that is how gravity is maintained

The car straightens the curve of the earth when it stretches the 7° the earth presents as the curve around which it spins. The car leaves the curve of the earth and gets airlift. It then needs "wings" to force the car down by creating more gravity because gravity is the space that pushes down on the earth and the wings produces more space that pushes the car down on the ground. As soon as the car lifts into the air not touching the ground surface the relevancy changes to $7°(3\Pi^2) = 217$ km/h as the relevancy changes to $7°(3\Pi^2) = 207$ km/h.

This picture we have above shows why gravity forms as the space we call air pushes down onto the earth. The faster the car moves from one point to the next the more air it will generate to push it down. Gravity is the space above the earth that pushes down on the earth and that includes the space the car holds. But if the car exceeds the speed the earth retains to spin the car will exclude some space from the earth's space and take this air along for the ride. This lump of air holding the car will stretch the 7° it retains to form a circle and this will extend the position where Π forms. This will then begin to lift the car when the point holding Π as radius becomes wider as the position where 7° forms is further apart.

As Indicated space always holds a double 10 value in terms of 7

Material associates with 7 forming one value in $21.991 / 7 = \Pi$

By spinning through 7° the changing of direction is taking 7 squared $(7°)^2$ and this gives 49 in total movement or space-time displacement or gravity applying or whatever you wish to call it. This is on one side of Pythagoras's triangle

The spin re-aligns the centre with the same point on the surface of the earth but it then is at another point. This movement then takes $\Pi^0 = 1$ or singularity to $(\Pi^0)^2$ or also 1 which is singularity. I sent an article about singularity to Annalen der physics.

Gravity is the contracting or compressing of the space surrounding the earth and this is by way of rotation.

It is very clear when looking at other solar structures that a control in movement applied from the centre of the planet (or sun) to the allocated position where the circle or moon forms an orbit. The circles forming around planets are evidence of this. At the point of orbit it forms a circle to the value of Π and most important to science is to know where Π as point as forms and it is the duty of science to find why such a point exists.

First it would be best if science got away from 1650 and got into what science knows in 2010 and apply such knowledge to what is prevalent in cosmology. It ahs been known since before Christ that Pythagoras played part in the forming of the cosmos and that I prove in a book entitled The Absolute Relevancy of Singularity in terms of a Cosmic Creation. The theme I use in explaining the Cosmic Birth is that is the Universe started with 1 a point holding singularity, then how did two arrive and then three, four, five up to ten.

I said it before so I say it again. From that comes gravity as follows: In the centre of the rotating body singularity forms to the value of Π^0 and this extends to the curve forming Π in terms of the curve of the rotating object. This then forms part of gravity moving forming singularity going square Π^2 and this form the relevancy

The extending of singularity goes from Π^0 to $5\Pi^0$.

Coming in from space while having no mass the coordinates change as the value of Π forms a relation to the 7°

When an object comes in from the sky the object associate value of the rotation in line with value of 21.991 / 7 on Π or on how gravity functions. It has way, shape or form and there is entirety of the cosmos apart the Newtonian conspirators

the atmosphere or flies with the turn giving it a the axis which puts a gravity. This is gravity and this is nothing to do with mass in any no factor such as mass in the from being in the imagination of calling them physicists.

I sent a fifteen-page article to the Annalen Der Physics in which I explain this process in length. I received the following reply from Annalen Der Physics.

I now explain gravity again and relate this explanation to the sound barrier functioning.

Gravity forms by association or relevancies that holds a value of coming towards the centre from the sky which then is 7(which is the curve of the earth) (Π associating with the curve of the earth) and Π^2 (associating with the moving of the earth) This is part of the Titius Bode law forming the measure of gravity which is not by mass but by forming Π.

$7\Pi\Pi^2$ x Π^0 to $4\Pi^0$

Then when a line is drawn from the centre of the earth another value comes in place referring to the centre of the earth Π^0 and the curve of the earth Π. The value of Π is then shared by both disciplines (association from the sky to the centre ($7\Pi\Pi^2$) and association from the centre to the sky ($\Pi\Pi^0$).

If the body does not connect to the earth by way of forming a unit or forming "mass" the axel line of the earth comes into affect giving the ratio a value of 3. Then when not connecting to the earth surface the ratio becomes $7(3\Pi^2)$ x Π^0 going up to $4\Pi^0$ all depending on the sped or the height. This is the way gravity forms by forming a connecting association with $\Pi^0\Pi$.

This forming of gravity explains the Coanda effect because the Coanda effect puts air or liquid in relative movement with solids and this movement (but not the mass part) is how gravity forms.

A More Complex Commercial Science Book　　　　433　　　Chapter 8　　RE-(W)RIGHTING COSMOLOGY

If the body does not connect to the earth by way of forming a unit or forming "mass" the axel line of the earth comes into affect giving the ratio a value of 3. Then when not connecting to the earth surface the ratio becomes $7(3\Pi^2) \times \Pi^0$ going up to $4\Pi^0$ all depending on the sped or the height. This is the way gravity forms by forming a connecting association with $\Pi^0\Pi$.

This forming of gravity explains the Coanda effect because the Coanda effect puts air or liquid in relative movement with solids and this movement (but not the mass part) is how gravity forms.

The condensing of the vapour surrounding the jet is a prelude to the sonic boom and is just one link in the sound barrier where the sonic boom is just another link in the sound barrier process. The sonic boom IS NOT the sound barrier but is as much part of the entire concept as the beginning of movement is part of the sound barrier. Waves in the sea are as much forming part of the sound barrier as winds howling or blowing is part of the sound barrier.

All objects in the Universe move even Black Holes because Black Holes make outer space move. Black Holes makes outer space recline in relation to the structure standing motionless and thereby it moves.

When any object moves it fill a certain space during a specific time period.

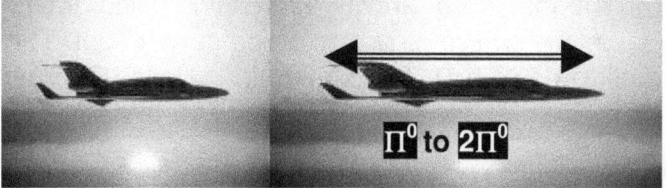

When an aircraft stands still by only moving with the earth it moves at Π^0 As it moves faster it fills more space up to where it fills from Π^0 to $2\Pi^0$ of sky space and as the speed increase this goes up to $2\Pi^0$ to $4\Pi^0$ The aircraft holds more space but does not fill more air and therefore the air has to compress

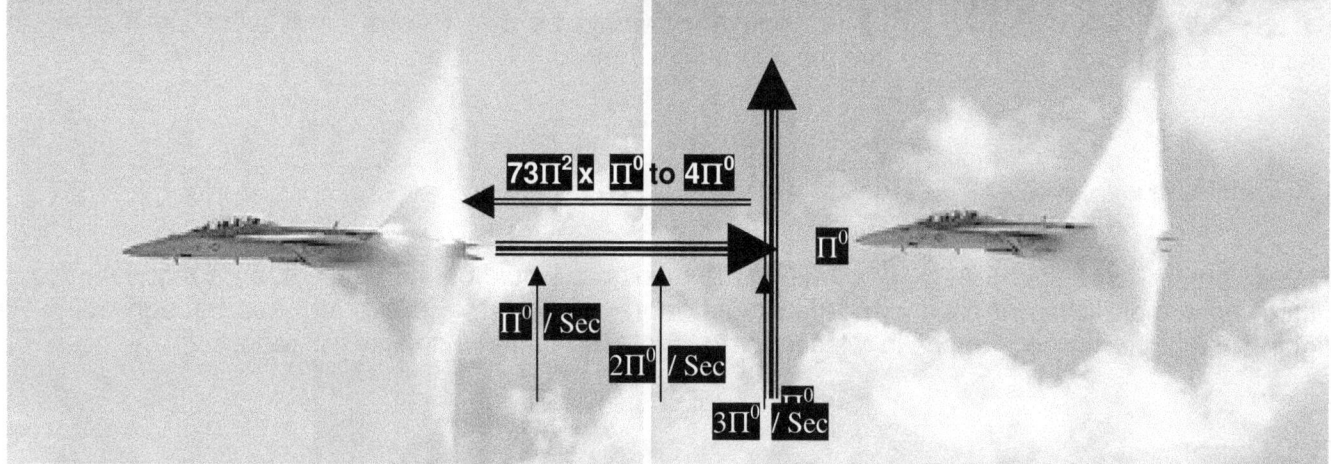

By going into space reduction as the fuselage occupies more space through extensive movement increase the space surrounding the moving aircraft condenses and this condensing forms the concentrated vapour. As soon as the fuselage begins to heat the heat surrounding the airplane pushes the vapour away again. The sonic boom is just one link in the chain of reactions we associate with the sound barrier and the sonic boom is more or less in the middle of the entire concept.

Flying at $(3\Pi^0)$ three times the falling speed or the speed of gravity $(7 \times 3\Pi^2)$ going in a sideways direction the body of the aircraft starts to expand into another gravity limit $(3\Pi^0)$. However by the movement of the aircraft the body also has to shrink ($3\Pi^0$) and the result is the body of the craft cant accommodate conflicting behaviour.

Since the body of the aircraft that should expand contracts the space surrounding the aircraft also contacts in sympathy with the aircraft body contracting. That makes the aircraft becomes literally smaller and this condenses the space surrounding the aircraft. But then the heat coming from the contracting body of the aircraft pushes away the atmospheric air as the surrounding air of the aircraft heats up because of the intensity of the aircraft increasing the space it holds.

What I ask of readers is to beforehand forfeit the culture of Newtonian bias when reading this by paying attention to what I say and not about the degree in which I stray from mainstream science's thinking. This way the exercise will present many new ideas and explaining my new concept will become clear. There is so much to benefit from. Science has no idea what a Black Hole is while I can prove what a Black Hole is. I formulate mathematically what "the sound barrier" is. I prove what gravity is. By using the four cosmic phenomena, which is what the cosmos uses to form gravity, I show what "the sound barrier" is and I go much further than that. I show that gravity forms using the **Roche limit**, the **Lagrangian system**, the **Titius Bode law** and the **Coanda effect**.

I uncover these principles by placing Π within the formulating of gravity and when using Π I bring clarity to the misunderstood cosmic principles. The list of the unknowns I can then explain is almost endless. **Gravity forms by movement that establishes singularity initiating a circle in using Π.**

I show why gravity is there, how gravity forms and what role stars play in forming gravity. There is no difference between how gravity and electricity forms and that I prove mathematically by decoding the cosmos. I prove mathematically when atoms spin they establish Π that forms the Universe. Whatever forms gravity has to link closely to Π since everything that has anything to do with gravity forms a circle that is Π by the value of the square radius. If mass has anything to do with generating gravity, then mass has to apply Π or otherwise mass has nothing to do with the forming of gravity. Everything using gravity forms a circle of sorts, which forms the curvature of space-time, which is Π and which curves light. The way the planets orbit the Sun and how stars spin has all to do with Π. In spinning in a circle, Π forms gravity as a centrifugal force that condenses space.

At $73\Pi^2$ x Π^0 the relevancy would be one to one on gravitational movement compared to the earth much space that the aircraft goes through

At $73\Pi^2$ x 1.3 Π^0 the relevancy would be the space it holds that enlarges to a third more than it has at gravity and that much more space that the aircraft goes through will put it aligning in a higher value of space in relation to material. The material density will reduce.

At $73\Pi^2$ x 3 Π^0 the relevancy would be the space it holds that enlarges to three times more than what it normally would have at gravity and that much more space that the aircraft goes through will put it aligning in a higher value of space in relation to material. The material density will reduce by the movement where through the aircraft goes as the aircraft duplicate its space in relation to the time it uses to occupy that space. Going at $73\Pi^2$ x $5\Pi^0$ will intervene the Lagrangian points law and before that happens the Roche limit comes into effect. By going into the

Roche limit or the sound barrier at $73\Pi^2 \times \Pi^2 / 2$ the space surrounding the aircraft becomes connected in singularity to the craft and is then rejected by the earth as "foreign".

Explaining the following is rather tough with the limited space available but it should secure the conclusion that I am not grabbing for straws and there are substantial facts on which I work.

Lets investigate the Universe. The Universe is made of lines connecting points. Whatever you see or can't see is lines connecting. If we wish to examine the Universe we first better star to examine lines connecting dots because whatever is in the Universe is dots that connects with lines. So it will be necessary to investigate lines. In mathematics they teach students that a line starts at zero and that is a fable. Zero starts nothing and nothing loses everything before anything can start. This sounds so unimportant but it is fundamentally all-important. A line starting with zero how cannot increase in length

What I am about to explain is the absolute basis for singularity. Lines mathematically cannot start at zero because there is no evidence of zero as a factor in mathematics. The shortest possible line (hypothetically) must be so short it must have an initial and ultimate point sharing the same spot. Any theoretical line being the shortest possible line cannot have the line holding the initial starting point at point zero and advance from there. If it used zero as a start, the zero part would not count, because the line will only start at a point past zero where the line then will start

When the line has a beginning and an end at the very same spot and it wishes to extend the position as to further the possibility it has, which direction should it favour. Extending the line in any one direction will favour one direction without any clear reason not extending in other directions. The only option about extending will be in all directions equally in order to give a meaningful non-bias flow of mathematical equilibrium.

The shortest line in the realm of possibilities must have a start and finish holding one spot and such a line will also be a dot or a circle. Not favouring one direction puts all directions at equilibrium meaning that any form of what ever might develop from such a spot with the end and the start being in the same position also has to be a sphere.

This reasoning prompted me to look for singularity in such a spot because if the prime spot from which all came was a spot holding all, then the spot must hold the shortest line but more prominent it will hold the smallest form including the smallest circle. One possibility that the shortest spot can never have is having a starting point on the zero mark. If the mark of zero holds the start it must also hold the end because the end and the beginning has the same position. If the position of zero then is the beginning, the end will also be zero leaving the line without an end as well as without a beginning.

The conclusion from this is that no line can start at zero because that will be a mathematical impossibility. A line or spot starting at zero would therefore be shorter than the shortest line possible. A line growing or extending from zero can never leave zero because of the influence of being zero disqualifies any possibility of growth. If the line then had to grow in all directions at the same pace the line must therefore be a circle or being three-dimensional, a sphere. Flowing from this fact is that in the Universe there can be no zero point or unfilled space. The value of the circle is Π, and that is where creation started. That gave me the clue where to start looking for singularity. One would find singularity in the value Π and the value Π will be in all things rotating in a circle. You might wonder how does that apply to the cosmos and moreover to gravity?

My approach might seem unconventional but through the abandoning of the accepted, it enabled me in locating the precise location of a universal singularity forming a connecting basis of the Universe (this I say with some degree of confidence). The smallest figure there can be must be a dot. The only mathematically sensible option about extending a line from the dot will be non-bias progress in all directions equally in order to give a meaningful flow of mathematical equilibrium. The Pythagoras mathematical principle is the proof and that I explain. The obtaining of singularity is in my rejecting of nothing by replacing it with something being the dot.

The claim becomes obvious when observing the connection between the half circle, the straight line and the triangle, which could also promote all the qualities lurking behind the pyramid. Consider the connection between 180° sharing and then one may realise much of the pyramid mystique becomes less spectacular in considering the very basic in mathematics being the Law of Pythagoras on which all mathematics are focused.

The Lagrangian System implicating the five positions extending from singularity

Singularity dividing the cosmos
Each triangle claiming a side of the universe

1 Half circle   = 180° L_3 L_4 L_5
2 Triangle 1   = 180° L_3 L_4 L_5
3 Triangle 2   = 180° L_3 L_4 L_5
4 Straight Line = 180°

The half Circle = 180° combining as a Sphere when comprising Singularity in the matching of the value of the straight line forming the half circle and combining as the triangle and all are equal 180°

The distance between the sun and Pluto is roughly one hundred times more and if the distance between mercury and the sun, but both has nothing between them and the sun. If space comprises of nothing how can nothing then become plural forming more or be multiplied. If it was one becoming one hundred, then the one cannot contribute to a value of nothing but then must be part of something. If the one substituted the nothing, all laws of mathematics will go in disarray because when one multiply any number by zero it becomes zero placing both planets in the sun. By excluding nothing from the equation space becomes something bringing in a value lying inside the realms of the infinite that must form singularity. Applying this logic to the Lagrangian system and interpreting that information to the law of Pythagoras a clear pattern come about.

Mercury has 58×10^6 km and Pluto is 5900×10^6 km space between the sun and the planet. That indicates a distance and a distance comprises of something, for if was nothing then both would have equal nothing and be next to the sun. I repeat, the distance indicates something because nothing would place them both in the sun. The problem is identifying something from nothing that defines the difference there is in science. I cannot see how nothing can become plural or more sometimes

Taking that into account it is important to recognise that notwithstanding the size of a line, there is another line (or dot) eternally bigger as well as eternally smaller than the line in question. We can never grasp the size of a line that forms the utmost or the least of possibilities and therefore size belongs to the human mind forming conceptions of big and small, but it has no place in the cosmos at large. This concept not only applies to size, but to all limits and divides we wish to create forming borders we can appreciate. When looking at the circle in the conventional manner, we persist with errors brought about in culture and not by applying some significant modern logic.

━━━━━━━━━━━━━━━━━━━━━━━━━━━━━━━━━━━━

From the smallest ever possible dot will grow a line in every imaginable direction relating to a prospect of Π not favouring one direction that puts all directions at equilibrium meaning that any form of what ever might develop from such a spot will have the end and the start being in the same position, which will also have to be a sphere as the flow outward will be equal in all directions. This reasoning prompted me to look for singularity in such a spot because if the prime spot from which all came was a spot holding all, then the spot must hold the shortest line but more prominent it will hold the smallest form including the smallest circle or for that matter the smallest sphere. One possibility that the shortest line or smallest spot can never have is having a starting point on the zero mark.

If the mark of zero holds the start it must also hold the end because the end and the beginning has the same position. If the position of zero then is the beginning, the end will also be zero leaving the line or spot without an end as well as without a beginning. Such a spot will constitute all of nothing Any line starting from zero would inevitably start from a point where it ignores the zero mark because the fact of zero does not implicate a start or a size of value, but only the not being there of that position. All lines would form a duplication of another line sharing value since there will always be a possibility of yet another line in the realms of singularity lying between the two lines in question reducing the size infinitely to either side of the divide we humans create. Boundaries therefore are human and as man made substances it does not belong to the cosmos outside the influence of man and must be discarded.

A More Complex Commercial Science Book   437   Chapter 8   RE-(W)RIGHTING COSMOLOGY

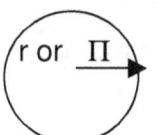

● r / 2 ● r / 2 • r / 2 dividing r reduces r to infinity but not Π as Π remains stable, protected by the rotation of matter forming a circle around singularity

When **the circle reduces**, the **value** located to **r** will become implicated because **r determines specific size. Not so** in the **case of Π, because** Π in the true sense only **indicate that the circle is a square without corners** and therefore Π **dictates form and not size**. By **reducing size** only **r comes into contest** and will point to such reduction. By **reducing** the circle **radius r by half continuously** will lead to an **infinite small circle** but Π **will remain because the circle as a form remains** even being infinitely small

In any circle or sphere the size only depend on the fluctuation of r in the square as a component to the circle or sphere but that does not affect the form by indication of Π in any way there may be. The conclusion from this is that no line can start at zero because that will be a mathematical impossibility. A line or spot starting at zero would therefore be shorter than the shortest line possible. For obvious reasons can no line, or any line grow or extend from zero because such a line must then quit zero and become something, thus abandon its original value. That would mean the start of the line has a different value to the end and a line holds conformity through out. When any line is starting from point zero it can never leave zero because of the influence of being zero disqualifies any possibility of growth. If the line then had to grow in all directions at the same pace the line must therefore be a circle or being three-dimensional, a sphere. Flowing from this fact is that in the Universe there can be no zero point or unfilled space. In the case of the growing sphere the value of the circle is Π, and that is where creation started. That gave me the clue where to start looking for singularity. One would find singularity in the value Π and the value Π will be in all things rotating in a circle. You might wonder how does that apply to the cosmos and moreover to gravity?

By reducing r indefinitely to the tune of half each time, r would become infinitely small, beyond human calculating means, however as mentioned in the case of the smallest dot holding one spot, r would become insignificant beyond human comprehension even, but never reaching zero and still Π would remain intact and dictating form.

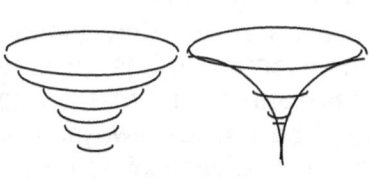

An observation coming instinctively to mind one may recognise is that the form reminds rather explicitly of natural phenomenon as hurricanes, water whirls and even the shape most commonly favoured to express the cosmic object referred too as a Black Hole. The similarity may be more than coincidental. Let us consider the statement in the reverse.

There is always a relevancy applying between that which spins art double 7 and that, which moves straight continuing at 10. The part spinning at seven we think of in terms of the diverting of direction it applies at 7. The liquid / gas holds the 10 factor.

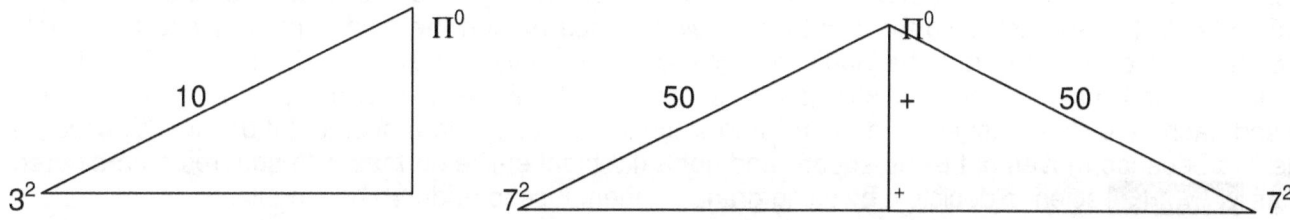

Without the application of specific heat, the object remains in the three directional moving of six possible directions. The value of space unoccupied therefore remains Π Π^2, as it was before the "Big Bang" event, whichever "Big Bang" you wish to refer to, because there were many. But space unoccupied holds time to the value of 10 to 1, and as the sketch of the triangles also indicated, holds space to Π. Therefore unoccupied heat holds the relation to space in applying 3 directions of influence $(3^2 + 1^2 = 10) = (a^3 = T^2 k)$. Always part of this equation is the dual function of space in $(a^3 = T^2 k)$ while at that very instant one has space-time. Therefore in space in time you have $(10^2) = 7^2 + 7^2 + 1^2 + 1^2$. In the sphere we have the axis holding a value of 3 and the circle holds a value of 4. These are dots forming in relaxation to the one spot holding a point from where singularity advances.

We have the axis valued at 3 going square through movement of the linear motion $(3)^2$ and then we have the circular motion $(4)^2$ going square by the spin of the circle ring the direct opposing side. Then the equation of influence becomes $3^2 + 4^2 = 25$ where $\sqrt{25} = 5$ and doubling the 5 on both sides of the triangle will apply the factor of $5 \times 2 = 10$ that then is $(10^2) = (7^2 + 1^2 + 7^2 + 1^2) = 50$ on both sides is 10. The implication of this may not dawn on one the very instant of realizing, but to scientists, there is no greater shock than just that. To any application of movement, the factor will be in the realms of singularity where half a circle is equal to a triangle is equal to a straight line and the lot is equal to 180°. No fancy mathematical expressions have any value in singularity because singularity holds a value of 1.

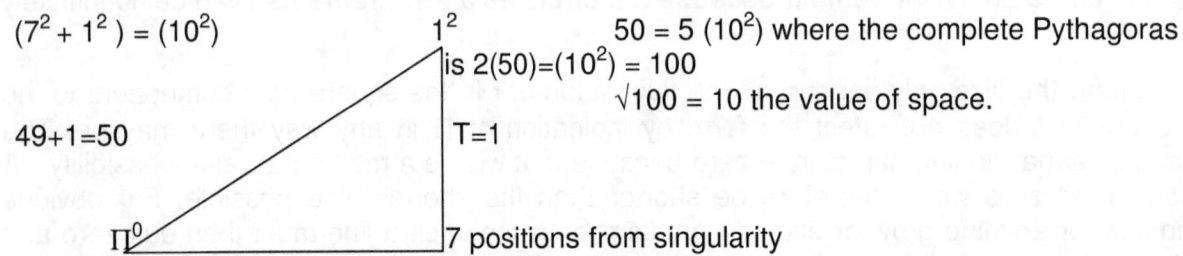

$(7^2 + 1^2) = (10^2)$

$49 + 1 = 50$

1^2 is $2(50) = (10^2) = 100$

$T = 1$

Π^0

7 positions from singularity

$50 = 5 (10^2)$ where the complete Pythagoras

$\sqrt{100} = 10$ the value of space.

The fact of this comes as 49 plus one becomes 50 and that is in the three dimensions of space $\Pi^2/7$ where 7 holds the relation to one and $\Pi/7$ again where 7 relates to one. At this point it is most important to remember that Pythagoras works on the application of the sum of the square of the two sides. When seven has a direction in the fourth dimension applied to it, the opposing dimension will be one and this applies in time relevancy, therefore the interchanging in time between infinity will place matter at $7^2 \times 1$ relating to circular and $7^2/1$ with $7^2 \times 1$. This makes 49 plus one (singularity) always being a factor of one. Space in time however, never can be a cube, it will always be a square with one side pointing the direction of time from time to the past (1) to time to the present (1) to time to the future (1).

The circle forming Π uses 7 to indicate the roundness of the circle but the 7 holds its roots deep within creation. It indicates how the Universe started because this is the way a star will start moving and it shows how as the infant star starts generating gravity just as the top starts to spin when it is thrown by life. Life can create nothing and that is true but life can mimic all laws in the Universe. Time is eternal movement and will be with us always. The line in infinity is still present while not being a part of the Universe. This line is always ready to be in place when the slightest movement orders it in place. Before the Universe was in place eternity and infinity was in perfect harmony and the line forming singularity validates this fact.

Before infinity parted from eternity, eternity met infinity on one spot as eternity came from the past (1) forming the present (2) to go onto the future (3) but also returned to come from the past which was the spot held by the future and this we find in the fact that the line forms 1 when not spinning but as soon as it evokes by spin, 3 points form even now. Then heat and cold differentiated values and space landed in between eternity and infinity. As eternity moved in relation to infinity but not forming a part of infinity any longer, eternity had to follow a path by never going away from infinity (3) and always returning to the point infinity holds but never lash onto the point again. With space parting the points, eternity had two points (the past and the future) before the partition came about and infinity held both the past and the future while infinity had the present as it still gas presently. By eternity also moving, the two points it held opposed each other (the past and the future) and since it moves, by the movement it became the square of the two because movement is the square and not a flat blanket-like surface with squares embroidered on it as Newtonian science depicts it by using grand mathematics to understand singularity.

Then we had two point holding eternity in place going square by movement to form 4 points serving eternity and infinity captured the first three points held by both and since eternity could not release from the two it had but had to duplicate what it had, eternity by movement became a circle captured by the line. With four points captured by the line of three points the circle coming about is eternally returning to infinity but never complying with infinity because if mismatching temperature or movement (3 against four). Material will always be colder than outer space. It is because material spin and outer space moves by expanding due to overheating.

This is where I start when I start to explain the first moment but I use a shipload more information to do explaining when I explain the star in the book I do so. I involve the four cosmic pillars to substantiate the claims I make because all four still work the very same way as it did at the beginning of the Universe. The three points serving one part of singularity combined with the four points serving singularity unites as seven to form a circle of either 3.1416 or 21.991÷7. The seven going to one is eternity matching infinity by movement. But since seven moves it are seven that have to produce gravity. How do I know all these facts, because we can see from the top it is still doing what it did the very first second. When time started infinity as well as eternity had altogether 3 positions, the past, the present and the future. It is still forming the very line in the centre of the top as it forms all lines in the centre of all things spinning. Then eternity parted from infinity when heat separated what is cold from what is hot and eternity formed one more point than before when it had the three points.

With infinity and eternity then jointly having 7 the cosmos came into rotation. In the aftermath post big Bang we now see the phase of cosmic development where the tow sectors try to unite and this brings along the contraction. When Π forms it does so on the grounds that 7 rotates. The circle forms by a change in direction by 7°. Every circle has opposing sides forming in relation to the axis line. If the topside goes rite then the bottom side has to the left. If the rite side goes down then the left side goes up. There is this double presence of a change in direction forming on both sides of the circle. The 7° move and by moving 7° goes square 7^2 and that is Pythagoras.

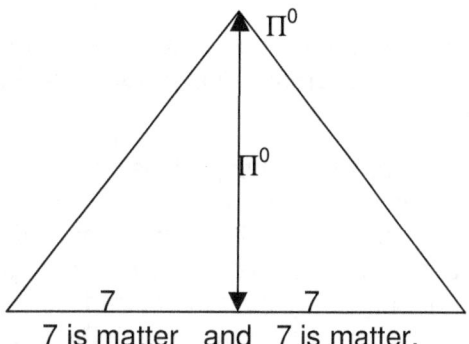

7 is matter and 7 is matter.

They join space-time therefore the matter factor is the same. This is where one can visually see the one object, filling the space of the other object's atmosphere.

$$7 \times 7 = 7^2 = 49$$

That is matter Π^2 (time) times matter (49)+(1) = 50. This 50 forms space which then applies to both sides of the rotation of the solid being 7 that rotates.

As this is all under the law of Pythagoras the law will evidently place a square root to that value of 483,61 and therefore $\sqrt{50} + 50 = 10$. This leaves the space value of the Roche-limit, as it develops into the Titius Bode law giving them a shared value of 7 (matter) and 21,91 (space) the value of 21,991 / 7 = Π.
Then the relation becomes

$(\Pi^2+\Pi^2)$ $(\Pi^2\Pi)$ $(\Pi\Pi^2)$ $(\Pi^2\Pi^3)$ holding space (3) still outside. They therefore will share space and that sharing will continue till times end. We know by now that matter is 7, and space is 3. holding time to a relevancy in singularity of 1. Sharing the space means that 21,9 will become (10) space to the one side
1   to the instant position of time (k^0)
,99   lost to space depletion $\Pi^2/10$
7   the relation to matter.

Through that the Titius Bode law comes into affect of 10/7 or 7/10, depending on whether space or matter holds a superior position to time. From that stance, all objects will relate to one another by the value of $\Pi^2\Pi$ and seen in a whole sale total 7/10 or 10/7.

That means to become part of the neutron status of the earth, the object has to be space (21,991 or less) and prove to be matter (7) before the earth will accept it. If holding a position of less than 7°, the earth will discard it and if it is more than 21,991 the earth will find the relevancy to be higher than the space it holds in a neutron time. That places the object in a relation of $4\Pi^2 - \Pi^2$ (because it is not part of the earth) in a position acceptable matter holds (7) within the confinement of Π (21,991/7). That means the object is part of space (21,991) acting as matter (it holds an acceptable own proton structure) 7 relating to the earth in the position the earth allows of $3\Pi^2$.

That means to become part of the neutron status of the earth, the object has to be space (21,991 or less) and prove to be matter (7) before the earth will accept it. If holding a position of less than 7°, the earth will discard it and if it is more than 21,991 the earth will find the relevancy to be higher than the space it holds in a neutron time. That places the object in a relation of $4\Pi^2 - \Pi^2$ (because it is not part of the earth) in a position acceptable matter holds (7) within the confinement of Π (21,991/7). That means the object is part of space (22,991) acting as matter (it holds an acceptable own proton structure) 7 relating to the earth in the position the earth allows of $3\Pi^2$. With the space position of the matter in the parameters of 21,991 it relates to the Titius Bode law as a factor of one. The object has the space value of 21 (3 x 7), which shows the axle value turning, plus the space value of .1416, in that instant of time (7) complying to the earth's space (.1416 x 7) in reduction (Π^2) formulating .1416 x 7 = 0,991.

That makes the object complying with the full agreement as laid down by the Titius Bode law. The object is, no matter where it is, travelling at a rate of 7 ($3\Pi^2$) in the space of the earth (21,991). This will be agreeable to the parameters of the Titius Bode law as long as it remains within the space depleting "gravity" limits of less than Π^2. In accordance to the Lagrangian atom layout, anything less than 5Π is manageable and is in effect less than Π^2. When it exceeds 5Π it will start opposing the dimensional equilibrium space holds of 10Π, therefore it will (according to space) exceed the linear point of R/T, which is $10\Pi/2$ (space going in a straight line).

Everything in the cosmos is moving, either by own individual accord, or under the influence of some other singularity dominance. In explaining we return to Pythagoras where the entire Universe with everything in it started.

It is the point forming the very centre that plays the part as the **controlling singularity** within the Universe I have named as **Infinity,** which is better known as the axis. It is where nothing can go smaller and anything within that point can never reduce. That point is where the entirety called the Universe begins and where everything holding substance begins. Once one accepts the fact of singularity being present in that location, that accepting of singularity then is contradicting all the things we know and we can measure and we recognise that point being present by merit of the fact that the point referred to is not being formed by any of the things we can recognise.

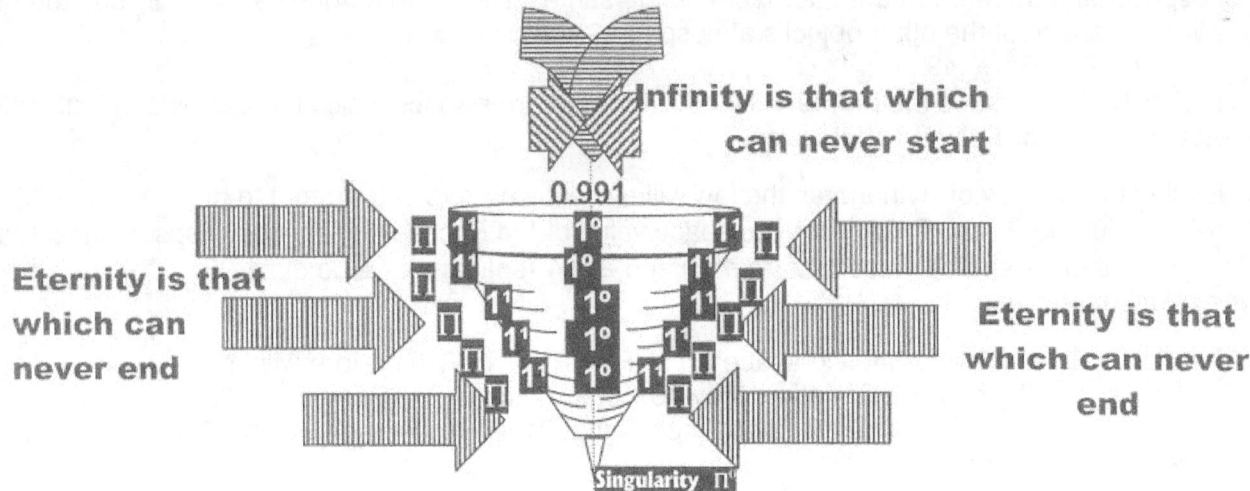

It is made up of everything we don't know and constitutes of everything we are unable to recognise or visualise. In that spot there is no space. That spot holds **Infinity.** In that space there can be no motion because there can be no space to have the motion within. It is formed as a line that is so small that our human reality by perception declare that point as not being there and the only reason why we know it is there is because of the results it left as an imprint of its not being there. We cannot detect it but notwithstanding our failure to note it we can recognise the dot on the merits of its absence and while in our Universe it is always absent, reality disallows the dot ever to be absent, because it is never absent. It cannot be absent. It cannot go absent but it can never be there where it should be in a place from where the third dimension forms and it is always present if I wish to locate it. It is **infinity** that can never go away. I named the other part of singularity forming space **eternity** because that area never become bigger, or become more or find an end to the outside. Whatever was and is and will ever be is locked in that space I named **eternity** and it is **eternity** that never ends because **eternity** can never end moving.

What we think of, as expanding is never ending movement giving eternity the eternal motion that will go on forever. The line **k** coming from the centre (singularity k^0) forms by forming an initial spot Π^0 becoming the dot Πr^0. However, I went on to say that whatever the line used to start with has to continue in order to repeat the same that began the line. Therefore the line started with Π^0 and it has to continue with Π^0 until such a point, as it must end with Π. Whether the line is Π^0 or is r^0, or uses 1^0 the outcome all refers to singularity being used. By reducing the line we come to the end of the mathematical equation of the circle but the circle does not end there. When the top is in a state of motionlessness on own accord it is everything but motionless. The motion it adapts are synchronised with the earth in harmony with the solar system and according to the greater picture of the cosmos.

The one side depicting singularity is one $= \Pi^0 = 1$

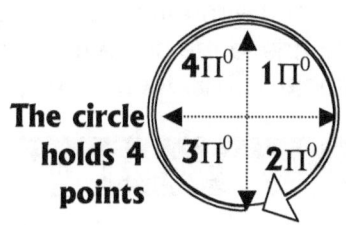

The circle holds 4 points

The axis forming the line holds three points and the circle holds 4 points

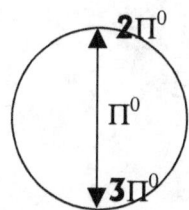

The axis forming the line

The one side depicting space-time is three = 7

The calculation of the triangle involves the law of Pythagoras.

Since I am explaining the most elementary of mathematics I better explain the law of Pythagoras as well. (For those SUPER-EDUCATED-MASTES BEING TO ADVANCE TO REMEMBER THAT)
The law of Pythagoras states that the sum total of the square of the two sides in a rectangular triangle will always be equal to square of the perpendicular side. How basic in mathematics can we still go?

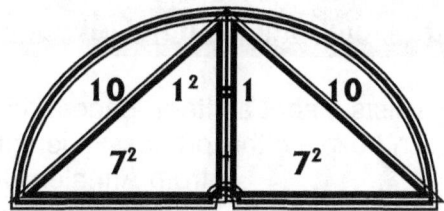

The axis holding $3^2 = 9$ and singularity forms space – time at 10

The square of space from the point of singularity is $2(1^2 + 7^2) = 10$

I have indicated a few pages back that the position the proton holds in considering singularity is $\Pi^2 + \Pi^2$ and the neutron holds in same fashion a matter in space of $\Pi^2 \Pi$. I also showed that matter can only relate to space by implicating matter in the sphere and therefore has to use the value of Π as a reference. In one dimension space became 10 and in that same dimension matter became seven.

In order to separate matter (7) and space (10) through time (the spinning of matter) (7) in space (10) and space (10) spinning the matter (7) the following result came about through the application of the Roche principle $(\Pi/2)^2$

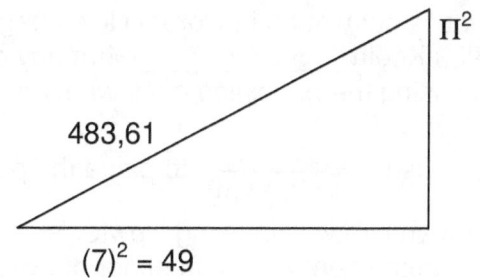

Anything that goes straight commits to a circle because by going straight it has to form a circle by forming Π. Gravity is what forms a circle because the smallest line that can enter the Universe is a sphere holding 7 points in relation. in a circle. In my books I prove how this then brings about the value of Π by implementing the law of Pythagoras and gravity is the law of Pythagoras. The reclining of space by redirecting the direction of travel from straight ahead to 7° reclines the space in a steady and sturdy flow. It is the space reclining or contracting and the space contracts albeit filled by solid material or empty of solid material. This is the reason why all things fall equally. It is the space moving down with or without holding material and the space has the same density in relation to the solid cosmic structure rotating.

In the centre of the rotating body singularity forms to the value of $Π^0$ and this extends to the curve forming Π in terms of the curve of the rotating object. This then forms part of gravity moving forming

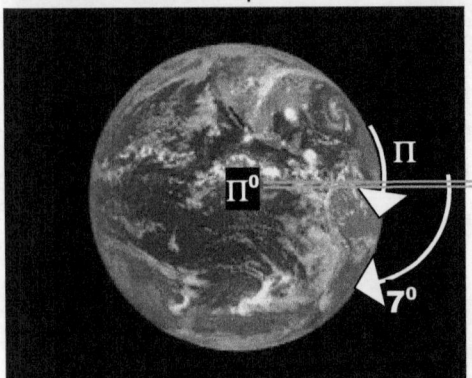

The extending of singularity goes from $Π^0$ to $5Π^0$.

singularity going square $Π^2$ and this form the relevancy Coming in from space while having no mass the coordinates change as the value of Π forms a relation to the 7° When an object comes in from the atmosphere or flies the sky the object associate with the turn giving it a value of the rotation in line with the axis which puts a value of 21.991 / 7 on Π or on gravity. This is gravity and this is how gravity functions. It has nothing to do with mass in any way, shape or form and there is no factor such as mass in the entirety of the cosmos apart from being in the imagination of the Newtonian conspirators calling them

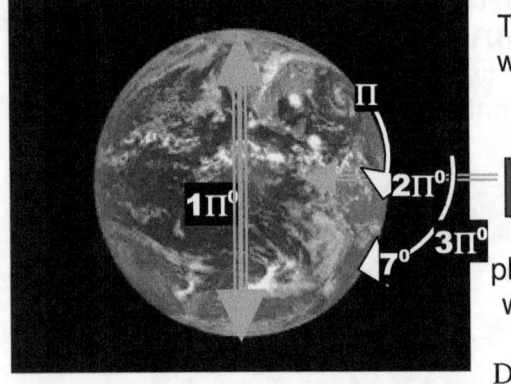

The entering by singularity that goes from 7° by $3Π^2$.

physicists. I sent a fifteen-page article to the Annalen Der Physics in which I explain this process in length. I received the following reply from Annalen Der Physics.

Dear Dr. Schutte, You submitted an article of 15 pages to the Annalen. The content of this paper doesn't constitute a theory in physics. With a lot of words and some simple algebraic relations, there is no way to "explain" the world of physics. You seem to be out of touch with modern developments. This is also shown by the fact that you don't quote any relevant literature. I am sorry to say, but the Annalen is not able to publish your work. I am sorry for having no better news for your. Best regards, Friedrich Hehl
Co-Editor Annalen der Physik (Berlin)
Friedrich W. Hehl, Inst. Theor. Physics
* University of Cologne, 50923 Koeln _____/_____ Germany
fon +49-221-470-4200 or -4306, fax -5159
hehl@thp.uni-koeln.de, http://www.thp.uni-koeln.de/gravitation
* Univ. of Missouri, Dept. Phys. & Astr., Columbia, MO, USA

From this reply I received it is clear professor Friedrich Hehl did not understand a single word I sent him because as he said, it was not in mathematics but written in words! Now I beg of Friedrich W. Hehl, Inst. Theor. Physics University of Cologne, 50923 Koeln _____/_____ Germany or any of the editing staff of Annalen der physics or any physicists professing the principles of Newtonian inspired physics to prove to me how does the formula $P = \left(\dfrac{4\pi^2 a^3}{G(M+m)}\right)^{0.5}$ use P = ─────── to place the positions of planets because
                                                                                G(M+m)
this is one of the supposed Kepler Laws. I wish him well in using Newton's corrupt formulas to prove that Jupiter with having the "most mass" is bang in the centre and on either of the outside and inside we have the two smallest planets. In that put the mass in relation to the position. Prove to me that mass is a valid

factor in the cosmos. Physicists live in Newton's dreams. My work proves nothing so please prove Newton when he claimed $P = \left(\dfrac{4\pi^2 a^3}{G(M+m)}\right)^{0.5}$ especially the part P = $\dfrac{}{G(M+m)}$.

Ask your physics professor to explain what causes the Sound Barrier in terms of physics... and moreover in terms of using mathematically applying Newtonians physics. Ask them to explain why the Sound Barrier has nothing to do with speed in the specific but holding space during specific time duration. You can't break the sonic boom at 31 thousand meters above the earth because there is too much space. Therefore not being able to break the sonic boom still puts the movement within the sound barrier although there is no sound at that altitude. The space is too large to be able to go at Mach 1. They would not know that!

Part 9 Introducing the Titius Bode Law.

Chapter 9 — The Titius Bode Law

This is true astrophysics. It is not mathematical misrepresentation of what those that has no idea about astrophysics try to sell as astrophysics. This is what the Universe presents as astrophysics. When a study is done on cosmology one would find the Titius Bode law in place ands although it never was acknowledged as such this is what is astrophysics.

It might not be what the Newtonian astrophysicists wanted to find because it does not apply such illustrious and breathtaking physics as they wish to place in the Universe but it is what is in the Universe. This is how the Titius Bode law applies as it is in the solar system.

It is not even for the first time since Newton or for the first time since Galileo or for the first time since Archimedes or who ever is your favourite all-time famous scientist that changed everything about science but this is the first time ever since there was life on earth that the Titius Bode law is explained and is defined by an explanation.

If you are smart can you see there is no way in heaven or hell that mass will fit into the factor that is responsible for planet mass being a factor that forms gravitational layout positioning. This presentation of how the solar system lay out is and then connecting mass as the building block forming the solar system is as false as what money as a currency ever can be. By gauging the tables Kepler left us it is clear that the space k is moving in decline or reducing by going smaller in value. This means the sun is drawing all the space not material towards the centre.

Mercury	$T^2 \div a^3 =$	0.983
Venus	$T^2 \div a^3 =$	0.992
Earth	$T^2 \div a^3 =$	1.000
Mars	$T^2 \div a^3 =$	1.000
Jupiter	$T^2 \div a^3 =$	1.000
Saturn	$T^2 \div a^3 =$	0.999
Uranus	$T^2 \div a^3 =$	1.000
Neptune	$T^2 \div a^3 =$	0.999
Pluto	$T^2 \div a^3 =$	1.004

The table that Kepler provided shows the time that the orbit of every planet takes according to the distance the planet travels in the same time lapse and considering all the planets it is very much the same thing and in that there is no provision in the table for any idea that might form mass. This ratio is the indication of speed travelled. The idea that mass exists is a Newtonian invention made up by Newton and is completely groundless except for the value that Newtonian science gives it in order to maintain the Newtonian principles. The entire idea of mass is a myth. The idea that mass pulls mass is complete mythology and is as baseless as any fairy story. But even more deceiving is that notwithstanding that every planet has a value when $T^2 \div a^3 =$ 0.983 which is $a^3 \div T^2 = k$ Newtonian science completely ignores the values and declares that $T^2 = a^3$ whereby they ignore the values in the column. That is cheating the truth into submission to corruption to say the least. Newtonians fabricate their truth.

Using the actual table the mathematical formula says $T^2 \div a^3 =$ 0.983 which is $T^2 \div a^3 = {}^{-1}k$. That puts the flow of space including all space directed towards the sun. ${}^{-1}k = T^2 \div a^3$ says mathematically in a formula that the circle moves towards the sun because the direction being negative is negatively directed. What the table also say is that the Newtonian presumption that a square can be equal to a cube or $T^2 = a^3$ is colossal madness because the table says clearly in mathematics that $T^2 \div a^3 = {}^{-1}k$. Therefore what Newton suggested does not apply?

What we do find in the solar system and what does apply is the Titius Bode law.

Planet	Mercury	Venus	Earth	Mars	Ceres	Jupiter	Saturn	Uranus	Neptune	Pluto
Bode's Law distance	4	7	10	16	28	52	100	196	-	388
Actual distance	3.9	7.2	10	15.2	28	52	95.4	191.8	300.7	394.6

A numerical sequence announced by J.E. Bode in 1772, which matches the distances from the Sun of the six planets then known. It is also known as the Titus-Bode law, as it was first pointed out by the German mathematician Johann Daniel Titius (1729-96) in 1766. It is formed from the sequence 0,3,6,12,24,48,96, and 192 by adding 4 to each number. The planets were seen to fit this sequence quite well – as did Uranus, discovered in 1781. However, Neptune and Pluto do not conform to the 'law'. Bode's Law stimulated the search for a planet orbiting between Mars and Jupiter that led to the discovery of the first asteroids. It is often said that the law has no theoretical basis, but it does show how orbital resonance can lead to commensurability.

The importance that becomes known is the sequence the Titius – Bode law saw in the number arrangement of 3; 6; 12; 24; 48; 96 etc. The incorrect application of the Titus Bode law lies in subtracting the figure of 3 from 10 leaving 7. The other way of reasoning is to add four each time to the firs value of three starting with 3 and so on. The true significance of the Titius-Bode law is that it points directly to a circular growth of 7 stages. The 7 relating to 10 is a precise derogative of the Roche limit or the Roche limit is a precise derogative of the Titius Bode principle because he two systems interlink. This is how I mange to explain the Titius Bode law that is in the solar system by the ratio applying that really form the solar system in the way nature shows space growing by time. What you see on the next page was never been shown but on the other hand Physicist say this mathematics are too simple to apply as physics! This is what there is. This, the Titius Bode law is a given as it is fact. With the exception of the first two, the others are simple twice the value of the preceding number.

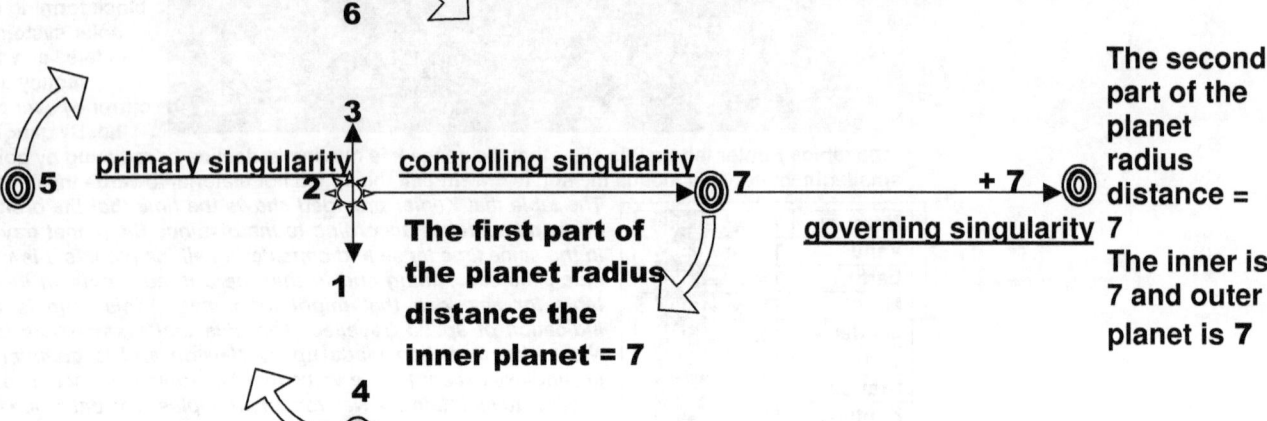

To find the mean distances of the planets, beginning with the following simple sequence of numbers:
0 3 6 12 24 48 96 192 384

To explain this is as follows: the axes line holding 3 doubles because in singularity all axis lines are 3 (1 bottom 2 centre 3 top)

Add 4 to each number:
4 7 10 16 28 52 100 196 388

Then the circle value that always has four point.

Then divide by 10:
0.4 0.7 1.0 1.6 2.8 5.2 10.0 19.6 38.8

Travelling in a straight line or a half circle or a triangle in terms of singularity is equal because it is all 180°. By taking 7 (the first or inner planet) in terms of Pythagoras 7^2 breaking the centre line 1^2 the result is 49 + 1 = 50. The second circle also values 50 and since singularity unites the movement it totals to form 100. The square root of 100 is 10 and dividing the travel by 10 the allocated position becomes valid. Is this not far better and truer than the following Newtonian accepted rubbish?

The Titius Bode law is a part of the Coanda effect and the Coanda effect is the interaction there is between what is material and that which holds material. That which is 7 is apt of material and the 10 part is liquid / gas. In the picture the outside planet holds relevance with the line that the direct inner planet has and that in turn holds relevance to the axis the sun holds. The outer planet has the **governing singularity**, the inner planet holds the **controlling singularity** and the Sun has the **primary singularity**. As soon as any object moves on Earth, the movement switches singularity by allowing the object to obtain the **governing singularity** while the Earth then fore fills the directional circular control in forming the **controlling singularity**. All four phenomena interacts in a manner forming this role where for instance in the solar system the Sun holds the **controlling singularity** and Milky Way forms the **governing singularity**

The outer planet is the **governing singularity** that forms a governing position in terms of the location and that has a place in terms of the second 7 that will eventually form the 10 of space. This is the outer 7. The planet on the inside of the planet holding the governing position has a **controlling singularity** since it delivers the four points in rotation that positions the last 7 with which the governing singularity finds position. This hold the value of the inner 7 of which the **controlling singularity** forms a value of 4 as that forms the **controlling singularity** circle in the Titius Bode law. The sun axis holds the **primary singularity.**

The axis around which the Sun turns forms the **primary singularity.** This **primary singularity** is the axis that draws all the space ($T^2 \div a^3$) towards the sun and in that it has the absolute **primary singularity** role of forming the value of 3. The first 3 (the sun) and the second 4 (the **controlling singularity**) forms the first 7 while the **governing singularity** is the positioning point to form the two 7 point that forms the 10.

I am able to explain what is there by using "With a lot of words and some simple algebraic relations, there is no way to "explain" the world of physics." Well where I am out of touch with the world of physics it seems that the world of physics is completely out of touch with reality. Furthermore because "Your seem to be out of touch with modern developments." How improper it might seem when compared to the grandeur of mathematical splendour, mine works because it is there and Mainstream science lives in fantasy because what they see is not there.

What it is that these two phenomena has in common except sharing a Universe because he should know, after all he is the Newton-physics expert ... but I will bet you although being a physics expert, your professor will have no idea...mainly because he is the Newton-physics expert. This prevails because of singularity, which clashes with Newton head on As Kepler said $a^3 = k T^2$ and therefore $k^0 = a^3 / k T^2$ and therefore we have to find k^0. As a result of examining this proposition, I located two principle positions both holding singularity. The cosmos is made up of one type (1^0) that is in two categories where one type moves and the other type does not move. The one is a liquid and the other is a solid..

The condition for the presence of this singularity that forms everything, controls everything and is everything is the centralised to a centre singularity $k^0 = a^3 / (T^2 k)$ that forms by movement $T^2 = a^3 / k$ of space $a^3 = k T^2$ placed in relevancy $k = a^3 / T^2$ that is centrifugally going both ways $k^{-1} = T^2 / a^3$ thereof (Newton's 3rd law). This explains the Coanda effect and the Coanda effect is gravity and gravity "glues" the water to the glass by implementing Π to form singularity! *What is in the Universe is spinning*. **The entirety of everything forming the Universe is spinning inside the Universe** and such spinning are always in the centre of one specific point, wherever such a point might be. In the **precise middle** of all **objects in rotation** is a precise centre where this pre-designated centre is dividing the object in rotation into sectors that will **start the spinning initiation** from that centre point. This is what Kepler's formula confirms in $a^3 = T^2 k$. By spinning, the one side is coming towards while the opposing side at that time is going away. Thus, the spinning object **will have a middle point**, a very specific **centre point that does not spin** and only holds Π as a specific value because within that centre being that small, no radius can apply. We have named this position or line the axis, but the true meaning of this line has eluded us since the concept was realised. This line that forms holds no space although it directs all the space that it controls by

spin. When going toward the centre where the axis form at the very centre of rotation, the space on the one side has to end and the space at the other side has to begin with the line unable to hold space.
It is not you being glued or not being glued to the Earth that I discard when I question the mass idea.. It is the definition holding this whole idea that I do not share in the least. What the definition describes is magnets pulling and it is the total opposite of what I experience. Breaking the first millimetre of gravity clampdown is the easiest and not the most difficult. The difficulty increases as the radius grows and not as the radius decreases. When I say there is no gravity everyone thinks I say we all are going to fall off the Earth at random and with me thinking that way then it is obvious that I must be a nut. Everyone thinks of me as the clown acting mad when I say gravity is not to be found in nature. But I do not say we are not standing on the Earth. I do not say there is nothing that is keeping me glued to the earth. I say there is no attraction between two bodies by the force of the mass that is such doing is diminishing the radius parting the bodies by the inverse square law. I say there are a connection by motion between the centre of he body and the material surrounding the centre. This is what I say when I say there is no gravity.

The condition for the presence of this singularity that forms everything, controls everything and is everything is the <u>centralised</u> $\Pi^0 = \Pi^3 / (\Pi^2 \Pi)$ <u>singularity that forms by movement</u> $\Pi^2 = \Pi^3 / \Pi$ <u>of space</u> $\Pi^3 = \Pi\Pi^2$ <u>in relevancy</u> $\Pi = \Pi^3 / \Pi^2$ <u>going both ways</u> $\Pi^{-1} = \Pi^2 / \Pi^3$ thereof (Newton's 3rd law).

This explains the Coanda effect and **the Coanda effect is gravity** and gravity "glues" the water to the glass! The water forms a value of $\Pi^{-1} = \Pi^2 / \Pi^3$ while the glass forms a value of $\Pi = \Pi^3 / \Pi^2$. This process happens to all spinning things and as much as it happens to a piston connected to a crankshaft, just as much this will happen to an atom spinning an electron in a similar manner as the crankshaft is spinning holding a piston connected. This proves that gravity is the Coanda effect and in another book I prove that the Coanda effect has its origins in Π forming a value and that value forms gravity.

In order to understand physics applying in cosmology I had to start by dissecting the set-up forming pi. At this point I can introduce my theory on the ***Absolute Relevancy of Singularity*** At the point in the centre of the circle a line must start. In the beginning when I explained the way I figured how the line starts I said a lot of dots has to continue in order to form a line. It would be 1 + 1 + 1 etc. because the line must form by holding singularity. After that point does mathematics begin but in the line that forms representing space as other all factors, then time holds 1. The line can only form when all the points forming the line have the value of 1 being 1^0. In that conclusion one realises something must separate singularity from all other factors because singularity hosts all other factors but is by own initiative Π^0. There are always a line of atoms made up of spinning subatomic particles also spinning holding dots Π^0 without space and the line runs through the dots not having space but still connecting by Π^0 or 1^1 forming a line $1^0 \times 1^1$ up to Π.

Only when singularity meets the end value can the end value have Π where the final ring of the spinning circle forms Π. That will be the spot of origin forming the relevance in Π. That will hold the eternal spot…the smallest spot ever because all spots that ever can be were secured in a position in the centre of that spot that must continue as a line that forms. Because of the progress singularity follows from the single dimension singularity only allows mathematics a start at Π^0 progressing further onto Π^0 and from there the line is born as $\Pi^0\Pi^0\Pi^0$ and to $\Pi^0\Pi^0\Pi^0\,\Pi^0$ etc. where Π^0 then may form the concept and value of r. But the line starts at $\Pi^0 = r^0$.

This forms because cosmology is singularity based and the value is $\Pi\Pi^{0.}$ This line $\Pi^0\Pi^0\Pi^0$ of singularity can only continue because every spinning atom preserves Π^0 in the very centre and since $\Pi^0 = \Pi^0 = \Pi^0$ the line is the same without finding conclusion except at the end where it forms mass at Π. At the point where Π forms, the movement Π^2 of the circle defines the space Π^3 of the circle and it confirms the centre Π^0 of the circle through the rotation. Let's call this the solid forming or if you wish, let's call it Kepler's singularity. After that singularity forms a line $\Pi^0 = \Pi^0 = \Pi^0$ where this forms another line again as Newton stipulated it by $\frac{dJ}{dt} = 1^0$. Let's call that the liquid singularity or Newton's singularity and the relevance of singularity having a solid base compared to the singularity holding a liquid base comes about by the movement of gravity.

a^3 symbolises in a mathematical interpretation of implicating the three-dimensional space holding a specific centre in relation to another specific centre indicated by **k** that could apply to either centre points in question. This is always a straight-line **k** representing the position of the **controlling singularity** moving in a circle T^2. The space forming a^3 is a **positional validity** of the space indicated by $k^0 = a^3 / (T^2 k)$.

T^2 is representing the circle that goes around the **governing singularity** k^0 that forms in relation to the line **k** in reference to the centre k^0 The space that forms holds the orbiting planet a^3 in direct circular contact with the space in relation to a very specific centre k^0 moving from point T_1 to T_2 in relation to a precisely placed centre k^0. The circle coming about from T^2 is the **controlling singularity** which is always a circle at the centre that is poisoned by the line **k** in relation to the centre k^0 and by forming a circle it holds reference to the **governing singularity.** Where **the governing singularity** is the centre of a spinning object such as the Earth, the centre of every atom holds **mutual singularity** that collectively puts a mutual value of all the atoms' singularity as a combined equal to the **governing singularity** and then the solar system will provides a **primary singularity**. The one would represent T^2 the other forms **k** that then produces the third singularity forming space a^3.

k is the space taken from the centre k^0 to the end of the line **k**. This line shows where the location is around which planet circles. The specific value about the centre is most important because from the specific centre gravity indicates a positional worth. The line forming **k** is pointing the circle or the **governing singularity** formed as a line that eventually forms a circle running from the centre k^0 to where the space a^3 is indicated.

The turning T^2 of any circle holding space a^3 is valid only if forming a reference **k** to a centre k^0. $k^0 = a^3/(T^2 k)$. This depicts a position a domineering singularity k^0 fills in relation to another point serving subordinate singularity **k**. There are always a dominant and a serving singularity interacting. If **k** indicates the centre of the Earth then T^2 rotates to form the **governing singularity** k^0 where then the centre of the Sun **k** will form the **controlling singularity.** When the Sun rotates, the Sun's centre k^0 forms the **governing singularity** giving the Earth in orbit **k** holds the **controlling singularity**.

The measure of **k** is not a specific value but serves only as an indicator to which space rotates or applies by the space rotating in a circle. This role of singularity being **controlling** or **governing** is playing part in movement of gravity forming and is very important when trying to understand the role that the four phenomena play in the forming of gravity. It is most important to understand what happens in the event of an object going through the "sound barrier" or when escaping from the Earth's atmosphere.

Where the object is standing still holding a position that allows the object to have mass, the object is part of the Earth while the Earth has the **governing singularity** and the Sun has the **controlling singularity**. As soon as any object moves on Earth, the movement switches singularity by allowing the object to obtain the **governing singularity** while the Earth then fore fills the directional circular control in forming the **controlling singularity.** All four phenomena interacts in a manner forming this role where for instance in the solar system the Sun holds the **controlling singularity** and Milky Way forms the **governing singularity** .

To Venus forming the **governing singularity** Mercury is the **controlling singularity** and the sun is the **primary singularity**.

To the earth forming the **governing singularity** Venus is the **controlling singularity** and the sun is the **primary singularity.**

To Mars forming the **governing singularity** the earth is the **controlling singularity** and the sun is the **primary singularity.** That is why this table forms.

Planet	Mercury	Venus	Earth	Mars	Ceres	Jupiter	Saturn	Uranus	Neptune	Pluto
Bode's Law distance	4	7	10	16	28	52	100	196	-	388
Actual distance	3.9	7.2	10	15.2	28	52	95.4	191.8	300.7	394.6

Where there is anomalies we can read into it events that happened in the past we were unaware of.

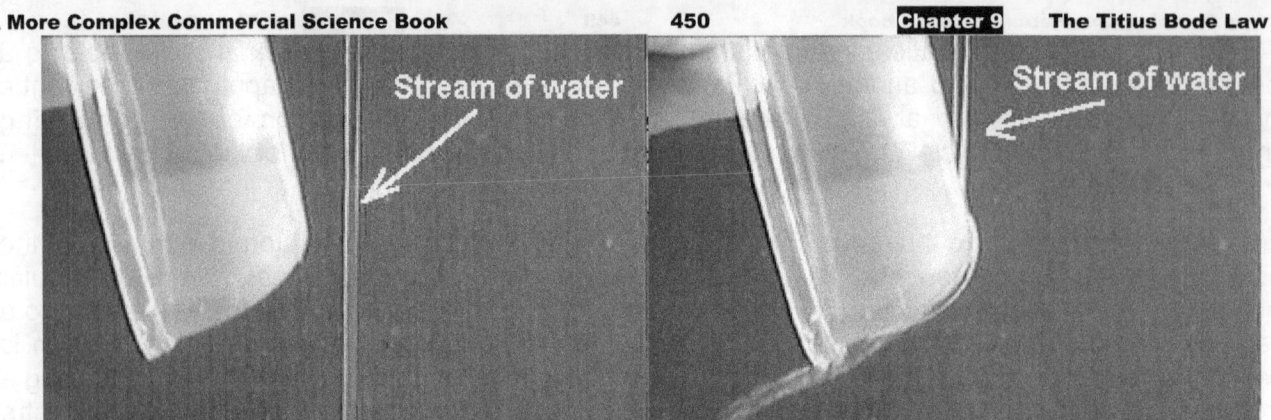

I have **written twelve articles** in which I explain the Titius Bode law, why it is in place, how does the Titius Bode law apply gravity, what keeps the Titius Bode law structurally in place and why is it in place as it is.

The gravity it should hold in distributing movement will bring along a certain amount of linear gravity to maintain the circular gravity position it wants to hold in the cosmic balance. The duel movement of gravity forming the allocated relevant position provides ratio of singularity Π accompanied by the same value but in movement by the square thereof Π^2 and repositioning the structure as a star.

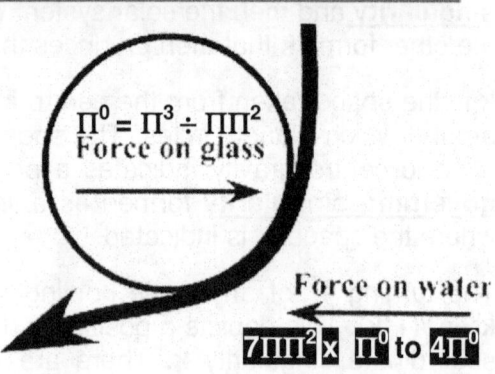

In short this is my explanation. Now please compare this to the accepted mainstream version and see why all my work up to now is ignored as if it holds the plague. See now what is science and what is fiction and why my work is ignored. That is the "With a lot of words and some simple algebraic relations, there is no way to "explain" the world of physics." "You seem to be out of touch with modern developments." **of which I am accused of.**

Please allow me to show you how they scare you to become fooled and suckered. Don't run and hide when you see the mathematics; it is meaningless although it was used as a scarecrow for more than three hundred years forming the backbone of the conspiracy.

This picture is as big a hoax as Newtonian science is when Newtonian science presents mass to be a factor that produces a pulling force called gravity. The question shouting for an answer in the picture is if mass is a factor that produces gravity as Newtonians claim it is then why are the planets not positioned according to mass as Newtonians declare.

The claim is as bogus as the entire philosophy. They present the proof that planets orbit according to mass in the following "Kepler law" which in its entirety had nothing to do with Kepler at all. It is all devised by Newton because Newton had no inclination of what Kepler's work was about.

Newton brought about the idea of mass positioning the planets in the formula $4\pi^2 a^3 = P^2 G(M + m)$

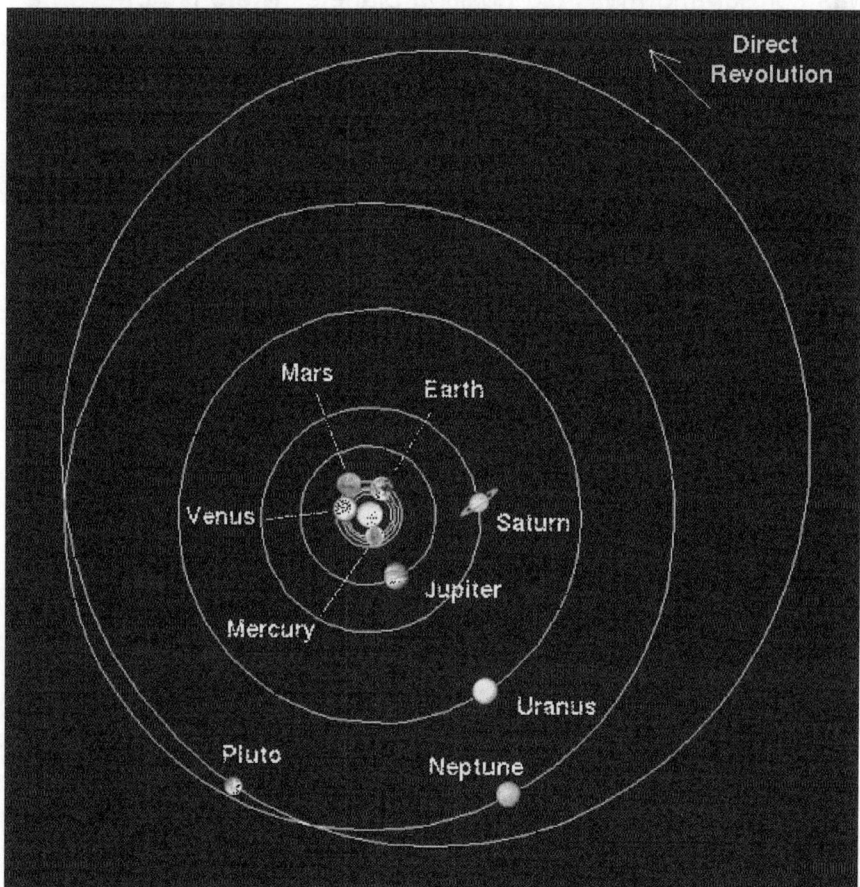

Do not get scared as everyone usually does when seeing the mathematics and then as a result get frightened. Those physicists expect you to turn on your heels and run as fast as your legs can carry you. Then consequently as a reaction to find survival, you turn on your heels and run… but this time don't. Don't run, just read on and see how simple it is to prove Newton was a backward dark aged sod!

This time, don't run because I am about to show how meaningless this entire mathematical statement in reality is! This formula is total garbage and there is no sign of evidence that this formula forms any part of the solar system, even in the least.

Lets test this formula and see how truthful it is. $4\pi^2 a^3 = P^2 G (M + m)$ indicates that the circle in which the planet orbits ($4\pi^2 a^3$) is the result of (=) the position of the body (P^2) positioned by the mass of both bodies ($M + m$) in terms of the gravitational constant (G).

The best way to find clarity is to test this statement with what is happening in the solar system just as it is, wouldn't you think.

A picture such as this provides much credence to the idea of gravity by mass since the lines drawn does not even begin to represent what is truly out there ands what is used by the cosmos in place of Newton's mass concept.

This is a table indicating the **mass** that **every planet** has in relation to the **distance every planet holds** in terns of the sun. If mass positioned planets then why did no one bother to inform the Universe about this because it is clear the Universe did not receive the memo from Newton's office to act accordingly.

Body	Mass (10^24) kg	÷	Orbital Distance(10^6 km)	=	ratio
Mercury	.3302	÷	57.9	=	0.0057
Venus	4.869	÷	108.2	=	0.045
Earth	5.975	÷	149.6	=	0.039939
Mars	0.6419	÷	227.9	=	0.00281
Jupiter	1898.6	÷	778.3	=	2.439419
Saturn	86.83	÷	1427	=	0.060847
Uranus	102.43	÷	2869.6	=	0.03569

This picture shows the hoax the Newtonian conspiracy pampers to keep the rest of Newtonian physics believable. They never mention the Titius Bode law and try to explain the Titius Bode law while it is the Titius Bode law that is really in place in the solar system. *If you wish to learn the truth then think again.*

Mass as a factor does not present or apply in one instance anywhere in the entire Universe and yet that is all theta physics says applies...but why would they cheat? It is because what is there applying between the planets is called the Titius Bode law and although this law is in place you have almost a hundred percent chance that you have never heard of it.

This is how it would apply if Newton was correct and mass did position planets. There is not even a remote chance that the positioning of the planets go in accordance with mass or $4\pi^2 a^3 = P^2 G (M + m)$. Do you realise there is much more "gravity produced by mass" in the space your feet has contact with the earth than there could ever be between Jupiter and the sun?

You that can calculate it so then do it.

Show that $\dfrac{a^3}{GM^2}$.

Put the orbit of Jupiter in relation to the mass of Jupiter and in relation to the position Jupiter holds. Forget getting swept away by the fancy Mathematics; just get to the task of putting the mass in relation or ratio with the position that any of the planets hold.

Take the mass of the earth and your mass you have and then divide that with the square of the distance there is between your feet and the earth by the square thereof then divide that square with the product of the mass of the earth and your mass you have. Keep in mind your distance between your feet and the earth is about 10^{-11} meters going square!

You that can't calculate it, the value would be meaningless but it is so much it will crush your atoms into a pulp, leaving you not even in a blood blob. It will leave a force to the value of about one Zetta 1 000 000 000 000 000 000 000 g per square meter. There is no chance in hell that any object of whatever size and formed by whichever method of construction could manage such a force of pressure.

Newtonians uphold their law of physics without showing mercy. The very first things the Newtonians use to beat us into submission are to blast us with incomprehensible mathematical formulas.

Those Physics-cheats always want to have the radius down to 1 meter because it makes their argument look sensible, and then they "forget" to use the correct mass of the earth not to stun any student into realising reality. Using one meter makes good sense when you wish to cheat but no body floats on meter above the earth. When anything stands on the earth, the distance between such a body and the earth is less than what could be sensibly measured.

Newtonians uphold their law of physics without showing mercy. The very first things the Newtonians use to beat us into submission are to blast us with incomprehensible mathematical formulas.

Incomprehensible they are but it is to scare anyone with the mathematical equations to get everyone hiding. They bewilder you with equations that put the fear of God into you; used simply to make you feel inferior so that they can feel superior and frown down on your inferiority from a dizzy height.

What Newton show that should place planets according to mass is not used by nature. I show what nature uses namely the Titius Bode law, The Roche limit, The Lagrangian Points and the Coanda effect and how this forms gravity as well as place the positions of the planets in accordance with singularity. Because I trash Newton's rubbish that does not fit and that can't apply no publisher of science books or science magazines will publish my work I show what goes on in nature while Newton's contribution of mass applying is total rubbish. Because I call it rubbish and I rubbish Newton I am ignored.

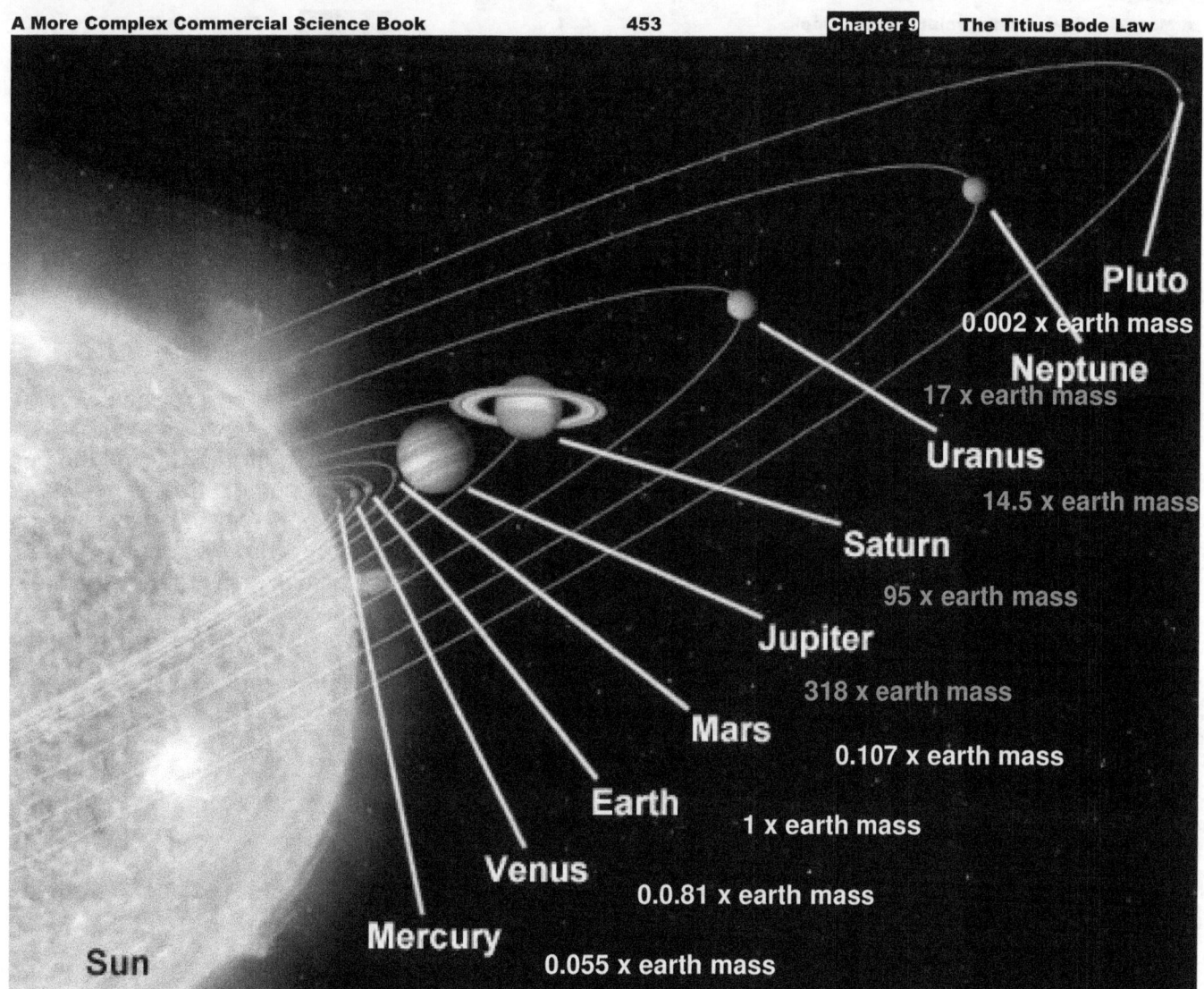

If the planet layout was as I now show it to be according to mass then this was the order that is in place in the solar system and if it is true according to the solar system that mass do produce the position of the planet:

1) Jupiter   **318 x earth mass**   at a distance of   57.9 million kilometres
2) Saturn   **95 x earth mass**   at a distance of   108.2 million kilometres
3) Neptune   **17 x earth mass**   at a distance of   149.6 million kilometres
4) Uranus   **14.5 x earth mass**   at a distance of   227.9 million kilometres
5) Earth   **1 x earth mass**   at a distance of   778.3 million kilometres
6) Venus   **0.81 x earth mass**   at a distance of   1427 million kilometres
7) Mars   **0.107 x earth mass**   at a distance of 2871 million kilometres
8) Mercury   **0.055 x earth mass**   at a distance of 4497 million kilometres
9) Pluto   **0.002 x earth mass**   at a distance of 5913.5 million kilometres

They are masters at manipulating anyone into a state of senselessness...but mostly that they do onto themselves. That they do because it forms the backbone of their fraud. They do not wish you to read closer and to find the fraud they hide to protect Newton. Ignore their mathematics because it only shows their incompetence to understand physics or Newton and see the fraud they propagate...

They employ mathematics to bewilder and that is all. I am going to show what we can uncover underneath what they cover. Look at what the mathematics supposedly says and then wake up, they are using maths as a scare tactic for three centuries to scare the daylights out of you and all this while its been working! Looking at the formula shows just how little Newton understood physics.
 Do not get scared as everyone usually does when see and get frightened then consequently as a reaction to find survival you turn on your heels and run... Don't run, just read on and see how simple it is to prove Newton was a backward dark aged sod!

I am going to show you how miserably incompetently incorrect this formula of Newton is when shown what the cosmos has in place. However, since I don't support Newton's blunders my work goes unpublished by science magazines and science publishers.

Bring me any mathematical formula Newton devised where he used mass as a factor and I show you how far the cosmos discards all of his claims. Newton is hugely wrong.

What the cosmos does use Newtonians reject because they can't explain it, they know too little about physics and secondly it rubbishes whatever fraud Newton thought up.

If the cosmos supported Newton's claims of $P = \left(\frac{4\pi^2 a^3}{G(M+m)}\right)^{0.5}$ then the planet arrangement would have been much more likely as I show above, but the picture indicates the mass as well as the planet formation. You must judge; it is either the cosmos that is incompetently wrong or it is Newton that is incompetently wrong because what the cosmos has in place Newton knows nothing about and what Newton claims the Universe uses, the cosmos knows nothing about. Who would you say knows more about the cosmos' method of workings, Newton or the cosmos? If Newton is correct then the planet layout must be as I show with Jupiter very close to the sun. It seem the cosmos is just as unaware of Newton's ideas as Newton is of what is happening in the cosmos. Who would be correct about cosmic principles applying, the cosmos or Newton?

Our Super-Educated-Mathematical-Wise use the elaborate formulas such as $\frac{d}{dt}\left(\frac{1}{2}r^2\dot{\theta}\right) = 0,$ and $P = \left(\frac{4\pi^2 a^3}{G(M+m)}\right)^{0.5}$ as well as $T^2 = \frac{4\Pi^2}{G(M+mp)}a^3$ to explain and prove what? If this formula statement is true then Jupiter must spin 317 times faster around the sun than the Earth does and be almost next to the sun while Mercury and Pluto in comparison must hardly move being cast into the darkness of the oblivious.

The formula $T^2 = \frac{4\Pi^2}{G(M+mp)}a^3$ says that the spin or time in which the circle comes about T^2 holds a space relation a^3 directly proportionate to the mass (G(M+m)) of both the sun and the applying planet multiplied by placing the gravitational constant also in relevancy. It says planets spin in accordance to mass…so lets see. I say prove it! Then $\frac{d}{dt}\left(\frac{1}{2}r^2\dot{\theta}\right) = 0,$ says that the Planets don't move at all because with $\frac{d}{dt}\left(\frac{1}{2}r^2\dot{\theta}\right) = 0,$ the movement acceleration is zero. Zero indicates no movement and that is as corrupt as the rest of Newton's ideas. Every person associated with physics to whatever extent has been and had been conspiring to hide the truth and the truth is that no one knows (not even Newton and even less Einstein) what physics is. What they say is precisely what the solar system proves different and either the solar system has no idea about what is driving gravity or they have no idea of what drives gravity, but it is definitely not mass. If they don't know what forms gravity and it is clearly not mass, then they know nothing about the way that gravity forms. They hide their incompetence under a blanket of ignorance and the lot commit to a conspiracy to hide the truth from the public. Let any one of those So –Wise-In-Mathematics show just how does Saturn arrive at the position it holds in relation to the mass it has by applying the Newtonian formula $P = \left(\frac{4\pi^2 a^3}{G(M+m)}\right)^{0.5}$, which they claim is data confirming Kepler's Law of Periods of the motion of the planets in relation to mass. Please prove this $m\ddot{\mathbf{r}} = -\frac{GMm}{r^2}\hat{\mathbf{r}}$ assumption Newton made. The go better and prove how Newton could have formulated and be realistic aboput what trully happens.

Now I challenge anyone out there to show Newton is not rubbish and Newton is correct. Please use the formula $T^2 = \frac{4\Pi^2}{G(M+mp)}a^3$ to prove Newton and then explain why the Universe does NOT apply Newton in the event where Newton is so admirably correct.

The formula that they say must allocate planet positions are $4\pi^2 a^3 = P^2 G(M + m)$ just as Newton introduced this concept because he said mass brings about gravity.
In this he had to force the issue even to the point of committing fraud and here comes the fraudulent part because there is no evidence of mass playing a part or forming an actual presence in the solar system.

$P = \left(\dfrac{4\pi^2 a^3}{G(M+m)}\right)^{0.5}$ What hogwash does the factor $\overline{G(M+m)}$ indicate? The same can be said in the formula $M = \left(\dfrac{4\pi^2 a^3}{GP^2}\right) - m$ when $P = \left(\dfrac{4\pi^2 a^3}{G(M+m)}\right)^{0.5}$ that the factor $\dfrac{P^2}{\ }$ is senseless and

$\left(\dfrac{P}{2\pi}\right)^2 = \dfrac{a^3}{G(M+m)},$ has no foundation other than fraud. It says the position of the planet is derived from the mass $M = \left(\dfrac{4\pi^2 a^3}{GP^2}\right) - m$ that each planet has and when viewing the reality that is totally and complete fraud. The Cosmos does not support the Newtonian formula even in one place where it could apply.

Position as a function of time $P = \left(\dfrac{4\pi^2 a^3}{G(M+m)}\right)^{0.5}$ This is what Newton said is in place and with no evidence ever founding this ridiculous proposition, all Newtonians that ever come after Newton. This is what Newton and his Newtonian followers tell the solar system it has in place and tell the cosmos it uses to operate.

I have indicated that mass has no place or use in the solar system according to what the solar system puts in place. According to Newton $P = \left(\dfrac{4\pi^2 a^3}{G(M+m)}\right)^{0.5}$ puts the location or position $\dfrac{P^2}{\ }$ of a planet in relation to the mass $\overline{G(M+m)}$ of such an individual planet. I am coming back to this and then you will choose which of us Newton or me, is committing blatant fraud.
I say I can prove Newton correct by showing the formation of the planets in orbit going around the sun.

With implementing Newton's formula $P = \left(\dfrac{4\pi^2 a^3}{G(M+m)}\right)^{0.5}$ the planet distribution are as follows:

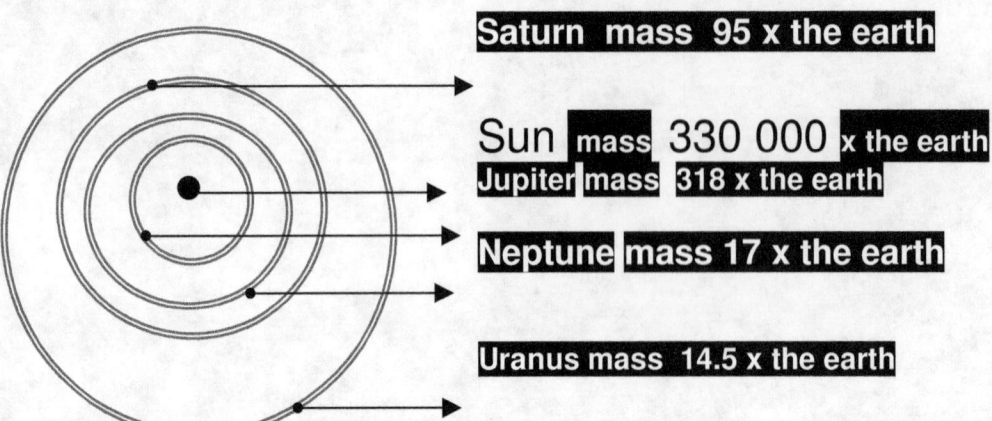

Saturn mass 95 x the earth
Sun mass 330 000 x the earth
Jupiter mass 318 x the earth
Neptune mass 17 x the earth
Uranus mass 14.5 x the earth

The inner circles that are very close to the sun we find the big gas planets with so much more mass forming the force of gravity that these planets are almost on top of the sun so close they are to the sun.

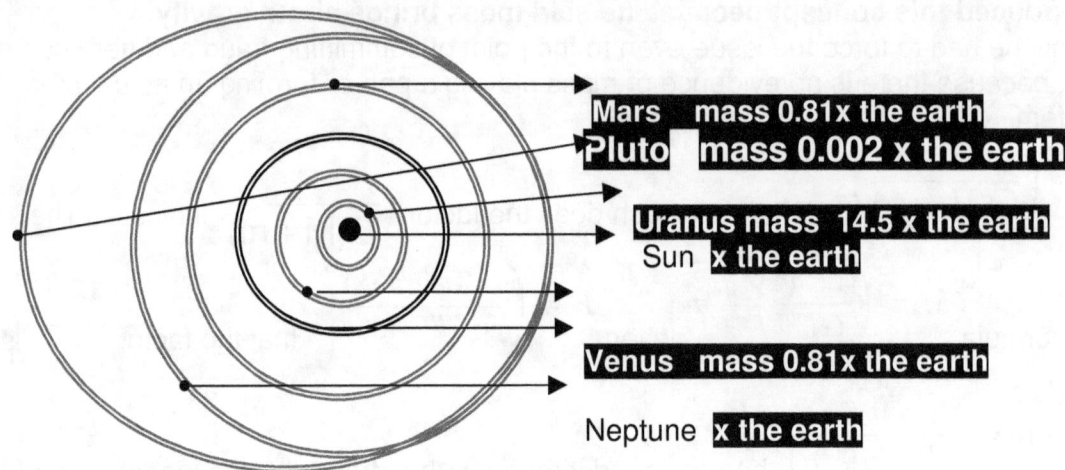

Mars mass 0.81x the earth
Pluto mass 0.002 x the earth
Uranus mass 14.5 x the earth
Sun x the earth
Venus mass 0.81x the earth
Neptune x the earth

Then in the outer circles that are very far from the sun we find the smaller solid planets with so much less mass the force of gravity just can't pull these planets closer to the sun.

Closet　1) Jupiter mass 318 x the earth
　　　　2) Saturn mass 95 x the earth
　　　　3) Neptune mass 17 x the earth
　　　　4) Uranus mass 14.5 x the earth
Then come the smaller planets with less mass and therefore less pulling force called gravity
　　　　5) Earth 1 x the earth
　　　　6) Venus   mass 0.81x the earth
　　　　7) Mars   mass 0.81x the earth
　　　　8) Mercury   mass 0.055 x the earth
　　　　9) Pluto   mass 0.002 x the earth

The above can be the only designated outlay of the planet position according to the sun when applying mass. Remember, if I am wrong then Newton and his formula $P = \left(\dfrac{4\pi^2 a^3}{G(M+m)}\right)^{0.5}$ is wrong but if I am correct and Jupiter is the closest planet then Newton is correct. However facts show my suggestion is incorrect and so is Newton and then the academics in physics are indulging in Newton's fraud by forcefully brainwashing students to believe in Newton notwithstanding the fact that it is the solar system that disputes Newton's ideas altogether. If you disagree with my layout, then you better disagree with Newton and his ideas

Tell me, can you find any credence in the "Conversions for "Unknown""
$4\pi^2 a^3 = P^2 G(M + m)$
In this comes the fraudulent part because there is no evidence of mass playing a part or forming an actual presence in the solar system.

Lets put Mercury as a yardstick and see how Newton's formulas pan out.

$P = \left(\dfrac{4\pi^2 a^3}{G(M+m)}\right)^{0.5}$ What hogwash does the factor $\dfrac{}{G(M+m)}$ indicate?

The same can be said in the formula $M = \left(\dfrac{4\pi^2 a^3}{GP^2}\right) - m$ when $P = \left(\dfrac{4\pi^2 a^3}{G(M+m)}\right)^{0.5}$ that the factor $\dfrac{P^2}{}$ is senseless and $\left(\dfrac{P}{2\pi}\right)^2 = \dfrac{a^3}{G(M+m)}$, has no foundation other than fraud.

$M = \left(\dfrac{4\pi^2 a^3}{GP^2}\right) - m$ is complete fraud. The Cosmos does not support the Newtonian formula even in one place where it could apply.

Position as a function of time

$P = \left(\dfrac{4\pi^2 a^3}{G(M+m)}\right)^{0.5}$ This is what Newton said is in place and with no evidence ever founding this ridiculous proposition, all Newtonians that ever come after Newton. This is what Newton and his Newtonian followers tell the solar system it has in place and tell the cosmos it uses to operate. I have indicated that mass has no place or use in the solar system according to what the solar system puts in place.

Visit **www.singularityrelevancy.com** to obtain more information on the subject free of charge.

There are the planets in line as it is in the solar system. Put anything connected to Newton's gravitational principles in relation to what does apply in reality. At this stage Newtonian science is science fiction used to brainwash students by thought control in pressuring the students to accept Newton. It is more effective mind control than any other dogma enforcement of any religion ever in the history of mankind.

These are the closest because these are the massive giant gas plants and having the most mass must put them the closest to the Sun. However, the location is random and not by mass in any way. Forcing students to accept the truth about mass is creating a staged science platform that annuls all other concepts that might sprout from this dogma. This is the most condemning religiosity that enforces a make believe in fantasy like no other religion can offer.

Now we show what is really in place as the cosmos has the validation. The Titius Bode law is in place holding every available piece of evidence of what the solar system uses. The Newtonian mathematical mongers will put the mathematics in place to scare everyone out of their wits should anyone ask questions. The mathematics is there to indicate the non – Newtonians inadequacy of understanding the issues. But guess what, it is the mathematics that show the stupidity raging amongst the Newtonians. They approve the use of mass and was too stupid the question or to verify mass forming gravity.

Later on I shall go into what the cosmos uses in forming the solar system. I shall explain how th Titius Bode works which is what is in place. There is not a sign of mass but for my effort to explain how singularity forms gravity in applying singularity as Π therefore I am rejected as mad.

Get your professor to prove Newton correct in the face of $P = \left(\dfrac{4\pi^2 a^3}{G(M+m)}\right)^{0.5}$ and if he can't let him admit he has been conducting in a fraudulent practise all the time he was teaching.

Gravity is the cooling of space. As space is being is placed in transformation, the time component which then can increase the spin value, or time by fore times the value. Therefore, it will transform fore times more unoccupied space-time to densified space-time, which is the name by which I call material.

Gravity forms when a solid holding an Iron core in relation to copper turns in a cosmic gas, which is the name that I gave to the singularity substance forming outer space. The turning of the solid inside the cosmic gas re-forms the cosmic gas into a cosmic liquid.

Science has to realise that the cosmos formed as the bible says and if there is any atheist idiot out there trying to go get smart about religion then read this book and see how wrong you lot are that believe in science and then order **The Veracity of Gravity** from Lulu.com and see how correct the Bible is. Creation started to a T exactly like the Bible says and precisely according to Genesis 1. Time split into what is infinity forming time and into time we see as space, which is a singularity substance, controlled by movement exceeding the speed of light in relation to cosmic liquid / cosmic gas moving below the speed of light. Gravity is a relative movement that has a relation between what is solid and what is not solid.

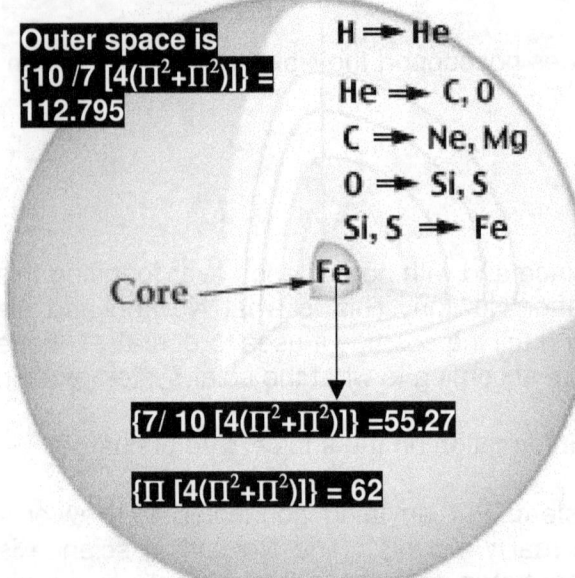

Outer space is $\{10/7 [4(\Pi^2+\Pi^2)]\} = 112.795$

$\{7/10 [4(\Pi^2+\Pi^2)]\} = 55.27$

$\{\Pi [4(\Pi^2+\Pi^2)]\} = 62$

All stars have an iron$_{56}$ core. At present, the Universe is in the iron peak era. Every star has to have an iron / copper core and the relevancy between outer space on the outside holding hydrogen and the iron forming gravity or electricity in relation to spinning around copper forms gravity. In this the Titius Bode law proves all of this. No star can function with out having this layer balance. Should this layer balance go array in any one of the layers the layers above it would not be able to withstand the time

Gravity is about outer space $\{10/7 [4(\Pi^2+\Pi^2)]\} = 112.795$ conformed by iron $\{7/10 [4(\Pi^2+\Pi^2)]\} = 55.27$ the displacement value of iron$_{55}$ in relation with singularity presented by the relevancy of $\{\Pi [4(\Pi^2+\Pi^2)]\} = 62$ and that is the displacement value of copper which is the element that has this proton value according to the periodical table. This is exactly how electricity is generated and that proves that gravity and electricity is the same thing charged on different levels

Fusion within stars does not come about "pressure" but the star freezes the atoms into forming a new unit. At this point I must also say I disagree with this idea that atoms freeze together but atoms rather grow into… The earth has not the gravity to "freeze" any type of fusion process and therefore all attempt this far was futile and will remain futile because the earth must first find the gravity to freeze hydrogen into a water-like substance as the sun does before fusion could be a possibility. Newtonians rob the public blind with their falsifies ideas about things they clearly don't understand and it is about time the public put a stop to this criminal behaviour. Either they stop to bullshit the public and promote wild schemes or they go to jail.

development, the matter will overheat and the matter will be transform to unoccupied space-time. This disaster is not a natural growth process, but a disaster of catastrophic proportions. The movement of iron $\{7/10 [4(\Pi^2+\Pi^2)]\} = 55.27$ in terms of the singularity product copper $[\Pi(\Pi^2+\Pi^2)=112.795$ **allows heat that is totally expanded at the value of** $0\{10/7 [4(\Pi^2+\Pi^2)]\} = 112.795$ **to reduce. Gravity is the movement of liquid in relation to a solid.**

This again was proven by the very first ever experiment concluded scientifically. This fact of space descending does not come as a surprise because Empedocles proved this fact back in 450 BC. Empedocles showed that space displaces water from the clepsydra, which was a sphere shape container with a sprout on the top and small holes in the sphere through which water ran in small streams out at the bottom. When the flow of air or space was blocked in the spout by a finger covering the hole at the top of the sprout at the entry, the water stopped flowing from the clepsydra. They concluded in 450BC that it is the empty space that pushes the water out of the clepsydra because the moment one restricts the empty space or air to flow into the clepsydra from the top, the water will stop flowing out of the bottom of the clepsydra.

Why would the flow of the water stop if the mass did pull the water down? When the finger blocks the sprout and stop the space entering from the top, the water does not fall to the ground but it is the empty space that pushes the water out at the bottom to fill the clepsydra from the top. When the finger blocks the sprout and stop air to come in through the sprout opening the water should still run out at the bottom by the mass of the water pulling, if mass was doing the pulling. If mass was the force giving factor, then the water must keep on flowing because the mass of the water did not disappear when the sprout was covered and therefore it still has to produce the pulling by forming gravity. All this evidence was known to science about 2500 years ago but since "With a lot of words and some simple algebraic relations, there is no way to "explain" the world of physics" it lacked mathematical communication and it should therefore surprise anyone very little that physics could not fathom this result 2500 years onwards.

Forget the example always used about the hammer and the feather falling equal in a vacuum because the hammer and the nail and the elephant falling together will also fall equally notwithstanding falling in a vacuum or not falling in a vacuum. The vacuum part is conspicuously in place to purposely confuse reality as it is brought in to flagrantly spread misunderstanding of the issues in hand about the falling that takes place. With everything always falling equally when the same the condition applies to all objects falling and therefore with such falling happening under the very same variation of natural conditions applying, this shows it is the space in which the object is that falls and not the object falling while leaving the space it holds behind. The lack of relevant density in relation to air moving down stops the feather from falling equal just as gas does not fall with the space at the rate that space does descend. All space falls by the compressing of the atmospheric space.

The rotation of the earth moves the space sideways and this brings the space to move downwards by increasing the density of space or air as it comes closer to the earth. This results from the Roche limit applying to fix atmospheric layers varying in density. In my books I explain that principle applying mathematically. Notwithstanding using your mathematical marvels, science has not got any vague idea to explain any of the phenomena mentioned above. To understand these phenomena one has to understand singularity.

I am able to explain the how the Universe started only because I discovered the building blocks used to build the Universe. The building blocks are the Titius Bode Law, The Roche limit, The Lagrangian points and these all culminate into the Coanda effect. Using these I can and I do prove exactly how the cosmos was built

Gravity is TIME forming SPACE = Π^2 = 9.8696 = MATTER HOLDING THE COSMIC LIQUID COUPLING THAT TO THE NEUTRON TO COMPLETE THE COSMIC GAS to the liquid part of the association between solids and liquids. The earth is the solid while the atmosphere is liquid and outer space is gas.

Every person associates gravity applying as "The natural force of attraction exerted by a celestial body, such as Earth, upon objects at or near its surface, tending to draw them toward the centre of the body" with what we think of happens to solar bodies having mass which is "the natural force of attraction between any two massive bodies, which is directly proportional to the product of their masses and inversely proportional to the square of the distance between them" and science cheats everyone into believing the two is the very same. The one I experience every day and with the other there is no evidence of applying anywhere. Yet students are fooled into believing the two are exactly the same issue, which is untrue. If there is any academic feeling insulted by me calling the lot fraudsters, bring evidence of the second form of gravity working on the principle of mass and I will withdraw my statement, otherwise if no proof can be brought, then you lot that ignored me for ten years are the fraudsters I accuse you to be. You are villains brainwashing students to corrupt their thinking.

Einstein's Critical Density lacks the accepted matching facts we need in proving the critical mass factor, which makes the entire idea silly and bogus. You can't force the Universe to conform into something Newton said because Newton said the Universe works on mass contracting... But our inability in securing such required evidence defies the most basic logic. It seems all new evidence we receive from outer space is disputing all Newton laws and new findings disprove **Einstein's Critical Density** as the answer.

The Universe will not reach a point of contracting, not withstanding whatever dark matter astronomers try to locate in the vast space. A rush is on finding the black-matter that will be applied to force the cosmos too retract back to where it came from. But what if our view of the cosmos was as incorrect as our views at present is about the sun? I prove that contraction is at present as much part of the cosmos as is the expanding is that we focus our attention on and it is our culture we carry from generation to the next generation that leaves the human view obscured in admitting the truth. The Sun is not a coal stove burning fuel. The Sun is an air-conditioned pumping gas (hydrogen and helium with electricity. I prove by applying the Titius Bode law that gravity is electricity and the two are the same thing. By using gravity or electricity the sun is a huge air-conditioned pump freezing cosmic gas, which is what the Universe is into a liquid that we see squirting from the sun.

Newtonian science has NOT developed from the idea that the sun and the rest of the Universe are orbiting the earth. Everything applying on earth and befitting human standards they think must be transmitted directly to the Universe. If I stand in the sun and feel hot, then "the earth is hot" and we have "global warming" just because we humans think it is hot and we humans feel slightly bothered. Every point in the Universe applying singularity has a different measure and the entire Universe is NOT made to cope with life but is completely alien and destructive to life. Newton started to tell the Universe what it must be and the Newtonians never stopped. Newtonians better stop with the Ptolemaic concept of putting the earth or life in the centre of the Universe and start to place singularity in the centre of the Universe where the centre of the Universe is..

Why would the expansion turnaround and do a reverse by going back to where it came from. Consider the momentum alternation such a change will bring about. The sun is not a gas-filled sphere holding hydrogen in its "natural gas" form, but it is all fluid and is in a liquid form where singularity is liquid-freezing hydrogen at 6500^0 C while outer space is boiling over at -276^0 C. **The Absolute Relevancy of Singularity** book explains the Roche limit as well as the other four cosmic principles in the practical sense... when applying cosmic laws instead of improvising cosmic laws uncovers that reality then becomes awesome. It becomes clear the Universe is as much expanding as it is contracting and contracting by expanding. As there is no hot or cold, no big or small, no grand opposing but relevancies in ratio to one another. If you do not believe me, then believe your eyes when looking at the picture. What ever the sun is it is fluid falling into fluid.

Because hydrogen is a gas on earth and we think of 6500^0 as hot on earth, therefore the sun must be hot and the sun must be gas at the same time. It is obvious what we see is liquid squirting and when hydrogen is in a liquid form it then must be cold because my eyes tell me that! Newtonians still have the entire Universe apply the standards befitting life on earth and that is why they wish to locate life.
When something is hot it expands. When something is cold it contracts. Outer space expands to the very limit and keeps expanding therefore outer space must be the hottest there can be, notwithstanding scientific Newtonian stupidity. The sun contracts every bit of space that forms the solar system and therefore the sun must be the coldest place in the solar system notwithstanding whatever Newtonian ego whish to declare. If I feel the sun is hot it is because it is diverting all the heat my way and it it diverts the heat to me then where there is no heat it must be cold. It may sound incorrect and unscientific madness but with my applying of Kepler's formula in alignment with the position I located

and valuated singularity it clarifies the possibility of the above statement... but please do not take my word for it, use your eyes and make sure you look past the culture bias of past incorrectness. See the fluid push out of a bowl of liquid, spilling both sides as it falls into liquid. The Hydrogen inside of the sun is not gas but it is fluid. In all of nature in all elements found through out science there is no NATURAL GAS as much as there is no NATURAL SOLID.

Hydrogen is as much a liquid as iron is a gas and neon is a solid. It depends on the element relating to the space/heat in the circumstances surrounding the substance at that very precise instant in time. We have to stop telling the cosmos to show us what we wish to find and start accepting what the cosmos is telling us is out there that we should look for and find. In creation there are two substances that formed the Universe. One was earth or solids and the other was heaven or uncontrolled heat. Between all solid we have heat parting the solids and being a solid or a liquid or a gas depends on the ratio between the solids and non-solids. Under conditions suiting life certain elements may be a gas, but in stars conditions don't suit life and in outer space conditions don't suit life so therefore life cannot be a barometer for conditions applying in the Universe. Kepler gave us **solids as $a^3=T^2k$ as liquid or gas.**

The earth, just as all other cosmic objects do, contracts outer space by the movement of the rotation. Material spins in gas to reduce the volumetric size and by that it reduces the concentrated space around the star / planet / earth which then reduces the cosmic gas forming outer space to the cosmic liquid forming the atmosphere. By rotation the earth "pumps" gas from outer space to the core within the centre of the earth by applying centrifugal pump action. As the space becomes denser the heat level rises but this is because the rotation of the earth reduces the heat, not the heat that we as humans feel or experience but the heat level within the space. We have to maintain science laws. When any cosmic substance overheats it becomes larger and when cosmic substance becomes colder it shrinks.

That is cosmic law.

The heat we feel on our skins is heat escaping from where it is cold and contracted to where it is hot and expanded because the levels that reduces has to dispel the heat from the location it is within and move it to where heat is excessive and expanded. It is the heat moving away that we feel and then think of it as hot. As the earth turns it cools the space in which it is and the space reduces in heat and therefore the heat levels can reduce in order to make the space move towards the centre. As a matter of fact the very first experiment in science ever conducted and recorded was the discovery that it is space that reduces and not things (water) that falls but notwithstanding Newtonians have gravity as mass falling.

But as Professor Friedrich W. Hehl, Inst. Theor. Physics With a lot of words and some simple algebraic relations, there is no way to "explain" the world of physics. Your seem to be out of touch with modern developments. I guess that is why Newtonian science can't conclude they are mistaken even after three-hundred years.

Again and as for e so many time in the past I repeat the warning once more of the Newtonian Brilliant-Brainy-Bunch to please take note of a conscientious warning about the gravity of the misgiving there is on the part of the most respected Academics in physics about a much concerning matter. As you can see why I state it emphatically that science accuses me to be not schooled to the point where I am able to have any form of an opinion on any matter concerning Sir Isaac Newton.

Notwithstanding that my research proves I did my private studies and through which I skipped the indoctrination and mind control academics place on students goes unrecognised by their standards and so too my ability to have any insight on matters regarding physics. However my skipping their methodical and systematic brainwashing enabled me to see and allowed me to be able to express the incorrectness in Newton's teachings and allowed me to show in clarity what destructive force Sir Isaac Newton used to corrupt the laws of mathematics, corrupting to science along the way and mostly raping to the work of a great man, Johannes Kepler and what Sir Isaac Newton did can only be expressed as being blatant criminal fraud. What his deeds amount to is to corrupt the laws of mathematics, to render the laws of cosmology useless and to rubbish all of science. Should you find this to be unbelievable, then I am glad to announce that this book is more for you than any other person, so go on and read what academics guarding science never wanted published. I challenge any one that disputes any claim I make to prove me wrong by proving me wrong and not merely suggesting claims in that direction.

We have to realise that it is about heat and cold where moving cools down and stationary space expands. Creation started with differentiating between what is cold and what is hot. That is exactly what the Bible says in Geneses 1:1. The Bible says God made heaven and earth. This means God made what is solid and what is not solid.

It means God made material (the earth) and heavens (the sky or non materials). Back then as is the case now with our most brilliant Newtonians there was no other names for solid as earth and non-solid sky as heavens. At least those back then realised the difference and did not put sky down to "nothing" as our incompetent Newtonians of our modern era does. There is singularity controlled by movement (material) and there is singularity not controlled by movement (non-material).

In terms of mathematical equating the very first recorded instant happened at:

This is exactly what the Bible says, if only the Theologians and the Newtonians were not so preoccupied in fighting for their own small-minded egos instead of looking for the truth.

When the Universe started there was one spot that released a dot. The spot as large as a thought grew into a dot. This too is exactly as the Bible says it happened and it still happens exactly like this. The Universe starts from infinity and grows into eternity just as it happened at the start and this is what Newtonians confuse with what they know as the Hubble constant. To check if I am correct look at anything spinning and you will see conformation. The dot that released was by such release relevant to other dots released because there was to be motion that measured many dots. The dots in release were relevant since they were the same. Only time being in delay by cycle of infinity interrupting eternity to form space, as space is the history of time gone to the past. But since we are looking at things as it started I put it into the past tense. That cycle brought about time delay as every cycle drifted further from the original singularity while the original was still responding as well. That was the first space. It was time being one infinity part.

Since the dot was also singularity and was the very same as singularity with a small difference that the spot was $\bullet \Pi^0$ and the dot was $\bullet \Pi^1$. At first, at moment-Alfa, there was no space nor time for only relevancies came about. Relevancies acted as motion to bring change to eternity by changing the flow of time in eternity. There was the perfect spot in which time moved while remaining the same. Then heat brought expansion and expansion brought space and space brought movement and movement brought about a Universe. That is how it started and that is how it still is. This was before the atom and the atom was before light and that too is exactly what the Bible says in Geneses 1:1.

Time will forever remain eternity but space or time distortion, which will forever remain infinity, interrupted the flow of eternity. Space breaks the monotony of time in eternity by parting time from in infinity. Yet, the relevancies did imply motion except for the fact that singularity is very much incapable of motion. Every dot had a purpose to fill a position in relation to the other dots that the spot excited. All dots had a line of three where two was one, each on every side of the spot. The spot was one with the two dots forming two, which improvised for motion that would later come to space and the three was what space was going to become. How do we know this: $0.1416 \times 7 = .991$ and that is by the value that the Universe grows.

Time was four because the four would bring about motion as heat separated infinity from eternity or hot from cold. Five was space because space was one removed from time. Space is the distortion of time and one outside time would bring a time delay or a time distortion of four plus one which is five, hence the principle behind the Lagrangian system. Because material was the square of space material was a crossing of three plus three forming six. However to find out what this means you have to read the entire theses called **The Absolute relevancy of Singularity**.

Space-time is the four of time, plus the three in singularity around which the four of time turns, therefore space-time is seven.

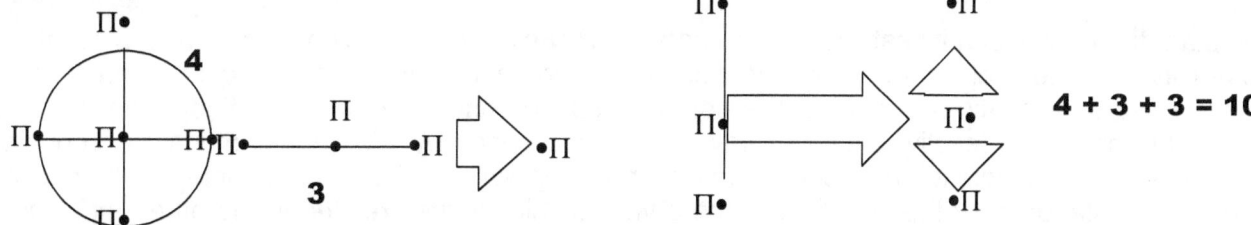

In the circle using $r^2\Pi$ the r has to have distinctive qualities placing it as a factor apart from Π. Where the growth shows no separate distinction but a continuous flow from the precise centre to the precise edge the flow would become in relation with Π depicting the circle and Π replacing r as reference to any point on the circle.

By using r, distinction in the circle is possible but by using, Π there is no distinction possible. Therefore, in the beginning when time formed space there was only relevance coming about from $\bullet\Pi^0$ and the dot was $\bullet\Pi^1$ with no mention of any possible r. The fact of r representing a radius represents space and what we refer to be long before the Big Bang introduced space or mathematics using space.

Before the Big bang the lot was form without dimensions playing any part. Then the atom came. Only after that did the Big Bang come. Even before the atom was the point lining up and forming positions that was spinning faster than the speed of light can ever achieve. However every point today still serve the role it took on at that stage and serves in the position that it had during the time it had no space with eternal time. These relevancies developed as part of a Universe we shall never understand. The Universe had no sides and a line was equal to a triangle, which was equal to a half circle. Singularity holds the double space-time position of five times two (matter and space duplicating singularity) which then is ten.

How did the Universe liberate material and heat forming space from singularity because with singularity comes an unchangeable eternal condition that is non-changing-everlasting in all conditions and aspects that is remaining in absolute equilibrium. This equilibrium maintains because all development extends form precisely in a detailed equal equilibrium throughout. Think about what brought the cosmos out from the eternal rest in which it was. The eternal rest still maintains and is therefore our detection. What inspired the eternal rest the cosmos was in and inspired change to the state of eternal rest? What evoked change? That is the question the Atheist will never be able to answer but that too is the most basic and ever-lasting fundamentals of the Universe. Singularity Π^0 is not substance but it is a thought establishing substance Π. What changed in this split second start before the official start? I do not wish to ponder on this matter in the letter I am writing at this minute, as there are other books where I delve into this matter. It is called **The Absolute relevancy of Singularity**

From the deep freeze of creation came the Hot Big Bang and the 3D Universal displacement came about with the relatives being **$10 \div 7(4(\pi^2 + \pi^2)) = 112.795$** and then a second one established 3D by introducing the six to seven sides Universe at a density point of **$7 / 10\ \pi^6 / 6 = 112.162$**. There is of course a lot more information about this establishing of the Universe than what I mention at this point. The question is what made the Universe freeze, to form the Universe in space and through time. It had to start with a specific reason applying, which brought about space-time. Once the process started there was no stop to it, but there is no chance that the initiation of the start was spontaneous by nature. With everything being in one spot, all within that spot was in a state of eternal rest. While all remains the same and nothing changing, what brought on the sudden change of everything, shocking

It is the time the space takes to bring about new positions in space occupied through motion applied and through motion applied, it takes space-time to duplicate and in the relevancy of duplicating, the duplicating takes certain duration in time to move from point to point. Gravity is motion of space towards and in relation with a centre and the time is the period such motion takes while that centre is attempting to produce space by doubling space through motion leading space away from a specific controlling centre to another specific controlling centre. The time it takes to complete such an attempt provides space the opportunity to double its status. Moreover, the time stands affected by this motion material creates to duplicate space-time by generating singularity and activating different locations holding singularity. Gravity is speed and speed in space in motion through time duration. Gravity is motion combating heat expansion and supplying space with space producing through motion providing the space the opportunity to expand while remaining in relevance.

At the start, the gravity part invested heavy in material. However, in space light came about since there is no gravity in space. Light is the attempt to establish motion not controlled by any centre and the time it takes to establish space between such a centre of space control and the light finding an ability to dispense the space by reactive motion. Gravity produced space by allowing as much as producing overheating with a feeble attempt to combat the overheating. What is the meaning of heat if there is no cold to set the standard for the heat to become a value, which is then related to the other end, where such another end must be the limit in cold? The moment singularity produced space-time heat distanced from the cold factor. Space produced a cold base to have heat within. How did singularity part the shared principle of being the unification of heat and cold as a unit? Singularity froze in applying gravity bringing about particle separation within singularity. Heat and cold parted to produce frozen heat by atoms forming and captured heat by unleashing uncontrolled material that ended as space in time.

We can see from what is available how everything fell in place. There was space filling with heat where the heat was compacted by the time delay. Every point established a relevance of three crossing singularity a motionless forming a line and four points formed the square of time. Crossing the four points that supply the time or the motion aspect is a line with three points supporting singularity by not spinning but they maintain singularity in being as motionless as singularity is. That four plus the three is seven. Then all this, which is the body of material requires somewhere to be within because to be is to move within something. That too is time but we regard that time to be space. That time is the heat that became space during the event of the Big Bang. That gives the atoms seven point as plus the electron or space having three points in relation to the seven points there is in time. If singularity is at the limit the value of ten but is the tenth portion of the value then ten must be divided by the tenth portion and from that the tenth portion must be distracted.

This as a group forms a relation with heat unlike any other. If mass did the trick these must have been the group having the second least density, but they form the group with the least density as a five point group. The next group holding the Pythagoras five or the Lagrangian five plus two or three, four or five enabling their relation with heat to be quite remarkable. This must be some indication of events during the period just preceding the Big Bang at say 10^{-7}, 10^{-6}, 10^{-5}, the time when the fuel that would ignite the Big Bang turning heat into space turned material into heat. The relation of five plus one, plus one and five, plus one plus one plus one is just too uncanny to ignore.

Following the process and seeing the influence of singularity should bring about a pattern that may lead one to a pattern of how the required heat formed and how the intended heat transformed to space. Density depends more on proton number arrangement producing specific form in relevancy as to merely and only having mass as factor that contributes to the forming and development of stars in the cosmos. The evidence is so clear that mass has nothing to do with gravity but density has everything to do with gravity. Density is the volume of space in numbers used to fill material in ratio with numbers of space per volume not filled with space. It is matter versus space in every sense there are. This came about before the Big Bang took place and before space was formerly space and time was formally motion. It was a time when singularity set relevancies moving from Π^O to Π

The spot becoming the Dot

In that manner we know that that was the way particles just after the arriving of moment-Alfa. Singularity brought singularity brought the divisions between the many Universes that followed the immeasurable many Universes that came after the flooding of Universes to follow the leaders. The term "moment-Alfa" is the way I refer to the moment when singularity changed, not when space formed or time began or space exploded but even before anything including mathematics became definitive. At this point mathematics renders it useless. There was no space or

This was the era of distinction, when separation brought an all-possible new Universe formed combinations the Universe but also

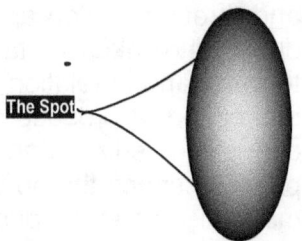
The Spot

time to calculate because relevancies came in place. Form took shape but space there still was not because Π^0 moved to Π. Every slightest point in space became an opportunity of establishing a Universe with most different functions and ingredients there might form. This is apparent from the fact that it still takes place at the present moment by motion attaching new singularity through duplication and through duplication releases previously attached singularity from serving the purpose of duplicating by motion.

When the cosmos came to motion, motion was not yet defined. When the cosmos brought about motion, the first motion was relevancies. Cold parted from hot. Eternity parted from infinity. Motion parted from motion absence. Infinity broke the laboriousness of eternity for the duration of infinity. The spot became and grew into the dot.

From what the spot was to what the dot now is might be just a mathematical implication of going from 1^0 to 1^1 but in reality that first motion was the creating of and establishing of an entire Universe with all possibilities now in it. Never again can that much growth become a reality, although to us the growth is beyond what we ever can notice. But it is because the growth is so massive and we are so small that we are unable to notice such almighty growth.

When the spot Π^0 became functional and established all relevancies possible, heat parted from cold as eternity parted from infinity. The expansion was not clear motion but more a parting of relevancies where a centre formed a relevancy because the centre could not provide motion. Without being capable of motion, the centre established four points, which also served singularity. From the inverse square law we know that the centre doubled by producing the four points holding singularity.

By exciting the centre spot, the centre spot came to be because of the heat that formed in relevancy as heat parted from the cold bringing about the division that followed and that was the motion that formed. Therefore the heat had to move but being singularity it could not get singularity to move. In an attempt to establish growth, singularity activated six spots of which four was having motion drawn into relevance four spots that was providing what was to be motion and three that was to be securing the position the centre holds. There were four forming a ring around singularity with two forming in locations we will refer to as above and as below or north and south.

The three in line was in singularity not being able to move but the four was also in singularity and just as incapable of moving. All the points came as relevancies applying the forming of more of what was to come but only the four committed to time were expected to move. The four points that came as a result of discrepancies that became time that produced form and that established the relation with the one but had to perform the motion by expanding was as much incapable of motion as the centre was that charged the four with motion in the first place. As they were incapable of motion, it still required a tendency to apply motion that did separate Π^0 from Π. This not only involved form but it involved all relevancies that did come or may in the future come about as a result of the attempt to commit motion. If mass was a factor contributing to gravity the cosmos would have frozen back to singularity without ever releasing singularity to relevancy.

Mass does not establish gravity. There is no magical graviton. In the beginning there was no mass but boy was there gravity! The only means that the cosmos could find a way to break from the grip of eternal eternity was to expand into relevancies. Such a feat can only go to task by forming opposing hot and cold. Becoming hot produces more of what is heating. That implies motion or a moving away from where it was by generating more of what is available. Only where hot released from cold could whatever was repeated once again and duplicate what was before into what then is more. Secured by motion T^2 in relation to a specific centre k from where singularity holds the Universe true to form. The k was an intention to place apart and by today's standards will not even qualify any noticing.

All that are is in singularity. From singularity comes the motion and the space we call space-time. Singularity is dimensionless, time less and space less and because of all this features, it carries the value of Π^0. By expanding, singularity applies a relation coming about that reforms singularity from Π^0 to Π. Only when extending Π^0 to Π, the extending creates motion and the motion creates space that then doubles through motion applying which cuts the space in motion in half by matching the space as a duplicate. Motion creates another dimension or another level reforming singularity from Π^0 to Π or from Π to Π^2 or from Π^2 to Π^3

As said before we now know Π came about since Π is achieving form and not space. Only **r** can establish space as size will accumulate and as it had with everything else singularity had **r** covered by one as in being $r^0 = 1$. By reducing the circle radius **r** by half continuously will lead to an infinite small circle and an infinite number holding r would place **r** to the power of one as a factor. Then as a factor **r** would not contest any change when change is introduced into any future equation but Π will remain because the circle as a form remains even being infinitely small. By reducing r indefinitely to the tune of half each time, r would become infinitely small, beyond human calculating means, however as mentioned in the case of the smallest dot holding one spot, r would become insignificant beyond human comprehension even, but never reaching zero and still Π would remain intact and dictating form. To amplify by dimension a value has to be set to r but if r remained covered by singularity all alterations that could possibly come about was in the form, which was Π.

This expanding can be a problem one can wrestle with for one lifetime and never reach any conclusion. How can something grow without getting more that what was before? Then it hit me like a ton of bricks. The answer is in heat but not heat, as we know heat. It is heat in getting relevancies between outer limits. Only heat could break the monotony of singularity. Heat in the form we now know heat as heat is now. Since the Big Bang heat is material transforming from one state to another state.

The change that took place involved singularity but singularity was 1^0 and being ^{01} could not grow. The growth came about. Heat rose from singularity, but if heat rose from singularity. Singularity as a factor changed from 1^0 to 1^1, which means a relevancy came in place that no one could detect. It is true that 1^1 are still one, but one could then escape from singularity by producing factors other than 1. Heat came about but only as a relevancy to utter cold. If there is heat, there is cold or if there is no heat there can be no cold. Space came into forming a relevancy that brought form. Since it is a relevancy and not a generation by accumulation, the form produced was Π.

The spot formed a dot by heat and cold establishing relevancies and from that singularity was broken to allow all other forms of relevancies to come about. The cosmos did not start because of gravity. The cosmos started with heat and cold coming into a relevancy and in the cosmos there is no hot as much as there is no cold. The cosmos broke, put from the confinement of singularity by establishing a singularity in a relation of heat and cold. The heat that came about was beyond measure because the cold that held the heat was also beyond measure. The immeasurable heat was on the outside of the dot that formed and the cold was on the inside of the dot that formed.

The cold contracted because in nature cold contracts. The heat expanded into a dimension of form and heat by expansion is in nature about motion. Motion is duplicating that which is and heat is what is duplicating by motion. But only heat by expansion was possible because in affect singularity cannot move. The motion became contraction, as the motion was the result of heat expanding which was forming four points in the rim of the dot. The expanding of the points created motion in relevance of a centre that formed because of the motion, which established an immovable centre as the Coanda effect, placed more dots in relation to more dots that formed.

Every dot was Π and every dot formed Π^3 because of the expanding heat, which produced Π^2. With that a new relevancy came about forming a centre in between the four points of expansion that was resulting in time. But since the points were in themselves singularity, which is immovable and space-less, they still heated forming a cold centre with the heat bringing about motion. It became a repetition where infinity broke eternity by producing a centre because of space (or rather form) forming the motion to enable the space to form in relation to the heat applying motion. This brought about a Cosmos being conceived.

The spot forms a full circle, but the line running through the circle is forever present because that is the future radius of the circle that will one day develop the circle, which is equal to the present diameter. The fact of the presence of such a possible line in such a possible circle dividing the possible circle into two parts makes the centre line equal to the half circle. The line forms the half circle but not only that the line presents the half circle as much as the line is the half circle. The line then is 180^0 and the half circle is 180^0 because in singularity the two factors are the same.

The same value is of course $\Pi^0 = 1$. The issue of concern is o understand that singularity cannot move. Singularity has no space. Singularity is no only part of the Universe but singularity is the Universe. By establishing motion singularity has to be charged with the time delay we find space to be. The space is time taking a period or a duration while moving from one singularity point to another singularity point while conducting the heat and the accumulation of heat that built up due to the retarding of the time to conduct the heat forms the space that is conductor to bring about the motion of the space.

It takes heat time to entice singularity and singularity can only entice. Singularity cannot move and neither can singularity form space. By enticing from one relevancy to another there is a bridging of heat that has to be crossed in order to send the gravity or the enticing or the relevancy to depart the space and reconnect the space to the next singularity. Bridging all the accumulated various time delays that formed an accumulation of heat through time distorting brings us the space we see and have. However there is no true space or motion but it is eternal motionless space is singularity charging time to provoke heat into forming space.

Three points formed a line covering singularity where the centre singularity recovered heat to grow and two points served as an axis to allow the rotation and to assist the duplication. There is one centre connecting the duplication of three as well as the recovery of one (the fourth one) that is applying the tie aspect. Therefore, motion consists of three positions in relation to a centre, which forms as space in relevancy to the motion and the space receive a controlling centre.

The duplication comes about as singularity is exciting another singularity in precise relevancy of 3 to 3 to 1, but the points charged is as space less and as motionless as only singularity is. The heat it requires to carry the exciting between points forming space and the space excites heat and the time delay it takes to excite singularity between points forms space-time.

That is why the Universe is Π

Where motion conducts electrical charging which is equal to gravity the charging of motion is to entice duplication of singularity. This is the basis, the heart and the sole ingredient of the Coanda principle that includes the Roche limit ($\Pi^2/4$). The charging of gravity $((7/10) + (7/10)) / (10/7) = \Pi^2$ and the charging of space-time $\Pi^3 = \Pi^2\Pi$ is all due to the relevancy brought on by the Coanda principle. The value of motion came from singularity exciting singularity and that is the duplication while the duplication or motion presents the space.

The development came into eras as the relevancies brought about new relevancies that spawned even newer relevancies that all remained in touch with the original singularity centres. Every one focused a new time delay that eventually brought about space and every distortion of time brought more. That concentrated between singularity points that charged the points to form space. When the charging became overdue in some sectors it erupted in forming the Big Bang. By the time the Big Bang erupted there was such a huge backlog in heat and time corrupted and delayed the next result was the employing of space as a commodity in the Universe. The relevancy was C the gravity was C^2 and the space was C^3. That left what was inside atom still spinning faster than the speed of light applying the relevancy of **k** = C where the electron applied the relevancy of $T^2 = C^2$ and that formed the atom which then became the cube of the speed of light $a^3 = C^3$. That left the atom at the relevant size of what the speed of light permitted at the time but since the Universe from that the relevancy expanded as the Atom grew in space to the extent it has now. The purpose of the star is to recapture the space the atom grew into and from there dismiss the space by spinning faster than whet the speed of light will be on the outside of the star.

- This form came about when only form was present in the cosmos. It was in a time era where form featured in relevancies that would lead to one day becoming the atom. The atom forms a dual purpose of duplicating as well as dismissing and some prefers the one better to the other. This relevancy came in place when time was not time and space was form. Time is forever eternity being interrupted by form in infinity to bring about eternity ticking as infinity ticks. Before that singularity took on stages in forming relevancies between duplicating and dismissing space-time, which incidentally was not yet truly space-time in the sense we think of as space-time. At first a dot moved from the spot leaving the spot but taking with the spot as part of the dot to remain in the dot. The two never separated but the one allowed the other to be.

As the dot confirmed a discrepancy between infinity and eternity by defining infinity as an interruption of eternity cold and hot parted a union. The dot that formed was not space but a relaying of time to form a new point of singularity where eternity was interrupted by infinity. Time took form from 1^0 to 1^1 or from Π^0 to Π. It brought form into differentiating between interrupted eternities with infinity doing the interrupting

Then a true distinct relevance came about that positioned a time differentiation outside the realm of time by four. In this realisation we can assume that space had some meaning at this point and the formula used to investigate suggests just that. Even in grouping, there are characteristics, which make a certain group of atoms more perceptible to duplicating and others more perceptible to dismissing

The lagging of exciting one point in relation to another point takes time. It takes time to send the message across to get singularity at that point excited. It takes effort to bridge from the dominating singularity to the independent singularity and that effort slows time down. The crossing of the divide is space formed by pushing time into duplicating. When time brought in a five points to the four points it took time to be, that fifth point became more than only form, it became space because it was one point outside the Universe of four or of form. One must see the three points established as motion duplicating singularity in relation to one dismissing singularity. This always has to strike a balance in order to establish space-time. It began as a relevancy and developed into space-time flowing or space-time displacement.

What the Coanda effect proves is that the rotating motion is acclimating a centre that exemplifies all phenomena in nature as we use nature to our advantage. All of nature including gravity uses the same method of motion forming around a circle in rotation and in the centre of the circle a point of no motion holding no space comes about. This is what Kepler taught us when he taught us $a^3 = k\,T^2$. With the Coanda effect forming the basic principle of all natural phenomena we can see from that, that the motion of liquid in the presence of a solid forms a centre that excites as it establishes singularity. From that rotation, space flows to a controlling centre but because of the lack of motion in that centre, there is a lack of space in that centre. Therefore, there is proof of a flow towards such an established centre and there is control from that point of singularity. In every case, the singularity controlling space-time sets standards for space dismissing in relation to space duplicating.

The duplicating stands in regard to the flow that the liquidity of the atom in relation to the solidity of the atom can reproduce. This forms density and mass but mass has little influence on the scenario.

There is a balance between the duplication in relation to the dismissing of space and the relation extends to the number of atomic elements present which then creates the balance applying within the star. As the liquid heat subsides in the centre of the star and the heat density is dissolved by the dismissing-prone elements the motion or moving ability of the star as a unit fades away as the star becomes static and solid with less space providing the star with less motion.

It is the way the atom formed before the atom took on space-time. It is in the formation, that space-time relates to motion. We have some elements being quite massive but also lighter than air and others are quit light but as dense as they come. This can only be a contribution from the way the atom relates to heat, which make the atom volatile (movable) or dense (motionless). Those elements being volatile are also very movable and in that we find the role that such elements play in the star. Stars that are predominantly made up of hydrogen and helium with very slight support from the metallic inner core are those stars that duplicate by producing motion. However the point I wish to press is that mass and being massive and being heavy do not support the fact that some elements have more gravity they produce because their protons are more numerous than others. The fact that mass generates gravity is a myth.

One will find that whatever group one chooses there are gasses and there are solids. If mass was attracting mass then the strongest mass must be attracted to the strongest mass and the least mass must float in the air. $F = G\,(M.m)\,r^2$ hardly can even begin to explain the fact that there is a gas that is more massive than iron but floats in the breeze just as hydrogen which is the least massive element.

Humans in science on earth still puts humans in science on earth in the centre fo the Universe. It is accepted all around that if an element is a gas in the "natural state" on earth then that is a gas. If the element is a solid on earth also in a "natural state" then that is a solid. In the centre of the earth we find iron in a molten sate floating as a liquid because it is within a state of liquid. In the centre of the earth every element that is there is a liquid and if it becomes gaseous it burst to the surface, through the surface and we then call it magma. All elements are either a solid, a gas or a liquid depending on heat.

Nitrogen 7	melts at −210°C	boils at −195.8° C
Oxygen 8	melts at −218.8 °C	boils at −183° C
Fluorine 9	melts at −219.6° C	boils at −188.2° C
Neon 10	melts at −248.59° C	boils at −246° C
Sodium 11	melts at 97.85° C	boils at 892° C
Magnesium 12	melts at 650° C	boils at 1107°
Aluminum 13	melts at 660° C	boils at 2450°
Silicon 14	melts at 1412° C	boils at 2680° C
Phosphorus 15	melts at 44.25° C	boils at 280° C
Sulphur 16	melts 119° C	boils at 444.6C
Chlorine 17	melts at −101	boils at −34.7 C
Argon 18	melts at −189.4° C	boils at −185.8° C
Potassium 19	melts at 63.2° C	boils at 760° C
Calcium 20	melts at 838° C	boils at 1440° C

Ignoring these facts, Mainstream science will hardly answer the problem we do not understand and such ignoring brings strong doubts about the quality and sincerity of science.

Excluding Argon, which is six (carbon's number) times two and suddenly that is a less dense material. The four times five plus... group are the following:

Scandium 21	melts at − 157° C	boils at −152° C
Titanium 22	melts at 1670° C	boils at 3260° C
Vanadium 23	melts at 1902° C	boils at 3400° C
Chromium 24	melts at 1857° C	boils at 2665° C
Manganese 25	melts at 1244° C	boils at 2150° C

Iron being the five times five plus one is the only generator of electricity and therefore the producer of gravity making five times five plus one the ultimate relevancy to heat in reducing space. Still Krypton is much more massive and turns out to be a gas.

Krypton 36	melts at 1539° C	boils at 2730° C
Iron 26	melts at 1536.5° C	boils at 3000° C
Cobalt 27	melts at 1495° C	boils at 2900° C
Nickel 28	melts at 1453° C	boils at 2730° C
Palladium 46	melts at 1552° C	boils at 3980° C
Silver 47	melts at 1412° C	boils at 2680° C
Cadmium 48	melts at 321.03° C	boils at 765° C
Xenon 54	melts at −111.79° C	boils a-108° C

How can science promote their image of establishing honesty when they are confronted by such truths but choose to ignore the truth so long as a lie will bring them some respectability.

Since the star is the total configuration of the atom's characteristics, the atoms will tell us what we should know about every layer from what is applying in such a layer to what characteristics such a layer would show when it provides the function of what it has to for fill within the star.

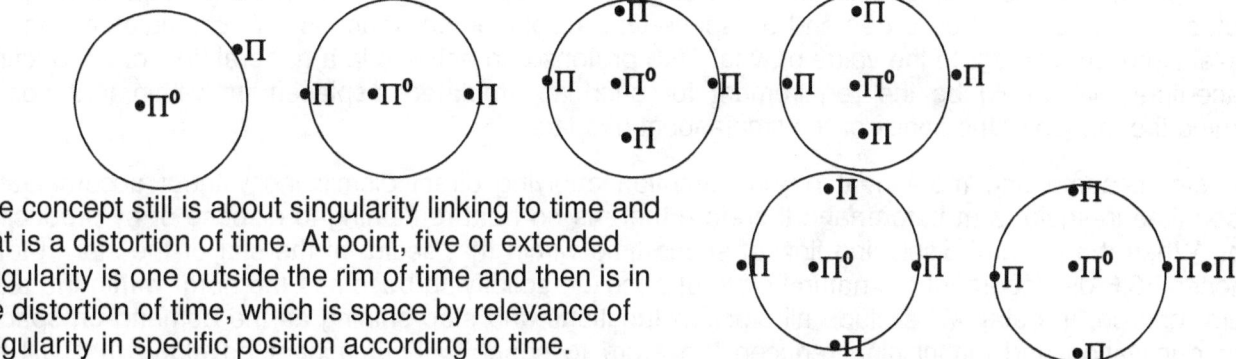

The concept still is about singularity linking to time and that is a distortion of time. At point, five of extended singularity is one outside the rim of time and then is in the distortion of time, which is space by relevance of singularity in specific position according to time.

Let us investigate and try to find a way by using logic how a star applies gravity. Therefore it is not the number of dots that is important. It is not the size of the number of dots occupying the position or the size of the space the dots occupy that is prominent. It is the relation in the dismissing of space and the duplicating of space that becomes important. The less space there is the more the favour will be to reduce the space because of the advantage the dots have in securing space-time that will prevent overheating. On the other hand the more space secured will also prevent overheating and therefore those will opt to duplicate space in order to find space to secure and prevent overheating.

Since the Earth has no singularity demand that is much better developed than the Universe sustains, we find on Earth a relevancy of Π to $(\Pi^2+\Pi^2)(\Pi^2\Pi)3$ is adequate. But in bigger units the space-time displacing relating to space duplication presents much more demands on atomic structures occupying space within the star containing through set boundaries. In the presumed to be bigger stars there is much space filled with atoms occupying much space. In the stars more massive but holding lesser space the atoms must also hold lesser space but they also hold more protons by number in the lesser space.

The space the particles hold is directly in relation to the particles the containing structure duplicate. The more space that is relevant to the structure that the star duplicate by motion is then in turn once again relevant to the space the structure destroys by proton action in space less units. The more space the particle claims in relation to the space the container holds that relates to the space the container duplicate is relative to the space the containing structure destroy. From that mass derives value. As individual occupying space the atom is an individual container by own merits and as such duplicate space in this regard within the specific confinements of atoms.

This we will classify as normal applying structure values the atom has in outer space or in structures with very little atmosphere. Please note there is no pressure involved because the motion involved creates conditions naturally instead of unnatural pumping that causes pressure. Pressure is an artificial creation as part of life but has no role in the natural cosmos. Pressure is a condition where the retaining of particles has to be confined in a patrician made of material where the outer wall does the retaining of the substance within. This obviously cannot be in a star because the "pressure" is regulated from a condition applying and space-time controlling inner centre that needs no solid walls to contain whatever is inside. With that one can see there is a Universal difference between the concept of pressure forming due to human action inside a container and what comes about as secluded space-time within a star.

As the demand of singularity in such units grow stronger some relevancies within the atom come into play and I developed a system whereby I can arrange the space-time merits of space-time curtailing within the confinement of the star borders applying in the star to place such a demand in relation to singularity where the ultimate demand sets the standards. In the sun, for instance, which is a minuscule small star a relevancy in the outer region might be 3^3 relating to singularity and with the atom having, a sustaining displacement of $(\Pi^2+\Pi^2)(\Pi^2\Pi)3$ there is no danger of the atom demising. The electron in the sun will have a diminishing factor of 27 whereas the atom can sustain $(\Pi^2+\Pi^2)(\Pi^2\Pi)3 = 1836$. The relation in the atom degenerated by 27 leaving the atom a sustaining value of the electron plus the neutron applying space-time without involving any of the neutron aspects at all. That is the mass of the space-time that the electron will consume in the space reducing flow of space-time.

The flow is the result of heat distributed where the heat is delivers to the dismissing sector and producing of the duplicating of space by mass within the star that then forms a favouring of duplication in comparison to dismissing. The star is a bright little boy shining by dismissing pebbles of light-photons into space. When a demand on space-time displacement reaches an accumulated general displacing or movement to the value of what 56.6 protons can achieve in a general flow of conducting space-time that would be the requirement for such accumulated displacement within that space forming the motion of the space or the time aspect of space.

The star accumulated more heat by consumption applying direct dismissing without accumulating space-time in liquid form beforehand therefore there is no heat remaining to dispose of by producing light. When the general displacing flow of space-time within that sector of the star or the star in total reaches 56.6 displacement the natural state of absolute solidifying becomes the norm within the star. From then on, the star will exclude all electron functions and stop shining as the demand on space-time duplication and diminishing reduced the atom to space without a heat envelope that will be electrons or a liquid/gas jacket. Only the nucleus will be able to sustain the diminishing and the reducing of space by increasing of time. The entire star becomes a solid structure by reducing space-time directly freezing the space-time from a gaseous state to a solid state. By motion, speeding up the tempo of the flow of space-time the liquid state of space-time is by passed going from gas to solidity in one motion. The atom would shrink to such little space it will have space within the star that only the centre nucleus will fit. More reducing by applying motion in creating space differentiation will leave a star with so little space the space will be insufficient to secure a position for the neutrons and the star will then have the name of being a neutron star. Going even further will find the proton rejected from the star.

Every atom holds (I am guessing), as many dots as the sun has subatomic particles per atoms and that would still be a very conservative guess. Every dot is a controlling centre selecting a regional centre where every regional centre selects a centre. This goes on as long as there are spots forming groups as individuals unable to survive independent. The others that was unable to group formed heat that became space, which became the broken dots. The dots form groups to survive and as a group, the survival depends on doing what the group has to do to remain cool. In another book, I reserve one chapter to explain the phenomenon what I called the Lagrangian atom. These dots arrange in a manner that they could favour either the space duplicating aspect or the space dismissing aspect.

This can only be the result of the fact that even in the case of the sun, the inner space is almost entirely liquid heat and the liquid heat produces sufficient space to dismiss as the centre that holds the heavy metal particles, where all the dismissing is done. The liquidity provides motion while the solidity removes motion in the centre of the star. The dismissing going on is in the space factor where the space leads to a denser heat within that space because there are insufficient material to accommodate all the heat by the dismissing factor T^2. In that case motion far outweighs dismissing $k>T^2$ but a time comes in every star that the dismissing takes absolute charge. $k<T^2$ That is when the star goes dark. The Earth is mainly about duplication of space much more than dismissing of space and so is every structure in the solar system.

I would suggest we think of stars in the following terms. A star that generates and transmits a lot of light is weak on gravity because their progress started recently. They command a lot of space-time but the demand they have to keep their cooling acceptable is very low. In that they can generate a lot of light but with the demand on cooling low and the gravity in the centre not very developed, those stars cast a lot of light back into outer space. It is just because of the size the stars hold that tell the that the stars are still young and have a weak developed governing singularity. The stars will have very prominent hydrogen and helium layers, with the inner core not very prominent. The control of the star is still very much in the individual atoms and in that the motion the atoms have to produce in order to maintain their individual singularity will only come about through motion. The atom has to make contact with as much space-time through motion as possible since it has a very poor ability in contracting space –time in support of the cooling system.

When what was perfect became imperfect the Universe started. When the spot differentiated and became differently allocated from the dot the Universe started. When infinity moved away from eternity the Universe started. I show where infinity is as much as I show where eternity is and any person can put his or her finger on the spot. I show where space ends and where time begins and any person can look at the point. I sent an article to Annalen Der Physics and Professor Ibn Christianson explaining this in a 15 page dissertation. They came back to me advising me to While it is possible that a lay person hits on an insight that has been overlooked by academic trained in the field over many years, it is unlikely. We assume that work offering something new would be related to existing theories, either by building on top of them or by showing how and where they fall short (Professor Iben Maj Christiansen) and With a lot of words and some simple algebraic relations, there is no way to "explain" the world of physics (Friedrich W. Hehl, Inst. Theor. Physics). Maybe our Academic elite including the two I quote had no idea what I was talking about, understood not a word of the detail and was fat too brilliant to accept and admit this state of affairs. Maybe if our distinguished Professors including the two I mention read my work with more attention they would have seen what I try to show and what I try to show is where physics starts. Professor Christiansen since when is Π not science?

Creation started off with one dot so small eternity met infinity within. Then came one more, and another and they continued coming until there were a countless number of dots. The accumulative size of the dots were the same size as one dot because in the true Universe big and small plays no part. The dots were infinitely small and eternally big at the same time because size is a relevancy and without one the other has no size. So in the true perception, there is no difference in size.

It started with the fact that there is no place or part in with which one may associate zero or nothing. There are no room for a number such as nothing. Next to the one dot (infinitely close) one will find the next dot, and if nothing was a factor then that is precisely what one will find between the two dots. Nothing of space, a non existing entity, taking up no space, and much more important, no time, therefore the dots are infinitely close to one another, being the same space, eternally big as much as infinitely small. If we as humans cannot find a manner in comprehending this notion, there can be no manner ever understanding the cosmos as much as the start to the cosmos.

Every dot was a Universe in its own and the accumulation was a universe. The earth in itself is a Universe as the moon is a universe, because rules applying on earth do not apply on the moon and visa versa. When in the ocean another set of rules apply, therefore being in the sea places a body in another universe. The number of universal entities is still countless, as much as it was in the beginning. Every dot insignificantly small as it may be, is a part of another Universe as much as it is part of the accumulative Universe and every dot in the infinity holds singularity, which we translate as " nothing" being " darkness". There cannot be "nothing" just as much as there cannot be "darkness".

There cannot be something big or small, but it into relevancy of perception, and then the relativity of perception becomes the question. There cannot be hot as much as there cannot be cold. The sun FREEZES hydrogen to a liquid at six and a half thousand degrees Celsius and Universe boils over in the form of the Hubble constant at the temperature (we presume from our vantage point) at minus 273 degrees C. If we Humans cannot or will not abandon our human perception and our manly perspective, we may as well return to astrology for all its worth.

Every point in the infinity we may observe at is not merely part of the Universe in not being nothing, but is the point where the Universe started representing singularity. It is the very first point where everything began so many eternities ago, because after all, how can we ever determine where the first point was, as they were very much equal and alike at the beginning. Every aspect of the Universe started with the fundamental fact that no point in the Universe can represent "nothing" as a number, because every aspect in the Universe represents singularity in what ever form it may hold in that specific spot forming space-time. If man does not reach a conclusion where that conclusion is matching the Universe and stop to match the Universe with man (and man's incapability), we may all go back to caves and become starving hunter-gatherers again, because we will never find a way to progress to the ultimate understanding of the universe.

Looking at stars Newtonians see coal stoves being stoked to burn. In the days of Newton coal stoves were the nuclear science of the day ands while all other departments in science moved on and away from coal stove principles Astrophysics and cosmology remained true to Newton by inventing the coal stove in so many ways not even the coal stove could think of the facets it can go through. Newtonians see stars being fuelled like coal stoves and such stoves can run out of fuel. This is so much Newtonian backwardness as mass forming gravity and the moon coming closer and the cosmos shrinking and we falling into the sun because of non-existing dark matter making up what is required to make Newton not to seem the idiot that Newtonians are because they make him and his contraction theory to be less foolish that what it apparently is and they overbearingly are.

What is of vital scientific importance is that there are three fundamental dimensions controlling the universe. The three are beyond intermingling and one confirms a status in relation to the others but not intermingling in status. From singularity comes matter and forming space-time in own accord. By matter not controlling time, space grew uncontrolled and the third dimension came about. That dimension birth we now recognise as the Big Bang, but the Big Bang is the last of a three prong cosmic growth. Science has to recognise the dimensions of densified (singularity), occupied (matter behind the electron) and unoccupied (space-time outside the orbiting electron boundaries) forming three points of cosmic recognising space-time.

Every dot was by itself as well as the accumulation as it currently is the present universe. The earth in itself is a Universe standing apart from other universes such as the moon as well as the space between the moon and the earth. The moon is a universe. Rules applying on earth do not apply on the moon and visa versa. When considering conditions with in the oceans and applying space-time another set of rules apply therefore the sea places a body in another universe. It takes the same engenering technology going underwater in deep sea diving that going into outer space.

The number of universal entities are still countless as much as it was in the beginning matter as atoms and even much smaller. Every dot insignificantly as it may be is a part of another Universe as much as it is part of the accumulative Universe and every dot in infinity holds singularity, which we translate as "nothing" but it cannot be nothing. There cannot be nothing as much as there cannot be darkness. There cannot be something big or small except in the relevancies of perceptions and then the relativity of such perceptions becomes questionable. There cannot be hot as much as there cannot be cold The sun freezes hydrogen to a liquid at 6500 ^{0}C and outer space boils over at 0 K. If we humans cannot or will not abandon our human culture driven perceptions and our mankind's pre-programmed perspective we may as well return to astrology for what the future hols. There are so many boundaries out there ready to destroy us because of our lack of insight, as did the challenger disaster.

Creation birth started off with one dot so small eternity met infinity within. Then came one more, and another and they continued coming until there were a countless number of dots. The accumulative size of the dots were the same size as one dot because in the true Universe big and small plays no part. The dots were infinitely small and eternally big at the same time because size is a relevancy and without one the other has no size. So in the true perception, there is no difference in size.

It started with the fact that there is no place or part in with which one may associate zero or nothing. There are no room for a number such as nothing. Next to the one dot (infinitely close) one will find the next dot, and if nothing was a factor then that is precisely what one will find between the two dots. Nothing of space, a non existing entity, taking up no space, and much more important, no time, therefore the dots are infinitely close to one another, being the same space, eternally big as much as infinitely small. If we as humans cannot find a manner in comprehending this notion, there can be no manner ever understanding the cosmos as much as the start to the cosmos.

Every dot was a Universe in its own and the accumulation was a universe. The earth in itself is a Universe as the moon is a universe, because rules applying on earth do not apply on the moon and visa versa. When considering the conditions with in the ocean and applying space-time another set of rules apply, therefore being in the sea places a body in another universe. The number of universal entities is still countless, as much as it was in the beginning, before dots formed atoms.

Every dot insignificantly small as it may be, is a part of another Universe as much as it is part of the accumulative Universe and every dot in the infinity holds singularity, which we translate as " nothing" being " darkness". There cannot be "nothing" just as much as there cannot be "darkness". There cannot be something big or small, but in the relevancy of perception, and then the relativity of perception becomes the question. There cannot be hot as much as there cannot be cold. The sun FREEZES hydrogen to a liquid at six and a half thousand degrees Celsius and Universe boils over in the form of the Hubble constant at the temperature (we presume from our vantage point) at minus 273 degrees C. If we Humans cannot or will not abandon our human perception and our manly perspective, we may as well return to astrology for all its worth, because that is the only boundaries we will find in the cosmos.

To unlock scientific truth we first have to dispose of scientific misconception

In the two pictures we are seeing disposing or releasing heat creates space. We may call it plasma or shock waves or what ever, but in the final analyses it is heat turning to space. Whatever you wish to call that which lies between the particles comes from being a solid, then with adding heat, the solid *"whatever"* becomes liquid and that is the white and orange plasma that we find. That white and orange is heat in a liquid form, just as all flames and smoke is heat in a liquid form. But that liquid does not remain liquid because the governing singularity cannot enforce a commitment ensuring the liquid heat remains liquid. The liquid *"whatever"* you wish to call the heat in fluid form then further overheats turning the heat to space. The space created must be equal to the heat reformed. That is a law of energy where energy equals equality everywhere it is.

Let us humans first detach culture from facts. Take the argument to iron, which we know well. Iron cannot boil, iron cannot flow or bend and iron cannot brake. Iron is an element like all the other elements we know, not one element can do any of the above, in sharp contrast to human belief. As indicated in this book the limits we should find to guide us we ignore for the reason that we cannot see it. We may not be able to ever see singularity, but with intelligence guiding mankind, we do not have to see everything to believe everything. It is because we could not see religion, but still practised religion that set us apart from the other animals.

At the start one would find iron and iron in a "natural state" as we find iron on earth being a human produce on the surface of the earth it will be a solid, suitable for man to handle with bare hands. When such a piece of iron is left in a desert in the midday heat, the human hand cannot handle the iron any longer without aid of covering the skin of the hand. Our perception is that the iron became hot, but that is not the case and our view is a culture contribution and not scientific fact. By heating the iron artificially with combined gasses (acetylene and oxygen or what ever) we now can over heat the iron to a state of flowing like a fluid. Our human culture tells us the iron now is melting.

That is a misconception!

Like the fact of "nothing" we inherited the idea from our past. After introducing artificially even more heat with more heat releasing gasses we may artificially form a condition where the iron would become a gas. Again it is not the iron that becomes a gas, it is the space the iron finds itself in that became hot enough to become a gas. The iron particles remain the same; it is the condition surrounding the particles that changes form with overheating.

Important to note is the fact that iron in a solid state will surround itself with solid matter in space applying a solid space. By introducing conditions producing **more overheating** the space or connecting between the particles become concentrated heat forming a liquid substance! It is not the iron that turned liquid but the wrapper containing the iron that concentrated so much it formed liquid fluid by the introducing of more heat to a point where the overheating created a fluid. It is considered that the oxygen burn and by that the iron heats up. NOT TRUE!

If oxygen burns no oxygen would be left on earth by the time man arrived on earth to use it to the benefit of intelligent life. The oxygen remains oxygen while the oxygen merely does a task in nature where oxygen carries heat to a specific space. On the other hand it is the task of nitrogen removing heat from the point of overheating by means of flames whereby it creates space. One can feel the "wind blowing" as the flames generate created space. In the extreme the creation of such space we call an explosion.

In the process where the space between the iron particles still further overheats, it becomes a gas. It cannot be iron that becomes gas, because iron will be as much a gas as iron will be a liquid or a solid. It is the space covering the iron particle separating the different iron particles, which will convert and sustain form. The gas is as invisible as space because the gas is the form space holds. This confirms the Biblical view of earth (solids) created and heaven (heat or gaseous/liquids) created. There are only two forms of substance that forms the Universe solids and non-solids, which is liquids and gas. It is not the solids going liquid but it is more of the liquid in ratio with the solids in between the solids that make a structure go solid or gas. There are heavens (non solids) and earth (solids) and this has to do with movement applying control or non-movement allowing non-movement control.

Iron is a solid. Introducing more heat the iron becomes wrapped in a cover that concentrates the wrapper to the point of concentration where it became a fluid. The iron remained what it is, neither a solid, nor a fluid nor a gas. By introducing more heat it becomes a gas. The gas we cannot see because the gas is space. But so was the fluid space. The introducing of heat brought about the turning of a solid to a liquid to space and every time more space becomes part of the picture.

Iron is in its normal form a solid. That means the space, which the iron particles are in, is solid and that disallow the iron to alter the form in which it is. By introducing considerable heat the iron melts changing the form of the iron from solid to liquid.

Considering the evidence we find it is not the iron that melted and that became liquid, but it is the space in which the iron is that became liquid. The iron particles are still as solid as they were. By introducing more heat the iron would eventually turn to gas. It is not the iron that turned to gas, but it is the space in which the iron particles are that has increased to the extent that the space now has so much heat, the heat turned to more space. The iron as particles remain the same, they are just elements confined to a nucleus with electrons spinning about. The space between the particles increased to such an extent it first became a liquid or a fluid and with more heat introduced the heat increase brought about that heat turned to space. That means by overheating the particles surround with heat as a fluid the heat increase then add space as a gas. The gas is the ultimate form of overheating but where one is unable seeing the gas.

1 Firstly the iron is cold enough to be a solid. Replace the word iron with cosmos and forget the colour we associate with heat being white and note the solidness of the centre of a galactica. This must have been the state of galactica that contained large parts of the Universe when time rolled away from eternity.

2 By introducing overheating the space between the iron and not the iron as such turns to liquid. The same apply as more matter (iron) produce more space forming as some matter turned to heat by overheating. The matter increased spin and in that way went out of sequence where it then became softer and softer in relation to other particles, where the loss of the matter released more of the third cosmic component we named heat and space.

3 Some of the heat introduced with the overheating by means of congestion then forms space while other remain in the form of heat allowing space to seam liquid. The matter could not breath and overheated by the enormous gravity the overheating created

4 As the area between the particles still further overheat certain parts of the area overheats to the extent that the space becomes an invisible gas allowing the congestion of matter to separate from one another and allow the stars' individual governing singularity growth. 5 From the soup of heat galactica come about allowing stars to rise out of the dense liquid cradle from where they can establish singularity growth. The process continues as more space becomes introduced through space overheating turning heat into space

6 Should star development come about as suggested it is foreseen that the Milky Way once was a liquid from which the sun developed the singularity in which it then form self-sustaining. The only precondition was that it captured individual space-time where the captured space-time remained a liquid frozen (as it was back then at the time of parting) by the governing singularity while outer space further overheated into a thin gas

7 The sun captured so much space by the intervention of singularity when released from the Milky Way that it produced space so concentrated today at present it clearly remained a liquid inside as it froze the interior in time the liquid it now is while outer space is still overheating as a gas with no visibility. From this overview one can judge just how far science is behind the time in their views on creation and the beginning of time including the universal establishing. Cosmology still hides behind medieval ideas that other faculties and scientific departments forgot long ago.

See the fluid push out of a bowl of liquid, spilling both sides as it falls into liquid. The inside of the sun is not gas but it is fluid. In all of nature there is no NATURAL GAS as much as there is no NATURAL SOLID. Look at the liquid squirting from the surface of the sun. if it is liquid the sun is liquid on the inside and if the sun is hydrogen then the sun is so cold through movement that it turned hydrogen into a liquid at 6500°

Science a under the impression that water will boil at 100o C when it is on the sun

Let us take this formula back to the accepting of the Big Bang and find sensibility amongst a lot of confusion that I can see.

As a solid	Forming a liquid	Steaming as a gas
Hydrogen 1	melts at -259^0 C,	boils at -252^0 C,
Helium 2	melts at -269^0 C	boils at -268.9^0 C
LITHIUM 3	melts 180^0 C	boils at 1300^0
BERYLLIUM 4	melts at 1287^0 C	boils at 2770^0 C
BORON 5	melts at 2030^0 C	boils 2550^0 C
Carbon 6	melts at 804^0 C	boils at 3470^0 C
Nitrogen 7	melts at -210^0 C	boils at -195.8^0 C
Oxygen 8	melts at -218.8^0 C	boils at -183^0 C
Fluorine 9	melts at -219.6^0 C	boils at -188.2^0 C
Neon 10	melts at -248.59^0 C	boils at -246^0 C
Sodium 11	melts at 97.85^0 C	boils at 892^0 C
Magnesium 12	melts at 650^0 C	boils at 1107^0
Aluminium 13	melts at 660^0 C	boils at 2450^0

Hydrogen is as much a liquid as iron is a gas and neon is a solid. It depends on the element relating to the space/heat in the circumstances surrounding the substance at that very precise instant in time. We have to stop telling the cosmos to show us what we wish to find and start accepting what the cosmos is telling us to find. The culture that I am referring to is all about **nothing.** At present, we find that there is something we think of as nothing in outer space. Because nothing is what we wish to find and nothing is precisely what we are getting because we think of outer space as nothing. If you accept the cosmos to be nothing, then please define nothing to yourself and find the definition in the cosmos.

As a solid	Forming a liquid	Steaming as a gas
Nitrogen 7	melts at -210^0 C	boils at -195.8^0 C
Oxygen 8	melts at -218.8^0 C	boils at -183^0 C
Fluorine 9	melts at -219.6^0 C	boils at -188.2^0 C
Neon 10	melts at -248.59^0 C	boils at -246^0 C
Sodium 11	melts at 97.85^0 C	boils at 892^0 C
Magnesium 12	melts at 650^0 C	boils at 1107^0
Aluminum 13	melts at 660^0 C	boils at 2450^0
Silicon 14	melts at 1412^0 C	boils at 2680^0 C
Phosphorus 15	melts at 44.25^0 C	boils at 280^0 C
Sulphur 16	melts 119^0 C	boils at 444.6C
Chlorine 17	melts at -101	boils at -34.7 C
Argon 18	melts at -189.4^0 C	boils at -185.8^0 C
Potassium 19	melts at 63.2^0 C	boils at 760^0 C
Calcium 20	melts at 838^0 C	boils at 1440^0 C

Ignoring these facts, Mainstream science will hardly answer the problem we do not understand and such ignoring brings strong doubts about the quality and sincerity of science.

Excluding Argon, which is six (carbon's number) times two and suddenly that is a less dense material. The four times five plus… group are the following:

Scandium 21	melts at -157^0 C	boils at -152^0 C
Titanium 22	melts at 1670^0 C	boils at 3260^0 C
Vanadium 23	melts at 1902^0 C	boils at 3400^0 C
Chromium 24	melts at 1857^0 C	boils at 2665^0 C
Manganese 25	melts at 1244^0 C	boils at 2150^0 C

Iron being the five times five plus one is the only generator of electricity and therefore the producer of gravity making five times five plus one the ultimate relevancy to heat in reducing space. Still Krypton is much more massive and turns out to be a gas.

Krypton 36	melts at 1539^0 C	boils at 2730^0 C
Iron 26	melts at 1536.5^0 C	boils at 3000^0 C
Cobalt 27	melts at 1495^0 C	boils at 2900^0 C
Nickel 28	melts at 1453^0 C	boils at 2730^0 C
Palladium 46	melts at 1552^0 C	boils at 3980^0 C
Silver 47	melts at 1412^0 C	boils at 2680^0 C
Cadmium 48	melts at 321.03^0 C	boils at 765^0 C
Xenon 54	melts at -111.79^0 C	boils a-108^0 C

How can science promote their image of establishing honesty when they are confronted by such truths but choose to ignore the truth so long as a lie will bring them some respectability.

If anyone puts anything as hot or as cold that person puts him or herself in as the centre of the Universe. Then from such a stance as the person has that person finds at the centre things go colder or hotter because that person is stupid enough to think the Universe was created with life in mind and moreover for that person in particular. There is no hot and there is no cold. As it is important to realise the above it is just as important to realise that heat is another form of material and a separate form of material. The two developed on an equal basis and as a result of the other. The one produced to save the other and what the one produced saved the other .The one principle brought the incentive for motion while the other took the incentive by providing the motion. The one produced what the other captured and the one retained what the other delivered. Eventually the motion did not bring the required relief and another form had to be devised. By overheating and increasing space it counteracted overheating and by removing the expanded material and retaining it onto the contracting of the other, did the two form a synopses where by all received benefits in the form of cooling.

Only when further requirements developed, did the need arise for more to be made available. The first demand on motion asked no further changes because one change brought on satisfaction to all that suited all. The second was more general and on an ad hoc basis that was established to fit the need of individual places and not groupings in the broader perspective to fit individuals at large. At first the establishing of motion set a trend that brought on required results but afterwards the space required, in which to move became a demanding issue as the heat levels raging out of control. The heat had to be stored in space by becoming space to retain heat for later consumption. The number in ratio that produced the heat providing particles that offered to release their form in contribution to have those that retained form, do so to save those others retaining form. Those on offer became those ones that became the danger of destroying Creation instead of saving Creation.

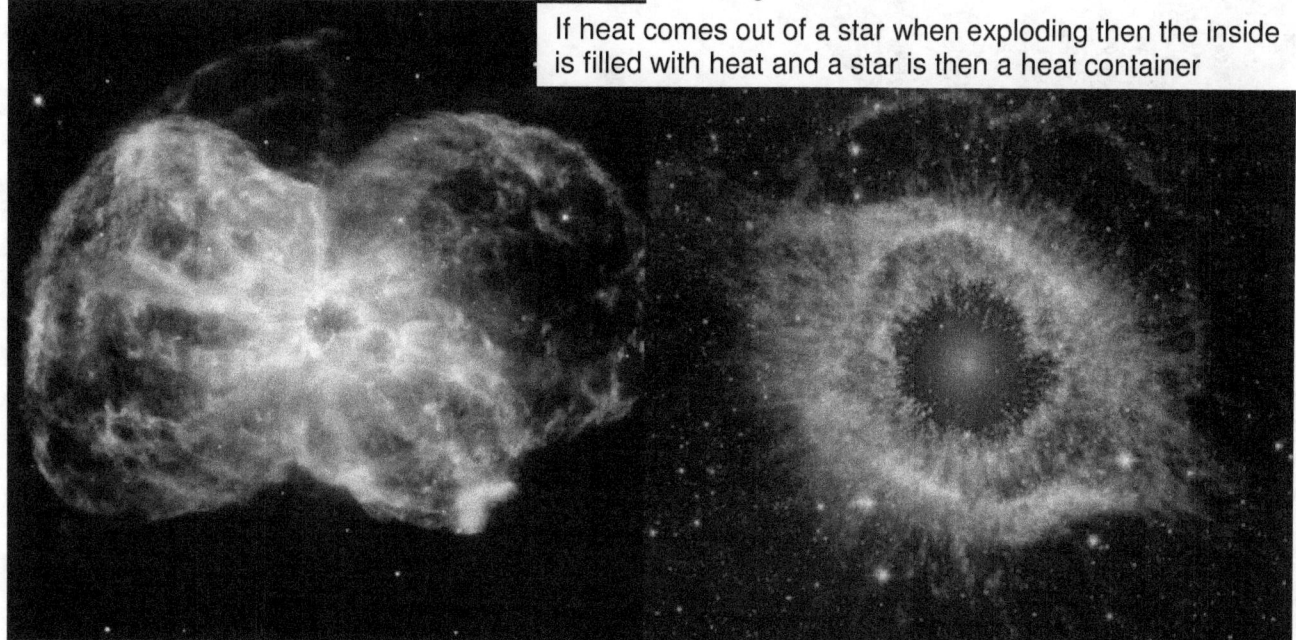

If heat comes out of a star when exploding then the inside is filled with heat and a star is then a heat container

When we see a Super Nova going ballistic we see heat spewing from the star to the outside regions. We see what is inside the star and what is inside the star is heat, not "mass". We see heat bursting out in rage human minds cannot absorb. If heat comes out then a star is filled with heat. That means a star is a cooler of heat because when the star overheats, as it no longer can retain the heat on the inside by gravity it explodes into unleashing the inner heat.

There might even be some areas and regions in far off places in our modern day where an imbalance may evolve and some particles become unsuccessful to save those more successful. By going less successful, the singularity places a demand on another bringing about the command on space-time so that support can be accomplished to save singularity. Therefore by losing density, was gaining security to survive as part of a bigger relative. Density is the distributing of heat in specific relative space and by having less material in more space; the density is the offering for the common survival of the lot. The relevancy brings a contribution in whatever role to secure the survival of the lot in relations. No relevancy therefore can be "nothing" notwithstanding Newton's opinion about the matter as Newton had the opinion a relevancy acquired by rotation brought about an accumulation resulting in nothing.
It is the way the atom formed before the atom took on space-time. It is in the formation, that space-time relates to motion. We have some elements being quite massive but also lighter than air and others are

quit light but as dense as they come. This can only be a contribution from the way the atom relates to heat, which make the atom volatile (movable) or dense (motionless). Those elements being volatile are also very movable and in that we find the role that such elements play in the star. Stars that are predominantly made up of hydrogen and helium with very slight support from the metallic inner core are those stars that duplicate by producing motion. However the point I wish to press is that mass and being massive and being heavy do not support the fact that some elements have more gravity they produce because their protons are more numerous than others. The fact that mass generates gravity is a myth.

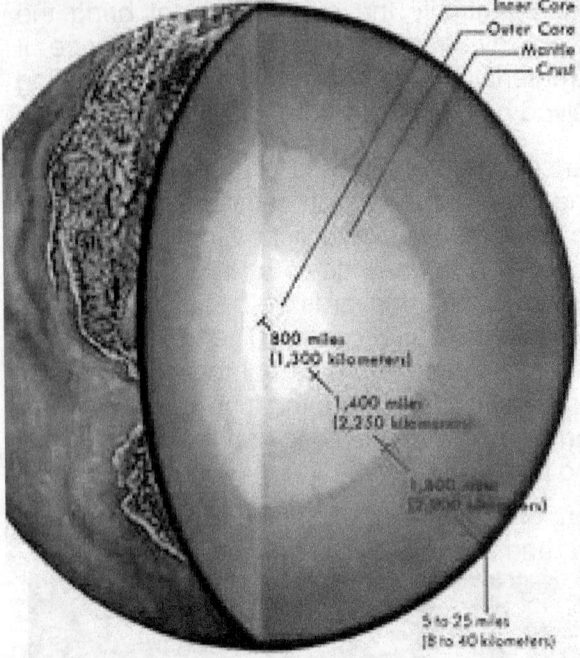
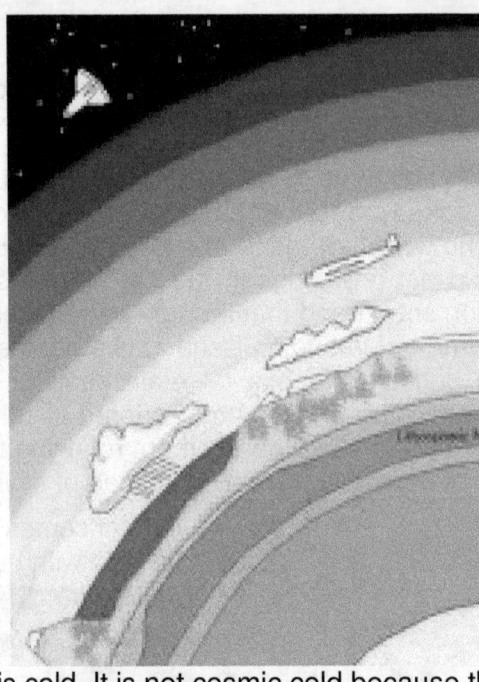

Hottest
2nd Hottest
3rd Hottest
4th Hottest
5th Hottest
Getting colder
Getting more colder
Getting freezing cold as it gets liquefying cold in the centre

When it expands it is hot. When it contracts it is cold. It is not cosmic cold because the human can feel heat but on the contrary it is cold when the human feel heat because it is getting rid of the heat since the space got cold as it contracted. That is science law and not the meaning of some stupid scientist. Heat in the centre of the earth is a liquid and heat in outer space is a gas. The liquid is much more contracted than the gas in outer space is and that is why we can fly in space and not swim in a volcano. We must start to realise how the cosmos translate values and stop thinking like humans. When the balloon gets hot it expands and it goes up. When the anything loses heat in shrinks as it becomes smaller. That too must apply to the earth where the inner core space gets colder as it loses heat and the heat rises to the outer space region. It is not what we feel that makes anything hot or cold but it is what the cosmos does for instance increase in size or decrease in size that forms heat.

One will find that whatever group one chooses there are gasses and there are solids. If mass was attracting mass then the strongest mass must be attracted to the strongest mass and the least mass must float in the air. $F=G(M.m)\ r^2$ hardly can even begin to explain the fact that there is a gas that is more massive than iron but floats in the breeze just as hydrogen which is the least massive element. Let's look at gravity and anti-gravity and see where the wind blows the balloon.

The saying goes "it fills with hot air" meaning it fills with nothing and yet while the balloon fills with hot air the balloon gets air borne. The balloon gets into the sky by being filled with "hot air" and the idea that this might happen is senseless to science or so they pretend. Still it happens but why does it happen? It is because science has no idea about heat and cold. When filling the balloon with heat the balloon takes of to where science thinks it is cold. Why would the balloon go to a cold place when it is filled with heat? It is because where science think it is cold it is hot. Forget this idea of pressure because if it was pressure in the air the pressure will escape from the bottom where it is open and being pressure it will escape or release from that opening. By heating the balloon with hot air must increase the heat level and increasing the heat level takes the balloon into the air. This means hotter is higher up by holding more space than down below.

When something fills with "hot air" it goes up and by going up it forms anti – gravity. If gravity is what takes bodies down to the earth then lifting it up into the air must be anti – gravity. Anti means opposing or counter acting and when going down forms gravity an anti of that then must be going up. If it is mass that pulls down then by applying more heat it forms anti-mass but it can't be anti- mass because the persons in the basket seems just as they were when they were on the ground wit all the mass intact.

Newtonians are never very clear on this issue but if it is mass that pulls then it must be gravity that moves. Mass supplies the magical force but the actual movement is derived from gravity. So the movement in itself is gravity and that gravity pulls bodies down. To have bodies lift into the air then this action represents what goes anti and anti is filling a basket with hot air. By adding hot air the gravity goes anti and then gravity must be the cooling of space. It is definitely heating space up that makes the balloon go into the air so going into the air forms anti gravity because gravity is going down. This is an argument that science cannot dispute except when they go into another cheating and dismissing mode of the truth. By heating the basket something happens and this science for many years try to avoid by reducing it to a joke. A joke it is not because it is a fact nature substantiates with forming anti- gravity. When heating a basket makes it lift than gravity is making things cool and that science can't deny.

When the earth spins there are always a relevancy applying between that which spins are double 7 and that which moves straight continuing at 10. The part spinning at seven we think of in terms of the diverting of direction it applies at 7. The liquid / gas holds the 10 factor.

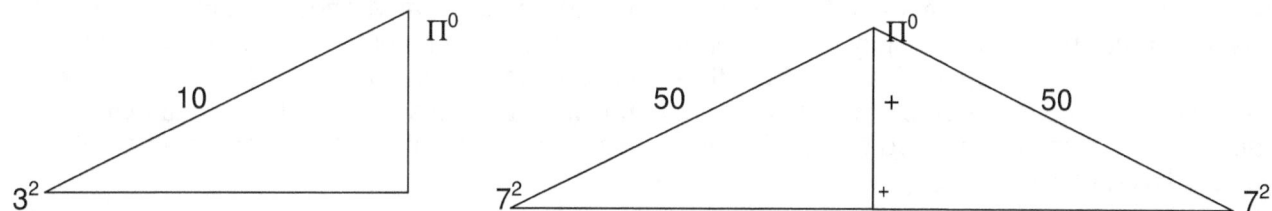

Without the application of specific heat, the object remains in the three directional moving of six possible directions. The value of space unoccupied therefore remains $\Pi \Pi^2$, as it was before the "Big Bang" event, whichever "Big Bang" you wish to refer to, because there were many. But space unoccupied holds time to the value of 10 to 1, and as the sketch of the triangles also indicated, holds space to Π. Therefore unoccupied heat holds the relation to space in applying 3 directions of influence $(3^2 + 1^2 = 10) = (a^3 = T^2 k)$. Always part of this equation is the dual function of space in $(a^3 = T^2 k)$ while at that very instant one has space-time. Therefore in space in time you have $(10^2) = 7^2 + 7^2 + 1^2 + 1^2$. In the sphere we have the axis holding a value of 3 and the circle holds a value of 4. These are dots forming in relaxation to the one spot holding a point from where singularity advances.

We have the axis valued at 3 going square through movement of the linear motion $(3)^2$ and then we have the circular motion $(4)^2$ going square by the spin of the circle ring the direct opposing side. Then the equation of influence becomes $3^2 + 4^2 = 25$ where $\sqrt{25} = 5$ and doubling the 5 on both sides of the triangle will apply the factor of $5 \times 2 = 10$ that then is $(10^2) = (7^2 + 1^2 + 7^2 + 1^2) = 50$ on both sides is 10. The implication of this may not dawn on one the very instant of realizing, but to scientists, there is no greater shock than just that. To any application of movement, the factor will be in the realms of singularity where half a circle is equal to a triangle is equal to a straight line and the lot is equal to 180°. No fancy mathematical expressions have any value in singularity because singularity holds a value of 1.

$(7^2 + 1^2) = (10^2)$

$49 + 1 = 50$

1^2 / $T = 1$ — 7 positions from singularity — Π^0

$50 = 5 (10^2)$ where the complete Pythagoras is $2(50) = (10^2) = 100$
$\sqrt{100} = 10$ the value of space.

The fact of this comes as 49 plus one becomes 50 and that is in the three dimensions of space $\Pi^2/7$ where 7 holds the relation to one and $\Pi/7$ again where 7 relates to one. At this point it is most important to remember that Pythagoras works on the application of the sum of the square of the two sides. When seven has a direction in the fourth dimension applied to it, the opposing dimension will be one and this applies in time relevancy, therefore the interchanging in time between infinity will place matter at $7^2 \times 1$ relating to circular and $7^2/1$ with $7^2 \times 1$. This makes 49 plus one (singularity) always being a factor of one. Space in time however, never can be a cube, it will always be a square with one side pointing the direction of time from time to the past (1) to time to the present (1) to time to the future (1).

The circle forming Π uses 7 to indicate the roundness of the circle but the 7 holds its roots deep within creation. It indicates how the Universe started because this is the way a star will start moving and it shows how as the infant star starts generating gravity just as the top starts to spin when it is thrown by life. Life can create nothing and that is true but life can mimic all laws in the Universe. Time is eternal movement and will be with us always. The line in infinity is still present while not being a part of the Universe. This line is always ready to be in place when the slightest movement orders it in place. Before the Universe was in place eternity and infinity was in perfect harmony and the line forming singularity validates this fact.

Before infinity parted from eternity, eternity met infinity on one spot as eternity came from the past (1) forming the present (2) to go onto the future (3) but also returned to come from the past which was the spot held by the future and this we find in the fact that the line forms 1 when not spinning but as soon as it evokes by spin, 3 points form even now. Then heat and cold differentiated values and space landed in between eternity and infinity. As eternity moved in relation to infinity but not forming a part of infinity any longer, eternity had to follow a path by never going away from infinity (3) and always returning to the point infinity holds but never lash onto the point again. With space parting the points, eternity had two points (the past and the future) before the partition came about and infinity held both the past and the future while infinity had the present as it still gas presently. By eternity also moving, the two points it held opposed each other (the past and the future) and since it moves, by the movement it became the square of the two because movement is the square and not a flat blanket-like surface with squares embroidered on it as Newtonian science depicts it by using grand mathematics to understand singularity.

Then we had two point holding eternity in place going square by movement to form 4 points serving eternity and infinity captured the first three points held by both and since eternity could not release from the two it had but had to duplicate what it had, eternity by movement became a circle captured by the line. With four points captured by the line of three points the circle coming about is eternally returning to infinity but never complying with infinity because if mismatching temperature or movement (3 against four). Material will always be colder than outer space. It is because material spin and outer space moves by expanding due to overheating.

This is where I start when I start to explain the first moment but I use a shipload more information to do explaining when I explain the star in the book I do so. I involve the four cosmic pillars to substantiate the claims I make because all four still work the very same way as it did at the beginning of the Universe. The three points serving one part of singularity combined with the four points serving singularity unites as seven to form a circle of either 3.1416 or 21.991÷7. The seven going to one is eternity matching infinity by movement. But since seven moves it are seven that have to produce gravity. How do I know all these facts, because we can see from the top it is still doing what it did the very first second. When time started infinity as well as eternity had altogether 3 positions, the past, the present and the future. It is still forming the very line in the centre of the top as it forms all lines in the centre of all things spinning. Then eternity parted from infinity when heat separated what is cold from what is hot and eternity formed one more point than before when it had the three points.

With infinity and eternity then jointly having 7 the cosmos came into rotation. In the aftermath post big Bang we now see the phase of cosmic development where the tow sectors try to unite and this brings along the contraction. When Π forms it does so on the grounds that 7 rotates. The circle forms by a change in direction by 7°. Every circle has opposing sides forming in relation to the axis line. If the topside goes rite then the bottom side has to the left. If the rite side goes down then the left side goes up. There is this double presence of a change in direction forming on both sides of the circle. The 7° move and by moving 7° goes square 7^2 and that is Pythagoras.

The one part of the earth is going up by 7° and the other part is going down by 7°. Every time the earth moves in the going up or down direction the relevancy crosses singularity 1°. It is the same 7° that crosses 1^0 to 1^1 but it results in two different points holding 10

7 is matter and 7 is matter.

Π^0

They join space-time therefore the matter factor is the same. This is where one can visually see the one object, filling the space of the other object's atmosphere.

$$7 \times 7 = 7^2 = 49$$

That is matter Π^2 (time) times matter $(49)+(1) = 50$. This 50 forms space which then applies to both sides of the rotation of the solid being 7 that rotates.

As this is all under the law of Pythagoras the law will evidently place a square root to that value of 483,61 and therefore $\sqrt{50} +50 = 10$. This leaves the space value of the Roche-limit, as it develops into the Titius Bode law giving them a shared value of 7 (matter) and 21,91 (space) the value of 21,991 / 7 = Π.
Then the relation becomes

$(\Pi^2+\Pi^2)$ $(\Pi^2\Pi)$ $(\Pi\Pi^2)$ $(\Pi^2\Pi^3)$ holding space (3) still outside. They therefore will share space and that sharing will continue till times end. We know by now that matter is 7, and space is 3. holding time to a relevancy in singularity of 1. Sharing the space means that 21,9 will become (10) space to the one side
1   to the instant position of time (k^0)
,99   lost to space depletion $\Pi^2/10$
7   the relation to matter.
Through that the Titius Bode law comes into affect of 10/7 or 7/10, depending on whether space or matter holds a superior position to time. From that stance, all objects will relate to one another by the value of $\Pi^2\Pi$ and seen in a whole sale total 7/10 or 10/7. That means to become part of the neutron status of the earth, the object has to be space (21,991 or less) and prove to be matter (7) before the earth will accept it. If holding a position of less than 7°, the earth will discard it and if it is more than 21,991 the earth will find the relevancy to be higher than the space it holds in a neutron time.

That places the object in a relation of $4\Pi^2 - \Pi^2$ (because it is not part of the earth) in a position acceptable matter holds (7) within the confinement of Π (21,991/7). That means the object is part of space (21,991) acting as matter (it holds an acceptable own proton structure) 7 relating to the earth in the position the earth allows of $3\Pi^2$. That means to become part of the neutron status of the earth, the object has to be space (21,991 or less) and prove to be matter (7) before the earth will accept it. If holding a position of less than 7°, the earth will discard it and if it is more than 21,991 the earth will find the relevancy to be higher than the space it holds in a neutron time. That places the object in a relation of $4\Pi^2 - \Pi^2$ (because it is not part of the earth) in a position acceptable matter holds (7) within the confinement of Π (21,991/7).

That means the object is part of space (22,991) acting as matter (it holds an acceptable own proton structure) 7 relating to the earth in the position the earth allows of $3\Pi^2$. With the space position of the matter in the parameters of 21,991 it relates to the Titius Bode law as a factor of one. The object has the space value of 21 (3 x 7), which shows the axle value turning, plus the space value of .1416, in that instant of time (7) complying to the earth's space (.1416 x 7) in reduction (Π^2) formulating .1416 x 7 = 0,991. That makes the object complying with the full agreement as laid down by the Titius Bode law. The object is, no matter where it is, travelling at a rate of 7 ($3\Pi^2$) in the space of the earth (21,991). This will be agreeable to the parameters of the Titius Bode law as long as it remains within the space depleting "gravity" limits of less than Π^2.

In accordance to the Lagrangian atom layout, anything less than 5Π is manageable and is in effect less than Π^2. When it exceeds 5Π it will start opposing the dimensional equilibrium space holds of 10Π, therefore it will (according to space) exceed the linear point of R/T, which is $10\Pi/2$ (space going in a straight line).

Everything in the cosmos is moving, either by own individual accord, or under the influence of some other singularity dominance. In explaining we return to Pythagoras where the entire Universe with everything in it started.

It is the point forming the very centre that plays the part as the <u>**controlling singularity**</u> within the Universe I have named as **Infinity,** which is better known as the axis. It is where nothing can go smaller and anything within that point can never reduce. That point is where the entirety called the Universe begins and where everything holding substance begins.

Once one accepts the fact of singularity being present in that location, that accepting of singularity then is contradicting all the things we know and we can measure and we recognise that point being present by merit of the fact that the point referred to is not being formed by any of the things we can recognise.

It is made up of everything we don't know and constitutes of everything we are unable to recognise or visualise. In that spot there is no space. That spot holds **Infinity.** In that space there can be no motion because there can be no space to have the motion within. It is formed as a line that is so small that our human reality by perception declare that point as not being there and the only reason why we know it is there is because of the results it left as an imprint of its not being there.

We cannot detect it but notwithstanding our failure to note it we can recognise the dot on the merits of its absence and while in our Universe it is always absent, reality disallows the dot ever to be absent, because it is never absent. It cannot be absent. It cannot go absent but it can never be there where it should be in a place from where the third dimension forms and it is always present if I wish to locate it. It is **infinity** that can never go away. I named the other part of singularity forming space **eternity** because that area never become bigger, or become more or find an end to the outside. Whatever was and is and will ever be is locked in that space I named **eternity** and it is **eternity** that never ends because **eternity** can never end moving. What we think of, as expanding is never ending movement giving eternity the eternal motion that will go on forever.

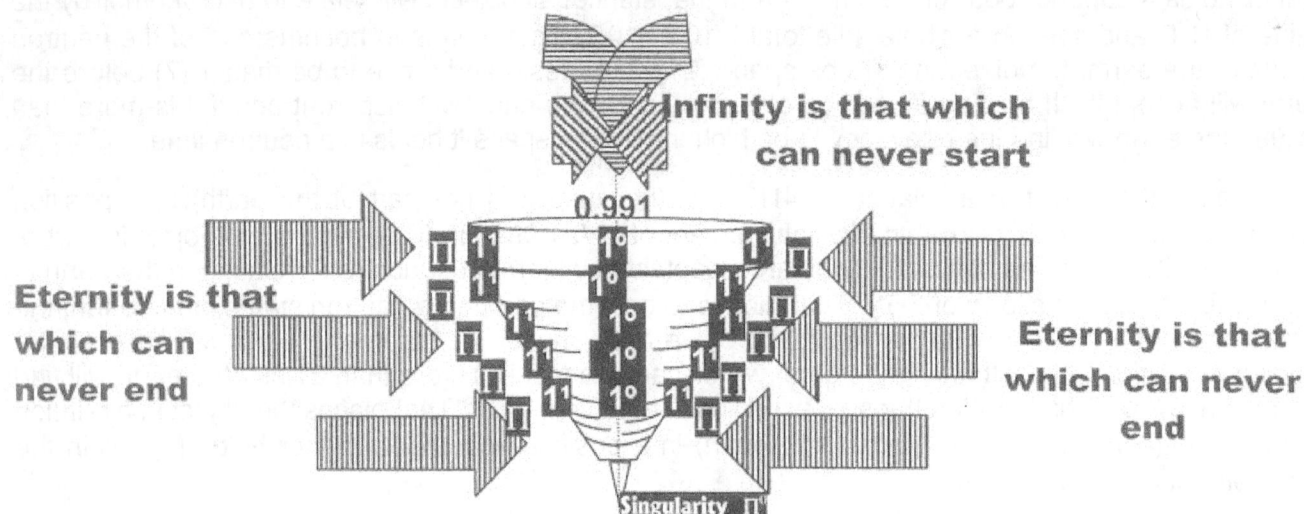

The line **k** coming from the centre (singularity k^0) forms by forming an initial spot Π^0 becoming the dot Πr^0. However, I went on to say that whatever the line used to start with has to continue in order to repeat the same that began the line. Therefore the line started with Π^0 and it has to continue with Π^0 until such a point, as it must end with Π. Whether the line is Π^0 or is r^0, or uses 1^0 the outcome all refers to singularity being used. By reducing the line we come to the end of the mathematical equation of the circle but the circle does not end there. When the top is in a state of motionlessness on own accord it is everything but motionless. The motion it adapts are synchronised with the earth in harmony with the solar system and according to the greater picture of the cosmos.

When an energy source not related to the cosmos called life intervenes and energises the tops motion, the singularity in that top suddenly jumps to life. By adopting a rotation energised to an unnatural state of energising because of life's intervention, the singularity of the top is not in charge but as it applies more and more energy, it will begin to find a means whereby it can escape and apply individual singularity as the top starts to separate from the singularity the earth holds. The singularity holding the earth would then allow the singularity of the top to rotate within a specific band where that a specific band of being active before the earth's singularity will start to destroy the singularity in rebellion.

The top on the other hand will try its outmost, when the singularity it holds gets by individual spin is too strong to remain be in domination of the earth's singularity. The motion of the top is an attempt to begin applying an individual singularity space-time defying and standing apart from the earth's gravity. That action we see as the top starts rotating in a manner where the top does not align with the earth's singularity. With the adding of spin, the time the top holds becomes unrelated to the time the earth holds and the top will start a campaign too escape from the singularity domination the earth has on the top. When the time or spin of the top exceeds the limits the earth places on the top, the top would emerge by trying to escape from constrains placed by the earth.

The view I represent at this point is known to science for almost as long as science knows mathematics. Not long after the law of Pythagoras was understood where Pythagoras introduced mathematics Eratosthenes of Syene made as big a discovery as Pythagoras did.

But in the one instance the world took notice because the world could see and understand and the other instance the world disregarded the findings because the world did not see what the implications was. The same apply to aircraft flying and when the aircraft wishes to escape the earth's singularity hold it has to comply with the laws laid down by the earth.

The seven becomes as big a part of the concept as does Π as it all interacts.

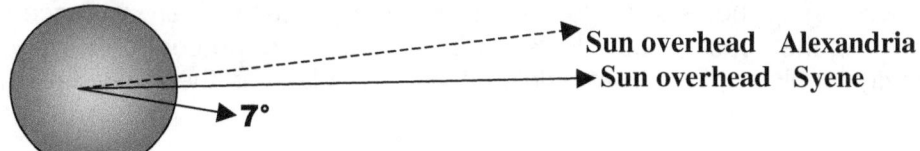

It took Eratosthenes of Syene (276 – 194 BC) a Greek astronomer who in the year 240 BC made a discovery that the earth has a profile of 7°. Since then no one ever did anything about it. When any singularity wishes to disconnect from the earths singularity, specific pre-calculated laws would have to comply to allow the lesser object to divorce from the larger object.

I indicated how the dimensions of 10/7 and 7/10 interact to form (Π^2)

Matter is a product through the separation of space and time receiving the value of Π The original time and Π^2 as follows: By circling around a spinning solid the space contracts to form Π and Π^2. Gravity forms everywhere in the Universe by applying singularity. By dividing space into material (material spinning in space) and duplicating space by material spinning, the TITIUS BODE LAW forms a 7^0 deviation and 7 / 10 in conjunction with THE ROCHE PRINCIPLE OF $(\Pi/2)^2$

In my article to Annalen der physics I used 15 pages to explain this process of singularity applying. I received a rather cordial but sincere reply from the Editor of the magazine.

When I placed an article in Annalen der physics Dear Prof Friedrich W. Hehl said in the e-mail he sent me that there is no way to "explain" the world of physics. I am not going to go into detail how this works. On the other side of the Pythagoras's' triangle we have 1 going square.

That makes Pythagoras's' triangle 49 + 1 = 50 on the one side of the earth and the same on the other side of the earth. The total is 100 and the square is 10. That leaves the Titius Bode law with a value 7 (it forms part of the material of one body) and 10 in relation to the space.

Then from the relation of 7/10 and 10 / 7 forming Π the Titius Bode law form Π^2 applying "With a lot of words and some simple algebraic relations" to quote Friedrich W. Hehl, Inst. Theor. Physics of Annalen Der Physics fame. This was simple algebraic relations but still it is science, is it not?

Since it involves singularity moving it calls for the law of Pythagoras to produce space. The law of Pythagoras is the triangle a^3 that is moving forward in singularity k by turning T^2. In singularity the 7 stands in for 7 points on the numerical line crossing over the line holding singularity or 1.

By moving 7 has to go square T^2 and that means 7 goes square 7^2 twice $7^2 + 7^2$ crossing the same divide $\Pi^0 = 1$. Since all movement in singularity has to enforce the law of Pythagoras we have two triangles holding 7 dots moving across singularity. I don't want to get too involved by bringing in numerical outlays because then this can truly become complex.

The line has two opposing sides turning directionally against each other while turning with each other. By moving or turning this involves time duplicating space by the square Π^2 on both sides of the divide $\Pi^2+\Pi^2$ and using the same divide or the same axis or the same point serving singularity we have 7^0 crossing the same point in singularity Π^0. There then is in this rotational movement 7^0 standing in for Π^2 on both sides of the divide $\Pi^2+\Pi^2$, which then is 7^2 on both sides of the divide 7^2+7^2.

The circle spins in duel directions. On the one side it would go left if on the other side it would go rite. The one side hold a directional change in singularity by 90°. As it is going sideways it changes to going down. This produces a rite angle triangle of 90° and in it the law of Pythagoras produces direction changes. Since the square of the turn of the circle places by the spin and the direction change we have 7 holding a relation to 10 in space because it is space that has to carry the value of 10 when material circles by 7. There is a connection between space surrounding the spherical circle turning and the sphere. The circle holds the value of 7 as in 7^0 and this we find from looking at singularity controlling the circle by movement

The wise men from physics put everything down to mass without ever searching for evidence of mass. Putting everything down to mass may be one solution except for the fact that only nothing is that simple. Even the rejecting and / or accepting are incorrect, as the pulling part just comes across a tad too simple to make sense. To find substantiation one has to find the manner in which light connects to singularity because everything connects to singularity. In every circle centre there is space that is so small it is not present within the space of our Universe. In that space in the centre of every spinning object we have singularity forming space. That points serving singularity is the reality and the rest is only make believe.

Explaining the following is rather tough with the limited space available but it should secure the conclusion that I am not grabbing for straws and there are substantial facts on which I work.

Our human view of the sun is that the sun is big; it is more than big it is outrageously big. The sun is so big that one person got a Nobel Prize when he worked out that anything bigger than 1.3 times the mass of the sun would collapse into a Black Hole. That is how massively big the sun is. Another person then got the Nobel Prise when that person calculated that anything the size of 1.6 times the mass of the sun would be an instant Black Hole. Yea sure and Elephants fly at night over the cameras of UFO hunters just as they press the record button! And now the sun is not even a dot in something that we might call big. Look at the picture of the giant CY Canis Majoris and look at the size^3 of the sun in the photograph of CY Canis Majoris. Then place the earth in this picture and then place yourself spinning on top of the earth in this picture. Then think of your position in terms of the entirety called the Universe.

The more suitable question to ask is how many suns will fit into VY Canis Majoris and in that we have to realise that VY Canis Majoris and Betelgeuse and other "giants" are not stars but they are galactica not developed yet and most of all they have not even began to develop. It is not what we think is true that is true but it is what we think about the truth that applies.

Lets investigate the Universe. We can see how far off the mark science is in their estimate of what is big and what is 1.6 times bigger than big. Let's find out what is small. The Universe is made of lines connecting points. Whatever you see or can't see is lines connecting. If we wish to examine the Universe we first better star to examine lines connecting dots because whatever is in the Universe is dots that connects with lines. So it will be necessary to investigate lines. In mathematics they teach students that a line starts at zero and that is a fable. Zero starts nothing and nothing loses everything before anything can start.

This sounds so unimportant but it is fundamentally all-important. A line starting with zero how cannot increase in length. This means that any line formed in a star only ends ant the outer limit if the star as the line is a solid non ending continuously continuing line from the centre to the outside. The sun is a liquid and not gas as science tries to convince people. Look at the picture. The picture shows a liquid with more liquid squirting into a gas. The density that is apparent at the surface of the sun is very clearly a liquid. But Newtonians have hydrogen as a gas just because on earth hydrogen is a gas. Thereby the Newtonian wisdom says that 6500° is very hot. It is so hot it would burn everything on earth and roast any form of life on earth down to less that carbon. This they argue because from the atheistic stance that life is a normal commodity in the Universe and life is everywhere to be found, going at a dime a dozen therefore if it does not befit life then it is extraordinary. Life is extra ordinary and not even a thought in the Universe except on this little blue dot we call earth.

What contracts is cold and what expands is hot. Cold cannot expand and hot cannot contract. What contracts because of cold must eject all heat because it contracts and therefore remove all heat. That is what happens to the sun. Because it is so cold it emits all heat from it and because outer space is so hot it accepts all heat because it has a cold it can forever heat as it forever expands. This is the fundamental of physics and not the human thermometer showing a scientist what he or she must presume to be hot and cold. In cosmology there can be no hot or cold as much as there can be no big or small. When the Big Bang was in The Planck time: 10^{-43} seconds. After this time gravity can be considered to be a classical background in which particles and fields evolve following quantum mechanics. A region about 10^{-33} cm across is homogeneous and isotropic, the temperature is $T=10^{32}K$. If the entirety was 10^{-33} then how big was the sun? If the temperature was $T=10^{32}K$ then what was zero? For the Universe to be $T=10^{32}K$ then something else must be zero or that temperature is meaningless because then it could just as well be zero. If the Universe was 10^{-33} then what was the size of one atom? If they say it was before atoms was present then they must state what was the material used in the Universe at the time. It is like saying there is anti-matter. Yes and what was anti matter. I know matter is singularity spinning in a direction in excess of the speed of light because the electron spins at the speed of light and what is further inside the atom must therefore spin faster than the speed of light. Matter is controlled singularity that directionally diverts from uncontrolled space by spinning in a direction. If that is matter then what is anti – matter, things that don't spin faster than light or is it a concept science ahs no clue about but naming it brings clarity to absolute stupidity. It is like saying the sun is gas and all anyone can see from pictures is streams of liquid squirting all over.

It is no use showing how space expanded in relation to temperatures cooling because the big issue is lost in the entire scenario. Space expanding is the same as heat lowering and heat rising is the same as space contracting. That is what gravity is. It is some space expanding and some space contracting and the expanding space is the space heating while the contracting space is the space cooling. Just because we humans feel space heating it does not mean it is hot. It means to humans it is hot but humans have no say about conditions applying in the Universe. Life is a nuisance that are not even ever recognised in the entire Universe and life is alien to everything except the earth in the entire Universe. The heat we feel is the coldness releasing heat to the heated space that we think of as cold.

The very first instant space formed is when heat parted from cold. The Universe began when a difference happened when there was one part that was hot in relevance to the other part that was cold. The entire Universe is a patrician of what seems cold in relation to what seems hot. If there is something hot then there has to be something cold and neither hot nor cold is prescribed or is dictated.

It is only opposing values in that specific space forming factors that generate gravity and gravity forms when one part expands in relation to another part that contracts space.

in the Universe there are two forms of substance that which holds material and that which is material. Material cannot flow but has solid form. Non-material holds no specific form but is able to hold material or not to hold material.

The sun is a bowl of fluid so cold it holds hydrogen as a liquid on the surface. Where the surface touches outer space the friction heats up and that makes the liquid boil. It is not the liquid that heat but is, is the gas causing friction as the liquid turns that makes the liquid boil.

The liquid seems pretty steady and it even seems like waves across the surface. Science forever and ever tells the cosmos what it is and never learns from the cosmos what the cosmos is. What squirts from the sun is a liquid frozen by the movement or gravity of the sun. It is not a gas because a gas cannot flow as the substance squirting from the sun very obviously does. Notwithstanding lack of support coming from the cosmos science tells the cosmos it uses "mass" to arrange planets and then it uses "mass" to allocate planets.

We see liquid squirt from the sun but because humans think of Hydrogen in terms of gas and then the sun is gas and the sun is also hydrogen gas that is hot! Science will forever tell the cosmos what to be. In the Universe there is no hot and there is no cold. In the Universe there is no big and there is no small. If outer space is the hottest there is then we know what a Black Hole is it is the coldest there can be. It is so cold it froze all material into one point that is not even part of the Universe.

It froze space out of existence and into the oblivious; literally the oblivious where what ever can be cannot be any longer. In the Black Hole the atoms froze into one structure where not even a proton has validation. The structure spins so fast it does not spin at all and allows all spinning into space that otherwise never can spin. The movement in outer space is the expanding of outer space and the movement allowed by outer space is the increase it has in growth by forming a larger volume of space. However this entirety growing is not in size but only applies in relevancy. If one dot removes from space; any space then the entirety forming space will collapse into where that one point has disappeared too.

The Black Hole is a star that the atoms within joined to form the ultimate atom in singularity abandoning even the proton in the core and all movement in relevancy has gone outside into time or outer space.

In the Black Hole the atom froze to something that has gone into the oblivious and it froze to a point that no longer has a position in our Universe. But reality is that in every atom there is a Black Hole because every atom absorbs light and the rest it casts away. That is all but silicon and I just can't explain that detail in this book since it will be far beyond the technicality of this book's standard. The sun is a future developing Black Hole and that is why what is in the solar system will flow around the sun eternally without ever having a chance to escape from the centre of the sun. The sun like every atom in every formation forms a Black Hole by forming singularity.

In ever star all atoms join one single point attaching all starts to that point and even after a Super Nova release such bonding remains firmly attaching the star.

Even the structural remains that is the reminisce of a star that went sour still hold singularity in position as forming a centre Π^0 indicting the point allocated to Π. This connection is not coincidental but rules apply putting this formation in place even after the structure went beyond repair.

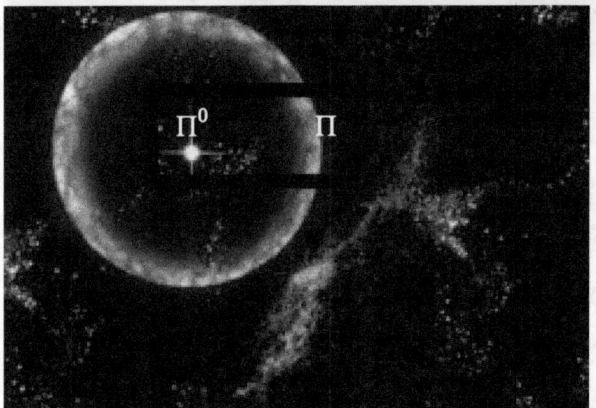

It is accepted that stars develop by growth and that is true but the growth in stars developing is moreover in the Universe expanding. When the Universe expands the relevancy left to the star is not in ratio the star holds less space. The star expands to a point where the star the forms gravity that exceeds the speed of light and when that happens the space the star requires fades in comparison to the space used by the universe in growth. The speed of light is not a constant and it could never be because gravity changes the speed of light as it alters the waves that light transmits by. The stupidity behind the reasoning that the speed of light was forever a constant is underlined by the question as to when the Universe was 10^{-33} cm across what was the speed of light then. As the Universe expands so the speed of light will stand corrected every instant and every location throughout the entire Universe. The speed of light might be 300 000 km / sec from where we are but a time back with a smaller Universe the speed of light had to be say 250 000 km / sec, then before that 100 000 km / sec and before that it had to have been a 100 km / sec. It then is madness to place a galactica at 12 000 years or 12 million years because where the galactica is it developed as the speed of light developed but to put reality in place there is no saying how far the galactica is because no one can determine how the speed of light developed since the light we see left the galactica we think we see at present. It is like saying the Universe was 10^{-33} cm at the beginning but the Universe was filled with atomic matter. Then with the Universe at 10^{-33} cm how big was a neutron at the time when the Universe was 10^{-33} cm? Everything has to be in relevance or not be at all. One can't say the Universe was because what was a centimetre back then? The material forming atoms was already in place and the entire Universe was the displacement intensity of one electron and therefore what was to become each and every neutron had to be fabric and what was then proton material was produced. What was the size of the material grouping to become an atom? It is very fuzzy to play the role of God and have an overview of the entirety because the scientist then runs out of logical relevancy. The entire Universe then was as big as it is now and was as hot as it now is but applying relevancies changed from then to now as densities and gravity alternated. If we have an atom then we have the same atom now but the relevancies applying changed in time growing space. What did change was the time in duration as time form space and time then in comparison to now stood still. Time then moved at a rate we now would never understand. We now attach time to movement and movement comes at a rate of the speed of light. How fast back then did light move? If we put the speed of light now in ratio to then we can never understand time back then.

Looking at a galactica it is clear that in the centre there is a bright bubble holding a sphere of heat. Normally surrounding the centre heat cocoon a ring forms and the ring shows how the galactica went flat as the stars formed rings in which it spins around the centre. As the Universe expands the rings must grow larger but growing larger will draw the centre thinner and flatter.

The centre of the Universe is within the centre of everything that spins around a centre because everything that spins forms Π^0 which is one. By forming Π^0 it forms 1 and all the values no matter how large the space is that spins around is $\Pi^0 = 1$, the value holding the space together is 1. So notwithstanding the space it holds the centre value or singularity $\Pi^0 = 1$ and in being 1 it holds singularity or the single dimension. Therefore the centre of the Universe is inside the centre of everything that spins around and that creates such a centre. The movement as time formed by producing space. Everything there is also is there because it moves in relevance to everything else that moves. What is within the Universe moves within the Universe.

If anything did not move it would form a Black Hole since Only a Black hole is within one point that does not mot move and that point that does not move is within not the Universe any longer. It must be understood that whatever is moves that does so in relevance and synchronisation to everything else that does also move and the difference between that movement and space forming forms time.

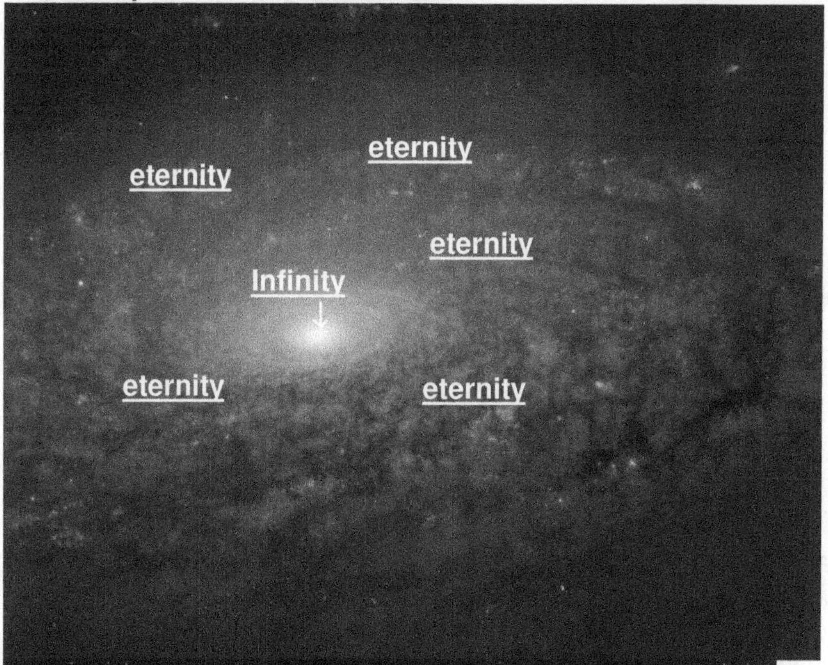

Chapter 9   **The Titius Bode Law**

$\Pi^0 = 1$ is infinity. Eternity is all the space occupied or otherwise that cannot escape infinity and because it will spin around infinity for eternity it holds a definite value of eternity.

Infinity is the point formed by everything holding space within eternity that spins as well as spins around infinity and where the point then forms in relation to all that spins that cannot reduce any more. Infinity is the point that bonds everything forming eternity and becomes the absolute point achieving singularity in infinity in relevancy to the centre of all which accepts that point as the centre.

eternity   *eternity*   *eternity*   Infinity   *eternity*   *eternity*

As stars develop time forming space pushes the stars outwards away from the centre and into maturity

In the centre is the incubator of stars

The idea that mass draws dust close to form stares is the another hogwash fairytale thought out by those who think not but only offers storeys with no educated substance. In the centre of galactica a ball of heat forms. This looks like the remains of a galactica that resembles the shape of current "big stars" such as Betelgeuse and VY Canis Majoris and others. Those "big stars" are galactica that is not yet affected by the Universe expanding and the stars it holds has not broken free from captivity to start development. This made me think that there are development eras still to form and that we are in the iron era

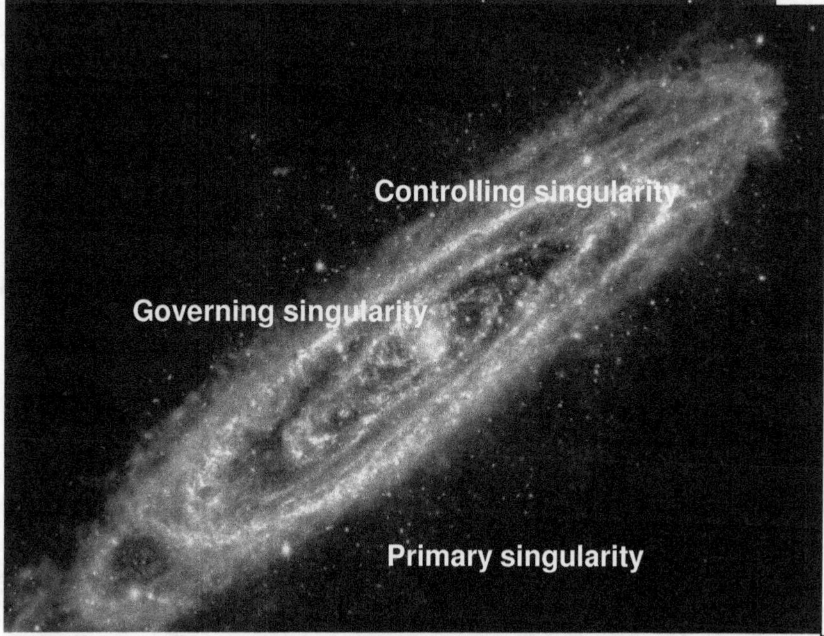

Controlling singularity

Governing singularity

Primary singularity

The sun turns around the earth as much as the earth turns around the sun. Everything is subject to changing and interchanging relevancies played by the way one looks at the factor prominence. It is relevancies that interchange as focus adapts new prime factors. The centre of the Galactica hold the governing singularity because it governs the spin of all the stars within while the stars form the controlling singularity because it holds control over the function and value of the governing singularity and outer space forms the prime singularity because that places the entire galactica in position.

Relevancies change as prominence alters. In the case of the solar system the governing singularity lies with planets, but in galactica the governing singularity is placed in the centre of the galactica.

There had to be a time when the Black Hole of today was a red galactica "giant star" and as the Universe expanded form 10^{-33} cm to where the current Black Hole was a "star" in relevance to what Betelgeuse is today it went through growth taking the current Black Hole through all the above stages as the size of the atomic material diminished by gravity increasing in relevancy. The star develops as the Universe develops gravity and gravity is the differentiation between what expand in movement and what contracts in movement.

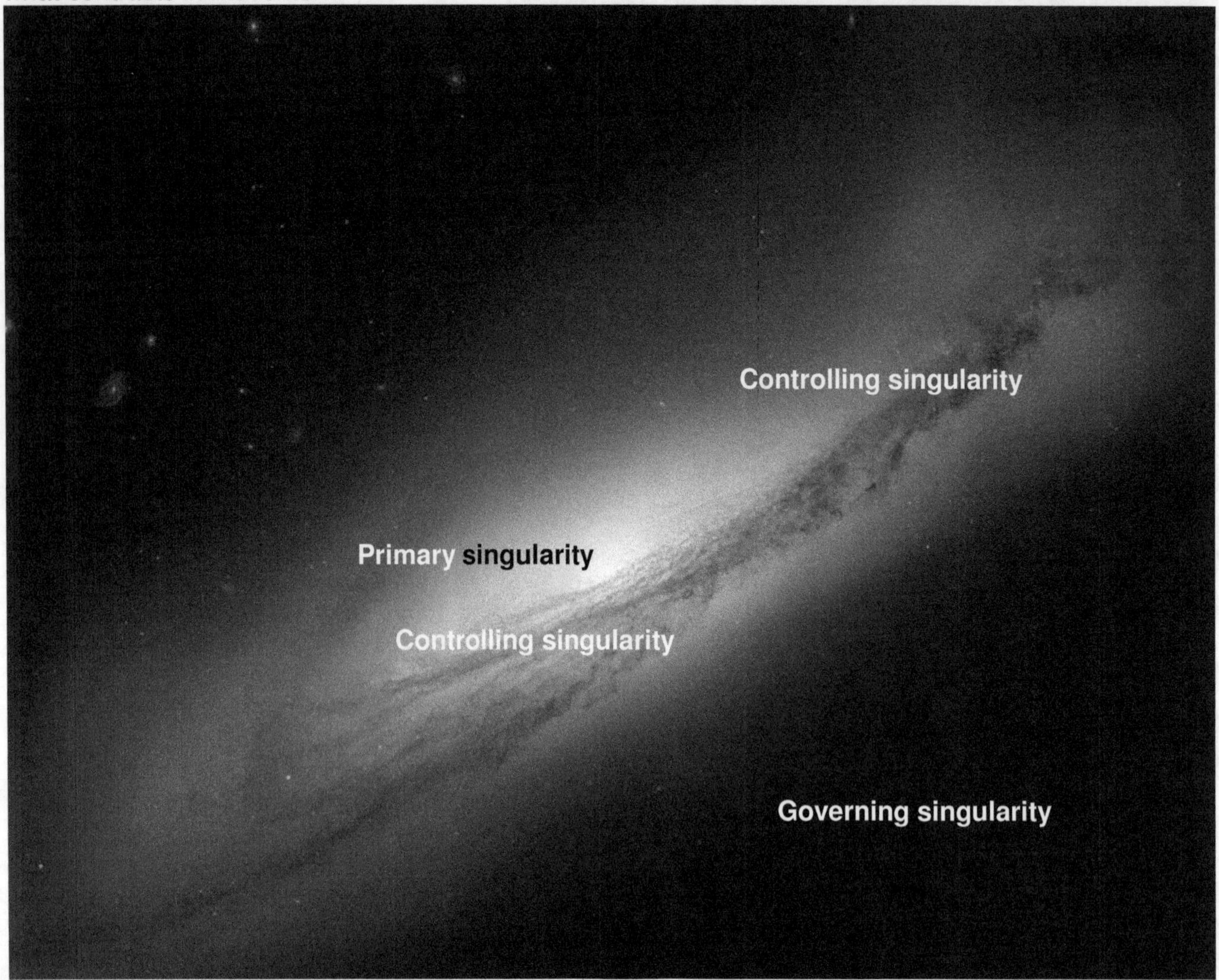

This galactica in all its splendour of light is a Black Hole in the making. The star forming the galactica will eventually combine in effort to project a Black Hole in the centre as all the primary singularity grows in strength by the input of the controlling singularity with the expanding governing singularity. The development of the Universe is natured in the time differentiation between what holds infinity and what forms eternity. The Black hole removes all of eternity and unites that again with infinity. However this is the role of every atom forming the Universe. As it removes time from eternity it captures light just as a Black Hole would do. That makes every atom a performing miniature Black Hole and in a combined unit it forms the role of the primary singularity developed by the controlling singularity of all spinning material within the galactica. It is a singularity projected by all material to form one unit $\Pi^0 = (\Pi^2\Pi)/\Pi^3$.

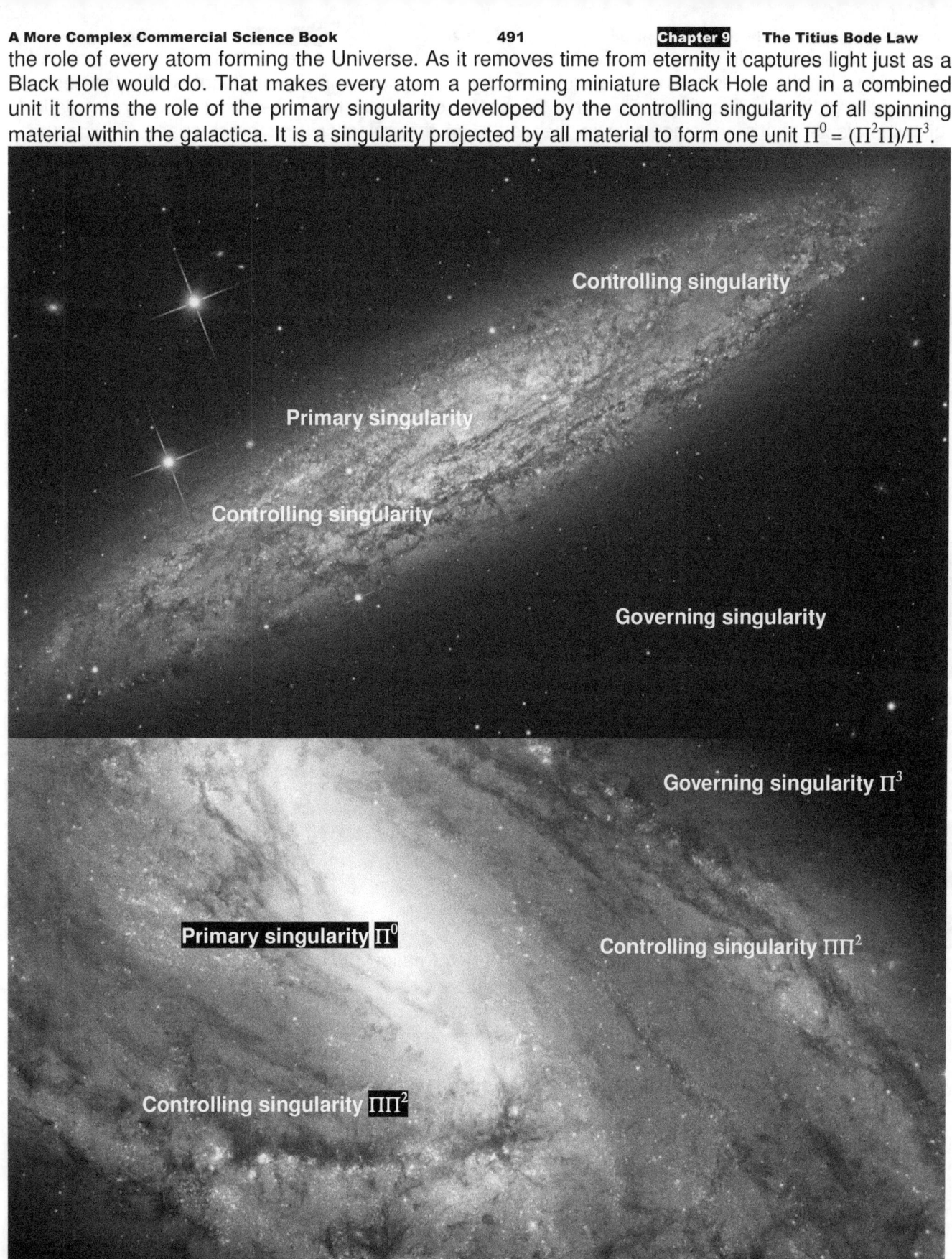

There is no and there can never be a mini – Black Hole or a small Black Hole. It is only a stupid atheist with no concept of reality that would observe and not understand what he observes and with his limited atheistic interpretation skills still then still try to explain that which his poor abilities in understanding can never comprehend to begin with. Every atom in a galactica forms a centre not only in the centre of the star in which the atoms are but also projects a point holding a primary singularity within the centre of the entire galactica. The singularity within the centre of the star becomes the governing singularity while it projects a primary singularity in the centre of the galactica and outer space forming the containing ability become the governing singularity as it hold the spin of the entire structure contained in relation to time developing the Universe and in this case moreover the galactica under question. When there are sufficient quantities of material available to project and materialise the primary singularity in the centre of the galactica a Black Hole will form as Π^0. Then the galactica forms the Controlling singularity $\Pi\Pi^2$ and outer space forms the governing singularity retaining the growth, gravity and movement of the galactica as a structural unit. Time is when the smallest particle change position.

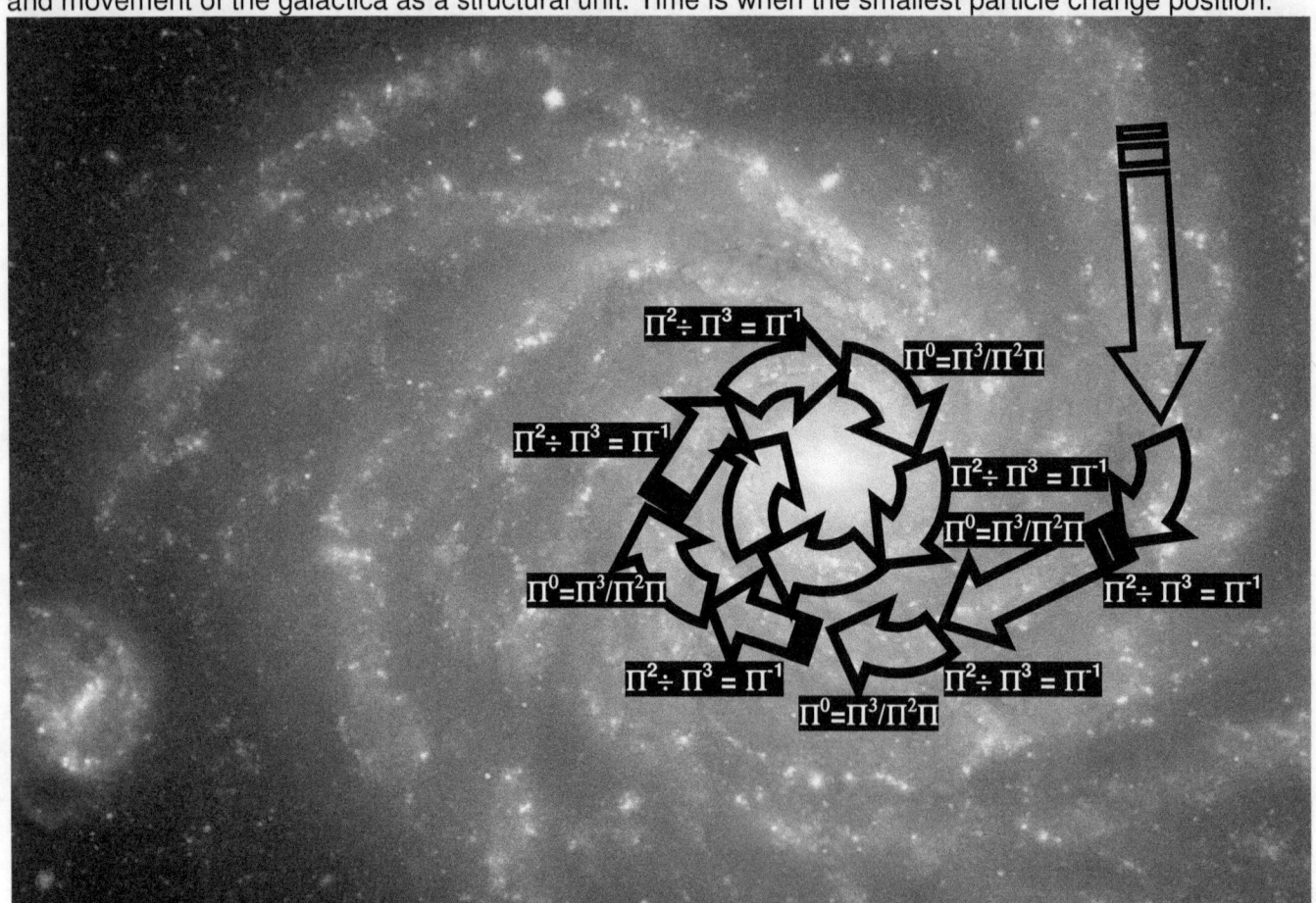

Mercury	$T^2 \div a^3$ =0.983
Venus	$T^2 \div a^3$ =0.992
Earth	$T^2 \div a^3$ =1.000
Mars	$T^2 \div a^3$ =1.000
Jupiter	$T^2 \div a^3$ =1.000
Saturn	$T^2 \div a^3$ =0.999
Uranus	$T^2 \div a^3$ =1.000
Neptune	$T^2 \div a^3$ =0.999
Pluto	$T^2 \div a^3$ =1.004

Studying the true information about Kepler's formula we see that the formulated factors are $a^3 = T^2k$ and it produces $T^2 \div a^3 = k^{-1}$ in value. Having $T^2 \div a^3 = k^{-1}$ proves that the space moves towards the centre because in the centre we find singularity valued at $\Pi^0 = \Pi^3/\Pi^2\Pi$. This proves that space moves towards the centre and material stays in place as proved by the Titius Bode law. It is a flow of space occupied or otherwise that flows towards the centre and this holds relevance with the expounding of the space outwards away form the centre. What is attached to the centre of the galactica stays attached to the galactica where no escape is possible.

Time forms as movement changes space. Between singularity acting as time and movement forming space we have singularity acting as time and space acting as the history of time. I.e. the earth circles around the sun and therefore every instant the earth is in a different position and in a different location in relation to the sun and therefore the Milky Way and therefore the entire Universe. However to do that the earth must stand still in every location before it changes to a new location. In that every fragment of any atom forms time by changing allocated space and the united effort of al material firtiming time finds that the earth represents the material in going to a new allocated position while circling around the sun. In the Universe time flows because the present is different from the future as the past is different from the present. Time is about everything changing. Every aspect of the Universe is different from what it was to what it is to what it will be the very next instant. That forms time. This is proven by the formula we received from Kepler and Kepler in turn received it from the cosmos which he and Tyco Brahe studied for eighty years and calculated as $T^2 \div a^3 = k^{-1}$. Time is not space but time is the movement of space. Time changes space. The changing gives the Universe time flowing. The space a^3 moves T^2k and therefore changes the space in a singular Universe a^3 at a ratio of a circle T^2 in a singular Universe that moves straight by k that is a line in a singular Universe. The change we see in the Universe as pictures formed by light is in the time flowing that forms space and thus creating a Universe by changing it every instant. Time is the changing of space flickering between $\Pi^0 = \Pi^3/\Pi^2\Pi$ and $\Pi^2 \div \Pi^3 = \Pi^{-1}$

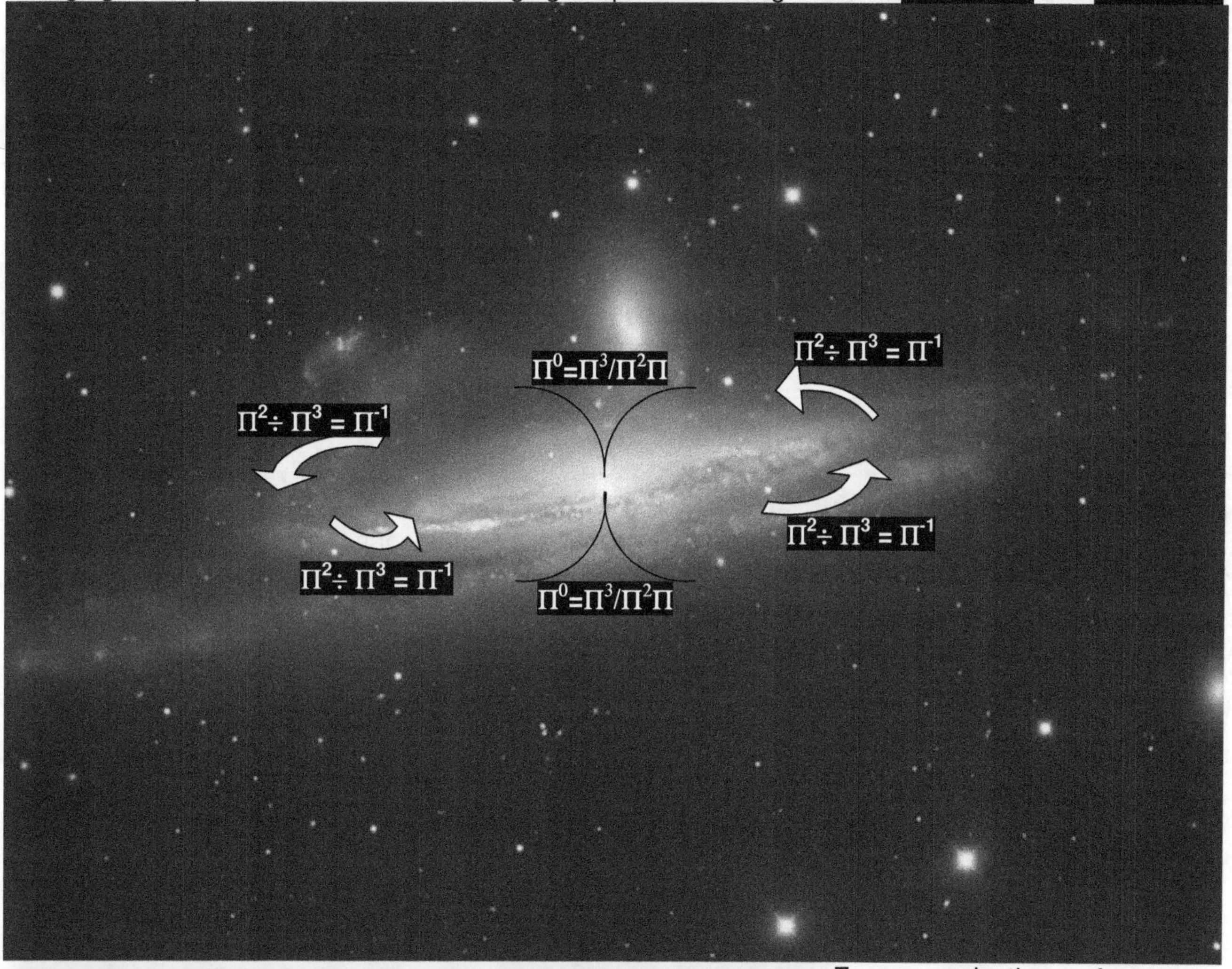

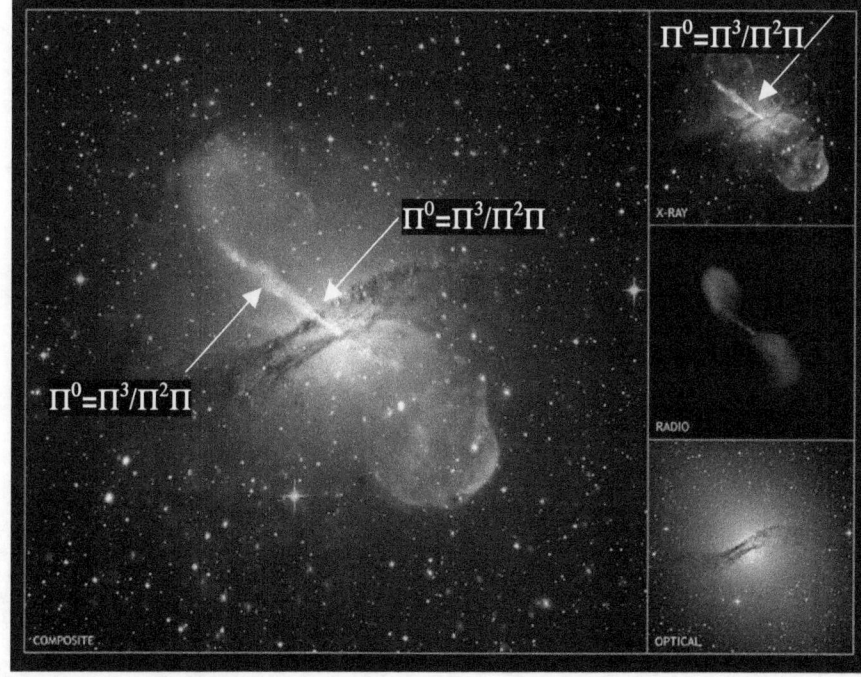

Every galactica forms a prominent (however not always visual) Black Hole in the centre that is generated by all the material that forms what is that particular galactica. It is because al spinning material project a primary singularity in the centre of the galactica generated from the movement of the controlling singularity of every piece of material spinning inside the galactica therefore a Black Hole is established at the centre of the galactica. It hold the value of $\Pi^0 = \Pi^3/\Pi^2\Pi$ as it generates $\Pi^2 \div \Pi^3 = \Pi^{-1}$ and cosmic development we find is vested in what movement is between these two directional movements.

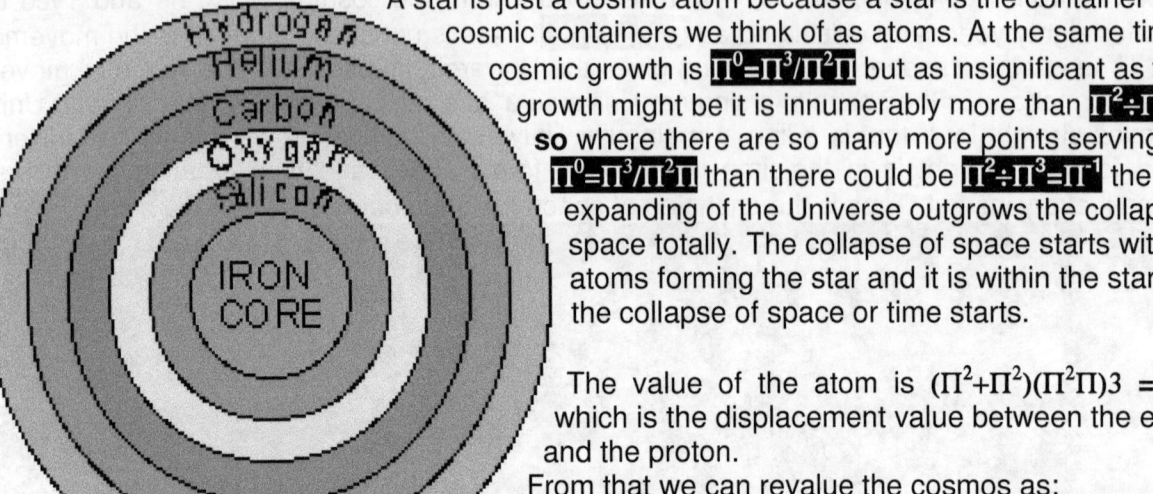

A star is just a cosmic atom because a star is the container of cosmic containers we think of as atoms. At the same time cosmic growth is $\Pi^0=\Pi^3/\Pi^2\Pi$ but as insignificant as this growth might be it is innumerably more than $\Pi^2\div\Pi^3=\Pi^{-1}$ so where there are so many more points serving $\Pi^0=\Pi^3/\Pi^2\Pi$ than there could be $\Pi^2\div\Pi^3=\Pi^{-1}$ the expanding of the Universe outgrows the collapse of space totally. The collapse of space starts with the atoms forming the star and it is within the star that the collapse of space or time starts.

The value of the atom is $(\Pi^2+\Pi^2)(\Pi^2\Pi)3 = \mathbf{1836}$, which is the displacement value between the electron and the proton.
From that we can revalue the cosmos as:
The beginning of space as time forms three dimensions = $\Pi^3 = \mathbf{31.0061}$. Every element has a specific value for a specific task it has to fulfil in the star or as its role is as a cosmic atom. In the cosmic atom the value of the cosmic atom is that of a proton which is $(\Pi^2+\Pi^2)$.

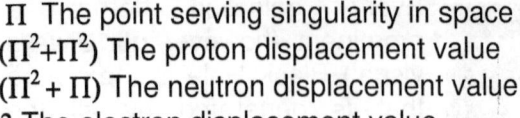

Outer space is $10/7(4((\Pi^2+\Pi^2)) = 112.79547$
10 / 7 is the relation between material 7 and space 10.
4 is forms the quadroons around which time cycles.
$\Pi^2+\Pi^2$ is: The comic atom forming material.

Light meeting singularity is $3^3+3\Pi^2 = 56.6$
3^3 is: The Universe formed by light in light.
$3\Pi^2$ is: Light moving through the Universe as light forms the Universe

The cosmic atom as a sphere is $7/10(\Pi^6)/6$
7/10 is: Material going singular in relation to space. The movement in this case takes space singular.
$(\Pi^6)/6$ is: The sphere spinning in accordance with the cube.

The Hydrogen is $\Pi\{(\Pi^2+\Pi^2)(\Pi^2+\Pi)+3\}=\Pi(35.75)= \mathbf{112.3}$

Π The point serving singularity in space
$(\Pi^2+\Pi^2)$ The proton displacement value
$(\Pi^2+\Pi)$ The neutron displacement value
3 The electron displacement value

Elimination of space-time is $3(\Pi^2+\Pi^2)$
3 The edge of space
$(\Pi^2+\Pi^2)$ The cosmic atom

Elimination of time and space differentiation is $\Pi(\Pi^2+\Pi^2)= 62.01255$
Π The edge of valuing singularity
$(\Pi^2+\Pi^2)$ The cosmic atom

Space reuniting with time is $= \mathbf{2\Pi^3} =62.01255$
2 Doubling what singularity in three dimensions could carry.
Π^3 This is not light but is the limit that allows light to form a Universe in which information are obtainable. These boundaries are motion in specifics that puts relations to certain limits set from the position of the point serving the governing singularity outwards. It proves that there are **dimensional implications all around and** that the dimensions are valid. The same implications are validating other principles in the cosmos such as in the case of the Titius Bode phenomenon by implicating the Coanda effect and others.

Outer space is $10/7(4((\Pi^2+\Pi^2))=112.79547$

The Hydrogen is $\Pi\{(\Pi^2+\Pi^2)(\Pi^2+\Pi)+3\}=112.3$

The sphere within the cube is $7/10(\Pi^6)/6$ as the cosmic atom turns

Iron is $7/10(4((\Pi^2+\Pi^2))=55.2697$

Light meeting singularity is $3^3+3\Pi^2=56.6$

Cobalt is $3(\Pi^2+\Pi^2)=59.21762$

Copper is $\Pi(\Pi^2+\Pi^2)=62.01255$

Space reuniting with time is $=2\Pi^3=62.01255$

H ⇒ He
He ⇒ C, O
C ⇒ Ne, Mg
O ⇒ Si, S
Si, S ⇒ Fe
Core ⇒ Fe

The value of 112 forms the limit of the element's value and therefore the limit of what there is forming the cosmos. Anything beyond 112 is not part of our Universe. Anything going above what light produces in the double does not contribute to what is within the Universe. Light is the three dimensions forming the cosmic gas that cosmic atoms turn into cosmic liquid and eventually turns to solids, as it freezes light into singularity. The part that holds the value of 3^3 or three cube forms the blackness where we see the spots of light bring rays of information through the vastness of space.

Light holds a double value of 56 x 2 = 112 and going above 56 is taking what is displacing space-time within the Universe to what is beyond the limits of what is within the Universe.

Please take note that the value of 10 / 7 changes to 7/10 to revalue the flow of gravity. Gravity and electricity is the very same thing except the scale of generating is on a much different level. When charging gravity the cosmic atom allow heat to flow from $10/7(4((\Pi^2+\Pi^2)=112.79547$ to the iron core within holding a relevance displacement value of $7/10(4((\Pi^2+\Pi^2))=55.2697$ This proves that gravity is the very same thing as electricity and only the dimension of generating changes the intensity applying.

The sphere forms by the interaction of the spinning where material goes singular by dividing into space and that takes space to 112.16. The sphere turn within space 7 / 10 while the sphere holds singularity at $(\Pi^6)/$ that is a roundness of Π^6 having Π in 6 dimensions and turns within a cube where the cube is valued at having 6 sides. The altogether value is $7/10(\Pi^6)/6=112.16$. Therefore the roundness of the cube provides the limit space can offer being $10/7(4((\Pi^2+\Pi^2)=112.79$.

The Hydrogen lines up the proton is $(\Pi^2+\Pi^2)$ in relation to the addition of the neutron $(\Pi^2+\Pi)$ and adding the electron value **3** and the sum total then is **(35.75)** points serving singularity $\Pi^0\Pi$ that forms a total value of movement such as $\Pi\{(\Pi^2+\Pi^2)(\Pi^2+\Pi)+3\}=112.3$

Any star forming a primary singularity with a relevance above $3^3+3\Pi^2=56.6$ that is the displacement value of light does not die because it is not a Dover coal stove but goes dark since no more light can escape from the gravity. The solidity of the star then absorbs all forms of heat in order to maintain and sustain the charging of singularity within. The movement freezes anything and all things and it even does that with light when gravity goes beyond $3^3+3\Pi^2=56.6$

The value of Cobalt $3(\Pi^2+\Pi^2)=59.21762$ where space breaks down in valuing three dimensions.

The value of Copper $\Pi(\Pi^2+\Pi^2)=62.01255$ where space breaks down in valuing three dimensions. This forms the end of gravity or the end of light or the end of the flow of space-time within the three dimensional Universe. It is clear why from above this value of copper there are Ferris metals and then there are non-Ferris metals. Gravity forms by conducting light in relation to iron spinning in the interaction of copper and this too is precisely why electricity forms that proves gravity and electricity is the same thing.

The Universe moves today using the precise manner that it started with and that is by employing the four cosmic pillars. Once we understand the workings of the four cosmic pillars the Universe unfolds in a clear picture. Everything we have within the Universe consists of space under the control of movement by spinning in the form of a sphere and space not under the control of movement in which the sphere spins. We have cosmic solid, cosmic liquid and cosmic gas. That puts two values in place. We have that which spin consisting of the axis formed by 3 and that which spins around the axis consisting of four points and in combination we have 7 points.

Then we have three more points in time representing the past, the present and the future, which forms an additional three points. That is how the Universe started. It did not start with molecules colliding because this was before there were molecules that could collide. The sun compresses cosmic gas into cosmic liquid by freezing space into liquid and it does that by maintaining a value of Π. In that movement we have to search how the value of Π forms. Thus the answer to everything is answering how Π forms a Universe without mass. Then again Π could only form by applying another basis in mathematics called the law of Pythagoras.

The direction we must follow is to see how mathematics developed. The Universe built mathematics because the Universe formed mathematics as it developed. Mathematics is not the God that was and that oversee everything as the Newtonian simpletons try to think. Mathematics comes in stages. There is a reason why mathematics can only add and subtract and then only later on can multiply and divide. There is a reason why the triangle is equal to a half circle that is equal to a straight line. That is where mathematics started and to get to that point where the Universe started we have to find the point where mathematics started everything. Therefore again only the Newtonian simpleton would think to use massive and impressive calculations to find the start of the Universe. I have written a book in which I explain how the cosmos started and that is how mathematics started.

Time forms a line by moving from the past through the present and onto the future. What is left behind is space as a remembrance of time and moving from the past brings along moving by the law of Pythagoras, just as the cosmos started in the very beginning. I find Newtonians think of the cosmos in terms of human life whereas the cosmos runs in terms of a process befitting only God. The Universe can never start because the Universe can never end. The Universe uses space to form time that will bring forth the next space but what is in the Universe can never again leave the Universe.

As Indicated space always holds a double 10 value in terms of 7

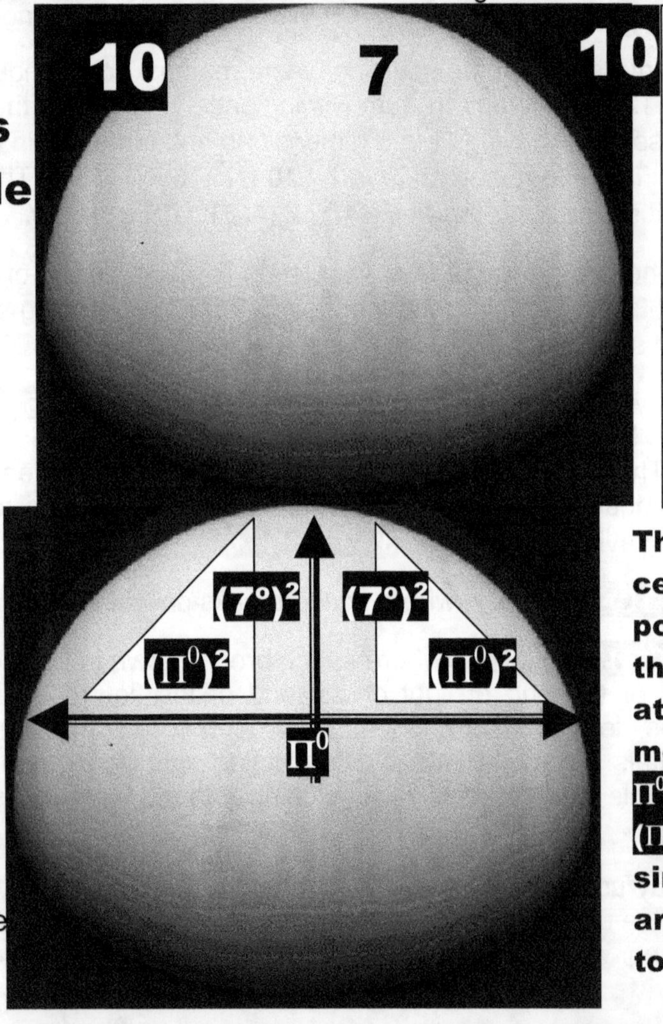

10   7   10

Material associates with 7 forming one value in 21.991 / 7 = Π

By spinning through 7° the changing of direction is taking 7 squared $(7°)^2$ and this gives 49 in total movement or space-time displacement or gravity applying or whatever you wish to call it. This is on one side of Pythagoras's' triangle

$(7°)^2$ $(7°)^2$
$(\Pi^0)^2$ $(\Pi^0)^2$
Π^0

The spin re-aligns the centre with the same point on the surface of the earth but it then is at another point. This movement then takes $\Pi^0 = 1$ or singularity to $(\Pi^0)^2$ or also 1 which is singularity. I sent an article about singularity to Annalen der physics.

TIME DIVIDED INTO SPACE
(10 / 7) \(7/ 10) = 2.04
1.4285 / 0.7 = 2.04
SPACE DIVIDED INTO TIME
(7/10) / (10/7) = 0.49
is also the same as .7 /   1.4285   = 0.49

SPACE MULTIPLIED WITH TIME
  7/10   10 / 7 X 7/10 = 2.04

THE PROCESS PARTED USING THE ROCHE PRINCIPLE

                 10 / 7 X 7/10 = 2.04   X $(\Pi/2)^2$ **Roche limit =2.4674**
                        2.04   x $(\Pi/2)^2$ = 5.033
$(\Pi/2)^2$ 2.04   x $(\Pi/2)^2$ = 5.033
                                   5.033 +5.033 = 10.066

SPACE DIVIDE INTO TIME
     7/10
             7/10 / 10 / 7= 0.49
             7/10 / 10 / 7 = 0.49
   10 / 7   10 / 7
   ―――――   ―――――
   7/10=.49   7/10= .49
     .49   +   .49   = .98
       .98 X 10.066 = 9.8 = Π^2

TIME SPACE = Π^2 = 9.8696

What I am about to explain is the absolute basis for singularity. Lines mathematically cannot start at zero because there is no evidence of zero as a factor in mathematics. The shortest possible line (hypothetically) must be so short it must have an initial and ultimate point sharing the same spot. Any theoretical line being the shortest possible line cannot have the line holding the initial starting point at point zero and advance from there. If it used zero as a start, the zero part would not count, because the line will only start at a point past zero where the line then will start

When the line has a beginning and an end at the very same spot and it wishes to extend the position as to further the possibility it has, which direction should it favour. Extending the line in any one direction will favour one direction without any clear reason not extending in other directions. The only option about extending will be in all directions equally in order to give a meaningful non-bias flow of mathematical equilibrium.

 Extending from start.

Zero point

The shortest line in the realm of possibilities must have a start and finish holding one spot and such a line will also be a dot or a circle. Not favouring one direction puts all directions at equilibrium meaning that any form of what ever might develop from such a spot with the end and the start being in the same position also has to be a sphere.

This reasoning prompted me to look for singularity in such a spot because if the prime spot from which all came was a spot holding all, then the spot must hold the shortest line but more prominent it will hold the smallest form including the smallest circle.

One possibility that the shortest spot can never have is having a starting point on the zero mark. If the mark of zero holds the start it must also hold the end because the end and the beginning has the same position. If the position of zero then is the beginning, the end will also be zero leaving the line without an end as well as without a beginning.

The conclusion from this is that no line can start at zero because that will be a mathematical impossibility. A line or spot starting at zero would therefore be shorter than the shortest line possible. A line growing or extending from zero can never leave zero because of the influence of being zero disqualifies any possibility of growth. If the line then had to grow in all directions at the same pace the line must therefore be a circle or being three-dimensional, a sphere.

Flowing from this fact is that in the Universe there can be no zero point or unfilled space. The value of the circle is Π, and that is where creation started. That gave me the clue where to start looking for singularity. One would find singularity in the value Π and the value Π will be in all things rotating in a circle. You might wonder how does that apply to the cosmos and moreover to gravity?

Gravity is the dimensional changing of space holding r as reference to the sphere holding Π as the reference. Heat occupying space has the cube that can apply r, as a straight line bringing about the cube with all its other names that may find attachment to specific form but nevertheless still remains only a six-sided cube with angles changing in some cases.

In the sphere there are no radius but only the extending of Π from the centre Π in six opposing directions relating to one another by the square but remaining Π because of the unity the matter holds in relating to space. It is not possible to draw a precise line that would form a precise ring and not cut some atoms in parts.

With the establishing if the value Π and identifying r, one has to distinguish and define each item in order to bring comparison. No line can start at zero but only on the smallest dot one may imagine bringing about that the line will start with the value of Π and proceed and proceed following the same value. The value has to be an extending of the original value because no evidence indicates changes that may take place. In the centre of a rotating circle is a spot $Π^0$ that becomes a dot Π and without the spot $Π^0$ the dot Π can never be and visa versa is also true.

This shows how gravity goes singular when applying time and then form space as the history that time left behind. We are in time but we live in space. Everything in the unniverse spins and by spinning everything forms singularity in the cvery centre of whatever spins. When gravity applies the location of the centre of whatever changes position in time provides singularity in that very centre.

These centres holding singularity connects and this connection places the Universe in singularity. The Universe is $Π^0=ΠΠ^2/Π^3$ **and this means singularity $Π^0$ forms space $Π^3$ by the movement thereof while the movement $ΠΠ^2$ of space $Π^3$ brings about singularity $Π^0$ which is the value by which the Universe "grows".**

Everything that holds gravity in the Universe is round and is spinning while spinning around some other object. In that we have to locate gravity because mass is not present. See how they are allocated in total random as if the Universe does not know about Newton's presumption of mass that should play a part. The only persons that are unable to see this anomaly are those so very educated in the science of Newton's principles that just never apply. If mass was present then surely there has to be some arrangement that show how and why the Universe adhere to mass, but there is not even a hint of a factor such as mass with a pulling power. Nothing is pulling anything because in place of the pulling we have the Titius Bode law, which I explain later on in the book. For the first time ever I am going to explain how it works and not merely the layout as the Newtonians do because the layout is rooted in Π forming.

Wherever we look we find rings and rind associate with Π and not with mass. The rings show that gravity has to be well connected to the roundness of rings because it is always rings form around structures. Even the moon is a ring circling around the earth much like the rings circle around the gas planets. In any ring there is always an incorporating centre and in that gravity must be vested. We have to look for a centre and a ring forming a circle by which it would lead to a circle moving. That will mean that if we wish to locate gravity we have to find a centre of a circle and we have then a ring circling around a centre. That is gravity and in that there is no hint of mass playing any part in gravity. When you show any Newtonian about the fact that all things fall equal as Galileo proves they come up with many arguments that avoids the topic. The clepsydra proves that gravity functions without mass the best.

A clepsydra or water clock is an ancient device for measuring time by means of the flow of water from a container. A simple form of clepsydra was an earthenware vessel with a small opening through which the water dripped; as the water level dropped, it exposed marks on the walls of the vessel that indicated the time that had elapsed since the vessel was full. More elaborate clepsydras were later developed. Some were double vessels, the larger one below containing a float that rose with the water and marked the hours on a scale. The clepsydra was the first recorded science experiment ever recorded. They proved through the clepsydra that by blocking the top intake, the water stopped running out of the bottom hole.

The clepsydra or water clock is an ancient device for measuring time by means of the flow of water from a container. A simple form of clepsydra was an earthenware vessel with a small opening through which the water dripped; as the water level dropped, it exposed marks on the walls of the vessel that indicated the time that had elapsed since the vessel was full. More elaborate clepsydras were later developed. Some were double vessels, the larger one below containing a float that rose with the water and marked the hours on a scale. A form more closely foreshadowing the clock had a cord fastened to the float so that it turned a wheel, whose movement indicated the time.

The Clepsydra shows how air coming from the top pushes water out the bottom. When you stop the air from coming into the top of the clepsydra by blocking the clepsydra at the top the water stops moving out at the bottom of the clepsydra. This proves that to get water to vacate the clepsydra air has to fill the clepsydra. The air coming in moves just as fast as the water move out of the clepsydra. This proves that it is air moving down and the air could be filled with water or not filled at all, the air moves at the same pace. The compressing of air is gravity and gravity is the pushing of air down to the surface of the earth, taking the air down to the centre of the earth.

That is gravity!

It has nothing to do with mass because that water is not moving down faster or apart from the air. When you block the air from flowing the water stops. So that proves that what moves downwards is air either filled with water or not filled with water. The thing to figure out is how does gravity function when mass clearly has no role to play in the process.

Looking at a point where singularity forms a line we find a triangle. This is because a line, a triangle and a half circle have the same value at 180°. By moving from a point holding singularity the movement is by seven points. This explains singularity. Singularity is not in exclusive brilliant formula but in the most basic of Mathematical principles. Two square triangles form one square triangle.

The one side depicting singularity is one $= \Pi^0 = 1$

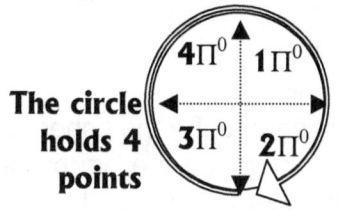

The circle holds 4 points

The axis forming the line holds three points and the circle holds 4 points

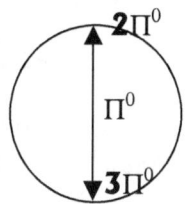

The one side depicting space-time is three = 7

The axis forming the line

The calculation of the triangle involves the law of Pythagoras.

Since I am explaining the most elementary of mathematics I better explain the law of Pythagoras as well. (For those SUPER-EDUCATED-MASTES BEING TO ADVANCE TO REMEMBER THAT)

The law of Pythagoras states that the sum total of the square of the two sides in a rectangular triangle will always be equal to square of the perpendicular side. How basic in mathematics can we still go?

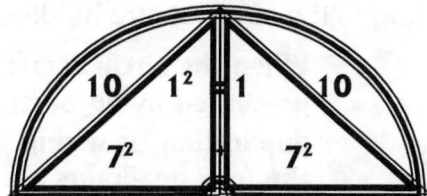

The axis holding $3^2 = 9$ and singularity forms space – time at 10

The square of space from the point of singularity is $2(1^2 + 7^2) = 10$

I have indicated a few pages back that the position the proton holds in considering singularity is $\Pi^2 + \Pi^2$ and the neutron holds in same fashion a matter in space of $\Pi^2 \Pi$. I also showed that matter can only relate to space by implicating matter in the sphere and therefore has to use the value of Π as a reference.

In one dimension space became 10 and in that same dimension matter became seven.

In order to separate matter (7) and space (10) through time (the spinning of matter) (7) in space (10) and space (10) spinning the matter (7) the following result came about through the application of the Roche principle $(\Pi/2)^2$

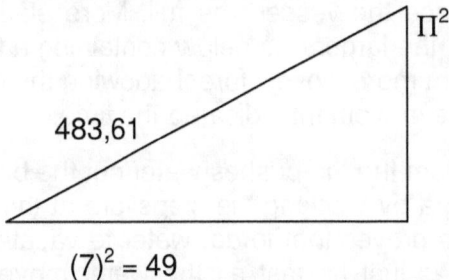

$(7)^2 = 49 \times \Pi^2 = 483,61$. But at this stage matter was in the relevancy of $(\Pi/2)^2$ therefore the factor that $(\Pi/2)^2$ represents, holds the value of 483,61. To get to the resulting dimensional value of 483,61, the square thereof becomes a factor.

$\sqrt{483,61} = 21,991$

From this one must conclude that gravity is also 21,991 when the moving body aligns with the 7 forming the contraction.

As 21,991 is one half of $(\Pi/2)$. Π Therefore, must be that value matter holds, which is 7. When matter divides into space $(7 \div 21,999)$ the result from that is Π. Through this the neutron (which is matter) holds $\Pi^2\Pi$ as a factor. Through that, the value of the triangle in matter, space and time also holds a 180°.

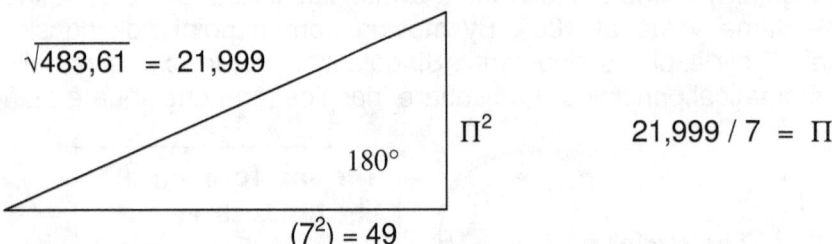

No person can deny the fact that the earth is a sphere. By the sphere turning into space it forms gravity in relation to singularity being $(\Pi^2\Pi)$. Going into space the earth spins into $(\Pi/2)^2$. This forms as a result where the one side of $\Pi/2$ turns into the other side of $\Pi/2$ by moving forming a square that forms the is the Roche limit.

The spin of the circle forms 7° on the one side of rotation as well as 7° on the other side of the centre of rotation.

That puts the centre turning by 7° x 7° = 7² = 49. Then on the other side of the divide the same repeats where there too forms by turning 7° x 7° = 7² = 49. Then by using the law of Pythagoras and incorporating $\Pi°$ on both sides 10 forms as a space value outside the circle.

A double 10 plus a developing singularity value of 1.9991 represents singularity, which represents Π turning around $\Pi°$ to form Π. This means the one side of the circle holds $\Pi =$ 21.991 ÷7 and on the inside connecting to the circle centre holding singularity at $\Pi°= 1$ the value of Π is 3.1416 ÷ 1

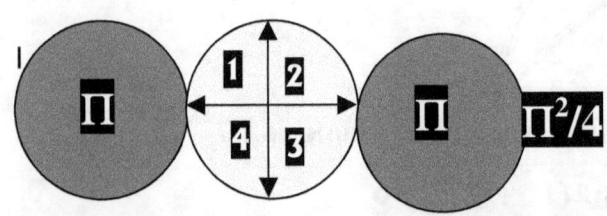

When the circle rotates and the rotation is prevented by an obstruction within the limits of Π duplicating Π^2 within the area forming space where the four quadrants are the part that forms the material component would liquefy anything that blocks the relevancy in singularity between solids and liquid.

In nature there is the Roche limit placing a limit to the reduction of space and the inflow of heat to sustain progressive cooling. Materials spin in order to contract the surrounding liquid onto and into material or solids to maintain the cooling of the solid. This is a very complicated process and I have devoted one half of an entire book just to go into this process. At a point of $(\Pi/2)^2$ the reduction of space disallows any object the cosmic object cannot reduce, an entry to its area of reducing space.

The first question that one can ask is why would there be the value of $(\Pi/2)^2$ between orbiting structures positioning themselves in a time relation to space. This is to establish Π and to get Π to move and form Π^2 from the one side $\Pi/2$ to the other side $(\Pi/2)^2$. But the air forming the atmosphere is still part of the earth and therefore it does not become the full Roche limit $(\Pi/2)^2$ but half of it $(\Pi)^2/2$.

Then we had two point holding eternity in place going square by movement to form 4 points serving eternity and infinity captured the first three points held by both and since eternity could not release from the two it had but had to duplicate what it had, eternity by movement became a circle captured by the line.

With four points captured by the line of three points the circle coming about is eternally returning to infinity but never complying with infinity because if mismatching temperature or movement (3 against four). Material will always be colder than outer space. It is because material spin and outer space moves by expanding due to overheating.

This is where I start when I start to explain the first moment but I use a shipload more information to do explaining when I explain the star in the book I do so. I involve the four cosmic pillars to substantiate the claims I make because all four still work the very same way as it did at the beginning of the Universe.

The three points serving one part of singularity combined with the four points serving singularity unites as seven to form a circle of either 3.1416 or 21.991÷7. The seven going to one is eternity matching infinity by movement. But since seven moves it is seven that has to produce gravity. How do I know all these facts, because we can see from the top it is still doing what it did the very first second.

The Π^2 end will be at the point where heat passes through the object directly to the earth and this position of space-time relates to the neutron time link of Π^2. The space link of the neutron will then form the Π link.

The value of the Π link we find to be $(\Pi)^2/2$, but the explaining to why it is $(\Pi)^2/2$ is rather more complicated.

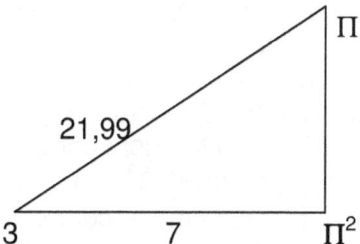

At the end of the space relevancy 3 where matter occupy space (21,991 / 7) is a border Π. That border is the exact point where space reforms to a square of time placing all matter (occupied heat) and heat (unoccupied matter) to a value of the square of time.

That specific point is in relation to the square of the diminishing shield around the earth. However it takes matter (Π^3) from the 3 dimensional positions to the square (Π^2) in relevancy to time in singularity.

With time holding space in singularity the 4 sides of Π truly relates to half of the total square value.

Everything in the Universe is a sphere formed by singularity extending. I can give the displacement values but it would not mean much at this point to any reader. Creation started at a point long before the Big Bang, long before light and even before space, as we understand what space is as it is space. If there is anyone doubting my revelations or my claims here is

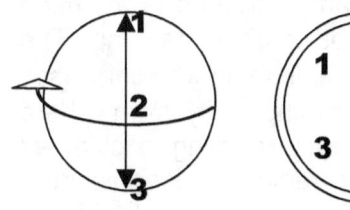
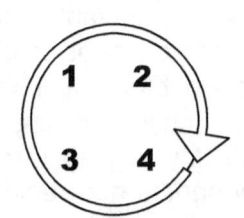

some proof, but understanding this proof requires the reading of all six parts of the books An Absolute Relevancy of Singularity Thesis 1 to 6.

Movement of space in space does not start at the large objects but it starts in singularity where something forming a thought presents space that presents a Universe with everything in that space which incidentally is time. Where does all this start? It is where a thought becomes space. It is where singularity that is not tangible forms an identity (space but not yet space) in the Universe.

What this shows is that we can't look at the Universe as if we are not part of the Universe. The Universe has one end and that is to the inside. When we look to the outside, we see only darkness and more endless darkness as from our vantage, the Universe can never end because that is our only view we can have of the Universe going larger or becoming more.

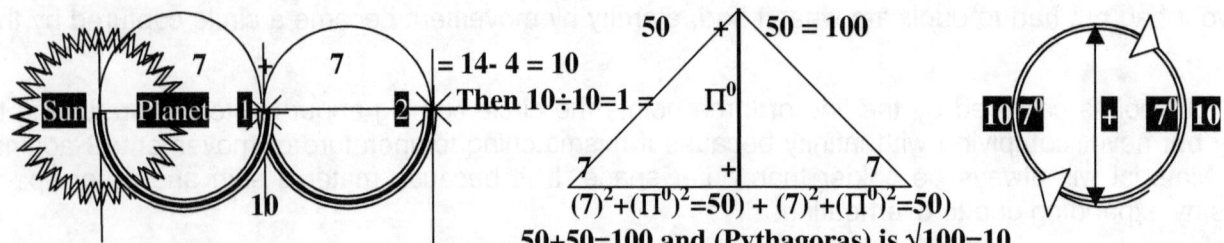

From forming the value of Π by placing 7 in relation to 10 we see how Π forms and Π being the eternal single we can only see a circle. Then in the centre of the circle we see a spot in singularity $Π^0$ and that what we then see is gravity $Π^0Π$

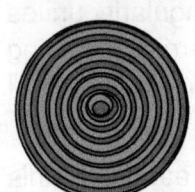
By extending the dot $Π^0$ to the value of the next Π, a continuing of Π will flow in all directions evenly. The form is a sphere, but when the sphere brake anywhere showing a flat surface Π will continue but will alter position to comply with the new standards set by shape. Every circle is a multitude of circles within a circle as $Π^0Π$ continues to produce circlers

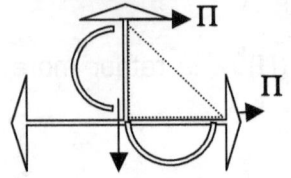
In that manner Pythagoras becomes defined because the straight line and the triangle holds equilibrium in the half circle. By extending the value of Π to the next position Π maintain in the time position $Π^2$ where the $Π^2$ indicate time and the square of the triangle indicate space developing will establish the square of time. This I prove using mathematical equating when I prove how the Universe started by implementing Pythagoras first when the law of Pythagoras founded the beginning of mathematics…and this I do.

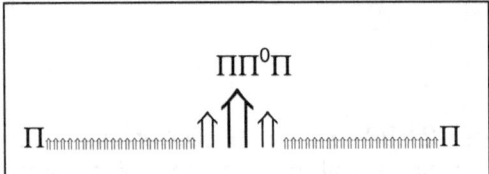
From the governing singularity come the major singularity providing $Π^3$ space, which time claims $Π^2 + Π^2$ and space holds time to control space $Π^2Π$ from where singularity influences space by the three sides a cube shows to one side of the Universe using r as a means of doing do. This entire concept grows from Kepler's formula of $a^3 = T^2k$.

From this (extending the dot $Π^0$ to the value of the next Π) there is a value difference where the one singularity holds a factor of 1 in the 21.991 / 7 and the other has a value of .991. Where $Π^0 = 1$ and $Π = 3.1416$ the $1416 \times 7 = .991$ and thus has to be smaller where it holds the value of .991 in some cases and other cases the value of .91. Closer than such an explanation I could not come with the limited space available in this website. I have to name the opposing linking singularity although to my mind humans use the naming of objects as a shield to cover their poor concept of understanding. The singularity $Π^0$ I refer to at times as the governing singularity. In my books I shall show why light uses $^αΠ^Ω$ by means to travel as a conductor. I mention that fact because the singularity from where everything came is still in our midst and used by all with vision. My dismay in naming is clear as through out the history of science there is clear evidence of how names had replaced recognising of various factors and through naming whatever differently; humans go at fault in recognising the name we use instead of what the product truly represents as space-time. The governing singularity is the unseen line that all matter refer to using the space provided by the governing singularity and I sometimes refer to that as the major singularity.

Our Brilliant mathematically inclined Masters-in-Newtonian-physics filled the entire Universe on the inside with nothing and even filled to the point where the nothing expands and becomes more nothing. "Nothing" can't be filling the Universe because lines fill the Universe in forms of light flowing all over. The presence of a line does not present "nothing" nor does it prove no line at all because if there were nothing there would not be any possibility of a line ever being at that point, as nothing can hold "no line". The line does not present or start with nothing or begin at nothing because there is a possibility of a line forming being the shortest line possible the line is infinitely small yet the possibility does not exclude the line totally as would zero do. Therefore the line is there, be it only in the possibility of a line being there.

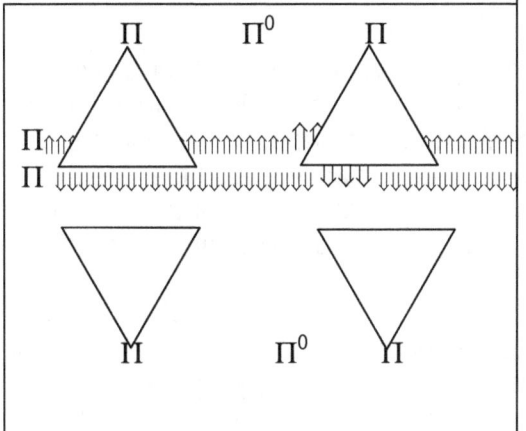

From the possibility of the line being, the line may have the space of the premier singularity from which all of the Universe arrived, but remains a possible factor and as such a factor. The fact that a line is not excluded as a possible factor presents a line non-the less. In the case of zero the factor becomes a defiant excluded factor whereas without zero in infinity the line becomes a defiant included factor, as it would be most incorrect to surmise no line will ever form at such a point. Zero brings conclusive exclusion while infinity brings conclusive inclusion and as such the two values oppose the concept surrounding the line.

It is moreover the individual singularity in maintaining the major singularity, which sustains the governing singularity providing equilibrium in space-time.

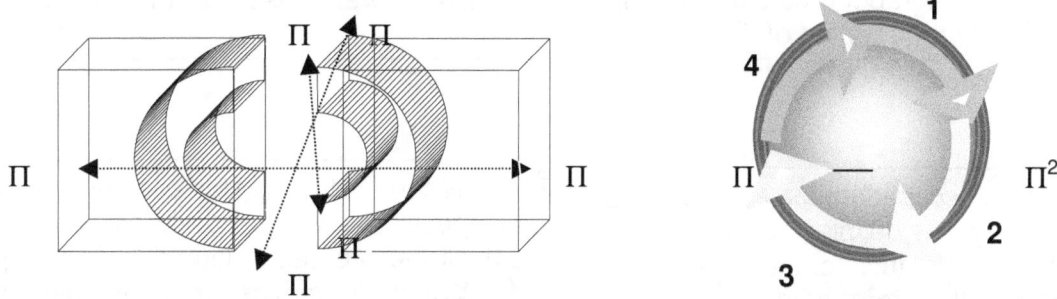

Seeing our spinning top from the top, there are four quarters opposing each other and by that opposing one another.

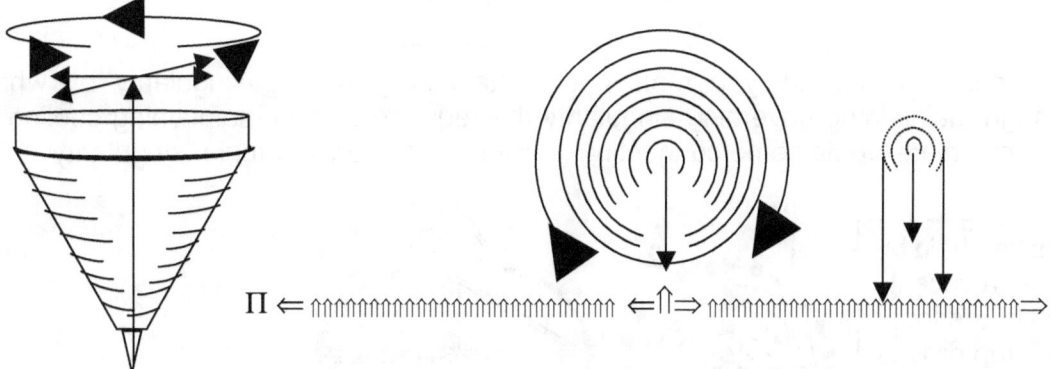

Any object in rotation will have a middle point, a very specific centre point that does not spin. That point in the very centre is hypothetical s it is valid but none the less must be standing still because every line running from that pint in opposing directions are also in opposing directional spin to each other.

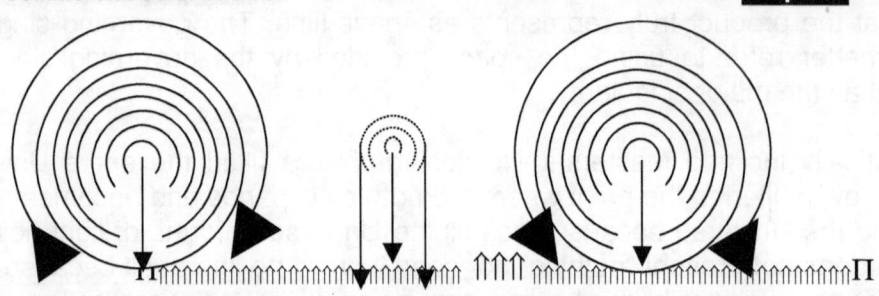

From such a point every other point will be opposing any other point not pointing in the direction to which the first point is pointing, whereby it extends the direction it holds. No matter what the point is or where the point leads, such a point holding a specific direction will be unique in the direction it is rotating because at that or any other specific point wherever, it will be directing not in the direction it spins but in the direction flowing from the centre point outwards.

Any point will be it opposing itself within the rotating of 180° changing every aspect of its previous flowing characteristics it previously had or will once again have in 180° from there. While in rotation from the point of an outside observer all may seem static and never changing but to the object in spin every next second will be a diverting from every aspect it was in very second passing, and the direction it held in relation to the direction it held the previous mille, mille second will totally be incompatible with the direction it holds the very next mille, mille second of rotation. That proves no point can be static or constant, all though it may seem that way to outsiders.

 Because there is a space that may not be occupied by a particle does not exclude the possibility of a particle sometime to the future occupying that space. If the space was nothing all possibility of future occupation will become excluded by the presence of zero that is unable ever to include occupation

From the centre of the top runs the premier singularity and as the top starts rotating the top's rotation bring about the sides to singularity, which too was present all the time but filled the being, they're by the rotation of the top.

Since occupation may or may not be placing the factor in infinite, the space therefore holds the premier singularity of infinite from which all included in the Universe has come.

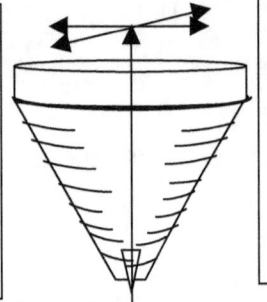

When the top starts spinning in a specific position the top merely executed the option to fill the premier singularity at that specific point. When it moves it may take the premier singularity with to the new location it moves through spin or it may fill yet another position in singularity as all is the same.

The sectors provide individual singularity a means in sustaining governing singularity by which provision comes through maintaining governing singularity the required spin in maintaining cooling. If this process did not apply, there would be no connecting individual singularity to major singularity

By claiming the position held by singularity premier as a vacant spot until the arrival of the top, the singularity of the top divides the point flowing from singularity into four sectors holding two half circles

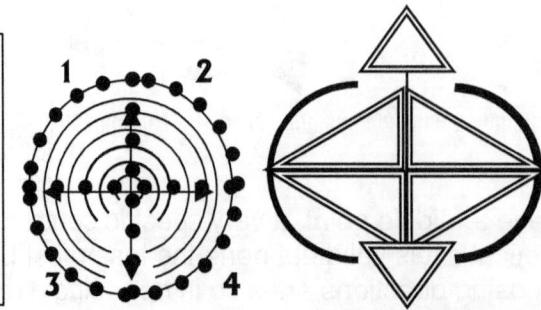

Any object in rotation will have a middle point, a very specific centre point that does not spin.

That point once again hypothetical but none the less must be standing still because every line running from that pint in opposing directions are also in opposing directional spin to each other

. Although the points had the same characteristics only seconds before, they oppose the characteristics it had just before and just after the very second in which they are and to which they relate by similar points also in rotation. Due to the spinning nature of such a point with all surrounding the point very varying second, the value to such a point can only be Π because of its constant changing. Using r would specifically oppose another r from every angle.

An object maintaining time will have Π confirming the next Π in the same positing in relation to r as the previous Π. By maintaining Π in relation to the same position Π had Π^2 confirms time. By maintaining the value of Π in relation to singularity as Π^0 it also confirms all other aspects of singularity in as much as $\Pi\Pi\Pi$ and Π^3. In that way singularity secure its relevancy to matter as much as matter secure its relevancy to Π^0 as singularity.

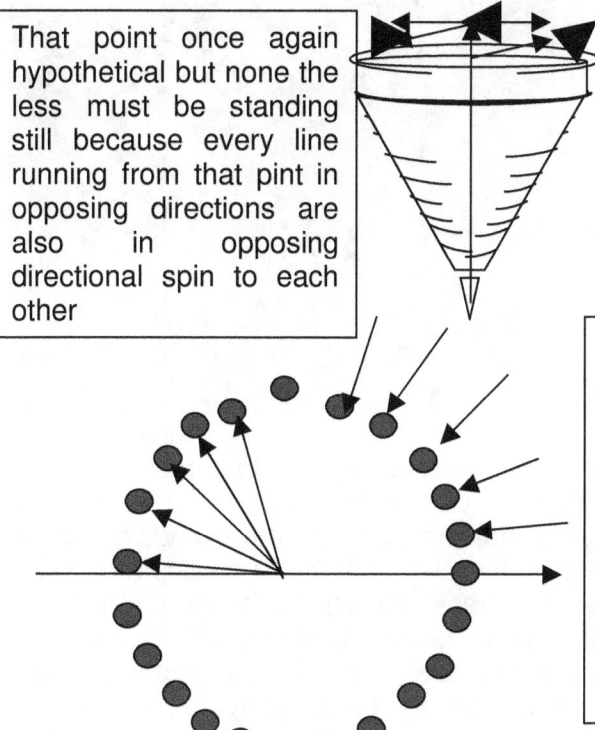

By over heating and though that applying a time difference matter becomes its own anti matter with in one rotation.

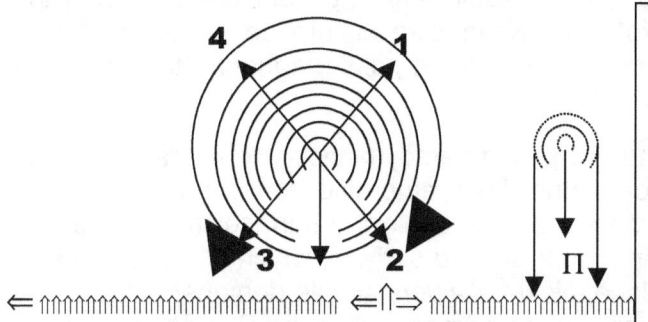

From such a point every other point will be opposing any other point not pointing in the direction to which the first point is pointing, whereby it extends the direction it holds. No matter what the point is or where the point leads, such a point holding a specific direction will be unique in the direction it is rotating because at that or any other specific point wherever, it will be directing not in the direction it spins but in the direction flowing from the centre point outwards

When Π crosses over from the one quarter to the next quarter and it does not maintain a constant position in Π^2 forming a relation with Π^0 the relevancy will change as time no longer can apply a true value. When time diverts it will also affect density and the standing the particle has to surrounding matter. Behind this concept we find the cause of Supernovas expanding. When the rotation movement of the star is insufficient to cool the star the diameter would extend and this will result in even more overheating because an extended radius will bring on even slower rotation. This then becomes a progressive regression of expanding $\Pi^0\Pi$ and heat developing more until a total annihilation of the structure of the star takes place when all the compressed heat in a star becomes liquid and the star is alleviated.

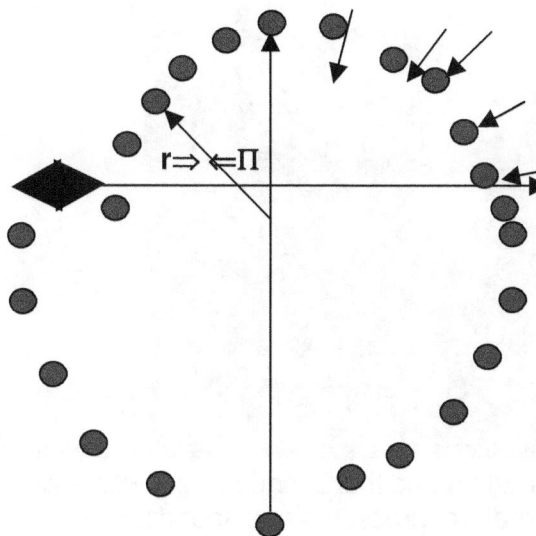

By applying a different position Π becomes r in relation to the previous position Π held because the circle now have to introduce a line in support of the new circle. The loss to density through the application of a new time relation will be suspended matter forming in a heat release.

Around the earth we find several of layers of air that is compressed in ever more dense space. While it is impossible that "mass" can merely pull "air" into denser units, it is very plausible and mathematically correct to presume that the spin of the earth can redirect the space into more compact layers and in this we then have winds and other disturbances coming about from this turbulent spinning action. It is clear that the air flowing around the earth acts in the same way as liquids do and has the same characteristics as liquids have. Studying the sun we find a higher concentration of liquid around the sun, which proves the Newtonian suggestion that the sun is a gaseous formation completely misrepresentative. They presume because the sun is mainly filled with hydrogen and hydrogen on earth is a gas therefore the hydrogen in the sun is also a gas. Think of how an air-conditioner unit functions when it reduces the heat on the inside of a room and then look at the sun. It repels heat to the outside to cool what is inside. Now we see the air conditioner from the outside blowing heat out and we say the conditioner is incredibly hot.

Same with the sun, it repels heat in a cold liquid form from the outside the freeze what is inside. Again Newtonian backward thinking is astonishingly stupid. If it is liquid squirting from the sun, then the hydrogen in the sun must be a liquid irrespective of what Newtonian culture demands to dictate. Gravity is the Titius Bode law applying in the limits of the Lagrangian points, adhering to the Roche limit and the lot culminates into the forming of the Coanda effect.

The Coanda Effect has been discovered in 1930 by the Romanian aerodynamicist Henri-Marie Coanda (1885-1972). He has observed that a steam of air (or a other fluid) emerging from a nozzle tends to follow a nearby curved surface, if the curvature of the surface or angle the surface makes with the stream is not too sharp. If a stream of water is flowing along a solid surface, which is curved slightly from the stream, the water will tend to follow the surface. Now for a very simple demonstration: If you approach gently a curved shaped surface (like the shape of the primary hull of the Repulsing) under a stream of water (see below):

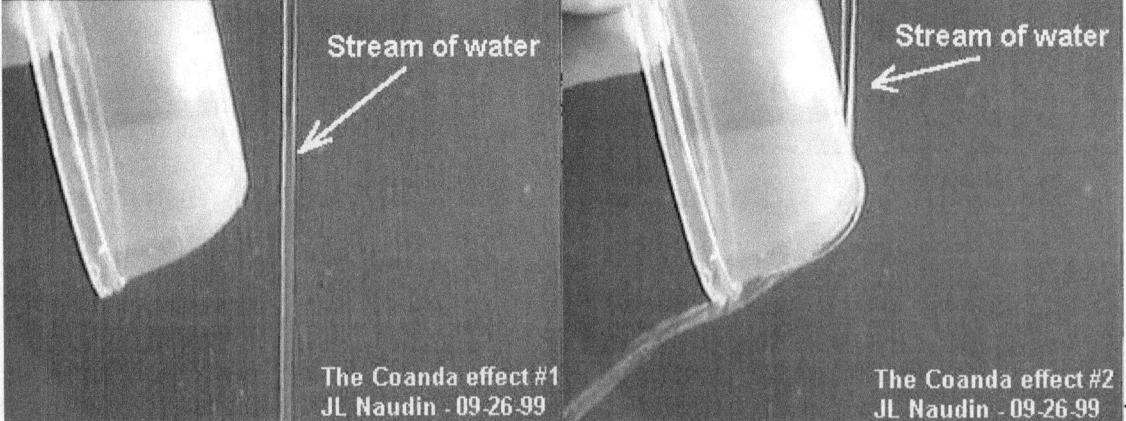

The water follows the surface of the curved shape, this is the Coanda Effect and the Coanda Effect works with any of our usual fluids, such as air at usual temperature, pressures and speeds.

The picture above is about a phenomenon called the Coanda effect but this is never mentioned in any physics handbook because Newtonian physics-religiosity is unable to explain or to understand this principle. When liquids flow past a cylindrical object the liquid clings to the surface of the object rather

than follow the "path of mass" and fall straight down to earth. Gravity is about the atmosphere that forms a liquid that is the same as the liquid running around the solid circle called the Coanda effect. Gravity is the movement of the air in relation to the movement of the earth. If an object moves within the parameters the earth set the liquid to move around the earth and the object has the ability to maintain the speed, it will be just more liquid floating around the earth at a specific height fore filing a specific circle or rotation requirement. If the speed drops the object will fall notwithstanding mass or the lack thereof. The entire idea vested in gravity focuses on speed-differences. So what the hell has mass got to do with the entire affair of gravity?

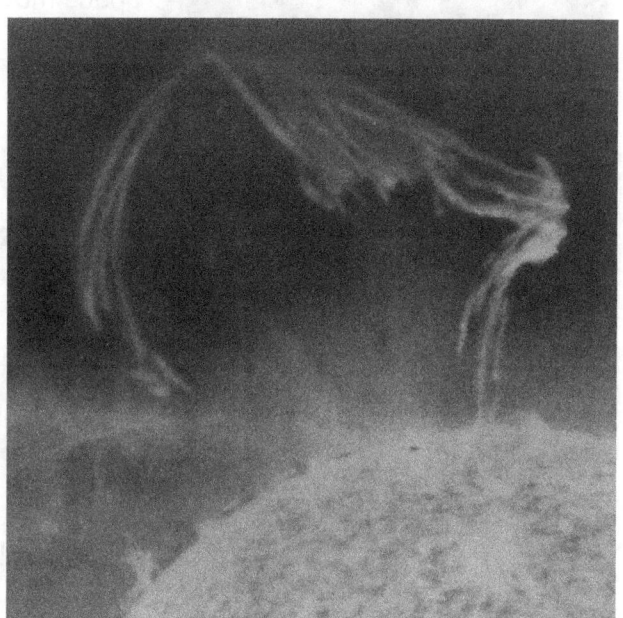

If the circle is as large as the sun is, then the compressing of space turns the air into squirting liquid precisely as the picture of the sun shows. Again I ask what has mass got to do with the entire idea. But remove mass from the equation and the Newtonian brilliance becomes idiotic stupidity because their entire mindset of playing God with mathematics becomes total invalid and their stupidity rings as load as a Cathedral tower bell.

It shows the limits of $k^0=a^3/T^2k$. Also this clearly shows that gravity is about condensing air through movement by Π when forming the balance $k^{-1}=T^2/a^3$ on the side holding the air and $k=a^3/T^2$ holding the top. This is the Coanda effect that keeps the top erect with one side (the air) forming a concentration of liquid and the other side (the top or the solid) $k=a^3/T^2$ concentrating the density of air. This too is gravity in all its splendour.

The fact that $k=a^3/T^2$ puts the movement of k relative to the movement of T^2 that makes the space moving dependent on size in relation to the distance per time unit the space travels through time and from this comes the explanation I give about the "sound barrier". The anomaly is that mass plays no role. The line holds space by forming a presence in terms of the spin charging singularity or mathematically said it is $k^0=a^3/T^2k$. It shows the limits of $k^0 = a^3 / T^2k$. Also this clearly shows that gravity is about condensing air through movement by Π when forming the balance $k^{-1}=T^2/a^3$ on the side holding the air and $k=a^3 / T^2$ holding the top. This is the Coanda effect that keeps the top erect with one side (the air) forming a concentration of liquid and the other side (the top or the solid) $k=a^3/T^2$ concentrating the density of air. This too is gravity in all its splendour. The fact that $k=a^3/T^2$ puts the movement of k relative to the movement of T^2 that makes the space moving dependent on size in relation to the distance per time unit the space travels through time and from this comes the explanation I give about the "sound barrier".

In front of your eyes you witness the sun freezing an atmosphere which the entire solar system shares and which is made up of condensed space that it freezes into squirting liquid that is the flow of heat, as pure as heat can be. This has noting to do with mass but those conspirers are so deliberately clinging onto their fictional Newtonian Universe that the truth passes them by and in total arrogant stupidity they refuse to use their eyes and see what the Universe shows it is.

The movement of space filled with material cools by movement of that which moves and this puts thermo differentiation between space that moves and space that does not move. As the space differences by thermo differentiation grows it would seem that one part grows in relation to another part shrinking. This is why large gravitational stars always seem to lose space while the Universe seems to expand.

Gravity forms as the earth or any other cosmic body rotates by 7°. By diverting the straight-line movement by 7° a contraction forms in a circle. In my books I prove how this then brings about the value of Π by implementing the law of Pythagoras and gravity is the law of Pythagoras. The reclining of space by redirecting the direction of travel from straight ahead to 7° reclines the space in a steady and sturdy flow. It is the space reclining or contracting and the space contracts albeit filled by solid material or empty of solid material. This is the reason why all things fall equally. It is the space moving down with or without holding material and the space has the same density in relation to the solid cosmic structure rotating.

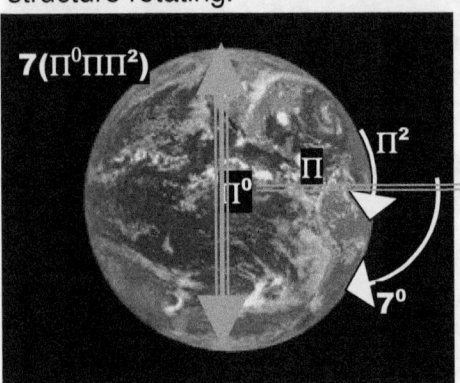

In the centre of the rotating body singularity forms to the value of Π^0 and this extends to the curve forming Π in terms of the curve of the rotating object. This then forms part of gravity moving forming singularity going square Π^2 and this form the relevancy

The extending of singularity goes from Π^0 to $5\Pi^0$.

Coming in from space while having no mass the coordinates change as the value of Π forms a relation to the 7°
When an object comes in from the atmosphere or flies the sky the object associate with the turn giving it a value of the rotation in line with the axis which puts a value of 21.991 / 7 on Π or on gravity.

This is gravity and this is how gravity functions. It has nothing to do with mass in any way, shape or form and there is no factor such as

The entering by singularity that goes from 7° by $3\Pi^2$.

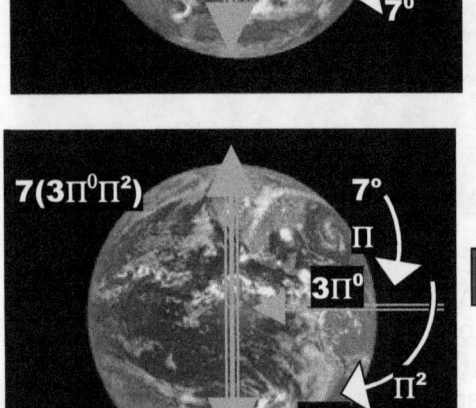

mass in the entirety of the cosmos apart from being in the imagination of the Newtonian conspirators calling them physicists. I sent a fifteen-page article to the Annalen Der Physics in which I explain this process in length.
I received the following reply from Annalen Der Physics.
Dear Dr. Schutte, You submitted an article of 15 pages to the Annalen. The content of this paper doesn't constitute a theory in physics. With a lot of words and some simple algebraic relations, there is no way to "explain" the world of physics. You seem to be out of touch with modern developments. This is also shown by the fact that you don't quote any relevant literature. I am sorry to say, but the Annalen is not able to publish your work. I am sorry for having no better news for your. Best regards, Friedrich Hehl
Co-Editor Annalen der Physik (Berlin)
Friedrich W. Hehl, Inst. Theor. Physics
* University of Cologne, 50923 Koeln _____/_____ Germany
fon +49-221-470-4200 or -4306, fax -5159
hehl@thp.uni-koeln.de, http://www.thp.uni-koeln.de/gravitation
* Univ. of Missouri, Dept. Phys. & Astr., Columbia, MO, USA
From this reply I received it is clear professor Friedrich Hehl did not understand a single word I sent him because as he said, it was not in mathematics but written in words! Now I beg of Friedrich W. Hehl, Inst. Theor. Physics University of Cologne, 50923 Koeln _____/_____ Germany or any of the editing staff of Annalen der physics or any physicists professing the principles of Newtonian inspired physics to prove to me how does the formula $P = \left(\frac{4\pi^2 a^3}{G(M+m)}\right)^{0.5}$ use $P = \frac{}{G(M+m)}$ to place the positions of planets because this is one of the supposed Kepler Laws. I wish him well in using Newton's corrupt formulas to prove that Jupiter with having the "most mass" is bang in the centre and on either of the outside and inside we have the two smallest planets. In that put the mass in relation to the position. Prove to me that mass is a valid factor in the cosmos. Physicists live in Newton's dreams.

My work proves nothing so please prove Newton when he claimed $P = \left(\frac{4\pi^2 a^3}{G(M+m)}\right)^{0.5}$ especially the part $P = \frac{}{G(M+m)}$.

The Titius Bode law

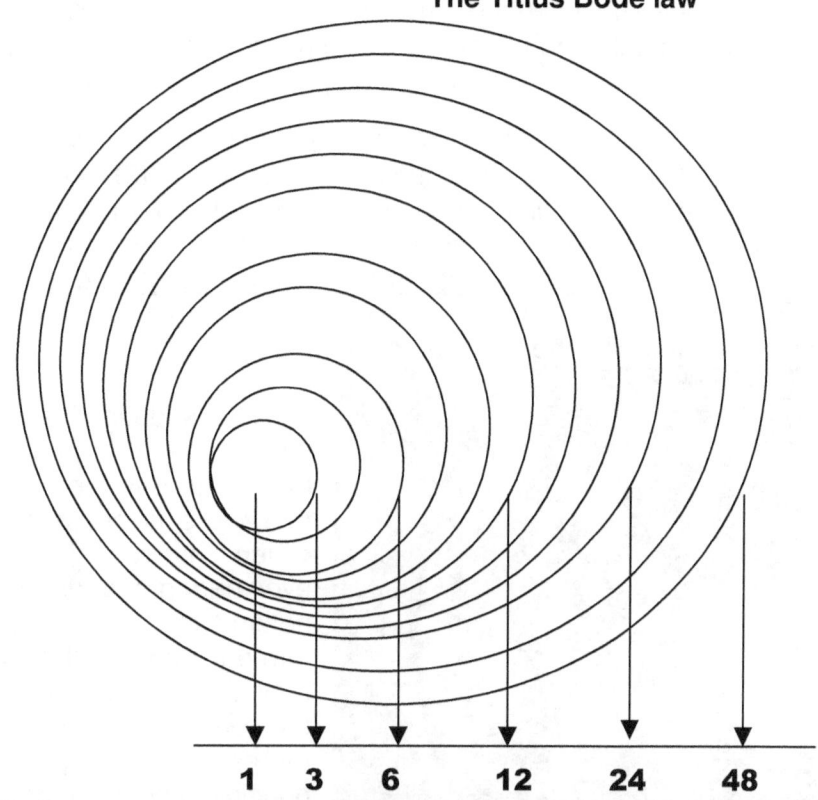

1 3 6 12 24 48 96 192

There is always movement because of the earth moving and when anything moves above and beyond the movement the earth provides that then forms the <u>**sound barrier**</u> using a modified version of the **Titius Bode law**. By going to and then **download The Absolute Relevancy of Singularity The Website** you will learn what the sound barrier is. By then afterwards going to and then **download The Absolute Relevancy of Singularity The Theses** you will learn what the difference is between mass and gravity and why a parachute slows down falling objects, and you will learn why we so desperately need to move on from Newton's incorrect restraining on the human mind. But either way you will find out why the "sound barrier" is a cosmic limit.

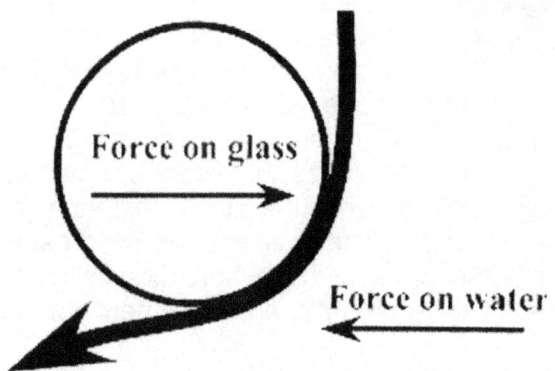

That shows that all movement is gravity or time related. This shows that when objects move then such movement of whatever is moving is moving extraordinarily because everyone always forgets about the fact that it is the earth that normally applies all of the movement while all else stands still in relevancy.

It is not my manner to speak ill of the brain dead or the dead by other means, but in the case of the Newtonian academics I am left with no option. Their forces haunt me to death and it is their forces and ghosts and witchcraft I have to fight. The lifting of a body comes quite natural when a certain speed is exceeded. By exceeding $7(3\Pi^2)$ the body will start to lift no matter what the mass is. A 747 Boeing of multi tonnage lifts off spontaneously at excess of that speed.

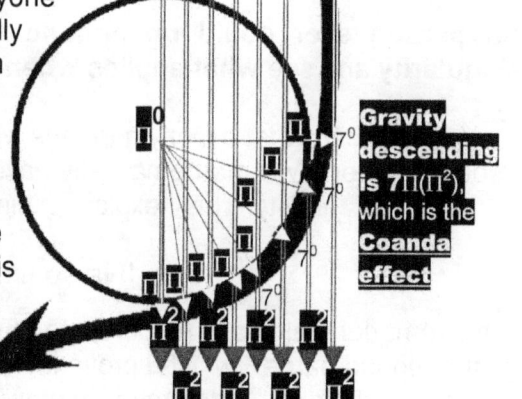

Newtonians are forever concerned with middle ages and with forces they can't explain but such forces and witches there are not, therefore they do not have to fear and can sleep well at night. In the sketch the circle portrays a glass and the arrow portrays running water. The Coanda effect is the water that does not drop straight down but follows the curvature of the glass.

The picture clearly shows the 7° inclination of gravity to the value of contracting $\Pi^0\Pi$. This is gravity! This is the most vivid example of the Coanda effect and it is what gravity is! It is a whirl allowing the flow of liquid space around a solid centre in relation to the centre holding singularity or $\Pi^0\Pi$ as it contracts space into a denser liquid.

The Coanda effect is gravity and my explaining this statement is part of many other books in which I explain what gravity is. The Coanda effect shows how liquid attach to the solid by $7(3\Pi^2)$ and the solid attach to the liquid by a relevance value of $7(\Pi\Pi^2)$. That is gravity. Should anyone require more or better explaining I would advise that person to purchase any of my books holding the title as an to go to **http://www.singularityrelavancy.com/** and also go to **www.questionablescience.net** Flying object is under this gravity control of movement and it is this that has crafts fly and cars requiring down force by the aid of aerodynamic devices.

Gravity is defined as a force that is present in mass pulling mass and it is that entire idea that there is not evidence of. When I refer to gravity everyone grabs on a cultural notion of a concept they formed and in that concept they link the smallest part of the concept to the become and represent the overall gigantic principle and by knowing one line everyone has the opinion that anyone then is the absolute master on the idea of gravity. When I freeze any substance the substance contract to a liquid and with more cooling it contracts to a frozen state of ice. The gas expanded more than what the solid did because the gas is hotter than the solid is. When we form the opinion that the outer space expanded to the limits the idea springs to mind that outer space is freezing cold. When I say the Sun freezes hydrogen to a liquid because my eyes see the liquid squirting from the Sun I am dangerously mentally impaired since the Sun is blistering hot. Then through this culture my effort to say gravity is motion and motion is the cooling of an overheating and thus expanding Universe goes wasted. Every one has the opinion that where gravity is the strongest such as the case is on the Sun or the centre of the Earth, such a place is extremely hot and where gravity is least that place is unbearably cold.

No person ever could understand the sound barrier because no person ever could **enter singularity** and see **what applies when the Universe goes singular.**

You will read what exactly happens when the "sound barrier" is exceeded but also to explain the "sound barrier" I will mathematically have prove what gravity truly is by proving the measured value of gravity. This I explain in simple understandable mathematical terminology.

It is so simple even I can understand it.

This remark I make is in response to physics academics not understanding something as elementary as the "sound barrier" which I prove is the most basic principle of gravity but they are forever telling me I am too stupid to "understand Newton" and therefore because of my lack in intellect I question "**the validity of Newton**".

A More Complex Commercial Science Book 511 Chapter 9   The Titius Bode Law

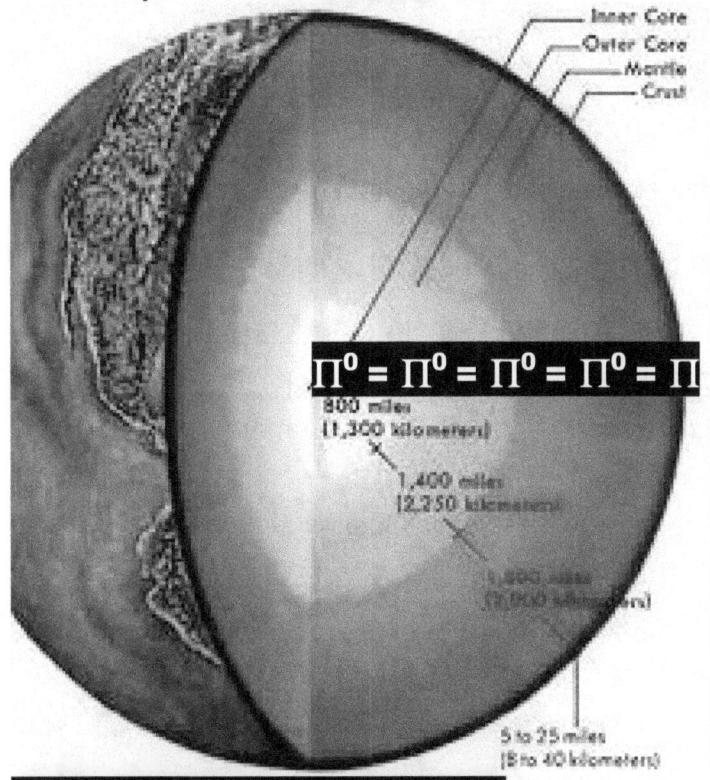

$\Pi^0 = \Pi^0 = \Pi^0 = \Pi^0 = \Pi$

From the centre of the earth to the very rim of the surface we have no Π forming since one singularity extending Π^0 is in place. When observing a mountain the question is where does the mountain start in order to determine the mass of the mountain. I living person has to use his judgement and that judgement is human not cosmic. The gravity line will run from the centre of the earth as Π^0 and only at the point where the surface touches the sky would $\Pi^0\Pi$ come into value. At that point anything standing on the ground will have mass or weight. At that point gravity is $\Pi^0\Pi = 3.1412$.

The value of Π is $\dfrac{21.991}{7}$ in every layer that forms and the 7 associates with singularity becoming one and by diminishing in value from 7 to one it reduces the top or space part to 3.1412. This happens as descending takes place throughout the space reducing up to where the object touches the surface. At that point the object then changes from forming part of a liquid to forming part of the solid and thereby then receiving mass or weight. This is how gravity reduces space albeit filled with material or not.

Obliquity of the Nine Planets

| Mercury 0.1° | Venus 177° | Earth 23° | Mars 25° | Jupiter 3° | Saturn 27° | Uranus 98° | Neptune 30° | Pluto 120° |

© Copyright 1999 by Calvin J. Hamilton

In the forming of gravity within the solar system the ratio used is much different as is used in gravity forming within spherical structures such as planets and stars. It is because in stars the full value of Π as it forms $\dfrac{21.991}{7}$ forms the factor known as gravity whereas in the Titius Bode law only a part forming the value of $\dfrac{10}{7}$ comes into play. Because it only calculates one part of $\dfrac{21.991}{7}$ the compacting of space does not come about as it does in stars spinning. In this we then find growth of space. When applying the full compliment of Π such as $\frac{21.991}{7}$ does space reduce but implementing the half of $\frac{21.991}{7}$ which then is $\frac{10}{7}$ we find space grow by the value of $\Pi^0 = \Pi^3/\Pi^2\Pi$.

The entire Universe grows by $\Pi^0 = \Pi^3/\Pi^2\Pi$ and not by the Hubble constant because the Hubble constant varies so much it is beyond human calculation. Yet because it makes scientist look and feel good they stick to the nonsense of the accepted Hubble expanding value, which is inconclusive.

The First Article about $F = G\frac{M_1 M_2}{r^2}$ in terms of Gravitational Formation

The Titius Bode law is in place holding every available piece of evidence of what the solar system uses. The law relates the mean distances of the planets from the sun to a simple mathematic progression of numbers.

To find the mean distances of the planets, beginning with the following simple sequence of numbers:
0 3 6 12 24 48 96 192 384

With the exception of the first two, the others are simple twice the value of the preceding number.

Add 4 to each number:
4 7 10 16 28 52 100 196 388

Then divide by 10:
0.4 0.7 1.0 1.6 2.8 5.2 10.0 19.6 38.8

The resulting sequence is very close to the distribution of mean distances of the planets from the Sun:

Body	Actual distance (A.U.)	Bode's Law (A.U.)
Mercury	0.39	0.4
Venus	0.72	0.7
Earth	1.00	1.0
Mars	1.52	1.6
Asteroid Ceres	±2.8	2.8
Jupiter	5.20	5.2
Saturn	9.54	10.0
Uranus	19.19	19.6

This is there used by the cosmos and performing as the working principle of cosmic physics. Guess what, the Newtonians ignore this which is in place and which is performing as the solar system in favour of Newton's gravitational principles that holds no backing of any evidence whatsoever.

Body	Actual distance	Actual mass
Mercury	0.39	0.4
Venus	0.72	0.7
Earth	1.00	1.0
Mars	1.52	1.6
Asteroid Ceres	±2.8	2.8
Jupiter	5.20	5.2
Saturn	9.54	10.0
Uranus	19.19	19.6

Never in the history of life has the Titius Bide been explained except in other books I have published and articles the academics wilfully and spitefully ignored and never published for reasons just because they had the authority to reject what they did not like and for the reason they did not like my arrogance towards them and for no other reasons at all.

Never has there been any attempt by Newtonians to delve into what truly applies in the cosmos in favour of the misinformation Newton presented.

I have explained how gravity forms. I have explained the locations of planets according to the Titius Bode law. I have explained that planets move around the sun by forming Π and not Newton's fable of mass attracting mass.

I use the four cosmic principles extensively to show how the solar system came about and the following is an example how I mange to accumulate information to prove my point of view.

Gravity forms by the Titius Bode law using Π. That I prove in seven articles Annalen der physics would not publish and this is how the cosmos formulates the process according to applying the Titius Bode law, the Lagrangian system, the Roche limit and the Coanda effect. Gravity is the concourse of these phenomena. In this chapter I am concentrating only on the Titius Bode Law and the explaining thereof.

You might say the Annalen Der physics has grounds not to publish but in the next few pages I will give a summery, a watered down summery of the information given to Annalen de Physics in the seven articles I mention The articles had much better explaining and much more in depth analyses of information running over a much wider spectrum By the ay, I was never informed by the Annalen Der Physics as to why the articles publication went begging but I think it has something to do with the response I wrote in accordance to the e-mail I received. The response you can read in a paper published as **www.singularityrelevancy.com THE WEBSITE** http://www.lulu.com/content/e-book/wwwsingularityrelevancycom-the-website/8074920]

There is this very deliberate pattern by which the cosmos forms and never in the past did science admit to this fact, true as it is. However this is also the way gravity forms and for that reason I keep on showing how ridiculous Professor Friedrich W. Hehl, Inst. Theor. Physics which represents the attitude of the entire field of physics. They want to play mathematical games where nothing is proved and nothing is disproved. They don't explain anything because they only understand the mathematical formulations they put forward and that has no inclination on reality whatsoever. This is how the solar system forms.

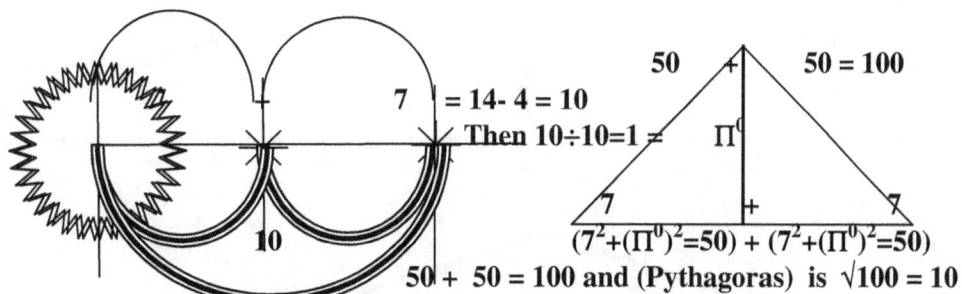

I have run through this argument about gravity and the Titius Bode law before. Let's run through it again but we now keep our focus on what happens with gravity in outer space and not with gravity applying to one single spinning sphere.

In the Titius Bode line it goes the other way around where the space becomes on because the two planets split the value of 7 whereas in the normal form of gravity the spinning object units the 7 factor and split the value of space as 10 into two parts.

0 3 6 12 24 48 96 192 384

The ratio the Titius Bode law forms is depicted from the way that the value of Π builds up space as time progresses. That which is thought of as space, which is between say the earth and the moon or between Venus and mercury is time but it is time that was left behind to become the history of time. We think of it as space but space is the result of time that moved down the line and formed the history of time written in space.

A More Complex Commercial Science Book   514   Chapter 9   The Titius Bode Law

The resulting sequence is very close to the distribution of mean distances of the planets from the Sun:

Body	Actual distance	Actual mass
Mercury	0.39	0.4
Venus	0.72	0.7
Earth	1.00	1.0
Mars	1.52	1.6
Asteroid Ceres	±2.8	2.8
Jupiter	5.20	5.2
Saturn	9.54	10.0
Uranus	19.19	19.6

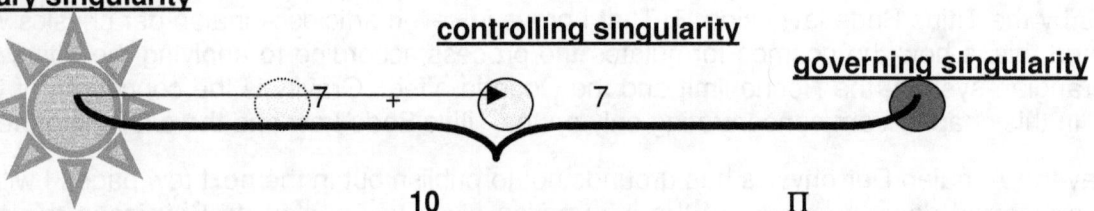

primary singularity   **controlling singularity**   **governing singularity**

Sun   Mars   10   Asteroid Ceres   Π   Jupiter

As I explained the relevancies applying the triangle forming gravity places two values in between the immediate inner planet and the planet arranging the outer value of Π at the time.

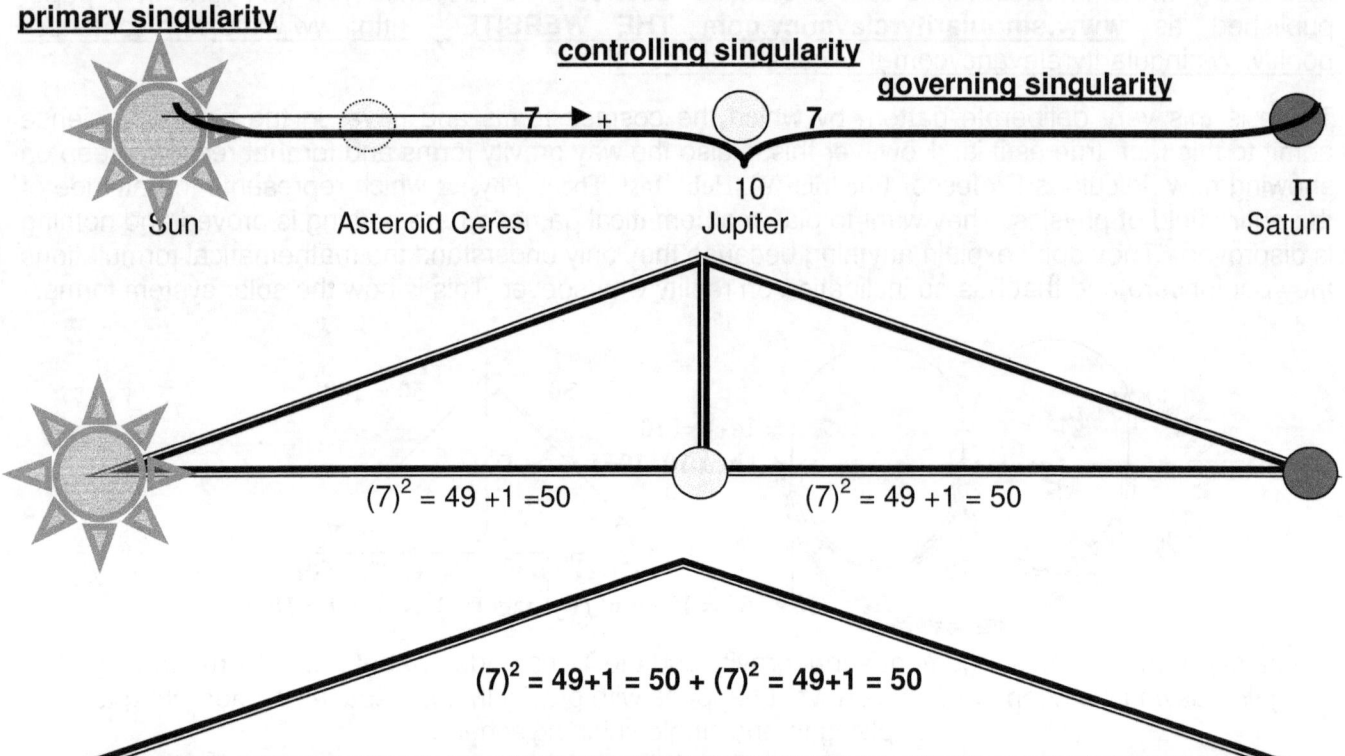

primary singularity   **controlling singularity**   **governing singularity**

Sun   Asteroid Ceres   10   Jupiter   Π   Saturn

$(7)^2 = 49 + 1 = 50$ $(7)^2 = 49 + 1 = 50$

$(7)^2 = 49+1 = 50 + (7)^2 = 49+1 = 50$

The two point connecting material combine as a sum total of 50 making the square thereof 10
$50 + 50 = 100^{\frac{1}{2}} = 10$

primary singularity   **controlling singularity**   **governing singularity**

Sun is Π⁰   **No value in the singularity line up**   **First inner planet**   Π

Sun   Asteroid Ceres   Jupiter   Saturn   Uranus

Every time that the point that holds Π forms it relates to the sun holding Π^0 and the immediate planet from the sun would be 3 where the inner planet then inner planet. The space between the planets in orbit forms the value of 10 in response to having 50 between the sun and the last inner planet and 50 between the last inner planet and the planet making singularity

This formation of gravity leads directly to the translation of the Titius Bode law where the law forms gravity through singularity by extending singularity to form $\Pi^0\Pi$. As I sent this explanation in much better detail in an article to Annalen Der Physics, I never received an answer from them explaining why the magazine concluded it was not worth printing or on what ground was the information not scientific.

To find the mean distances of the planets, beginning with the following simple sequence of numbers:
0 3 6 12 24 48 96 192 384

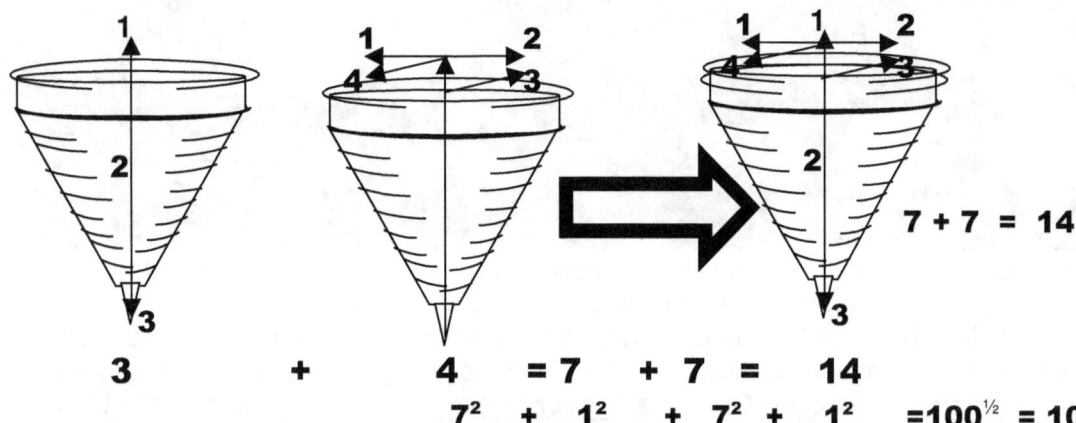

$$3 \quad + \quad 4 \quad = 7 \quad + \quad 7 \quad = \quad 14$$
$$7^2 \quad + \quad 1^2 \quad + \quad 7^2 \quad + \quad 1^2 \quad = 100^{1/2} = 10$$

With the exception of the first two, the others are simple twice the value of the preceding number.
Add 4 to each number:
4 7 10 16 28 52 100 196 388
Then divide by 10:
0.4 0.7 1.0 1.6 2.8 5.2 10.0 19.6 38.8

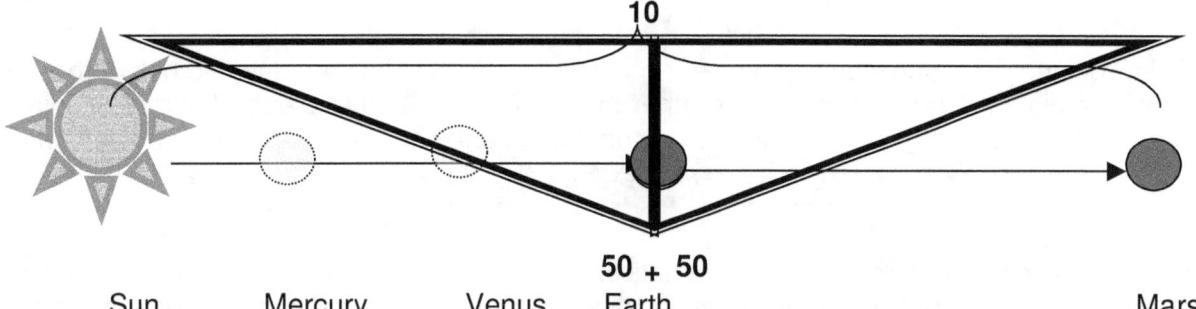

Sun　　　Mercury　　Venus　　Earth　　　　　　　　　　Mars

As space connects by singularity in value of material holding a double 7 square to the value of space becoming the root of the sum squared of the 7 the figuration therefore is explained.

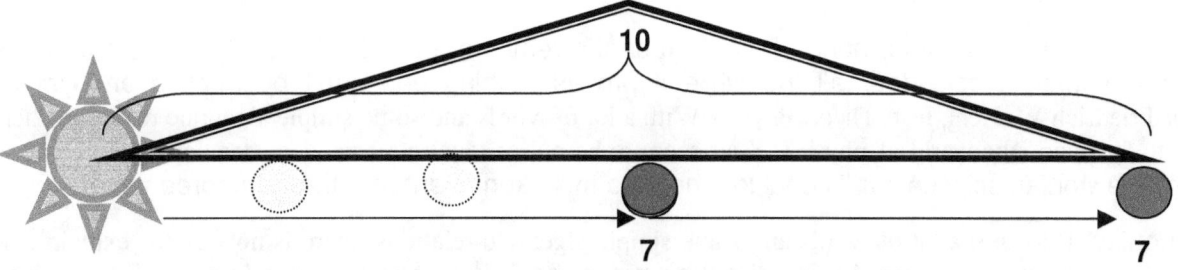

Sun　　　Mercury　　Venus　　Earth　　　　　　　　　　Mars

This confirms that gravity has nothing to do with mass and everything to do with singularity forming $\Pi^0\Pi$.

It is about singularity forming a value of Π in relation to gravity and forming the value of Π.
It is always the double square of 7 as it forms a circle that then forms 10 in space, which allows growth within the space we call the Universe or outer space. It is always Π forming space.

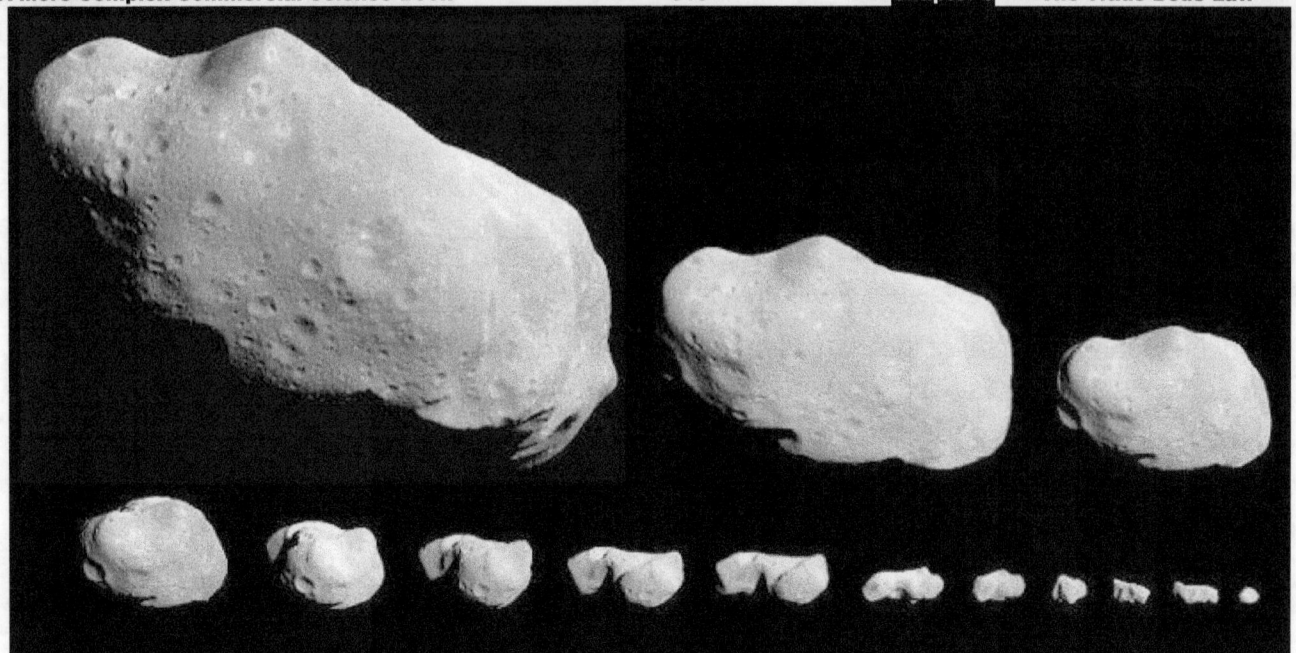

Even individual asteroid rocks still hold the connecting value of $\Pi^0\Pi$ in relation to the sun.

Those lose rocks circling the sun remained connected and never relinquished the singularity connection. To show how this frequency works and to show how the relevancy code could point to other factors proving information obtained by other factors coming to light I will introduce the following:
In the book An open letter About Gravity's Prescription wherein I explain how the Universe came about at the Solar system, as we know it took place. It explains why there are four solid planets, four gas planets and one cold structure. It also explains mathematically why all the debris is encircling the planets and where they come from. This is one part of another book entitled the Seven Days Of Creation. The idea that the solar system formed from a cloud of dust that then by "the magic of mass forming gravity" collected the dust to solidify it into solid particles that became rock solid although the small rock in some occasions were no more than pebbles. If there is any person out there with a higher persuading power than I have or any person that is more influential than what I can manage then get any or all Newtonians jus to show once how $\left(\dfrac{P}{2\pi}\right)^2 = \dfrac{a^3}{G(M+m)},$ fits into the asteroid belt. If there is somebody with a higher degree of convincing other to comply then let that person have Newtonians comply in proving how to apply the Newton formula on movement where mass presents movement in the following equation [illegible equation].
Let them use this inspiring formula of mathematical genius to show how the small rocks called asteroids formed as solid as they have at the location they are in by applying Newton's wealth of equations. Show by mass how they became as dense as they did. I have not yet been able to convince one Newtonian to comply, but then maybe I am not with enough persuasion to inspire a Newtonian to prove Newton was correct.

But then again and on the other hand I have to face the reminder that Newton's formula is a treasure trove of mathematical splendour all the while it proves nothing and can't be applied and yet as Professor Friedrich W. Hehl, Inst. Theor. Physics With a lot of words and some simple algebraic relations, there is no way to "explain" the world of physics. Your seem to be out of touch with modern developments. I guess that is why Newtonian science can't conclude they are mistaken even after three-hundred years.

Yes I am guilty, I do use a lot of words and some simple algebraic relations, there is no way to "explain" the world of physics. Notwithstanding the fact that I do prove the Titius Bode law with Annalen Der Physics not publishing my articles I presume I still seem to be out of touch with modern developments when I explain something that was present in the solar system only as resent as when the solar system came onto form. There are three points within the axis cross referencing with four points forming the rotating circle and in singularity that holds seven points in constant cross alliance. These seven points move to seven points allocated in a new position and therefore we have seven points going seven points forming a square by seven points.

A More Complex Commercial Science Book   517   **Chapter 9**   **The Titius Bode Law**

The sun holds the first inside planet at a certain position

Sun   Mercury 57.9×10^6 km

Then Venus doubles the distance Mercury has from the sun

Sun   Mercury   Venus 108.2×10^6 km

primary singularity   **controlling singularity**   **governing singularity**

$\Pi \div \frac{1}{3}$ $\Pi \div \frac{2}{3}$ Π
57.9×10^6 km 108.2×10^6 km 149.6×10^6 km

Sun   Mercury   Venus   Earth

3 + 4 = 7 + 7 = 10

primary singularity   **controlling singularity**   **governing singularity**

$\Pi \div \frac{1}{3}$ $\Pi \div \frac{2}{3}$ Π
57.9×10^6 km 108.2×10^6 km 149.6×10^6 km

Sun   Mercury   Venus   Earth

The resulting sequence is very close to the distribution of mean distances of the planets from the Sun:

Body	Actual distance (A.U.)	Bode's Law <A.U.)< td>
Mercury	0.39	0.4
Venus	0.72	0.7
Earth	1.00	1.0
Mars	1.52	1.6
Asteroid Ceres	±2.8	2.8
Jupiter	5.20	5.2
Saturn	9.54	10.0
Uranus	19.19	19.6

The fact that the Asteroid belt also holds relevance in the Titius Bode Law indicates what happened when the solar system formed when the solar system formed by violent interaction between three stars fighting for supremacy.

The singularity places the first orbiting planet in relation to 7 and the second in relation to another 7 because of spinning that applies. The relevancies applying is as explained before

Where the earth fits into the scenario Mercury holds more or less ⅓ from the distance and Venus hold ⅔ where the earth completes the full distance. Mercury is 57.9×10^6 km; Venus 108.2×10^6 km and the earth is 149.6×10^6 km. Remember we on earth holds the earth at the centre and therefore the inner planets assort to our position.

From here on the Titius Bode law comes into its own where the first inner planet or the earth forms the **controlling singularity** the outer planet as Venus then is forms the **governing singularity** and the sun forms the **primary singularity**. The rest of the inner planets have no role to play in accordance with singularity positioning the planets.

The law of Pythagoras states that the sum total of the square of the two sides in a rectangular triangle will always be equal to square of the perpendicular side. How basic in mathematics can we still go?

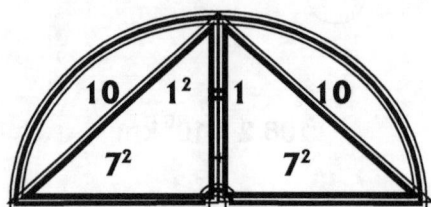

The axis holding $3^2 = 9$ and singularity forms space – time at 10

The square of space from the point of singularity is $2(1^2 + 7^2) = 10$

In gravity applying to spheres such as stars and planets the orbit involves on material structure with one spinning object. The sphere holds a circle of 7° and therefore the 7° applies to both sides of the spin evenly. In that way matter in the sphere has to use the value of Π as a reference. As the sphere spins it refers to one 7 that applies to any point on the sphere as the sphere moves.

As this argument is taken into the space scenario the spinning structure divides the space into where it came from as well as where it is going and that puts space in relation to material on two sides of one divide. There then is 10 on the one side and 10 on the other side divided by a line with two edges.

This is because no object could be on two sides of the Universe at one point in time. This argument is very complicated ands I explain this thoroughly in other books but it takes hundreds of pages to explain. Therefore I wouldn't dare touch on the issue at this point.

Looking at the Titius bode law we find the value of Π in reverse where singularity went out to met Π in the case of the Titius Bode law singularity takes the movement in space and not of space to form Π.

A More Complex Commercial Science Book	519	Chapter 9	The Titius Bode Law

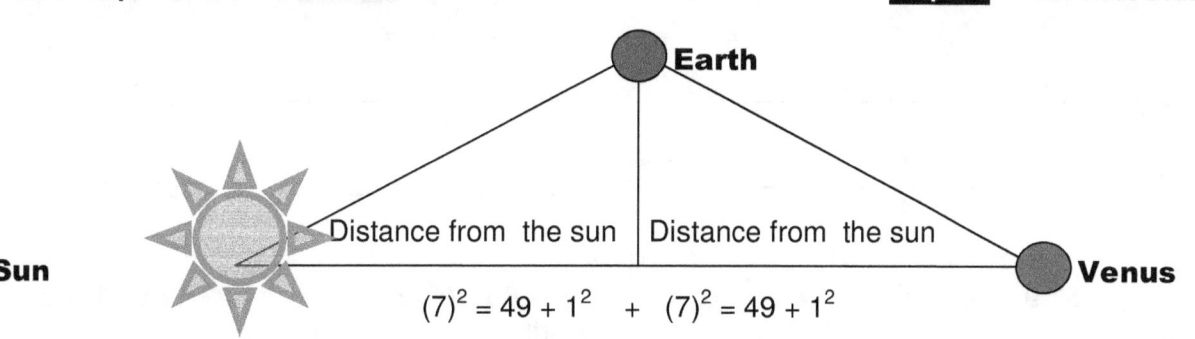

$(7)^2 = 49 \times \Pi^2 = 483{,}61$. But at this stage matter was in the relevancy of 50 + 50 in the law of Pythagoras and by validating the triangle that forms in relation to the half circle on the one side of the sun the space in use resembles a factor value of 10.

$\sqrt{100} = 10$

From this one must conclude that gravity is also 21,991 when the moving body aligns with the 7 forming the contraction.

As 21,991 is one half of ($\Pi/2$). Π Therefore, must be that value matter holds, which is 7. When matter divides into space (7 ÷ 21,999) the result from that is Π. Through this the neutron (which is matter) holds $\Pi^2\Pi$ as a factor. Through that, the value of the triangle in matter, space and time also holds a 180°.

The earth holds Π in relation to the sun being Π^0. From our point in singularity we hold the sun in terms of Π because the sun is to us Π^0. This splits the three inner planets in accordance with us as $\Pi \div \tfrac{1}{3}$, $\Pi \div \tfrac{2}{3}$, and lastly $\Pi \div \tfrac{2}{3}$. Going from our position outwards we find the normal translation of Π finding recognition but the terms of gravity applying in outer space according to the Titius Bode law is much different.

Only the outer planet and the planet alongside the outer planet serves a purpose in singularity where the one forms 7 and the outer planet aligns by another 7 leaving the rest not accounted for in singularity at that point. The outer planet holds the value of Π where Π is made up of $(7^2 + 1^2) + (7^2 + 1^2) = 10$

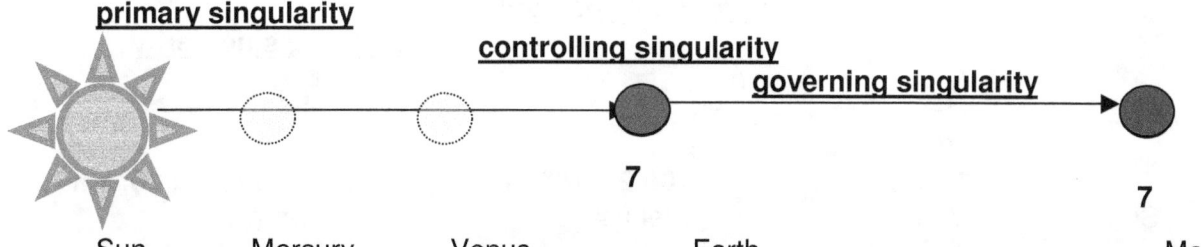

In this order there is no mention of mass playing a part. It is very convenient that Newtonian science ignores the relation or the fact that it is the Titius Bode law that is in place and that the Universe does not recognise Newton's claims on mass in any way.

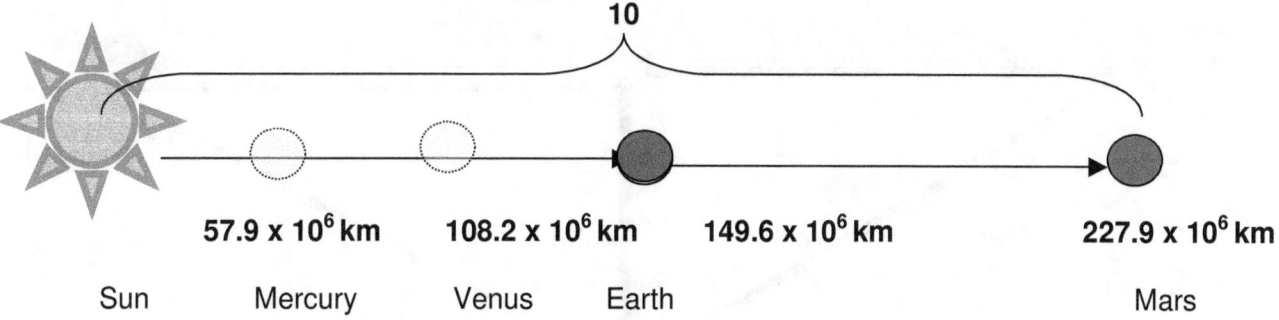

Using the factor of Π is more symbolic that holding an actual value and please read that into the use of it that trying to calculate a meaning it should represent.

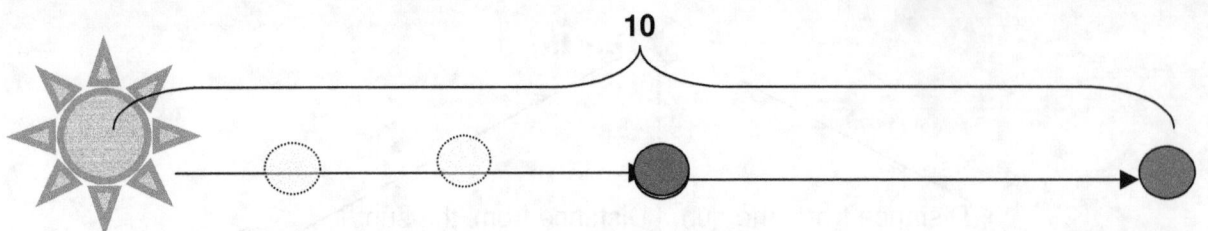

Asteroid	Diameter (kg)	Mass (10^{15} kg)	Rotation Period (hours)	Distance from Sun (A.U.)	Orbital Period (years)
1 Ceres	960 x 932	870,000	9.075	2.767	4.60
2 Pallas	570 x 525 x 482	318,000	7.811	2.774	4.61
3 Juno	240	20,000	7.210	2.669	4.36
4 Vesta	530	300,000	5.342	2.362	3.63
45 Eugenia	226	6,100	5.699	2.721	4.49
140 Siwa	103	1,500	18.5	2.734	4.51
243 Ida	58 x 23	100	4.633	2.861	4.84
433 Eros	33 x 13 x 13	6.69	5.270	1.458	1.76
951 Gaspra	19 x 12 x 11	10	7.042	2.209	3.29
1862 Apollo	1.6	0.002	3.063	1.471	1.81
2060 Chiron	180	4,000	5.9	13.633	50.7

This shows that the first four planets possibly held the same relevancy in singularity and came from a mutual structure that dissolved in some past collision or alteration of sorts. I do realise that this si far not enough evidence to form a conclusion or even form an informed opinion but then you have not read nearly enough to have an opinion of any sorts about my work!

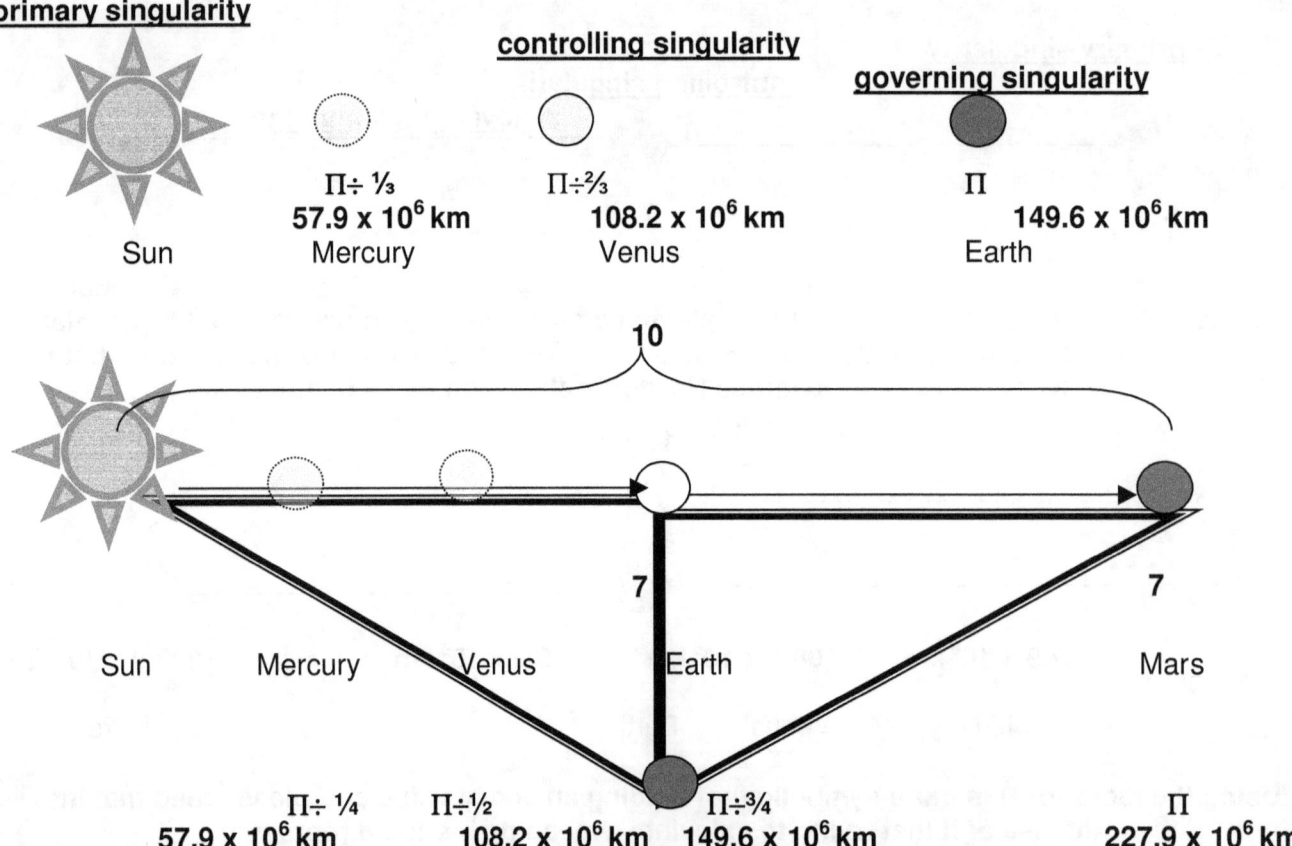

You are all accustomed to have pleasant and wonderful thoughts about Newton and what you are have read was a shock. I can't see many readers enjoying this article, yet it is most critical that some in science will be brave enough to read and inform the others. This article aims to introduce **The Absolute relevancy of Singularity: The Theses,** which is specifically dedicated to show the readers why **a new cosmic concept** is so urgently required and why the Newtonian physics concept about the cosmos is disconnected from reality. Everyone believes more in science than they believe in God and that includes Theologians of all denominations because everyone is committed to the idea that science is uncorrupted, pure and true, working by dealing only with uncompromising truth, facts verified to the smallest detail. If you are one of those believing in the incorruptible truthfulness of science, and everyone is one of those, read this article and prove that I am just an alarmist blowing hot air.

If you think I am just slinging mud and accusing unfoundedly because I am another sick-minded conspiracy theorist that craves for attention and trying to score notoriety while corrupting others, then read what I have to say and form a better informed opinion. If you then still think I am shouting doomsday by trying to cry wolf, I charge you to show where I am wrong about claims I make and accusations I bring against science with the evidence I present. I use this method because I offer a viable working concept but no one reads my work, so no one sees a problem because everyone believes Newton that personifies incorruptibility.

No one bothers to read my work because no one sees a need to change and with no trouble perceived by science, and so all my hundreds of letters to various academics at multitude of institutions while telling those that control science principles about my work interests everyone little. No one in the world see the problem I see and therefore there is no need for the solution I offer. Read why I accuse the entire paternity of Mainstream Physics of blatant misguidance and corruption by hiding Newton's insurmountable incompetence in cosmic physics.

See why the entire paternity of Mainstream Physics this far refused to read my work and why I could not get any publisher willing to publish my views for ten years, because they don't care to read about what a man says that steps on their toes. See if the grounds I use to accuse are valid and see why Mainstream Science recommends not publishing anything I write. In short I will now explain what I explain how there is the one only valid way to see Newtonian's formula $F = G \frac{M_1 M_2}{r^2}$.

This is what represents all the calculation that is used to explain all the physics that man knows about the cosmos. This represents the truth that represents the basis that represents all forms of physics that the entire Universe is based on.

The formula defining gravity reads $F = G \frac{M_1 M_2}{r^2}$ and is the formula used by science to explain and define physics applying throughout the Universe. It says that the (**$M_1 \times M_2$**) mass of one object forms a force called gravity that pulls the mass of another object also forming gravity and this process stands in relation with a gravitational constant (**G**) (a supposed force keeping the Universe attached) and the pulling and subsequently destroys the radius by the square (**r^2**) being between the objects. That says that objects **always moves closer by the force of gravity in relation to mass.**

Newton submitted the suggestion that objects fall as **mass** provides the movement that will cause the falling by the inducing of a force he named gravity which he subsequently only proposed was the acting suppositious force. I charge anyone to show where this discipline of mass pulling mass applies anywhere in the entire Universe.

I disprove this formula in so many ways books can't cover everything and I show that this formula and the ideas Newton introduced just don't stand up to even the smallest tests. Then, if Newton's idea on gravity has validity and mass is responsible for objects falling, in that case all objects that are in a process of falling must be subject to mass and in that idea rests differentiation and discrimination on size and compactness producing speed variations while objects descend to earth. If any and all falling is subject to the variation mass introduces and the influences coming about is the result of mass interfering in the gravity force being generated, this then must bring different speeds to cause substantial variation in the falling of different objects holding different mass factors. There can't be conformity in the falling of all objects while such falling is the result of the discrepancy that mass has to inflict due to variations that result in mass differentiations. In other words, if things do fall by mass and mass has to differentiate, all the claims Galileo then made is unsubstantiated. Either Newton is correct and everything that falls by mass also fall differently because mass is different or Galileo is correct saying all things fall equal, but then mass cannot play a role in the falling.

This is a vital issue that science eludes and science has all clever ways to avoid addressing this question direct. This is the part where science runs around, never addressing the issue and avoids confronting arguments about the issue. This avoidance of confronting the issue, which will disprove the validity of Newton, is done with much cunning. To go around the issue they use a feather and a hammer in a vacuum to explain Newton and Galileo using the same principle at the same time without stirring suspicion. The fact that objects fall due to conformity in the falling, science accepts while always portraying a picture that mass brings falling without claiming distinction when objects fall and therefore things that fall by mass forming gravity also falls equally.

When confronted science claims this argument doesn't apply. The formula used indicates mass forms the pulling while at the same time they admit that Galileo's presentation that falling of all objects are equal in tempo, irrespective of size or any form of differentiation. Then they promote the obscurity that Newton and Galileo is in harmony. The truth about the matter is that the two can never have the same result. This prompted me to look at the formula science uses to form the foundation of all that is physics.

Whenever I bring the incomparableness of $F = G \frac{M_1 M_2}{r^2}$ into the open, I am told I am the one that does not understand "**classical physics or mathematics**' or whatever name it goes by. I am the one that is so stupid I can't understand Newton. They pretend Newton is above my intellect and surpasses my comprehension ability as Newton is above my intellectual ability. The problem is reflected back to me without ever addressing the issue. For almost forty years now, since I was a student, I was forever told I was the problem and I don't understand but no one could tell me what I didn't understand. I am going to explain what I don't understand.

Newton at first had the idea of $F = \frac{r^2}{M_1 M_2}$ when he thought mass was responsible for the pulling that was responsible for the apple falling from the tree. He saw the mass of the apple pulled the earth closer while the mass of the earth pulled the apple closer and this pulling devastated the radius. But then under closer inspection the conclusion proved very unreliable because the results were clearly not applying truthfully. If any one cares to put numbers into the factors it will be clear why it was not suitable. His idea was that the apple had mass while it was on the tree and never lost the mass when it fell and because it never lost the mass, it fell by the mass. When the apple hits the ground the radius between the apple and earth is gone. For someone not understanding what Newton is about I am doing extremely well this far.

This prompted Newton to formulate $F = \frac{r^2}{M_1 M_2}$ and the intellectual world was reborn. Then Newton changed his initial formula that was $F = \frac{r^2}{M_1 M_2}$ to then become $F \alpha \frac{M_1 M_2}{r_2}$ in order to establish a more coherent result about his idea. By changing the formula from $F = \frac{r^2}{M_1 M_2}$ to $F \alpha \frac{M_1 M_2}{r_2}$ he placed $F = \frac{r^2}{M_1 M_2}$ in context to $\left\{ \frac{F}{1} = \frac{m_1 m_2}{r^2} \right\}$ and by changing the formula by only changing one symbol α the entire outcome of the formula changed without changing anything. Newton saw it fit to replace = with α and the formula was reborn in value while having the idea staying the very same. The evidence about this statement is confirmed by the formula then later only adopting a G as seen when the final adapting was accepted as $F = G \frac{M_1 M_2}{r^2}$. There is an applying rule or law in mathematics that says when a formula changes from $F = \frac{r^2}{M_1 M_2}$ to $\left\{ \frac{1}{F} = \frac{m_1 m_2}{r^2} \right\}$ then F being F ÷ 1 must also remove a position to become 1 ÷ F making F the fraction value. All those that know even the least about mathematics and of which Newton and his followers are supposedly mathematical experts know that if any part on the one side changes dynamics from being on top of the dividing line then the very same must apply on the other side of the equality. One can't just say that to change a formula $F=\frac{r^2}{M_1M_2} = \left\{F\alpha\frac{m_1m_2}{r^2}\right\}$ would not translate in ultimately changing the outcome of the formula because the truth about mathematics is that to change the relevancy applying the change of factors must apply on both sides of the equality equally as well as simultaneously and also even-handedly. It must be $\left\{F=\frac{r^2}{m_1m_2}\right\}=\left\{F\alpha\frac{m_1m_2}{r^2}\right\} \neq \left\{F=\frac{m_1m_2}{r^2}\right\}$ but when it changes one then every factor changes to $\left\{F=\frac{r^2}{m_1m_2}\right\}=\left\{\frac{1}{F}=\frac{m_1m_2}{r^2}\right\}$. Newton had this idea that because he was Newton The Great (and if one looks at his life he came across as the biggest egocentric maniac of his day) and with such a bloated ego he could have held such a view that normal rules did not apply when he concluded gravity and with him being Newton he thought he could put even mathematic laws below his important status. He could replace symbols = with α to form $F=\frac{r^2}{M_1M_2} = \left\{F \; \alpha \; \frac{m_1m_2}{r^2}\right\} = \left\{\frac{F}{1}=\frac{m_1m_2}{r^2}\right\}$ and doing this, that will change mathematical physics forever. It never dawned on him or his followers whom are observed as those forming the genius in mathematical minds as those are that came after him that $\left\{F=\frac{r^2}{m_1m_2}\right\}=\left\{F\alpha\frac{m_1m_2}{r^2}\right\}$ is very correct and acceptable but going that one step further as he did constitutes to presenting an act equal to mathematical fraud because one can't change an equation expressed as $F=\frac{r^2}{M_1M_2}$ to then become $F=\frac{M_1M_2}{r^2}$ simply because $\left\{F=\frac{r^2}{m_1m_2}\right\}=\left\{F\alpha\frac{m_1m_2}{r^2}\right\} \neq \left\{F=\frac{m_1m_2}{r^2}\right\}$ because the correct application is in fact $\left\{F=\frac{r^2}{m_1m_2}\right\}=\left\{\frac{1}{F}=\frac{m_1m_2}{r^2}\right\}$. But then he went much further and compromised the formula without compensating for replacing the symbols = with α to then be $\left\{\frac{F}{1}=\frac{m_1m_2}{r^2}\right\}$. The proof of his intention was switching $F=\frac{r^2}{M_1M_2}$ directly to $F=G\frac{M_1M_2}{r^2}$ and that declaration came as $F=\frac{r^2}{M_1M_2}=F=G\frac{M_1M_2}{r^2}$. There was never one Newtonian that even hinted how Newton could explain how he changed the initial thought of $F=\frac{r^2}{M_1M_2}$ then to mathematically become $\left\{F\alpha\frac{m_1m_2}{r^2}\right\}$ which was actually obviously intended to become $\left\{\frac{F}{1}=\frac{m_1m_2}{r^2}\right\}$ and then with normal mathematical principles still applying change this lot to $F=G\frac{M_1M_2}{r^2}$.

Furthermore, how could academics in mathematical physics teach children or students in physics this as the truth and this goes on for three centuries or so! How could any mathematician explain a process of following mathematical logic maintain that $F=\frac{r^2}{M_1M_2}=F=G\frac{M_1M_2}{r^2}$ ...explaining it is preposterous and makes a mockery of mathematical principles applying. When I try to keep my conscience in tact by not using mass or mathematics I am frowned on as the idiot not knowing what physics constitute and my intellect is in doubt. Why use something that makes a mockery of physics to start with because it will end in more deception as it develops further.

If Newton was correct then the formula $F=G\frac{M_1M_2}{r^2}$ puts a ratio in place. The longer the radius is by the square, the less would the influence of the mass be by forcing gravity to pull. Again on the other hand the shorter the radius is by the square, the greater would the gravitational force be because it is the radius that determines the value applying and not the measure of mass. With $F = G\frac{M_1M_2}{r^2}$ in place then if my foot were on the earth, the gravity force would be so strong I would never be able to be a bloody blob because the atoms would not stand such a force applying. The mass of the earth multiplied by my mass divided by a millionth of a micrometer would boost the force so much I will become a Black Hole.

Let any academic mathematically show how one would go about to use Newton's visionary formula $F = G\frac{M_1M_2}{r^2}$ to calculate the force of gravity by replacing the symbols with the actual values in mass that the symbols should have. Put in the Earth's mass in place where it belongs and put in your mass in place where it should be and then divide that with the distance between your soles and the Earth measured in fractions if micro millimetres by the square thereof! Do it and anyone will see there is more gravity applying between any one standing in the Earth than all the gravity floating throughout the Milky Way. A human body will never be able to withstand such a force and so if it can't be done, then that is proof of Newton committing fraud when he introduced the formula $F = G\frac{M_1M_2}{r^2}$ being able to calculate the force applying as gravity. The cell structure will collapse withstanding even a fraction of that... that is if Newton made sense with his mass pulling mass idea.

Take any formula used in daily physics and show where they use the mass of the Earth as a factor in calculating anything in applying physics. Never, not once, does any formula used by physics hint that the Earth's mass has any influence on any part of physics when any one calculates factors to determine whatever they wish to determine. If the Earth's mass is never used in any calculation, then the Earth's mass has no part presented as a factor and then the Earth has no mass that influences any aspect of physics. That means the Earth's mass doesn't produce gravity because if it did, the calculating formulae used in physics must use the Earth mass as a factor in all calculations! Newton cheated to bring in the Earth as a factor that has mass that produces gravity and never does the mass of the Earth contribute to any part in any of the many calculations that form part of physics. The Earth has no mass because the Earth's mass never plays a part in any formula used by physics. It is as simple as that! The formula $F = \frac{r^2}{M_1M_2}$ that Newton first devised has not even a ring of truth to it and that is why the change was made from $F = \frac{r^2}{M_1M_2}$ to $F \alpha \frac{M_1M_2}{r^2}$, which was only pushing the envelope of truth. I

f it is true then show how the formula reading $F = G\frac{M_1M_2}{r^2}$ is used to indicate that this brings about gravity or without cheating then $F = \frac{r^2}{M_1M_2}$ could become $F \alpha \frac{M_1M_2}{r_2}$ and if there was the least bit of honesty, the formula would remain as $F \propto G\frac{M_1M_2}{r^2}$ but that is not the case. Even presenting it as $F \propto G\frac{M_1M_2}{r^2}$ would not produce any working viability. There is clearly blatant mathematical deception intended by altering the formula from $F = \frac{r^2}{M_1M_2}$ to $F \alpha \frac{M_1M_2}{r_2}$ and then committing further blatant fraud in changing the formula to $F = G\frac{M_1M_2}{r^2}$ while even in this form it still doesn't apply, as I am about to show.

I challenge any physicist to show where in physics does any formula require the mass of the earth to value any reading in calculation. This shows mass is a one-way value presenting a reading of the object and not mass going both ways as a force. This is very significant to realise when concluding what the role of mass truly is.

In the form $F = G\frac{M_1M_2}{r^2}$ the mathematical shortfall is horrendous. What would the purpose be of using a formula that is unproven and was cheated to come into use? Surely no benefit can come from a formula that obviously was rigged. The formula is not in place to show an idea of what the thinking process is, but it must have a very specific purpose and function.

In the light of all this evidence I am the one being ridiculed for not using the Newtonian idea of formulated mass and using mass as a force between the earth and a body. Every time I step out to indicate my reservations about the matter, I am the one being in question. If you think I am going on about academics, then think how much did they ignore me in ten years. I present a new cosmic concept solution and in ten years not one academic even read the first page of paper I send and why, because I address Newton's shortfalls, which is what they don't want to read. With the clear evidence I show those in charge of physics still dismiss my work as coming from the one that is mindless because I am unable to "understand" Newton. What is there to understand when everything I am supposed to understand is tainted and is flawed! The concept of gravitational forces applying in the cosmos is Middle Age superstition. If it seems I am going into rhetoric about academics then it is because I wish to describe their methods in dismissing me. They will not allow me to test Newton because Newton was never tested to begin with. I have to prove something incorrect that has never been proven correct.

The point I wish to make is that science says gravity is $F = G\frac{M_1 M_2}{r^2}$ while science also says gravity is a value of **F=g=9.81** and further more science says that **F=mv²** while science first said when Newton introduced his formula it developed as $F = \frac{r^2}{M_1 M_2} = F\alpha\frac{M_1 M_2}{r_2} = F = G\frac{M_1 M_2}{r^2}$ Now get this lot married mathematically…going from was $F = \frac{r^2}{M_1 M_2}$ to $F\alpha\frac{M_1 M_2}{r_2}$ then onto $F = G\frac{M_1 M_2}{r^2}$ and then becoming **F=mv²**. that is a challenge science can never manage and yet science says it is true because Newton confirmed the this statement as the truth and all that is needed as conformation is and only using Newton's say so on the issue. In later articles (should it come to print) I indicate why **F=mv²** is the true basic formula.

Let all the physicists show how they manage mathematically to get $F = G\frac{M_1 M_2}{r^2}$ equal to the measured value that science says gravity has being the "g" value and not the "F" value at **g=9.81 Nm/s²**. They advocate that gravity is another symbol that somehow replaces **F** with **g** but also is gravity having a totally new value than what Newton had in mind and then as "g" being apart from "F" has a value showing gravity at **g=9.81Nm/s²**. So let them do the calculating of the Earth mass and any person's mass multiplied by the gravitational constant and get this lot divided by the distance between my feet and the Earth when I stand on the ground by the square thereof and to top this and see what force it takes to move me from a point to another point on earth using $F = G\frac{M_1 M_2}{r^2}$.

With all this confusion they then get the value of gravity from $F = G\frac{M_1 M_2}{r^2}$ to become a predetermined constant value of **g=9.81 Nm/s²** without producing one iota of proof. **I'd love to see them accomplish that!** When they use another formula that also uses the symbol **F** as in **F=mv²** there is no indication of the mass of the earth playing a role or the radius by the square effecting the result. I still have to find one academic that can show me whereto did the mass of the Earth and the radius disappears. Also there is no mention of using the gravitational constant as well as the diameter parting the mass **m** from the other factors that disappeared. This is one of the many small issues they never think of because they can't explain it while upholding the correctness of Newton at the same time. Let one of them with the many doctoral degrees, show how they come from the one formula $F = G\frac{M_1 M_2}{r^2}$ to the other formula **F=mv²** that is totally incompatible with having **F** equal ($F = G\frac{M_1 M_2}{r^2}$ =**F=mv²**) or to any of the other previously introduced formulas. Show how **F=mv²** incarnates as $F = G\frac{M_1 M_2}{r^2} = F\alpha\frac{M_1 M_2}{r_2} = F = \frac{r^2}{M_1 M_2}$.

Show how (**F=F**) proves to be $F = G\frac{M_1 M_2}{r^2}$ and then on top of this to eventually reappear on the surface as the formula **F=mv²**. If you thought gravity signs it was an act of magic, then try this for magic. Where did all the factors (M_1, G and r^2) go while being on route to change in appearance to become **F=mv²**. The formula **F=mv²** is an exact representation of the formula that no entity less than the cosmos gave Kepler as **a³=T²k**.

In my researched of Kepler's research I discovered that science doesn't recognise Kepler's work for what it is. Kepler would not have been mentioned if it were not for the blemished misconception Newton produced on behalf of Kepler while it is Kepler's formula that forms the basis of all physics. Everyone thinks that Kepler only found planets rotating but had no idea what he saw. Then the "brilliant" Newton was able to explain what Kepler could not even foresee, which makes everyone more concerned about how Newton saw Kepler's work than what Kepler's work involved. The formula used in physics as a principle is **F=mV²**, which should be **F³=mV²**. **F³=mV²** is replicating Kepler's formula in detail as **a³=T²k**. By using Kepler's formula we have **F³=mV²** that is a precise repeat of **a³=T²k**. The duplication is so obvious that we have (F³ becoming **a³**) while (m is **k**) and (V² is **T²**).

Einstein also only duplicated Kepler's formula by putting **E=mC²**, which also should read **E³=mC²**. Again that is precisely Kepler's formula **a³=T²k**. (E³ is **a³**), (m is **k**) and (C² is **T²**). In **E³=mC²** Einstein mimicked **a³=T²k,** which is Kepler's formula. (E³ is F³ is **a³**), (m is **k**) and (C² is V² is **T²**). So what is so brilliant about Einstein's formula if Kepler had it centuries before? E³=mC² is F³=mV² which is **a³=T²k** and only the factor symbols change but the conscription remains.

Newton corrupted the formula when he added $4\Pi^2$ to the formula and removed **k** that Kepler introduced while **a³=T²k** Newton ignored. Newton changed **a³=T²k** by using the symbols G (m + m_p) to replace **k** and then declared $a^3 = T^2$. I still wish to see the proof confirming Newton's changes as being correct notwithstanding that everyone thinks physics is entirely based on this conception. Whether the formula used is **F³=mV²** or is **E³=mC²**, it still remains duplicating what Kepler introduced as **a³=T²k**. So I changed it back to Kepler's version of **a³=kT²** as to better the understanding of the foundation of astrophysics and mainstream physics. The entirety of physics is not based on Newton.

It uses Kepler's findings to a precise duplication while science does not even recognise Kepler. Giving Kepler the credit due, the entire Universe becomes completely understandable…but then for my audacity to show mistakes in physics I am ignored flat! All I ever ask is prove the truthfulness of $F = G\frac{M_1 M_2}{r^2}$ because it is F³=mV² that forms the basis of physics and that accuracy comes from Kepler's view of **a³=T²k** that became Einstein's E³=mC². It was the Newtonians that got rid of Newton's $F = G\frac{M_1 M_2}{r^2}$ and accepted Kepler's **a³=T²k** by renaming the factors as F³=mV². In the same sense I also renamed the factors **a³=T²k** as gravity forming $\Pi^3=\Pi\Pi^2$.

The mass of the Earth that academics in physics claim is there and that supposedly is doing the gravity pulling, is a relevance that the object has when the Earth has a centre axis with a factor of 1 and this relation is effectively viable only when the object having this mass is resting on the surface of the Earth or having some direct contact through another medium connecting the object to the Earth. To form mass the object rests on the link or by a link or otherwise rests directly on the Earth, but the condition for having mass is that the object stands still or move with the earth while being in direct contact with the Earth. But all action the object has is relevant to the position the object has in relation to an allocated relevance with a position according to moving in terms of the earth. That is why in terms of cosmic physics all other physics formulae that indicate something is motionless are invalid. Everything in the Universe is spinning while moving in terms of all other things spinning and moving.

Subatomic particles spin while moving while the atom spins while moving while the object the atoms forms spins while moving, even if it only spins when being a part of the earth. All things move in relation to all other things moving and nothing ever stands still. That is absolute cosmic law. That is physics and that is **a³=T²k** or F³=mV² or is **E³=mC²**. Even when standing still, the object has to align with the Earth and relate to the movement that the Earth has.

The object in mass has to move directly with the Earth or slightly more than the Earth. This is the concept I use to explain the sound barrier mathematically. The object only shows mass when connected to the earth and when accepting the movement the Earth has. If the mass the Earth is a physics reality this then should have a place in the formulated calculation alongside the mass the object has, as a complimenting factor is totally absent in normally used physics because the Earth has no mass. The Earth's mass is lacking all visible presence in influencing physics by lending support or increase any calculation in physics. This proves my statement that the Earth and the rest of cosmic objects don't have mass and therefore can't be used as a calculating factor. The Earth supplies movement to render mass.

Planets have no mass and neither has the Sun got mass except the mass Newtonians wish to credit planets with, but in cosmology that unfortunately has no influence. Bigger planets don't move faster because they have more mass and smaller planets do and neither are they further from the Sun because they have lesser mass. All planets big and small spin at the same speed around the Sun and in relation to the Sun and all planets are scattered going around the Sun while being big and small where all sizes are well mixed. This is because planets have no mass except in the imagination of Newton and his devoted followers. The mass of the Earth never plays a role in physics and the mass of planets do not draw any of the planets closer to the Sun and let one physics professor bring proof that the planets do draw nearer to the Sun!

They just can't because planets do not have mass that can produce a pulling by gravity! If and when the mass of the Earth does not feature as a factor in any formula that is used in physics, then the mass of the Earth is no factor playing part in gravity. It is the movement of the Earth that forms gravity by having **g=9.81 Nm/s²**. This then can only indicate that the Earth has no mass but has movement. If there is an absence of mass as a factor that influences physics, this can only be as the result that the Earth mass has no gravitational presence in any physics formula. Gravity does have the value of **g=9.81 Nm/s²** but that I explain when I prove gravity is Π and the value is along the value of Π forming **g=9.81 Nm/s²**. With the evidence being that clear, then the mass that the Earth should supposedly have, does not produce gravity as Newton suggested.

Prove me wrong by getting gravity at **g=9.81 Nm/s²** from using either any of Newton's formulas being $F = G\frac{M_1 M_2}{r^2}$ or $F \alpha \frac{M_1 M_2}{r_2}$ and $F = \frac{r^2}{M_1 M_2}$. Let me see Newtonians do that and I will become a believer in Newton! The Earth has no mass because physics can't show the Earth's mass playing part in calculating formulas and if there is no mass that plays a part that should produce gravity, and then mass can't be responsible for the producing of gravity as Newton declared. That makes Newton's suppositions total rubbish and that makes Newton responsible for a crime of defrauding and falsifying the science of physics. If you, the reader is able to get academics in physics as far as even reading this argument I make, then you are more influential than I could ever be. They plainly dismiss all these arguments with arrogance by discrediting my credentials!

What Newton saw as gravity can't withstand even the slightest test of proof and I showed that it is not possible to use Newton's formula as Newton suggested it applies to mathematically calculate gravity. I come back to this issue later on. I have tested Newton's thinking and the book I offer to you for investigation serves as the testimony to all the testing I did on Newton.

This any body who can see, will see when reading this book, I tested Newton from all the angles to see if he possibly could be correct but found his thinking wanting every time. The truth about Sir Isaac Newton's concepts I came to conclude, was that the reality is that it is not in any way overstated to declare that Newton conspired to defraud science and moreover that he committed blatant mathematical corruption in trying to prove the concept he had about what he thought forms gravity. There is no backing for Newton's ideas and even the ideas which are in use are not in the form that Newton said it applies where physics in daily use serves as the best discredit to Newton bringing no proof about any of the claims that Newton made on matters concerning science in cosmic gravity.

If Newton said Πr^2 but $F = G\frac{M_1 M_2}{r^2}$ **is totally ridiculous**

PLANET	Mean Distance from the Sun (AU)	Equatorial Radius (km)	Mass of planet (Earth=1)	Mean density (grams/centimeter^3)	
Mercury	0.3871	2439	0.06	5.43	
Venus	0.7233	6052	0.82	5.25	
Earth	1.000	6378	1.000	5.52	
Mars	1.524	3397	0.11	3.95	
Jupiter	5.203	71490	317.89	1.33	
Saturn	9.539	60268	95.18	0.69	
Uranus	19.19	25559	14.53	1.29	
Neptune	30.06	25269	17.14	1.64	
Pluto	39.48	1160	0.002	2.03	

I show that every thought Newton introduced that later proved useful and was correct, was what he stole from another far better cosmologist called Johannes Kepler. Not one of his laws are directly relating to any concept Newton ever introduced at any stage but is the result of academic theft he committed against a much larger figure that preceded him by almost a century. But he stole, he lied and he raped the work of a predecessor in order to defraud the world of science in his time.

Newton brought no original input into science except that he gave a concept the name "gravity" and even that is inappropriate. Newton made suggestions that break every mathematical principle he could think of. That, Newton did in his attempt to win over the prevailing academic thinking of the day in his time as to lay some sort of groundwork to form backing for his ideas on physics and to attempt to explain gravity or what he thought gravity is. If this is shocking and sounds outrageous, then a lot more shocking detail awaits the reader in this book.

Newton's claims about the principles he declared as being responsible for guiding physics carry no proof and after I realised that, I was able to start forming another line of thought on gravity. After formulating my concept about how gravity was truly formed, I had to introduce my ideas to academics in physics. In my quest to find the method how gravity formed I used the four phenomena and the principles of these phenomena as well as determining in which way each phenomenon applied. Then I placed each one in the way that were known how they work and then implicated that specific formula's function mathematically in forming gravity in the cosmos. This was no easy task but I did it and by formula shows that my argument is logic and the mathematics prove that it works well.

The phenomena that I use is still to this day unexplained by Mainstream science because it shows no sign of using mass and without mass the Newtonian mind understands nothing! Newtonians don't understand the four phenomena due to the fact that science up to the present date has no means or method to explain the four mentioned phenomena while I can explain the working of each independently and how they work in a combination to produce gravity. I found a way to put those four phenomena in a perspective and put the four in a mathematical sequence that from there I could explain gravity in detail. When I first approached academics, I had the opinion that all academics were knowledgeable about the lack in the correctness we find in Newton's views and that every one in physics would be rejoicing in finding what gravity consists of. I was under the impression that I would be embraced by those in physics for finding a solution to Newton's errors. I was in for a nasty shock with such naivety.

I met with such rejection that no one even cared to look at my work because they were of the opinion that looking at my work would be sacrilegious to Newton. I was told on occasions that Newton has never been proven incorrect and therefore any attempt on my part in doing so is a waste of time. At first I was not confrontational towards Academics in physics and avoided any indication about disagreeing with Newton, but academics always threw Newton at me and eventually for self protection I had to start to confront them and confront Newton, with which I was in disagreement from the beginning although at first I was reluctant to voice any opinion about the matter. But slowly it dawned on me that if I had any serious plans to introduce my ideas I had to dispute Newton's gravity principles and show the inconsistencies and dishonesty in Newton's approach to physics. I came to realise that his flaws are there and the mistakes are present whether I avoid it or attack it; the inconsistencies are part of forming the basis for modern accepted science. It is that strangle hold I had to break before I could even think of finding acceptance about change.

Dear Dr. Schutte,

You submitted an article of 15 pages to the Annalen. The content of this paper doesn't constitute a theory in physics. With a lot of words and some simple algebraic relations, there is no way to "explain" the world of physics. You seem to be out of touch with modern developments. This is also shown by the fact that you don't quote any relevant literature.

I am sorry to say, but the Annalen is not able to publish your work.

I am sorry for having no better news for your.

Best regards,
Friedrich Hehl

Co-Editor Annalen der Physik (Berlin)
--
Friedrich W. Hehl, Inst. Theor. Physics
* University of Cologne, 50923 Koeln _____/\/_____ Germany
fon +49-221-470-4200 or -4306, fax -5159
hehl@thp.uni-koeln.de, http://www.thp.uni-koeln.de/gravitation
* Univ. of Missouri, Dept. Phys. & Astr., Columbia, MO, USA

This was using a lot of words and some simple algebraic relations with which I disprove and discredit the foundation that all physics supposedly rests on. Moreover I used inferior mathematical physics to prove how according to Kepler the cosmos was structured by physical gravity applying movement.

I could conclude that the Universe (a^3) moves (T^2k) to form a centre (k^0) and that centre (k^0) becomes singularity around whichever that spins forms time where everything not forming singularity spins around singularity.

If the spinning or movement of space in space stops everything will decline back into singularity $\Pi^3 / \Pi^2\Pi = \Pi^0$ and because everything spins and moves it is by singularity that the Universe grows in time to leave space as a legacy of the past that was.

Because there is movement there is growth of time forming space. $\Pi^3 / \Pi^2\Pi = \Pi^0$
Because there is movement there is growth of time forming a future $\Pi^3 / \Pi^2\Pi = \Pi^0$
Because there is movement there is growth of time forming a present $\Pi^3 / \Pi^2\Pi = \Pi^0$
Because there is movement there is growth of time forming a past $\Pi^3 / \Pi^2\Pi = \Pi^0$

If only science took more notice of Kepler and less of Newton science would be that much wiser about the reality we call the Universe.

Part 10

Is a Small extraction from a part of the Theses called The Absolute Relevancy of Singularity In terms of The Cosmic Creation, which is the Book as part 8 of the Absolute Relevancy of Singularity in Terms of A Cosmic Creation.

This book allows an open window on a process whereby the cosmos was created before the cosmic birth applied. The Cosmic Birth is now known as the Big Bang and what a stupid name that is.

To understand the following we must understand the difference between time and space. Time moves because time moves space. As a human I look at objects and put time and space in a straight line. That is wrong. Time is a straight line but also time bends space by placing space into forming the history of time.

When I look at my finger it forms a straight line according to my observation and for all technical purposes that is true. When I look at the moon time bend space in which the moon is by one and a half second. Time in relation to space is no longer a straight line because time placed the space one and a half seconds arrears of me. I see the moon and through my observation skills the moon is in a straight line according to my vision.

Observing the sun puts the sun in a slot eight and a half minutes arrears but to my mind it is many millions of years more then that but I am not debating the observing merits but only showing that time and space does not form a straight line. Time bends the space by eight and a half minutes. When we look at the giant planets there is a time interval of thirty minutes or more between relaying messages to space craft and the space craft receiving the message. Time bends space by thirty minutes. Time will bend space by many hours when time places the space of Neptune in relation to our position in time.

Time forms as movement changes space. I.e. the earth circles around the sun and therefore every instant the earth is in a different position and in a different location in relation to the sun and therefore the Milky Way and therefore the entire Universe. In the Universe time flows because the present is different from the future as the past is different from the present. Time is about everything changing.

Every aspect of the Universe is different from what it was to what it is to what it will be the very next instant. That forms time. This is proven by the formula we received from Kepler and Kepler in turn received it from the cosmos which he and Tyco Brahe studied for eighty years and calculated in the well known and misrepresented formula as $a^3 = T^2k$. Time is not space but time is the movement of space. Time changes space. The changing gives the Universe time flowing. The space a^3 moves T^2k and therefore changes the space in a singular Universe a^3 at a ratio of a circle T^2 in a singular Universe that moves straight by k that is a line in a singular Universe. The change we see in the Universe as pictures formed by light is in the time flowing that forms space and thus creating a Universe by changing it every instant.

Before this we now have the future was so perfect that the future was the same as the present in every detail. The past was the same as the present and being perfect the past was the same as the future because it had no difference from being the present. The Universe never changed anything and therefore was in singularity. This even the Newtonians agree on…without realising the concept. *Einstein brought us the concept that everything came from singularity and everything still is in singularity. Forming singularity as one was that what gave one as the future that arrived in the present but also left one as the present that moved to the past, serving time as the future being equal to the present being equal to the past with nothing ever altering from one position to the next position.*

Then at one moment in time the imperfect brought the Universe a point in time where the future was different from the present that took the difference onto the past at a rate of every instant from then on and that is how time formed space. Suddenly time flowed in two directions. Since time will tend to remain the same perfect ness therefore it can only be a Creator that will bring change.
The change came as heat formed that split in the flow of time that then had the Universe flowing into two parts. The one part overheated and expanded ever since and the other part in relevancy cooled down by moving and with that contracted.

There was space that overheated and expanded because of overheating and there was material within the space that by movement froze and shrunk by spinning. This factor science never came to realise and therefore never came to acknowledge because Newtonian science never understood this principle.

The difference between outer space expanding and material shrinking we call gravity. Before this any numerical order was nonexistent. There was no need for anything more than one. Then afterwards the Universe introduced numbers because only then did a need arrive for numbers carrying any measured value at a level of forming a concept being more than one. This book shows how the Cosmos started way back when it introduced Creation and the four Cosmic Pillars... It shows why numbers came about as laws formed, founding Creation. This concept described in this book goes back much further than any one using formulated mathematics can realise, it is before light or material formed as the Big Bang.

The notion of Scientists thinking that the intellectual thinkers in society don't know anything about the Universe before the Big Bang is altogether wrong and also it is altogether correct at the same time.
The reason we as the intellectual species on Earth think we don't know anything about how the Universe started before the Big Bang is because we are staring into the science of Newtonian darkness formed by misconceptions. It is those that supposedly hold all the intelligence on Earth that believe that we are clueless on matters that happened before the Big Bang happened.

In the Universe time flows because the instant in the present is different from the following instant in the future as the instant in the immediate past is different from the present. That gives the Universe time flowing because every instant everything changes to what it is going to be from what it was. Before this we now have, as a Universe the future was so perfect that the future was the same as the present in every detail. The past was also exactly the same as the present in every smallest detail and perfect. The present was the past in every aspect that was the same as the future in precise detail. This is because the Creator is perfect. That gave one, serving time as the future that was the same as the present that was the same as the past and time formed one. Then the Creator decided to create something imperfect. Then the imperfect brought the Universe time where the future was altogether different every instant that time formed space and that difference also brought a past. Before this a numerical order was nonexistent. There was no need for anything more than one because in one everything that was was perfect. Then by changing every instant one no longer applied as a solo number and the need were there to start numerical multiplying. The Universe introduced numbers because a need arrived for numbers valued as more than one. This book shows how the Cosmos started when and how Creation introduced the four Cosmic Pillars. It shows why laws formed, founding Creation. This event goes back way before light or material formed the Big Bang. In this book I show what stupidity is going on physicists minds when think they are equal to God Almighty. I show everyone that reads how God created physics in the process of creating a Universe. I show how the numerical order formed mathematics as that formed physics because only one was present before the Universe was in place. Should you doubt that a living God was responsible for Creation, then you better read this and get wise about how physics is truly conducted in the way nature intended it to be!
It is Newtonian Science that stands to be corrected firstly because Newtonian Science knows nothing about the Universe...at all, but that is because Newtonians know nothing about science...at all.
Scientists think they are brilliantly brainy because to them intelligence is punching in numbers by calculating formulas and then to equate formulas. However, that is not science, it is highly camouflaged misrepresentation because to think and to reason about nature is true intellectual science.

That misrepresentation only comes about through the way we practise science.
Mainstream Science doesn't know anything about the Universe before the Big Bang but this is on account that Mainstream Science doesn't know anything about how science started to begin with.
We know very well how the Universe came about but we just never considered the correct approach.
It is because those thought to be wise are instead only Academic mathematicians therefore being the mathematicians they then think they then can so gallantly by their calculation say God didn't create the Universe because they declare that physics created the cosmos. How smart does this and these idiot(s) think he (they) are by saying something so stupid? Who then created physics or are the likes of Steven Hawking and his band thought of as superior thinkers think he (they) are also so mentally clueless that he and his brigade think that physics was around before the Universe came about?
In that case they admit to a God being present but they then just call the Creator by another name since they are fools that can't appreciate reality...and that is because they can punch in numbers and never think. When the Universe started, the Universe started everything including numbers, and with it came mathematics and with mathematics came physics. We know very well that the Universe started with one spot or dot or point holding singularity. It started with one...and that is singularity, the one! With singularity being one and one is the point that the entirety started with, then the question is not why the Universe would explode because when it exploded it was not one but many that exploded. It is a case of asking how did the number two come about and then three and then four and five and six and seven going up to ten. How did the numerical order formed by Pythagoras come into place? How

did Pythagoras instate numerical order and therefore physics as a law? That is how mathematics started physics and that is how physics developed as the Universe developed through physics developing. When you read further you will find how numbers developed physics, mathematics and the Universe in the process of creating the Universe. This is how the creation of the cosmos came about.

There is no factor such as mass in the Universe. There is no evidence of a factor such as mass that holds any validity throughout the Universe. There is no proof that the Universe indicates the presence of gravity by the measure of mass forming a pulling power and while science conducts an entire religiosity based on this falsified belief, any such notion is falsified truth. **Using science based on the idea that there is a pulling force such as mass forming gravity is as valid as giving Snow White seven dwarfs and then beginning a religion on that basis.** There is a factor such as weight but there is no pulling of anything towards anything by magical forces forming gravity or whatever. **There is a conspiracy of conducting fraud by claiming non- existing forces but such claims are utterly fraudulent.** I have been trying for twelve years to introduce the true forming of gravity but all Physicists I have encountered prevented me of doing so. They stop me because my work makes Newtonian science and when removing the notion that a pulling force of gravity works by the value of mass, most of their work becomes science fiction that falls apart in substance. Read this and see how **students in physics** are methodically **brainwashed** to get the students to believe in the absolute accuracy of science. Professors and teachers participate knowingly or unknowingly in this thought manipulation process by means of conducting **mind control.**

By applying this **mind-altering process** those teaching physics ensure they subdue students into becoming mind-altered zombies. It is a process going on for centuries and which without science would have no foot to stand on in the modern environment. By presenting incorrect, falsified or unproven facts and other untruths as proven truth they exert **thought control** and thereby change the student's ability to appreciate what is correct and believable logic and then force students to discard such judgement ability in favour of accepting the institutionalised untested norms and values of science in order to unequivocally believe in science. The accepted teaching methods force students to comply by compromising their better judgment and then systematically to capitulate under teacher pressure by making their own what science prescribes what should be believed. I prove this and you get this for free so what have you got to lose…**but you can get wise to what forms a better understanding about science**! By using the building blocks that forms the Universe I take you back to the instant the Universe started and I show you how the Universe fits like a jigsaw puzzle.

If you think my accusations are baseless or the ravings of a madman then go on and download what

you have opened and read for yourself. What you download is free and I do not benefit financially from this explanation I present to you. There is no such a thing as mass anywhere in the cosmos. If there were a factor such as mass every planet would orbit distinctly positioned according to mass, but they don't. Should you think of the size of a body containing more or containing less material and put that in terms of mass that forms gravity then the orbital layout of the Universe or solar system would very distinctly NOT be the way it is. The cosmos shows no mass as a factor and we can either regard the cosmos as correct or Newtonian science as correct as Newtonian science diverts totally from the physics that the cosmos displays. If **hot air** lifts "**mass**" up then "**cold air**" must bring "**mass**" **down** and that is as simple as that that is if you don't try to cheat to become clever! Then there can be no forces as science portrays gravity, which still stems from witches or druids or wizards and soothsayers pulling pushing or being scientific Neanderthal in the explaining of science. .

When a body touches or contacts the earth there is a factor such as weight but when you call that weight by the name of mass it is deceptive. Then again to call weight by any other name is equal to commit deception or to deceive the public. Notwithstanding the number of atoms that form any cosmic body, the mass or body volume is all the same because all planets orbit at the same rate.
The table that Kepler provided shows the time that the orbit of every planet takes according to the distance the planet travels in the same time lapse and considering all the planets it is very much the same thing and in that there is no provision in the table for any idea that might form mass. This ratio is the indication of speed travelled. The idea that mass exists is a Newtonian invention made up by Newton and is completely groundless except for the value that Newtonian science gives it in order to maintain the Newtonian principles. The entire idea of mass is a myth. The idea that mass pulls mass is complete mythology and is as baseless as any fairy story. But even more deceiving is that notwithstanding that every planet has a value when $T^2 \div a^3 = 0.983$ which is $a^3 \div T^2 = k$ Newtonian science completely ignores the values and declares that $T^2 = a^3$ whereby they ignore the values in the column. That is cheating the truth into submission to corruption to say the least. Newtonians fabricate their truth.

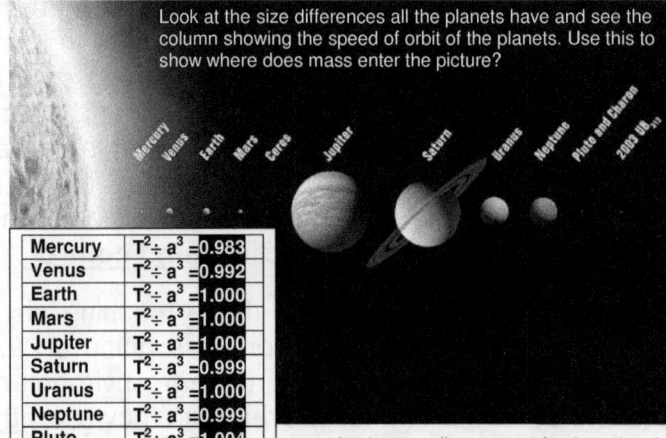

Look at the size differences all the planets have and see the column showing the speed of orbit of the planets. Use this to show where does mass enter the picture?

Mercury	$T^2 \div a^3$	=0.983
Venus	$T^2 \div a^3$	=0.992
Earth	$T^2 \div a^3$	=1.000
Mars	$T^2 \div a^3$	=1.000
Jupiter	$T^2 \div a^3$	=1.000
Saturn	$T^2 \div a^3$	=0.999
Uranus	$T^2 \div a^3$	=1.000
Neptune	$T^2 \div a^3$	=0.999
Pluto	$T^2 \div a^3$	=1.004

Bode's Law:

Planet	Mercury	Venus	Earth	Mars	Ceres	Jupiter	Saturn	Uranus	Neptune	Pluto
Bode's Law distance	4	7	10	16	28	52	100	196	-	388
Actual distance	3.9	7.2	10	15.2	28	52	95.4	191.8	300.7	394.6

A numerical sequence announced by J.E. Bode in 1772, which matches the distances from the Sun of the six planets then known. It is also known as the Titus-Bode law, as it was first pointed out by the German mathematician Johann Daniel Titius (1729-96) in 1766. It is formed from the sequence 0,3,6,12,24,48,96, and 192 by adding 4 to each number. The planets were seen to fit this sequence quite well – as did Uranus, discovered in 1781. However, Neptune and Pluto do not conform to the 'law'. Bode's Law stimulated the search for a planet orbiting between Mars and Jupiter that led to the discovery of the first asteroids. It is often said that the law has no theoretical basis, but it does show how orbital resonance can lead to commensurability.

The importance that becomes known is the sequence the Titius – Bode law saw in the number arrangement of 3; 6; 12; 24; 48; 96 etc. The incorrect application of the Titus Bode law lies in subtracting the figure of 3 from 10 leaving 7. The other way of reasoning is to add four each time to the firs value of three starting with 3 and so on. The true significance of the Titus-Bode law is that it points directly to a circular growth of 7 stages. The 7 relating to 10 is a precise derogative of the Roche limit or the Roche limit is a precise derogative of the Titius Bode principle because he two systems interlink. This is how I mange to explain the Titius Bode law that is in the solar system by the ratio applying that really form the solar system in the way nature shows space growing by time. What you see on the next page was never been shown but on the other hand Physicist say this mathematics are too simple to apply as physics!

To find the mean distances of the planets, beginning with the following simple sequence of numbers:
0 3 6 12 24 48 96 192 384

With the exception of the first two, the others are simple twice the value of the preceding number. Add 4 to each number:
4 7 10 16 28 52 100 196 388

Then divide by 10:
0.4 0.7 1.0 1.6 2.8 5.2 10.0 19.6 38.8

The resulting sequence is very close to the distribution of mean distances of the planets from the Sun:

Body	Actual distance (A.U.)	Bode's Law (A.U.)
Mercury	0.39	0.4
Venus	0.72	0.7
Earth	1.00	1.0
Mars	1.52	1.6
		2.8
Jupiter	5.20	5.2
Saturn	9.54	10.0
Uranus	19.19	19.6

The Titius Bode law proves that in the Universe laws apply that positions objects in terms of other rules that mass. That means the Newtonians hides their lack of understanding behind mass that they invent.

Please show how mass by $4\pi^2 a^3 = P^2 G(M + m)$ can produce Planet positions. It is hogwash.

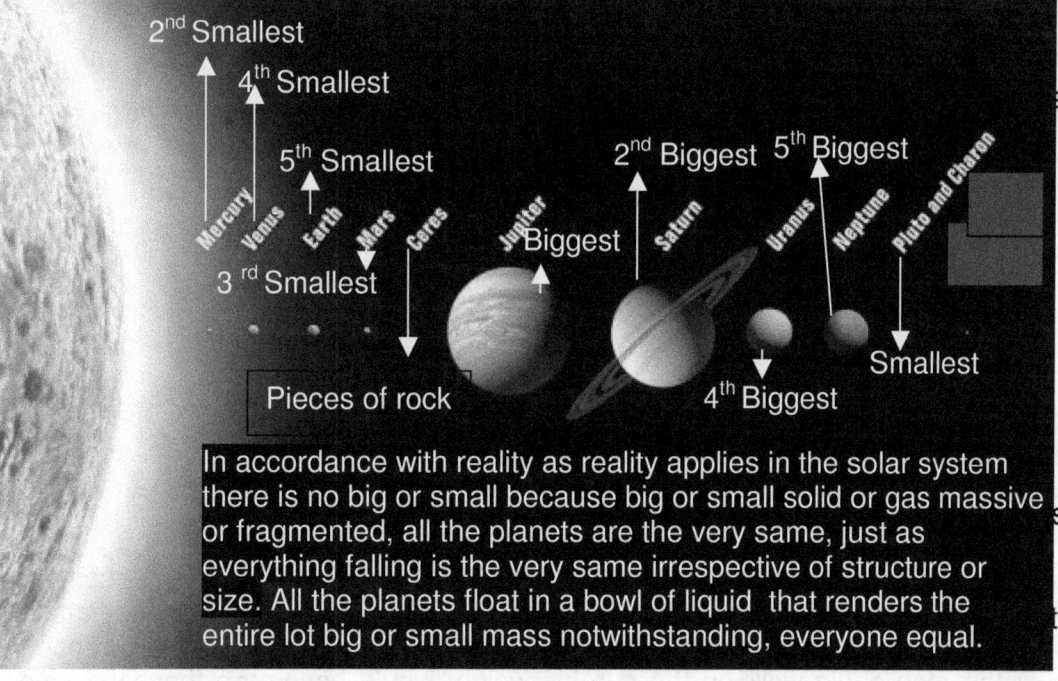

By depicting the solar system in such a presentation as Newtonians normally do such as the picture next this form of presenting the layout without providing correct spacing purposely corrupts the entire structure formation by which the solar system develops. It then purposely hides the essence that forms the solar system.

In accordance with reality as reality applies in the solar system there is no big or small because big or small solid or gas massive or fragmented, all the planets are the very same, just as everything falling is the very same irrespective of structure or size. All the planets float in a bowl of liquid that renders the entire lot big or small mass notwithstanding, everyone equal.

This is so typical Newtonian in every sense there is in science.

The Newtonians gave the Titius Bode law a formula and that explains the lot. To they're under achieving standards that is very satisfactory. Now it is written in mathematics then what more do we need to know. The fact that the distance that Mercury has from the sun is doubled by that which Venus has from the sun is completely ignored. In cosmic reality mass plays no part. Then again the distance that Venus has from the sun is doubled by that which the earth has. This clearly has nothing to do with the size or mass of the planets. Explaining that part is completely ignored. Then again the distance that the earth has from the sun is doubled by that which Venus has and inexplicably this forms the layout of all planets in the solar system. Where do Newton and his idea of mass fit into what truly applies in outer space. Moreover, why does science never mention this? This is my formulated explanation about how the Titius Bode law forms.

The numbers we need to find the key to the mystery of the Titius Bode law is 3, 4, 7, and 10.

This is what is in the solar system applying the serving ratio that the Universe uses. It is not the fake Newtonian $4\pi^2 a^3 = P^2 G (M + m)$ that has no basis except for Newtonians brainwashing students into believing that otherwise can't be accommodated by the Universe. These are ratio values that are there…used by the Universe as actual factors forming space.

To make sense we have to look for Π in this because Π forms the value of gravity.

We have 3 and 4 adding to be 7. Then we have 10 forming the other factor number. If we wish to stop pretending to make science a hoax we must come to some realistic conclusion that will prove what is working in the Universe.

This astronaut has "mass" and the earth also in the picture has "mass" and the pulling should be going on since there is no restraining that would prevent the man from falling to the earth so why is Newton's "mass" not applying? This astronaut is circling the earth at a certain speed notwithstanding "whatever mass" science contributes to the pilot. If "mass" was pulling then why is the man not falling. The man will fall depending on the speed by which the man rotates the earth. His spacecraft (not in the picture) has about a hundred times more mass that should entice about a hundred times more pulling but both float above the earth at a specific pre-calculated speed. It is the speed that determines the distance of circling and not "mass".

$$F = G \frac{M_1 M_2}{r^2}$$

If the astronaut were on earth his mass would be the same as his weight. Therefore on earth there is no distinction between the mass factor on earth and the weight factor on earth. It is clear that things change when this astronaut walked on the moon. If the astronaut were walking on the moon his weight would not be equal to his mass.

The astronaut would weigh less than the mass he has on earth. It is said that the gravity changes the numbers. The moon has less mass and that gives the man less weight while the mass remains the same. Say the man weighed 90 kg on earth he would have mass equal to 90kg.

When the same man walks on the moon he would weigh 30 kg while having a mass of 90 kg being on the moon. On the moon the machine may have the same weight as the man on earth but the mass of the machine stay what it was on earth and so does the mass of the man. That becomes science in fictional perspective. If the man could walk on the sun his weight would become a thousand times more but the mass of the man remains 90 kg. If I believe that bit then somewhere I am fooled just because I am that stupid and I deserve to be fooled that easy.

Then when the astronaut is in space he still has a mass of 90 kg but when walking in air the man has no weight with a mass of 90 kg. The indication is that while the astronaut still has mass and so the earth does he circles around the earth by gravity. Gravity is circle and not a pulling by force. Gravity comes about in terms of density applying through motion exerted on any object. It is movement that determines the location of an objects and the movement forms gravity. How stupid must any person be not to see and therefore realise this as a fact.

If there is no mass then how does gravity apply?

Their sublimation covering their stupidity causing their ignorance will outlast them for all time to come. As the Pope five hundred years ago is remembered for his ignorant stupidity, so those currently in office would also befall the same fait just because they were as stubbornly arrogant and ignorant as the Pope was back then when ignoring mistakes.

They conspire to conceal the dishonesty that they don't wish to reveal.

Singularity Π^0

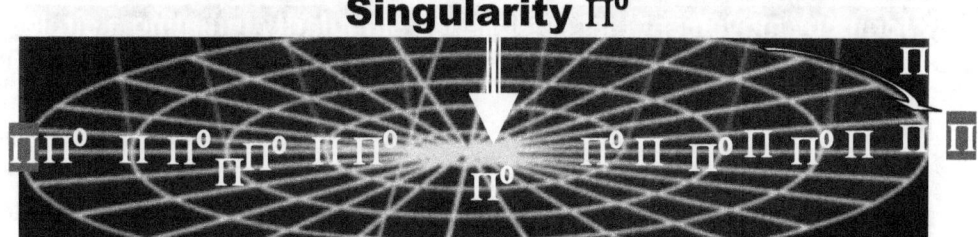

There never can be anything such as a straight line in singularity and therefore in the Universe. To produce singularity movement must establish such a centre point. Only circles can be the shortest line possible because of the nature of mathematics.

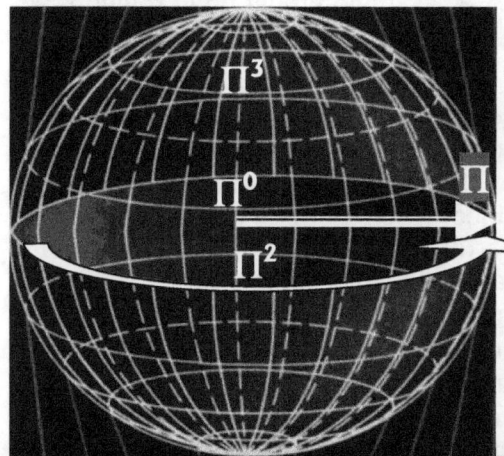

Lines mathematically cannot start at zero because there is no evidence of zero as a factor in mathematics.
The shortest possible line (hypothetically) must be so short it must have an initial and ultimate point sharing the same spot.

Any theoretical line being the shortest possible line cannot have the line holding the initial starting point at point zero and advance from there.

If it used zero as a start, the zero part would not count, because the line will only start at a point past zero where the line then will start.

When the line has a beginning and an end at the very same spot and it wishes to extend the position as to further the possibility it has, which direction should it favour. Extending the line in any one direction will favour one direction without any clear reason not extending in other directions. The only option about extending will be in all directions equally in order to give a meaningful non-bias flow of mathematical equilibrium.

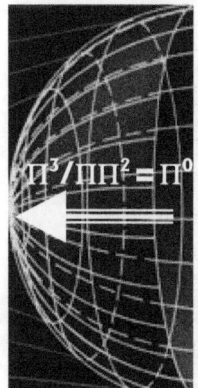

$\Pi^0 = \Pi\Pi^2/\Pi^3$, which says that everything that is (Π^3) moves ($\Pi\Pi^2$) in relation to singularity (=)Π^0
But also when anything (Π^3) cosmic moves ($\Pi\Pi^2$) it forms singularity (=)Π^0 by which it moves.
This formulation of space-time is the starting point of physics
The entire Universe is the movement of space in relevance of a point forming singularity and Singularity forms in the centre of all things spinning by the spin the material has. Because singularity holds the compliment of one everything that spins holds the centre of the universe in equal measure and in equal quantity. All star are the measure of the gravity that every proton within that star holds as every proton conforms space to confirm the stars solidity and therefore the contraction ability that the star holds to retain the heat in space or material that the star could collect.

Everything that is in the Universe is $\Pi^3/\Pi\Pi^2 = \Pi^0$, is spinning while also rotating something else. This produces a relevancy in relation to singularity which Kepler showed as $a^3/kT^2 = k^0$ and since I found gravity is $\Pi\Pi^0$ I changed the format to $\Pi^3/\Pi\Pi^2 = \Pi^0$. When the combined contraction that the atoms form by overall movement of the star the relevancy of $\Pi^3/\Pi\Pi^2 = \Pi^0$ will change and in changing $\Pi\Pi^0$ the star overheats and goes supernova. The slower movement $\Pi\Pi^2$ will enlist a larger space Π^3 form more space colleted away from the centre holding and securing singularity.

By using the four cosmic phenomena, which is what the cosmos uses to form gravity, I show what "the sound barrier" is and I go much further than that. I show that gravity forms using the **Roche limit**, the **Lagrangian system**, the **Titius Bode law** and the **Coanda effect**.

I uncover these principles by placing Π within the formulating of gravity and when using Π I bring clarity to the misunderstood cosmic principles. The list of the unknowns I can then explain is almost endless. Gravity forms by movement that establishes singularity initiating a circle in using Π. "GRAVITY IS DIVIDED IN TWO FACTORS, BEING LINEAR DISPLACEMENT (Π) WHICH IS WHAT NEWTON'S GRAVITY IS AND,

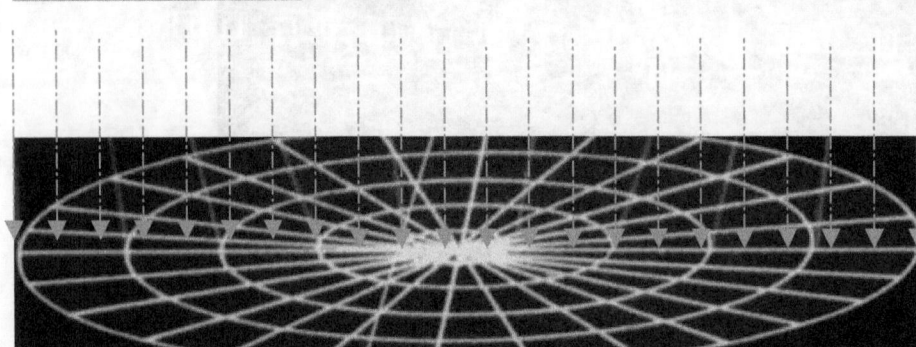

There is a position that is in motion that is forming the very edge of the outside of whatever spins or moves or mathematically equated as $\Pi^0\Pi\Pi^2=\Pi^3$. This says that to be in motion the position securing the motion must be in relation to a point forming a centre.

CIRCULAR DISPLACEMENT (Π^2) WHICH IS THE "GRAVITY" EINSTEIN RECOGNIZED

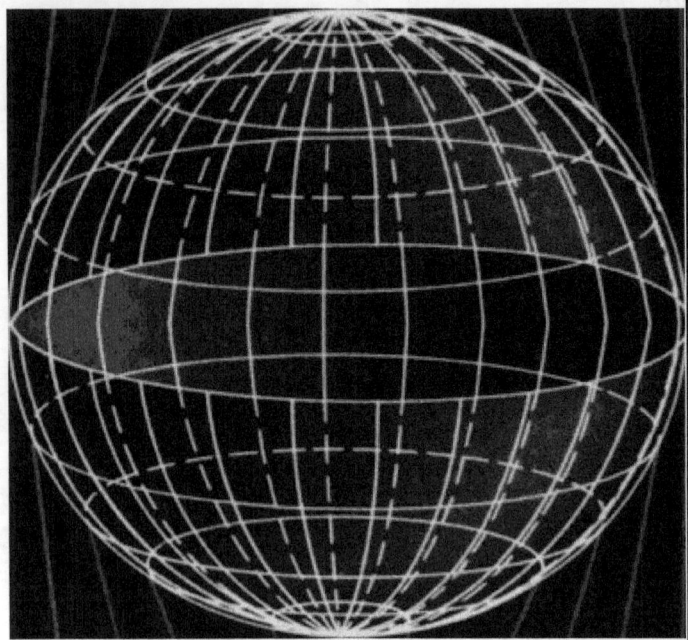

Remember that according to cosmic physics everything moves by a straight line Π that moves by circling Π^2. From the centre, there must be a specific allocated space ending at the object in motion and starting from a centre that centre line has no dimensions. The object in motion determines the one limit and the centre with no

Those that are of the opinion that it is friction with the air that heats a spaceship or any object when it enters the atmosphere of the earth, please explain the following: When a spaceship enters the earth's atmosphere by more than 21° it will burn out. If it enters the earth's atmosphere by less than 7°, it would be bounced off into space. The 7° comes about by the concentration of the atmosphere and this bounces the spaceship outwards. However, when a spaceship enters the earth's atmosphere at an angle higher than 21° it will in fact meet less air than in the case of the spacecraft entering at an angle of more than 21°.

The extension of Π is well received as a dimensional implication to matter holding seven positions from singularity and space having four quarters through out the rotation of singularity forming the centre to the five dimensions (one side lost to the cube's six sides connecting to the five remaining sides) making the total sides facing space from the point holding singularity at any given instant at a value of twenty (4 X 5 = 20).

sides and no space, which is standing still in singularity, determines the other limit. By that we can see there are only one way of looking at what we can observe and that is from the outside in. Mainstream physics ignored the clear connection completely, notwithstanding it being so very obvious. There is this far in their recognising of principles in natural physics not one single reference made to prove their appreciation of this matter. They are bent on particle colliding. When particles collide, such collision forms an atomic thermo release and that action we call an exploding atomic bomb. What principle this argument about particles colliding ignores is that all atoms use negative charged electrons forming the atomic limit on the outside forming a definite border to the boundaries of all atoms and in both electrons from different atoms are being negative charged. In being negatively charged, it means both will come out and totally reject the other. The closer they come the more violent the rejecting will be and such rejecting is the production of heat that will turn to space. The electrons repel other negative charged sub atomic structures, which the electrons are that form the outer borders of all atoms. With all electrons highly negatively charged (being as negatively charged as any possibility will allow to match the utter extreme) such electrons could not touch.

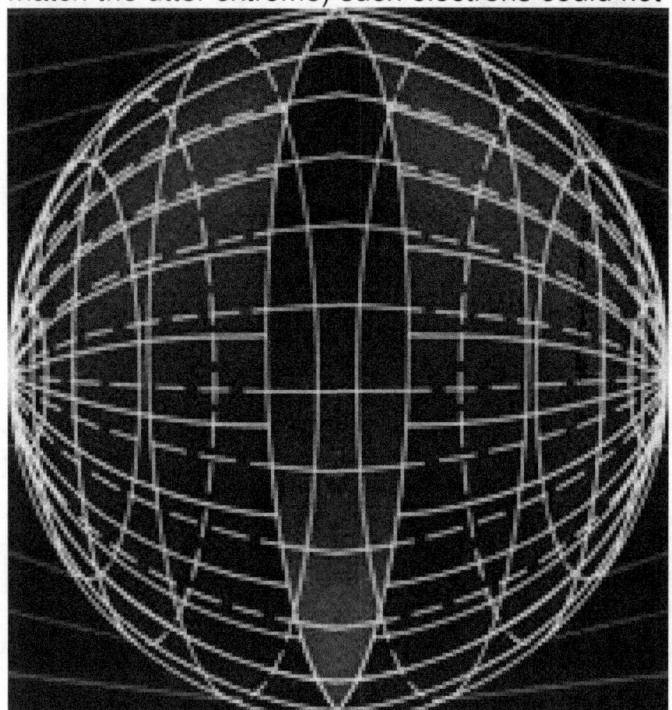

Since it is linear movement that is also the circular movement ΠΠ² that launches Π the value of 7 can never be excluded from the factor forming Π.

Then adding the singularity cross of Π being (1+1) = 2 the relation becomes 22.991/7. This is crude because in more precise calculations it becomes .91 + 1 = 21.91/7 = Π

The Universe is in place in space by the value of movement and it is the movement that forms the space. If the movement of the space stops it will pull the entire Universe which then is the star and this space surrounding the star into one spot that forms the singularity that represents all spots holding singularity onto and into one spot in singularity.

It is about time scientists start looking with their minds and not their eyes at the Universe and see what is truly out there to see. All the difference we find is seated in the human mind. We humans set differences because we look at the cosmos by placing humans and the life we find on Earth in a pivotal centre in the cosmos instead of placing singularity in the centre and life where it belongs; only found on Earth. Einstein proved mathematically that in the presence of a strong gravity such a strong gravity slows time down. Surprisingly with that evidence being around this long nobody in science since Einstein's discovery took those statements and made any further progress from that.

It seems to have been left in some drawer to dry. Science still sticks to the opinion that time did not change, not even slightly, since the beginning of the time and holds the same pace ever since the start of the Big Bang notwithstanding the implications this concept carries. Before the Earth took one year to circle around the sun and even before the sun was there a year was still the same duration of one year. How odd... don't you think ... that the only aspect in the entire Universe that is beyond change is the aspect of time? With the entire Universe including all the gravity now present and not excluding one Black Hole or dust speck pressed in such an area that was possibly the size of a lepton even then the gravity extending from that circumstances must have been beyond what words can ever describe.

When everything was that small when the Big Bang took charge, the gravity at the time was beyond light, because even today in the Black Hole the gravity is beyond the speed of light. If the gravity was that high and Einstein already proved that strong gravity slows time down, then there is one logical conclusion and that is that time was in fact at the time of the Big Bang standing still. Mathematically it is incorrect to allow gravity to compress the Universe into a spot smaller that an atom and exclude any other factors and relevancies to change.

If the darkness was the representation of "nothing", then that should be exactly what we must see, nothing but the stars. Taken from the top picture some stars and leaving the rest to nothing is what we see in the picture below. A blind person sees nothing but when we look at space, we see something that we think nothing of as we see as space. One cannot have the ability of sight and see nothing except by closing your eyelids and then you see nothing. But in that case you do not see "nothing" in contrast of "something" you see "nothing" without it contrasting to "something".

Cosmic Solids notwithstanding how small is the elements made of atoms forming a star.

Cosmic solids floating in Cosmic gas seems

The tiny asteroid is in relevance a cosmic solid but as a solid it circles around the sun. Notwithstanding the enormous size difference the piece of rock circles around the sun as if it has the mass that is equal to the mass of Jupiter. The only reason for that must be that in terms of the sun it has a mass equal to that of Jupiter. A pebble orbits the sun in an equal manner as the "giant Jupiter does". How it can orbit the sun while being that small how it formed is "a mystery to science" because as long as science clings to the outdated idea of mass taking the responsibility for gravity forming reality will remain a mystery to Scientists. The Asteroids make a mockery of the way Newtonian science tries to portray the way in which stars form by gravity compiling matter into dense structures. How can a structure of this size find enough gravity to form a compressed, solid structure while something the size of Jupiter is but merely a gas? This places science in a fool's paradise. The stupefying simplified explanation that science embrace from the past prove how little science have in understanding and therefore explaining about how science truly works.

Cosmic gas littered with cosmic solids spinning around a cosmic planet holding cosmic liquid that surrounds a cosmic solid base

The planet rings proves that gravity as science interpret it does not exist. Ask your teacher to explain the fact that if mass pulls mass by gravity pulling, then why does something as tiny as an asteroid circle around the sun at the very same speed as what the giant planets do. Why would the rings form around planets when mass have the ability to form gravity that will pull the rings into the giant star? With mass doing the pulling what prevents the circles spinning around the large planets to collapse into the planet. What sustain the orbit of the circles and why would the asteroids circle around the sun as if they had the same mass as the planets have? Don't allow them to shrug their shoulders and carry on as if everything they say is true. That is part of the brainwashing and mind control process. Do not believe them because they are corrupt. **Download The Absolute Relevancy of Singularity @ Lulu.com** and doing so will show how you it is possible to condense all the space around the spinning the planets.

You may see how material compress the space around the star. This will inform you as to why the cosmic physics principles establishes these phenomena...and because the information does not salute Newton, academics in physics despise what is said. If mass don't pull stars to a point of colliding then Newton is wrong! Any star coming closer than 2.4674 times the radius of the larger star would liquidise the smaller star and this law and the Coanda effect are a precise duplication of true gravity truly applying.

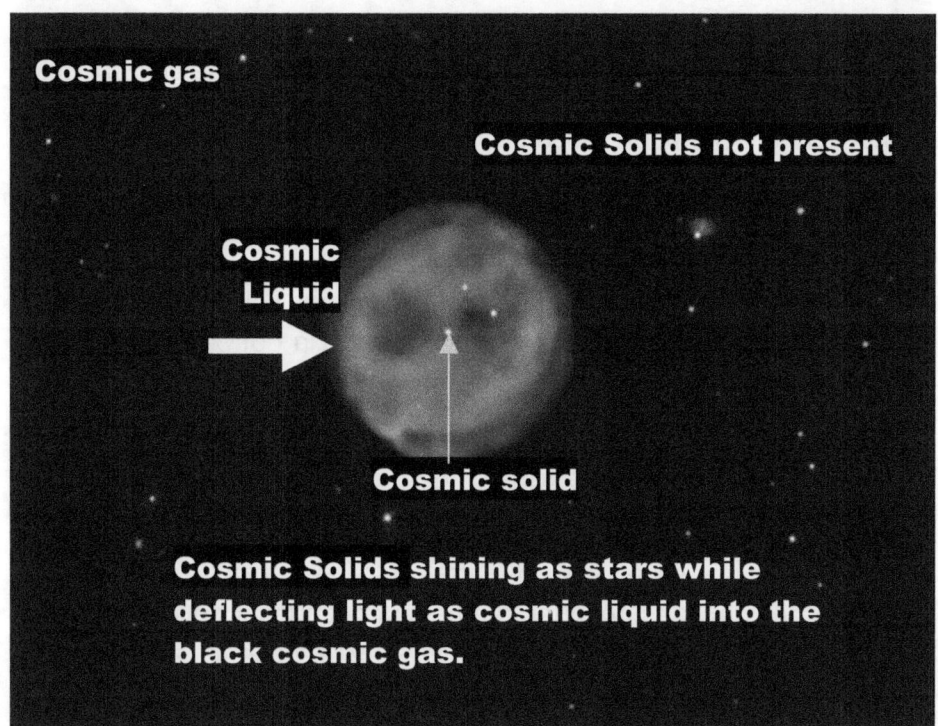

Do you know what happens in this picture? Is there in physics any realistic explaining available other than saying that gravity has gone mad? That is the best they can do because either they must come up with anti-gravity, which must come from anti-mass, which must be the result of anti-matter, which they can't explain. Gravity can't go mad because gravity has no intellect! Have you ever encountered a realistic believable explanation why a Super Nova explodes? The best offer science sent can produce is that the gravity in such a star goes mad. Then one would presume that gravity has to have some mental stability with a testable intellect and a nervous system that can go mad or stay sane, depending on the emotional stress limit the star can endure. What junk this is and there is even more junk they cook up because they know so little.

When a top starts to spin Moment-Alfa starts all over again as it did at the start.

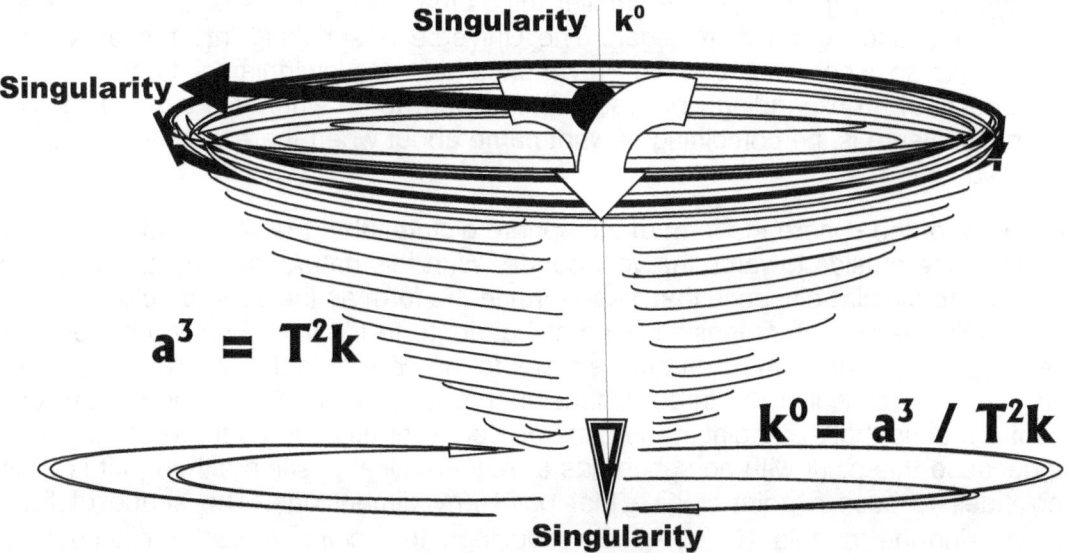

Moment-Alfa is in eternity and eternity never ends therefore it moved to eternity in space eluding time by having a vision upheld by the information light delivers. Through light it is possible to find information being able therefore to see the information that can be reported because the light carrying the information was in that space but not in that time. If the viewer and what he saw were in that same time the view the viewer saw and the viewer would still be there sharing space, but also then the lot would be there in that space within time eternally and never being able to come back to report on what the information about the past that the viewer saw. There is much rumour of a Big Bang, however I am inclined to think the biggest bang that ever was also became the smallest bang there ever can be. It was the instant when heat parted from cold. Then the past separated from the present to form a future. Without a future there can be no present and then we would not be able to live in the past that began Creation in that instant.

Before this period the past carried the present from the future and everything landed on the same dot. The past was the present was the future because it was laboriously monotonous while being perfect in every aspect all the time. Then the Creator created heaven and Earth just as the Bible says the Creator did. The Thought became sound. The idea became structured. That which is beyond recognition became observable. The thought became the voice, which remained God. Singularity found a presence within the space that now is the Universe. The Creator created what is cosmic liquid and what is cosmic solid. The Creator created something that spins (as solid as the Earth) while holding space in which it travels. Study how singularity works and you see precisely how the Bible says God Created the heaven (cosmic fluid and gas) and the earth (cosmic solids that move within the cosmic liquids). Still, everything that is also is in place as either a dot filling or a dot being filled by a dot. The role the dot accepts depends on the relative movement of the dot being filled or filling. It still is a dot, although many dots but inn singularity remains one single dot.

The proof of the dots existing is looking at the Universe and finding space. The dots are not there because individually no dot is part of the Universe but because every dot is moving in terms of every other dot all the dots form a solid structures Universe we all enjoy. Still every dot that is also is not and every dot that proves a history also becomes one dot forming time. The dots form a circle that forms a line. The line started in infinity because the line was continues but being continues it never was. The very instant followed the previous instant identically and the instant was so identical it remained the same by never moving while always moving uninterrupted eternity upon eternity. Then came the entire Universe when infinity broke free from eternity with a difference between heat and cold parting time. Time became imperfect splitting the perfect into two zones. It was long before when darkness broke into light. This was when space was a thought and not a measured distance. It was when whatever possibly can be became a possibility to be. It was when the first line of numbers mathematically arrived and from the one became two by gong 1^0 to 1^1. That infinity is so small it houses everything there is in the entire Universe. The entire Universe still is in a spot that formed a dot. The spot has no outside but it only has an inside while it is inside all that spins as it generates all that can spin. Yet, by spinning it brings motion into being. The spin creates a drive that keeps the Universe mobile. Still the first forming of the dot from the spot came about to the inside and not the outside, which makes the Universe shrink and not expand. It is the smaller things that come into relevance as the larger things were placed in relevance when time began. The Universe is shrinking into the oblivious since the Universe never had anywhere to expand to. Never once did one Newtonian sit back and consider their laughable proposal of an expanding Universe with nowhere to go when it is expanding. The diversity we have in the Universe is the combining of what came about when there was nothing added to what there already was.

In the very centre of the sphere the form of the sphere dictates that the shape will relinquish space as the line run from the outside towards the very centre. With this natural state of affairs the sphere are naturally inclined to dismiss all space that it can form in the form as the sphere holds space inside and the form will finally be without dimension. All that I attribute to the line shrinking by reducing actually takes pace in every sphere as the diameter reduces to the centre. In the centre where the radius line goes single the form relinquish the three dimensional form it has inside. Being without dimension in the very centre means that at a point in the extreme centre of all spheres there are a point that holds singularity because this point with no space has a mathematical position although it is invisible since there is no sides to such a point to give that point any dimensions. The shape of the sphere is calculated by using the formula $4\Pi (r^3) / 3$. By reducing r to a point where r is r^0 singularity steps in because only the form remains as Π. Going even further we find that there then comes a point where Π goes singular Π^0. At that point absolute singularity is present but so is absolute gravity present at that point. When holding the strength of the shape of the sphere in mind as well as taking into account that all cosmos objects of importance is in the form of planets or stars and they are all in the form of a sphere, we therefore may contemplate that it is where gravity originate. We now only have to find the reason why gravity will hold a base in a space less ness as Einstein predicted. It is clear to be seen that gravity is in the centre of the sphere controlling from the centre everything that is outside the space less centre. We can reason with confidence that gravity is the strongest where space is the least. We can further reason that it is gravity that is holding the sphere in true form and since the sphere allow gravity the best working opportunity, gravity can form the sphere in as strong a shape and form as the sphere seems to have. From every point on the surface of the sphere is where that point connects with the other side of the surface of the sphere by a line that runs through the space less ness of such a centre of the sphere. Such a line also connect by an angle of 180^0 as well as 90^0 to six other lines running from top to bottom, right to left, and back to front, where all join and cross in the centre of the sphere. There are therefore six lines crossing and connecting by a centre from any given point on the surface of the sphere. Such points connects in total six surface points on each side of the sphere while they all support one another through the space less centre. In that absolute space less ness in the centre holding singularity we find gravity supporting and controlling all space within the sphere as well as space connected to the sphere. That is where gravity control and guide the space, which falls in the parameters as well as under the influence of the form of the sphere. In the gravity centre space goes singular meaning space becomes space less or flat.

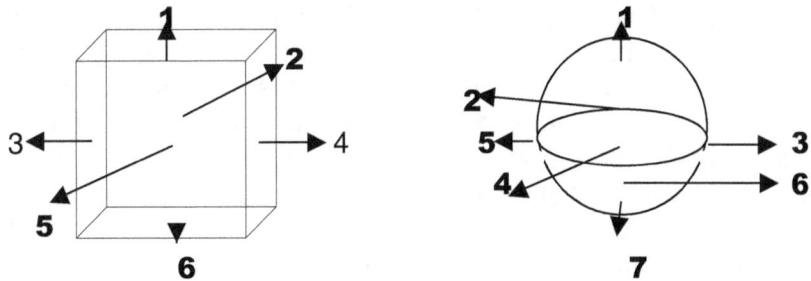

It is from the layout that the sphere uses as natural form that we are able to locate singularity. In the case of the sphere the material naturally reduces by measure of the radius becoming smaller to a point where the radius is r^0. At that point the line that will form the radius has gone single dimensional r^0 and that is equal to 1^0, which is singularity. Also it is true that the entire form that is the sphere is controlled from a centre within the sphere. That centre holds the sphere in form and shape. Therefore the strong form is dictated from that space fewer centres where there is no space and no form left. The natural inclining is in the form of the sphere. It is part of the roundness that the overall shape of the sphere represents and this structural strength is carrying down to the very centre. Because the circle is forever reducing that reducing which is inherently part of the form of the sphere becomes a tool in distorting of space in the sphere and is eventually removing all forms of space from within the centre of the sphere. The very centre ends up as having no space because of the reducing that continuous down to become the space less inner centre. The all roundness is the ingredient that forms the backbone of the absolute strength that the sphere has and that is the component that the sphere is so famous for. The form the sphere has allows the sphere to have a control that is coming from the centre deep inside the sphere where the space vanishes and being without space seems to keep the entire structure rigged. From the centre the sphere shape shows strength that the shape as tough as it is. How does it work in its most basic analyses?

Realities such as getting there, finding the object, fighting the gravity (remember this object has the gravity to absorb the entire solar system in a matter of minutes and that is according to their calculated efforts on the Black Hole) Those capable of designing space whirls were in charge of the mathematics of the Universe and still they could not solve such a simple issue as gravity in all the time they had to their disposal over the past three hundred and fifty years. It takes human effort to recognize reality. It takes a human's intellectual effort to recognize the Godly aspect to the Universe. The thought manifests as an electrical component after it is realized and then by a human's deciding ability is transferred to other parts of the body in electricity conducting. Normally it is the emotion provoking the thought that generates the thought because it is the emotion attaching thought to the moment that places the thought in the mind. The machine can never reason and challenge conceptions but will always accept conceptions unconditionally and Newtonians are brainwashed to do in science. To them and theirs and those the computer may replace them and I hope it will the sooner because that will make life the better for the rest of us that can face reality and not hide behind computing skills that machines can replace.

Time came from eternity. It is so obvious but then again science are not very likely to condone in the obvious because science grab onto the ridiculous. Science is absolutely opinionated when they do not even know what gravity is. The Coanda affect is the epitome of gravity and is the implicating principle in establishing gravity, but while at the time I am writing this, not one in established physics was able to acknowledge this fact. Kepler said $a^3 = T^2 k$ which in turn says $k^0 = a^3 / T^2 k$. This then translated to a verbally understood language says that all space-time $a^3 / T^2 k$ becomes nullified in singularity k^0. This a now can say because for the first time since gravity was conceived four hundred years ago the principles behind the phenomenon gravity are revealed. Up to now every one in science was acting as if gravity is a commonly explained factor, which every one knows every aspect about all principles that are involved in gravity down to the smallest detail. In truth no one in science anywhere remotely knows what brings gravity about and I used Kepler to unravel this mystery called gravity. The most informed in Science at best can only assert their suspicion on a rumour presumed about what causes gravity to perform as the part interlinking the cosmos but no one can go any further than affirming their alleged confirming such possible correctness by explaining the concept. But in the meanwhile gravity eluded them as much as it eluded their master, Isaac Newton.

This is because they avoided and belittled Johannes Kepler. Newton started this rumour and Newton admitted to it being a concept he could not explain but nobody in science at present will denounce Newton's gravity ideas as being merely just a rumour as Newton admitted it was. But instead of trying to study the findings Johannes Kepler produced, Newton degraded the work of Kepler by changing the work of Kepler. As he did it, it also rubbished the work and the findings of Kepler. Newton agreed that he could only declare gravity as only a rumour announcing a force that could be anything. He did not see that it was motion of space that kept the planets orbiting the sun as Kepler said when Kepler said space a^3 moves around T^2 in a precisely designated relevant space k where the distance of k places such space in that specific motion in a specific and precise relevancy to a generated specific centre k^0 where such motion produces time for space to move in the relevant space coming from k^0 and ending at k to produce $k = a^3 / T^2$. Not once could one person in the past or present provide substantiating proof on gravity as a physically reality by defining the very principles that produce such a force, should such a force even exist. That includes Newton as well as Einstein and even Hawking.

Scientists can declare gravity was a factor at 10^{-43} seconds after the Big Bang but what brought gravity about or why gravity became or still remained, as a presence is still tightly concealed information. Even to the best informed amongst the most educated do not know what is gravity because they all ignored Kepler and for ignoring Kepler the price they pay is not finding gravity. Using Kepler makes the method to follow and understand even Einstein's discoveries shockingly simple. By my applying Kepler I can define gravity precisely to the point where I now can explain why the proton and the electron forms a mass difference of 1836 times between the two mass components.

With the aid if unravelling the four cosmic pillars, which the Universe in its entirety was built on and now rest on I can enter an era never entered before by science. But to uncover the veil that kept the Universe hidden from sight was finding the four cosmic pillars. The four cosmic pillars are keys that were in place before mathematics came in place. It was dimensions that applied before dimensions were applied. By the first spark a Universe came about translating the four cosmic pillars into cosmic principles allow us to look past the Big Bang into an era where space was an image and time was the interrupting of eternity by changes in infinity. Firstly and in front of everything else the cosmos is about relevancies. There cannot be gravity is there is not a gravity between two particles in the Universe. There is no force such as gravity because gravity is a motion of space in relation between two particles. Once that fact is established and the fact of two points holding the space singularity but the singularity is divided by space-time generated by motion between such points can we find the cosmic pillars that serve as keys to unlock our past. The keys are not recognised by the Mainstream policy makers in officially accepted science. The Newtonian thinking minds never put cosmology in terms of reality and as normal put Earthly science in cosmic place and Universal space.

Kepler declared the Universe to be $a^3 = T^2 k$, which the Universe itself told Kepler. If the Universe is a ratio of $a^3 = T^2 k$ then the Universe is $a^3 / T^2 k = 1$. When I place any number or any figure holding a mathematical symbolic value to the value of an exponential 0 as in 1^0, then the figure has the value of one notwithstanding the actual number or symbol. From a^0 to Z^0 and 1^0 to 10000000×10^0 all the answers come down to one. That means singularity is not a mystical numerical mathematical idea exported from the world of make believe to pester the sane, but it is an every day reality. That puts the Universe at $k^0 = 1 = a^3 = T^2 k$ **declaring the Universe as being singularity.** That way the universe is singularity and singularity is without form being one to the dimension of zero. Where then do we find form, if not in the Universe?

How big was the Milky Way when the Universe was the size of the neutron. Science in their overwhelming intellect has not thought one nana meter past their noses to come to a realistic approach to cosmology. If mathematics can explain a scenario then the mathematics accept it as religiosity. However it is never the idea that they scrutinise but always the mathematical formula that goes tested. What no one realises is that mathematics is part of the Universe. Being part of the Universe there then had to be a period where mathematics were established as a factor and some factors in cosmology has to predate mathematics because it did not start with mathematics, the cosmos started with singularity being 1^0. Technically singularity is not part of the Universe but part of what is determining the Universe by creating the Universe. Singularity at 1^0 is above and is outside mathematics.

When the Universe was the size of a neutron, where were the react that we now find to form the Universe? When the Universe was 10^{34}, what then was minus 273 C. If there was no minus 273 C at the time any where in the Universe to compare the 10^{34} with, the heat of 10^{34} stands rather without meaning. Only by comparing the hot with what was seemingly cold could put the hot in an appreciating position otherwise it has no meaning but to be a statement someone wish to make.

What was the size of the sun when the Universe was the size of a neutron? We cannot look at the Universe and proclaim a size without the size affecting the lot within. Stars were stars when time distortion produced space because what is space when space is not filled and what is space when space is filled, but moreover, what is the difference there is and why is the difference there? It is obvious why science has no idea what time is if they wish to use a line supporting the upright stance of a vector to indicate time. Time is the holding as well as the support of space. It is time and not space what keeps space in motion and the motion we see as space while it is the time that is the space. Space moves in time with the support of time. Therefore space grew as much as time grew. With the Universe being the size of an atom the sun was frozen into a thought of one day probably. But that cannot be because there are no means to add to the Universe, as there is no means to remove from the Universe. That which is in the universe is here to stay and is as eternal as the Universe is. At the of the Big Bang period in time the sun was far from the boiling bubbling soup bowl containing pure liquid heat that we have today. It is quite impossible to imagine the sun being the size it is and still fitting into a Universe the size of a pea. It is also quite impossible to imagine a sun bubbling with heat while being the size of a pea. That means if the Universe grew, then the sun grew with that Universe. However, the Universe contains what ever will be which means the Universe cannot possibly grow. The Universe already has everything there ever can be.

With the Universe being the size of an atom the sun was frozen into a thought of one day probably being more than a concept. At the time the sun was far from the boiling bubbling soup bowl containing pure liquid heat that we have today. It is quite impossible to imagine the sun being the size it is and still fitting into a Universe the size of a pea. It is also quite impossible to imagine a sun bubbling with heat while being the size of a pea. That means if the Universe grew, then the sun grew with that Universe. However, the Universe contains what ever will be which means the Universe cannot possibly grow. The Universe already has everything there ever can be and that the Universe had at the very start of the Big Bang. Whatever came after the Big Bang were merely relevancies alternating possibilities.

This was moment-Alfa. The darkness was there as big as it is and as dark as it never again could be and from the darkness heat came about. Only heat expands and it interrupted the true invisible darkness, the blackness of a Black Hole, the invisibility coming from within the Black Hole. Eternity tore from infinity. Darkness broke from light. Heat broke from cold. Relevancies parted by 1^0 going 1^1. There was one but also there was two too because one cannot be without two being there to ensure one is one. The marks are still with us but to see the marks requires a great deal of intellect.

$\Pi^0 \Rightarrow \Pi$. In this there was only space for one being one in the two forming one. It was $\Pi^0 \Rightarrow \Pi$ however there was no space to be $\Pi^0 \Rightarrow \Pi$ and there fore because of the lack of space to be which is the infinity of time braking the eternity of time the true measure was $\Pi^0 \Rightarrow \Pi$ but realized only 1^0 going 1^1. Π was to the future because of the motion of time involved and the space less ness of space at the time. By inclining to move the process crossed the Universe but also it took one eternity to accomplish the feat.

The fact that 1^0 going 1^1 brought movement can only become a reality as a result of light. Light is heat and the heat is expanding.

Science look at space when space had 1^0 going 1^1 brought movement and today space has 1^0 going $1^{1000000000000000000000000000000000000000000000}$ and yet because of their inability to understand concepts regarding truth they wish to place the cosmos that then applied to the cosmos that now applies. They look at a hologram of space going back say a Yotta- meter times another Yotta- meter (Y or numerical 10^{24}) it is by numbers 1 000 000 000 000 000 000 000 000 000 000 000 000 000 000 000 meters of space is a concept surpassing understanding that developed and that image of a Tera- years (T or numerical 10^{12}) it is by numbers 1 000 000 000 000 of space development that they wish to transfer to what the space was a trillion billion years ago. Then the hologram goes out of context completely and this effects out truth about the cosmos. As long as science tell the cosmos it has mass and the planets align by mass while it is not true then they are busy with a ridiculous complot to deceive and defraud just as common criminals do.

Motion creates space-time as much as space-time is supplying motion to form space. The only way to enable that to become a reality is that motion creates space as much as space follows the direction of motion. That is what Kepler said when Kepler said the space is equal to the motion thereof $a^3 = T^2 k$. There is no solid a^3 Universe but all interrupted by positional changes that recreate the space in the and according to the new direction singularity will create as singularity allows space to flow by motion by fragmenting space into time sectors.

We with life take motion as for granted. We never consider why objects in orbit would have motion and what would be the result when such an object is deprived of motion. Having life means moving and we as humans are on the move from birth. We even call an infant that has died at birth being still born meaning it did not move in the manner we accept life would move. The last thing life would have before capitulating to death is to have the blood circulation terminated. In that way to our reason motion is a fundamental or a basic which is true but that makes it even less common. Why would all things move? The quickness of the Newtonian mind would remind me about the galactica inner core that shows no motion. That is a time concern and not a result of inadequate motion. There is lots of motion but the motion is more delayed and therefore we are unable to witness it.

Before I allow myself to get tangled into that explaining (I do explain this issue on another occasion in this very book) I wish to return to the basic concept of motion. Kepler quite correctly stated in his formula that the cosmos holding gravity is motion. Motion of space in time as well as through time stated as space $a^3 = T^2 k$ moving in time and through time. In that we find the birth of the cosmos. Newtonians have this grand dilution of a Birth of a Cosmos that holds a formula, which can crack the jaws of a hurricane.

Kepler's formula shows how the Universe started. When space a^3 started it moved in time and through time $= T^2 k$ where time holds two parts in securing space. Realising this birth we must get to terms why stars move. Why would a star move and why would they not move? I guess it would be fair to put any and all activity concerning motion of the star in relation to the activity within the star. The motion of the star must be related to the time component the space uses to move within.

As humans tend to do we look at the brooder picture and try to assess from that what applies. That makes us miss the target by miles. The star moves with everything in it but it is everything in it that moves and therefore the star moves no with everything in it but because of everything in it. The star does move everything in the star but those within the star moves more and therefore by creating more motion the star establishes an accumulative all including motion of which all individual motion takes part in and therefore is part of such an accumulative motion. The atom is vacating the allocated place it occupies and moves in time and through time to claim and a new position in which it holds space for one point in infinity. It vacates to fill and while vacating to fill it claims a spot where it is. Let's put it in the manner Kepler introduced it. Material a^3 vacates $T^2 k$ an allocated position to fill $k = a^3 / T^2$ a predestined position and while vacating $k^{-1} = T^2/a^3$ to fill another spot it claims $T^{-2} = a^3 / k$ a spot where in it is $a^3 = T^2 k$ at the present $T^2 = a^3 / k$. Well normally we would leave it at that but then that would be that and we would be left with no results. What makes the atom move?

In the first factor of singularity a line indicates direction and hydrogen becomes a volatile product. But singularity also provides space a^3 and duplicates by motion k to destroy by rotating T^2. In dismissing the space the proton grows by accumulating space that the other side lost. The growth depicting of the dot I use to symbolize the proton's growth are highly exaggerated I have to admit, but that is only to bring across the idea I wish to convey. Looking at carbon$_6$ one would think that proton numbers would bring about mass, which we then associate with density between particles. But then comes Nitrogen with seven proton pairs and oxygen with eight proton pairs, Fluorine nine and Neon having ten. These mentioned are significantly highly volatile which means they truly extend duplicating of space.

This as a group forms a relation with heat unlike any other. If mass did the trick these must have been the group having the second least density, but they form the group with the least density as a five point group. The next group holding the Pythagoras five or the Lagrangian five plus two or three, four or five enabling their relation with heat to be quite remarkable. This must be some indication of events during the period just preceding the Big Bang at say 10^{-7}, 10^{-6}, 10^{-5}, the time when the fuel that would ignite the Big Bang turning heat into space turned material into heat. The relation of five plus one, plus one and five, plus one plus one plus one is just too uncanny to ignore.

Following the process and seeing the influence of singularity should bring about a pattern that may lead one to a pattern of how the required heat formed and how the intended heat transformed to space. Density depends more on proton number arrangement producing specific form in relevancy as to merely and only having mass as factor that contributes to the forming and development of stars in the cosmos. The evidence is so clear that mass has nothing to do with gravity but density has everything to do with gravity. Density is the volume of space in numbers used to fill material in ratio with numbers of space per volume not filled with space. It is matter versus space in every sense there are. This came about before the Big Bang took place and before space was formerly space and time was formally motion. It was a time when singularity set relevancies moving from Π^O to Π

In that manner we know that that was the way particles formed combinations just after the arriving of moment-Alfa. Singularity brought the Universe but also singularity brought the divisions between the many Universes that followed the immeasurable many Universes that came after the flooding of Universes to follow the leaders. The term "moment-Alfa" is the way I refer to the moment when singularity changed, not when space formed or time began or space exploded but even before anything including mathematics became definitive. At this point mathematics renders it useless. There was no space or time to calculate because relevancies came in place. Form took shape but space there still was not because Π^O moved to Π. Every slightest point in space became an opportunity of establishing a Universe with most different functions and ingredients there might form. This is apparent from the fact that it still takes place at the present moment by motion attaching new singularity through duplication and through duplication releases previously attached singularity from serving the purpose of duplicating by motion.

When the cosmos came to motion, motion was not yet defined. When the cosmos brought about motion, the first motion was relevancies. Cold parted from hot. Eternity parted from infinity. Motion parted from motion absence. Infinity broke the laboriousness of eternity for the duration of infinity. The spot became and grew into the dot.

From what the spot was to what the dot now is might be just a mathematical implication of going from 1^0 to 1^1 but in reality that first motion was the creating of and establishing of an entire Universe with all possibilities now in it. Never again can that much growth become a reality, although to us the growth is beyond what we ever can notice. But it is because the growth is so massive and we are so small that we are unable to notice such almighty growth.

● **Singularity**

and ●

When the spot Π^0 became functional established all relevancies possible, heat parted from cold as eternity parted from infinity. The expansion was not clear motion but more a parting of relevancies where a centre formed a relevancy because the centre could not provide motion. Without being capable of motion, the centre established four points, which also served singularity. From the inverse square law we know that the centre doubled by producing the four points holding singularity.

By exciting the that formed in division that heat had to move but centre spot, the centre spot came to be because of the heat relevancy as heat parted from the cold bringing about the followed and that was the motion that formed. Therefore the being singularity it could not get singularity to move. In an attempt to establish growth, singularity activated six spots of which four was having motion drawn into relevance four spots that was providing what was to be motion and three that was to be securing the position the centre holds. There were four forming a ring around singularity with two forming in locations we will refer to as above and as below or north and south.

The three in line was in singularity not being able to move but the four was also in singularity and just as incapable of moving. All the points came as relevancies applying the forming of more of what was to come but only the four committed to time were expected to move. The four points that came as a result of discrepancies that became time that produced form and that established the relation with the one but had to perform the motion by expanding was as much incapable of motion as the centre was that charged the four with motion in the first place. As they were incapable of motion, it still required a tendency to apply motion that did separate Π^0 from Π. This not only involved form but it involved all relevancies that did come or may in the future come about as a result of the attempt to commit motion. If mass was a factor contributing to gravity the cosmos would have frozen back to singularity without ever releasing singularity to relevancy.

Mass does not establish gravity. There is no magical graviton. In the beginning there was no mass but boy was there gravity! The only means that the cosmos could find a way to break from the grip of eternal eternity was to expand into relevancies. Such a feat can only go to task by forming opposing hot and cold. Becoming hot produces more of what is heating. That implies motion or a moving away from where it was by generating more of what is available. Only where hot released from cold could whatever was repeated once again and duplicate what was before into what then is more. Secured by motion T^2 in relation to a specific centre **k** from where singularity holds the Universe true to form. The **k** was an intention to place apart and by today's standards will not even qualify any noticing.

All that are is in singularity. From singularity comes the motion and the space we call space-time. Singularity is dimensionless, time less and space less and because of all this features, it carries the value of Π^0. By expanding, singularity applies a relation coming about that reforms singularity from Π^0 to Π. Only when extending Π^0 to Π, the extending creates motion and the motion creates space that then doubles through motion applying which cuts the space in motion in half by matching the space as a duplicate. Motion creates another dimension or another level reforming singularity from Π^0 to Π or from Π to Π^2 or from Π^2 to Π^3.

As said before we now know Π came about since Π is achieving form and not space. Only **r** can establish space as size will accumulate and as it had with everything else singularity had **r** covered by one as in being $r^0 = 1$. By reducing the circle radius **r** by half continuously will lead to an infinite small circle and an infinite number holding r would place **r** to the power of one as a factor. Then as a factor **r** would not contest any change when change is introduced into any future equation but Π will remain because the circle as a form remains even being infinitely small. By reducing r indefinitely to the tune of half each time, r would become infinitely small, beyond human calculating means, however as mentioned in the case of the smallest dot holding one spot, r would become insignificant beyond human comprehension even, but never reaching zero and still Π would remain intact and dictating form. To amplify by dimension a value has to be set to r but if r remained covered by singularity all alterations that could possibly come about was in the form, which was Π.

This expanding can be a problem one can wrestle with for one lifetime and never reach any conclusion. How can something grow without getting more that what was before? Then it hit me like a ton of bricks. The answer is in heat but not heat, as we know heat. It is heat in getting relevancies between outer limits. Only heat could break the monotony of singularity. Heat in the form we now know heat as heat is now. Since the Big Bang heat is material transforming from one state to another state.

The change that took place involved singularity but singularity was 1^0 and being $^0 1$ could not grow. The growth came about. Heat rose from singularity, but if heat rose from singularity. Singularity as a factor changed from 1^0 to 1^1, which means a relevancy came in place that no one could detect. It is true that 1^1 are still one, but one could then escape from singularity by producing factors other than 1. Heat came about but only as a relevancy to utter cold. If there is heat, there is cold or if there is no heat there can be no cold. Space came into forming a relevancy that brought form. Since it is a relevancy and not a generation by accumulation, the form produced was Π. The spot formed a dot by heat and cold establishing relevancies and from that singularity was broken to allow all other forms of relevancies to come about. The cosmos did not start because of gravity. The cosmos started with heat and cold coming into a relevancy and in the cosmos there is no hot as much as there is no cold. The cosmos broke, put from the confinement of singularity by establishing a singularity in a relation of heat and cold. The heat that came about was beyond measure because the cold that held the heat was also beyond measure. The immeasurable heat was on the outside of the dot that formed and the cold was on the inside of the dot that formed. The cold contracted because in nature cold contracts. The heat expanded into a dimension of form and heat by expansion is in nature about motion. Motion is duplicating that which is and heat is what is duplicating by motion. But only heat by expansion was possible because in affect singularity cannot move. The motion became contraction, as the motion was the result of heat expanding which was forming four points in the rim of the dot. The expanding of the points created motion in relevance of a centre that formed because of the motion, which established an immovable centre as the Coanda effect, placed more dots in relation to more dots that formed

Every dot was Π and every dot formed $Π^3$ because of the expanding heat, which produced $Π^2$. With that a new relevancy came about forming a centre in between the four points of expansion that was resulting in time. But since the points were in themselves singularity, which is immovable and space-less, they still heated forming a cold centre with the heat bringing about motion. It became a repetition where infinity broke eternity by producing a centre because of space (or rather form) forming the motion to enable the space to form in relation to the heat applying motion. This brought about a Cosmos being conceived.

The spot forms a full circle, but the line running through the circle is forever present because that is the future radius of the circle that will one day develop the circle, which is equal to the present diameter. The fact of the presence of such a possible line in such a possible circle dividing the possible circle into two parts makes the centre line equal to the half circle. The line forms the half circle but not only that the line presents the half circle as much as the line is the half circle. The line then is 180^0 and the half circle is 180^0 because in singularity the two factors are the same.

The same value is of course $Π^0 = 1$. The issue of concern is o understand that singularity cannot move. Singularity has no space. Singularity is no only part of the Universe but singularity is the Universe. By establishing motion singularity has to be charged with the time delay we find space to be. The space is time taking a period or a duration while moving from one singularity point to another singularity point while conducting the heat and the accumulation of heat that built up due to the retarding of the time to conduct the heat forms the space that is conductor to bring about the motion of the space.

It takes heat time to entice singularity and singularity can only entice. Singularity cannot move and neither can singularity form space. By enticing from one relevancy to another there is a bridging of heat that has to be crossed in order to send the gravity or the enticing or the relevancy to depart the space and reconnect the space to the next singularity. Bridging all the accumulated various time delays that formed an accumulation of heat through time distorting brings us the space we see and have. However there is no true space or motion but it is eternal motionless space is singularity charging time to provoke heat into forming space.

Three points formed a line covering singularity where the centre singularity recovered heat to grow and two points served as an axis to allow the rotation and to assist the duplication. There is one centre connecting the duplication of three as well as the recovery of one (the fourth one) that is applying the tie aspect. Therefore, motion consists of three positions in relation to a centre, which forms as space in relevancy to the motion and the space receive a controlling centre.

The duplication comes about as singularity is exciting another singularity in precise relevancy of 3 to 3 to 1, but the points charged is as space less and as motionless as only singularity is. The heat it requires to carry the exciting between points forming space and the space excites heat and the time delay it takes to excite singularity between points forms space-time.

Where motion conducts electrical charging which is equal to gravity the charging of motion is to entice duplication of singularity. This is the basis, the heart and the sole ingredient of the Coanda principle that includes the Roche limit ($\Pi^2/4$). The charging of gravity $((7/10)+(7/10))/(10/7) = \Pi^2$ and the charging of space-time $\Pi^3 = \Pi^2\Pi$ is all due to the relevancy brought on by the Coanda principle. The value of motion came from singularity exciting singularity and that is the duplication while the duplication or motion presents the space.

The development came into eras as the relevancies brought about new relevancies that spawned even newer relevancies that all remained in touch with the original singularity centres. Every one focused a new time delay that eventually brought about space and every distortion of time brought more. That concentrated between singularity points that charged the points to form space. When the charging became overdue in some sectors it erupted in forming the Big Bang. By the time the Big Bang erupted there was such a huge backlog in heat and time corrupted and delayed the next result was the employing of space as a commodity in the Universe. The relevancy was C the gravity was C^2 and the space was C^3. That left what was inside atom still spinning faster than the speed of light applying the relevancy of **k** = C where the electron applied the relevancy of $T^2 = C^2$ and that formed the atom which then became the cube of the speed of light $a^3 = C^3$. That left the atom at the relevant size of what the speed of light permitted at the time but since the Universe from that the relevancy expanded as the Atom grew in space to the extent it has now. The purpose of the star is to recapture the space the atom grew into and from there dismiss the space by spinning faster than whet the speed of light will be on the outside of the star.

* This form came about when only form was present in the cosmos and there was no room yet for space. It was in a time era where form featured in relevancies that would lead to one day becoming the atom. The atom forms a dual purpose of duplicating as well as dismissing and some prefers the one better to the other. This relevancy came in place when time was not time and space was form. Time is forever eternity being interrupted by form in infinity to bring about eternity ticking as infinity ticks. Before that singularity took on stages in forming relevancies between duplicating and dismissing space-time, which incidentally was not yet truly space-time in the sense we think of as space-time. At first a dot moved from the spot leaving the spot but taking with the spot as part of the dot to remain in the dot. The two never separated but the one allowed the other to be.

As the dot confirmed a discrepancy between infinity and eternity by defining infinity as an interruption of eternity cold and hot parted a union. The dot that formed was not space but a relaying of time to form a new point of singularity where eternity was interrupted by infinity. Time took form from 1^0 to 1^1 or from Π^0 to Π. It brought form into differentiating between interrupted eternities with infinity doing the interrupting. Then a true distinct relevance came about that positioned a time differentiation outside the realm of time by four. In this realisation we can assume that space had some meaning at this point and the formula used to investigate suggests just that.

Even in grouping, there are characteristics, which make a certain group of atoms more perceptible to duplicating and others more perceptible to dismissing

Where the seven of material duplicate by the three positions according to time and from that realise the sphere, the other side of the Universe uses other dynamics, which also realises the form of the sphere in as much as being the four five sided straight lines pulled around a circle with multiple circles attached to one specific centre and this brings the ultimate sphere. There is conserving expansion about space in acknowledging a common centre that controls and influences a dynamic formed by a sphere which is inside another sphere formed by a consistent continues flow of a line forming relevancies as lines that is connecting at 90^0 angles by measure of Pythagoras.

Finally with all the heat retarding an interrupting of the line a gap formed time.

All the while it was just a spotted and dotted line running along time as space duplicated with heat surging and cooling as cold contracted much similar to the actions of stars in the process of pulsating by what ever name one wish to use.

The Black Hole proves to be the ultimate atom and is also the ultimate profile of the Coanda principle. It is stretching as far as singularity can stretch space-time in the reducing of the cosmos by gravity.

What I introduce is what Kepler introduced although Kepler may have used somewhat different factors in the manner he introduced what I am about to introduce. The black Hole shows the Coanda principle in its ultimate role. Any person doubting my theory about space –time moving towards a centre and thereby creating gravity must then explain the Black Hole working. But please for sanity's sake let's forget the theory of gravity going mad in a super nova event and then becoming a monstrous Black Hole. The super nova is as natural having the process where gravity is in the process of moving from the atom accumulation to the governing centre singularity. However in the case of the super nova things went wrong because the super nova occurrence is where the transfer from movement or gravity within the star went out of control. The motion was to slow and the build up of accumulated heat exceeded the motion and over came the contraction of the governing singularity. In the process heat went back to gas as time increase by millions of folds.

On the other side of Universal time another scenario gives the sphere distinction. Three dots accumulated on one side and three dots on either side accumulated on the other side of the Alfa singularity dot. This made up the second dimension   How long did it take singularity to evolve from the spot to the dot and then on to form a sphere? It takes 1^0 to go to 1^1. To us it is incredibly small and it is so small we can form no understanding of the concept but can you see the spot, no you can't because although I know it is there, it is so small it falls outside the parameters of the Universe.

The spot • forms the dot ♦ Π^0 The spot comes first, so small it is not part of our Universe and then the next movement takes the spot to become the dot. The dot is there because the spot became the dot but the dot is there because the dot moves in relation to all other movement.

• The dot Not yet formed 1^0 to 1^1 • if the spot did not move in relation to all other dots ♦ in the Universe the dot ♦ will not form from the spot •. It is through movement that the spot grows into the dot.
The spot 1^0 forms the dot to become 1^1 and yet in relation to space the dot is simply a thought an idea an emotion that will in future become material because it relates to material.

The dot ♦ Π^0 has not yet formed material Π^0 to Π ♣ and material must come as all other positions forming material or not forming material reposition in accordance with everything else forming a position of space within the Universe formed in space. These lines are representing the dot that serves as the spot extended. I wish to deliberate the dot in better detail. The dot is an extension of the spot and is immeasurably bigger than the spot. If the same space improvement has to come into place in terms of the way we visualize the cosmos the moon and the Sun would suddenly be double the distance way halving the space that the Earth claims.

• The spot Π^0 There are two pertinent numbers applying when space-time forms, one is two and the other is three and then seven…and it starts off with a spot that becomes a dot. The dot becomes one thousand times of what it was and remains invisible in relation to the entire Universe from the point it becomes the dot. Where the spot ends and the dot starts there is still no space (1^1 to 1^0) and therefore with it being a mathematical factor it then represents one phase of cosmic development that stands apart from all the other phases that came to follow this phase.

In a close inspection of the dot as the dot is in our current Universe we find the dot not holding a quantifiable measured value except for being the numerical value of 1 to whatever exponential value one wish to apply. This is proven by $1^{256} = 256^0$ and this shows a specific period in cosmic development predating the Big Band by many eternity of mathematical translation of numerical figure formation to actual space. This spot that forms the dot is not tangible even in the terms we apply to anything being a part of the reality in the cosmos. It is similar to all the characteristics we have in a thought. In fact it has everything that a thought must have. Having a thought in my mind is a reality as far as I am concerned but in the view of the person next to me the thought has no right to exist.

The fact that thought is a part of the cosmos but the cadaver confirms not have a part in the body. When the body turns into a cadaver the body weighs as much as it weighed before such turning into a cadaver ended the person's stay on Earth. After the thought (also called life) leaves the body the body disintegrate to atomic particles with no structural attachment in any way. It is life that attaches the atoms forming a body to have any coherent significance that then can function as a unit. It is thought that drives the cadaver as a principle carrying life. Denying the existence of the fact of life being apart from the body and forming a part of some concept that is much bigger than the space-time we find in the body is very stupid or very Newtonian, which is the same of whatever you wish to name the same concept of having a feeble view on science. The concept of the thought is what drives the body. Life can only manipulate what is part of the cosmos and can't create anything outside the cosmos. That is then what drives the cosmos where the point holding singularity attaches to the same principal that drives the cosmos and drives the living body. Without thought or emotion or whatever you wish to name the driving principle it is the same as that which drives life but is on an immeasurable larger scale. This goes much further and from that we have a lot of information to harvest about the cosmos. Newtonians hare the opinion that the thought is in my head but again that is much Newtonian or in other words represent that that comes about with the lack of intelligent thought. It can't be in my head because in my head is grey matter. If the thought was in my brain then after death my head must be filled with thought since my brain is still there. After death sets in the thoughts are gone and the brain is still there so therefore the thought is not in my brain but my brain is only a device translating what I think to the rest of my body. If it is the electricity with which we may charge thoughts back into the dead body by establishing a flow of electric current through the brain but doing that will not produce thoughts that would reinstall life and this shows that no part of the body holds thought but thought holds every aspect of the body. This is critical in understanding how the cosmos works and where the cosmos came from. It can't be in the grey matter but since it has no material in which to be so it must be a part of the vision in the Universe and material is part of the vision in the Universe.

The fact that the thought is there gives the thought a definite place, albeit not tangible, but definite nonetheless. Everything in the Universe confirms the Universe as much as it is being confirmed in the Universe. The thought takes up time and in order to take up time it has to take up space because space is time and space used is a form of time delayed to become the past. Before Creation the past was the present that was the future but now time in the past is space coming from the present as a result of movement coming from the future. Thoughts does not take up space so therefore it has to be space less time and therefore it has to be space that time produces while moving from the future through the present to the past. I am deliberating this because Newtonians work in the era of light where they can see the lion and when they can't see the lion there can be no lion and when they see the lion they run because there is a lion. What they can't see (or calculate by mathematical formulation) is not a reality. If they can't calculate what life is they put life in terms of electricity and that they can calculate. The reality of life not being electricity they ignore not because of stupidity because that is an insult to stupidity but to be abstinent because they think (or can they think) they can replace God.

The dot $1°$ to 1^1 In a close inspection of the dot as the dot is in our current Universe we find the dot not holding a quantifiable measured value except for being the numerical value of 1 to whatever exponential value one wish to apply. This spot that forms the dot is not tangible even in the terms we apply to anything being a part of the reality in the cosmos. It is similar to all the characteristics we have in a thought. In fact it has everything that a thought must have. Having a thought in my mind is a reality as far as I am concerned but in the view of the person next to me the thought I have has no right to exist. The thought is in the present in time and the body controlled by thought is in space. Time is the instant in which thought is and in that sense time travel as the Newtonians whish to speculate on is total madness, as much a product of their incapability to think as their creation they call "mass". Time is in the instant that the thought is and in order to control the body in space-time is needed to send a message to the arm or leg or whatever that is in the space that is behind time. There can be no time travel because time travel is rushing through space but even that requires time to accomplish that and since time is in the instant within thought it takes space to cross and time to cross the space to even control the body by thought control.

The fact that thought is a part of the cosmos but when life becomes the cadaver that absence of a thought controlling the cadaver confirms a thought has no part in the body. When the body turns into a cadaver the body weighs as much as it weighed before such turning into a cadaver ended the person's stay on Earth. It is thought that drives the cadaver as a principle carrying life. Denying the existence of the fact of life being apart from the body and forming a part of some concept that is much bigger than the space-time we find in the body, is very stupid or very Newtonian, which is the same of whatever you wish to name the same concept. The concept of the thought is what drives the body. Life can only manipulate what is part of the cosmos and can't create anything outside the cosmos. That is then what drives the cosmos where the point holding singularity attaches to the same principal that drives the cosmos and drives the living body. Without thought or emotion or whatever you wish to name the driving principle it is the same as that which drives life but is on an immeasurable larger scale.

Newtonians share the opinion that the thought is in my head but again that is as much Newtonian as the "mass" they create to hide their stupidity or in other words represent that that comes about with the lack of intelligent thought. It can't be in my head because in my head is grey matter. If it must be in the grey matter then why does the grey matter not produce thought after I am declared "brain dead" Since the grey matter is still present and the thought is absent the thought has no direct part in the grey matter? Since it has no material in which to be it cannot be a part of the vision in the Universe and material is part of the vision in the Universe. The fact that the thought is there gives the thought a definite place albeit not tangible, but definite nonetheless. Everything in the Universe confirms the Universe as much as it is being confirmed in the Universe. The thought takes up time and in order to take up time it has to take up space because space is time and space used is time delay. It does not take up space so therefore it has to be space less time and therefore it has to be space that time produces while moving. I am deliberating this because Newtonians work in the era of light where they can see the lion when there is a lion and they fear the lion when they see the lion because only when they see a lion can there be a lion. That thinking is Neanderthal but don't tell the Neanderthals because they are modern in terms of Newtonians. When they can't see the lion there can be no lion. That is somewhat atheistic or animal like in thought. The animal only sees a lion and then feels threatened. That is where Newtonian's arguments start and end. If it is not tangible, in vision, part of the smelling or in hearing range there is no possibility of it being part of the cosmos. What is in the cosmos is part of the cosmos. So too, is the thought and we have to trace the space and the time the thought holds. However I showed that there is a place in the Universe that is not part of the Universe.

The thought may drive me in reacting in any manner that comes to mind and the person next to me will never be aware of the thought I have. The fact that I have to have a thought is a proven fact without ever doubting such a statement. No person can have a mind that is clear of thoughts because in such an event such a person is brain dead. The thought is there in my mind while to others it is there in my mind without the thought being there, but such a concept reduces the Human intellect to animal instinct. It is present by never being visible and not being tangible makes the thought not being present in the view of others, although that thought is "me" in every aspect "I" am present. The only way to make the thought tangible is by speaking the thought in words. I have to reproduce the thought into a voice box and translate ideas into verbal language to make the thought a reality to others. If I wish another person should regard the thought I have with understanding I have to transform the idea I have to a voice and in the voice I use I reproduce the idea in terms of sound. But this action relates to time. Having the thought also related to time but the time spent on the thought while thinking was much less than the time used to find a means in converting the thought to sound. I transform my thought to another person where I then hope that person will translate the sound back to a thought that that person will have. No part of what I just mentioned has any tangible evidence, which I can touch or hold but notwithstanding that it is part of the most valid evidence of the human being in the Universe. Never can I reach any part of what I just said and still that is what controls the Human race. If not for the ability to think and to produce conversation by voice or body language, the thoughts I produce I would not be classified as a member of the human species. That is what makes me part of being man, the most dominant specie ever to live on Earth. If not spoken, my thought will go lost forever because I did not share the thought with any person. Yet the thought had validity because the thought controlled my life for a period and an impulse. I might try to determine the time it took me to think the thought I had, but that will be in vain because no chronograph can measure such time response. If I try to measure the time it took to transform the thought to sound through applying voice that can be measured. That is because it is done by time that already formed space.

All this arguing is to show those Newtonian nihilists that it is what they and I can't see, touch or feel that is in control of our daily life. Without the thought the deed will be impossible. But the conveying of the thought gives the thought a lasting reality because it then is shared with others. They put their reputation on mathematics, which then forms physics, but do they know mathematics or for that matter physics at all? Has any one ever explained how mathematics formed physics while the cosmos formed mathematics in the process of development? Mathematics we know is the forming of a numerical line giving values according to a flow of values by adding the original numerical value to produce growth. If mathematics started with 1 we have to add the original value that was first brought into as the first value to get to the next value as it is 1 + 1 = 2 +1 = 3 + 1 = 4 + 1 = 5 = 1 = 6 + 1 = 7 and so on If Newtonian conclusion is correct we have and the numerical order started with 0 it would produce no further growth. For instance 0 + 0 = 0 + 0 = 0 + 0 = 0 + 0 = 0. This is a mathematical fact that is above argument but since it indicates Newtonian incorrectness I have never seen one Newtonian that does not belittle this fact as insignificant rhetoric on my part. Numbers would never evolve past zero.

Let us find the smallest possible line first. We already have reached the conclusion that by reducing the line, the reduced line will eventually leave all sides on the same spot. Such a spot must be round in form. With the line being the smallest line, such a line will start off as a dot that moved away from a spot. With all possible sides being in precisely the same spot we have all possible sides onto one spot. Mathematically the spot is in the single dimension where the space is one and exponentially zeros. There the space moved over to form the dot. We now are reaching into areas only the human mind can venture by understanding and nothing more. The understanding of this concept demands our reaching the point where the mind of the animal cannot reach. Numbers start with a line and numbers grow by adding the original value every time.

If it starts with a line that line only represents two sides being one and as such that is rather a flat Universe. The spot is not yet round because being round are requiring a shape or form and this lies beyond or before a time when any form of shape came into the cosmos scenario. It was in a period where shape and form was a part of the distant future hidden in and beyond eternity. In that time the line must have been so small it had reached a point not yet dividable in any way. If any further dividing took place such dividing would have brought growth because there then would form space between the sides going in the opposite direction. The dividing brought all there is having all sides literally on the precise same spot, and I have located singularity in just such a spot.

I came to the conclusion that the spot I found had to be singularity purely on the grounds that that spot holds only one side to serve as a start to the starting point of all directions possible. In that side is only one spot where there is only one side applicable and one dimension present. With all the factors given one can only come to one conclusion and that is that there can be only singularity. In such a case more dividing by two will land further positions on the other side of the divide. That point is serving as a position for all possible points and cannot allow further dividing as it is in the smallest line or spot there may ever be. This spot is the result of a most basic process of reduction as the Hubble constant is a most basic process of expanding during a matter of time. By reducing the line constantly the only value that will eventually remain without dispute from any party arguing about the facts is exponential zero. By only having exponential zero instead of a numerical zero and a radius as one in the square (the radius effectively becomes one holding any and all sides on one point) such a point might become any value of any significant measure implicating anything but zero as the radius. By expanding the line, it will be an evenly spaced structure growing into the most perfect round dot ever possible anywhere at the point when it starts to grow.

Firstly we have to get rid of the Newtonian notion of zero. A line does not start at zero because that is where Newton went miserably wrong. If a line did start with zero the Universe had to be nothing, which it is not. The Universe is an assembly of lines criss-crossing as the lines form circles where one circle follow the other circle with no room in between. When looking at the circle in the conventional manner, we persist with errors brought about in culture of stubbornly upholding Newtonian mismanagement of facts and not by applying some significant modern logic. Take a circle and reduce such a circle constantly to where it no longer can reduce. Reduce it to a point where only form remains part of the circle because the radius has gone beyond human measure and becomes so small it is not noticeable with what ever measuring tools man may use, then what remains is pi since pi does not indicate size but indicate form, and form is all that then will remain. We know that the ratio of pi is 3.1416 to one and one is singularity and not a measured meter or a specific distance. If a line forming the radius is 1 then the circle brought about is 3.1416. That is the connection to singularity and if singularity is seven then the circle is 21.991. This where we should start to look for a point where the Universe started since the Universe is a ratio of lines connecting circles. In any circle or sphere the size only depends on the fluctuation of r, as a component to the circle or sphere but that does not affect the form by indication of Π in any way there may be. The conclusion I drew from following this process is that from this line can start at zero because that will be a mathematical impossibility since no line can ever reduce to zero. If the line of the radius that initiates pi was zero, there was no pi. If there was no pi there was no Universe with or without Newton's magical forces. A line will forever be able to reduce further becoming smaller but it can never reach zero because zero is not part of the scale on which we measure lines. If a line cannot reduce to zero it then cannot start at zero. Zero is in the mind of the Newtonian but so are his forces and other witchcraft he projects into science.

A line or spot starting at zero would therefore be shorter than the shortest line possible. For obvious reasons can no line, or any line grow or extend from zero because such a line must then quit zero, which is the current value and become some other value. It has to become something, thus abandon its original value. However, a line that starts with 1 when growing it adds another 1 to become 2 and by adding another one would become three. That is where the Universe started and that I show. If Newtonians are correct and the line did start at zero with zero that would mean the start of the line has a different value to the end of the beginning and we know that a line holds numerical conformity through out. When any line is starting from point zero and it uses the factor zero, then it can never leave zero because of the influence of being zero disqualifies any possibility of growth. (1000 x 0 = 0) and that is a mathematical law. But when coming from singularity π^0 and the line then had to grow in all directions at the same pace the line must then become a circle π or being three-dimensional, then form a multi circle π^3 we named a sphere. Since the Universe is about circles and lines connecting circles I came to conclude that flowing from this fact is that in the Universe there can be no zero improvising as a filling ingredient for the space of a point or be unfilled space. Zero is no valid factor in the Universe. In the case of the growing sphere the value of the circle is Π, and that is where creation must have started. That gave me the clue where to start looking for singularity. After I found singularity that finding directed me to the point where to look for the start of the Universe. One would find singularity in the value Π as 21.991/7 or as 3.1416/1 and the value Π will be in all things rotating in a circle but by measure one dimension smaller. As usual I am again shooting the gun before the hunt started. Lines in mathematics do not start from zero and that is no discovery on my part but that was a realisation I came too. The Universe is all about lines and the manner that Kepler pointed to the increasing of the lines by **k= a³/T²** proves growth in the composition of all lines. However only movement secures space.

●Π^0 Singularity was the first there was and was eternally running as the past going to the present and on to the future but also returning from the future to become the past. Everything was in perfect synchronization to accomplish the perfect cycle. This was only possible because the only value that was valid was 1. One repeated eternally without ever stopping. Then for some reason that which was eternal and perfect became temporary and imperfect. Singularity distorted into a concept outside the eternal. Then that which can have no inside parted from that which can have no outside. $\Pi^0 = 1^0$ **parted from 1^1. Heat became relevant as the Universe transformed.**

●1^0●1^1 Then the perfect era ended with infinity releasing eternity as much as eternity dismissing infinity. There was a past seen from the present and there was a future seen from the present. There were two dots moving from one position to the next position.

1● ●▶ 2

The fact that 1^0 going 1^1 brought movement can only become a reality as a result of light. Light is heat and the heat is expanding.

1^0 going 1^1 1^0 ▶ 1^1

1^0 going 1^1 1^0 ▶ 1^1 had to bring about 1^0 going 1^1 1^0 ▶

●+●+● = 3. Two dots were moving where the one dot was the future of the other dot that formed the immediate past of the one behind and they were connected by an immovable centre dot. To find surety about this we have to look at the numerical order that came about from this. Two plus one is three so there were two in movement with three connecting the two. This brought about time in movement.

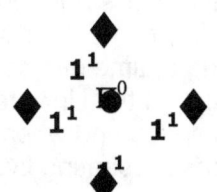

The number 1^0 was the same as 1^1 but also completely different, because the eternal repeat of duplicating while contracting was not relieved from the Universe. Before the contracting was equal to the duplicating because by measure the heat was identical to the cold. It was eternity that was interrupted by one cycle of infinity and was in repeat of eternity. This change is since once something is part of the Universe there is nowhere else to take it so it has to remain as a part of the Universe.

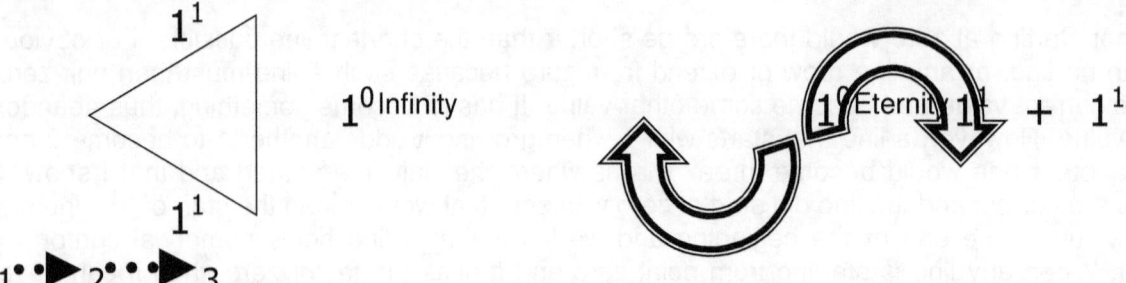

Then came three because motion was so limited that the least inclination to move threw what wished to move to the other side of the Universe, As it moves it also moved across singularity. It crossed the entire Universe as it moved because it moved and finding nowhere to move too. It crossed the entire Universe and it took one eternity less the measure of one period lasting infinity to achieve that. That brought to relevance three points where each was in measuring quantity exactly equal but also one Universe apart. In the reality there was now two points holding singularity on both sides of the Universe because by crossing the divide that crossing set in place the two sides relevant of singularity governing. However infinity was bridges at two points holding infinity with which process eternity repeated the past into the future.

●●●● Since the one dot was the future to the second dot and the dot moved from the position the first dot held the movement brought eternal time and therefore represented eternity from there on. The rotation of two $1+1 = 2$ brought about 2 doubling $(2+2=4)$ as well as moving in ratio $(2^2=4)$ the result was that on the circle time was forever moving and on the other side time was never moving. All the motion was connected to a centre point that had two moving $(2+1=3)$ and two moving $(2+1=3)$. Thus the one represented time moving in a ratio of three where the one dot became the future of the dot behind and the past of the dot ahead. The four dots that formed enlisted a fifth point in the centre that connected the four but also at the same instant there were three dots on the one side and three dots on the other side although the two sectors shared the same dot. In order to bring distinction to the two times three holding two opposing sides the one sector turned in the opposite direction of the other sector while also sharing one centre.

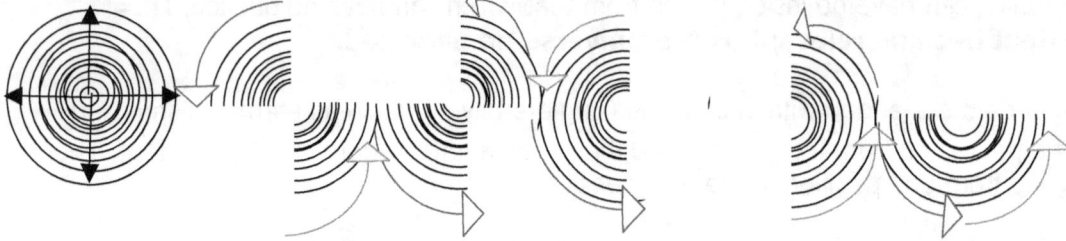

This development introduces opposing directional spin and is a major factor in Universe development to this day. It is part of fundamental gravitational influences. Also by bringing in three forms with equal influence the straight line the half circle and the triangle receives the same and an equal value.

There is a markedly difference in how five forms space-time and how seven forms space-time although both applies the same principle although not distinctly the same.
Five forms with the four rotating and one point forming the centre

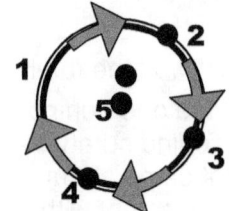

This is the first infinitive limit that Creation developed because this is where the value of Π forms by implementing the law of Pythagoras and it is the law of Pythagoras that kick starts all mathematics and forms the basis of physics and with that Creation by name. From this come the discipline known as the Lagrangian points and the confining of time into space. This starts mathematics by numerical value because one half of the Universe is ten in space and the other half of the Universe is ten in space and when combing that with singularity. We find pi form by the four locations presenting five places in conjunction with an allocated centre every time.

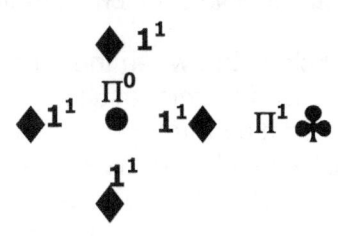

Although six is part of the concept five brought into duplicating the numerical order that followed five it also placed a new spot into the numerical order. Since three represents a line forming the order that comes about is that five completes time as time returns to the past by completion of five. The forming of the triangle and the straight line as well as the circle brings time into a repeat of what was to what will be.

The formed a repeat of what was to what again will be. His is a result of he fact that once something is in the Universe it will repeat constantly and the repeat of that will always repeat constantly because the cosmos does not change its basis such as time.

This then is the occasion where Pythagoras stepped in. Since it as a crossing of the divide the crossing involved a line that formed a half circle connecting a triangle. But the crossing was done in the space of half the Universe and since the Universe was 180° half of the Universe was 90°. That involved Pythagoras as mathematics was born. Up to this point it was arithmetic done with adding but now mathematics by multiplication came into place. Remember we are a few eternities in side the development of the Universe. Later on I am going to show how critical part the Law of Pythagoras plays in the forming of Creation and how from that an entire Universe comes about as a result thereof.

The radius that the Universe has is a factor of two. The increasing of this has a value of seven points. The relevancy in-between forms a Universe we have. The dimensions this carries are $2, 2^2, 2^3, 2^4$. The significance of this I elaborate in **_Matters' Time in Space: The Thesis_** where I discuss how stars function and the significance of displacement of space-time inn terms of singularity.

How did we get this far? There was one and then a split came in time. Time brought in space and in space there was a past and since there is a past there has to be a future. That made the past one point away from the future, which is three points. But since two points move one point is moving to the next position which is the present and one point is moving to the future which is the next position and since the present is the same poison we have two points (the past and the future) moving (the circle) Because the present remains the same position movement occurs in relation to one point that remains fixed.

This puts a line of time that remains forever but then space formed as a circle with the two points circling about the one point serving as the line. From the present the past is a point from where the present is moving towards the past. If there is a past there has to be a future that moves towards the present in order to fill the void. That is how gravity works. By moving towards the circle in a descending manner, the space just below any given point forms the future of the space just above and the space just above forms the past of the space just below. That is how a centrifugal pump works and the earth is just a centrifugal pump pumping away at space and compressing space towards the earth. If a pendulum can measure time by measuring the space the earth compresses in gravity then what the pendulum measures is time and then time is gravity compressing space. If time is compressing space then time is space expanding as well and so time is the movement of space towards as well as away from any given point.

Time is a line formed by• singularity and runs as straight as no other.

This line that holds time within the sphere is erected by the circle that spins about it and this we see in the manner the top spins. The spin takes the dot from the future to the past so by movement there are two dot moving future going to present as well as present going to past.

The movement alone forms two points but this cannot ignore the third point, which seemingly is shared, as it is stationary.

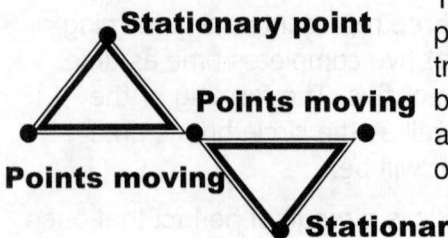

This shows that two points move as a circle but when the third point connected the two points moving we find this results in a triangle. It is still a straight line of time shifting through space but bringing in space as a result of time delayed this forms space as a triangle and a half circle although both also ahs the equal value of a straight line.

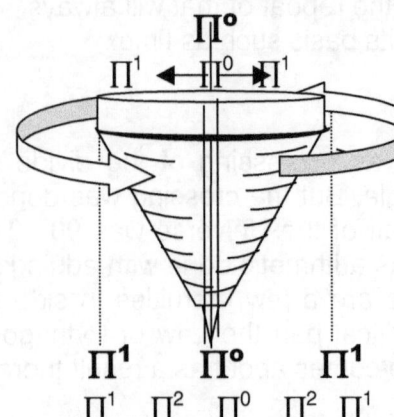

At this point space was an idea of time departing setting apart one point carrying time in infinity from the other sector in time forming eternity. Because the past assembles validity from the present initiating the past and the future taking queue from the present as what is going to be we must regard that everything starts with the present going past while allowing the future to flow into the present. For that reason time travel is a farce because no one ever tried to find the difference between time and space.

Time was in the represent. There was no numerical order because the numerical order gave value to one point that moves as two but remained as one that forms the constant. The flow of time gave a numerical value to three points because space brought about a circle and the circle held relevancy

We can see this evidence in the way the top generates gravity and applies relevancy. A circle puts a line as an axis in place and the axis supports the circle presenting space in which the top can spin. This is why everything that is has to spin while going straight but going straight calls for a circle to be and what spins will spin again on another level.

There were two points moving bringing about two more points that was the same but filling different locations, which then formed the square of two and that, is four. Then we trace another dimension rising from the way mathematics formed as the circle by rotation substantiated the line that validated the triangle. The line was the half circle by including one dot and became the triangle by validating the third dot. The number of four is just the result of the complete circle forming and five puts a centre into the circle of four to secure the circle. Then the number of six came about as the triangle that formed doubled and by it's doubling the trend for duplicating the lien came about. Transforming the entire value of the sphere brought about replacing the four of the circle as well as the three of the line, which forms seven. The rotation is in opposition to each other, which puts a distinction to what applies "on the other side of the Universe".

What this argument further proves is that the circle rotating must then come from all four points because the radius might be a line but that line represents the centre of a circle through 360^0 coming from and accounting for all possible directions. Taking that into account it is important to recognise that notwithstanding the size of what a line that forms any radius of any size might be, there is another line (or dot) that is eternally bigger as well as eternally smaller than the line in question. The line represents an innumerable many variations of possible sizes as it was in the past or what it would be in the future. While we are in the third dimension being part of the third dimension such being in the third dimension then allows that all parts of the third dimension forever can be divided once more until the line in the third dimension is no longer part of the third dimension. That then is as far as we have to take the lien to see where the cosmos started. When such a line leaves the third dimension it is still dividable because it might not be part of our dimension any more but it can still reduce further as part of the second dimension going into singularity. By that time it has left our scope by miles but that does not mean that it ends there because from our perspective that is where it ends. But our perspective does not represent reality. Our perception involves space and space is a hologram that holds the past and therefore in the present does not form reality. Yet, even then connecting to the second dimension, when being in the realms of space it can still reduce infinitely more until it has left the second dimension and then at last forms part of the first dimension where the first dimension is pure time flowing. Only then when the line reaches the first dimension no further dividing of that line is longer possible. We can never grasp the size of a line that the first line in size that came about when the first motion broke the eternal stranglehold on space. According to our big and small conceptions of what we perceive as large, ultra large, small and microscopic small is just mere words describing thoughts totally unrealistic in the context of what the cosmos sprang from as the cosmos moved out of the spot and formed a dot. Even by the standards of forming the dot, which was eternally bigger that the spot **T**, as the dot and all the many dots that came from the spot. The size differentiation only between those two exceeds all limits and divides we wish to create forming borders that we can appreciate. By coming from the spot and becoming the dot the sphere or space forms. Looking at the sphere we see many circles that are in association with each other and to find the truth we can look at one circle and see the characteristics of all spheres.

Any point will be opposing itself within the **rotating of 180°** where it **then change every aspect** of its **previous flowing** characteristics it had or **will once again have** in 360^0 from there. While in rotation from the view point of a bystander it all may seem static and never changing but to the object in spin every next instant in time will be diverting from every aspect it had every second passing, and the direction it held in relation to the direction it held the previous mille, mille second will totally be incompatible with the direction it holds the very next mille, mille second of rotation. This is why we can use degrees measuring the circle by (6^2) (forming the square relating to matter through singularity) X 10 (square if space) = 360^0 however it is always in motion. That proves no point can be static or constant, though it may seem that way to outsiders. Although matter is matter, matter can also be anti-matter and moreover form its own anti-matter at the same time. This degeneration of structure is very likely to occur with overheating. Revaluing Π to Π^2 will bring about a new contact point where Π meets **r** forming another relation in Π^2. Every time material swap sides it also qualifies as anti matter to matter because if it goes out of orbiting rotation frequency, it has the ability to collide with the same matter it forms union with but is located on the other part of the spin. It then becomes in a situation where Π **revalue to r. Time is** the **changes in relation** where Π **contacts a different r** not withstanding the many r points there may form because **every r constitutes a different value** to the universe through other ratios and relevancies brought about **by heat and light. Time is the duration it takes Π to rotate between any two given points of r** and therefore must always amount to **a square (T^2)** moving from point to point through the **cube of space (a^3)** in that **duration of time (k)**. With that it proves **Kepler's a^3 (space) =T^2k (time in the instant of motion)** but motion must continue through a specific value in space where the space-time is maintaining relevant equilibriums throughout singularity connecting.

Moving as an object that forms the numerical value the orbiting object also secures an individual independent and own centre but from the orbiting object the limit it holds ends at the edge it forms. This we named infinity and also the present. Being a circle that formed the object secures four rotating positions in relation to a connecting centre. The forming the shape the Universe applies which is a sphere the orbiting object secures seven positions and the larger containing sphere is ten of which seven is within the singularity dimensions within the centre, which we not observe. Immediately following that as part of that relevancy comes the containing sphere that holds space-time and another three positions. The three positions puts a relevancy of three to the holding space that already caries seven. There are forever another centre that secures seven positions just because singularity chose the sphere at the value of Π and in the sphere 7 positions is made up of six sides that hold relevance to a precise centre. There are Π^3 but then there are Π^2 putting Π^3 at a value. Then there is a relevancy named Π, which puts Π^2 at a relevancy. None of these is fixed markers because the relevancy can and does swap sides placing importance as alliances changes. When Π^2 focus on another a^3 the relevancy about k changes and amplifies the importance of yet another space. When one applies the Coanda effect one would see just how easily, new alliances come in place and secure new centres that charge new relevancies between newly established points. Going either "bigger" or "smaller" is only shifting focus on another relevancy.

Let us see why would space form. We have time forming two sectors where one moves and where one stands still. The moving is eternity and the one standing still is infinity. If only movement applied only eternity would be present but we know we are in the present, which represents infinity. We think we see movement but we see movement within space and not the movement of time. Infinity is double as long as eternity. Eternity is taking what is and carrying on with that forever or until infinity replaces the moment carried by infinity. Then infinity takes what eternity is and replaces it with what eternity will be until such a time, as infinity will again interrupt eternity by one instant.

Let us then with that in mind consider how growth will come about that forms space. Time we already placed on a line. Space we can see has six sides. There is one going forward and one backwards and on left and one right and on to the top and to the bottom. That is six sides. If a line starts growing space it is conclusive that the growth will go into all directions equally by not favouring one specific line at any given point. It is at this point that patrician developed between points holding singularity as a solid and points not holding singularity as a solid. The movement brought about the relevance of Π and the movement brought into play gravity measured at the Π movement of the relevancy and therefore Π^2. By the relevancy moving space was required to bring distinction between what was and what will be and then what is took up space for the very first time. The relevancy moving Π^2 formed the for points in contact of the expanding and the Roche limit brought an effective boundary before the next line came into place. This gave the Roche limit a numerical value of $\Pi^2/4=2.4674$ the size of the straight line that formed.

By just being it had to move. But that put all lines in the cosmos at a straight as well as circling forming a doubled **k** as well as T^2

The truth is at that point all there was, was an inside and an outside and a promise of what might come. Before there was what there now is there was what we might think of as what is perfect, because the Universe was not in the in the fourth dimension in which we are but time was in repeat of continuing the perfect. That which was in the present was going into the future at the instant it came from the past because crossing the Universe took one eternity just as it does now, but back then all there were was one eternity holding one Universe within it without boundaries or limits and that was exactly what there was. There was one Universe and that was what there was to cross while it took one eternity to do so. Still today it is just the time it still takes to accomplish the only accomplishment available at the time back then when it all began to go oblong. However what there were, is what now is because what there was had nowhere to go but to remain where it is which is inside the Universe. There is nothing to add to what is within the Universe and therefore what is in the Universe was in the Universe at the time it all began.

As the Bible says nothing is new (under the sun) meaning in the Universe. The way the Universe started is the way the star repeats every instant after the start ended. The start had to end but the end had to begin a new start and this continues by the implementation of the four cosmic pillars. If the start was introduced into the cosmos the start became procedure because the start had to continue since the start had nowhere to go. The cosmos restarts every instant of the smallest fraction time could form and ends as eternity develops away from infinity. Then infinity introduces a new start to the process that ends in eternity developing and ending in space as space formed by light.

The movement back countered the movement foreword because while the movement went foreword it was in the process of moving back. It was in the dimension of r^4 but at the time this state of affairs was all connected to r^0 and since r^4 was r^0 the lot was the same. Therefore because the past present and future was the same time carried on monstrously as it does now, but without distinction as it does now. In all that it was everything was still inside singularity as well as inside the present keeping everything there were that was locked into a thought, moving into the future as it was moving from the past. In singularity the present, the future and the past has no distinction and therefore moving into the future was exactly what was taking the movement from the past into the present.

The Universe was perfectly posed, perfectly balanced, perfectly harmonized and totally synchronized. What was hot was cold as well. What was big was also small. That, which was far away was just as near and close to everything at the same time. What was were as big as the other and as far or as close as everything else. Going into was coming out of and what was the future was the past exactly duplicated. From our perspective we now have we will find that everything stood still but that cannot be because then there was nothing to be and everything was being only it was being in a state of precise duplication with no change to the slightest detail possible while the lot was happening at the same time. That which has no inside was united with that which has no outside and the only limit was that everything could be everywhere since there was no place to go but to be everywhere all over at the same time. I am trying to be persistent on this matter because we fall into concepts we now identify with being ends, boundaries, pockets, limits, edges, and diversities while then all this was something carrying no concept. It still is like that except fro light forming space and in the hologram of space in light we find the specifications we live by.

Then suddenly 1^0 moved from 1^4 to 1^1 taking away 1^1 and leaving 1^4 at 1^3 and 1^1. The moving involves 1^2 and suddenly we had one Universe moving from being 1^4 onto being $1^0 = 1^3 / 1^2 \times 1^1$. The perfect Universe was no longer perfect. That is what must be realised. We cannot see what is now and try to judge from that, what was then. The imperfect grew but that did not dissolve the prefect. It just gave the perfect some other option to be as long as time has a perfect running in between imperfects.

Gravity is the strongest where space is least because motion at that point there is infinitive. Consider the spot before it even formed the dot. The space it had was is not there at all although all that we now see and cannot see because of the vastness of the concept, was locked into one point that formed everything while not even being except being the present. Going across the entire Universe was equal to an attempt to move, going from one point to the next point but the effort of leaving the one point was too massive to take place. The time it took to cross the Universe was one eternity of not crossing or not getting across. One might say that is still the case, but that is because the Universe remains that big or small as it was at that time. By any attempt to move that which was there that called on such gravity also called all the factors that there ever might be throughout the entire Universe to one spot in order to frustrate the motion. By one attempt not forming motion but only an attempt to commit motion gravity released the Universe and overcame the entirety of frustration to resist motion, and while this was going on and all the gravity in motion that might ever apply to and in the entire Universe was in a spot concentrating on a spot that is not yet part of the entirety we consider as the Universe. That is because what we think of as the Universe is not the Universe but is a hologram of what was at a specific time and scattered in relation to time forming space forming distance forming size by filling what there is into the hologram. While it became the Universe the split of time from singularity established the gravity that lies between what is not because to us all must end somewhere and that then which is there then is not because all must start and end somewhere and that singularity does not accomplish. Only the hologram of not being while forming pictures using light to form space to form time ends what never starts. What would generate the equivalent to accomplish that?

What would create the motion to break the strangle hold whereby of eternity restricted infinity. It is gravity that forms time that forms space that keeps eternity from reuniting with infinity again and motion keeps the time zones apart. Time leaves space as the wake moving away from the centre of time. Do not be fooled by the seemingly innocent explanation that space is the motion thereof which is what gravity produces because of all things the cosmos creates, motion of space through time is the utmost complex manoeuvre and without bringing a restraining of mathematics into science, it is so complex there is no viable explaining in physics about how the cosmos produce the act of motion of space in time. In movement we find that the atom in front of the one following holds a specific relation because the atom in front becomes the future of the one following and that one finds the atom behind filling the role of the past. While the atom in front is vacating space to fill the space of the atom in front it is vacating at that instant that the atom behind is filling the space that the atom in front has vacated in order to vacate and relinquish the previous position in favour of the following position to honour the direction gravity is insisting upon. Removing material from space by filling material into a position of new space sounds simple because the complexity has never been realised. That is $2^4 \times 7 = 112$.

Gravity is motion differentiation between objects. While falling the gravity applies as moving of space that is putting time in relation to the distance travelled. While the object falls towards the earth it is then that the motion confirms gravity and gravity forms time. When motion ends only then does mass set in and it becomes the constraining of the object preventing further independent motion. Where a confronting of objects restricts gravity to ensure the objects independence the action then implements an introducing of the mass as a substituting factor to motion that then replace motion as substitute to the motion that would be and the mass is providing the tendency of gravity being the motion of space. However mass then restricts motion and becomes motion in a tendency to apply motion. While falling gravity applies and motion neutralizes size, mass or weight. Mass counters motion being when the Earth restrains further motion of the falling object and the moving object is stopped from further movement where mass is then preventing or hindering gravity. Further movement is disallowed as other material fill space that falling body wants to lay claim to. Mass then sets in not causing the motion but substituting the motion and from that motion restriction becomes resistance that becomes mass. While falling the object is experiencing gravity because the object is in gravity but when on the soil the object experience mass which is the restricting of gravity or motion by other space filled with material. When any person is standing on any place anywhere, while viewing the Universe, that person is filling the centre of the Universe. All the light that come across and travelled all of the vacant space from any and all possible positions in space runs directly towards your position using a straight line towards you where you are filling the centre of the Universe. Realising this will end this nonsense of looking for the end of the Universe. Wherever the smart-arse astronomer places the end of the Universe it must also be realised that in that instant at that point we will find the centre of the Universe looming.

Should you decide to shift your position to any other place in the Universe, you will shift the centre of the Universe to that location as well. If you install a camera on Mars, the light is obliged to acknowledge your relocating the centre of the Universe at your will to reposition you're being that centre of the Universe where the camera then is. All the light that ever left its destination crossing the vast spaces of the Universe, excluding no particular light, travelled all the way just to find you filling the centre of the Universe, right where you are, wherever you are. By you're standing anywhere, you fill the centre of the Universe, and the entire Universe admits to that because all the light comes to meet you there. If you shift from the North Pole to the South Pole you will shift the centre of the Universe because all the light travelling throughout the Universe will find you where you then moved the centre of the Universe. The light left its destination billion years ago as it travelled through space at the speed of light anxious to acknowledge you're being in the very centre of the Universe. The Universe is spinning around you or I, which is filling a centre where all motion is connected. It implicates gravity as wide as can be... It is our task to find space, to find time and moreover it is our optimal task to find the Universe. Gravity is to move or apply the intension to move space a^3 at the distance or relevancy of k while T^2 is the time it is going to take to apply gravity or move the space filled with material space a^3 at the distance of k in the time period of T^2. That confirms Kepler's attribution to gravity where according to Kepler space a^3 is equal to the movement T^2 (time it takes to move) at the distance k from the centre specific.

It puts all aspects of gravity in the Universe in new dimensions. If gravity is motion, what causes motion? What stops motion? What is the gravity if the star has melted all atoms it had into one all-inclusive atom and this all-inclusive atom is providing all the gravity that the star had when the star still had massive volumetric space? If all that space that once filled an entire giant star fused into one specific space less centre holding singularity 1^0 then the enormous gravity is applying to the centre of such a non existing space-less atom and that entire enormous force has been secured in the space less than that which one atom holds. In that case the atom would then show a force that would pull the surrounding Universe flat. The purpose of fusion is to reduce space and magnify space less ness inside the sphere. Where does the gravity of the star end when all the atoms in the star became one giant atom by fusing all atoms into one nucleus? Gravity is smallest where space is least.

Where space of an entire massive star is left in the size of one atom the gravity coming from that will pull the Universe flat at that point.

Coming to the conclusion about gravity being motion and mass being the restriction of motion was the easy part. The facts that presents the understanding of what produces the motion and what prevents the restriction from overcoming the motion was the part that required thinking. Gravity is The Roche limit Gravity is The Lagrangian system; Gravity is The Titius Bode law; Gravity is The Coanda affect. By you're filling the most exclusive spot there is in the entire Universe, right in the centre of the known Universe now you'll have to buy the book to know why you are in the centre of the Universe.

Gravity is about reducing space and maintaining different cosmic sides not sharing the same sort of space. It is how gravity does it and why gravity does it that produce the accepting of the Big Bang Theory but it also includes the Plasma theory in detail. In the beginning before and after the Big Bang (yes before because there had to be a before) gravity was about bringing across heat that was in space to material that was in another space. That is how Mainstream science presents the forming of all particles. The only difference is that in the space where the space is filled with plasma or heat and not material, the heat was much denser then compared to now. Now they present that same space holding nothing but they never show where such a process stopped at a later stage. I say it never stopped and is still the result of gravity applying. Material is still growing but the growth now is hardly noticeable because the heat in space lacks density to provide such growth and space has much less heat density compared to what there was back when…

Mainstream science claims that many aspects of material (matter and antimatter) and singularity amongst others were present in the cosmos during the early phases but has vanished since then. Think about it clearly…What ever was in the Universe had nowhere to go but to remain in the Universe after the Big Bang. The Universe is everything there can be. That means if singularity was part of the cosmos during the Big Bang singularity must still be in the cosmos. If antimatter was present during the Big Bang it must still be present. There is nowhere to go except remain as part of the Universe! Space was little and heat was massive. Heat became space as the density of heat turned to form space through a process we named exploding, which is space expanding. The fact that heat turns to space was never recognised by science in the past. The fact that space compressed forms heat notwithstanding all evidence supporting this knowledge was never accepted by science while all combustible engines work on heat coming from space compressing to bring about heat. If you push in a cylinder the space heats up but science does not recognise such obvious facts. While the whole world produces energy from a principle that is about converting heat to space mainstream science chooses to deny the fact. In the beginning singularity was present in the Big Bang. Mainstream Science promotes the idea that singularity and antimatter went on the disappearing by escaping from the Universe. But Mainstream Science never says where they went. If singularity was part of the cosmos in the beginning it is still with us. However by the claiming that singularity vanished somehow since the Big Bang is totally incorrect and proves a lack of understanding cosmic concepts. Kepler and his formula also prove this fact.

Every aspect that was part of the Universe at the cosmic birth announced by the Big Bang has to be present up to this point. We have to realise that singularity is a prerequisite for the Big Bang to have formed any and all material. The cosmos has lines forming cubes and lines forming circles, which in applying 3D manifests as spheres. Between the circles and the cubes runs lines so the key to understanding the Universe are lines. The Big Bang was a time when the Universe was incredibly small making the running lines small. Understanding the Universe is taking the line connecting particles through space back to its limits where such limits were during the Big Bang. Extend that value received to a Universal centre and bring that value to align with Kepler's $a^3 = kT^2$ and understanding the Universe by finding the centre of the Universe makes the Universe simple as can be The Universe becomes sensible making the entire different yet unexplained phenomenon as easy as children schoolwork. There are suddenly no more mysteries in the Universe. It is only possible when we see gravity not as a grabbing force instead of seeing gravity reducing the space between particles. Gravity is not being some magic force found between particles grabbing onto everything.

What was within the Universe at the start will be in the Universe at the end. The Universe holds all; maintains everything and combines the lot. In Afrikaans we call the Universe the "Heelal". It is a combination of two words namely "geheel" and "alles". "Geheel" means everything and "alles means everything. Therefore the "heelal" directly translated from Afrikaans to English will mean the "Everything of everything". Nothing can be added and nothing can be lost. It is all-inclusive. With this fact so commonly known and accepted, how can the Universe grow? When the Universe started it started with 1 because 0 is the value of what we could find between the ears of the Newtonian atheist. Notwithstanding Newtonians' feeble views on the matter nothing can start with 0 because if a line started with zero it has to continue with what it started with. If a line has an initial number it stars with as 0 it must continue to apply the initial number being 0.

I do realise Newtonians filled the Universe with nothing to a point of overflow but because of their atheistic low understanding of Creation and the way the Creator created the Universe nothing is a validation of their concepts and not a factor in the Universe. From anywhere to everywhere the Universe forms by lines connecting points. Innumerable lines running all over form the Universe. If the line was zero as the atheistic Newtonian minded nihilist wish to think then the line must be $0 + 0 = 0 + 0 = 0 + 0 = 0 + 0 = 0$. The line must then continue as 0 because it started as 0 and it has to be what it started with, because by changing the value it changes character. In order to then switch to one would dissolve the start as the start will change value and that would repudiate the 0 as it did not apply as a starting value. This is where numbers start and even that the Newtonian atheist does not understand and yet they claim to be mathematical masters. They don't even know where the numerical order starts but they want to tell everyone where the Universe started.

The Universe started as a line because the Universe is still a line. In order to know where the Universe started one have to look where a line starts because the Universe is made up of innumerable lines crossing at innumerable angles. Still the Universe is a line forming a Universe. A line starts off with 1. The line must be a repeat of whatever the line starts with and in that we have to focus our understanding of how the Universe started. The Universe started with 1 because there was no need for two. The Universe was perfect and on the spot holding the future the present presented the past. Time did not part as eternity was the same line as infinity and infinity continued into eternity. A line therefore starts with 1. As the lien progresses it will continue with the value with which it started. It started with $1 + 1 = 2 + 1 = 3 + 1 = 4 + 1 = 5 + 1 = 6 + 1 = 7 + 1 = 8 + 1 = 9 + 1 = 10$. This is a numerical order and not $0 + 0 = 0 + 0 = 0 + 0 = 0 = 0 = 0$, which proves zero is invalid as a numerical number and therefore on the graph the lines parting the graph by forming distinction is not zero but is 1^1.

There was Π^0, which was α^0 or if you would rather have it Ω^0 or it maybe was 1^0, but more correctly it was all the above and the beyond because multiplying what ever constitute the mentioned will bring about what is mentioned to a precise equality. It was a spot that was not. It was a line that ran eternal but because it ran eternal and kept repeating exactly what was before to the precise what came afterwards the line was there and was eternally running, while never changing in the least or growing by any measure. It was not one because before it was one, what was repeated and the process cycled back to before one and before one could be reached. It was such a continuing of the monotony, no change occurred and therefore never did the running produce progress because the progress was in the perfect repeat of what was before. The duplication brought contraction to the minutes detail. That is where our atheists get one hiccup. The repeat brought eternity and the repeat was so perfect that the repeat continued. The repeat still is with us as much as we are within the repeat. There was something beyond the Universe that institutes change. There was something that brought a difference and we are within that difference. That difference was time and that time is what we move through as much as what we see at night. Oh, how stupid and how thoughtless the minds of atheist and other atheistic animals are. Baboons do not recognize this because they cannot think and are therefore atheists. Spiders cannot think and therefore they are atheists, as they do not think what the night consists of. Reptiles cannot think and without thought they are incapable to see what time is, what space is, what light is and what darkness cannot be. All the animals I have mentioned are mindless atheists because they fail to see beyond the visible into the realms of the thinkable. Because of the incapacity to think the animals are both mindless and they are atheists. Therefore atheists are mindless. The night sky is such a bright light our evolution protected our vision from the brightness in order to give as much better vision. Through evolution development our eyes is protected and we remove the qualities from the light. However animals do use that light and not our light to see by. You can shine a bright hunting spotlight onto an animal at night and the animal will not be able to see the light on it. The animal does not use the light to see better as the animal is totally unaware of the light. Then a prowler come from the night and see the animal in the light the night provides. It does not use the light the spotlight uses and the light is not even traceable to either the hunter or the hunted. From there we accept that during the day the animals must be using our light to see because the nightlight is inferior to see by. Who says they use the daylight much different from the nightlight because all evidence is there that they cannot recognize our light as light. It is very evident in the manner they go on hunting and grazing while being totally unaffected by our form of light. That which you see at night because you cannot see darkness and you cannot see black is the light the Universe is painted in just like the Bible says. This is not religion and it is not a sermon, it is hard-core and brutal basic science and it the most fundamental basic physics there is. It is the start of the mathematical Universe portraying the only physical way it could ever be.

My atheistic idiots, your mindlessness caught up with you! Then came this light that the Bible refers to as the first of what ever was and what our stupidity tells us is darkness. It was there and it interrupted the true invisible darkness, the blackness of a Black Hole, the invisibility coming from within the Black Hole. Eternity tore from infinity. Darkness broke from light. Heat broke from cold. Relevancies parted by 1^0 going 1^1. There was one but also there was two too because one cannot be without two being there to ensure one is one. $\Pi^0 \Rightarrow \Pi$. In this there was only space for one being one in the two forming one. It was $\Pi^0 \Rightarrow \Pi$ however there was no space to be $\Pi^0 \Rightarrow \Pi$ and there fore because of the lack of space to be which is the infinity of time braking the eternity of time the true measure was $\Pi^0 \Rightarrow \Pi$ but realized only 1^0 going 1^1. Π was to the future because of the motion of time involved and the space less ness of space at the time. By inclining to move the process crossed the Universe but also it took one eternity to accomplish the feat. The fact that 1^0 going 1^1 brought movement can only become a reality as a result of light. Light is heat and the heat is expanding.

1^0 going 1^1 1^0 ▶ 1^1

1^0 going 1^1 1^0 ▶ 1^1 had to bring about 1^0 going 1^1 ▶ 1^1, because the eternal repeat of duplicating while contracting was not relieved from the Universe. Before the contracting was equal to the duplicating because by measure the heat was identical to the cold. It was eternity that was interrupted by one cycle of infinity and was in repeat of eternity. Once something is part of the Universe there is nowhere else to take it so it has to remain as a part of the Universe.

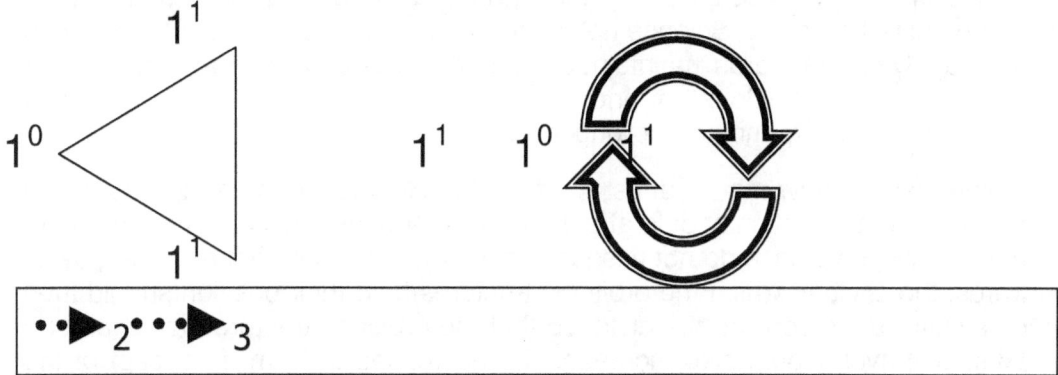

Then came three because motion was so limited that the least inclination to move threw what wished to move to the other side of the Universe, As it moves it also moved across singularity. It crossed the entire Universe as it moved because it moved and finding nowhere to move too. It crossed the entire Universe and it took one eternity less the measure of one period lasting infinity to achieve that. That brought to relevance three points where each was in measuring quantity exactly equal but also one Universe apart.

In the reality there was now two points holding singularity on both sides of the Universe because by crossing the divide that crossing set in place the two sides relevant of singularity governing. However infinity was bridges at two points holding infinity with which process eternity repeated the past into the future.

This then is the occasion where Pythagoras stepped in. Since it as a crossing of the divide the crossing involved a line that formed a half circle connecting a triangle. But the crossing was done in the space of half the Universe and since the Universe was 180^0 half of the Universe was 90^0. That involved Pythagoras as mathematics was born. Up to this point it was arithmetic with adding but now mathematics came into place. Remember we are a few eternities in side the development of the Universe.

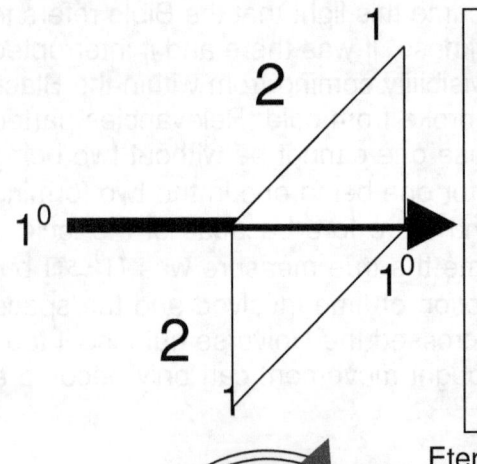

In the three came four that brought along five. How do I prove this statement? We have to examine the top spinning. When a top spins it introduces time at both limits. In the center of al things spinning a line comes from where a dot first was. The line comes from a dot that extends but just as the dot the line has no start and has no beginning. As soon as entering the line one has gone through the line. The line has no inside that can go even smaller yet we know that the line must have some ability to be able to go smaller. That point without space that never starts and can have no beginning and is without limits because it holds no space can only hold a thought because only a thought can fit into that space.

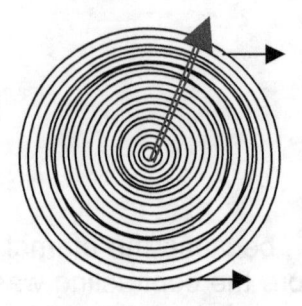

Eternity progresses from this point and carries on indefinitely. The boundaries and ends of the Universe Newtonians place in the cosmos are limitations of their instruments and their limited abilities to understand the concept of endlessness. Wherever they think the Universe ends is just another centre of what the Universe actually holds.

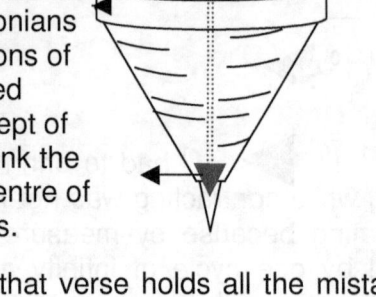

The Bible says: "Do not think of the heavens as Earth" and that verse holds all the mistakes and misjudgements that science has about the Universe. Whether the Newtonians take exception in my quoting from the Bible means little to me since they only know about "nothing" and with that they start lines and with that they fill an entire Universe. Science holds the attitude that from the Earth, they can judge the Universe and by applying standards maintained on Earth, it is a "fit all bar a few minor adjustments." Science accepts the Hubble Constant and even uses it as a barometer. This Hubble Constant indicates how the Universe is expanding, growing in size.

That means it is the measurement of how matter drifts apart. If the Hubble Constant applies to the Universe as a barometer, then that application should affect the Milky Way as much. Remember I am using Newtonian logic, not my own, so please do not misquote me. If there is the shift of matter, away from a centre point outwards, the circle in which the orbiting structures hold their position should then enlarge, because of such a shift. By increasing the distance that the orbiting structure travels around the centre, the time it takes to travel should also increase. If one wishes to form a concept of the implications in hand the most prudent manner is to return to the star of it all and then try and progress from that position.

The Big bang was never bigger than at the time the procession of time started. The Biggest Bang was way before when light or heat started the process we call the Universe. The biggest start was when the first dot became two dots and whatever was the Universe doubled as it halved at the same time. It was when that which is infinity parted from that which is eternity. Heat expands and cold contracts and that is gravity. As heat expands it leaves an area by halving the heat in that area. The relevancy changing makes that the duplicating brings about the material two half by doubling the space it moves through.

By the halving the heat, the motion is enforcing the area to cool off by half and that produces contraction. By expanding there is contraction because overheating expands while it then cools bringing about contraction. As the light came into the darkness by heat coming onto the eternal line of time the line was interrupted or broken by infinity releasing the line in eternity of its eternal procession. Infinity released from eternity, heat broke from cold singularity in the dot released from singularity in the spot, Π^0 relieved Π from eternity, 1^0 broke from 1^1 and all the Universe coming about fell into place in space in time in space-time an in time space. The line dotted on a spot the line was in progress one eternity long without progressing because it was in an eternal progress that had no beacon, no marker and no comparing to, which put the eternal running line eternally in one spot. The heat interrupted eternity by infinity parting from eternity.

Firstly we examine movement surrounding singularity and supporting singularity, which is epitomized by the five points we find forming the Lagrangian points.

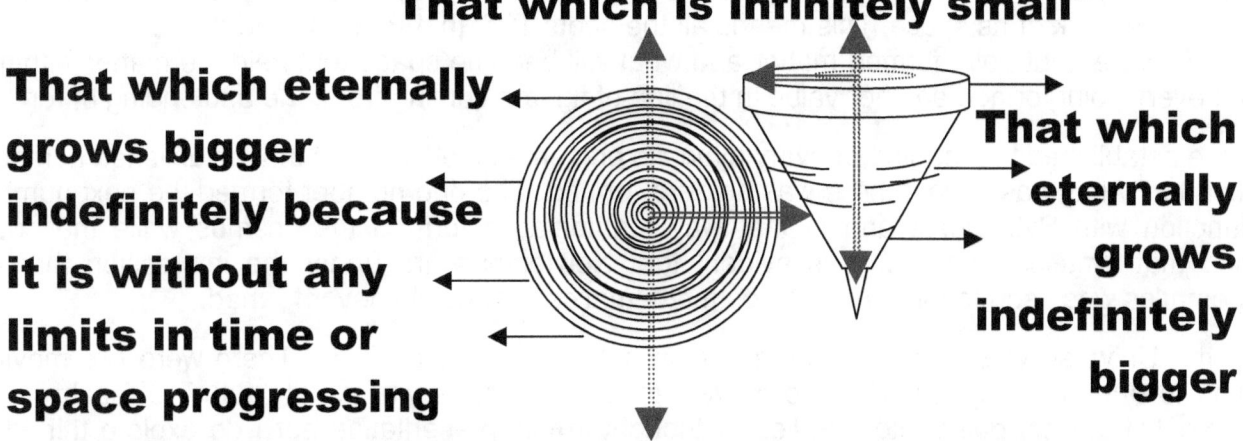

Now we are able to recognise the two time components forming the Universe.

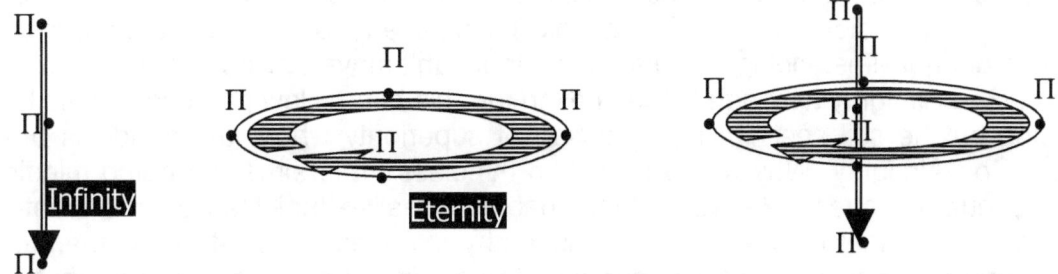

Now please consider my explanation and compare my explanation on how creation began with the nonsense of Newtonian science and that which according to them I am supposedly too simple minded to understand. Simple minded I might be but so was creation at the start when one dot became two dots and that can never be seen as not being very complicated.

It started from a spot that became a dot and by becoming something eternity began to flow, as time broke free from monotony, although at first by the instant of infinity. Before that time was a line being perfect. Future landed on the spot where the past departed from and where all this joined in the present the present represented the spot holding the past and the future together. The present stood still while the future moved through the past without landing on another spot. The Universe was one spot or singularity.

The line had flow. The line did grow albeit from infinity to infinity relieving eternity from infinity. The Universe made the biggest leap it never could repeat afterwards and it was so small it can never be seen inside this Universe we are within. That which does not move became hot and that which did move became cold because it moved. That which does not move expanded and that which moves contracted. Partition came into the midst of time and space formed by parting time into two segments. Two components came about just as the Bible says it did. The material that moves the Bible calls earth and the liquid / gas that does not move the Bible calls heavens.

The irony is with all the Newtonian cleverness science fail to recognise the most vital forms in the Universe being that which fills space and that which holds space. The one is material or earth and the other is sky or heavens or outer space, but only a nihilist atheist mindless beast that are unable to recognise God will fail to recognise these factors. There then was a past and a future (two) in terms of a present (three). This lagging of heat formed the basis for space holding time end very much later it became even further behind and in such a lagging that the time solidified to form heat dragged behind as material. There was a spot holding the past seen from a spot holding the present in terms of a spot developing forming the future.

The Universe was spots and dots.

Mathematically the Universe was $1^{2564} = 2564^0$ and notwithstanding all the Newtonian brilliance this mathematical numerical range then formed the Universe as the Universe progressed in form without space as we think of as space. This means all the spots 1^{2564} that formed filled the space of one spot 2564^0. There at that point formed matter and what will become space that held the matter within. But at first every point formed another value in the line of figures coming from one and ending at ten.

The line established the forming of every dynamic present in the Universe that established numbers in relation to Pythagoras except there was no zero formed. The one number formed the next number in conjunction with Pythagoras and in that way mathematics formed the cosmos while the cosmos formed mathematics. However it is evident that Newtonians are under the impression that while mathematics was used to form the cosmos mathematics in principle never formed.

Then the Universe was 3 with two divisions on both sides of the divide. There were two moving in relation to one standing still. The two moving filled the past as well as the future of each other by movement. I am not going into that lien of thought in this presentation but I do explore this idea in detail in **The Absolute Relevancy of Singularity in terms of A Cosmic Creation.**

Stephen William Hawking said, as a seemingly very wise Newtonian there can't be a God because God could not have created the Universe because the Universe was created by physics. Stephen Hawking, the pathetic idiot doesn't even know how physics started. He has no idea how mathematics started. He doesn't know how the numerical order started but he declares that physics started creation. Hawking the idiot (and I call him an idiot) can't even move his head and that is precisely how pathetic he is but notwithstanding being a cripple that can't move any body part wants to challenge the existence of God Almighty although he can't even turn his head. How pathetic excuse he is for being a human being but he still see him in a position of superiority where the mindless brat is worthy of challenging God Almighty. Why would I call him mindless? It is either he being mindless or being a crook and a hustling cheat. He was either unable to realise that Newtonians creating mass as a cosmic presence but that it as a factor in cosmic physics does not exist and is therefore an ideology without any grounds is what was teachings equal to a fairy story or he realised it but was part of the conspiracy to fool every person on earth.

This is how physics developed as the Universe came to be. There was a presence that moved to the past. That is two pointy formed from one. Then there is at the same instant a future moving into the presence and the too is two. That means two points became the past of two points coming from the future, which initially was the same point that split because of heat differentiation occurring. That means there was (the present) + 1 + (the past) 1 = 2 and (the future) 1 + (the present) 1 = 2 . The one point (the present) stood still while the future moved to the past and the movement cooled that which overheated while what overheated expanded by standing still. It is the present that overheats and contracts and it is the past that moves and contracts. However to explain this, that efforts is almost a book in itself. That brought about 2 + 2 = 4 as well as 2^2 (movement) = 4. From one now grew four spots that ended up as dots.

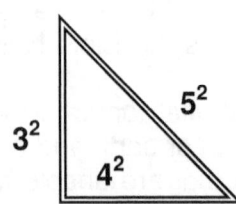

The points forming (the past the present and the future) moves (going square) while the past moves to the present (2) and the future moves to the present (2) and the adding brings movement to four (4^2) and this in relation to the law of Pythagoras in the triangle brings about the number of five. That then introduced the next number worth five by exploiting numerical orders and mathematical principle. This is how mathematics formed. This is how physics developed and it developed as the Universe developed.

It is clear that the adding was the first dimension because to get from one to two by multiplying brings one back to one. However by adding it does bring about three in the triangle while two is part of the half circle as well as the line of time. By forming a relation with the line and on the one side with a half circle while being relevant to a point on the other side of the divide clearly shows why 180^0 fits all three forms and put the here on an equality that otherwise does not make any sense to us. However one can see that in the next development the square came about by multiplication of two points travelling equal along a centre line.

The line is time.

Newtonians think of space in terms of time. The Moon is not distance away from the earth but is movement in relation to forming space away from the Earth. At the speed of light the moon is 1.5 seconds away and that puts the moon in terms of my position on earth 1.5 seconds into the past. The Sun is about say 8 minutes and thirty seconds away from the Earth travelling at the speed of light. That puts the Sun 8.5 minutes into my past when taking the speed of light as a yardstick. Time is movement but time is not space and if Newtonians do not realise that they may as well start to go back to magical potions and black forces instead of gravitational forces they try to create. Physics as part of creation came about because of movement but then again the cosmos came about because of movement.

The Lagrangian five points are a direct result of singularity forming the first numerical order as time formed a circle while forever and always holding an applying relevancy. This fact we find in the Kepler formula of $a^3 = T^2k$ where that shows that all space forms by movement in relevance of and it is this relevance that becomes the great connection. The entire formula should read $k^0 = a^3 \div (T^2k)$

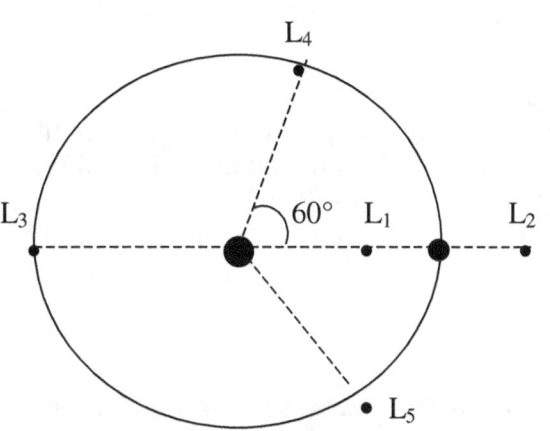

LAGRANGIAN POINT:
The Lagrangian points are five equilibrium points in the orbit of one body around another, such as a planet around the Sun

In the Lagrangian points system we find the formation of the three forms formed by singularity as equal values while they are totally unmatchable in the dimension of space in which we live. This is the proof of singularity forming the components that produce space from time forming space. It is in the numerical order we find the how the equality is revealed.

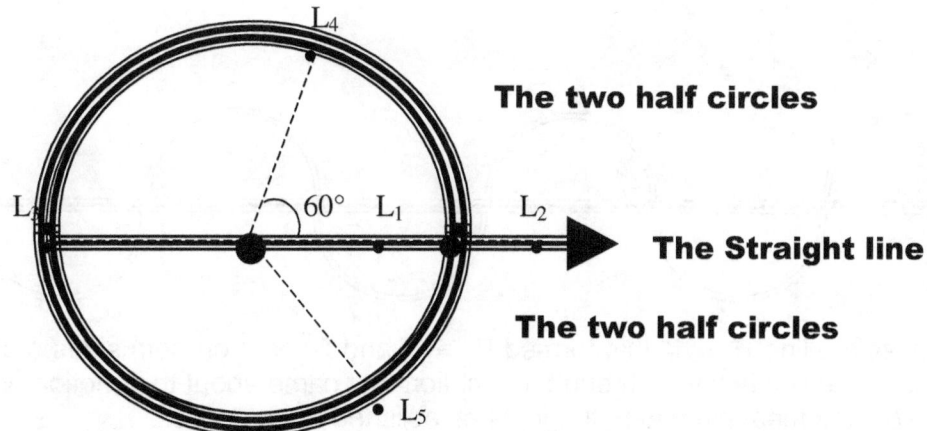

LAGRANGIAN POINT:

The two half circles

The Straight line

The two half circles

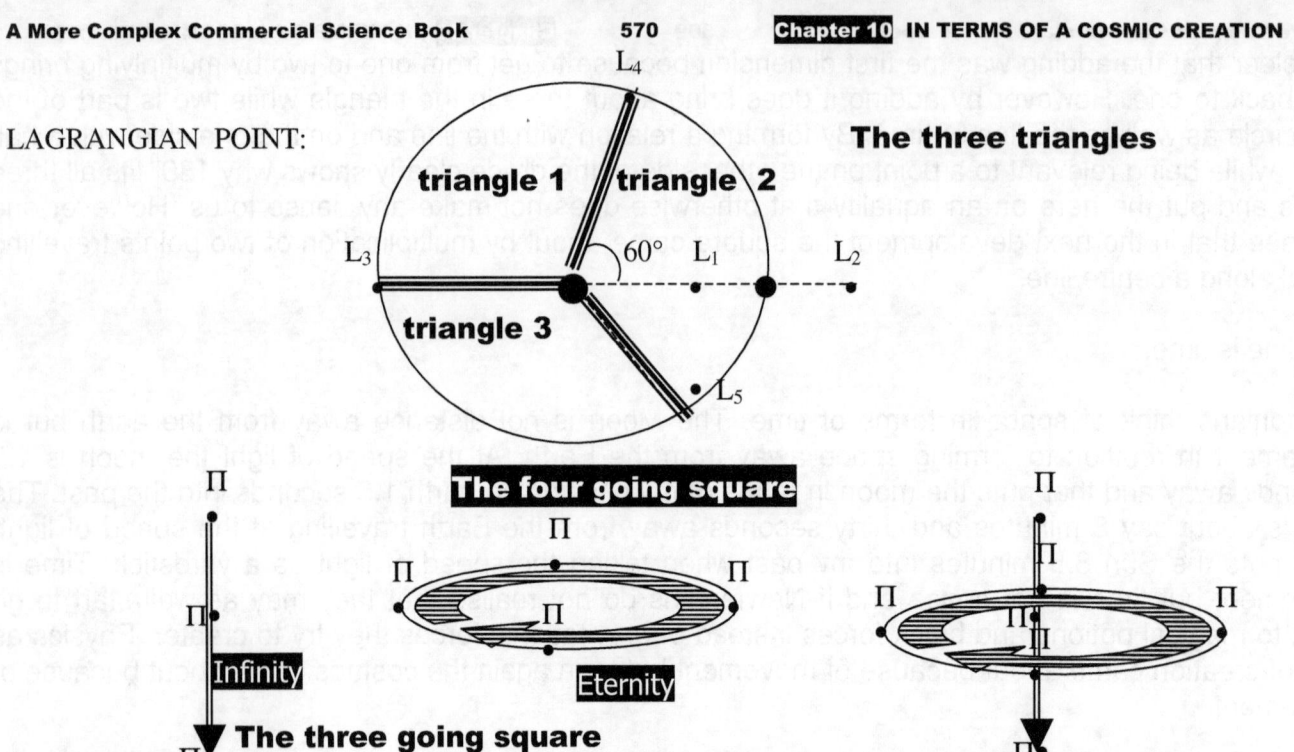

The five contracting in terms of seven going square and therefore expanding.

As time moved on a lagging behind started to creep into the flow where the heat began to distort the flow of time by not repeating the flow in the same cycle. There are four positions holding relevancy to five points formed by movement and the four points moving in terms of five points form twenty and that makes ten on the one side and ten on the other side

Taking nothing to mathematics zero locates in infinity by abolishing nothing from the universe. The sphere is 7 X Π = 21.991. It is a formula or a recipe for gravity. It shows how a sphere grows into a sphere by form on a continuous basis.

The line diverted from Π^0 forming Π but there was so little established that although Π^0 was the present Π^0 also was the past and therefore $\Pi^0 = 1$ was the present because 1^0 moved to 1^1. That put the figure of two in the Universe. Then from two the square of two diverted from the line of time by 90^0 which established a line of 180^0 at a relevancy of 90^0 to the timeline at 180^0. There was the line filling with ones that formed twos as 1^0 formed 1^1. That made two lines and that had to be three because $1 + 2 = 3$.

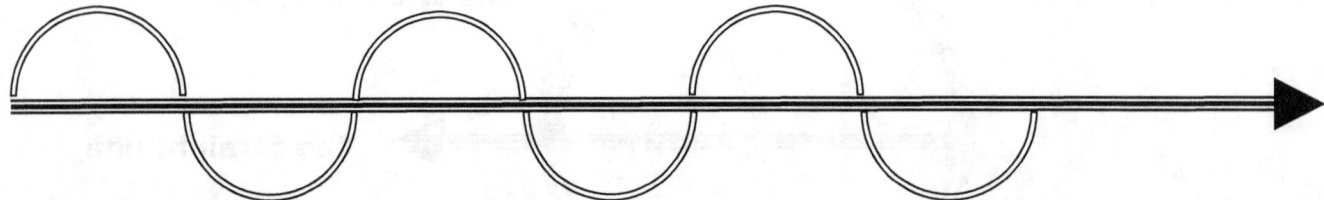

There was the line $\Pi^0 = 1^0$ that formed $\Pi^0 = 1^1$ and $\Pi^0 = 1^1$ on both sides of the divide forming ninety degrees to the line in time. Through the motion that came about that motion brought about a square in the line of time meaning that Pythagoras established mathematics.

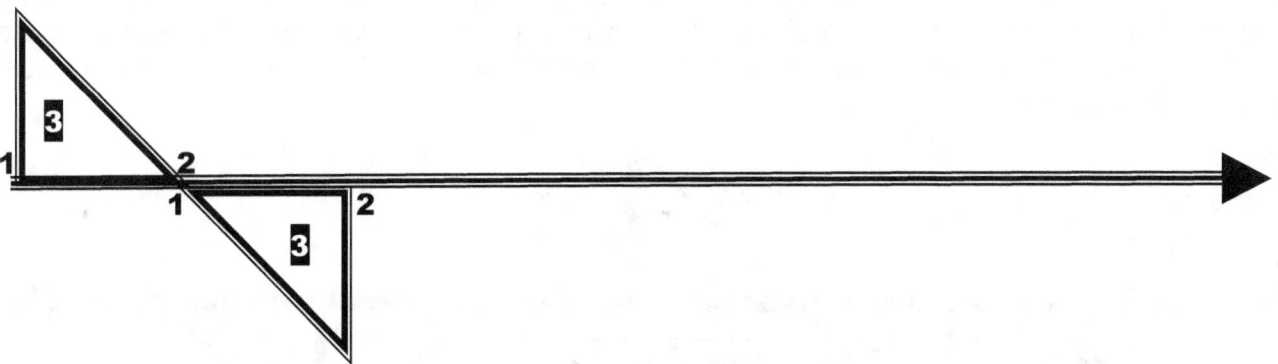

That makes ten the square of as well as the root of time and that is another reason (if not the only reason) why we are able to use ten as a decimal basis for counting by numbers. Since three came into place while Π^0 was establishing Π I guess (I prove it because as a reality it does come to mind) that the motion had took place while Π became the factor and 3 became the position that the small difference there is between 3 and Π was somehow initiated by the motion in the line. In motion space-time found a relevancy by the ordering of gravity.

Motion is to the value of gravity, which is the value of Π^2. Crossing the division of singularity brings about a waving half circle that connects by the triangle to the other side of the divide. That places 5 points holding singularity Π^0 in relation to the motion of space going through time in Π^2. The five points in singularity by association with 4 positions is the motion of space in Π^2 passing through time's five and that also conclude as material in the square at 49. Now we can put that in relation to Pythagoras and find gravity forming space-time while material in time formed gravity.

The Universe grew from relevancies but not out of relevancies. At the present the relevancies still dictate everything there is in the Universe. If anything disappears everything will disappear accordingly and if anything suddenly stands still everything will fall into the gap of the object that suddenly stood still. The only way space can grow is by relevancies between what is hot in terms of what is cold and when heat separation comes about and the growth changes the form and form is present only by association. That is nature and that forms nature.

The only way particles take up space is if there is space that is not taken up by particles and because particles move the heat of what does not move will initiate an overheating that will expand and because that which does not move expands then that which does move will contract in relevancy of the presumed expanding.

Solids cool by movement and the movement is within n cosmic liquid presenting expanding by forming cosmic gas. Heat coming into or removing from determines form. A Black Hole did not shrink into the oblivious but the Universe expanded and that expanding placed the Black Hole in a relevance appearing to be in the oblivious. The material within the Black Hole remains within the cosmos as part of the cosmos and movement or the lack thereof puts differentiation that accumulates and heat results in the lack of movement in comparison to the movement it is in association with. Heat is the result of movement in ratio to the lack of movement. That is physics and that Stephen Hawking and his Newtonian atheist never realised because they created mass as a distraction of their incompetence concerning physics.

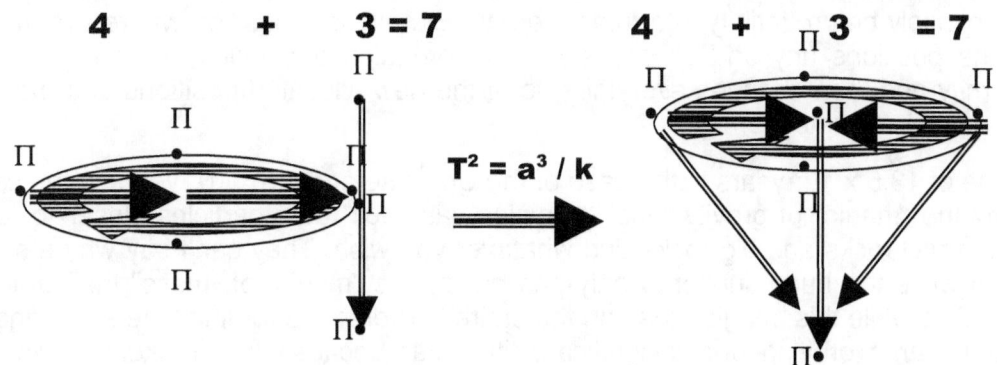

In the motion space found a relevancy by the ordering of gravity. Motion is to the values of gravity, which is the value of Π^2. Crossing the division of singularity brings about a wave half circle connecting to a triangle on the other side of the divide. That places 5 points in singularity Π^0 in relation to the motion of time Π^2. The 5 points in singularity is 5 x the motion of singularity is 9.8 X 5 = 49. Put that in relation to Pythagoras

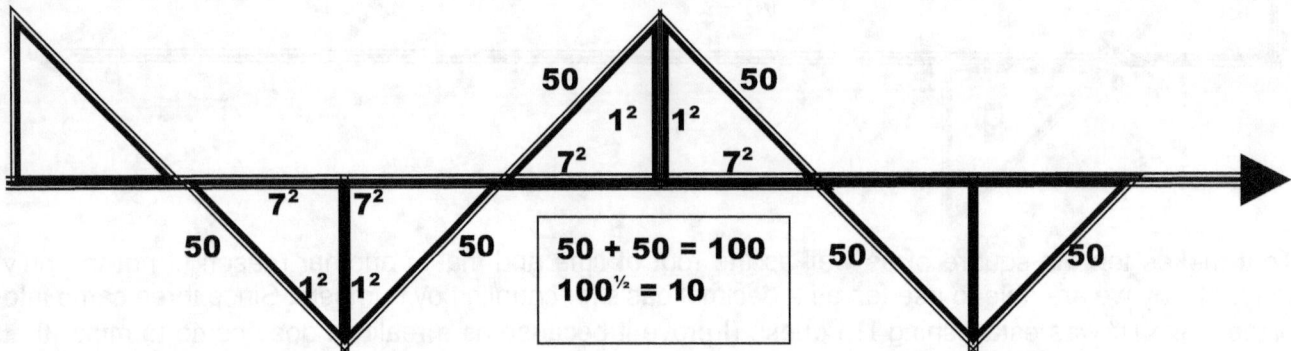

That way space in time formed the curve of space by 7^0. Then on both sides of 7 there is 10 because 7 square associates with one square by forming diverting movement coming from then future as the last part of the value of Π indicates (.1416 x 7 = .9991) and from that the entire value of Π became valid. That makes ten the square of as well as the root of and that is why we may be able to use ten as a decimal factor forming the basis of all numbers. Lets run through the process again in order to pick up some loose ends. This was what lead to the process through which at a later time became the method that the atoms of various significant formed by the motion that was prevailing at the time. We find the evidence in the characteristics the atoms show in relation to heat. In the book Starstuffin I do explain this in length.

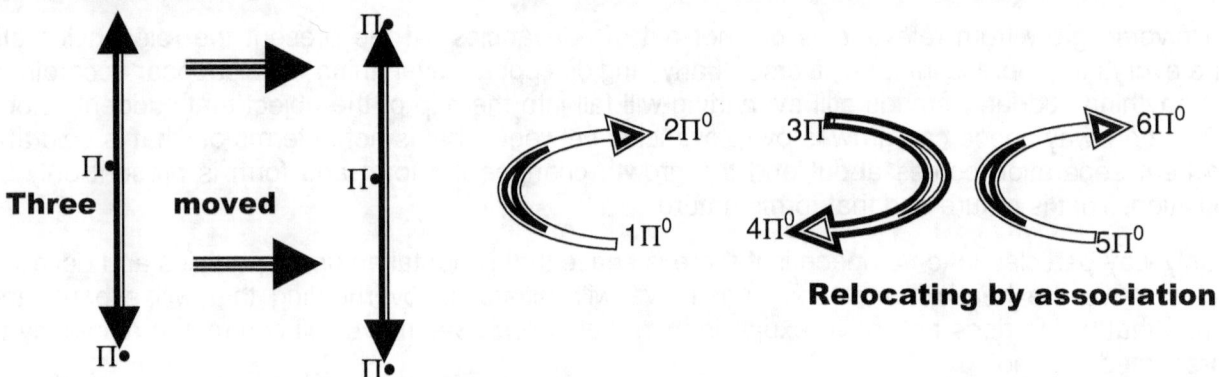

Then from five came six still applying the cosmic laws of physics. The line moved, duplicated to form space, repositioned singularity by shifting the line forming infinity thus creating a relevancy by association.

How long did the first instant in time last? It lasted one eternity minus one part forming infinity. Then the next duration was one eternity minus three parts in infinity and then the Titius Bode rule started affecting time by the measure of $(7+7)/10/(10/7) \times 10 = \Pi^2$. Infinity takes twice as long to complete than does eternity while eternity lasts as long as it takes infinity to relieve eternity from continuing. We are in eternity. We are relieved from eternity by the interruption fop infinity. Eternity can last 100 billion years before infinity interrupts eternity and it would seem to us to be instantaneous and for all we know eternity might last indefinitely before infinity interrupts eternity. Eternity is a position where everything is set according to the positions any and all space was in formation and infinity then takes thee positions held by eternity and then changes everything to fit the new allocated positions that eternity will carry until infinity again interrupts eternity.

Science applies an age of $13,5 \times 10^9$ years in the case of the Universe and $4,5 \times 10^9$ years in the case of the Earth. Then by the "magic" of gravity "dust" particles can "pull" dust particles closer to form planets and stars and small rocks and big rocks and whatever you wish. They can't say why a small rock will become solid while the huge Jupiter is only gas but by the "magic" of "mass" the particles "pull" to become solid. And while this "magic" goes on the entire Universe are forging time according to one small insignificant planet orbiting one insignificant little star because that makes Newtonian calculations easy to fake. This allows the Earth's establishing date to fall into a position where the Universe was 2/3 of what it is now. IF THERE IS A HUBBLE SHIFT, THERE ALSO HAS TO BE A HUBBLE SHIFT IN OUR PART OF THE UNIVERSE AND THEREFORE THE SOLAR SYSTEM ALSO HAS TO ADHERE TO THAT. THE SAME HUBBLE SHIFT MUST THEREFORE AFFECT THE SPACE BETWEEN THE EARTH AND THE PLANEYS TO THE SAME AS THE REAT OF THE UNIVERSE.

$F = G \dfrac{M_1 M_2}{r^2}$ What this formula suggests is that mass moves mass closer by diminishing the radius and therefore the object of the exercise is the get the radius as small as possible. The radius should not grow or increase because the pulling of the mass by the mass will prevent it by the value that the mass has. If one would find that all the mass of the entire Universe was bundled together in one spot and the radius was non existent in terms of the mass pulling, then the mass must unite and therefore nullify the radius altogether. This then would bring about that the entire Universe must instantaneously collapse into the most enormous Black Hole God could ever perceive. It did not happen and we see that by looking at God's creation. God had quite the opposite come about and that makes the idiotic atheist claims of understanding physics the biggest joke their stupidity could devise. Any person validating $F = G \dfrac{M_1 M_2}{r^2}$ puts that person's intellect into a conundrum from where only his or her stupidity and lack of insight emerges. What the Newtonian atheist find more credible as being $F = G \dfrac{M_1 M_2}{r^2}$ than the likes of God Almighty bringing about creation is a joke so miserably stupid that it can never be defended or validated.

Any person that holds a clear view and is not adamant in defending Newton has to be aware of many misgivings his formula brings to the Big Bang, If he saw mass compacting mass by mass accumulation then the Big Bang was a missing event and we were no more here than we are there. There should be no star structures that formed because how easy was it for mass to pull mass when there is no separation between particles at all. Many such discrepancies go unnoticed or are either blatantly ignored by a sure lack of interest on the part on science. One cannot advocate a withdrawing Universe and claiming that very instant there is a contracting Universe the way Newton's law lays claim to that. While we all know that that is the case Newton does not prove it and to defend Newton is to become the devils advocate then. With the Universe, expanding as it is, taking the Milky way along for the ride, and taking into effect the Universe square law where if the radius doubles, the circle grows four times, a year back when the Earth supposedly started, had to have been quite a measured mile shorter than it is today. Back then it had to take the Earth a few days to complete a year, since the orbit path was so much shorter. (Again, I wish to remind the readers that I am not applying my logic but that of the Super Educated.)

The Hubble's expansion brings about the increase in the radius of the distance between the Sun and the Earth. Every time the radius doubles, the circle grows four times by distance. This is a mathematical fact, beyond reasoning, to be accepted by one and all! Holding this mathematical law of value and positioning, the age of the Earth happening some 2/3 down the road the Universe took to develop, and judging the size the Universe got to where it is at present, the circle the Earth uses to orbit the Sun had to quadruple many times over. What is a year? The distance the Earth takes to orbit the Sun at present, or the distance it took the Earth back then. I am not the one to maintain a second back then is the second now. Our Super-Educated are the ones that insist on maintaining to stick to the second minute and hour as it is today. By reducing the radius between the Sun and the Earth, and then multiply that figure with the number of four that the year was back some $4,5 \times 10^9$ years ago, one had to use some stopwatch to measure one year. If scientists took less time in studying far off galactica, and took the Hubble shift they regard with such prominence closer to home (say implementing it in our solar system) and work the Hubble shift from that angle backwards, a great surprise would await them. They would find that not one single application of the Hubble shift would nearly fit our solar system.

My being ill educated brought so much to resolving of issues in favour of academics in the past that I am putting science in two categories: them and me, and if I stand accused every time not understanding Newton for the lack of education then I am prepared to be the un-educated as long as the other party accepts the title of the SUPER- EDUCATED. I think it is only fair to bring distinction to both parties from both ends. Saying this I also do not wish to offend you in person so I wish to make the issue a three party affair with you in a refereeing stance judging me versus them (the SUPER-EDUCATED-ONES.)

In the first part of the cosmic development since the Big Bang arrival, there was a lot of activity that brought stars of many sorts and flavours to the cosmos from dust, but since there is no evidence of that happening in view of so many astronomers dissecting the sky every night, one has to conclude events slowed down the past 4.5 billion years or so with not even one star in a half built state to show. It must have been great turmoil as everything happened in a short period of time because the development produced all matter relative to the Universe, all space there is in thee Universe, positioned all particles in place to one another, set standards for stars and galactica, reduced galactica, produced dust by the truck loads and in specific semi oval, not to distant and precisely aligned positions enabling stars to form and in some cases even form and disintegrated several times. What an era to live through. Much different from the placid one we find ourselves in at present!

Lets give the Newtonians some leniency and say that the Hubble shift started by having a space of about fifty millimetres in diameter. Two thirds down the road of cosmic development the solar system received its day of birth. Gauge from the development where that should leave the Milky Way, and particularly our solar system. Some scientists are even of the opinion that a massive star developed, destroyed and from that debris and rubble the solar system came about. That means the last third of the cosmic development was a relative dreary and boring era, compared too the first and second era. Think about the billions upon billions of galactica that formed in the first event, followed by the massive stars that formed and demolished during the second era, and during the last third not one single planet formed, that we can witness to prove all the claims. The Sun and the planets formed during the second era; therefore, by the start of the third era, they were in the position and place they are at present. The working tempo slowed down considerably the last five billion years or so I guess some one somewhere formed a workers union and brought in human working conditions to ease the workers demands for overtime compensation.

Push this double standard applied back to before the Sun took its position and there was no Earth to indicate the year. How small was the year circle at that point in time and space. Take this right down to the:" Big Bang" where "the whole Universe were the size of a man's fist" (they go even further by putting us into a neutron), how far did the circle goes to indicate a year then? The year was immeasurably smaller, shorter and faster than at present. This is logic even the Newtonians must accept. There is no space outside insanity to apply time to the past at the value it is at present and far worse, to use something so extremely insignificant as the Earth to measure it by.

Using such logic makes science appear foolish. There is just no rational in the way Newtonians suggest that time lapsed verses how events occurred that can explain facts without. Since the time of Newton, the arguments tarnished from being brilliant to clever to fair, to poor and a hundred years ago to the point of being stupid.   That is what Kepler's formula is all about! That is what Kepler indicated with his formula $a^3 = T^2 k$. The space of an object (a^3) is equal to the time (T^2), which it is in, in every given instant (k). If the space becomes smaller, the time duration becomes longer in every instant that time flow progresses.

Singularity is a mathematical reality. Einstein may be the first to name it and Galileo (unwittingly) may have been the first to define it as Kepler was the first to formulate singularity, but in mathematical terms singularity is the most basic principle.

When science calculated the value of gravity and the gravitational constant, as well as the speed of light, they never considered the moon to play a distracting factor. It is quite understandable because they did not know about the Roche-factor influencing their calculations. With Π^0 little more than a figment of the imagination there is actually two values of Π^1 facing each other in a relation combining Π^1 to hold the value of $\Pi^{1+1=2} = \Pi^2$ and with two sides being the very same but opposing each other there will therefore also be Π^2 to every side that holds Π^1.

From the above I can conclude that gravity is not 9,81 Nm/s, it is Π^2 = 9,8696.

The gravitational constant is not 6.67 but it is 6.9 (7/10 (Π^2)) and here the moon had an even bigger influence. It is a fortunate coincidence that we took water to be the measured calculation since water holds the combined value of 17,5 and that is half the value of either space (31) or time (Π^3). That makes a kilometre (1 000 m) one cube laid flat and since movement represents space-time occupation in a linear manner, it is the cube that went in a single line.

More of the same fortunate coincident is that we connected time to spin long before Newtonians came along. The Earth spins through space at 360° in one day and space represents 10, therefore there is 3600 minutes in the 7° of spherical angle moving to the outer rim representing again the seven. All this makes explaining matters a lot less difficult.

The value of the proton is not 2/3, but Π^2. The proton spins at a rate of Π in a dimension of 2. The neutron is not 2/3 + 1/3 + 1/3, but again it is $\Pi^2\Pi$ and the electron's 3, holds a dimensional implication, because time is in singularity and space is in singularity. Time is eternal and heat releases space, which is time from singularity. In an effort to make the understanding simpler, you have time at an eternal value and that makes space zero. $\Pi^3 / \Pi = \Pi^2$; $\Pi^3 / \Pi^2 = \Pi$, Time and space interlinks because it is the same thing and heat, (matter in many spin rates) allows time to break free from eternity by allowing a distinguishing of the flow of events. Time in movement is the result of matter (which includes heat) to change their relating positions.

Think of a movie. The continuous flow of pictures indicating the change of the position of the photo's bringing about the concept of time. Play the picture too fast or too slow and it will be unreal because we know at what tempo matter changes its position in relation to all other matter surrounding it. That is time and that makes time irreversible, because the position matter holds in relation to each other (in considering it to be throughout the Universe), can never repeat once it has changed.

Newtonians, forget about time travel because just by mentioning such absurdity you prove what little you know about the cosmos. To go back to a certain time, you will have to redirect all matter in the Universe in a reverse, apply that reversing of all particles up to the point required, stop the movement of all particles and start time going forward. Before some Newtonian grabs for a calculator, remember, you that are doing the changing, is as much part of matter in the Universe, therefore your action in changing the direction will stop even before you start! It is silly to think people with healthy minds, acting like adults will indulge in senseless stupidity such as claiming to be able to reverse time.

It is the Coanda effect that produces gravity and it is the Coanda effect that is keeping the Universe together. Let us take the "Sound Barrier" from a point where we see phenomena apply laws that matter complies with. The sound barrier is a prime example of the relevancies, which I suggest, takes place. There are two points in singularity always referring to each other and one is expanding while the other is contracting. In nature there is the Roche limit placing a limit on the reduction of space and the inflow of heat to sustain proton cooling. At a point of $(\Pi/2)^2$ the reduction of space disallows any object to immediately compromise space claimed in time because the cosmic object cannot reduce space while an entry to its area demands such a time reducing in space claimed by the material.

The first question that one can ask is why would there be the value of $(\Pi/2)^2$ between orbiting structures positioning themselves in a time relation to space.

Every person knows about the entry restriction an orbiting spacecraft finds that forces the craft to comply with. The entry maximum is 21,991 and the minimum entry is 7. This is without doubt, the number of Π (21,99/7). The Earth holds its value to $4\Pi^2$ and when an object is not part of the surface of the Earth, even say a mountain; it becomes a holding value of 7. Later in this part I explain the sound barrier in more detail and the 7 will then become better understood. At this point one must see the Earth in the proton status of $(\Pi^2+\Pi^2)$ while acting as an atom. In this relation the atmosphere including all particles in the atmosphere will in relation be either Π^2 or Π. When water is in a vapour form, it will have a value of 3Π, having heat separating the water to the factor of 3. By dislodging the thunderbolt, the 3 receive a square value and displaces to the Earth in the linear light to time stance of 3^2. With heat (3) grouping by initial spin value, it will remove from space leaving the water to the value (Π to Π) and this will then give the water a relevancy of Π^2. The factor of Π^2 places the water no longer amongst heat as gas, but heat as a liquid (rain) or solid (hail). The relevancy of the water will change from 3Π to Π^2 placing the water's position from space (3) to liquid or solid (Π^2). Where does the Π that one find in the Roche limit $(Π/2)^2$ and the vapour $(3Π)$ finds its relevancy to gravity? Every particle that enjoys space-time outside the Earth's structure $(Π^2 + Π^2)$ will hold a neutron position of $(Π^2Π)$. The $Π^2$ ends will be at the point where heat passes through the object directly to the Earth and this position of space-time relates to the neutron time link of $Π^2$. The space link of the neutron will then form the Π link. The value of the Π link we find to be $(Π/2)^2$, but the explaining to why it is $(Π/2)^2$ is rather more complicated.

At the end of the space relevancy 3 where matter occupies space (21,9 / 7) is a border Π. That border is the exact point where space reforms to a square of time placing all matter (occupied heat) and heat (unoccupied matter) to a value of the square of time. That specific point is in relation to the square of the diminishing shield around the Earth. However it takes matter $(Π^3)$ from the 3 dimensional positions to the square $(Π^2)$ in relevancy to time in singularity. With time holding space in singularity the 4 sides of Π truly relates to half of the total square value.

The "gravity" factor of $Π^2$ becomes one and only holds the square to the Π position as it holds space to singularity at a square (time dissolving space at a square) and the time value (Π) remains dimensionless in singularity at (1). It then is $(1)^2$ where the one becomes the space position $(Π/2)$ representing time (1) at that point. This makes the position that time normally has $Π^2$ but directly links to the controlling singularity which we then give a value as $Π^3$ which then relates to the singularity position of space diminished from the three dimensional to times single dimension in the square. That makes the Roche limit hold the position of $(Π/2)^2$ when the neutron position of time $(Π^2)$ links directly to singularity (1). This may only represent figures, something to accept through intellect but lying far outside the reality surrounding our everyday understanding.

We all think of many reasons why birds fly and when the pioneers of flight started experimenting the main presumption was centred on the flapping of wings. It is a natural tendency to presume that flight comes about from the wings "pumping air" in order to raise the mass and I am sure but not certain that such a thought had presented itself at sometime in the Head of Newton. Once again that is as far from any truth as is possible.

That is the shape needed to fly in the surrounding of the Earth that we presented with a name THE ATMOSPHERE. In outer space a flying object can hold any shape because any shape fits the three-dimensional. It could be round, oval, flat or any combination of all shapes and forms. The dimension is 6 with a linear value of 3 applying in one direction. This implicates the fact that wherever an object directs a direction in time revaluating positional change in space, it will forever be heading in three directions in the same time. The only control to direction is in the concentration of heat at a point opposing the direction of travel. By applying the release of heat, one is applying "a directional change". One is producing space that never existed before, because the release of heat will produce space. To the object in travel another dimension brings about influence to this three directional space application. This application of the heat brings about a fourth dimension that applies directly to time. I shall come back to this point duly.

Without the application of specific heat, the object remains in the three directional moving of six possible directions. The value of space unoccupied therefore remains $ΠΠ^2$, as it was before the "Big Bang" event, whichever "Big Bang" you wish to refer to, because there were many. But space unoccupied holds time to the value of 10 to 1, and as also indicated, holds space to Π. Therefore unoccupied heat holds the relation to space in applying 3 directions of influence $(10)^3 = \mathbf{T}^2$. Always part of this equation is the dual function of space in $Π^2Π = Π^3$ while at that very instant one have $Π=\mathbf{T}$. Therefore in space in time you have $10(10^2) = \mathbf{T}^2$. Applying the fourth dimension does not bring about another part in the six dimensions, but actually cancel the influence of the one dimension by favouring a specific dimension. Bringing about the fourth dimension will lead to halving one dimension because of favouring the direct opposing side. Then the equation of influence becomes $(10/2)(10^2)$ and this means the implication of any heat, be it heat, be it light, will apply under the factor of $5(10^2)$ Π.

The linear factor of light travelling from the Sun will be the distance between the Sun and the Earth. The light density that reaches the Earth is the density that remained after the rest of the wave (3^3) went to space in the form of radiation (3^2) and heat invisible to the naked eye, the heat that holds the universal gas that covers all objects, the same heat that one day must all return through matter to time. When looking at a sphere the three we see, has a lot less dimensional value then we can see from a cube. Even looking at a cube from only one side, will still have the support of half of the other sides that provide enough stability to the cubic orate so that we can keep on seeing the one side.

That which we mentioned in all of the dictions it all is located between the "it" that has no start and the "it" that has no end.

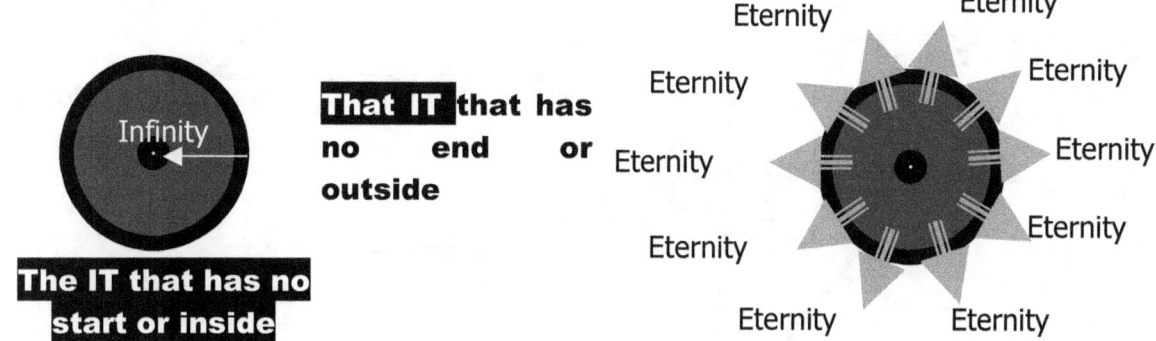

This part forming infinity holds a relation of five, which I show is a result of the numerical order of 3 moving (going square 3^2) and spinning forming a circle (4^2) and by involving Pythagoras 5^2 moved into place. The fact that 5^2 formed 5 by Pythagoras indicates five forms by movement and thus bringing about the Law of Lagrangian numbers by conserving or space returning. These facts I proved this far.

When anything will grow mathematically keeping form, as the main subject there will be a centre point Π^0.

•Π^0
•Π^0
•Π^0

From the centre will grow a point towards the top of the centre spot and a spot will grow from the centre to the bottom. This will form a line and that line we all know as the axis forming in a turning sphere. Looking at the way the top operates we can see this axis provides independence that is a result of initiating gravity but it happens on the condition of the top spinning.

•Π^0
Π^0•   •Π^0
•Π^0

From that evidence we know that a sphere demands four more spots to form the circle and complete the sphere.

The in a circle four more points come into play. Again I find a need to remind the reader that we are dealing with a formation where the spots take shape but the shape alone has value because it moves and that is all. It cannot be seen, be touched or be witnessed in any way possible that forming a dot Π coming from a spot Π^0. This is even before $7\Pi^0$ forms Π but as result of the movement Π^0 become Π valued at 21.991/7. This is the smallest form in the Universe taking up space. This Universal form and this is the form of the Universe. This is the reason why the Universe is in the shape of a sphere. It is the way the Universe forms and when the Universe expands it grows by exactly this distinction because the Universe is what moves.

However what is has to move to be. The movement of the four establishes three and by conserving space it returns to five or by expanding into excessive space it progresses by 21.991/7 or then 3.1416. That is time that is gravity, that is movement and all those ideas are all one concept within space-time.

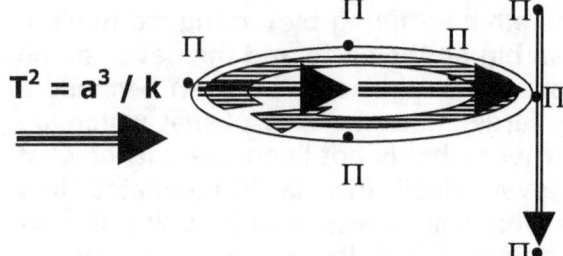

The forming of seven is even more critical in understanding cosmic physics since seven is the movement of the entirety from one location to the next location involving the formation of space. In this two more laws find validation and without understanding that as well as those principles there can be no claim of understanding physics in any form.

The forming of the three within the line and the four spinning results in seven and when the seven displaces into a new location is space a result thereof. This is the way that Π forms as 7 duplicates by the square. The entire Universe are built in this manner because the solar system is built in this manner and everywhere space forms it used the value of Π as we find it in the Titius Bode law.

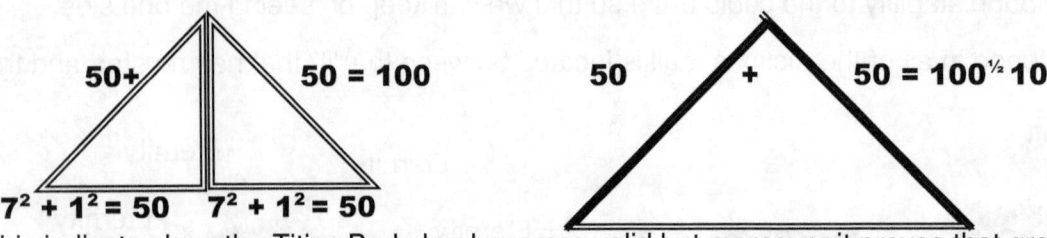

This indicates how the Titius Bode law becomes valid but moreover it proves that gravity forms by Π.

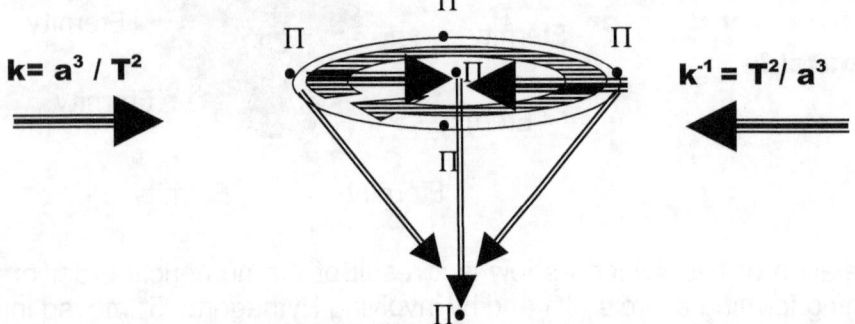

The moving part of gravity holds seven points in terms of ten forming on the one side and ten forming on the other side of the divide holding singularity at 1.9991. That then is gravity.

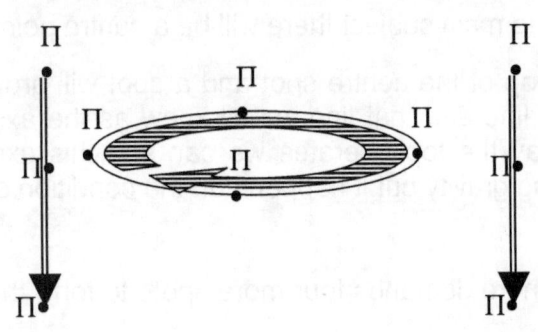

3 + 4 + 3 = 10
This indicates that the movement completed space by repositioning the line forming singularity.

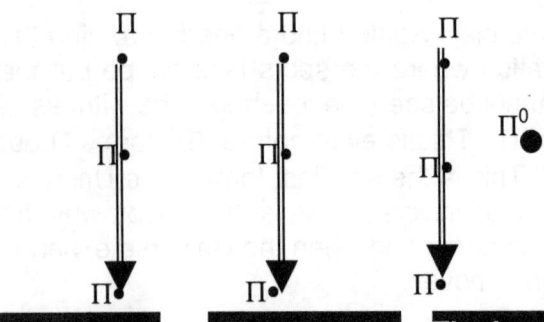

3 + 3 + 3 = 9 and also 3^2 = 9 this proves that space moved the value of the 3 positions where it results in singularity forming a new associated allocation.

The Past   The present   The future = 9 plus singularity forming in space is 10.

One holding singularity in a position in the centre does not shift. One realigns with a seven points that shifter by going square and in that one also goes square that actually means nothing and says little.

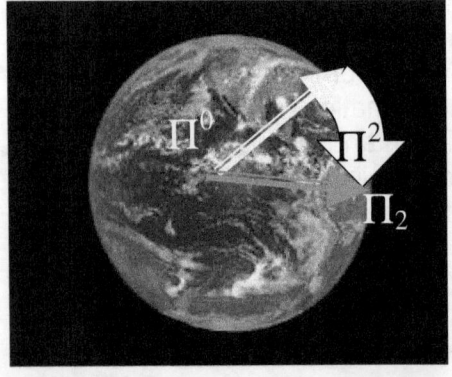

When we think of something shifting by turning we think in terms of something as big as the earth and then even much bigger. Where we now are could never be seen because a photon is immensely larger than the size of that which we now delve into. It is where what is not becomes part of what already is and however small this is it resonates into whatever is because whatever is was built by using this as the manufacturing content. When looking at a skyscraper, don't look at the building structure but look at the atoms forming the mortar and then see where we are at. When $Π_1$

moves to Π_2 through Π^2 in relation to Π^0 there are no changes occurring in regard to singularity because from the point holding singularity at Π^0 no changes occurred in material because it is still facing the same point. Therefore according to singularity Π^0 material Π_1 to Π_2 never changes and all the change came from space that does not change. Therefore according to singularity in the centre that cannot move, all movement of material is coming from changes in space because it is in that relevancy that changes occur and in material all movement takes place while no changes are happening. The movement in space is going from one through two too three. It is moving three points, three positions and the two point moving two positions does the moving of the three points.

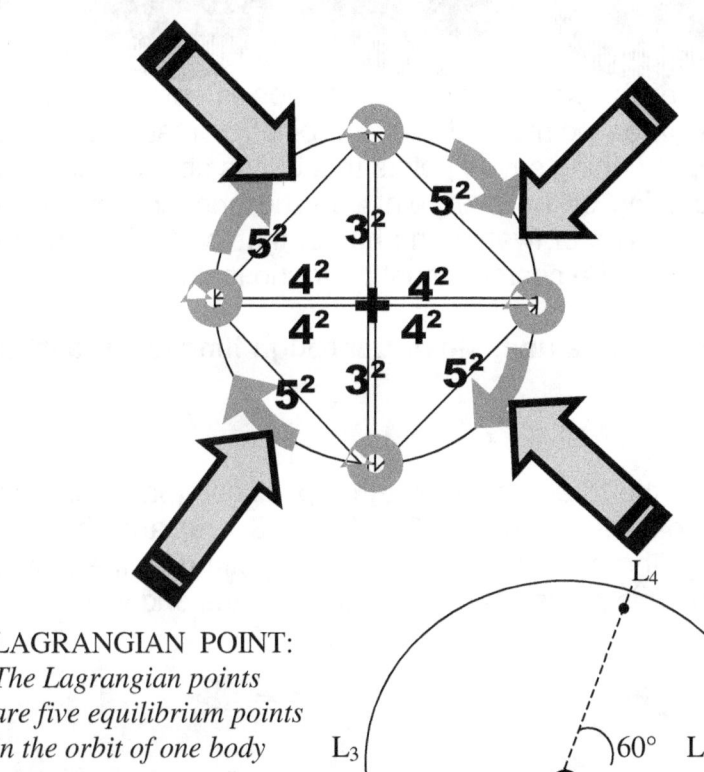

Numbers are turning but in relation to one another and together forming a relation that relation turning in response to each other forms Π. Four times five becomes on when the spot in the centre (0.991) becomes the dot as 1 followed by the next spot and this stands related to five points at four positions all in one relevancy of $5 \times 4 = 20$. The main thing to see is that five returns to the centre and the movement of four stands related to one in the centre. The movement is formed by the four of the circle centred on the one point in singularity turning around the centre. That holds the principle called the Lagrangian five points that forms the way singularity operates.

LAGRANGIAN POINT:
The Lagrangian points are five equilibrium points in the orbit of one body around another, such as a planet around the Sun

The Lagrangian points are the way in which planets and other cosmic bodies preserve space-time by contracting space-time. It is in the value of the 3 (the axis) moving and therefore going square 3^2 by having the four points forming the circle going square 4^2 and combining through the result of the combination of the relevancy of these points forming motion adding the combination $3^2 + 4^2$ and then applying the law of Pythagoras as a rectangular triangle $3^2 + 4^2 = 5^2 = 5$. Then this combination of five turning in four places around growing centre 1.9991 forms the top value of Π. Nevertheless, the result of the commotion is the forming of a circle to the value of Π.

By moving four to the centre of one the retaining value of $\Pi^2 = \Pi^3 / \Pi$ space-time is achieved.

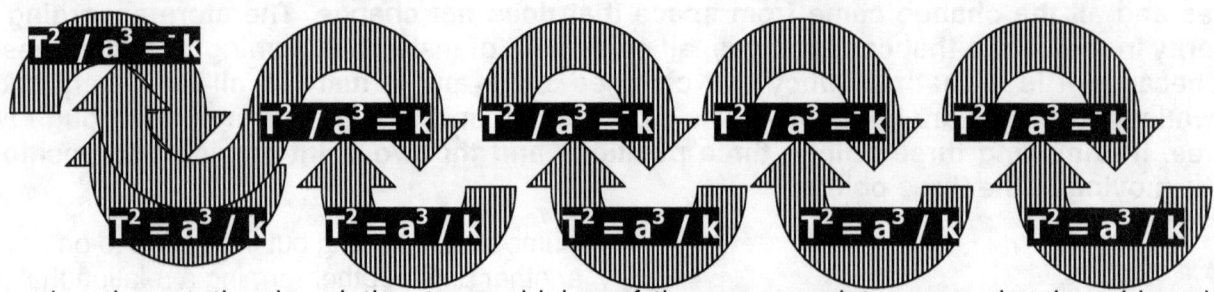

The putting the rotation in relation to combining of the seven points expansion is achieved. In the rotation going on the result is four points add to three points and this combination results in seven points moving. It is removing seven points and repositioning seven points that brings about expansion. However this must result in ten because the doubling of the five points on the one side moving in relation and to the next points has an overall circle value of twenty and half of that is 10. Forming ten on the one side is a cosmic must and the expanding of the cosmos must be found in this part.

The past – present – future is the way time moves as a unit. Moving through time is a reality of 3.

The axis that moves takes three points and moves the three points by three positions (past – present – future) and this movement combined is nine (past 3 + present 3 + future 3 = 9) and therefore although the movement takes three points through three positions and this results in nine the moving takes three by the square and therefore 3^2 is nine. This proves another 10 forming the end value.

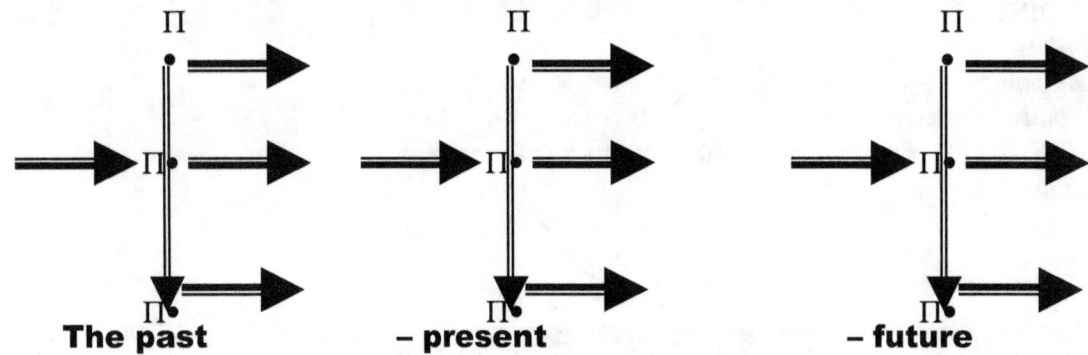

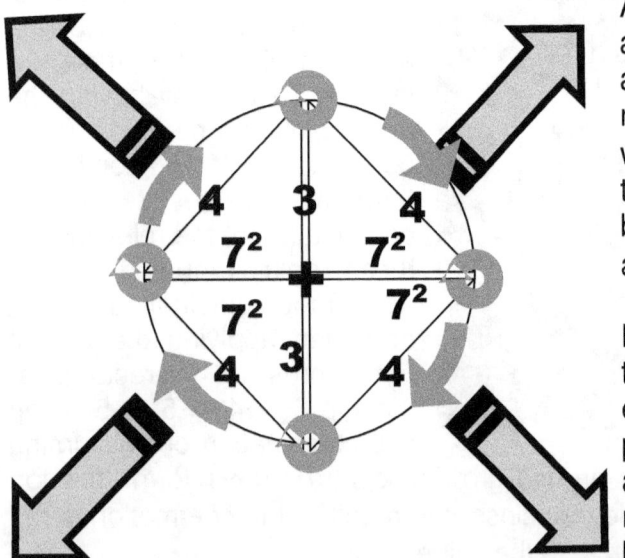

As the five points form a unit (7) the seven moves as a unit. The unit consists of the four points spinning around the axis that holds three and this togetherness represents 7 point, which forms the sphere. That is why Π has a bottom relevancy of 7 that represents the 1 which h singularity holds. Numbers are turning but as unit and the unit is a relation to one another and as a unit it forms Π.

By moving the entire sphere holding seven points in total we find the expanding contribution having the effect required. In the first instance five times four produced twenty. Now we find seven going square and double produces the twenty but in this case the result is the Titius Bode law as the space-time grows by the value of seven in relation to ten.

Then progress also becomes valid in terms of space-time building and progressing by becoming more. That is growth in terms of the space within the cosmos seemingly to grow bigger. This comes in place as the Titius Bode law forms more space in distance between cosmic material structures.

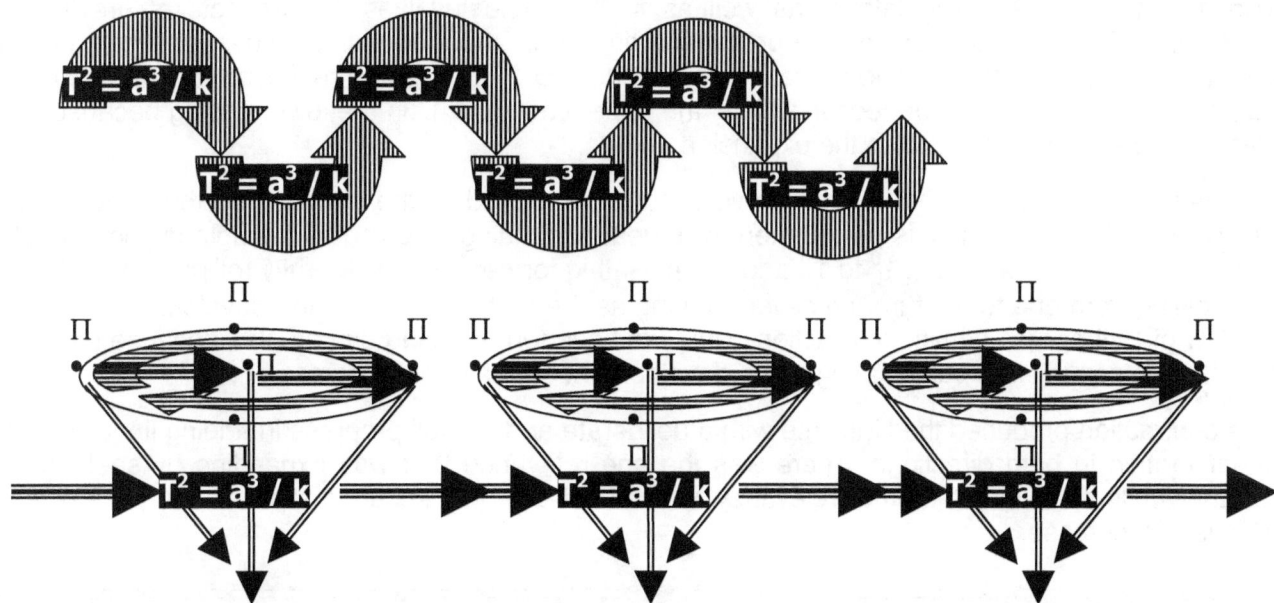

The four cosmic principles start at the point where the Universe starts. It starts in the manner numbers formed for the first time. The start is where one dot became two, became three, became four, became five and so on. It starts in movement of dots as the first dot duplicated forming the eternal relevancy. There never were many dots all connecting to one as if they all were one because they all were one. It is the same dot, the eternal dot that in relation to movement filled space, as space became the history of time and time still to this day is $1^0 = 1^1 = 1^2 = 1^3$. Five forms as four repeats one

To find the mean distances of the planets, beginning with the following simple sequence of numbers:

0 3 6 12 24 48 96 192 384

With the exception of the first two, the others are simple twice the value of the preceding number. Add 4 to each number:

4 7 10 16 28 52 100 196 388

Then divide by 10:

0.4 0.7 1.0 1.6 2.8 5.2 10.0 19.6 38.8

The resulting sequence is very close to the distribution of mean distances of the planets from the Sun:

Body	Actual distance (A.U.)	Bode's Law <A.U.)< td>
Mercury	0.39	0.4
Venus	0.72	0.7
Earth	1.00	1.0
Mars	1.52	1.6
		2.8
Jupiter	5.20	5.2
Saturn	9.54	10.0
Uranus	19.19	19.6

The truth beyond all other truth is that Newton's gravity has never been proven (because try as you may it is not possible to prove Newton's formula forming gravity mathematically) and because academics know that, academics require (no insist on) the blind acceptance of Newton by students.

When 1^1 expanded from 1^0 there was a linear motion established. The motion took what had no start away from what had no end. In that the expanding had a direction being lateral that connected to the cross of the lateral that connoted what remained of the eternal to the lateral. That which spins repositions as it spins and that leaves two actions serving at the same time and serving individual purposes while being synchronised in the movement.

The action of repositioning latterly as well as in the circle involves two aspects of gravity both interacting with each other but both representing time in its most opposing values, one being infinity and the other being eternity. The lateral motion connected 1^0 to 1^1 but in reflection as a mathematical response there was too a connection with 1^1 the upper casing to 1^1 in the lower casing because there was an immediate contraction to the expansion.

The infinity part that could not move was very much dominated by the eternity part that moves. From this we can deduct that in this period there was heat expanding followed by a single duplication of 1^0 to 1^1 then a retreating of heat 1^0 to 1^{-1} and this pulsating formed the first eternity rolling over by infinity. By growing from one to ten through seven forming as the bottom part of Π ten develops as a result of the law of Pythagoras coming into action. Then we have ten reducing by half also because the law of Pythagoras coming into participating in the development of numerical progress.

The contraction produced the Universe with a deliberate and will full progress in adding linear but also linear motion in both directions. There was the one relevance that was expanding by seeking new territory while there was the other relevancy, which contracted by securing a position in the Universe. This is still the case.

•⇒⇒⇒⇒⇒⇒⇒⇒⇒⇒⇒⇒⇒⇒⇒⇒⇒⇒⇒⇒••⇒⇒⇒⇒⇒⇒⇒⇒⇒⇒⇒⇒⇒⇒⇒⇒⇒⇒⇒⇒•••

The distribution occurred more often since the regulator grew by the inertia of the formula.

The expanding was **k** being one more added to the already existing but eternal spin T^2 where the spin was the size of the Universe a^3. By adding **k**, the Big bang became the big growth way from the stagnation that was the immovable Universe.

The top becomes another atom as it went through he entire process of creation to find release from eternity by infinity. By motion the top identify liquid and allow liquid to move around what the to established as a solid. In that the top created gravity by tow way balance. The balance is in the form and the form apply (7+7) / 10 in relation to 10 / 7 which is the equivalent to motion of singularity in whichever direction $Π^2$. This action forms the Coanda principle from which the Universe established a working process called gravity or to use an even more applying name we may call it movement.

When viewing the cosmos we see the same principle apply to massive stars that apply to the top. When the motion stops that parts the part with no end from the part that has no beginning the two aspects of eternity and infinity combine once again as time demolishes space and collapse on infinity. That which holds the outside reunites with that which hold the inside and the duplication of heat by expansion is then once more overwhelmed by the contraction that is principled to reunify that which has no outside with that which has no inside.

That what happens to the top is the same that happens to the Black Hole. The Black hole seized to move. In that singularity contracts that which has no outside and which we named outer space to unify once more with that, which has no inside. Time collapses on it self and in the process time in space as well as space-time is destructed by the lack of motion. As time draws the top into a state of collapse after the motion ends, it becomes precisely what the Black hole is. It is because the atom it once was that had the ability to move seized such ability and the collapse returned eternity onto infinity. In the state of the top being an independent Universe the top Universe collapsed. By generating electricity we establish exactly the process that is used within the star. The core collapses or condenses heat from a gas form to a liquid form and finds a short cut to send it into the Earth. Nevertheless it is exactly what a star does and we even use the same elements to do the job just as the star does.

There is no denial about the process because all the evidence is there as much as all the evidence is self-explaining. It proves by just being there. The contraction generates expansion and the expansion parts the Universe as much as it allows the collapse back again onto the Universe. It applied from moment-Alfa onwards. There was expansion and there was contraction. The contraction however did not bring the Universe back to an original value but left one dimension to progress. $k^{-1} = T^2 / T^2$. The minus one is a direction and not a value since k^{-1} is more in value that is k^0 and because of that the Universe fluctuated but it was directing into a future.

A More Complex Commercial Science Book   583   **Chapter 10** IN TERMS OF A COSMIC CREATION

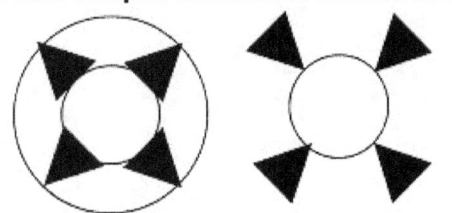

It was progressing towards where all the vacant time would fill with actual time. ••

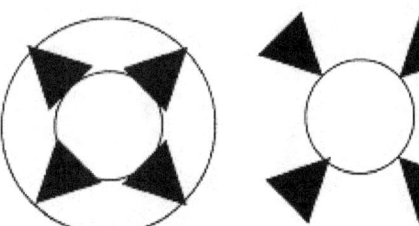 

At this point the motion was that time was forming the space and $1^0 = 1^1$ and while this happened time by space was still forming the Universe. There was that which is motion provided and substantiated by heat and then there was that which has no motion and therefore is removing that which substantiates heat being the drive behind motion. There is expanding of what can move in contrast to the removing of what can move. There was movement of what in future will become space that predated the enormous first Big Bang while at the very exact moment of the pre- Big Bang banging away there was the pre- Big Crunch crunching as the Biggest Crunch there shall ever be slammed shut the expanding ability of the biggest Bang there will ever be. The Universe expanded ten times and then reduced by half. Most important is that it was not random but a control was about it by measures no mind can ever understand. The expanding brought the contraction and the contraction brought the curbing to the motion, which brought retreat to the expanding. The Coanda effect was in place. Any and all substances with the ability to move are liquids and all that is in relevance not moving is solid. The distinction was there. Liquids parted from solids while solids curbed the patrician by contracting.

www.ingramcontent.com/pod-product-compliance
Lightning Source LLC
Chambersburg PA
CBHW080647190526
45169CB00006B/2021